Innovations in Agriculture for a Self-Reliant India

About the Editors

Dr. P K Ghosh, PhD topped in M.Sc. and PhD, with First Class First (90%) & Gold Medal in GBPUA&T, Pantnagar, He started his career as scientist in Indian Council of Agricultural Research (ICAR) in April 1993, completed 28 years' service (11 years as Research Management position and 14 years Professor/Principal Scientist) and is presently working as Founder Director and Vice Chancellor ICAR-National Institute of Biotic Stress Management (NIBSM)-Deemed to be University, Raipur, Chhattisgarh, India. Earlier he demonstrated leadership as National Coordinator, National Agricultural Higher Education Project-World Bank aided project, ICAR, New Delhi (2017-20), Director, Indian Grassland and Fodder Research Institute, Jhansi (2012-17) and Head, Crop Production Division Indian Institute of Pulses Research (2009-2012). His pioneering works on climate change, carbon sequestration, Resource use efficiency, conservation agriculture crop diversification, soil water conservation, Integrated Farming System (IFS) and coordination of higher education in 14 agricultural universities in India to bring excellence in quality of PG education program have been considered as outstanding contributions benefitting large numbers of farming community, scientists and students across the globe, as result he has been identified in November 2020 as one of the top two percent of agricultural scientists at global level based on the database analysed by Stanford University, USA. With his initiative PG program on biotic stress management at NIBSM has begun in 2020 affiliated with IARI, New Delhi. He has implemented National Initiative on Fodder Technology Demonstration throughout the country through 100 KVKs, established AdarhCharagram and grassland in 6 states in 657-acre area and promoted 17 Gousala in UP and Rajasthan for ensuring fodder security and livestock productivity. He is recipient of 19 national awards including M.S. Randhawa Memorial Award for best Administration, Excellence in Science Award and Sardar Patel Outstanding–Best ICAR Institute Award for leadership and also Fellow of National Academy of Sciences, India and National Academy of Agricultural Sciences (NAAS).One of his international book entitled "Carbon Management in Tropical and Sub-Tropical Terrestrial System" (published by Springer) is being referred globally and nationally to address Carbon Neutrality, which is one of the important points of SDG 2030. His another book "Grassland: A Global Resource Perspective" has become very popular to the International livestock and grassland working groups. He showed scientific leadership as Chairman by organizing 23rd International Grassland Congress (IGC) held first time in India with 450 international delegates and still acting as executive member of IGC continuing committee representing South-East Asia. He served as sectional president, Agricultural section under the general presidentship of Hon'ble Prime Minister of India in the centenary year (2013) of Indian Science Congress. Dr Ghosh holds many important positions nationally and internationally, few of them are Member, DST task force on Climate change on forestry, Chairman, National fodder planning committee, Editor-in-Chief, National Academy of Science, India, Facilitator, National Conference on Doubling Farmer Income (DFI) and Chairmen/Coordinator/Convenor of many national seminar/workshop/ conference/brainstorming, Member, Gujarat Institute of Desert Ecology, Member, Task Force on DUS guideline, Member, Board of Management /Academic Council/PG faculty of 9 Universities including Country expert/coordinator of many international projects funded by CGIAR/FAO/World Bank. Considering his academic, administrative and research excellence he was selected as co-opted member of DFI Committee, constituted by PMO and contributed to bring 14 volumes on DFI which are available in website of Ministry of Agriculture and Farmers Welfare and many actionable points and recommendation are being executed in different states. He also formulated four policy papers for the country and delivered 33 lead /Keynote Lectures in National and International conferences. He published 20 books and 161 research papers in the high impact factor journals with the total citation of 7103, h-index of 39 and i-10 index of 83 and total publication of 283.

Dr. Prabhat Kumar is National Coordinator, ICAR-National Agricultural Higher Education Project, KAB-II, Pusa, New Delhi. He obtained his B. Sc. in 1995 from CSAUA&T, Kanpur and M.Sc (Floriculture and Landscaping) in 1997 from PAU, Ludhiana and Ph.D (Floriculture and Landscaping) in 2002 from IARI, New Delhi. He worked as Assistant Professor, GBPUA &T. Pantnagar from 2003, and in the year 2012 selected as Associate Professor at HNBGU, Srinagar. Later he was selected as Senior Scientist at IARI during 2013. Since then he has worked as senior scientists and further got promoted to Principal Scientist. He has more than 17 years of experience in teaching, research and extension. He guided 10 M.Sc. and 01 PhD. students. He has handled more than 04 externally funded and 6 institute research projects and published 79 research papers, 35 popular articles, 06 book chapters, 11 technical bulletins and 03 books. He is associated in the development of 4 varieties namely; Pusa Deep, Pusa Bahar Pusa Sinduri and Pusa Shanti. He is fellow of Indian Society of Horticulture Research & Development in 2011 and awarded Himadri Young Scientist Award in 2012 for his outstanding contribution in Horticulture, Fellowship of Indian Society of Ornamental Horticulture in 2011 and 'Confederation of Horticulture Association of India, in 2012 for outstanding contribution in floriculture. He is also recipient of Dr.Ambedkar young fellow award in 2010 by SARC, Meerut. Established Model Floriculture Centre (MFC) under Horticulture Technology Mission.

Dr. Debashis Chakraborty has obtained his M.Sc. and PhD. degrees in Agricultural Physics from the Indian Agricultural Research Institute (IARI), New Delhi. He has joined as Research Scientist (ARS) in 2001 at IARI, and is presently the National Fellow of the ICAR with a programme on regional root-zone soil moisture mapping using satellite information. His research interests are soil-plant-water relations, geoinformatics and big data analysis, primarily in areas of alternate tillage practices, and remote sensing of soil moisture. He has worked as Post-Doctoral Fellow at Rothamsted Research, UK, with IRRI, Philippines and CIMMYT, India on a number of programmes. He was principal investigator in DST and ICAR sponsored projects, and worked with University of Sydney, Australia under the DFAT Fellowship. He was awarded NAAS Associateship in 2011. Dr. Chakraborty is a faculty, Post Graduate School, IARI since 2001, and is primarily involved in teaching soil physics and geoinformatics courses. He has guided six students published 90 peer-reviewed publications (h-index 25) and a book on Geographical Information System.

Dr. Debashis Mandal received M.Sc (Soil Science) from GBPUA&T, Pantnagar in 1994 and the PhD in 1999 from IARI, New Delhi. He currently holds a renowned post of ICAR National Fellow since 2014 at the Faculty of ICAR-IISWC, Dehradun. His research interests are assessment and monitoring of soil erosion, soil sustainability with emphasis on land degradation, reclamations processes and land degradation induced losses of productivity and ecosystem services. He has handled 20 research projects. His praiseworthy work on dissemination of soil and water conservation technologies for self-reliant (*Atmanirbhar*) rural India has been documented as an important success story. His contribution in conflict management towards water resource development has been highlighted in a peered reviewed journal which ultimately helped in achieving more income and production per drop of water. He has worked as a visiting scientist at CMASC at Ohio State University, Columbus in 2013 and conducted research with Prof. Rattan Lal, the world food prize winner for the year 2020. Debashis has around 65 peer reviewed articles in scientific journals, organized six conferences exclusively on Farmers First for conserving soil and water of the entire country, 10 books. He is recipient of 12 awards such as ICAR-National Fellow, Dr. B.C. Deb Memorial Award (2012) by Indian Science Congress Association (ISCA). Department of Science and Technology, GoI, Dr. K.G. Tejwani Award (2010-11) for Management of Natural Resources by Indian Association of Soil and Water Conservationists, Dehradun, Fellowship by Indian Association of Soil and Water Conservationists, Dehradun in 2018, and Policy Paper 88 of NAAS on Mitigating Land Degradation due to Water Erosion (2017).

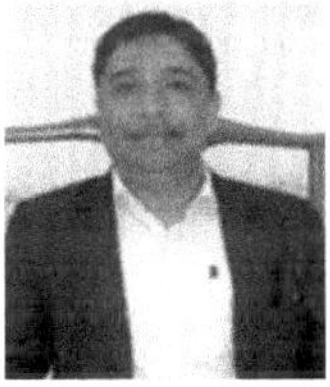

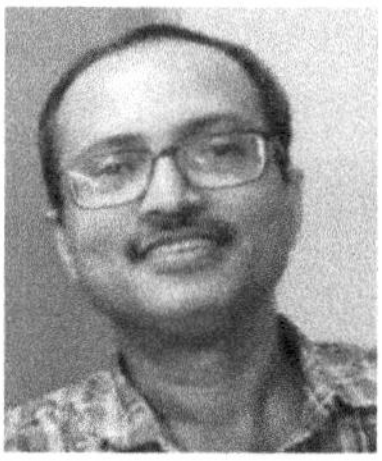

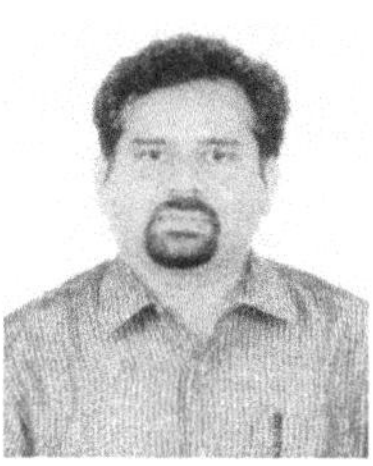

Dr. P.N. Sivalingam obtained his B.Sc (Ag) from TNAU Coimbatore, M.Sc (Ag) from GBPUA&T, Pantnagar and PhD. from IARI, New Delhi. Presently, he is working as Principal Scientist (Agricultural Biotechnology) at ICAR-NIBSM, Raipur. Earlier he was working as Scientist (Plant Biotechnology) at ICAR-CIAH, Bikaner for eight years and as Senior Scientist (Agricultural Biotechnology) at ICAR-NIBSM, Raipur for three years. He has made significant contributions for the last 13 years in the field of molecular characterization of plant pathogens affecting yield and quality of tomato, chilli, potato, grain legumes etc., Characterization of genetic resources of bioactive compounds rich under-utilized arid horticultural crops such as *khejri, ber, lasoda*, established tissue culture protocol for multiplication of date palm, molecular understanding of abiotic stress in *Ziziphus nummularia*, identification of non-host resistance mechanism in mungbean and tomato to begomoviruses and biotic stress induced promoters. He was a visiting scientist to Noble Research Institute, Ardmore, Oklahoma, USA to understand complex nature of gene expression. He has received six national awards from IARI, ICAR, New Delhi, NASI, Allahabad and Professional Societies. He had published more than 35 research papers peer reviewed journals. He is a member of National Academy of Science, India and reviewer of more than seven international journals and three national journals.

Innovations in Agriculture for a Self-Reliant India

P.K. Ghosh
Director and Vice Chancellor
ICAR-National Institute of Biotic Stress Management
Baronda, Raipur, Chhattisgarh-493 225, India
Former National Coordinator
ICAR-National Agricultural Higher Education Project
KAB-II, PUSA Campus, New Delhi-110 012, India
Former Director, ICAR-IGFRI, Jhansi, Uttar Pradesh-284 003, India

Prabhat Kumar
National Coordinator
ICAR-National Agricultural Higher Education Project
KAB-II, PUSA Campus
New Delhi-110 012, India

Debashis Chakraborty
Principal Scientist and National Fellow (ICAR)
Division of Agricultural Physics
Indian Agricultural Research Institute
PUSA Campus, New Delhi-110 012, India

Debashis Mandal
Principal Scientist and National Fellow (ICAR)
Indian Institute of Soil and Water Conservation
218, Kaulagarh Road, Near SBI Bank, Dehradun
Uttarakhand-248 195, India

P.N. Sivalingam
Principal Scientist
ICAR-National Institute of Biotic Stress Management
Baronda, Raipur, Chhattisgarh-493 225, India

CRC Press
Taylor & Francis Group
Boca Raton London New York

CRC Press is an imprint of the
Taylor & Francis Group, an **informa** business

NEW INDIA PUBLISHING AGENCY
New Delhi-110 034

First published 2022
by CRC Press
2 Park Square, Milton Park, Abingdon, Oxon, OX14 4RN

and by CRC Press
6000 Broken Sound Parkway NW, Suite 300, Boca Raton, FL 33487-2742

British Library Cataloguing-in-Publication Data
A catalogue record for this book is available from the British Library

Library of Congress Cataloging-in-Publication Data
A catalog record has been requested

ISBN: 978-1-032-15698-9 (hbk)
ISBN: 978-1-003-24538-4 (ebk)

DOI: 10.1201/9781003245384

DEDICATED TO
THE HON'BLE PRIME MINISTER
OF INDIA

नरेन्द्र सिंह तोमर
NARENDRA SINGH TOMAR
D.O. No. 123 /AM

कृषि एवं किसान कल्याण,
ग्रामीण विकास और पंचायती राज मंत्री
भारत सरकार
कृषि भवन, नई दिल्ली
MINISTER OF AGRICULTURE & FARMERS WELFARE,
RURAL DEVELOPMENT AND PANCHAYATI RAJ
GOVERNMENT OF INDIA
KRISHI BHAWAN, NEW DELHI

सत्यमेव जयते

20 FEB 2021

Foreword

Agriculture plays pivotal role in Indian economy by securing direct and indirect livelihood of most of the Indians. Indian Agriculture has witnessed great transformation with technological and policy interventions and the established notion of 'ship to mouth' has been changed into export surplus nation. However, we should not be complacent with past achievements as the food demand is increasing with increase in population and livestock. By the year 2050, country's food demand would be 333 million tonnes annually. Hence, our quest for new innovations should continue in light of national priority in a sustainable way. In this endeavour, future research needs to essentially concentrate on climate-resilient agriculture in order to ensure safety and security to our future generations on this planet.

In order to be *Atmanirbhar* in *Krishi*, we need the most productive and appropriate institutional mechanisms to lower down production costs, boost productivity and encourage better marketing linkages to increase farmers' net income. This will bring a paradigm change at the grassroots level. Innovations by agricultural start-ups in the form of products, services or applications can be meaningful solutions across agricultural value chain. Efficient use of these talent pools will act as the key driver for improving competitiveness in the agriculture sector.

Ministry of Agriculture & Farmers Welfare, Government of India, appreciates publication of such book which is timely and pertinent i.e **'Innovatlons in Agriculture for A Self-Reliant India'**. It gives me great satisfaction to note that this book focuses on the understanding and requirements of state-of-the-art research, trade and policy institutes, FPOs & use of modern ICT which will play pivotal roles in making India a US$ 5 trillion economy. The editors appropriately highlighted new innovations that will reignite agriculture as a career-option for the youths and repose India as a world leader in agricultural innovation.

I congratulate the editors for their diligent efforts in bringing out this comprehensive book.

Narendra Singh Tomar

Office: Room No. 120, Krishi Bhawan, New Delhi-110 001 Tel.: 011-23383370,23782691 Fax: 011-23384129
Resi.: 3, Krishna Menon Marg, New Delhi-110 001, Ph.: 011-23794697 / 98, Fax: 011-23794696

Preface

'........ *Atmanirbhar* is seen to be an answer to the revival of a Covid-impacted economy'

– Oxford University Press

India's journey from a food-grain importing country to a country with abundance in production and as one of the largest exporters of major agricultural commodities have been phenomenal. The passage to this glorious feat has witnessed a great synergy among the scientific tools and methods, institutional participation, policy initiatives and a strong will power of the Government. Food grain production in India has been increasing every year, although its contribution to the gross domestic product (GDP) declined due to growing share of manufacturing and services to Indian economy. The trajectory of Indian economy saw a sharp decline in 2020 due to Covid-19 pandemic, and a never-seen-before contraction of country's GDP. However, despite all other sectors had huge slumps, agriculture managed to clock a positive 3.4% growth rate analogous to 5.7% Gross Value Added (GVA) in the first quarter of 2021.

In the backdrop of Hon'ble Prime Minister of India's call for an *Atmanirbhar Bharat* (self-reliant India), and Government of India's bold and elaborate plan for a resurgent economy, agriculture and allied sectors must find the centre stage again, but this time with renewed form and vigour. It should be an all-new agriculture with three commitments: sustaining the production, protecting the environment, and doubling the farm income. It is indeed a daunting task, but when all is said and done, innovative tools and technologies appears to be the only solution. India is blessed with immense diversity, potential and human capital in all sectors, including agriculture. The country has already been moving in the path of progress in agriculture, but in *Atmanirbhar Bharat*, the Indian agriculture must be independent and self-reliant in all senses with sheer determination and self-esteem.

With over 40% of employment, agriculture and allied sectors necessitate detailed and categorized analyses of all the available new and novel technologies to explore the potential in augmenting the output. This is only possible through engaging scholars from different fields of agriculture, horticulture, dairy and animal husbandry to gather information on tools and technologies available on the horizon and their full prospective utilization. The necessity of bringing together all innovations in the field of agriculture and allied activities in the form of a book has long been felt. This becomes all the more necessity in the wake of country's pledge to be self-reliant in agriculture as livelihood, driver of food security and an industry, and therefore contributing to a bigger and more important role by the country in the global economy. There could be plenty of information available everywhere, but organizing these into a single book provides a ready-to-hand document for all, and especially to the people engaged in farm-entrepreneurship and policy organizations. Though the chapters have been constructed diligently and with utmost care, there could be mistakes and errors most unintentional, for which we apologize. We welcome suggestions which we will try to address in the future editions.

We take this opportunity to thank all the contributors for their sincere efforts for timely submission of manuscript in the amidst of covid-19 pandemic situation. Most importantly, the real motivation for drafting this book originated from the vision of a new '*Atmanirbhar Bharat*' envisaged by the Hon'ble Prime Minister of India; we are truly indebted to him for this inspiration. Therefore, we, the editors **dedicated this book to Hon'ble Prime Minister of India**. We owe a great many thanks to New India Publishing Agency (NIPA) and CRC Press for bringing this book to light. We hope that this book will lead in the country's quest for emerging as '*Atmanirbhar*' and be exemplary to the rest of the globe.

Editors

Contents

Theme 3: Innovations and Strategies for Abiotic Stress Management

Theme 4: Innovations and Strategies for Biotic Stress Management

Theme 5: Innovative Technologies for Agroecosystem Development

Theme 6: New Science and Speciality Agriculture

Summary

Climate change, food safety, and economic uncertainty in agriculture, and the unprecedented global disaster like COVID outbreak with dire social and economic impacts call for achieving the self-reliance and self-sustainability as a country. India had three greatest challenges ahead of the 21st century: (i) sustaining food and nutritional security, (ii) adaptation and mitigation to climate change, and (iii) sustainable use of like water, land and energy resources. By the year 2050, annual food grain production needs to increase to 333 million tonnes (Mt), which is possible only through use of innovative technologies. As per NITI Aayog, the policy think tank of the Government of India, the India stands 48th position in Innovation Index globally among 129 countries during 2020, compared to 52nd position a year ago. There is a huge potential for innovation as India has high human capital and demographic advantages to realise the capital potential. Several innovations in agriculture aim to make India self-reliant *viz.*, sustainable use of natural resource base, diversification in cereal based systems, establishment of zone-specific seed hubs, development of multiple stress-tolerant crops and animals, biofortified high-yielding crop varieties, target oriented programmes on developing climate resilient nutri-cereals, integrated farming system for small farm holdings, solve the crisis and ensuring their income, promoting secondary agriculture, alternative source of income organic and specialized farming, group farming approach, integrated aquaculture farming, promotion of high-tech precision farming, branding of local farm products (vocal for local), technology-enabled supply chains, wasteland utilization, achieving land degradation neutrality, agro-forestry and carbon farming, real-time market information and technology enabled crop insurance, developing water harvesting capacity and irrigation infrastructure, location-specific farm mechanization to engage youths in agriculture, rural entrepreneurship, strengthening food processing micro-industries in villages or sub-urban areas, seed export and regulatory system, feed and fodder sector, better survey, surveillance and quarantine systems, extension system with more public-private partnership and digital advisory services, micro-enterprise development in agriculture and allied sector, skill development and vocational training to farmers and youths for a strong marketing system, efficient utilization of government schemes, stronger linkages with international organizations, on-farm processing of agricultural products, innovations in post-harvest technology, solar and wind energy utilization in farming, bio-tech crops, agro-processing and service sector, contract farming, government interventions and support, bridging gaps in production and productivity in eastern India, promote horticulture on small lands ensure farmers the infrastructure and other facilities, facilitate agro processing for small farmer's produce, vertical farming creating sufficient cold storage facilities, and real time market information. In research on agriculture, innovations on climate resilient crop varieties, genetic engineering and gene editing, water resources and irrigation management, precision farming, artificial intelligence and machine technologies, diversification to higher-value crops, promotion of secondary agriculture, increasing shelf-life of perishables, and adding value to farm produce by branding them are needed. Similarly, innovations are required on modern techniques like digiculture, unmanned aerial vehicles, artificial intelligence for crop and soil monitoring, predictive agricultural analytics, supply chain efficiencies, agricultural product grading, smart agriculture, drone-based technology, agriculture machinery and implements, climate technology, vertical farming and data science. The list is exhaustive. However, the NITI Aayog predicts five most important innovations, which are expected to shape Indian agriculture industry as digital innovation, climate risk management strategy, developing the start-up ecosystem, farmer producer organisations and water management initiatives. To accommodate the large number of innovations spreading across all spheres of agriculture and allied sectors, eight broad groups are identified, which are briefly discussed below.

1. Crop Improvement

Several crop varieties have been released to make India self-sufficient in cultivation. Challenges remain to develop climate-smart, multiple stress resistant, bio-fortification varieties especially for protected cultivation, by utilizing genomic tools. The seed sector needs strong public and private partnership, which would cater to the needs of the farmers more efficiently with surplus seed availability. There is a need to innovate in all the components of seed chain. Harmonization of rules and regulations, liberalization of licencing system for quality assurance, strong policy support towards technological interventions, database development and reforms in the public sector are some of the areas which call for innovative approaches. Crop diversity was largely lost and only 12 plant species and five animal species are providing more than 75% of all human food items. Among ~7,000 species cultivated and domesticated by humans for food, feed and fodder, only 30 species currently cover global agriculture, indicating critical need of crop diversification. The demand for edible oil in the country is rising and domestic production has not been able to fulfil the gap forcing the country to import over 70% of its vegetable oil requirements. Cultivating oil palm with assured market price, promotion through independent regulatory board, enhanced processing facilities, supply of good planting material and crop insurance may help in meeting the domestic demand to a large extent.

2. Abiotic Stress Management

Population in India is likely to reach 1.8 billion by 2050, most of the agriculture products and inputs need doubling to meet the demand, without further degrading the available natural resources. Land degradation and water scarcity are the most pressing issues, which must be addressed through basic and applied research supported by increased funding, thus evolving to new knowledge and technologies. Better understanding of hydrological cycle, improved agricultural practices and development new tools are to be used for mitigating the impacts of chemical pollutants. Precision irrigation technology should be adopted to enhance irrigation efficiency for reducing over-exploitation of groundwater, and to enhance net irrigated area. India needs a second green revolution, which has to come through advanced technology application in arid regions of India. This requires enhancing the component diversity at farm level, and the creation of agricultural infrastructure facilities in the arid region. Innovations on climate -smart agriculture and structural reforms to combat climate change are required in the present context. Locating soil organic carbon enhancing spots, identifying higher biomass producing genotypes and factors governing carbon sequestration through the National Agricultural Research System are keys to mitigate the climate change. Managing Green House Gases emissions from agriculture and adopting climate-smart agriculture for improving farm productivity and farmers' livelihoods accord to Paris agreement, and can be achieved through IT-based sensors and communication technologies for real-time monitoring of crops and soils. Over the past 30 years, more than 2.5 million people and nearly USD 4 trillion have been lost to natural disasters in India. According to a World Bank report, nearly 75% of the losses are attributable to extreme weather events and climate change threatening to push an additional 100 million people into extreme poverty by 2030. It makes long-lasting impacts on the country's economy, and on agriculture. To mitigate and manage these losses various innovative methods have been evolved, capacity building of local people found to be the most important among these.

3. Biotic Stress Management

In India, crop yield losses are estimated as 37, 26, 20, 6 and 8 percent by weeds insects, plant pathogens, rodents and others, respectively with an annual loss of USD 31 billion. The important issue is strengthening pesticide regulation mechanism in India, as use of spurious pesticides will fail to control pests, and could be worse for the environment. Non-availability of branded pesticides and microbial insecticides on credit basis and lack of awareness among large numbers of farmers could be reasons for use of spurious pesticides.The size of the spurious pesticide market in India is as high as 30% by volume and 25% by value, and is growing at 20% per year leading to a loss of nearly 10.6 Mt of food grain production. Emerging of new approaches like nanotechnology molecular mechanisms (RNAi, CRISPR/Cas9) and identification of highly virulent biological control agents has resolved many concerns prevailing with the earlier strategies for pest management. The new Pesticide Management Bill, 2020 seeks to promote use of safe and effective pesticides. To achieve this, forecasting/forewarning techniques, innovative agronomic practices, precision pest management technologies, nanotechnology based formulations, innovative biological pest suppression, novel biotechnological interventions, strengthening bio-safety and bio-security issues, remote sensing and artificial intelligence usage in biotic stress management and reforms in the pesticide management could be new and novel innovations. On weed control, important emerging areas viz., usage of infrared, thermal energy, microwaves and radiations, hot water, steam, hot air, lasers, robotic, flame, remote sensing etc are evolving. Precise crop health monitoring is in great demand, so as to adopt precise management against the losses caused by pathogens. Assessing disease risk at regional and global scales is of great interest for on-farm management and assessment of agricultural policies and practices. However, most of the pest and disease models do not cater to meet the need of users, and therefore cannot be utilized to their potentials. Models to evaluate the impact of plant diseases on crop production under current and future climatic conditions are increasingly sought for. Another area of major concern is the biotic stresses in the emerging production systems like protected cultivation, organic farming and conservation agriculture. Web- and IoT-based forecasting systems, forecasting based on pathogen detection and phenotyping of resistance, hyperspectral remote sensing for disease detection, chlorophyll fluorescence imaging and expert systems - machine learning applications on detection and forecasting are expected major areas to grow in near future.

4. Agroecosystem Development

Indian agriculture is undergoing transformation to integrated farming systems (IFS) for making farm self-reliant and doubling the income, addressing the issue of climate change, improving resource use efficiency and rural bio-entrepreneurship. It is essential as the per capita land availability reduced from 0.5 ha in 1950-51 to 0.15 ha in 2020, therefore increasing production from a unit land possible through IFS. The IFS is the potential option, where 70% of the resources would be generated and recycled in the farming systems. Innovations on energy self-sufficient IFS, farm development card and development at commercial scale rather than sustenance level and development of location-and situation-specific are most required. Innovative use of software/ IT intervention should be initiated; system must be in force to sell local farm produce to global markets and strengthening of marketing channels. Hill agriculture needs special attention particularly in north-eastern region (NER), drought and floods being two critical factors revolving the livelihood and economy of the region. Indian Council of Agricultural Research Institutes and Central and State Agricultural Universities are working tirelessly to this effect. Although much emphasis is needed on seed production strategy, target-based programmes on nutri-cereals, IFS for small holding farmers, developing water harvesting technique and efficient water use strategy to build up soil fertility, promotion of high-tech automated-, specialized- smart-farming, promotion of maize as food-feed crop, rice-fallow-oil seed crop rotation, organic farming, promoting secondary and specialty agriculture, location-specific farm mechanization strategy processing, value addition and packaging and use of different apps in agriculture. Nutritional security in NER will further be achieved through better horticulture livestock and fisheries. Prevention of loss of biodiversity and integrated mountain development will support the economy on sustainable basis to achieve the vision of self-reliant north east with a focus on self-reliant India. The coastal areas of the country occupy about 10.8 million ha of land and are spread over the 7517 km long coastline. The coastal ecosystem is highly fragile and vulnerable to climate change, and endangered by the sea level rise following global warming. Since the problems of coastal agriculture are widely variable, there cannot be a universally applicable way to build self-reliance. Coastal agriculture contributes largely to country's food production, particularly of rice and fish. Many areas like sub surface drainage for reclamation of saline waterlogged soils, alternate land use systems, bio-drainage to combat water logging and secondary salinity, increase in cropping intensity in coastal saline soils, research on salt-affected lands and mangrove ecosystem and brackish water aquaculture, and the use of digital science and renewable energy resources needs innovative approach. In coastal areas, land shaping is an innovative technology to address land degradation (salinity), drainage congestion and scarcity of fresh water for irrigation and higher carbon sequestration due to more plant cover. Major constraints for adoption of land-shaping techniques are marginal land holdings, high initial investment and presence of acid sulphate soils etc. Integrated coastal resource management and proper land-use planning therefore need to be implemented. Planning and implementation of the policies need farmers' active involvement. Bundelkhand in central India is another important region blessed with plenty of natural resources. The major issues are extreme weather, water scarcity, low fertile soils, prevalence of mono-cropping low livestock productivity, unregulated mining, subsistence-like farming, lack of industrial development and support from off-farm employment poor infrastructure to support tourism and inadequate implementation of schemes and multiplicity. Focus needed on agriculture and allied sector includes watershed technology IFS, crop diversification, organic farming, agro-forestry, fodder technologies, soil testing, integrated nutrient management, micro-irrigation, breed improvement of milch animal, use of digital science and strengthening the seed sector. This region needs appropriate management of existing resources of soil, water, minerals and vegetation, emphasis on natural resources management with scientific and social planning of water use by different sectors. Some of the useful interventions could be cluster-based farming for organic and specialised agriculture, promotion of boundary plantation for natural resource conservation, and increase in tree cover in the region, facilitating credit and financial assistance and market infrastructure for agroforestry produce to bring it at par with agricultural produce for all policy matters, providing insurance for tree-based farming in line with crop insurance and subsidy for drip and micro irrigation facility, generation of alternative sources of income like secondary agriculture, local artisans, utilising the ITKs and establishing small scale agro-based industries to provide jobs to redundant agriculture labourers in agricultural slack season and reduce migration. Agroforestry is the complementary ecosystem for agricultural crop, research and developments, region-specific models, area under agroforestry, initiative of government on policy aspects and challenges elaborated extensively.

5. Technological Innovations

India is now transforming from an agrarian economy to an industrial economy. Even with a food grains production of 285.21 Mt and advancement in technological sector, INR 92000.00 crore worth food grains are wasted annually. Food processing sector contributes 14 percent of GDP of the country and generates employment. It is a key sector for doubling farmers' income. Advanced processing technologies such as edible coating (Waxing), irradiation (Cold sterilization), non-thermal and high pressure processing, pulsed electric field, manothermosonication, ultrasound, cold plasma, cryogenic grinding, modified atmospheric packaging (MAP) technology, hurdle technology, nanotechnology and bio-preservation are being implemented in the food processing with newer innovations. Better public-private partnership, advanced agro-processing centres, stronger farmer producer organizations and skill development centres and improved agri-business incubators are necessary for further adoption. The strengthening and expansion of secondary agriculture through rural industrialization by developing processing technologies and new products, the demand push for under-utilized crops have been created which may help a large number of farmers. Innovative post-harvest technologies, post-harvest goods and services for small-holder farmers and marketers will allow them to join in modern value chains, improve their potentials to explore new markets and help to reduce food waste while increasing their incomes. There is a necessity to devise more efficient processing techniques to add value to the crop as well as minimization of post-harvest losses. Space technology in agriculture is providing large-area synoptic coverage with repetitive, objective and scientific datasets to generate information on crops and natural resources on a real-time basis. This aids in manipulation of crop production and management practices improving the resilience of farming against weather aberrations. Advanced estimates of regional agricultural output are critical to anticipate and mitigate the course of food shortage from a sudden crop failure. Pests and disease assessment, agricultural drought monitoring, mapping of saline, alkali and salt-affected soils, and soil mapping of the whole country are excellent applications areas of space technology. Further identification of areas and planning for horticultural development, assessment and mapping of land degradation and soil erosion, disaster monitoring like locust incidence, crop insurance and potential fishing zone mapping can be successfully implemented by using the space technology. The digital farming solutions can lead to future smart agriculture for enhanced productivity through applications of robotics, drones, automated and semi-automated guided vehicles, artificial intelligence, machine learning, deep learning, big data, IoT and cloud-based computations. In India, 62 Mt of municipal wastes are generated each year, of which 75% is recyclable but only 30% is recycled. Processing, treatment and disposal technologies available with industries, academic institutions, research agencies and other platforms must be integrated to convert this large volume of wastes into wealth. Conversion of wastes into manure, energy, recycled materials and other value-added products are the real pathways towards a successful 'waste to wealth' programme. This need innovations on mobile phone-based applications (apps) and websites, start-ups and non-governmental initiatives, community participation combined with engineering and machineries for utilizing the wastes, growing awareness among citizens and innovative solutions to improve the efficiency of waste management and make value-added products. Hi-tech agriculture is the fast developing areas for innovations and it revolutionised in the field of agriculture. It needs greater attention particularly in India to harvest its full potential.

The technology innovation heaven could be in the field of post-harvest and marketing system. This will likely save 30% losses of farm produce. Loss of food relates also to scarcity of water, wastage of inputs, labour and land, and the increase in the GHGs emission. Majority of loss and wastage is occurring at the production (harvest) and post-harvest stages. Even though the use of machineries increased in the past few decades in India, problems associated with improper drying, storage and delivery exist. Efficient post-harvesting technology and management are necessary to counter this loss, especially in roots, tubers and oil-bearing crops (>25% of total loss), followed by fruits and vegetables. Farmer Producer Organizations (FPOs), Self Help Groups (SHGs) and Farmer Interest Groups (FIGs) can play major roles. Establishment of traceability, supply chain and cold supply chain management, entrepreneurship and skill development programmes, quality and safety management systems, technical interventions and novel technologies, smart farm practices includes drone-based crop protection systems, crop maturity sensors, robotic harvesting systems, heavy metal detection sensors, non-contact quality analysis devices, mobile applications for crop nutrient information all requires new innovations and implementation of recent development to minimize food losses. Technologies like controlled atmosphere storage, ethylene controlling technology, low-temperature drying by heat pump, vacuum impregnation, vacuum frying, non-thermal inactivation, 3D printing, vacuum packaging, intelligent packaging system and nitrogen flushing are to be modified as per local conditions for industrialization. Future post-harvest management should be focused on new cooling systems and temperature control, sustainable packaging to reduce food wastes, time-temperature indicator technologies, active, smart and intelligent packaging systems, use of nanotechnologies in different stages of post-harvest management, information technologies like genomics and computer modelling etc. Efforts are also to be made in the upgrading of local knowledge on food supply chains and food uses, strengthening supply chain integration, transferring superior technologies, building skills and capacity in the food supply chain, and improving small and medium-scale focused entrepreneurship opportunities.

6. Development in Allied Agriculture Sectors

There is an urgent need to infuse new ideas and innovations for rejuvenating the livestock and fodder sector. This sector is contributing nearly 4.9% to the national GDP. However, main challenges are improved breeds of livestock, need-based stocking and stall feeding, and innovations to augment the fodder availability collection and conservation of crop residues, popularization of agro-climate based fodder crops, year-round fodder production, utilization of orchards and non-competitive lands and non-conventional fodder resources, hydroponics, bale and feed block making, azolla production, leaf meals, fodder banks and forages from forests. Innovations on participatory fodder seed production, use of digital science and apps in technology up-scaling, public-private-partnership (PPP) for rejuvenation of community grassland, involvement of FPOs, NGOs, and their linkages, capacity building and women empowerment.

Although a large number of new and potential innovations and practices have been developed, lack of adequate publicity and awareness inhibits their quick adoption. Self-reliant India- the dream in livestock and fodder sector is expected to gain momentum through modern breeding programmes and feeding and health approaches to all categories of livestock, maintaining feed and fodder database involving various working groups, Tapping export potential of Indian livestock products, entrepreneurship development, commercialization and value addition to farmers groups and other private stakeholders, Application of ICTs and modern tools for extension services including vaccination, addressing the issue of insurance of milch animals.

Aquaculture including fish production has increased to 6.2 Mt in 2017-18 from 1.9 Mt in 2000-01. About 88% of the farmed fish comes from freshwater aquaculture. In spite of its rapid growth, aquaculture today is beset with multiple issues impacting production and productivity such as disease, absence of species diversification and genetic improvement, poor brood and seed, high input cost, lack of access to institutional credit, credit guarantee and insurance, environmental sustainability etc. Major limiting factors are (i) In the marine sector: overcapacities in territorial waters, weak regulation, inefficient management and prevalence of traditional fishing practices, inadequate infrastructure; (ii) In inland sector: tenure and lease rights, use of obsolete technologies for harvesting coupled with low capital infusion and inadequate infrastructure. India has immense potential in terms of resources and youths. Innovations on strategic implementation of policies, budget allocation, linkages among institutions, smooth credit flow, research, innovations, development of new technologies, community-based extension activities, entrepreneurship development, better market infrastructure information system, proper regulations and skilled manpower are keys to become self-reliant in fisheries sector.

7. Agricultural Marketing and Extension

India is leading in production of many agricultural crops, but finding a market for surplus and getting fair prices have always been major challenges. Efficient regulated system to help the farmers from not getting manipulated in the market is still lacking in spite of many Govt. initiatives. To get a faster growth, higher investment in infrastructure viz., opening up of railways infrastructure to 100% foreign direct investment, rural electrification, driving the construction of new airports and sea ports across India *etc.* Creation of adequate infrastructural facilities in a liberalized and market-driven economic environment is necessary particularly in rural areas for minimizing economic disparities between rural and urban areas. Small farmers cannot hold stocks for long. Hence, the value of storage facilities at the FPOs level must be enhanced. FPOs will need large working capital to give advances to farmers against their produce as collateral. Therefore, handholding by the government in this area is required to strengthen the market infrastructure using advance technologies. Physical markets with modernized facilities and services both for the farmers and the buyers will create competitive trade environment. It is mandatory for maximum use of ICT in agricultural marketing, modernization in the marketing facilities for agri-logistics, develop the modern pack-houses for handling huge horticultural produce, establish linkages for small and marginal farmers with domestic or international markets etc. Farmers have knowledge of only 40% of the developed farm technologies. Access to information and improved communication is a crucial requirement for sustainable agricultural development. Rural populations still have difficulty in accessing crucial information in order to make timely decisions. To enhance the information base of the farmers, mobile apps in technology transfer, need of mobile ICT interventions for agriculture, agro-information services through mobiles, mobile based projects' overview for rural India, engaging farmers and enriching knowledge such as Agropedia (ICAR initiatives) are disseminated to farmers for enhanced usage.

8. Agriculture Education

The Government of India established Department of Agricultural Research and Education in 1973 to strengthen the linkage between the centre and state

Governments and to deal with the international agencies. Agriculture growth rate is 4% and to keep up with the growth, technology and innovation needs proper agricultural education system in the country. The Government reforms are necessary on the organizational level, e-Governance in agricultural universities, financial management system and prioritizing and monitoring system of education. High demand for competent human resources in the area of veterinary and animal sciences, dairy technology, food technology, horticulture, fisheries and agri-business management need to be created. Enhancing the quality of education management, entrepreneurship development and self-employment orientation in agricultural education, periodic revision of curriculum, distance education of agriculture and training, developing locally competitive human resources in advanced areas of science, etc are the most important areas to augment agriculture education system. Moreover, the education on agriculture should be modernized to attract the youth.

Theme 1: Agriculture Sector in India Challenges – Now and in Future

1

Indian Agriculture: Issues, Challenges and Priorities

P.K. Ghosh[1], P.N. Sivalingam[1], Debashis Chakraborty[2], Debashis Mandal[3] and Prabhat Kumar[4]

[1]ICAR-National Institute of Biotic Stress Management, Raipur, Chattisgarh
[2]ICAR-Indian Agricultural Research Institute, New Delhi
[3]ICAR-Indian Institute of Soil & Water Conservation, Dehradun, Uttarakhand
[4]National Agricultural Higher Education Project (NAHEP), ICAR, New Delhi

Introduction

Over the past decades, agriculture sector played a crucial role in enhancing the economic growth of India, and will continue so in future. It employs 55% of the workforce, thereby contributing to 17.1% of the country's Gross Value Added (GVA) for the year 2017-18 (MoA & FW, 2019). Agricultural production in India grew at an average of 3.6% annually since 2011, thanks to the policy decisions on inputs including improved fertilizer delivery systems and high yielding varieties. Release of improved varieties in cereals, pulses, fruits, vegetables and ornamental flowers as well as high-yielding breeds in the livestock and fisheries sectors has increased the diversification. Availability of a range of nutrition rich food commodities played crucial role in reducing undernourished population from 24% in 1990-92 to 15% in 2014-16 (OECD, 2018, FAO statistics). India is currently the largest product export economy of the world. India is leading exporter of rice globally, amounting to 8.34 Mt in 2019, and currently holds >85% of global basmati rice exports. The country has shown massive resilience to very recent global outbreak of COVID-19, and managed to tackle the social and economic challenges with a spirit of self-reliance. The '*Atma Nirbhar Bharat Abhiyan*' by Government of India with a special economic package of Rs 20 lakh crore to enable the resurgence of the economy through three mantras, 'innovation, integrity and inclusion' as suggested by none other than the Hon'ble Prime Minister of the country. It is therefore imperative that the agricultural sector too needs new initiatives to continue with the success, and realize the potential of self-reliance.

The Green Revolution

The 'Green Revolution' in India began in the mid-1960s marking a transition of Indian agriculture from a food-deficient country to one of the leading agricultural exporters. Genetically improved high-yielding varieties of rice and wheat were introduced along with adoption of new agricultural practices and technologies resulting in phenomenal increases in their production. The improved rice and wheat varieties, developed by International Rice Research Institute (IRRI), Philippines and the International Maize and Wheat Improvement Centre (CIMMYT), Mexico had several traits and specific genes responsible for short stature, photo insensitivity and high yield potentials. All these led to increase in yield and cropping intensity culminating to doubling of the production (e.g., 50 Mt wheat grain productions in 1950 to 79 Mt in 1964 and to 95.1 Mt in 1968). Recent estimates show that the production of wheat grain would reach to a record level of 1072 Mt in 2020 (FAO, 2020). Success of the Green Revolution clearly shows the importance of science, government policies, farmers' enthusiasm and assured and remunerative marketing for successful dissemination and adaption of technologies. The role of Indian Council of Agricultural Research (ICAR) in collaborating with leading international organizations and the determination of the Government of India especially the erstwhile Ministry of Food and Agriculture in implementing the programmes were phenomenal.

Green Revolution has negative consequences also like loss in genetic diversity fatigue in soil and depletion in groundwater resources, environmental consequences through indiscriminate use of fertilizers and pesticides, and increase in pest-vulnerability to crops, it was indispensable to counter the food shortage and brining self-sufficiency in food supply in a country with more than a billion population. Although India in contemporary times has gained self-sufficiency in wheat and rice production (FAO, 2018), issue of food security is predominantly existing in the country. A 'second' Green Revolution is awaiting since the agricultural landscape has largely changed over the years, natural resources is getting scarcer and climate change is threating agriculture like never before. This evergreen revolution, as Dr. Swaminathan (2006) puts in, 'involves integration of ecological principles in technology development and dissemination'. This resonates with the call from Hon'ble Prime Minister on India's 75th Independence Day, to move from 'food security' to 'nutritional security' and doubling farmers' income by 2022.

The thrust of the green revolution was majorly on fine cereals namely, rice and wheat. However, the interventions mainly benefited irrigated regions and resourceful farmers, bypassing the larger segment of rainfed farming and resource poor farmers. The current phase of agricultural development aims at greater inclusiveness and equitable transformation, whereby all farmers are able to reap economic benefits. The Doubling farmers' income (DFI) Committee is guided by the fact that more than 85 per cent of the farm holdings are small and marginal and are economically challenged. Currently, the agricultural extension services are available on a pluralistic platform. DFI Committee proposes to transform agriculture extension as an important engine for doubling the farmers' income by promoting new generation farmer collectives, such as Commodity Interest Groups (CIGs), Farmer Producers Organizations (FPOs) and Village Produce Organizations (VPOs) – in the form of Cooperatives, Companies and Societies etc.

Issues and Challenges in Agriculture

1. Food grain surplus

Total production of foodgrains reached a record 295.67 Mt in 2019-20, with maximum share of rice and wheat. Similarly, horticultural production also reached an all-time high of 320.48 Mt. The With an impressive food grain surplus in the country, it needs to critically assess whether the country needs to produce further. Managing this surplus has become a challenge now. There is a potential to shift some areas under rice and wheat to other crops which are high in nutrition and remunerative as well. This will necessitate in bringing essential changes in current agricultural set up, and offer high potential in achieving the doubling farmers' income.

2. Nutritional security

Although being surplus in rice and wheat as chief food grains, the country is facing deficit of pulses and of oilseeds. There has been a big gap in demand and supply of major pulses in India. Short supply led to increase in price, adversely affecting the nutritional security of a large number of citizens. However, government has acted swiftly following an unprecedented crisis in 2015-16 production due to erratic rainfall (pulses in India is predominantly rainfed). It increased the minimum support prices (MSP) and procurement from the farmers at the MSP, leading to a steep jump in pulses production (PNI, 2020). However, pulses are still out of reach to a large section of people, and still non-remunerative to the domestic farmers. The availability of fruits, vegetables, milk, meat and fish has increased due to large scale production gains over the decades, but affordability to a vast majority, including large number of farmers too, remains an important issue even today. India ranks 94 out of 107 countries in Global Hunger Index 2020 with 14% of undernourished population, which is indeed a serious obstruction on the path of development. Bio-fortification of crop plants (nutritionally enhanced food crops with increased bioavailability) is an upcoming, promising, cost-effective and sustainable approach to address wide spread malnutrition problem. In this regard, ICAR has developed and released many biofortified varieties of across cereal, millets, pulses, oilseed, vegetable and fruit crops (Yadava et al. 2017). The biofortified varieties with enriched protein content, essential vitamins and minerals are 1.5 to 3.0 times more nutritious than the traditional varieties.

3. Market instability

The agricultural sector often faces challenges despite provision of high subsidies and minimum support price, and a positive and favourable policy approach by the government. Since the input market is highly regulated, the output prices do not keep pace with the rising levels of inflation and as a result, the farming sector's profitability is reduced to a significant extent. Agricultural marketing infrastructure in India lacks farmer-friendly logistics and results in a huge difference in prices paid to the farmers and the prevailing market prices for agricultural products. Even though farm inputs are subsidized, agricultural labor poses a huge problem with rising wage rates and unavailability of seasonal work force in rural India. All these are making farming a non-profitable, risky and unattractive option. Additionally, the large migration of work force during COVID-19 outbreak indicates an imbalance in labour availability in the country for agriculture.

4. Trade imbalance

Share of India's agricultural exports and imports in the world was 2.27% and 1.90%, respectively in 2017 (WTO Trade statistics). Agricultural exports marginally increased from 9.4% of country's agriculture GDP in 2017-18 to 9.9% in 2018-19; During similar period, agricultural imports declined from 5.7 to 4.9%. However, India still lagging in terms of exports, although being one of top agricultural producers in the world. India ranks 34th in wheat, 14th in vegetables and 23rd in fruits in spite of being 2nd largest producer in wheat and fruits, and 3rd in vegetables. Therefore, a time bound action plan is need of the hour for promotion of the trade.

5. Ensuring rural prosperity

Nearly 58% of India's population lives in villages comprising of 120 million farm families, 85% being small and marginal farmers. Agriculture and allied sectors had 4% growth in GVA in FY 2020-21 (IBEF, 2020). This accounted for 15.2% of country's economic growth, which is much higher compared to 4.7% by the industrial sector in 2019-20. Latest available estimates of average income of an agricultural household were 6,426 with a wide regional disparity (NSSO Situation Assessment Survey, 2013). A high 22.5% of farm household income was below poverty line in the same year. A non-agricultural worker's income was more than 3-times of a farmer, which was ironically the same at in 1983-84.

6. Small and marginal farmers

The farmer producing organizations (FPOs) are the legalized form of farmer-owned (70-80% small/marginal farmers) institutions, registered under company's act and governed by a set of state laws and by-laws. Around 6000 FPOs are currently existing in India, of which 3000 are registered as farmer producing companies

(FPCs) and remaining as cooperatives or societies (NABARD, 2020-21). These companies are mostly registered in four states (Maharashtra, Uttar Pradesh, Tamil Nadu and Madhya Pradesh) with a fairly low representative in dairies and non-farm activities like food processing, weaving etc. The policy paper issued by NABARD lists seven major issues and challenges *viz.*, lack of technical skills and awareness, inadequate professional management, weak financial condition, poor access to credit, lack of risk mitigation mechanism, limited access to market, and inadequate access to infrastructure. Most importantly, farmers are unable to act as manager in understanding resource optimization techniques, identification of the best agri-practices and make their presence felt in organized markets. The paid-up-capital of these companies, except a few, is also bare minimum, which severely restricts them to undertake business activities.

7. Climate change consequences

The concentration of greenhouse gases like CO_2, CH_4, CFCs, N_2O and O_3 increased the ambient air temperature in India to 0.6- 0.8°C between 1850-2010 (Jayaraman and Kamal 2018). This resulted in reducing rainfall, crop production and productivity in many places (IPPC 2001; Mall et al., 2004). The water availability had been also declined during this period. The southern or eastern Maharashtra (Western India), Andhra Pradesh (Southern coast of India), northern Karnataka (South-western India), Odisha (Eastern Coast of India), Telangana (South eastern coast of India) and western Rajasthan are current major drought prone regions of India. The frequency of drought is set to increase between 2020 and 2049 (Vibha, 2017). The climate-induced changes are set to bring additional challenges to the farmer's ability to adopt coping and adaptation measures that are warranted.

8. Rapidly growing urbanization

With growing urbanization, less of the population is directly linked to agricultural production, while every individual remains intrinsically linked to agricultural output as consumers. However, projections indicate that until 2050, more than 50% of country's population will still be rural. Thereafter, if the trends continue, the urban population may start to overtake rural population. However, agriculture and allied activities will remain a focus area of the economy, not only from the perspective of employment to half the population, but also for nutritional and food security concerns of the other 50%. Rapid urbanization will accelerate the demand for higher quantity of nutritional foods, especially fruits, vegetables, milk, meat and eggs, which has to come from shrinking land and water resources. With growing income and living standards, especially of the middle class in India, the demand for animal origin foods (milk, meat and fish) is expected to rise sharply. In 2050, the projected consumption of milk is 209 Mt (Kumar et al., 2014). The poultry market (broilers and eggs) is projected to reach 60 billion USD from current 25 billion USD (RM Report, 2019). This demand can only be met by at least doubling the animal productivity, which is a serious challenge to the livestock sector. India's livestock productivity, however, is 20-60 per cent lower than the global average, posing a great challenge to meet the growing demand.

9. Degradation of natural resources

Since land is a limited and non-renewable resource, its ability to support the agriculture declines with continuous cultivation, even we use the best of technologies and management. Hence, it is imperative to not only preserve but also to enhance its quality to keep pace with the increasing demand of production.

10. Oil seed production

Despite being the fifth largest oilseed crop producing country in the world, India still imports nearly 60% of country's edible oil requirement. Annual edible oil consumption grew by 6-7 lakh tones (per capita increase is 6%) due to enhanced consumption driven by rising urban population, changes in consumption patterns, and failure of domestic production to keep up with the demand. Although India's edible oil imports reduced to 133 lakh tonnes in 2019-20 against 149 lakh tonnes in the previous year, it is still the largest importer of edible oils, putting a huge burden on the country's foreign exchange.

11. Issues on crops with special reference to maize

In maize crops, the most critical factors to realize its market potential are enhancement and diversification of germplasm using modern tools and techniques, development of diverse and productive breeds, fine-tuning of resource conservation techniques, and reduction of cost of cultivation by input use efficiency. More than 75% of maize cultivation in India is under rainfed conditions, severely limiting its productivity. On the other hand, high-yielding cultivars are yet to reach many farmers. The pulse crops need a dual approach of area expansion and productivity enhancement. Increase in area can be achieved through higher cropping intensity like cultivating in post-*kharif* rice fallows. Short-term gain in productivity can come from improved agronomic practices including use of quality seeds; while in the long run, better pulse varieties are the solution. An estimate shows average national productivity of pulses should be more than 1.2 Mg ha^{-1} and an additional area of about 3.5 Mha should be brought under pulses to realize this (Malo and Hore, 2020).

12. Genetic improvement

The Indian Council of Agricultural Research has developed and released more than 3500 and 1600 high-yielding and hybrids of field crops and horticultural crops for different agroecological regions of India. However, development of climate-resilient, biofortified and nutri-cereals crops requires more emphasis. In this direction, next-generation artificial intelligence can be explored to process the large amount of genomics and phenomics data for effective understanding of gene-environment interactions (Harfouche et al., 2019). Crop breeding and genetic improvement can be accelerated through different speed breeding models worked out in different crop plants (Fiyaz et al., 2020). Many crop wild relatives (CWRs) have been lost over decades. Wider gene pools of major crops including CWRs need to be sequenced, and the CWRs and the crop genetic resources (CGRs) can be explored through genomics assisted breeding to break the yield barrier.

13. Loss of produce

Insect pests, pathogens, nematodes and weeds cause 20-40% crop production globally (FAO estimates) with a cost of 220 billion USD. In India, 30-35% yield losses occur due to peat-diseases, with an annual estimate of 36 billion USD (Dhaliwal et al., 2015). India is the lowest in the world in pesticide consumption (0.5 kg ha^{-1} annually), against 5-7 kg ha^{-1} in the UK and 13 kg ha^{-1} in China. Current approaches like nanotechnology, molecular arsenals like RNAi, CRISPR/Cas9 and highly virulent biological controls can aid to better pest management. Introduction of exotic pests and diseases from other countries also pose serious challenges in country's fight against the yield losses. Invasive and transboundary insect pests like fall armyworm, fruit flies, bollworms, whiteflies, locusts cause great economic loss to agriculture in India. Use of artificial intelligence, remote sensing and machine learning especially in inaccessible areas must be successfully implemented. Regulation of biosafety on genetically modified crops against the pests needs revisit. Department of Biotechnology, Government of India have prepared a new draft on Genome Edited Organisms: Regulatory Framework and Guidelines for Risk Assessment 2020 for regulation of GMOs, and invited comments. Forecasting needs to be expanded to new crops and diseases. Similarly weed management requires massive testing of new approaches like use of infrared thermal energy, microwaves, lasers, hot water and air, steam, pneumatic control robotic, flame weeding etc. Their usage in the field of small and marginal farm needs further modifications. Further, post-harvest loss occurs during supply chain due to proper storage facility, mostly for fruits, vegetables and meat. Public-private partnership for cold chain systems using advance technologies can be the solution (Ndraha et al., 2020). Maintaining quality of the food is another challenge, as the usage of chemical preservatives is not preferred by consumers, and this has public health issues too. Many new technologies, but problems associated with proper drying, storage and delivery still exist. In future, new cooling systems temperature control, sustainable packaging systems, nanotechnology, IT etc are also make the losses minimum.

14. Under-productive livestock

Introduction of improved breeds and their crossing with indigenous stocks enhanced the production of milk, meat and other associated produce. However these cross breeds are susceptible compared to their native counterparts. Some of the deadly diseases such as black quarter, foot and mouth, *Hemorrhagic septicemia* anthrax, fascioliasis, brucellosis, tuberculosis, rabies etc are still affecting these animals, and causing mortality and less productive. It is happening even after vaccination programme in place. In most of the cases, animal feeding depends on open grazing; however, grazing land is decreasing. In India, maintenance of animals is mostly with conventional practices, which demand labour and is less remunerative. Fodder availability during summer months is difficult. It is essential to have drought and temperature tolerant fodder crops in India and utilization of arid and wasteland in different parts of the country.

Sustainable Development Goals and Indian Agriculture

The Sustainable Development Goals (SDGs) were adopted in September 2015 as a part of the resolution, 'Transforming our world: the 2030 Agenda for Sustainable Development'. Jaipur declaration, India is committed to achieve the 17 SDGs and the 169 associated targets, which comprehensively cover social, economic and environmental dimensions of development and focus on ending poverty in all its forms and dimensions. At the Central Government level, NITI Aayog has been assigned the role of overseeing the implementation of SDGs in the country.

Priorities in Agriculture

The priorities need to be in accordance and at pace with the global targets. The Government of India has effectively fast-tracked the initiatives to double agricultural productivity and incomes of small-scale farm producers to achieve in doubling farmers income by 2022. This is also the target-year set by the Development Finance Institution, well before the target-year set in SDGs. The Development Monitoring and Evaluation Office (DMEO) of NITI Aayog has carried out a draft mapping of the SDGs against nodal ministries, with information on related interventions and targets.

1. Protecting and empowering farmers

The Farmers' Produce Trade and Commerce (Promotion and Facilitation) Bill 2020 (now an Act) will allow intra- and inter-state trade of farm produce beyond the physical premises of markets (APMC), giving freedom of choice to farmers and traders to sell and purchase agricultural produce. Under extraordinary circumstances like war, famine etc., Government will regulate the supply chain of food items under the provision of Essential Commodities (Amendment) Bill 2020. The Farmers (Empowerment and Protection) Agreement of Price Assurance and Farm Services Bill, 2020 will empower farmers to engage with agri-business firms, processors, wholesalers, retailers, exporters, and shield farmers from market price fluctuations. These bills will raise price realisation by farmers, supply food to the consumer at competitive prices and to compete with import and push export, thereby advancing the idea of 'one nation one market'. Promoting contract farming by encouraging groups of farmers on one hand and processors, aggregators or large retailers on the other, to enter into agreement for crop production and purchase is another bold step.

2. Biosafety

Another priority should be the regulation on sale of spurious, substandard and misbranded pesticides in the country. India needs to make embargo or restriction within the country for emerging new pests and introduced pests. Advance technologies; remote sensing, Artificial intelligence, biopesticides are also emerging field, which will make further strengthen in the field of biotic stress management.

3. Public-private partnership

A public-private partnership (PPP) is a contractual agreement between a public agency (federal, state or local) and a private sector entity. Although the PPP is not a new concept, the level is very low. Public sector has greater focus on the benefit sharing while private sectors focus more on economic calculations. Certain areas of PPP at present are field trials, pesticide testing, germplasm evaluation, biosafety studies, animal feeding studies, ecological studies, etc (https://icar.org.in/files/Public-Private-Partnership.pdf). As any partnership requires patience and trust to succeed, PPP should move from 'partnership for benefit' mode to 'partnership for prosperity' mode.

4. Promote agroforestry

Agroforestry is an easy alternative to meeting the target of 33% forest cover from the present level of 25% of total geographical area of the country (ISFR, 2019). This will comply with the commitment of reducing GHGs, increasing share of non-fossil based power generation and additional C sinks through additional forest and tree cover by 203. Agroforestry can also play crucial role in achieving the targets of SDG1 (on ending poverty), SD2 (on ending hunger), SDG 6 (on protecting water-related ecosystems), SDG 7 (on access to sustainable energy to all), SDG 13(on combating climate change) and SDG 15 (on protect terrestrial ecosystems).

5. Improvement in livestock productivity

For improved productivity of the livestock, the use of bio-technological, molecular and nano-technology approaches in breeding, feeding, reproduction and health care need to be considered. With increasing demand for animal products and by-products, the production and productivity of livestock sector in the country need to be enhanced, where livestock is majorly with resource-poor farmers. Livestock also serve as financial instruments, by providing households with an alternative for cash savings, which can be sold and transformed into cash as and when needed. For some poorer households, livestock provide a means of income diversification to help deal with times of financial difficulty

Other issues and priorities

1. Increase the production of agricultural commodities through developing new sustainable technologies and infrastructure facilities in rainfed areas.
2. Promotion of water-saving micro-irrigation techniques and improving water-use efficiency through coordinating the irrigation and agriculture departments, private sector and farmers' water-user associations.
3. New innovations using digital platform for strengthening of the existing agricultural extension in India. New policies need to be framed on revamping the extension system towards translational research, by following farmer participatory approaches focussing the progressive farmers including farm youths for transfer of technologies in faster way.
4. Identification and exploitation of agroecological zones for product diversification that can help farming communities to increase their incomes.
5. Harnessing the potential of the genetically engineered crops in enhancing the income of the small and marginal farmers with proper regulatory requirements, and removing excess unnecessary restrictions.
6. Policy decision for increasing the capital investment in agriculture by both public and private institutions in the potential areas with clear objective of the realistic growth of agriculture to meet the emerging needs of the people.

Policy interventions with focus on doubling farmers' income

1. Linking Govt. schemes with modern technology with science base: Several initiatives like

- Skill India, Stand up India
- Swachh Bharat Abhiyan
- Start-ups/Incubators
- Doubling of farmers income by 2022 etc.
- Paramparagat Krishi Vikas Yojana (PKVY)
- Rastriya Krishi Vikas Yojna (RKVY)
- National Nutrition Mission
- National Mission on Organic Farming

Have enough potential to double the farmers income and raise the standard of living and bring the socio-economic change in the society particularly in transforming rural economy.

2. Protected cultivation, vertical farming and secondary agriculture: Protected cultivation and vertical farming have enormous potentials in India; however poly house or protected structures should be as per prevailing agroclimatic conditions to make it profitable. Secondary agriculture like processing, packaging, branding and value addition technologies need to be promoted through trainings to make these sustainable and profitable enterprises.
3. Quality assurance in agricultural education: With the establishment of National Agricultural Education Accreditation Board, it has now been possible to maintain the quality of agricultural organizations/institutes of agricultural education.
4. Formation of different village level centre/ association /group: There are about 85000 self-help groups (SHGs) involved in various activities. Women SHGs in India have risen to the extraordinary challenge of COVID-19 pandemic. They are meeting shortfalls in masks, sanitizers and protective equipment, running community kitchens, fighting misinformation and even providing banking and financial solutions to far-flung communities. A special mention is Koel Aajeevika Apparel Park, Palamu. So far, more than 19 million masks have been produced by some 20,000 SHGs across 27 Indian states.
5. Digital Science to educate farmers through KVKs: India has witnessed a rapid adoption of digital technologies in last one and a half decade. This growth may be attributed to internet penetration, availability of affordable data services, proliferation of feature phones as well as 'budget smartphones', and the rise of innovative digital services. India is the second largest internet user base with penetration of almost 50% today and its digital divide is narrowing fast. Digital technology holds tremendous potential to transform the Indian agricultural economy and impact the lives of Indian farmers. Almost 80% of 600 plus Agritech Start-ups in India are working on digital models which will play a pivotal role in transforming Indian agriculture. These models offer huge potential for global markets wherever there is high density of small holder farmers. Digitization of Agriculture will bring much needed efficiency and transparency to the sector for the benefit of farmers and consumers. Availability of accurate and almost real time data will enable multiple applications to solve for farmers' challenges. The most commonly used low-cost ICT individual and social media tools having huge potential to deliver relevant and actionable information to the students and farmers in time are Mobile Smartphone, Mobile voice calling, Video calling, Messaging, Facebook, Twitter, WhatsApp, Telegram e-mail

are. It has changed the landscape of global development. It is just not shaping communication but also the future, by influencing opinions and mobilising masses during this unrest situation. Globally, the market size of AI in agriculture stood at USD 852.2 million in 2019, and it is expected to reach USD 8379.5 million by 2030, with a CAGR growth of 24.8%. Precision agriculture and farm management is the largest category within AI and related technology segment, wherein predictive analysis is the fastest growing sub- category. Crop health monitoring will be the highest revenue-generating application globally. The Indian government is serious about leveraging new-era digital technologies for agriculture. In May 2019, NITI Aayog, the government's main think-tank, signed an agreement with one of the big software firms to develop a model for crop yield predictions using AI so that farmers can be provided real-time advisories. The number of people who will be using the Internet by 2020 is expected to be 700 million.

Finally, the issues and challenges todays Indian agriculture faces are loss of surplus produce, nutritional security, market instability, trade imbalance, ensuring rural prosperity, climate change consequences, rapidly growing urbanization degradation of natural resources, oil seeds import and under-productive livestock. All these challenges need to be addressed involving all stakeholders (public and private sector) in a time bound manner through holistic approach as our Government committed to zero poverty and hunger, combating climate change carbon neutrality, land degradation neutrality and quality education by 2030 as these are the major agenda points of sustainable development goal. Therefore immediate need is to go in for paradigm shift on adopting modern agriculture practices and new science *viz*; Digital and data science, Disaster management Climate smart agriculture, Waste to wealth, Post-harvest management, Hi-Tech Agriculture etc. to put forth Indian agriculture for food sufficiency along with nutritional security and income security of the small and marginal farms and to ensure ever green sustainable agriculture.

References

Dhaliwal, G.S., Jindal, V., and Mohindru, B. 2015. Crop losses due to insect pests: Global and Indian scenario. Indian J Entomology 77(2), 165-168.

FAO, 2020. GIEWS - Global Information and Early Warning System. Available at: http://www.fao.org/giews/countrybrief/country.jsp?code=IND. PNI, 2020. Press Network of India. Available at: https://www.pninews.com/msp-the-driving-force-of-pulses-production/

FAO. 1999. Agricultural Biodiversity, Multifunctional Character of Agriculture and Land Conference, Background Paper 1. Maastricht, Netherlands.

FAO. 2004. The State of Food and Agriculture 2003– 2004 (Food and Agriculture Organization of the United Nations, Rome).

FAO. 2013. Statistical Yearbook of the Food and Agricultural Organization, FAOSTAT. Agricultural Development Economics Division. p. 1-77.

FAO. 2017. The Future of Food and Agriculture – Trends and Challenges. Food and Agriculture Organization of the United Nations, Rome.

FAO. 2018. http://www.fao.org/3/ca6030en/ca6030en.pdf

Fiyaz A R., Ajay B.C., Ramya K.T., Aravind Kumar J., Sundaram R.M., Subba Rao L.V. 2020 Speed Breeding: Methods and Applications. In: Gosal S., Wani S. (eds) Accelerated Plant Breeding, Volume 1. Springer, Cham. https://doi.org/10.1007/978-3-030-41866-3_2

Government of Telangana. 2015. Task Force Report: Agricultural Challenges and the Way Forward. Submitted to NITI Aayog by the Government of Telangana, Agriculture and Cooperation Department.

Harfouche, A.L., Jacobson, D.A., Kainer, D., Romero, J.C., Harfouche, A.H., Scarascia Mugnozza, G., Moshelion, M., Tuskan, G.A., Keurentjes, J.J.B., Altman, A. 2019. Accelerating Climate Resilient Plant Breeding by Applying Next-Generation Artificial Intelligence. Trends Biotechnol. 37(11):1217-1235.

IBEF, 2020. Indian Brand Equity Foundation (An initiative of the Ministry of Commerce and Industry, Govt. of India). Available at: https://www.ibef.org/industry/agriculture-india.aspx

ICAR. 2011. Vision 2030. Indian Council of Agricultural Research, KrishiBhawan, New Delhi, 22 pp.

ICAR. 2015. Vision 2050. Indian Council of Agricultural Research, KrishiBhawan, New Delhi, 34 pp.

ICAR-CIRG. 2013. Vision 2050. Central Institute for Research on Goats, Makhdoom, Farah, Mathura (UP) India, 38 pp.

IPCC. 2001. (Intergovernmental Panel for Climate Change), Climate Change - The Scientific Basis, Contribution of Working Group I to the Third Assessment Report of the Intergovernmental Panel on Climate Change. In: Houghton JT, Ding Y, et al. (Eds.), Cambridge University Press, Cambridge, UK, p. 881.

ISFR, 2019. Indian State of Forest Report. Forest Survey of India, Ministry of Environment Forest and Climate Change, Dehradun.

Jayaraman, T., Kamal, M. 2018. Climate Change and Agriculture: Current and Future Trends, and Implications for India. Centre for Climate Change and Sustainability Studies, School of Habitat Studies, Tata Institute of Social Sciences, Mumbai, India, pp. 1-19.

Kumar, A., Joshi, P.K., Kumar, P., Parappurathu, S. 2014. Trends in the consumption of milk and milk products in India: Implications for self-sufficiency in milk production. Food Security 6(5): 719-726.

Mall, R.K., Lal, M., Bhatia, V.S., Rathore, L.S., Singh, R. 2004. Mitigating climate change impact on Soybean productivity in India: A simulation study. Agricultural and Forest Meteorology 121(1-2): 113-125.

Malo, M., Hore, J. 2020. Pulse production in India: major constraints and way forward. In: Fedorov, S. ed. Research trends in multidisciplinary research and development. Vol. 5. Weser Books, Germany, pp. 35-63.

Ministry of Agriculture and Farmers' Welfare (2015–16) State of Indian Agriculture 2015–16. Directorate of Economics and Statistics (Pocket), Department of Agriculture, Cooperation and Farmers' Welfare, Ministry of Agriculture and Farmers' Welfare, Government of India, New Delhi.

Ministry of Agriculture and Farmers' Welfare (2019) Agricultural Statistics at a Glance 2019. Directorate of Economics and Statistics,

Ministry of Agriculture and Farmers' Welfare (2019) Annual Report 2018-19. Department of Agriculture, Cooperation & Farmers Welfare, Ministry of Agriculture & Farmers' Welfare, Government of India, New Delhi, pp. 1–220.

Ministry of Environment Forest and Climate Change (2015) Development Goals in India: A Study of Financial Requirements and Gaps. Ministry of Environment, Forest and Climate Change, Government of India.

Ministry of Finance (2015–16) Economic Survey 2015–16. Ministry of Finance, Government of India, New Delhi.

NABARD, 2020-21. Paper on PFOs, Status and Issues. National Paper-PLP 2020-21.
Ndraha, N., Vlajic, J., Chang, C.C., Hsiao, H.I. 2020. Challenges with Food Waste Management in the Food Cold Chains. In: M. R. Kosseva & C. Webb (ed.) Food Industry Wastes Assessment and Recuperation of Commodities. Academic Press: Chennai, India. pp.467-483
R&M Report, 2019. Indian Poultry Market: Industry Trends, Share, Size, Growth, Opportunity and Forecast 2019-2024. Research and Market IMARC Group, 96p.
Swaminathan, M. S. 2006. An evergreen revolution. Crop Science, 46(5), 2293-2303.
UN (2013). World population prospects: The 2012 revision. Key findings and Advanced Tables. ESA/P/WP.227. United Nations, Department of Economic and Social Affairs, Population Division.
Vibha, D. 2017. Water and Agriculture in India: Background paper for the South Asia expert panel during the Global Forum for Food and Agriculture. Federal Ministry of Food and Agriculture. pp. 1-28.
Yadava, D.K., Choudhury, P.R., Hossain, F., Kumar, D. 2017. Biofortified Varieties: Sustainable Way to Alleviate Malnutrition. Indian Council of Agricultural Research, New Delhi.

2

Innovations in Agriculture An Overview

Jagdish Kumar[1], P.N. Sivalingam[1], Mallikarjuna, J[1], S.K. Jain[1] Sridhar, J.[1], K. Kiran Kumar[2], Sujay Anand[3] and P.K. Ghosh[1]

[1]*ICAR-National Institute of Biotic Stress Management, Raipur, Chhattisgarh*
[2]*ICAR-Central Citrus Research Institute, Nagpur, Maharashtra*
[3]*ICAR-Indian Institute of Pulses Research, Kanpur, Uttar Pradesh*

Introduction

India is one of the agriculturally important countries in the world and has strong net-work of agriculture education, research and extension. As per the historical record India is committed to thrive and excel independently. During 1965–66, India imported food grains from the US under PL-480 scheme, and the situation was "ship-to-mouth". This condition was also exacerbated with the growing population which made India to increase its food production at most priority with commitment. With the introduction of high yielding semi dwarf varieties of rice from IRRI, The Philippines and wheat from Mexico which combined with right policies of government to provide essential inputs and market facilities, credit etc. revolutionized the Indian agriculture. In addition to these innovative ideas and policies the food production reached to the level of self-sufficiency until today. For instance the food production during 1950-51 was 54 million tonnes and reached to 285 million tonnes during 2018-19. After the fourth five year plan, India made its stand of self-sufficiency even affected by any year of worst drought. Later, the focus had been given on agricultural research and education in the country for sustaining self-sufficiency in agriculture sector at all times of change of conditions within the country and also globally. Although more than 65 per cent of rural population is depend on agriculture sector, the average monthly income is about Rs. 6500. It reveals that under employment and large population in India depend on agriculture which is unsustainable (Rena, 2003). Therefore, it is needed to books the rural economy, create new infrastructure and other innovations in this sector. India has been emphasizing for innovative techniques. The main advantage for India is the richest natural resources and need only planning and execution on innovation front.

Self-reliant living means learning to grow your own, make your own, sell your own and bake your own in the rural homestead farming, urban homestead or mini-farms. Self-reliance is the way forward. There is no better time to do this, especially with the presence of climate change, food safety, and the gloom of economic uncertainty. In the context to make India *atmanirbhar*, for achieving self-reliance and self-sustainability in agriculture is easier and more cost-efficient compared with other sectors of the economy. Presently, Gross Value Add of agriculture sector is about 18 trillion rupees with the growth rate of approximately 2 per cent in year to year basis. a key economic driver, the sector needs to adapt to the challenges that it is facing today. India has three greatest challenges ahead of the 21st century and they are (i) sustaining food and nutritional security (ii) adaptation and mitigation to climate change and (iii) sustainable use of critical resources like water, land and energy. By the year 2050, annual food grain production would need to grow to 333 million tonnes. Even then also the contribution of key food grains in acreage terms in India is 15 per cent and the production contribution is mere 8.7 per cent. It clearly shows that the use of innovative technology is necessary to meet the forecasted demand in a sustainable manner and move Indian agriculture along the growth path and make India *atmanirbhar*. India needs to rapidly move towards 'innovation led agricultural growth'. This has to be achieved with speed scale and sustainability. On the basis of current situation, it is foreseen that India have to and being adopted advanced methods for agriculture. India's achievements in agriculture sector are unevenly distributed and have remained lower than the potential. Agriculture is not to be compared with achievements made in space, IT telecom, services, automobiles, medical science, transport etc though we achieved big compared to the situation in 1960s.

The decade of 2010-20 was declared as the Indian Decade of Innovation. We have benefited from our established 'Indian Agriculture Research System'. However we need to understand that any National Agricultural Research System (NARS) is activity based. Agricultural Knowledge and Information Systems (AKIS) are output based. National Agricultural Innovation Systems (NAIS), however are outcome based. We are being benefitted in this system available in India. Technological and institutional innovations are needed throughout the production marketing, policy research and enterprise domains. First and the most powerful way of getting 'more from less' is by using the power of new technology, such as information and communication technology, nanotechnology, space technology modern biotechnology, etc. However, a robust policy level innovation is a must to achieve this.

Several innovations in agriculture made India self-reliant by

- **Sustainable use of natural resource base**: Balanced nutrient use, INM residue management, efficient crops and water management, enhancing NUE, WUE, WP etc.

- **Crop diversification**: Diversification of rice, wheat, maize and cotton based system to reduce farmers' risks and enhance income and employment. Of late, enabling technology and policy support helped pulse production to reach near self-sufficiency. On the contrary, oilseed sector is still lagging behind and the current import is 67% of the requirement costing >70,000 crores INR annually. To reduce it to about 15% by 2022, area expansion (4.5mha), yield enhancement (0.33t/ha) and yield gap reduction need to be pursued. Rice fallows and half of the upland rice area measuring around 12-14 mha should be used for oilseeds and pulses.
- **Establishment of zone-specific seed hubs:** Seed is the basic requirement for a secure and sustainable farming. Non-availability or poor quality seed is the basic constraints to most farmers which affect their farming every year. Good quality seedling materials (propagules) for fruits and vegetables are also not available at the time of planting season. The start-up is also possible for the rural youths and farmers. Thus, there is an urgent need to strengthen existing seed production programme and extend it to field and fodder crops, livestock and even fisheries for self-reliant. This is an integral part of the Self-reliant farming. Interlinking of hubs can be established in a particular region in the country (Singh et al., 2019).
- **Development of multiple stress tolerant crops and animal variety/ breeds**: Flood and droughts occurs in same place in areas in a single cropping season. Again multiple diseases, insects cause huge loss to agriculture. CR Dhan 801 & 802 are the example of multiple stress tolerance which can tolerate both flood and drought to a great extent.
- **Biofortified high yielding crop varieties**: To overcome protein energy malnutrition and also Zn and Fe deficiency, there is an urgent need to develop and promote biofortified crops.
- **Oilseeds, pulses and vegetables in cereal based system**: Oflate, enabling technology and policy support helped pulse production to reach near self-sufficiency. On the contrary, oilseed sector is still lagging behind and the current import is 67% of the requirement costing >70,000 crores INR annually. To reduce it to about 15% by 2022, area expansion (4.5mha), yield enhancement (0.33t/ha) and yield gap reduction will be pursued. Rice fallows and half of the upland rice area measuring around 12-14 mha should be used for oilseeds and pulses
- **Target based programme on climate resilient nutri- cereals**: Millets are much more climate resilient (requires less water, thrives under harsh environment) and enriched in nutritional properties and minerals than the cereals. To overcome protein energy malnutrition and also Zn and Fe deficiency, there is an urgent need to develop and promote biofortified crops

- **Promotion of self-reliant and climate resilient Integrated farming system for small holders:** About 85% farmers in India are small and marginal. Also climatic anomalies like drought, flood, frosts, and cyclones are a recurrent phenomenon in India. Thus, IFS is the de-risking mechanisms not only enhance productivity, nutrition and employment but also reduce farmers risks due to climatic aberrations and provide assured income under all circumstances. Therefore, **Promotion of specialized farming for high income is the need of the hour.** Poultry farming, Commercial fish farming, ornamental fish, Dairying etc, should be sub-component of IFS. Crop varieties released through AICRP programme are evolved under different region and time likely to adapt variation in the climatic factors. Each region of the country should come out with some IFS model for replication in farmers' fields.
- **Promoting secondary and specialty agriculture**: Scented/aromatic products/products with GI Tags, Niche items etc. must be promoted for higher income.
- **Promoting alternative source of income**: Each farmer must have at least one alternative source of income either through a shop, processing unit, custom hiring, or any other service sectors.
- **Promote organic farming:** Productivity of all crops in India is much less than the world average productivity. Organic system of cultivation increases productivity in long run. It also improves the quality of produce and economic returns apart from ensuring sustainability of agriculture. Thus, organic system of cultivation needs to be encouraged.
- **Group farming approach:** can be a viable option for marginal (<1 ha) and small land holders (1-2 ha). These categories of land holders share about 86% of total 145.73 million operational land holders with an average holding size of 1.08 ha, and about 40% workers of total population in India. Pooling of cultivable land, invested funds and labour force would be required. This approach could provide small and marginal farmers, distressed farmers a for developing multi-enterprise-based farming and cluster-based promotion of agro-commodities higher economic returns, for example group farming outperformed individual farming in Kerala, south Goa, Odisha, and in many places. This may be promoted to other states as well. This will empower farm women also.
- **Integrated aquaculture farming:** This would be effective for resource poor farmers as well as unemployed youths and migrants to produce additional fishes from marine and inland aquaculture. Shrimp polyculture with brackish water fish, and crab-fish polyculture in coastal belt of India have been found to be very lucrative and promising. This will open up huge export earnings in aquaculture products.

- **Promotion of high-tech automated precision farming:** Target educated youths with IT knowledge. Moderately skilled agricultural workers with access to smart apps using smartphones or tablets also benefitted. Precision agriculture could be achieved with the use of advanced GIS/GPS and sensors can guide planting/irrigation, monitor yields, fine tune inputs and by improving yields as well as reduce the use of water and fertilizers.
- **Branding of local farm products** to be sold globally, the branding helping to tide over price spirals and making farmers quality conscious
- **Technology-enabled supply chain** through the use of RFID, advanced GIS/GPS, tracing and traceability systems
- **Wasteland utilization:** States like Gujarat have more than 25 lakh hectares of wasteland. This is non cultivable. It is highly saline and underground water is also saline. Such land has remained uncultivated for 1000 years. But modern technology now provides solution to this problem. Salinity resistant crops like baval, date palm, Jatropha etc can be developed. There is technology available for harvesting every drop of water that is received by rainfall. Micro irrigation is for efficient and economical use of water resources. The challenge is to develop these land resources. The constraint is its cost. One-hectare land would need approximately Rs. 2.0 lakhs. This is not within the means of a farmer. This is an area where participation by industry is essential. Private and public relationship can go a long way in utilization such land resources and provide more employment and income.
- **Achieving land degradation neutrality**: India has about 120.8 m ha of degraded land, which is not only contributing to low productivity and poverty but also to greenhouse gas emission and huge economic loss. UN General Assembly 2015 has call for achicving a LDN world by 2030 and India has committed to rehabilitated 26 m ha. This needs a special attention of all sectors
- **Agroforestry and carbon farming:** Integrated farming system incorporating agroforestry may help meet the future requirements of agriculture diversification. Inclusion of small farmers in Green India Mission by adopting agro-forestry in CDM projects through community and grid approach will be helpful to get extra income through carbon credits. At national level a simple carbon credit system should be developed especially for small and marginal farmers so that financial benefits can reach to them at the time of payments for irrigation, electricity, purchase of farm equipment and machinery etc. Tree based integrated farming system (agroforestry) should be promoted (a minimum 5-10% land of small and marginal farmer and 20-25% land of large farmers) to ameliorate the effect of climate change on agriculture. Accelerated Fodder Development Plan (AFDP) and proposed Fodder Mission need to include agroforestry for silvipasture and tree fodder production in all areas and especially in arid and semi-arid areas.

- Moderately skilled agricultural workers with access to smart apps using smartphones or tablets also benefitted. **Real time market information** by using mobile communications, voice based call centers and expert systems for real time price discovery, weather information and cultivation trends.
- Technology enabled **crop insurance**, where use of **real time data from weather stations** could be used to predict the rainfall.
- The `**collective intelligence** of the innovators must be used in enhancing the productivity of the workers in agriculture, while reducing or removing the drudgery in their fields.
- **Developing water harvesting capacity and irrigation infrastructure**: RWH and its efficient use must be promoted to reduce pressure on ground water and increase irrigated area.
- **Promotion of specialized farming for high income**: Poultry farming, Commercial fish farming, ornamental fish, Dairying, and so on. Each Region of the country should come out with some models for replication in farmers' fields.
- **Location specific farm mechanization strategy to attract and retain youths in agriculture**: Special emphasis on AI technology like use of drone for pest management.
- **Strengthening food processing micro-industries at the village or sub-urban areas:** Target in our country is to process about 25% of produce which is presently less than 3%. In USA as high as 65% farm produce are processed. Promotion of this particular enterprise will provide employment to farmers and rural youths. Scientific back-up is not lacking. Food processing is important in India because the country is leading in production of pulses, jute and milk, ranked second in production of vegetables and fruits, third in cereals in the world. There are about 39,748 food-processing industries under organized sector in India. Processing of several agri- and animal- based products can be developed at the rural level, and also for huge export earnings. The government schemes would be effectively utilized to initiate micro level industries by the farmers and rural migrant workers. The industrial extension team should give it priority. It can develop young entrepreneurs form farmers. These projects require low investment. With proper bank credit linkages, the rural scene can change rapidly. Along with this, there is a need to pay attention to the growing demand of services in rural areas and more rural youth to seize it. Similarly, strengthening quarantine system needs special attention to protect genetic materials and now allow spurious materials inside.
- **Seed export and regulatory system:** India can emerge as an important seed producer and a large exporter of seeds to many developing countries in

South and South-East Asia as well as Africa. The country can produce very competitively priced seeds for hybrid rice, hybrid corn, hybrid *Bt* HT cotton and several vegetables including tomato, potato and okra, provided we set our regulatory system right. To stop illegal spread of some GM crops our regulatory system should be very effective and simple. The issues related to the bottlenecks holding India back from becoming the seed capital of the developing world should be solved at government level. As livestock is also equally important component, quality seed production for fodders and forage crops need to taken with R& D.

- **Strengthening feed and fodder sector**: There is a chronic shortage (about 67%) of green as well as dry fodder in organized sector.
- **Strong survey and surveillance** for trans-boundary pests and diseases using modern tools preferably, remote sensing and drones etc.
- **Strengthening quarantine system** to protect genetic materials in the country and to prevent the introduction of exotic insect pests, diseases and nematodes which may cause damage to crops.
- **Robust extension system with public-private partnership**
- Strengthening village level group organizations viz. Farmer Producer Organizations (FPOs) and Self-Help Groups (SHGs) by providing more financial support and liberalised governance.
- **Micro-enterprise development in agriculture and allied sector** is a very good option. It has ample potential to provide employment and generating a stable income to farmers. This can be achieved by formalizing FPOs and SHGs in India. Farmers may form cluster-based cooperatives for crops and commodities viz. rice, jute, potato, mango, bamboo, chilli etc., which will create more employment, boost production, marketing and better livelihood to returnees and the famers.
- **Skill development and vocational training to farmers and youths** will accelerate agri-based business by the rural people in India. Youths should be encouraged to join the '**Make in India**' programme of the Government of India. For skill development, technical support should be obtained from National Skill Development Council and Agricultural Skills Council of India, National Institute of Food Technology, Entrepreneurship and Management and the Indian Institute of Food Processing Technology. As penetration of the mobile services and other digital infrastructure has increased, India has significant advantages, thus, mobile-based extension approaches (**digital advisory services**) to be scaled-up to connect with small and marginal farmers to provide real time solutions for their agri-needs, weather advisories, and also information about the input providers such as soil-testing labs, seeds, agro-chemicals and fertilizer suppliers etc.

- **Strengthening marketing system**: *e*-NAM, FPOs, MSPs for more crops and products, etc.
- **Efficient utilization of government schemes:** By developing efficient mechanism to transfer funds to farmers under the existing Government schemes, and new allocation of funds for agriculture and allied sectors towards self-reliant mission India announced by the Government in the pandemic era. The Government's financial package should be utilized for promotion of herbal plants cultivation, integrated beekeeping and value addition. The schemes should be implemented effectively such as *'Agri Infrastructure Fund'*, *emergency working capital, Pradhan Mantri Matsya Sampada Yojana, Animal Husbandry infrastructure development, Employment push through the Compensatory Afforestation Management and Planning*
- **Stronger linkages with international organizations:** The suggestions and policy recommendations of international organizations (FAO, WHO, IFPRI IMF, World Bank, UNDP, CGIAR institutes, IFAD, WFP etc.) should be considered for implementation of different schemes at the country level. This will strengthen Indian agriculture and to revive battered economy,and to make India self-reliant.
- **On-farm processing of agricultural products** using solar energy applications, compost preparation using agricultural wastes and residues, biopesticide from locally available plant resources, android based solutions for soil, water and nutrient management, sensor based grading and packaging of produces, drudgery reducing tools and implements for marginal and small farmers, cold chain transportation of agricultural produces, replications of cooperative model of AMUL dairy in village level.
- **Innovations in post-harvest technology:** Need innovation to mitigate food loss and wastage with timely intervention. The Government should encourage public-private partnership programmes, making traceability in post-harvest systems, utilizing novel technologies in reducing post-harvest losses.
- **Agricultural extension through digital advisory services:** Strengthening of the extension system in the country is urgently required. Adoption of the best practices is critical and digital communication has become necessary to bridge the gap in technology dissemination. As penetration of the mobile services and other digital infrastructure has increased, India has significant advantages, thus, mobile-based extension approaches to be scaled-up to connect with small and marginal farmers to provide solutions for their agri-needs, weather advisories, and also information about the input providers such as soil-testing labs, seeds, agro-chemicals and fertilizer suppliers etc.

- **Solar and wind energy utilization in farming:** This field need priority and huge initial investment for implementations. Government policy should be strengthened innovations for small and marginal farm level.
- **Bio-tech crops:** Bio-tech crops are part of new technology innovations in agriculture. They reduce cost and /or add value, and /or increase productivity. USA, China, Brazil and a number of other countries have gone ahead and have availed of their benefits. India should not lag behind. One reason is that they provide better productivity and reduction in cost and therefore increase incomes of farmers. This is also illustrated by adoption of *Bt* cotton and increase in cotton production and productivity. At present, the mechanism of introducing such *Bt* crop is a long drawn out process. It is important to come out boldly and move with great speed. If this is not done, one major danger is that, other countries may start exporting agriculture commodities of India at lower prices than at which we produce. This can finish small and marginal farmers. This threat is a genuine one and policy makers should not take it lightly.
- **Agro processing and service sector:** Another area which needs attention is the expansion of agro processing. Only 2-3% of fruits and vegetable produce are being processed now. Agro processing adds life and value to agriculture produce. Massive efforts are needed here. The industrial extension team should give it priority. It can develop young entrepreneurs form farmers. These projects require low investment. With proper bank credit linkages, the rural scene can change rapidly. Along with this, there is a need to pay attention to the growing demand of services, in rural areas and more rural youth to seize it.
- **Market infrastructure:** Israel has shown a way to the world that even on a desert land, with a low rainfall, a prosperous agriculture and agro processing industry can be built up. In fact, geographical and agro-climatic situation in Gujarat, Rajasthan and Israel are the similar. States like Gujarat can provide 100% requirement of country in crops like oil seeds and pulses, where today major imports are taking place. Israel technique, as it is known, is high-tech. It is based on soil and moisture conservation and continuous soil analyses to apply required nutrients. But crucial to success is the market strategy. Daily planeloads of products are sent to important capitals of Europe. The government provided great support in developing common brand for export of agricultural products and helped market efforts by setting up link centers. This is not difficult to replicate. It is necessary to develop an independent market identity for export of horticulture produce. This has been done for milk products under AMUL brand – but it did not extend to other agriculture produce. At present, there is under-cutting between the producers themselves, which is taken advantage of by importers.

- **Agriculture research institutes:** The scientists of premier institutes, the Indian Council of Agriculture Research (ICAR) must come out from their ivory tower. They must move to the village to understand real problems, understand why they the scientists gave failed. ICAR has been a mute spectator when *Bt* cotton seeds were sold at an exorbitant price. It did not do anything to develop *Bt* variety on its own initially. ICAR needs to be reoriented and made accountable to farmers.
- **Contract farming:** Contract farming is actually a contract to purchase agri produce at a predetermined price before the season commences. it is not a contract for taking over the farm. Farmers are to be provided a supply of quality raw material and a better price. The processor in turn stands to gain as he has a steady supply for processing ensured. Gujarat policy for contract farming provides for (i) tripartite agreement between the farmer, contract sponsor and APMC; (ii) A facilitator role for APMC; and (iii) A provision in the agreement to encompass all the details / specifications of contract.
- **Government interventions and support:** Indian agriculture needs diversification, innovation and adoption of practices for its sustainability. Indian farmers are also facing problems of high indebtedness due to low income, which is not enough to repay loans and feed family or live honourably. The high indebtedness need to be reduced by increasing the production and productivity of all crops and by creating opportunities of additional income from allied sector activities through better integration, and bringing out change in agricultural marketing practices and laws. It may be necessary to assure farmers with farm income and help them in their individual crisis and individual crop failures apart from supportive general crop insurance. India being a country of small farmers, it is necessary that they get the benefits of trade within the production and productivity of their small land holdings.
- **Fill gaps in production and productivity in eastern India:** This has been noticed by the GoI and various programmes at national level are being implemented since 2000 to bridge the gap. Priority need to manage the natural calamities in this region.
- **Promote horticulture on small lands.** Growing of horticultural crops such as leafy vegetables, vegetable crops and flower crops will boost the earnings of farmers from their land holdings.
- **Assure income to small farmers:** A farm income assurance scheme should be introduced either by way of remunerative price mechanism for all crop with procurement facilities at farm gate or through private market mechanism with subsidies payable directly to farmer. Some areas need attention like **Rescue small farmers from individual crisis, Ensure farmers the infrastructure and other facilities, Facilitate agro processing for small farmer's produce.**

- **Vertical Farming:**Vertical farming is a concept of commercial farming whereby plants, animals, fungi and other life forms are cultivated for food, fuel, fibre or other products or services by artificially stacking them vertically above each other. Vertical farming is a popular method of agriculture in urban areas which needs immediate attention. The idea defines the production of fruits, vegetables, medicinal producing plants and other plant products in the cities
- **Creating sufficient cold storage facilities:** By capacity enhancement of existing cold storages, and creating new infrastructures for storage of vegetables and fruits for year-round supply to market, and also to consumers.
- **Real time market information** by using mobile communications, voice based call centers and expert systems for real time price discovery, weather information and cultivation trends needs immediate attention.

Innovations in Research

To achieve self-reliance and self-sustainability in agriculture, new innovations in research and technological advances for increased productivity are required.

Climate resilient crop varieties: The impact of inevitable climate change could be seen not only on food availability but also on quality of food. Improved climate-resilient crop varieties will be important for the food and nutritional security in time to come. It will help in sustainable productivity against changing weather factors including rising temperatures and water scarcity, resistance or tolerance to the emerging of pests and disease while adapting to climate change. Breeding for improved varieties will require broadening the genetic base using wild species in crossing programmes. Conventional breeding should also be strengthened utilizing molecular assisted breeding and recently evolved speed breeding approach.

Genetic engineering and gene editing: Crop improvement through conventional breeding techniques has given good productivity gains over the years but to address the future requirements of food, genetic engineering will be one of the options to increase yield as well as tolerance/resistance to biotic/abiotic stresses. Already crops are being engineered to have more resistance to abiotic and biotic stresses, for varieties which photosynthesize better etc. Another recent tool is being utilised by scientists is CRISPR/Cas9 (gene editing) which allows to target and alter the genome of an organism by cutting out or replacing specific parts of a DNA strand's genetic sequence. The potential of gene editing into breeding programs need to explored and this may facilitate targeted crop improvement and could be utilized for improving nutritional quality as well as pest/pathogen resistance.

Water resources and irrigation management: Innovations or new mechanisms to enhance the productivity of irrigation water with the focus on "more crop per drop" are required to be studied. On-farm management of water, use of efficient delivery system such as drip/sprinkler irrigation and also piped irrigation can be explored. Drip irrigation can also be utilized for delivery of soluble fertilisers. Solar-powered products for irrigation have come up and need to be scaled up as they are useful to small farmers and in areas where electricity is a problem. Focus on development of crop varieties requiring less water should have our priority.

Precision farming: Precision farming is an approach which uses low-input with high-efficiency and sustainable agriculture to get increased yields. It has many benefits apart from increased productivity including less soil degradation, reduction in chemical usage, efficient water use, reduced cost of cultivation among others. Thus, farmers will use less quantity of inputs and target very specific areas. Introduction of soil health cards among farmers a right step for the application of nutrients based on requirements. Commercial farming as well as horticultural crops have a good scope for precision farming using information and communication technologies to get timely information. Further research based information required for appropriate application of precision agriculture based on location-specific crop, soil and weather conditions. It is also important to create awareness among farmers about the demerits of imbalanced doses of farm inputs like irrigation, fertilisers and pesticides.

Artificial intelligence and machine technologies: Artificial intelligence technologies with the use sensors can be used to detect diseases and pests in plants, nutrient deficiency in crops, can detect weeds and help in deciding which pesticide/nutrient application within the affected area can be taken up. This will help to prevent over application of various inputs.AI can also be applied to create seasonal forecasting models with accuracy which can predict weather forecast so that farmers can take decisions in time. Such forecasting is valuable for small farmers where information reach is limited. Wild animal repellent sensors can also be developed as wild animals causing enormous losses in crops in India. Drone technology has been starting to be used for pesticide applications with better coverage, aerial photography for crops stand and damaged areas, and even drones have been used to supplement the pollination in trees.

Agricultural diversification to higher-value commodities: Encouraging farmers to diversify to higher value crops may bring in higher agricultural productivity particularly in the rain-fed areas to reduce economic risk.

Promotion of allied activities in agriculture having high growth potential or secondary agriculture: Some agricultural activities have particularly high potential for growth. The livestock sector (primarily dairy) is a source of income for 70% of India's rural families. Growth in milk production has been more, but future domestic demand is expected to grow. Milk production is constrained mainly due to the poor genetic quality of cows, inadequate nutrients, health issues etc. Solution to these issues can give good growth in this sector. Other activities in agriculture including mushroom production, bee-keeping, herbal/medicinal plant cultivation and fisheries etc. have great potential for growth and can be undertaken utilizing less space in innovative ways.

Increasing shelf-life of perishables: Innovations are required in increasing the shelf-life of perishable farm produce such as fruits and vegetables which are produced in large quantities during the season, however, because of lack of proper storage facilities decompose early and go waste. Some innovations are going on for biodegradable coatings applied on fresh farm produce to slow the decomposition process and wastage can be reduced.

Branding farm produce to add value: Branding of local farm products need to be taken up to increase the reach of the product. It will help to fetch good price to farmers and also encourage farmers to maintain quality. Standardisation and promotion of indigenous technology knowledge (ITK) available in the villages such as nutritious landraces of the crops should be taken up.

Role of Private Sector

The private sector has also played an important role in the advancement of the agro-technologies, and now the same needs to be arranged in India. Even the government needs to actively participate and encourage farmers to use modern techniques with incentives. With the vision of **'Make In India',** government involvement and government adhered innovation and investment is increasing. Now, it is the modern agricultural implements and machine manufacturers that need to bring innovation and new agro-techniques to Modernised Farmlands of India. In the seed sector, the growing need for R&D along with developments in **agrochemicals** such as increasing use of **bio-fertilizers** and bio-release **smart fertilizers** has been developed. The farm implements sector, however, is yet to become a major market segment and tractors remain a major sector of investment. **The irrigation sub-sector** also needs to evolve for Indian farming with large investments.

Innovations on Modern Techniques

(i) Digiculture: Application of modern digital tools in agriculture for making more profitable farming. They are, Artificial intelligence in agriculture, precisely in crop forewarning systems and plant protection systems, Development of digital tools for precision agriculture for timely management of farm activities, App based field monitoring systems, App based mechanization in agriculture-Due to shortage of labourers in new India, everything we need to depend on machines for field activities, Computer operated drones for input delivery, application of ROBOTICS in field of agriculture- for big data recording and management digitally and transfer to computers (Shetty et al., 2014)

(ii) Modern techniques are the future of Indian agriculture. They capture complete information about the commodities (growing information, pre- and post-harvest, transportation, warehousing, etc.) and proactively advice farmers on sowing, pest control, harvesting, etc. The future is with Unmanned Aerial Vehicles (UAV) and Artificial Intelligence (AI). **Unmanned Aerial Vehicles (UAV)** can be used to monitor and observe important data like multispectral imagery, visual and thermal imagery, humidity, the weather condition at a resolution of up to 1 cm/ pixel and air pressure. This data can help the farmer make an informed decision. **Artificial intelligence (AI) technology used for (i) Crop and Soil Monitoring**– crop and soil health can be monitored with sensors.(ii) **Predictive Agricultural Analytics**–Farming practices due to lack of access to the scientific understanding of crop lifecycle, pests, quality metrics, and the latest micro-fertilizers. (iii) **Supply Chain Efficiencies**– Real-time data analytics can be used for the efficient and smart supply chain (iv) **Agricultural Product Grading—**Crop diseases or pest infestation in the crops can be accessed by an image on a farmer's phone and determine the product quality in real time, without any manual intervention. Farmers just need to know the operation of an app on their phone.

(iii) Smart agriculture: Smart farming and precision agriculture play a crucial part in sustainable agriculture, using big data and new technologies such as IoT, drone imagery and smartphone apps to harvest as much matter and products as possible, while consuming less energy and inputs (fertilizers, phytosanitary products, water). Such as popularization of soil less cultivation-incentives for small scale industries to produce vegetables hydroponically.

Innovations in Indian agriculture made India self-reliant by introducing several technologies (i) **Drone based technology** for are very much useful for farmers in monitoring crop growth and spraying pesticides over the farm field by using drone technology and taking pictures, gathering accurate crop data is an emerging trend in this sector (ii) **Agriculture machinery and implements** -has powered the growth of the agriculture sector. New machinery immensely increased the speed of agricultural activities. Threshers, cultivators, mowers, seeders, drillers, Drillers are emerging and widely used agriculture implements in India (iii) **Climate Technology**-Rural agriculture in India is heavily dependent on climate. Natural rain plays a crucial role in the health of crop yield. New technology developed will give early alerts about storms, floods and other natural agricultural hazards. It'll recommend the percentile of rain and send a warning about the famine conditions. (iv) **Vertical Farming**-Vertical farming is a concept of commercial farming whereby plants, animals, fungi and other life forms are cultivated for food, fuel, fibre or other products or services by artificially stacking them vertically above each other. Vertical farming is a popular method of agriculture in urban areas. The idea defines the production of fruits, vegetables, medicinal producing plants and other plant products in the cities and (v) **Data Science**-Collecting data for soil health, fertilizers, drones, seed quality and amount of rain will help to assess the quality and predict the quantity of crop yield. Data science will highlight the improvement necessary in the agricultural and motivate people to research and escalate to new technology.

Role of NITI Aayog

Earlier India was far lesser level in the innovation in the field of science and technology. As per NITI Aayog statement, India is constantly on improving side of innovation as innovation index, India stand 52nd and 48th position globally among 129 countries during 2019 and 2020, respectively. It has been well notified that the huge potential in India for innovation as India has high human capital. It predict top five trend of innovations expected to shape Indian agriculture industry **(i)** The **digital innovation** in agriculture has applicability in infrastructure development supply chain management and technology enablement of areas such as quality traceability, logistics and distribution and other areas of value chain (ii) **climate risk management strategy** for effective water management, adopting to rising temperatures, facing drought situations. Solutions for early warning system can play a vital role in estimating and minimising the risk due to erratic climate change events (iii) **developing the start-up ecosystem**- more incubation happening for developmental and early stage start-ups while more funding is likely to continue in mid stage start-ups. A large proportion of the Indian population is dependent on agriculture, the majority of which is made up of small and medium farmers. There is tremendous potential for agri-tech start-ups to enhance rural livelihoods and reduce costs associated with climatic uncertainties and traditional farming practices. To enable this, financial credits/ concessional loans have been provided as a part of *Aatmanirbhar* Bharat *Abhiyan* to boost crop management, repayment, efficiency of rural banks, and infrastructure development. Advancements in technologies which can improve the quality of agricultural produce such as IOT/ analytics/ block chain for climate intelligence, forecasting solutions, machine learning to identify crop stages, artificial intelligence to reduce crop wastage, soil health monitoring, plant image recognition, geospatial tracking, and sustainable packing, can help achieve this goal. (iv) **Farmer Producer Organisations (FPOs)**: number of farmers getting connected through FPOs. The FPO structure is currently in need of support services to enable them to secure business and market linkages. Other key issues such as better insurance terms, quality assessment infrastructure, precision agriculture solutions for better crop management, access to finance, IoT based applications etc. needs to be addressed for the success of the FPO's and. (v) **Water management initiatives** like watershed management, drip irrigation and water user association can play an important role in strengthening the agricultural sector. The Government bodies, private players and the growing agriculture start up community can together transform the sector's outlook. This aided by digital drive can start the new agri and technology revolution in the sector (Chand, 2019). The agri-innovation has been taken up right path in India to achieve its goal and we need more commitment for achieving the challenges ahead to feed the growing population. As per NITI AYOG, the new vision for agriculture innovation will focus for Growth to efficiency, Employment Generation, Food Security to Nutrition and Health, Shortage Management to Surplus Management, Input Intensive to Knowledge Intensive Agriculture, Climate Change and Sustainability, Production and Producers and Policy Interventions, Regulations and Reforms.

Conclusion

The agri-innovation has been taken up right path in India to achieve its goal and we need more commitment for achieving the challenges ahead to feed the growing population. Top five trends of innovations expected to shape Indian agriculture industry are **digital innovation, climate risk management strategy, developing the start-up ecosystem, Farmer Producer Organisations (FPOs) and water management initiatives.** The new vision for agriculture innovation must be focussed for growth to efficiency, Employment Generation, Food Security to Nutrition and Health, Shortage Management to Surplus Management, Input Intensive to Knowledge Intensive Agriculture, Climate Change and Sustainability, Production and Producers, Policy Interventions, Regulations and Reforms. There is a need to draw area production plans for animal husbandry and to grow crops to meet India's nutritional requirement considering agroecological zones and the changing climate. The aim should be to figure out a suitable basket of crops for each area. Integrating the efforts of these organisations to identify niche areas viz; value addition, rural infrastructure, logistics, warehousing, quality certification, etc can provide livelihood opportunities for migrant workers bearing the brunt of the Covid-19 lockdown. The effective utilisation of MGNREGA can also create adequate work days, with a thrust on rural infrastructure. The Cabinet's decision to allow farmers to sell directly through e-trading platforms to entities of their choice instead of being confined to state *mandis* is another game changer. This will enable contract farming and inter-state trade. However, breakthroughs are needed in areas of transportation, logistics, and digital access and payments to enable widespread prosperity. Thus, focus must be on production, distribution, and management of food i.e., the complete agricultural value chain.

References

Chand, R. 2019. Transforming Agriculture for Challenges of 21st Century. In: 102 Annual Conference, Indian Economic Association (IEA), 27-29 December 2019, AURO University, Surat (Gujarat), 23p

http://agrospectrumindia.com/analysis/83/180/innovations-required-for-successful-indian-agriculture-in-future.html

http://www.mashelkar.com/convocation-addresses/176-building-a-great-indian-agriculture-innovation-system

https://agfundernews.com/five-trends-in-agritech-innovation-in-india-to-watch-out-for-in-2017.html
https://geographyandyou.com/a-critical-review-of-the-green-revolution-in-india/
https://ruralmarketing.in/industry/agriculture/is-a-brighter-future-awaiting-indian-agriculture-sector-post-covid-19
https://vicepresidentofindia.nic.in/pressrelease/research-and-scientific-innovations-must-aim-benefiting-farmers-vice-president-delivers
https://www.ciiblog.in/technology/tech-innovations-have-the-potential-to-transform-agriculture-production/
https://www.pwc.in/government-reforms-and-infrastructure-development/agriculture/top-five-trends-indian-agriculture-is-expected-to-witness-in-2019.html
https://www.startupindia.gov.in/content/sih/en/bloglist/blogs/Innovations_Agritech_AGC.html
Rena, R. 2003. Green revolution: Indian agricultural experience – a paradigm for Eritrea. New Jersey, USA: Eritrean Studies Review, 4(1):103-130 (A Biannual Journal Published by the Red Sea Press).
Shetty, P., K. Manorama, M. Murugan and M. Hiremath. 2014. Innovations that Shaped Indian Agriculture-then and now. Indian Journal of Science and Technology. 7: 1176-1182.
Singh, J., V. Kumar and T. Jatwa. 2019. A review: The Indian seed industry, its development, current status and future. International Journal of Chemical Studies. 7(3): 1571-1576

Theme 2: New Trends and Future Research and Development in Crop Improvement

3

New Plant Breeding Techniques and Current Trends in Plant Breeding Research

Kumar Durgesh, Gyan, P. Mishra, Rekha Joshi and S.M.S. Tomar

Division of Genetics, ICAR-Indian Agricultural Research Institute, New Delhi

Introduction

India has progressed well on food production from 51 million tonnes (mt) in 1947, 286 mt in 2018-2019 owing to the largest contribution from wheat and rice (about 213 mt). But simultaneously our population has also grown to about 135 billion in 2018. Therefore, the impact of higher production is not realized in absolute terms. Indian agriculture has come a long way since independence achieving food security in major crops, declining rate of rural poverty, rise in agricultural incomes and nutrition levels have gone up. Higher production has also been recorded in sugar, cotton, oilseeds, fruits, vegetables, milk, fisheries and poultry. All this has made us a proud with achieving self-sufficiency in food security. However, on the socio-economic front, the picture does not become rosy as we have a lower strength of socio-economic development indicators even among developing countries. Still the reports of under nourishment, hunger and poverty rings the bell. However, the importance of agriculture cannot be negated because it provides sustenance to mankind, food and employment to all. Considering the emerging challenges of climate change impacting agricultural productivity and increasing variability, poor investment in research, skill development of farmers, post-harvest losses etc. New strategies and innovations in agriculture, particularly on the development of climate resilient varieties with high productivity and better nutritive quality are called for, which has to be led by structural reforms and infrastructural augmentation.

Early History of Plant Breeding in India

Prior to start of systematic agriculture, plant improvement was restricted to crop domestication to find out eco-geographical adaptations of various crop plants. Plant breeding in India started in response to the stoppage of supply of cotton from the U.S.A. in 1863 and 1864 on account of the American Civil War that the British industrialists began to consider India as a great source to which they might look for the large supplies that were so urgently needed for their Manchester and Lancashire textile mills (Mehta, 1986). The efforts to expand cotton cultivation in India by introduction of the exotic seeds from Malta and Mauritius and selections started in 1790 in Bombay province and continued till 1910 considering several other countries. The first varieties Co 1 and Co 2 were developed at Coimbatore, Tamil Nadu (India) from Cambodia cotton during 1913-1914 from the material introduced in 1907 from Thailand. Next to cotton, the improvement in sugarcane was also considered and as early as in 1827, thick, noble Mauritius canes were introduced to India however, these canes were not found suited to conditions in India, being poorly adapted and susceptible to diseases. The problem of finding suitable sugarcane varieties for India had to wait till the time newly established Agriculture College, Coimbatore came into existence. Several attempts of introducing thick canes grown in other countries were made but failed to find suitable canes, specially for the northern provinces. At this point of time interspecific crosses between the indigenous canes (*Saccharum barberi* and *S. sinense*) and the thick (noble) canes (*S. officinarum*) were made but the resulting hybrids were not found to be hardy enough for Indian conditions and the then the hardy wild relative of cane, *S. spontaneum* was utilized. The hybridization resulted into the noble cane 'Vallai'. Selection from these crosses finally gave superior varieties which were well adapted to Indian conditions. Thus, came the first variety Co 205 that made Co (Coimbatore) canes famous in India and abroad.

Similar attention was needed for wheat which was frequently of inferior quality not acceptable to millers and bakers in England. To bring reforms at this point of time was to apply science on a large scale to the study and practice of Indian agriculture. Henry Phipps of Chicago, an American made a huge donation to start an agricultural institution and thus an agricultural research institute (I.A.R.I., then called Imperial Agricultural Research Institute) was established at Pusa in Darbhanga (now Samastipur) district of Bihar (India) in 1905, which was moved to New Delhi in 1936. The breeding of crops, cotton, wheat and sugarcane started in the early years of plant breeding in India. Initially, thousands of wheat specimen were collected, classified and recorded huge data on inheritance of characteristics such as beardedness, grain colour, felting, grain consistency and shattering at Pusa during initial years 1906-1909. This made the basis of pure line selection possible. At Pusa, four varieties *viz.*, Pusa 20, Pusa 21, Pusa 22, and Pusa 23 were at first isolated from mixture following pure line selection. These varieties gave rise to Pusa 12 and Pusa 4 and later on discarded for various reasons.

Besides the cotton, wheat and sugarcane, the work on other crops also started in the beginning of the 20th century. The earliest rice breeding started in India by Parnell in 1915 who made single plant selection and subjected to progeny test testing and the rice breeding work was continued by K. Ramiah, who selected superior types like GEB 24 (Government Economic Botanist), PTB 10 and SLO 16. GEB 2 gave rise to H35 suitable in Andhra Pradesh. Rice improvement work was started by the Economic Botanist in Bengal in 1911 who continued his efforts till 1928 to study the crop which was grown under different conditions and seasons-upland, lowland and boro etc. A variety N22 popular in Tamil Nadu became popular in U.P and Gujarat also. Despite lack of Government support, work on sorghum also has been started by Dr. G.N. Rangaswamy during 1925 to 1928. Among *rabi* jowars M 35-1, a selection of the variety known as Maldandi, has continued to be popular with the farmers for decades in the northern part of Karnataka. Despite the importance of pulses in vegetarian Indian diet as the main source of protein pulses received very little attention of plant breeders. The I.A.R.I. at Pusa studied a collection of varieties of *arhar* (*Cajanus cajan*) and Bengal gram (*Cicer arietinum*). A wilt resistant arhar variety WR 15 became popular. In gram, varieties NP25 and NP58 were recommended and they found a place in some of the states. Urd (T7 and T17, a medium early variety T21) and mung bean (T44, a selection of which at IARI was named Pusa Baisakhi) varieties, described by Howard (1914) and Bose (1932, 1939) were given to farmers for cultivation. Kadam and associates (1943) developed an early gram variety of attractive grain size, *Chafa*, from the breeding work of 1938-1945.Work on oilseeds such as mustard (*Brassica juncea* var. T1193, RR9, Raya L18 and RT11), sesamum (introduced a few century ago in India), groundnut (TMV varieties, TMV2 and 11) and castor was also pursued and improved cultivars were given to farmers.

The population pressure kept on increasing in poor and developing countries during the turn of century, the food grains became highly deficient and had to depend on food imports to feed their population. There was also an atmosphere of gloom and despair with regard to the potential of many densely populated countries of Asia becoming self-reliant in food. Since the early days of the 20th century, classical breeding for manipulating different traits in plant has been a primary method for improvement and enhancing the ability of crop against different stresses. A dynamic and well-planned breeding program of varietal improvement and diversification of varieties was urgently needed to develop speedily the genotypes with both specific and general adaptation. It was also thought that the crop plants grown traditionally in absence of fertilizers and other inputs. In the beginning of 20th century, the research on several other crops such as rice, sorghum, pulses and oilseeds were also started.

Commendable achievements were made in wheat breeding but the menace of the three rusts hampered the rapid progress. It was the foresight of Dr. B.P. Pal and his ability to plan and implement programs of research which ushered the next era of wheat breeding which would combine high yielding ability and good grain quality with rust resistance and other desirable characters. He helped to develop a strongest national wheat improvement program in the world. Several excellent varieties of NP series such as NP710, NP809, NP852, NP839 etc and the wheat program simultaneously carried out at other stations also developed suitable varieties such as C591 and C36 from Punjab, K65 from Kanpur, Hyb65 and Hyb11 from Powekheda and C35-1 from Rajasthan.

Even in developing countries like India where more intensive work on wheat and rice improvement had been organized during 1950-1965. Traditional plant types had continued in the farmers' field. The varieties released showed only a marginal increase over those recommended a few years ago. It is remembered that in 1943 the north eastern states of India had suffered food shortage due to the devastation of rice crop by *Helminthosporium oryzae* causing famine. It was therefore, obvious that food security was a dominant item of independent India's agenda as it has undergone unforgettable agony due to severe famines. Some special programs were implemented by the Government *viz.*, The Grow More Food Campaign (1940s) and the Integrated Production Program (1950s) to focus on supply of food and cash crops respectively.

The availability of Norin10 in bread wheat and *Dee-Gee-Woo-Gen* in indica rice were dwarfing source for changing plant architecture using useful genetic diversity. The introduction of semi-dwarf, photo-insensitive and high yielding varieties of rice and wheat revolutionized the Indian agriculture. The first high yielding variety in rice was released in 1968 as Jaya. In wheat, Lerma Rojo 64A and Sonora 64 were introduced directly from Mexico, later Kalyansona and Sonalika were released for cultivation. The initiative of the IRRI and CIMMYT however, changed the entire outlook on the production potential of tropical and sub-tropical agriculture. Subsequently, the food grain production in the country dramatically increased to 108.46 MT in 1970-71, to 129.6 MT in 1982–81, to 176.4 MT in 1990–91. The prominent mega varieties of wheat, namely, HD2285, HD2329 and PBW343, WH 147, Lok1 under favourable condition (C36 and HW2004 under rainfed condition)and of rice, IR8, IR36, IR64, Swarna, MTU7029, MTU1010 BPT520 and Samba Mahsuri etc.

Wheat and rice are being grown under very diverse conditions in the country. Concomitantly, the pace of productivity and production was maintained by other crops such as maize and pearl millet. Therefore, the Green Revolution has changed the face of Indian agriculture and the country became self-sufficient in its food grain production. Thanks to genetic improvement of yield and quality coupled with advances in other areas of agricultural technology, such as mineral fertilizers pesticides etc., India has increased production and productivity in almost all crops of economic significance in an impressive manner. The population keeps on increasing and the pathway open for maintaining self-sufficiency are higher productivity and production. However, the use of conventional breeding techniques has made a significant contribution to lowering the cost of production without a concurrent reduction in yield by introducing genes for multiple resistance to pests and diseases and abiotic stresses. The improvement in crop yields (productivity) during the last 70-80 years of breeding has been achieved mostly through redistribution of dry matter between, the vegetative and the reproductive parts. This has helped to increase the harvest index of crop plants (Jain 1986).

The progress made in cereals, particularly in wheat and rice prompt us to think that fundamentally, there are no reasons why mutant genes for improved plant types should not be found in other crop plants for accelerating the process of their genetic improvement.

Present status

Agriculture has been the major livelihood source of Indian people from the ages. It is not only to sustain life but has been the mainstay of the Indian economy contributing about 15% of national Gross Domestic Product (GDP) and more importantly about half of the India's population is significantly dependent on agriculture and allied activities. The agricultural produce is supplied as raw materials to the industry and therefore, the agriculture becomes a source of employment generation. Now-a-days the agriculture has moved from traditional practices to fairly new technologies. The industry has seen an increasing number of traditional agriculture companies adopting farm automation into their processes. New advancements in technologies ranging from robotics and drones to computer vision software have completely transformed modern agriculture. In present scenario of agriculture innovation is becoming more important than ever before. The industry is facing huge challenges from rising cost of supplies, shortage of labour and changing consumer preferences. The agriculture technology has also seen a huge investment of more than 7 billion is last few years. The major areas of technological innovations are focused on vertical farming, automation and robotics, modern greenhouse practices, precision agriculture and block chain technology and more recently the CRISPR gene editing and artificial intelligence technologies.

The advantages of vertical farming are that it reduces the amount of land space needed to grow plants. However, this type of farming is associated with city and urban areas as it reduces the travel distance in supply chain and can thrive in limited space. The plants can be grown in hydroponics with assured water and nutrients. In recent times the greenhouse industry is also flourishing to produce vegetables, flowers and medicinal plants to compete directly with land based conventional food production. The greenhouses, highly technology based using LED lights, having controlled environment with tailored plants are emerging on large scale near cities. Successful greenhouses are fulfilling the demand for local food, whatever be the season. Farm automation is being practiced in crops by a large number of farmers across the country for the activities like ploughing, inter-culture, harvesting, threshing, planting of seeds and watering etc. An increasing number of companies are working on robotic innovation to develop farm machinery. Automation technology is beneficial and addresses major issues like a rising world population, farm labor shortages and consumer preferences.

The present-day agriculture is undergoing an evolution. The technology is becoming absolutely necessary as part of a commercial farm. New precision technologies are developing that allows farmers to maximize their yields by controlling ever variables such as moisture levels, pest stress, soil condition and micro-climate. Thus, by providing accurate technology for growing plants, precision agriculture enables them to increase efficiency and better returns. Another important aspect of present-day agriculture is the food safety. Food traceability has been on the focus of food safety with application of block chain. The food industry is extremely vulnerable, such as perishable food, for making mistakes that would affect the human lives likely with food borne diseases etc. Block chain's capability is necessary to analyze root-cause and track down the source of contamination and to solve food fraud, safety recalls, supply chain efficiency and food traceability of current food system. This will also bring transparency in premium products in the market.

The modern agriculture has turned totally scientific and using digital technology. This rise of digital agriculture and its related technologies have opened a wealth of new data opportunities. The latest techniques can monitor pant health, growth, soil condition, temperature, humidity etc. through remote sensors, satellites and UAVs within very short time. The use of this technology will allow farmers to gain better understanding of the situation on the ground. The data analysis will provide more information and the artificial intelligence can be used to achieve their goal of better harvest through making better decisions in the field.

Concerted efforts are to be made on generation and adoption of technologies during coming times. The vision of Indian Council of Agricultural Research is to encourage SMART FARMS with greater technological support to reduce post harvest losses, strengthen value chain and provide markets to increase the farmer's income.

Climate smart crops

Developing new genotypes of plants is on priority of Indian agriculture. Adaptation of crop plant to climate change requires the implementation a range of complementary strategies. Genotypes and production systems become most important to make agriculture resilient to climate change. Conventional plant breeding usually depends on selection in the target environment. The selected varieties do adapt well under the environment for which a variety is tested. The developed genotypes give higher yield only under selected optimal and favorable environment but yield less under unfavorable environments. However, climate smart genotypes require more proactive approach to climate adaptation. Genomics provide a key platform to understand the response of plants to the environment and breeding of better adapted crop varieties (Abbetron et al., 2015). The consideration of new niches of physiological and ecological adaptations becomes very important under this situation. For all this, we need to design breeding programs breeding for combining desirable alleles for specific traits, gene editing to deliver targeted alleles to provide required yield and thus ensuring the delivery of food and nutrition security.

Multiple disease resistance

Plants encountered pathogens including viruses, bacteria, fungi and nematodes frequently during evolution for growth and survival. In due course, plants have evolved acquiring structural, chemical, and gene-based defense mechanisms to combat pathogens of different nature. However, the pathogens also have an ability to develop new races to infect plants. Considerable efforts are required to determine the pathogenic mechanism and to identify host plant genes that confer resistance.

Generally, disease resistances in plants can be classified into two major categories: qualitative resistance conferred by a single resistance (R) gene and horizontal resistance or quantitative disease resistance which are controlled by polygenes of minor effect and are therefore called multigenic or genetically complex (Glazier et al., 2002). The qualitative resistance is simply inherited and mainly controlled by a single major gene and called race specific resistance. This type of resistance can be easily overcome by the evolution of new race(s) of concerned pathogens. The resistance controlled by many minor genes provides more protection to the host plant as the pathogen strains will not be under selection pressure. The loss of one or two genes from the cluster of genes may not leave the host completely susceptible (Poland et al., 2011). Since the quantitative resistance is conditioned by several genes, it is called race non-specific and characterized as broad-spectrum resistance and durable resistance. It is most desirable type of resistance in crops. One such example is a wheat gene *Lr34* providing resistance to many diseases may be due its linkage with other genes. Singh and Huerta (2003) characterized the genes *Lr34* and *Lr46* as slow rusting, which provide non-specific adult plant resistance but their effect is more reduced than that of race specific genes. Kranttinger et al. (2009) cloned the *Lr34* and it was shown that it is the same gene as *Yr18* imparting adult plant resistance to stripe rust, powdery mildew resistance (*Pm38*) and leaf tip necrosis (*Ltn*). A similar tight linkage or pleiotropic effect was observed between *Lr46* (showing partial resistance; Martnez et al., 2001) and *Yr29*, also slow rusting gene against stripe rust.

However, the genes responsible for a specific disease resistance are mostly minor with small phenotypic effect. The total phenotypic variations control by these minor genes is usually very low and identifying these genes is a cumbersome task. Identification of all the genes for the most prevailing diseases in different crops at genomic level is most fundamental issue of today's agriculture. However, many plants both in nature and agriculture are resistant to multiple diseases and this type of resistance is termed as multiple disease resistance. The multiple disease resistance could be under control of one gene (Wisser et al., 2005; Chung et al., 2010) but this phenomenon has not been explored well. Zwonitzer et al. (2009) identified QTL for resistance to several diseases viz., northern leaf blight, grey leaf spot and southern leaf blight in a maize inbred line population, which the evaluated and found resistance to multiple diseases.

Breeding for durable disease resistance is labor intensive and time consuming. The resistance controlled by a single gene (race specific straightforward type of quality) can easily be transferred. However, the single gene resistance is vulnerable and may be overcome in a short span of time. Therefore, a broad spectrum of resistance is preferred in a cultivar. To generate a crop cultivar with multiple disease resistance traits is difficult and time consuming. Transferring multiple genes into a cultivar using genetic engineering is likely to aid this process. The availability of biotechnological tools and molecular markers facilitate the introgression and pyramiding of multiple diseases resistance genes into cultivars.

Biofortification of crop plants

Recent advances in biotechnology specially in the area of genomics offer several opportunities to overcome constraints in the crop production. Accelerated crop improvement with respect to some traits for which conventional breeding could not solve, the molecular breeding could possibly help. Biofortification is a process of increasing nutrient density of food crops, increase micronutrient intake by undernourished and poor. It is important that the system continues with better facilities and in scientific way. Breeding fortified staple crops to have a higher content of micronutrients like iron, zinc and vitamin A should be on high priority. The availability of such crops is likely to improve poor people's nutrition status in a cost-effective manner. Researchers are working on biofortification and the success has been achieved in rice (Golden Rice), sweet potato and potatoes. Many other crops have been included for biofortification with particular micronutrient. Till, 2020 has developed more than 50 biofortified varieties of different crops including paddy, wheat, maize, lentil, mustard, groundnut and millets such as pearl millet, finger millet and little millet. The availability of biofortified varieties will help in dealing with nutritional deficiency in the country. Farmers are being encouraged to grow these varieties to enhance their income and socio-economic condition.

Three quality protein maize (QPM) hybrids *viz.*, 'Pusa HM4 Improved', 'Pusa HM8 Improved' and 'Pusa HM9 Improved' have been developed through marker-assisted introgression of recessive *opaque*2 gene into the parental lines of the original hybrids. However, selection of kernel modification (less opaqueness) among the *o2o2* segregants were carried out using standard light box text. Yellow maize though rich in carotenoids, is poor in provitamin-A (1-2 ppm). The natural variant of *crtRB1* gene was introgressed through MABB into the parental lines of the popular maize hybrids that led to the development of provitamin-A rich maize hybrid, 'Pusa Vivek QPM-27 Improved' which possessed 8-10 ppm of provitamin-A. *Opaque2* and *crtRB1* have also been combined using MAS and three hybrids, viz., 'PusaVivek QPM9 Improved', 'Pusa HQPM-5 Improved' and 'Pusa HQPM-7 Improved' rich in lysine, tryptophan and provitamin-A have been released for commercial cultivation. Alpha tocopherol that possesses vitamin-E activity in maize is low (6-8 ppm) in maize grains, and the same has been increased to 16-18 ppm by introgressing mutant version of *vte4* gene through MAS. Maize hybrids with high lysine, tryptophan, provitamin-A and vitamin-E and low phytic maize that increases the bioavailability of iron and zinc, has been reduced by 30-40% through introgression of *lpa1* and *lpa2* genes. These biofortified maize hybrids rich in nutritional quality are at par with the traditional hybrids for grain yield, thus assumes great significance in alleviating malnutrition through sustainable and cost-effective manner. Indian government has supported the work on producing nutria-cereals especially the millets towards meeting the nutritional gaps.

A list of biofortified crops released in India

Crops	Biofortified for	Content	Year	Name of variety / genotype
Groundnut	Oleic acid	~80%	2019	Girnar-4 and Girnar-5
Lentil	Iron	65.0 ppm	2017	Pusa Ageti Masoor
Lentil	Iron, Zinc	73, 51 ppm	2018	IPL 220
Maize	Provitamin-A, Lysine, Tryptophan	8.15 ppm, 2.67%, 0.74%	2017	Pusa Vivek QPM9 Improved
Maize	Lysine, Tryptophan	3.62%, 0.91%	2017	Pusa HM4 Improved
Maize	Lysine, Tryptophan	4.18%, 1.06%	2017	Pusa HM8 Improved
Maize	Lysine, Tryptophan	2.97%, 0.68%	2017	Pusa HM9 Improved
Maize	Lysine, tryptophan and provitamin-A			Pusa HQPM-5 Improved'
Maize	Lysine, tryptophan and provitamin-A			Pusa HQPM-7 Improved
Mustard	Erucic acid	<2.0%	2013	Pusa Mustard 30
Mustard	Erucic acid, Glucosinolates	<2.0%, <30.0 ppm	2016	Pusa Double Zero Mustard 31
Pearl millet	Iron, Zinc	73.0, 41.0 ppm	2017	HHB 299
Pearl millet	Iron	73.0 ppm	2017	AHB 1200

Crops	Biofortified for	Content	Year	Name of variety / genotype
Rice	Protein	10.3%	2016	CR Dhan 310
Rice	Zinc	22.6 ppm	2016	DRR Dhan 45
Rice	Zinc	25.2 ppm	2018	DRR Dhan 49
Rice	Zinc and Protein	20.1 mg kg^{-1} 10.3%		DRR Dhan 311
Rice	Low glycemic index			Madhuraj 55
Rice	Zinc	24.0 mg kg^{-1}		DRR Dhan 48
Rice	Zinc	22.8 mg kg^{-1}		Surabhi
	Zinc	27.4 mg kg^{-1}		Zinco Rice MS
Soybean	KTI	KTI-free	2018	NRC-127
Wheat	Iron, Zinc	40.0, 42.0 ppm	2017	WB 02
Wheat	Iron and Zinc	40.0, 40.6 ppm	2017	HPBW 01
Wheat	Protein, iron, zinc	12%, 42.1 ppm, 42.8 ppm	2017	Pusa Tejas (HI 8759)
Wheat	Protein, iron, zinc	13%, 43 ppm, 35 ppm	2017	Pusa Ujala (HI 1605)
Wheat	Protein, iron, zinc	14.7%, 46.1 ppm, 40.3 ppm	2018	MACS 4028 (d)

Hybrid technology

Prior to World War II, the most significant yield enhancement technology was the exploitation of hybrid vigour as in the case of maize coupled with good soil fertility (Swaminathan 1986). Heterosis has been a very important source of yield improvement in many crop plants, even in self-pollinated crops. Most of the hybrid work in India in the last few years has involved testing many pollen parents on relatively few male-sterile lines. Continued long-term improvement in pearl millet hybrid yields is dependent on the breeding of new male-sterile lines (A-lines) as well as pollen parents (R-lines).

The world's first commercial cotton hybrid, Hybrid-4 (H-4) a also known as sankar-4 was developed in 1970 by Chandrakant T. Patel, which revolutionized the cotton farming in Indian states of Gujarat and Maharashtra. A Chinese agronomist, Yuan Longping developed first rice hybrid in 1970. Rice hybrids are since then being developed and cultivated in Africa, America and Asian countries providing food for millions. India has also developed several rice like APHR1, APHR2, MGR 1 and KRH 1 which were released in 1994. The first super fine grain aromatic rice hybrid Pusa RH1 was released for cultivation in 2001 in the regions of Haryana, Delhi and Uttarakhand. Hybrids in other crops such as maize, both single and double cross hybrids (Ganga 101, Ganga 102, Ganga Safed, Trishulata, Co-6, COMH 6, COMH8, DMH 117, DMH 121 and PMH1) pearl millet (HB 1-released in 1965, ICTP8203, Pusa 23, BJ104 and HSB-67 etc.) and sorghum (10 hybrids viz., CSH26 to CSH35 under AICRP on Sorghum) played a very significant role in maintaining the pace of food production.

Varietal improvement adopting traditional breeding methodology continued and is still being followed. Recent example of release of Pusa Basmati 1121 (PB 1121) is a landmark Basmati rice variety having Basmati quality traits introgressed from traditional Basmati varieties such as Basmati 370 and Type 3 (Singh et al., 2002). It was released for commercial cultivation in 2003. It possesses extra-long slender milled grains (9.00 mm), pleasant aroma, and an exceptionally high cooked kernel elongation ratio of 2.5 with a cooked kernel length of up to 22 mm volume expansion more than four times, appealing taste, good mouth feel and easy digestibility. Owing to its exceptional quality characteristics, it has set new standards in the Basmati rice market. The cumulative foreign exchange earnings through export of PB 1121 since 2008 have been US$ 20.8 billion, which has brought prosperity to millions of Basmati farmers.

The trends of productivity over about 70 years have shown that Green Revolution has tremendously enhanced genetic gains in cereals such as wheat, rice, maize and pearl millet. The significant increase in productivity is considered as combined result of enhanced genetic potential of varieties and improved integrated management practices. To ensure food and nutritional security to the ever-growing India's population would be a great challenge. Use of technology and innovation in Indian agriculture would only be the way out.

Hybrid vegetable production under protected conditions

Vegetable cultivation faces various constraints in the production of disease free healthy and genetically pure seed for commercial cultivation. Due to the lack of proper isolation and attack by the insect-pests, and various fungal, bacterial and viral diseases, the quality seed production program under open field condition faces severe limitations. Thus, the hybrid seed production under protected conditions is aimed to fetch higher seed yield with better quality seeds. Viruses and their insect vectors are one among the most fatal causes limiting the quality seed production in a number of crops including vegetables. To prevent this, the insect vectors can be checked by growing the crops under protected conditions. By using this strategy, not only the quality but also the quantity of seeds produced per unit area can also be increased. For a number of crops, especially vegetable crops the seed production under protected conditions have proved highly economical and profitable. Another added advantage of this method is use of limited resources and manpower per unit area of cultivation (Singh et al., 2015). In the northern plains of India, crop cannot be taken throughout the year though the soil is fertile yet due to the extreme temperature (0-48 °C).

Similarly, in other parts of the country various biotic stresses limits the successful production of a number of vegetables like tomato, chilli, okra, cauliflower etc. under open field conditions. In trans-Himalayan cold-arid desert conditions during winter months, the temperature becomes extremely low (-5 to -30 °C) and growing any vegetable in such extreme climate under open field conditions becomes impossible (Mishra et al., 2010). In many countries like USA, Japan, China, Israel Morocco, Turkey etc. where climatic conditions restrict the choice for year-round outdoor production, vegetables are being exclusively cultivated under protected conditions. Protected cultivation is the most efficient means to overcome climatic diversity; thus, it is being used for high cost vegetable production, especially hybrid and colored vegetables (Capsicum, Okra, etc). Tomato hybrids when grown under both open-field conditions and under protected-conditions, gave significantly higher yield under protected conditions under cold-arid desert conditions of Ladakh (Singh et al., 2009a, b). India is becoming a major player of the greenhouse vegetable cultivation and total area under protected vegetable cultivation is approximately 10,000 hectares. India being a vast country with diverse and extreme agro-climatic conditions, the vegetables cultivation under protected conditions should be very effectively, economically and wisely used for year-round and off-season production of high-value, low-volume vegetables virus-free quality seedlings and also for the production of quality hybrid-seeds (Wani et al., 2011).

Accelerated breeding

Traditionally cultivated crops take a range of time period to complete their life cycle (attaining time from seeding to maturity). The crops can be divided into seasonal, annual or perennial depending on their maturity duration. A plant breeder has to restrict him for that period to take one generation for making observation recording data in their breeding experimentation. Rapid generation advancement in crop breeding provides opportunities for quick cultivar development by advancing generations in a shortest period. The accelerated breeding or speed breeding is becoming a powerful tool (Watson et al., 2018) to accelerate crop research and breeding that shorten plant generation time can help hasten the breeding and research. Fundamentally, the cultivar development takes relatively longer period. However, the current food situation demands the quick development of products in order to mitigate the natural vagaries of climate change influencing environment appearance of new races of diseases and pest infestation. Breeders are always in search of new methods and tools to develop varieties accordingly by studying merits and demerits of the new methodology (Forster et al., 2014; Lenaerts et al. 2020).

To advance generations in wheat breeding (Shuttle Breeding) was originally developed by the CIMMYT and was popularized by Nobel Laureate Dr. Norman Borlaug. In this system a different location with similar environment enabled an extra generation to be advanced each year, which permitted off-season breeding activities. The advantage of shuttle breeding is not only to advance the generations but to make selections under different ecological conditions, disease screening and facilitating crossing programme also. India is taking advantages of shuttle breeding at Wellington, Nilgiris, Tamil Nadu and Dalang Maidan in Himachal Pradesh; Aduthurai in Tamil Nadu for rice and Dharwad in Karnataka for chickpea.

However, rapid advancement of generation and molecular assisted backcrossing supported by speed breeding generally require initial investments in infrastructure and operational cost. It is therefore, essential to analyze both, cost and benefits, before adopting this innovative approach for rapid advancement of generations. As it is claimed now, that speed breeding may deliver rapid genetic gain and more responsive release of varieties are yet to be seen.

Role of advanced genomic tools

Design based crop improvement will be carried forward with the increased knowledge in the field of genetics and molecular understanding of the functionality of genes controlling target traits. Determining the best alleles and their combination in the genotype is being developed with respect to target production environment desired pest and disease resistance and nutritional quality. The genotype will be facilitated to be designed by identifying closest sequence to the desired product and then making the required changes by backcrossing or transgenic approach. The continuing advancement in DNA sequencing technology is becoming much more efficient and routine (Henry 2020). The application of molecular markers in selections, the use of genetic transformation, genome selection and the gene editing techniques will rapidly help in designing the plants.

Genetically modified crops

Biotechnological approaches of plant breeding have made a major impact on genetic modification of crop plants in order to enhance yield, disease resistance and quality. The best bet for alleviating the high incidence of hunger, under nutrition and poverty in an agriculturally important country like India, the efficacy of such as agriculture needs solid foundation by substituting the prevailing less productive technology by improved technology and technological innovations. From very long time, biotechnology was used as a tool for crop improvement, commercialization of genetically modified crops has been going on around the world for last 20-25 years. The countries like USA, Australia and Canada are growing genetically modified crops such as mustard, soybean and maize. The cultivation of GM canola since 1996 has changed socio-economic and environmental benefits to Canadians by earning foreign exchange even from India.

A phenomenal increase in area is noticed since the first Bt cotton variety was released. India is among the 10 countries with an area of more than 11 million ha exclusively under Bt cotton. Now even European countries have started cultivation of GM crops. Biotechnology is an important tool and a noble approach to harness its advantages and not a panacea as was initially understood. Researchers at Monsanto developed Bt cotton in 1996. Since then it has become a commercial crop in China in 1997 and the United States in 1996. A joint venture between Monsanto and Mahyco introduced Bt cotton in Indian in 2002. In 2011, India grew the largest genetically modified cotton at 10.6 million ha.

Besides, Bt cotton there are several other crops such as soybean stacked with herbicide-tolerance and insect-resistance, omega-3 rich soybean, drought tolerant maize and Golden rice have become important genetically modified crops. With wrong information spread by some NGOs world over on the potential contribution of technology, particularly the role of innovative bio-technologies deprived the farmers and general public.

Marker based genetic purity testing

Seed is the most crucial input in agriculture. The availability of genetically pure seed is the first and foremost need of the farmers. Till recently, public sector was the sole supplier of quality seed. With the intervention of private sector venture and the awareness of farmers, the demand of quality seed has been increased. However, the price of good seed has become an issue. Timely access to quality seed of improved, high yielding crop varieties at affordable price can be ensured by increased production of quality seed, reliable quality evaluation procedures and the application of appropriate technologies of seed quality maintenance. Therefore, holistic approach to produce must be followed. Standardization of cost-effective hybrid seed production technology is needed to produce quality seed. Standardization of protocols for variety identification and testing of genetic purity of seed using DNA markers should be practiced to have vigorous seedlings in the field. A model for regulation of seed dormancy and deterioration need to be adopted strictly. Molecular markers are constantly being used for both fingerprinting and genetic purity testing of various hybrid vegetables like tomato, beans, pepper and potato (Ansari, 2015). Genetic purity can also be tested using technique as reported by Mongkolporn et al. (2004). Purity of F_1 chilli hybrids was determined using two molecular techniques RAPD and ISSR. Four molecular marker systems—RAPD, ISSR, SRAP, and SSR were used to evaluate seed genetic purity of a hybrid cabbage cultivar 'Zaoxia 16' by Liu et al. (2007).

Doubled Haploid (DH) production

When haploid cells (gamete) undergo chromosome doubling, results in the formation of a DH which after selection may be used as a parental line. Before using them in the breeding program and their propagation through seeds, the fertility of DH has to be restored by way of chromosome doubling. The obtained DHs become homozygous for all the loci and they represent a new genotype/ variety (self-pollinated crops) or acts as a parental line for the production of hybrids (cross-pollinated crops). The production of pure lines using DH has several advantages over conventional methods; as it saves a lot of time particularly in biennial crops and in crops with a long juvenile period. For self-incompatible species, dioecious species and species suffering severe inbreeding depression due to selfing, DH could be the only way to develop inbred lines. This novel technique may be used to transfer the sterile cytoplasm of CMS lines into any desirable background in one step by avoiding many generations of backcrossing. A number of seed companies like Monsanto, Syngenta, Nunhems (Now Bayer), Namdhari HyVeg, have already made significant progress in this direction especially for a number of vegetable crops (Sadashiva et al., 2014).

Skill development

India has the World's largest scientific workforce in terms of science graduates being produced every year. Accountability at different stage of research is linked with excellence as it drives an individual or a group of workers to produce results. The technology produced in the laboratory must reach the field for proper application and adoption. The link between the extension worker and the farmers must be strongly built up to realize the gains claimed at the laboratory level. The link will also provide the knowledge to improve the efficiency of water and fertilizer management and to promote the economic sustainability of modem agriculture.

Major breakthrough in improving the productivity and profitability of wheat, rice maize, jowar and bajra was made because of availability of high yielding varieties and hybrids coupled with appropriate agronomic and plant protective measures. However, the condition of farmers has not improved significantly because of technology dissemination. Methods have been devised to deliver the technology and its adoption which will help to bridge the gap between potential and actual yields of major crops. When the first dwarf wheats were introduced in India in 1964, Dr. Swaminathan (1965) advocated to start with National Demonstration in small farmers' fields in order to introduce new wheat varieties to avail opportunities for yield enhancement.

Following National Demonstration opportunities were also created to link farmers and scientists together in a symbiotic partnership through a program of Whole Village Operational Research via Krishi Vigyan Kendra (KVK and Land to Lab program). At this point of time the best available scientific know-how to the small and marginal farmers enhance the rate of food production exceeding the population growth enhancing the image of India (Swaminathan 1993). However during the first decade of 21st century, the position in knowledge dissemination and skill training became dismal as revealed by NSSO surveys, although a significant but limited role was played by radio and television due to poor access at the farmer ends. Probably public agencies played a poor role in creating awareness among farmers as the farmers were dependent on input dealers for advice.

Very insignificant numbers of farmers were educated about World Trade Organization, crop insurance and technical literacy. For employment generation in agriculture, National Commission on Farmers (NCF) had suggested to operationalize *Krishi aur Udyog Vigyan Kendras* to give an income and employment orientation. It was also suggested by NCF to start *Gyan Chaupal* or Village Knowledge Centres based on the use of internet. NCF had also suggested the need of farmers to get advice on land use and anticipate meteorological and marketing trends. To get benefit of the knowledge in farm management, technology delivery system needed urgent attention, reform and revitalization. Presently, the government has taken initiatives, namely, *Make in India*, *Skill India*, *Start-up-India* and *eNAM* etc. to build new National Agricultural Innovation System that would achieve assured success.

Capacity building

The demonstration of proven technology is very important component of transferring to farmers for its adaptation and bringing into production system. Capacity building of young scientists on simulation modeling, high throughput phenotyping, greenhouse gases measurement etc. should be taken up through training programs. Simultaneously, if needed, the scientist must be trained abroad about advanced technology. Programs should be organized to encourage farmers to create awareness about drought preparedness, flood preparedness like drainage, water management including water harvesting, supplemental irrigation, and any abrupt climate changes occurring. The interventions in village Panchayats following participatory approach through the Village Climate Risk Management Committee. The Scientists should introduce to farmers the stress tolerant, (such as drought, heat, temperature etc) varieties and their method of planting etc.

Demand and supply chains

During the last 15-20 years, agriculture in India has undergone tremendous change in realizing higher production in crop plants, fruits, vegetables, milk, meat and edible oil. The country achieving not only sufficient food to realize self-sufficiency but certain commodities are in surplus. However, the resurgence of climate change has enhanced risk coupled with small farm holdings, depletion and degradation in quality of air, soil and water (short falls in rains, untimely and excessive rains floods and droughts) etc. A lot of agricultural products are being imported, an obligation of being a member country of WTO. Under these circumstances processed and convenient foods with increased urbanization and rise at income levels are replacing the conventional food products. This shift is being diversified and hence increasing demand for a large number of products.

The future of Indian agriculture under this challenging scenario, the Indian farmers have to become globally competitive with respect to the crop production with quality. Mechanism to bring raw produce or value-added products through different phases of production, delivery to consumers, a market focused collaboration may be developed and the holistic system research that moves from input to output research is needed to be developed from marketing and processing points of view. The public and private investors have to encourage contract farming to buy the raw materials from farmers which will lead to enhance the farmer's income and their livelihood so that the improvement become demand driven. The process of demand driven production will help in checking the migration to urban/cities due to their gainfully engagement. At the same time, it should be well recognized that market driven forces impact on poverty reduction through producers' income consumer prices, employment and wages may enhance the economy.

Women empowerment in India

Women empowerment in India is mostly dependent on many factors that include geographical location (urban/rural), educational status, social status (caste and class) and age. Therefore, it needs multifaceted approach which addresses the diversity of social structure. The policy decisions related to women empowerment have to be taken at administrative/political level. Policies on women's empowerment exist at the national, state, and local (Panchayat) levels in many sectors, including health, education, economic opportunities, gender-based violence, and political participation. However, there are significant gaps between policy advancements and actual practice at the community level. Women are put to a lot of discrimination, economic disadvantages, and violence against at the community level. As such, women and girls have restricted mobility, access to education, access to health facilities, and lower decision-making power, and experience higher rates of violence. Political participation is also hindered at the Panchayat (local governing bodies) level and at the state and national levels despite existing reservations for women.

The impact of the patriarchal structure can be seen in rural and urban India although women's empowerment in rural India is much less visible than in urban areas. This is of particular concern, since much of India is rural despite the high rate of urbanization and expansion of cities. Rural women, as opposed to women in urban settings, face inequality at much higher rates, and in all spheres of life. Urban women and, in particular, urban educated women enjoy relatively higher access to economic opportunities, health and education, and experience less domestic violence. Women (both urban and rural) who have some level of education have higher decision-making power in the household and the community. Furthermore, the level of women's education also has a direct implication on maternal mortality rates, and nutrition and health indicators among children. Based on the assumptions, that women differ from men in their social positions. These differences are of asymmetric, unequal power relations between the genders. Hence there is an urgent need to increase women's access to control over the strategic life choices that affect them and the access to several rights in the society.

Indian constitution provides reservation for women to enter politics at the Panchayat level. Due to vibrant women's movement in the last 50 years, policies to human rights for women in India laws on Domestic Violence Act and Right to Information Act have been enacted. However, the gap in policies and practice in women's empowerment exists. Despite the laws and initiatives by the civil society institutions, violence against women in India is widespread because the laws for sentencing of criminals are long arduous processes. But the policies relating to rights of women have positive trajectory and government is articulating many progressive measures to gender inequality (preference for son in the family) and improve economic and political situation. Research conducted by ICRW has indicated the discrimination against a girl child.

Another important aspect is the economic empowerment of women, which is complex and influenced by caste, class, religion and ethnic identities. Still the India's ranking in the world is dismal. However, the women workforce in some states and mega cities is far better than rural area. The diversity in the economic empowerment of women is affected by cultural, religious and ethnic identities, whereas the women movements assure economic, social and cultural rights in order to alleviate poverty and unemployment.

Public-private partnership

By encouraging public-private partnership in research and infrastructure development it is possible to enhance the agricultural production in the country. This is possible by undertaking case by case and developing models and tools to encourage Public-private partnership in agricultural research, infrastructure development such as roads, marketing facilities, storage facilities etc. The role of private sector in hybrid seed production of vegetables and maize is commendable and hope now they are venturing into hybrid seed production in rice too which is to appreciate.

Private sector investment

The governments of many countries have encouraged and supported private research and development by investing to achieve scientific endeavour. Initially, incentives were provided to public research institutions to boost research and privatized some institutions that conduct research on applied aspect. The system in public research focused towards basic research with objectives of stimulating private-sector innovations. The progress was monitored for effective implementation. Specific laws were enacted that allowed the private sector to conduct research supported by the liberalized input markets. Incentives were given in the form of specific subsidies with tax reforms for companies carrying out research and development in the field of agriculture and allied branches. They were encouraged to strengthen Intellectual Property Rights. Keeping all the points into consideration, India also need to change the research priorities and formulating policies to have positive change in:

(a) Public research and development should be equipped well to encourage basic research and the institutes that are engaged in applied research can be privatized. This step will help in reducing the number of research institutes under public sector. The institutes engaged in basic research can muster more funds for each institute;

(b) For popularization of agricultural, input industry should be totally privatized so that the development activities should be independent and encouraged; and

(c) Tax exemption for the income generated on technology transfer to farmers should be considered.

Policy intervention

The priority of the government is to enhance public sector investment in agricultural research and concerted efforts are needed to support the effective transfer of technology. At present only 0.6% of agriculture GDP is being spent on agricultural research and development, which is not adequate. To keep pace in agricultural productivity government needs to enhance the investment to at least 2 or 3 times from the present. The investment is insufficient compared to China, Brazil and Japan and even the under developed countries (Beintema and Stads, 2010). Therefore, an aggressive research and development investment strategy is indispensable for meeting the demands of present competitive world to have a reasonably good growth in agriculture, food security situation as well as sustained agricultural growth. Singh (2012) suggested to commensurate priority policy actions, such as: Integration of food security and sustainable/resilient agriculture into global and national policies, increase global and national investments in resilient agriculture and food systems and target programs and policies to assist vulnerable populations.

Water management policy and input subsidies

The agricultural production and productivity are basically based on natural resources like land, water and biodiversity. In recent times the water availability is shrinking, land is fatigued and biodiversity has been eroding. It is expected that by 2025 about 3 per cent of production will be at risk due to declining of water availability. The present situation and time demand that government should develop minor irrigation system, discourage subsidies on electricity as it is mining the water from deeper aquifers and adopt water saving technologies. The crops requiring more water to produce one unit of grain with less water to produce that unit should be considered under input management. Crops which mature in shorter duration must be encouraged. For minimizing over exploitation of ground water, there should be legally bound water management regulations, which can further enhance the area under irrigation. Subsidies on fertilizers need to be gradually phased out by emphasizing the adoption of eco-friendly nutrient management practices

Marketing infrastructure

India is booming record production of certain agricultural commodities. The pace at which the crop production is being enhanced, despite of bad weather and natural calamities in preceding years. However, the socio-economic condition of farmers do not show pleasing situation. This may be because they do not get the desired returns from the inputs they invest in production system. Frequently, farmers lose their produce by unavoidable factors during harvest, its transport and storage such as excessive rains, floods, post-harvest losses and crashing of prices. Since agricultural commodities are perishable in nature, their quality gets deteriorated continuously causing distress. Government has to plan to develop infrastructure through different schemes giving financial support, developing cold chain from harvesting to consumption. For value addition, a large number of cold storages with higher capacity and warehouses must be constructed, development of collection centres. This is particularly relevant for the high value segment that suffers from high post-harvest losses and weak farm-firm linkages. The excess food production of in some states can be diverted to production deficient states by an agreement between states. Creating better market facilities and providing market intelligence to farmers also can help reducing the post-harvest losses and enhance farmers' net income. In order to improve competition at retail level and to benefit consumers and producers, some innovative marketing mechanisms have been developed in some states, where the farmers sell their agricultural produce directly to consumers. These markets include *Apani Mandi* in Punjab and Haryana *Rythu Bazar* in Andhra Pradesh, and *Uzhavaar Sandhai* in Tamil Nadu.

Conclusion

The Mendelian and molecular genetics have facilitated the development of innovative approaches for achieving accelerated development of varieties which are economically and ecologically sustainable under the favorable environmental conditions. The modern tools and technology are helping to build an enduring global production and nutrition system to mitigate the problems arising from decreasing land availability, increasing population pressure, hunger and malnutrition. The farmers have to enhance their capacity to produce more from per unit of area of land without causing harm to the long-term production potential of the soil. Conventional breeding technologies have made a significant contribution in yield with low cost of production introducing genes for multiple resistance to pests and diseases, multiple tolerance to soil stresses, improvement in quality etc. The increasing availability of chemical fertilizers, water and other inputs had made it possible to evolve cultivars with a different kind of plant architecture marked by a high harvest index and response to increased plant populations. The kind of transformation achieved was accelerated after the discovery of genes having major effects on plant type. Following this accelerated improvement, a stage would soon be reached when breeders must select for increased dry matter production for there is a limit to which harvest index could be increased. An important objective for increasing crop yield in future must, therefore, be the search for genetic variability giving high rates of photosynthesis at high population density. Future food production will relay on the continued development of new crop varieties, novel crops and new type of plant-based foods. Potential crops will need research attention to be able to contribute to climate adaptation. A new biodiversity sources must be searched for new and useful genes for introgression into food crops.

Future strategies

Crop production in future will relay on the continued development of new and novel varieties and variety of plant-based foods. Potential and undcrutilizcd including non-traditional crops would need more research attention to fit the genotypes into climate adaptation. The basic need to develop the kind of varieties may get complemented by domestication of new crop species and the extensive use of wild relatives using pre-breeding precisely. Protected agriculture including vertical green houses are likely to produce more for better production and profitability of certain crops. At the same time a cautioned use of biodiversity without losing it in order to save water and other inputs. The farmers have to enhance their capacity to produce more from per unit of area of land. However, this has to be achieved without causing harm to the long-term production potential of the soil. Thus far plant breeders using conventional techniques have made a significant contribution to lowering the cost of production without a concurrent reduction in yield by introducing genes for multiple resistance to pests and diseases, multiple tolerance to soil stresses. Thank to innovation that we have been successful in producing sufficient to quench hunger in our country to a greater extent. New breeding technologies are to be adopted involving genomics such as genome editing and CRISPR/Cas9, which can efficiently accelerate breeding methodologies. To develop climate resilient genotypes, high throughput plant phenotyping when the genotype and environment interaction determine the plant performance and crop production needed to be followed. Data management and modeling supporting handling of data and digital farming must be improved for better communication and outreach to society at large.

References

Abberton M, Batley J, Bentley A, Bryant J, Cai H, Cockram J, Costa de Oliveira LJ et al. (2015) Global agricultural intensification during climate change: a a role of genomics. Plant Biotechnol. J. 1:1095-1098.

Ansari AM (2015) Molecular markers in vegetable improvement. Hort. Biotechnol. Res. 1:5-10. doi: 10.5455/jhcbt.2015-07-05.

Bose RD (1932) Studies in Indian pulses. No.5. Urid or black gram. Ind. J. Agril. Sci. 2:625-637.

Bose RD (1939) Studies in Indian pulses. IX. Contributions on the genetics of mung. Ind. J. Agril. Sci. 9:575-594.

Chung CL, Jamann T, Longfellow J, Nelson R (2010) Characterization and fine-mapping of a resistance locus for northern leaf blight in maize bin 8.06. Theor. Appl. Genet. 121:205–227.

Foster BP, Till BJ, Ghanim AMA, Huynh HOA (2014) Accelerated plant breeding. CAB Rev. 9:3-15.

Glazier AM, Nadeau JH, Aitman TJ (2002) Finding genes that underlie complex traits. Science 298:2345–2349.

Henry RJ (2020) Innovations in plant genetics adapting agriculture to climate change. Curr. Opinion Plant Biol. 56:168-173.

Howard A (1914) Some varieties of Indian gram. Memoirs of Dept. Agr. in India, Bot. Ser. 7(6):6.

Jain HK (1986) Eighty years of post Mendalian breeding for crop yield: nature of selection pressure and future potential . Indian J. Genet. 46 (Suppl.):30-53.

Kadam BS, et al. (1945) Chafa-a new variety of gram from Bombay. Ind. Farming, 6:444-446.

Krattinger SG, Lagudah ES, Spielmeyer W, Singh RP, Huerta-Espino J, McFadden H, Bossolini E, Selter LL, Keller B (2009) A putative ABC transporter confers durable resistance to multiple fungal pathogens in wheat. Science, 323:1360-1363. DOI:10.1126/science.1166453

Lenaerts B, Collard BCY, Demont M (2020) Review: Improving global food security through accelerated plant breeding. Plant Sci. 287:1-8.

Liu L, Liu G, Gong Y, Dai W, Wang Y, Yu F, Ren Y (2007) Evaluation of genetic purity of F1 hybrid seeds in cabbage with RAPD, ISSR, SRAP, and SSR markers. Hort. Sci. 42:724-727.

Martínez F, Niks RE, Singh RP, Rubiales D (2001) Characterization of Lr46, a gene conferring partial resistance to wheat leaf rust. Hereditas, 135:111-114.

Mehta TR (1986) Early history of plant breeding in India. Indian J. Genet. 46 (Suppl.):1 -11.

Mishra GP, Singh N, Kumar H, Singh SB (2010) Protected cultivation for food and nutritional security at Ladakh. Def. Sci. J. 61: 219-225.

Mongkolporn O, Dokmaihom Y, Kanchana-Udomkan C, Pakdeevaraporn P (2004) Genetic purity test of F1 hybrid Capsicum using molecular analysis. J. Hort. Sci. Biotechnol. 79:449-451.

Ortiz R, Trethowan R, et al. (2007) High ield potential, shuttle breeding, genetic diversity and a new international wheat improvement strategy. Euphytica, 157:365-38.

Poland JA, Bradbury PJ, Buckler ES, Nelson RJ (2011) Genome-wide nested association mapping of quantitative resistance to northern leaf blight in maize. Proc. Natl. Acad. Sci. USA ,108:6893–6898.

Sadashiva AT, Aghora TS, Rddy MK, Mohan N, Rao ES (2014) Growing interest in double haploidy for improvement of horticultural crops in India. Curr. Sci. 107:16–17.

Singh B, Tomar BS (2015) Vegetable seed production under protected and open field conditions in India: A review. Indian J. Agr. Sci. 85(10):1251-9.

Singh BK, Sharma SR, Singh B (2009b) Heterosis for antioxidants and horticultural traits in single cross hybrids of cabbage (*Brassica oleraceavar.* capitata I). Indian J. Agr. Sci. 79(9):703-708.

Singh R, Mishra GP, Ahmed Z, Singh SB (2009a) Effect of various greenhouse structures on yield and quality of different table tomato varieties under cold arid desert of trans-Himalayas. Indian J. Agril. Sci. 79(4):243-47.

Singh RB (2012) Climate smart agriculture for food security. NAAS News, January-March. 12:1-2.

Singh RP, Huerta-Espino J (2003) Effect of leaf rust resistance gene Lr34 on components of slow rusting at seven growth stages in wheat. Euphytica, 129:371-376.

Singh VP, Singh AK, Atwal SS, Joseph M, Mohapatra T (2002) Pusa 1121: a rice line with exceptionally high cooked kernel elongation and basmati quality. Int. Rice Res. Notes, 27:25–26.

Swaminathan MS (1986) Integration of the tools of Mendalian and molecular genetics in crop improvement. Indian J. Genet., 46(Suppl.):12-29.

Wani KP, Singh PK, Amin A, Mushtaq F, Dar ZA (2011) Protected cultivation of tomato, capsicum and cucumber under Kashmir valley conditions. Asian J. Sci. Technol. 1(4):056-061.

Watson A, Ghosh S, Williums MJ, et al. (2018) Speed breeding- a powerful tool to accelerate crop research and breeding. Nat. Plants, 4:23-29.

Wisser RJ, Sun Q, Hulbert SH, Kresovich S, Nelson RJ (2005) Identification and characterization of regions of the rice genome associated with broad spectrum, quantitative disease resistance. Genetics, 169:2277–2293.

Zwonitzer JC, Bubeck DM, Bhattramakki D, Goodman MM, Arellano C, et al. (2009) Use of selection with recurrent backcrossing and QTL mapping to identify loci contributing to southern leaf blight resistance in a highly resistant maize line. Theor. Appl. Genet. 118:911–925.

4

Innovations in Seed Sector and Its Regulations for *Atmanirbhar Bharat*

D. Vijay

Division of Seed Science and Technology, ICAR-Indian Agricultural Research Institute, New Delhi

Introduction

Agriculture is the backbone of Indian economy. Good harvest helps in moving the economy in the right direction. The agriculture sector accounts for 17% of US$ 2.6 trillion economy of India (Anonymous, 2019a) and employs 54.6% of the country's workforce (Anonymous, 2019b). India occupies first place in the production of pulses, jute and milk, and second place in rice, wheat, cotton, sugarcane, vegetables and fruits (FAO STAT, 2020). The agriculture export constitutes a fifth of total exports made by India and self-reliance in agriculture is required to become an economic superpower. The green revolution has paved the way from begging bowl to overflowing granaries in our country. The overall food production has increased from 52 MT in 1950-51 to 295 MT in 2019-20, which is a phenomenal growth of more than five times in seven decades span. Of the several factors that influenced the sea change in agriculture production and productivity, the quality seed of improved varieties occupies a crucial position. The green revolution was ushered in by the import of 18000 tonnes of seed of high yielding wheat varieties Lerma Rojo and Sonara 64 and rice variety IR-8. Till today, the quality seed alone contributes to 15-20% of total production (Chauhan et al., 2015), indicating its importance in agricultural production of the country. The problematic global scenarios like economic recession, climate change, water scarcity, population explosion, pollution or new pandemics drives the states to be more self-reliant. To be self-reliant is not to be altogether "self-contained". We should be opened for the best material available anywhere in the world, and simultaneously we should develop our infrastructure and industry to cater to the needs of our population as well as export to other countries. Seed sector is one of the potential areas for boosting self-reliance as well as has a lot of export potential. The government push for self-reliance under the present COVID-19 pandemic situation by infusing INR one lakh crore as agricultural infrastructure fund that will help in developing several start-ups. The success of this movement mainly depends on innovations in every sector, including the seed sector and simultaneously, there is a need to extend it to the regulatory mechanisms. This chapter will discuss the possible areas of innovations that are already in the pipeline and those to be targeted in the areas of seed production, quality assurance, marketing and regulation along with policy interventions needed to boost the seed sector.

Status of Indian seed sector

The Indian seed industry is the most vibrant industry with a cumulative annual growth rate of 10-14%, comprises both public and private sector companies. The national seed project (NSP) phase-I (1977-78), phase II (1978-79) and phase III (1990-91) implemented by Government of India has strengthened the seed infrastructure and paved the way for the development of seed industry in the country. Further, the collaboration between the public and private sector and the mentorship role played by National Seeds Corporation (NSC) strengthened the Indian seed corporate sector. The major thrust for the phenomenal growth of the Indian seed sector was received with the implementation of the new policy on seed development in 1988. This policy completely brought transformation into Indian seed industry by allowing the import of seed material into the country. This gave access to the best quality seed and planting material available anywhere in the world to Indian farmers. The huge investment in the private sector, including the initiation of multinational companies (MNCs), and intense research and development (R&D) for product development was ushered. The development of hybrids in several crops, including vegetables and commercial crops like cotton attracted a greater number of companies to invest in the R&D. At present more than 500 seed companies in India are providing a wide range of varietal choice to the farmers, and the seed business became farmer-centric and market-driven.

The Indian seed market value is US$ 2.6 billion and is expected to reach US$ 5 billion by 2030. India is occupying 5th position in the global seed market with approx. 5.8% share. India exports 33036 MT of seeds worth of US$ 101 million and imports 29456 MT of seeds worth of US$ 121 million of vegetables, flowers and field crops (ISF 2020). The trade balance of the USA, occupying the first position in the global seed market, is surplus for these three categories of seeds. Based on the latest five years data (2013-2017) available from International Seed Federation (ISF), India is having a trade deficit in vegetable, flower and food crop seeds; even though the import quantity is less than export quantity. This clearly indicates the need for targeting the crops of importance, particularly those which have export potential. The deficit is higher in the case of flower seeds (Fig.1) followed by vegetable seeds (Fig.2). In the case of food crop seeds, import and export are mostly balanced (Fig.3). The significant bulk of both imports and exports is of field crop seeds whose value is much lower than vegetables and flowers seeds.

The per tonne value of flower seeds imported is higher than vegetables and food crop seeds. The gap between import and export values of per tonne of seed is higher in the case of flower seeds (Fig.4) followed by vegetables (Fig.5) and food crops (Fig.6). Among different categories of seed groups, vegetables have average profitability of 20% against only 5% in field crops. The annual global seed trade is US$ 14 billion. The 50% of total export is combinedly by France, Netherlands and Germany whereas, 25% of the total export is by the USA alone. India's share in the global seed trade is only 1%. However, there is a huge potential to capture at least 10% share by 2028 (Kaundinya, 2020). India has become a member of the Organization for Economic Cooperation and Development (OECD) seed scheme in 2008. But the seed exports are staggering around 20,000 tonnes per annum, which can be improved by the participation of a greater number of seed producers. A total of 61 countries are participating in the OECD seed schemes, where 62,000 varieties consisting of 204 species were enlisted for trade. There is a considerable scope to enlist several important Indian varieties to enhance the trade under OECD seed schemes. India is also exploring the possibilities to get the EU-Equivalence to reach the high-value European markets and farms.

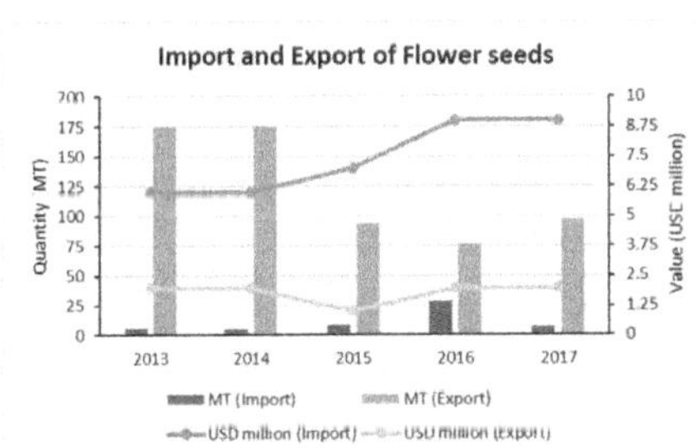

Fig. 1. Flower seed import and export in latest five years

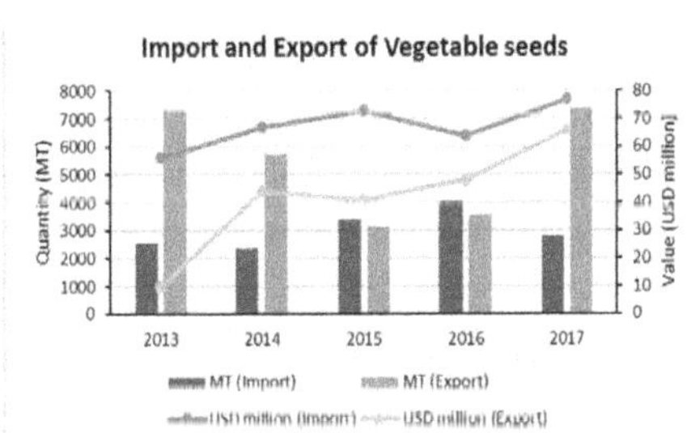

Fig. 2. Vegetable seed import and export in latest five years

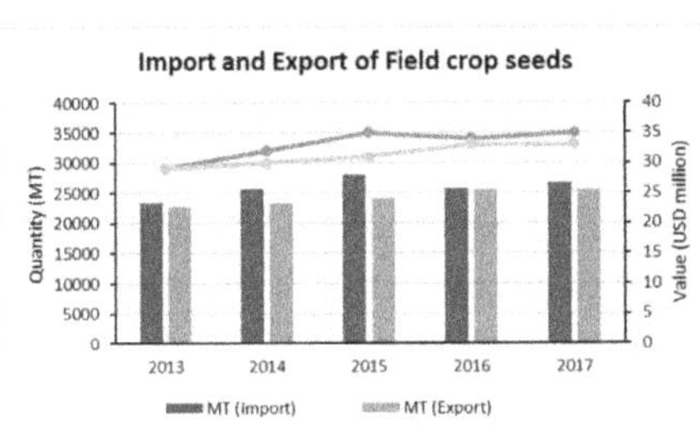

Fig. 3. Field crop seed import and export in latest five years

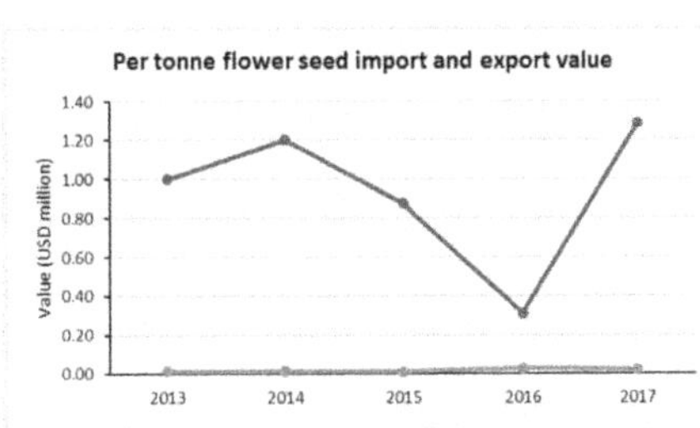

Fig. 4. Export and import value of flower seed latest five years

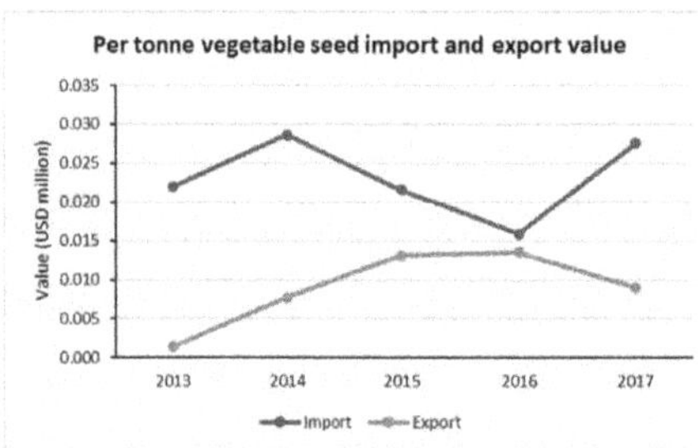

Fig. 5. Export and import value of vegetable seed

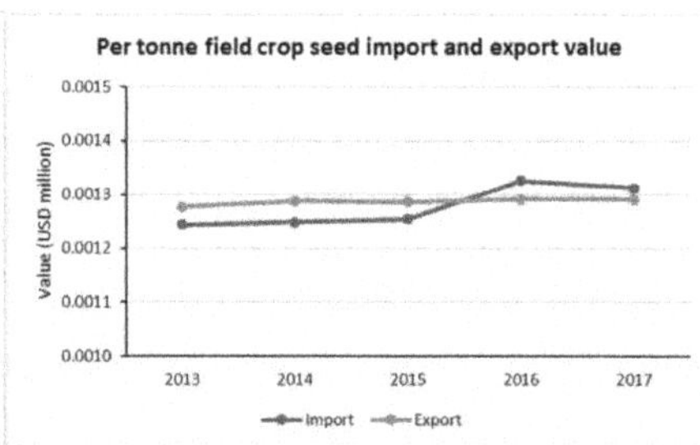

Fig. 6. Export and import value of field crop seed

Improvement needed over existing innovative models

Innovations are a continuous process and are essential to make a system more vibrant. The change in the requirements of the end-users acts as a guiding force for the innovations. The Indian seed sector is one of the most dynamic systems with equal participation by both the public and private sectors. The ushering of the green revolution in the 1960's by the import of huge quantities of seeds of dwarf varieties of wheat and rice is itself a disruptive innovation in the seed sector. Later in the eighties, the implementation of the new policy on seed development is another major innovative policy decision which changed the role of the private seed sector in India. Similarly, several innovations are either initiated or in the process of implementation to boost the present seed sector. These efforts need to be strengthened for having a positive impact on the Indian seed system.

(i) **Seed hubs:** Development of seed hubs is an innovative idea not only to meet the seed requirement but also to enhance production and productivity. The success of pulse seed hubs in recent past is a glorious example of this approach. The pulse seed hub project was initiated in 2016-17 under the centrally sponsored scheme of National Food Security Mission (NFSM) with an outlay of Rs. 225 crores with 150 KVKs of ICAR/SAUs for two years. A total of 3.57 lakh quintal of seed production was targeted for 2016-17 to 2018-19. The pulse production was increased from 16.32 MT in 2015-16 to 22.08 MT in 2018-19. The agriculture statistics before and after the project seed hub implementation increased the availability of seed by 16 lakh quintals that too from the public sector where the project was implemented (Fig 7). This helped in achieving the self-sufficiency in pulse crops. This innovative model has to be strengthened and extended for other crops like oilseeds where there is seed and production shortage.

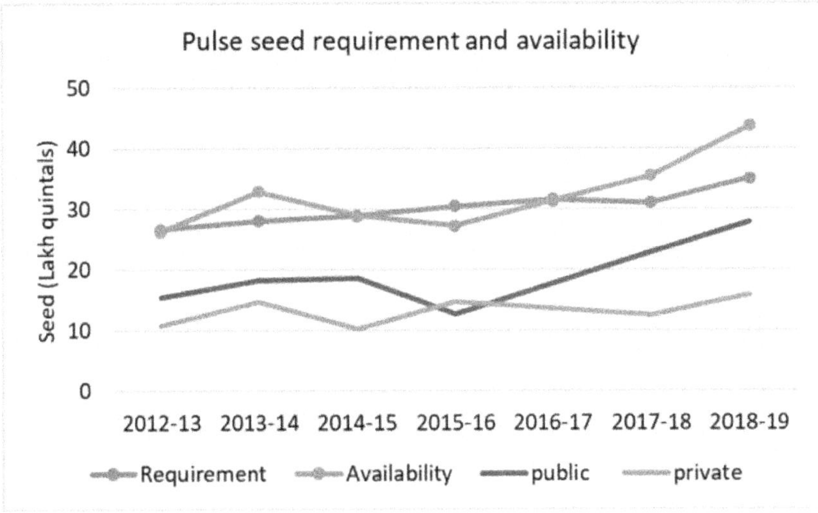

Fig. 7. Requirement and availability (from public and private sector) of pulse seed before and after implementation of seed hub program

(ii) **Seed banks:** This is a central scheme for establishment and maintenance of seed banks by the Department of Agriculture, Cooperation & Farmers Welfare (DACF&W) of Ministry of Agriculture and Farmers' Welfare (MoA&FW) started from 1999-2000 onwards through NSC, erstwhile SFCI, 18 State Seed Corporations. The main objective is to meet the seed requirements during natural calamities and unforeseen conditions. These seed banks can be strengthened and extended for different other purposes including conservation of native varieties, rare germplasms, speciality varieties and nutritionally superior genotypes. This program may be clubbed with seed village schemes to reap up better benefits for the self-sufficiency in targeted crops.

(iii) **Business class seed farmers (BCSF):** Indian agriculture needs to be developed as an industry. Individual farmers face a lot of difficulties in production, processing and marketing of agricultural produce, including seeds. Though with the initiative of public and private seed sector, many farmers produce seed by signing MoU, there is a need to develop a farmer-centric organization at farmers level to monitor seed production and marketing. Among the farmers, the progressive farmers can be grouped as business class seed farmers with the help of KVKs. They shall be trained and registered as a BCSF organization (BCSFO) to take up seed production in a competitive mode. These BCSFOs can have licence of newly developed varieties from research organizations for further disseminating to the farming community. The support from central and or state government shall be provided to BCSFO similar to FPOs. This can help in the fast spread of new varieties as well as make the farming community self-reliant. The development of these seed producing farmers companies will help towards increased seed exports along with self-reliance in the seed sector.

(iv) **VCU for all varieties:** The seed certification is restricted only to notified varieties, which undergoes the testing of value for cultivation and use (VCU) through AICRP trials. The private sector mainly sells the truthfully labelled seed of their varieties, which does not go through this VCU testing. The VCU testing helps in ascertaining the yield and also checking for pest and disease incidence at multiple locations. This test is essentially required to obtain EU equivalence, for pushing seed to European markets. This testing also helps in resolving the claims of various crop insurance schemes. In the new seed bill, it is proposed to make VCU compulsory for all the seed that is available for sale in the market. However, there is a need to develop well-established infrastructure and an independent agency, not involved in breeding and seed production, to monitor the testing. Registered third party agencies with qualified persons should be permitted to undertake VCU testing under the regulation of government organizations like ICAR which has a standard set up of crop-specific AICRP coordinating centres. This will boost up seed exports, assure the availability of quality seed to the farming community and helps in creating employment to the qualified agricultural graduates.

(v) **Seed traceability:** The quality of seed is very much essential to realize the yield potential of a variety. A lot of spurious seed is being marketed by various agencies which not only deprive the farming community economically but also reduce the food production of the country. Tracing the origin of the seed definitely helps in tracking the quality issues effectively. If the details of production and transport of seed lot are available, it can easily be checked, and low-quality lots either can be stopped from reaching the farmers or helps in resolving compensation claims efficiently. As desired by the government, there is a need to develop suitable software-based seed traceability mechanism in the country. Once it comes into usage, the seed quality assurance will be in place which augment the seed sector in the country.

Future innovations needed in the seed sector

With the changing needs of the society, following innovations are needed both in seed research as well as production and marketing sectors.

(i) **Identification of alternate production zones:** The seed production is concentrated in few pockets of the country like Hyderabad, Bengaluru, Eluru, Aurangabad, Nandyal, etc. due to several factors, more importantly suitability of climate is taken into account. However, with the changing climatic scenario, there is a need to identify new effective locations for various crops. The ICAR-Indian Institute of Seed Science, Mau through their coordinated trials identified alternate areas for hybrid seed production of paddy, sunflower, sorghum, maize, pearl millet, castor and jute. There is a need to effectively use these new areas for seed production and to extend them for other crops and areas. Similarly, the disease-free zones in the non-traditional production areas need to be recognized for export-oriented seed production. The identification of alternate production zones helps in sustaining the seed production within the country. A multi-disciplinary approach, including soil, seed, crop and protection scientists along with the meteorological and GIS data may bring required changes in seed production system.

(ii) **Custom seed production zones:** India is a vast country with variable climatic conditions as well as lot of seasonal variations simultaneously in different parts of the country. This diversity, along with low-cost human resources helps in developing custom seed production zones for other countries. This will not only help in creating seed export zones but also make farming a profitable venture. Proper guidelines and framework shall be developed to realize this idea into reality.

(iii) **Development of AI-based national seed grid:** Seed grid at the national level is visualized long back but could not be established. With the advent of artificial intelligence (AI) and penetration of information technology in the agriculture sector, it is far more achievable now. The national seed grid helps in connecting the seed production and marketing throughout the country and can be integrated with e-marketing. This may help in the reduction of the localized scarcity of seed availability in the country and effectively use the precious seed material. The varied agro climatic situation in the country leads to both excess and deficiency in seed production in different regions. An AI based online grid with live updating facility integrating both public and private seed sector along with transportation will help in boosting the seed sector as well as reduce the localized non-availability of seed material. This may work effectively even under natural calamities like floods and drought conditions. Initially, this can be implemented in the public sector seed production organizations like NSC, SSC and NARS institutes for certified and TL seed.

(iv) **Automation of seed testing**: Testing of seed quality is an essential step in the quality control system. It may be either through registered seed testing laboratories (STL) of state seed certification agencies, state agriculture departments or in-house seed testing labs of production agencies. In all the STLs, the seed testing is entirely manual, as a result the whole process gets delay. The automation of seed testing is possible with the development of AI-based machines. The germination testing, which is a prime test in seed quality assessment, can be automated, combining machine vision technology with artificial intelligence for bulk analysis of samples. Similarly, the new vigour tests can be integrated with the machine-based technology. The identification of inert matter, other crop seed and weed seed in seed purity testing can also be made through the development of large-scale databases for various crop species. Therefore, the automation of the seed testing helps in reducing the seed certification duration, which at present is one of the major hurdles during peak crop seasons. It also improves the accuracy and efficiency of seed testing laboratories.

(v) **Start-ups in seed delivery chain**: The seed delivery chain is crucial for the timely supply of this critical input of agriculture. At the same time, the timely availability of seed is essential for the food security of the nation. The encouragement for Start-ups in the seed delivery chain makes this process more effective and helps the farmers in choosing the best variety. The success of delivery chain Start-ups in other sectors may be exemplified in the seed sector also. The App-based systems integrating various production agencies and transport organizations helps in making this innovation possible in the seed sector. The seed delivery chain can be linked up to different government schemes and VCU database to identify the suitable varieties for farmers of other regions and supply their seed on a subsidized basis to farmers at their doorsteps.

(vi) **Digitization of seed procurement and supply chain:** The digitization of seed procurement process helps in easy identification of the material and speed up the seed certification process and minimize the involvement of middlemen. The digitization of the supply chain helps in the traceability of the seed production planning, sales order management, accounting, distribution of seed material at different levels of marketing, effective inventory control and human resource management. Although some kind of digitization has already been under practice in both the public and private sectors, they need to be integrated with several other AI-based technologies. A basic structure for the harmonization of activities across different stakeholders shall be developed. The new blockchain concept with decentralized and distributed ledger is the promising one for implementing in the seed sector.

(vii) **Integration of AI and sensor-based technology for seed quality assessment**: The adoption of new technologies for seed quality assessment *viz.*, e-nose, colour sorters, machine vision which are mostly sensor-based shall be integrated with the artificial intelligence for the seed quality assessment. The use of these technologies shall be standardized for all released varieties and rectify the deficiencies for easy adoption. The computer-based algorithms, electronic devices and seed science should come together to develop these kinds of innovations under a public-private partnership model. The sensor-based quality assessment enhances the accuracy of seed testing and helps in faster and efficient testing.

(viii) **Innovations in seed storage:** Storability of seed from production to next season sowing is an essential part of the seed system and is considered as a necessary evil. The loss of seed viability and vigour during storage is eminent. This component needs an innovative model consisting of identification of suitable locations with less humidity and low temperature using meteorological and GIS data, development of modified atmospheric storage units and customized transport systems. Therefore, an interconnected approach of research with industry is needed for its success. The development and growth of independent seed storage chains in a PPP mode will further strengthen seed quality and self-reliance in the seed sector.

Seed regulatory system in India

The seed supply system in India comprises both formal and informal sectors. The formal sector includes both public and private organizations and is controlled by several rules and regulations in the form of acts passed by the parliament. However, there is no control on the informal sector, which mainly operates at a local level in the unorganized way. This informal sector also comprises seed exchange among farmers. The seed, being the primary input of agriculture, is highly regulated in India as most of the farmers are small and marginal and are innocent. To protect the interests of the farmers and to make available the quality seeds in the market, several legislations were made. These legislations are oriented to regulate the formal sector from exploiting the farming community. Some of the critical seed legislations are, (i) Seeds Act (1966) and Seed rules (1968) (ii) PPVFR Act (2001), (iii) New policy on seed development (1988), (iv) Seed control order (1983), (v) EXIM policy (2002-07) and (vi) Biodiversity act (2002).

The Seeds Act was passed in 1966 and is the fundamental law that controls the seed quality through certification. However, only notified varieties are certified. In India seed labelling is compulsory, but not the seed certification. However, the label shall meet the Indian minimum seed certification standards. The un-certified seed with self-declared quality on the label is called truthfully labelled seed. The major part of marketed seed by the private sector is of truthfully labelled seed. Since the Seeds Act cannot control the seed quality of truthfully labelled seed, the seed control order was passed in 1983 under the essential commodities act of 1955 by declaring seed as an essential commodity. The seed control order governs the sale of seed through dealers and empowers the seed inspector to check the quality of the seed sold in the market and to seize the poor-quality seed. Of late, as per the obligations under WTO to give intellectual property rights for the crop varieties, the PPVFR Act was passed in 2001. Instead of joining the UPOV for the granting of varietal IPR to the breeders, India has developed its own plant variety protection under *sui-generis* system. The PPVFR Act of India is unique with the provision of farmers' rights and benefit-sharing clauses.

Along with the acts that regulate seed, several policies that direct the seed sector also played a crucial role in the self-sufficiency and industrial growth. Among the various policies, the New Policy on Seed Development (NPSD) of 1988 is one which played a critical role to make available the quality seed, present anywhere in the world, to the Indian farmers. The NPSD also brought a sea change in the Indian seed industry. With the permission to import seeds, the Indian seed industry has seen tremendous growth and ultimately the farming community got the advantage of availability of best varieties from different corners of the world. The biodiversity act gave the farming community the benefit-sharing for conserving the traditional and native genotypes and protected the Indian biodiversity from commercial exploitation without benefit sharing. Whereas, the EXIM policy has guided the industry for the import and export of seeds.

Policy interventions for self-reliant seed systems

The rigid regulatory system of seed should be eased towards encouraging the seed industry and simultaneously protecting the farmers' interests. Some of the following policy interventions are proposed for the development of self-reliant seed system in our country.

(i) Harmonizing the seed rules and regulations is the need of the hour. The seed act of 1966 is very old and needs immediate replacement. Although the new seed bill was placed in the parliament in 2004 until today, it could not be formed into an act. The new seed bill needs further deliberations as several new issues are emerging, and there are specific contentions between seed bill and the PPVFR Act 2001. Guiding and advising seed committees on scientific knowledge will be very effective for taking appropriate decisions. Harmonizing the procedures and standards in line with the international systems will help in ease of movement of seed to other countries and will make India, the seed hub of South Asia.

(ii) The seed quality testing and assurance system should be liberalized by licencing registered third-party entities to develop private STLs. This will help in effective implementation of quality control and make timely available of certified seeds for sowing. The private STLs will enhance the efficiency of quality control and also provides employment opportunities to qualified persons.

(iii) The emphasis on export-oriented seed production is needed. The scope of OECD seed schemes shall be extended to a greater number of varieties as most of the Indian varieties have potential to get exported to South Asian and African countries due to similar climatic conditions and food habits.

(iv) The development of seed hubs in flower, vegetable, and forage seed sectors will improve the availability of seeds of these species with in the country and also paves the way for their export to other countries. These seeds are of high-value and will shift the seed trade balance towards surplus.

(v) The seed chain has several loopholes which need to be plugged to make it an efficient system by taking appropriate policy decisions. The inefficient generation system of seed multiplication, will result in excessive production of higher-class seeds leading to wastage of precious resources, including human resources. Several orphan and neglected crops shall be brought under seed chain for their spread and availability. The seed multiplication of neglected crops like range grasses shall be linked to different government schemes for an assured market and development of organized production centres both at the public and private level.

(vi) Standard policy on GM seeds needs to be taken with prioritization of crops and traits so that the illegal spread of untested technology will be stopped. Specified production zones can be developed for GM crops for better containment with regulated seed supply and procurement system. Labelling of GM seed and their testing shall be pushed on a large-scale basis similar to registered human diagnostic centres with quality regulation. A single window platform shall be established in regulating the GM seeds for more efficient and time-bound action. The conflict of interest shall be addressed in the regulating system to have more transparency and confidence.

(vii) The informal seed sector shall be brought under quality assurance system, as more than 40% of total seed transaction in India is dealt under this system. The development of seed farmers' organizations and compulsory certification will ensure the quality of seed traded by the informal seed sector. However, the availability of the registered independent third-party certification and testing centres is essential to implement compulsory certification. Faster compensation provisions for financial losses at various stages due to mismanagement by these organizations needs to be put in place to protect the interests of the farmers. Efficient regulatory system to check illegal seed trade shall also be in place.

(viii) The adoption of producer compensation in place of subsidies will bring down the production costs and pave the way for increased export of seeds. This will also reduce the glut in the market during excess production and losses to the seed farmers. Simultaneous development of cold chains for seed storage will increase the shelf life and assured seed availability during calamities.

(ix) A reliable and robust database is required to integrate the IT and AI in the seed sector. Appropriate policy initiative is needed to collect the seed data by using various means including satellite-based remote sensing data and data collection by drones. The open-source data will help in third party verification of the success of different seed schemes and implementation of various programs through scientific analysis by academicians.

(x) Policy reforms in public seed sector are needed. Service nature of the public seed sector shall be kept in mind while formulating various policies. Mere duplication of private industry or privatization of the public sector will deceive the purpose of social responsibility and will be detrimental to Indian agriculture. The profitability and the public service, cannot be fitted into the same box. The policy shall demarcate the profit-oriented units and made them independent of public service units so that they can compete effectively with the private sector without any interference.

References

Anonymous. 2019a. India-Agricultural sector. https://www.export.gov/apex/article2?id= India-Agricultural-Sector.

Anonymous. 2019b. Agricultural Statistics at a Glance 2019. Directorate of Economics and Statistics, Department of Agriculture Cooperation and Farmers Welfare, Ministry of Agriculture and Farmers Welfare, Government of India. Pp. 301.

Chauhan, J. S., A. L. Singh, S. Rajendra Prasad, and Satinder Pal. 2015. Quality Seed: A Mega Factor in Enhancing Crop Productivity, p. 357-426. In: A. L. Singh, ed. Recent Advances in Crop Physiology, Volume II. Daya Publishing House, New Delhi.

FAO STAT. 2020. FAOSTAT statistical database.http://www.fao.org/faostat/en/

Kaundinya, R. 2020. India can become a hub for seed exports. The Hindu Business Line dated 23 July 2020.

ISF 2020. International Seed Federation. https://www.worldseed.org/

5

Innovations on Agrobiodiversity for Sustainable Agriculture in India

K.K. Gangopadhyay, Anuradha Agrawal and Gowthami, R.

ICAR-National Bureau of Plant Genetic Resources, New Delhi

Introduction

It is known and well documented that 'biological diversity' (i.e. diversity within species, between species and within ecosystems) has been declining at a much faster rate than earlier time in human history. The current rate of extinction is tens to hundred times higher than the average over the past 10 million years and it is accelerating (IPBES, 2019). The estimates of the number of species getting extinct every year vary from low (200 to 2,000) to high (10,000 and 100,000) (Bar-On et al., 2018). The delicate balance of clean air, water and biodiversity in nature which helps in survival on earth, is rapidly being destroyed due to humans' activity like growing of food, producing energy, dispose of waste and consume resources (IPBES, 2019). The current human population of the world (7.6 billion) represents only 0.01% of all living creatures and attributed to be instrumental in causing loss of 83% of all wild mammals and half of plants (IPBES, 2019). These unprecedented changes in important components of biodiversity are likely to continue or increase in the foreseeable future, unless we take active corrective measures (IPBES, 2019).

Whilst all biodiversity has intrinsic economic or social value, it is the 'agricultural biodiversity' ('agrobiodiversity' in short) that has the greatest importance for food nutrition and health security of humans. The FAO has recently defined biodiversity for food and agriculture (BFA) as a subset of biodiversity that contributes primarily to agriculture and food production (FAO, 2019). It includes the domesticated plants and animals raised in crop, livestock, forest and aquaculture systems, harvested forest and aquatic species, the wild relatives of domesticated species, other wild species harvested for food and other products. It also encompasses "associated biodiversity" of the vast range of organisms that live in and around food and agricultural production systems, sustaining them and contributing to their output (FAO, 2019). It comprises diversity of genetic resources (varieties, breeds), species used for food, fodder, fibre, fuel and pharmaceuticals, non-harvested species that support production (soil micro-organisms, predators, pollinators) and those in the wider environment that support agro-ecosystems (agricultural, pastoral, forest and aquatic) as well as the diversity of the agro-ecosystems (FAO, 1999). The genetic diversity ensures continuous improvements in food production, allows adaptation to current needs and ensures adaptability to future ones. The term of the also encompasses the many ways in which farmers use the natural diversity of the environment for production, including not only their choice of crops but also their management of land, water, and biota as a whole; that is the associated cultural practices and knowledge (Dulloo et al., 2017). Agrobiodiversity is essential for ecosystem services such as pollination, pest control, nutrient cycling and ensures greater adaptive capacity to global production challenges including climate change, emerging disease risk and changing consumption patterns (Dulloo et al., 2017). Using agrobiodiversity (from genetic to ecosystem level) produces a web of interactions, which make ecosystems more resilient, and helps in reducing unexpected risks (Bioversity International, 2019).

India's Agrobiodiversity Scenario

India is one of the world's most biologically, ethnically and culturally diverse countries, ranking second in food and agricultural production, globally. Although India's agriculture sector accounts for only 14% of the country's economy, but provides 42% of total employment (MOSPI, 2020). India is a mega-biodiverse country with only 2.4% of the world's land area and 7–8% of all recorded species in the world (MoEF, 2014). This includes ~11% species of the plant kingdom (47,513) and ~6.7% of the animal kingdom (1,01,167) documented in the world so far (Singh and Dash, 2014; Venkataraman et al., 2020). Of the 35 global biodiversity hotspots, four are present in India, represented by the Himalayas, the Western Ghats, the North-east, and the Andaman & Nicobar Islands (Mittermeier et al., 2011; Williams et al., 2011; MoEF, 2014).

In spite of the limited geographical area (328.73 million hectares (mha), the country supports approximately 16% of the world's human population and 20% of the world's livestock population. Nearly 45.3% (139.51 mha) of India's land is arable and under net sown area (GoI, 2020). Based on climate, soil, geology, vegetation and crops the country is distinguished into 10 biogeographic regions, 15 agro-climatic zones, 21 agro-ecological zones (AERs) and 60 agroecological sub-regions (AESRs) (MoEF, 2014; Srivastava et al., 2019). Nayar et al. (2009) identified 22 agrobiodiversity hotspots, while Singh (2015) described 21 National Agricultural Biodiversity Heritage Sites based on agricultural history/heritage, richness in cultivated plant species diversity, genetic diversity within crop / cultivated species and innovative agricultural practices developed by local tribes / communities' overtime for sustainability of agriculture and management of natural resources. The hotspots and/or agricultural biodiversity heritage sites are closely associated with the agroecological regions with some overlap. Thus, agroecological regions with distinct physiography, soils and bioclimate appear to have been the centers of agricultural biodiversity with crop domestication cultivation, adaptation and agricultural diversification/ development (Singh, 2017).

India has established five National Bureaux under the Indian Council of Agricultural Research (ICAR) to conserve and facilitate utilization of genetic resources of plants, animals, insects, microorganisms and fish. These are (i) ICAR-National Bureau of Plant Genetic Resources (NBPGR), New Delhi with >0.45 million accessions of plant genetic resource (ii) ICAR-National Bureau of Animal Genetic Resources (NBAGR), Karnal, which has a total holding of 1,23,483 frozen semen doses from 276 breeding males representing 38 breeds of cattle, buffalo, sheep, goat, camel, yak and horse, etc. (iii) ICAR-National Bureau of Agriculturally Important Microorganisms (NBAIM), Mau with a repository of 4,668 cultures (4,644 indigenous and 24 exotic accessions) (iv) ICAR-National Bureau of Agricultural Insect Resources (NBAIR), Bengaluru with 593 insect germplasm holdings (v) ICAR-National Bureau of Fish Genetic Resources (NBFGR), Lucknow with a repository of 2,553 native finfishes and Fish Barcode Information System with 2,570 microsatellite sequences. The respective Bureau coordinates the management of its genetic resource involving various stakeholders in its functional area (Mohapatra, 2016).

The immense diversity of traditional farming systems and practices in various parts of India contribute to the food security of hundreds of millions of people across the country. Today, India is the world's largest producer of milk, pulses, jute and banana, and ranks as the second largest producer of rice, wheat, sugarcane, groundnut, vegetables, fruit and cotton (FAO, 2019b). The country is also one of the leading producers of spices, fish, poultry, livestock and plantation crops. India ranks first in thc world in buffalo population, second in cattle and goat, third in sheep, fourth in duck, fifth in chicken and sixth in camel. Besides this, India also has amongst the world's largest diversity of domesticated animals, including cattle (42), buffaloes (15), goat (34), sheep (43) and chicken (19) (MoEF, 2014). Out of the 34,000 fish species reported globally (http://www.fishbase.org/search.php), more than 3,300 species occur in the Indian subcontinent. The database of National Bureau of Fish Genetic Resources (NBFGR) has 868 indigenous species; of which 877 are found in freshwater, 113 in brackish water and 1,878 in marine waters. Insects form nearly 75% percent of all animals, representing tremendous diversity (MoEF, 2014).

Concerns related to Agrobiodiversity

The unprecedented COVID-19 pandemic has brought to fore the importance of agricultural sector in ensuring food security during times of disasters and crisis. How food security is defined, which in turn becomes the basis of how agriculture programs and projects are designed to achieve it, raises an important aspect related to viewing agricultural systems as food systems (Ancog and Gregorio, 2020). Both globally and particularly, in India, it is well recognized that accelerated erosion of agro-biodiversity and its negative impacts on ecosystem services are threatening productivity and sustainability of agriculture and food systems (Paroda, 2018; Singh, 2019). Globally, humans have made conscious selections from an estimated 7,000 edible plant species that are either cultivated by them or harvested from the wild. Of these, only about 150 species have gained significant commercial importance; two-third of which produce 90% per cent of the world's food. But the intensity and options of continued human intervention have remained confined to just five staples, including three food grains – rice, wheat and maize, and two roots and tubers – potato and tapioca, which together produce two-third of the world's food (Paroda, 2016). It has been estimated that 75% of crop diversity was lost in the 20th century and only 12 plant species and five animal species provide more than 75% of all human food (FAO, 2015). Out of 3,00,000 identified plant species at world level, only ~7,000 have been cultivated and domesticated by humans for food, feed and fodder (FAO, 2019a). However, currently only 30 species form the basis of world's agriculture. Just three crops (rice, wheat and maize) account for about 60% of the calories and 56% of the proteins that humans consume directly from plants (FAO, 2015; 2019a). Climate change is predicted to reduce agricultural production by 2% every decade until 2050, with yields of major crops reducing by an average of 8% in Africa and South Asia (IPCC, 2014). Due to more frequent and unpredictable extreme weather events, small holding farming communities will continue to be the hardest hit.

As the future global food security is completely dependent on improved productivity and appropriate management of agrobiodiversity, two major international treaties of the United Nations (UN) draw attention of all governments towards it. First is the Convention on Biological Diversity (CBD). It specifically recognized the need for scientific research on genetic resources, *in situ* conservation and emphasized the need for utilization of genetic resources for sustaining food security at national level. Its Aichi Target 13 calls for maintenance of genetic diversity of cultivated plants, farmed and domesticated animals and of wild relatives, including other socio-economically as well as culturally valuable species, and to develop and implement strategies for minimizing genetic erosion and safeguarding their genetic diversity. Maintaining genetic biodiversity is also addressed under the UN Sustainable Development Goal (SDG) 2 related to 'Zero Hunger' that envisions eradicating hunger, achieving food security and improved nutrition, and promoting sustainable agriculture. Under this, Target 2.5 calls for conservation. of genetic diversity of seeds, cultivated plants and farmed and domesticated animals and their related wild species, including through efficiently managed and diversified seed and plant banks at the national, regional and international levels, and promote access to and fair and equitable sharing of benefits arising from the utilization of genetic resources and associated traditional knowledge.

The loss of agricultural biodiversity in global food production systems, as well as associated cultural practices and knowledge, is an issue of deep concern, particularly in centers of origin and diversity. The sustainable management of agrobiodiversity will determine our future, both in urban and rural areas. Agrobiodiversity plays an important role in achieving food security, eradicating hunger, improving human nutrition and provides essential functions in the agricultural landscapes. Keeping in view its enormous role, the 1st International Agrobiodiversity Congress (IAC) was held in New Delhi during November 6-9, 2016, which was inaugurated by Hon'ble Prime Minister of India (Paroda et al., 2017). While elaborating on the strength of India's agrobiodiversity and the associated knowledge, the Prime Minister called for intensifying research in agrobiodiversity, harmonizing global laws on conservation, and invited countries to develop a "shared vision" on the use of diversity to ensure food security and sustainability (Archak et al., 2017). The Congress provided opportunity for more than 60 countries and concerned international organizations to share the latest developments (Paroda and Agrawal, 2017). An important outcome of the IAC2016 was the 'Delhi Declaration on Agrobiodiversity Management' (see Box 1) which urges nations to accord top priority to the shared vision of agrobiodiversity conservation and their sustainable use towards achieving targets of SDGs relating to poverty alleviation, food and nutritional security, good health, gender equity and partnership by developing the necessary funding, legal and institutional mechanisms (Paroda and Agrawal, 2017; Paroda et al., 2018).

Box 1: Delhi Declaration on Agrobiodiversity Management	
1.	We call upon nations to accord top priority to the shared vision of agrobiodiversity conservation and sustainable use towards achieving the Sustainable Development Goals (SDGs) and the Aichi Targets of the Convention on Biological Diversity addressing poverty alleviation, food, nutritional and health security, gender equity and global partnership.
2.	We recognize the importance of traditional agrobiodiversity knowledge available with farm men and women, pastoralists, tribal and rural communities and its central role in the conservation and use for a food secure and climate resilient world. We, therefore, call upon countries to develop the necessary legal, institutional and funding mechanisms to catalyze their active participation.

3. We urge researchers and the policymakers to initiate, strengthen and promote complementary strategies to conserve agrobiodiversity through use, including greater emphasis on using crop wild relatives. We call for them to ensure a continuum between ex situ, in situ, on-farm, community based and other conservation methods with much greater and equal emphasis on each.
4. We propose that researchers employ modern technologies including, but not limited to, genomics, biotechnology, space, computational, and nano-technologies for genetic resources characterization, evaluation and trait discovery. The aim must be to achieve efficiency, equity, economy and environmental security through diversified agricultural production systems and landscapes.
5. We reemphasize the necessity of global exchange of plant, animal, aquatic, microbial and insect genetic resources to diversify agriculture as well as our food basket and to meet the ever-growing food and nutritional needs of all countries. To ensure this, nations need to be catalysed to adopt both multi-lateral (as envisaged in the International Treaty on Plant Genetic Resources for Food and Agriculture) and bilateral (as per the Nagoya Protocol) instruments to facilitate the exchange of genetic resources, while ensuring equitable access and benefit sharing opportunities.
6. Countries are also expected to harmonize their existing biosecurity systems, including phytosanitary and quarantine, and enhance their capacities to facilitate safe trans-boundary movement of germplasm.
7. We also expect that the governments and civil societies lay much greater emphasis on public awareness and capacity enhancement programs on agrobiodiversity conservation in order to accelerate its effective and efficient use.
8. We recommend the development and implementation of an Agrobiodiversity Index to help monitor on-going genetic resource conservation and management efforts, with particular emphasis on agrobiodiversity hot spots.
9. It is also urged that public and private sectors and civil societies henceforth actively invest in and incentivize the utilization of agrobiodiversity to mitigate malnutrition, increase the resilience and productivity of farms and farming households and enhance ecosystem services. Such efforts should lead to equitable benefits and opportunities, with particular emphasis on women and youth.
10. We urge countries to reprioritize their research and extension with increased investments to support the conservation and use of agrobiodiversity. Furthermore, we strongly recommend to create an International Agrobiodiversity Fund as a mechanism to assist countries and communities in scientific in situ and ex situ conservation and enhanced use of agrobiodiversity.
11. We urge the United Nations to consider declaring a 'Year of Agrobiodiversity' in order to draw worldwide attention and catalyse urgent actions for effective management of genetic resources by the global community.
12. Finally, we recommend that the International Agrobiodiversity Congress be held every four years, with Bioversity International playing the facilitator's role, to maintain the momentum gained in 2016 and continue emphasizing the need to implement the 'Delhi Declaration on Agrobiodiversity Management' and monitor the progress so made by the different stakeholders and countries.

Innovations in Agrobiodiversity for Sustainable Agriculture

1. **Broadening of food basket of the country by mainstreaming potential crops:** Cultivation of only a handful crops repeatedly in large areas has resulted in the loss of agrobiodiversity and as a result narrowed the food basket. India is a storehouse of a number of edible species with numerous benefits compared to the main cultivated crops such as millets (barnyard millet, finger millet, foxtail millet, kodo millet, little millet, proso millet), legumes (adzuki bean, fababean, moth bean, lentil, lathyrus, horsegram, rice bean, winged bean), pseudocereals (buckwheat, quinoa) which are yet to be scientifically researched, fully utilized and mainstreamed. Many of these crops are well-adapted to hot and hardy climates, easily cultivated in marginal lands with limited inputs and maintenance, have resistance to biotic and abiotic stresses, are part of the local diet and highly nutritive. Therefore, there is a need to bring other potential crops with high economic and nutritive value into the mainstream agriculture (Fig. 1).

	Activity	How to achieve	Expected output
1	Identification of priority species for mainstreaming	Based on previous research on underutilized crops done in the country by several institutes, superior species which has the potential to compete with the main corps should be identified for all the zones separately	New crops are identification for cultivation
2	Cultivation in the selective farmers fields	Progressive farmers with large land holdings should be selected for initial cultivation of selected new crops	Cultivation practice, post harvest handling, storage and processing
3	Value addition marketing	Instead of raw materials, value addition of final harvest with attractive post-harvest pacakging	Cost: benefit ratio estimation Ability to compete with main corps in markets Develop market-oriented strategies
4	Awareness on new crops	Success stories of the new crops cultivation should be disseminated to public through news, advertisements KVKs, farmers meetings, field visits with complete details	Mainstreaming of new crops

Fig. 1. Strategic plan for mainstreaming new crops for cultivation

2. **Measuring Agrobiodiversity Index**: The Agrobiodiversity Index is an innovative tool developed recently, which helps to gauge agrobiodiversity and identify concrete actions to help achieve diverse, sustainable and resilient food systems (Bioversity International, 2019). It not only measures species diversity, but also the potential contribution to healthy diets, sustainable agriculture and genetic resource management for future options. The index takes into account the state of agrobiodiversity in markets and consumption, in agricultural production, and in genetic resource management, looking at diversity in terms of species, varieties, functions, soil biodiversity and

landscape complexity. The data generated provides an indicator of the extent to which commitments and actions at national level support sustainable use and conservation on of agrobiodiversity for healthy diets, sustainable agriculture and future options. The index needs to be adapted to measure it at regional level (say agroecological zones), to determine the factors which require policy and structural reforms for long-term sustainability of agriculture.

3. ***In situ* and on-farm agrobiodiversity conservation as 'sustainable seed innovation'**: This approach comprises dual promotion of *in situ* innovation of plant varieties and agrobiodiversity conservation by the primary users (Kochupillai, 2016). Reframing conservation as sustainable seed innovation allows for a dynamic view of farmers' contributions. *In situ* agrobiodiversity conservation by farmers/pastoralists/fishermen not only provides immediate benefits, but also provides options for climate change adaptation which is critical to sustainable agriculture (Kopytko, 2019). Sustainability in this case implies not only agrobiodiversity conservation *per se*, but also insurance against crop failure and nutritional security by using the indigenous or traditional knowledge of small-scale farming communities (Kopytko, 2019). Traditional knowledge of what works under specific conditions, and continued access to genetic diversity provides a vital strategy for adapting to climate change (Smith et al., 2015). For farmers, having a portfolio of species and within-species diversity at on-farm helps them to withstand or recover from extreme weather events. Sustainable seed innovation within India receives support primarily from non-profit groups favoring open-source systems. Conserving natural and financial capital motivated farmers to adopt sustainable techniques, but attracting additional innovators requires development of new markets. India's Protection of Plant Varieties and Farmers' Rights Act (PPV&FRA) recognizes farmers as plant breeders but does not provide incentive to innovate sustainably. Currently, agricultural policies favour "progressive" farmers who may be following unsustainable practices incentivizes formal innovations, at the expense of sustainable innovations of farmers.

4. **Diversification through participatory approach of small holding farmers**: As extreme weather events become more frequent and unpredictable, small holding farming communities will continue to be the hardest hit. Diversification is one way to give farmers more options in times of need. With access to a diversity of crops and varieties, farmers are enabled to cope more effectively with the effects of climate change. But farmers do not always have the information or planting material to choose what diversity best suits their conditions. By harnessing the existing genetic diversity in the crops/animal breeds in different countries and proving to farmers directly would be an excellent solution. This diversity may be

in genebanks, breeding programs or in farmers' own fields. Much of the diversity that is available never reaches farmers' fields. This approach requires screening genetic materials regardless of their origin, to identify landraces or varieties that are better adapted to local growing conditions than what is available through commercial seed catalogues (Coto et al., 2019). The approach combines agromorphological and molecular characterization and farmers' evaluation to identify the best varieties. Farmers' participation in variety selection and breeding is central to this approach. Farmers test and evaluate the varieties on their own farm conditions, providing valuable feedback on their preferred traits. By involving them as 'citizen scientists', farmers increase their first-hand knowledge of useful varieties and traits. Scientists combine this farmers' generated data with agro-climatic and soil data, allowing them to identify which varieties are best adapted to local growing conditions, and to map traits that inform farmer decisions and breeding (Coto et al., 2019). Digital platforms and apps facilitate data collection and scaling, to reach thousands of farmers.

5. **Development of region-specific integrated farming system (circular agriculture) for small and marginal farmers:** In the country, about 85% of the operational holdings are small and marginal farmers. To attain food, nutrition and livelihood security, there is a need of integrated innovative approach for vertical expansion with less space, time with sustainable income round the year. Integrated farming system (IFS) is one such approach involving different components namely, crops, horticulture, dairy, fisheries, poultry, etc. with efficient utilization of available natural resources viz., land, water, nutrients, agrobiodiversity. It provides regular sustainable income around the year, decreasing cultivation cost through multiple uses of resources and providing much needed resilience for predicted climate change scenario, meets the basic needs of farm families along with nutritional improvement, raises the family income through optimum use of resources and proper recycling within the system, also to mitigate the impacts of drought, flood or other extremes weather events, Effective recycling of farm waste for productive purposes, community-led local systems for water conservation. The IFS models can fulfill the national mission of "Per Unit-More Income" among small and marginal farmers. Appropriate subsidies and incentives can be provided to farmers switching over to IFS. Capacity building and training especially targeted towards women and youth would be required for scientific management of IFS components like crop production, livestock, fish, mushroom cultivation etc.

6. **Development of strategies for conservation and enrichment of below ground biodiversity (BGBD):** Healthy and fertile soil is the basic requirement for sustainable agriculture. Soil harbors enormous below ground biodiversity which plays an important role in nutrient cycling, soil

health, soil carbon sequestration, soil texture and structure, mineralization and mobilization of nutrients, nutrient uptake by the plants etc. But other than soil health maintenance in terms of nutrient status, very little emphasis has been given for conservation of BGBD. In addition to other cultivation practices, major emphasis should be given to enhance the soil microbial biodiversity and effective soil microbes or microbial formulations should be identified for supplementing and enriching BGBD. Research required is identification of indicator species for assessing the soil biodiversity, artificial mass production of soil microbes or microbial formulations, multi-location field testing and validation of developed microbial formulations in soil health and soil biological diversity enhancement. The 'Soil Health Card' Scheme needs to integrate use of such microbial formulations and provide concurrent incentives/subsidies.

7. **Developing policy that explicitly acknowledges and integrates the role of agrobiodiversity:** Agrobiodiversity can deliver substantial environmental benefits with traditional crops and livestock varieties generally better adapted and more compatible with ecological conditions. The 2008 National Biodiversity Action Plan (NBAP), updated in 2014, prioritizes agrobiodiversity and calls for greater policy harmonization to avoid adverse impacts to biodiversity. However, agrobiodiversity is not integrated into some of the largest and most influential agricultural programs in India, such as the National Food Security Act or the Mid-Day Meals Program. The National Biodiversity Authority aims to coordinate efforts related to India's Biological Diversity Act (2002). The authority has sub-boards located in 29 states and has established nearly 38,000 Biodiversity Management Committees. National Biodiversity Authority has representation from agriculture agencies, forestry agencies, and environment agencies. This authority also needs to intensively integrate biodiversity concerns into agricultural policies and initiatives. Minor shifts in funding criteria for large central programs could significantly alter state-level activities. For example several benefits could be gained by aligning the National Food Security Act (NFSA) with the National Food Security Mission (NFSM), the National Nutrition Mission, and agrobiodiversity-related initiatives, particularly by including NFSM-advocated crops in the NFSA (e.g., pulses, which reduce the need for urea fertilization, fix nitrogen, and are generally high in protein and zinc). Likewise, numerous policies offer opportunities for very large benefits as a result of relatively easily achievable shifts as indicated here. To fulfill the obligation of Convention on Biological Diversity (CBD) Protection of Plant Variety & Farmers' Right Act 2001' (PPV&FRA) has been enacted in India to protect the rights of the farmers in respect of their contribution made at any time in conserving, improving and making available plant genetic resources for the development of the new plant

varieties. Moreover to accelerate agricultural development, it is necessary to protect plants breeders' rights to stimulate investment for research and development for the development of new plant varieties. Nevertheless, it facilitates to develop a database for utilization of indigenous plant genetic resources.

8. **Use of AI/ Blockchain technologies for Agro-Seeds Sector**: Data, including data associated with agrobiodiversity and locally relevant heterogeneous materials, is increasingly considered a tradeable commodity (Kochupillai, 2019). However, the transfer/sharing of such data needs to be regulated to facilitate its safe, beneficial, sustainable and equitable use in downstream research as well as in various emerging technological applications, the most prominent of which are artificial intelligence (AI) and machine learning applications. With growing economic and political relevance of data, several major ethical issues linked to its use are also surfacing. Platforms that adequately address these issues are indispensable to the meaningful and widespread adoption of (AI) applications that rely on data to design solutions for any specific sector/use case. Digital Ledger Technologies (DLT), including blockchain, can potentially help address several of these ethical issues, including issues of trust and privacy, security and "controllable" data sharing, fair, inclusive and equitable economic benefits for those sharing data, and traceability (for purposes ranging from economic benefit sharing to liability determination). Blockchains (in combination with other technologies like AI and Machine Learning) can potentially be used in the seed sector to incentivize the sustainable and equitable sale and use of agrobiodiversity and heterogeneous material in agriculture on one hand while also incentivizing benefit sharing and equitable downstream research with this biodiversity. Application of blockchain technology may support sustainable and equitable agriculture which (economically) benefits both small farmers and corporations, and help overcome issues that intellectual property law and biodiversity conservation regulations have together not been able to optimally resolve. Data platforms based on public or consortium blockchains with appropriate software architectures and consortium governance models have the potential to overcome issues of lack of trust (fear of biopiracy), that prevents farmers from sharing agro bioresources, and absence of means facilitating their traceability to source after any instance of sharing or sale. They empower data providers to collectively track, control and monetize the usage of their contributed data and assets. In particular, such systems can help by (i) maintaining an immutable record of transactions/transfers of agrobiodiversity from farmers to various categories of end users; (ii) collecting data about the agrobiodiversity transferred, such as unique cultivation methods, special characteristics etc.; (iii) ensure that the contributors and transformers of data can retain control over data user, how

and when it can be viewed or accessed; (iv) structuring the data to be used in associated solutions (such as AI and ML solutions) in a manageable form, and (v) facilitating automatic transfer of monetary benefits for the contributors of data, especially from their use in diverse applications, thereby effectively incentivizing continuous cultivation and engagement with agrobiodiversity (Kochupillai, 2019 and Kochupillai et al., 2019). Blockchain/DLT can help farmers engaged in traditional and sustainable agricultural practices using local heterogeneous seeds, and traditional soil and ecosystem management methods, manage and monetize their data/know-how while sharing these with AI/Machine learning apps and other users and stakeholders. This would significantly help incentivize and accomplish a goal that is of current and growing global relevance, namely, incentivizing research and *on-farm* innovation with agrobiodiversity.

Conclusion

Since the beginning of agriculture, human interventions in agro-ecosystems have helped in agriculture-led development leading to food and nutritional security of societies. Agricultural expansion and intensive cultivation have considerably reduced the existing agrobiodiversity. Tradeoff for increasing food grain production have contributed to decline and degradation in other ecosystem services like clean water, recycling and carbon foot-prints (Paroda, 2016). At the same time, everyone is concerned by the degradation of agricultural soils and environment, and unknown threats of climate change. The emerging scenarios have posed further management challenges of all sorts, especially the assessment of potential new genes to manifest adaptability and productivity under severe abiotic and biotic stresses. Most redeeming feature of the existing agrobiodiversity is that it offers great scope to adapt to or mitigate the climate change effects in future. Nonetheless, to achieve this, we shall need greater research and development efforts to harness available potential using new science before it is too late. In this context, new tools of research and accelerated germplasm use have been developed with increased sophistication and precision to attain both efficiency and targeted purposefulness. The cost of research has also increased tremendously, whereas the corresponding policy support and donor interests are diminishing, which needs to be reversed before it is too late. Thus, funding in research, capacity development and training, especially of rural youth and women, for enhancing their analytical and decision making capabilities, have emerged as additional challenges for managing the change relating to conservation and rational use of agrobiodiversity.

References

Ancog, R.C. and G.B. Gregorio. 2020. Impact of COVID-19 Pandemic on Agriculture Production in Southeast Asia: Reinforcing Transformative Change in Agricultural Food Systems (No. 2020-1). Southeast Asian Regional Center for Graduate Study and Research in Agriculture (SEARCA).

Archak, S., R.K. Tyagi, A. Agrawal and P.N. Mathur. 2017. Delhi Declaration provides a Roadmap for Agrobiodiversity Management. Indian J Plant Genet Resourc 30(1): 88-91.

Bar-On, Y.M., R. Phillips and R. Milo. 2018. The biomass distribution on earth. Proc National Acad Sci 115(25): 6506-6511. https://www.pnas.org/content/115/25/6506

Bioversity International. 2019. Agrobiodiversity Index Report 2019: Risk and Resilience. Bioversity International, Rome, Italy. https://hdl.handle.net/10568/100820 (accessed Oct. 10, 2020).

Dulloo, M.E., J.E.O. Rege, M.Ramirez, A.G. Drucker, S. Padulosi, N. Maxted, B.R. Sthapit, D. Gauchan, I. Thormann, H. Gaisberger, N. Roux. 2017. Conserving agricultural biodiversity for use in sustainable food systems. In: Mainstreaming Agrobiodiversity in Sustainable Food Systems. Scientific Foundations for an Agrobiodiversity Index, Bioversity International, Rome, pp. 103–140.

FAO. 2015. The Second Report on the State of the World's Plant Genetic Resources for Food and Agriculture. Food and Agriculture Organization, Rome.

FAO. 2019. The State of India's Biodiversity for Food and Agriculture. Rome. http://www.fao.org/3/CA3444EN/ca3444en.pdf (accessed Oct. 10, 2020).

FAO. 2019. The State of the World's Biodiversity for Food and Agriculture, J. Bélanger and D. Pilling (eds.). FAO Commission on Genetic Resources for Food and Agriculture Assessments. Rome, 572 p. http://www.fao.org/3/CA3129EN/CA3129EN.pdf (accessed Oct. 10, 2020).

GoI. 2020. Agricultural Statistics at a Glance 2019. Directorate of Economics & Statistics, Department of Agriculture, Cooperation and Farmers Welfare, Ministry of Agriculture and Farmers Welfare, Government of India. https://eands.dacnet.nic.in (accessed Oct. 10, 2020).

IPBES. 2019. Global assessment report on biodiversity and ecosystem services of the Intergovernmental Science Policy Platform on Biodiversity and Ecosystem Services. E.S. Brondizio, J. Settele, S. Díaz and H. T. Ngo (eds) IPBES Secretariat, Bonn, Germany. XXX pages.

IPCC. 2014. Climate Change 2014: Synthesis Report. Contribution of Working Groups I, II and III to the Fifth Assessment Report of the Intergovernmental Panel on Climate Change. [Core Writing Team, R.K. Pachauri and L.A. Meyer (eds.)]. IPCC, Geneva, Switzerland, 151 p.

Kochupillai, M. 2016. Promoting Sustainable Innovations in Plant Varieties (Vol. 5). Springer Berlin.

Kochupillai, M. 2019. Diversifying directions of knowledge flows using blockchain and AI to equitably monetize and (re)distribute value from traditional and sustainable farming systems (Working Paper) 2019. DOI: 10.13140/RG.2.2.17276.28805.

Kochupillai, M., J Köninger, N. Kopytko, J. Matthiessen, G. Radick and P. Rao. 2019. Incentivizing and Promoting Sustainable Seed Innovations in India: A Three Pronged Approach. doi: 10.13140/RG.2.2.12670.74568/1 (accessed Oct. 10, 2020)

Kopytko, N. 2019. Supporting sustainable innovations: An examination of India farmer agrobiodiversity conservation. J Environ Develop 28 (4): 386-411. https://doi.org/10.1177/1070496519870299 (accessed Oct. 10, 2020)

Mittermeier R.A., W.R. Turner, F.W. Larsen, T.M. Brooks and C. Gascon. 2011. Global biodiversity conservation: the critical role of hotspots. In: F.E. Zachos and J.C. Habel (eds.) Biodiversity Hotspots. Springer Publishers, London, pp. 3-22.

MoEF. 2014. India's Fifth National Report to the Convention on Biological Diversity. Ministry of Environment and Forests, Government of India, 100 p.

Mohapatra, T. 2016. Indian Agrobiodiversity System. In: Souvenir, 1st International Agrobiodiversity Congress, November 6-9, 2016, New Delhi, India, pp. 12-28.

MOSPI. 2020. Ministry of Statistics and Programme Implementation http://mospi.nic.in/?status=1&menu_id=15 (accessed Oct 10, 2020)

Nayar, M.P., A.K. Singh and K.N. Nair. 2009. Agrobiodiversity Hot Spots in India: Conservation and Benefit Sharing, Vol I & II, 217 & 307 p. Protection of Plant Variety and Farmers' Rights (PPV&FR) Authority, New Delhi.

Paroda R.S., R.K. Tyagi, P.N. Mathur, A. Agrawal, S. Archak, R.C. Agrawal and S. Dsouza (eds). 2017. Proceedings of the 'Ist International Agrobiodiversity Congress: Science, Technology and Partnership', New Delhi, India, November 6-9, 2016. Indian Society of Plant Genetic Resosurces, New Delhi and Bioversity International, Rome, 152 p.

Paroda, R.S. 2016. Agrobiodiversity needs dynamic change management. In: Souvenir, 1st International Agrobiodiversity Congress, November 6-9, 2016, New Delhi, India, pp. 1-11.

Paroda, R.S. 2018. Agrobiodiversity: dynamic change management. In: Reorienting Indian Agriculture: Challenges and Opportunities, pp.157-164.

Paroda, R.S. and A. Agrawal. 2017. Changing paradigms in managing agrobiodiversity through use: An appraisal. Indian J Plant Genet Resourc 30(1):5-12.

Singh, A.K. 2015. Agricultural Biodiversity Heritage Sites and Systems in India. Asian Agri-History Foundation, Secunderabad, India.

Singh, A.K. 2017. Revisiting the status of cultivated plant species agrobiodiversity in India: An overview. Proc Indian Natn Sci Acad 83(1): 151-174.

Singh, P. and S.S. Dash. 2014. Plant Discoveries 2013. New Genera, Species and New Records. Botanical Survey of India, Kolkata.

Singh, R.B. 2019. Agricultural Transformation – A Roadmap to New India. National Academy of Agricultural Sciences (NAAS), Delhi, India.

Srivastava, S.K., N.P. Singh, J. Singh, K.V. Rao and S.J.Balaji. 2019. Agriculture development-based mapping of agro-ecological sub-regions and its implications for doubling farmers' income in India. Curr Sci 117(2): 282-287.

Venkataraman, K., G. Sharma and D. Banerjee. 2020. Faunal Diversity of India. In: G.Dar and A. Khuroo (eds.) Biodiversity of the Himalaya: Jammu and Kashmir State. Topics in Biodiversity and Conservation, vol 18. Springer, Singapore, pp 71-92. https://doi.org/10.1007/978-981-32-9174-4_4

Williams K.J., A. Ford, D.F. Rosauer, N. De Silva, R. Mittermeier, C. Bruce, F.W. Larsen and C. Margules. 2011. Forests of East Australia: the 35th biodiversity hotspot. In : F.E. Zachos and J.C. Habel (eds.) Biodiversity Hotspots, Springer Publishers, London, pp. 295-310.

6

Innovations to Make India Self-Reliant in Edible Oil Production: The Story of Oil Palm

R.K. Mathur, K. Manorama, B. Kalyanababu, G. Ravichandran
K. Ramachandrudu and M.V. Prasad

ICAR-Indian Institute of Oil Palm Research, Pedavegi, West Godavari District Andhra Pradesh

Edible oil scenario

Vegetable oil or edible oil is the main source of energy in humans and it also supplies nutrients for growth and maintenance of human body. Edible oil demand in India has been increasing progressively since last one decade, primarily due to population growth, improved standard of living and also changes in food habits. Presently, the area under oil seed production is between 26 and 27 million ha with an average yield of 1000 to 1100 kg ha^{-1}. This accounts to 7.5 to 8.0 million tonnes of vegetable oil production locally, as against the current demand of 22.5 to 23 million tonnes per annum.

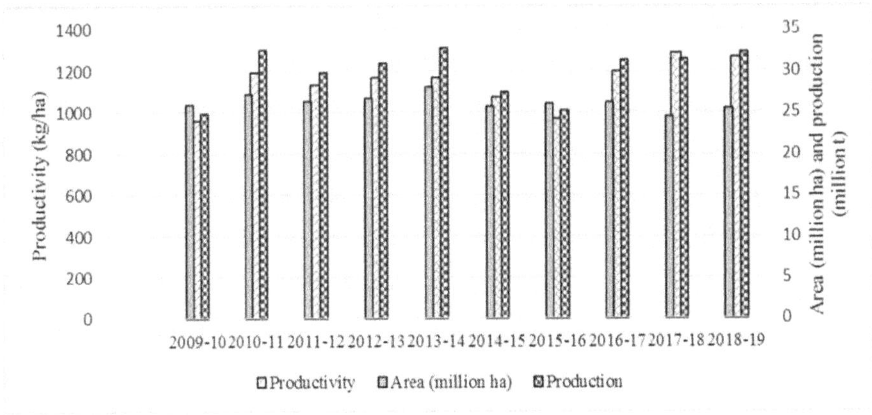

Fig. 1. Area, production and productivity of oil seeds in India in last 10 years (NOM, 2020)

India is one of the major growers of oilseeds. Its vegetable oil economy is fourth-largest after US, China and Brazil. It is the world's largest importer (ahead of the EU and China) and the world's third-largest consumer (after China and the EU). The demand for edible oil in the country is rising due to enhanced per capita consumption, growing urbanization and steadily increasing affluence levels. The growth in production of domestic edible oil has not been able to keep pace with growth of consumption and gap is being met by imports. The country relies on import to meet over 60% of its vegetable oil requirements; almost 60% of requirement is met through palm oil. Palm oil is cheap (due to high productivity per hectare) and it costs 20% less than most of the vegetable oils as well as versatile. The sectoral volumes of refined and expeller oils being consumed are in the range of 80 to 85 and 15 to 20 per cent, respectively. The major refined oils are palm oil, soybean, sunflower, cotton seed, rice bran, TBOs (Tree borne oilseeds) and others. The expeller oils being consumed are mustard, coconut, groundnut, sesame, linseed and others. The gap between demand and domestic production is being compensated by imports (60% of the demand) to the tune of Rs.70,000 crores. Low and unstable yields of most oilseed crops are mainly due to cultivation under rainfed conditions and it is the principal cause of this wide demand-supply gap.

Among the imported oils, crude palm oil (CPO) & crude palm kernel oil (CPKO) occupies the lion's share (nearly 60%) and it is followed by degummed soybean oil, RBD (Refined, Bleached and Deodorised) palmolein, crude sunflower oil and crude rape seed oil. The current productivity levels of oil seeds in India are very low in comparison with world average levels, viz., soybean – 29%, cotton – 71%, shelled groundnut – 72% and rapeseed/mustard – 52%. Government of India intends to increase the production of edible oils both from primary (soybean, rapeseed & mustard, groundnut, sunflower and safflower) and secondary (palm oil, coconut, rice bran, cotton seeds and TBOs) sources through various missions.

Table 1. India's import of major edible oils (Quantity in Lakh Tonnes)

Year	Soybean oil	Palm oil		Sunflower oil	Major edible oils	
		Crude	Refined		Crude	Refined
2013-14	13.5	51.3	25.4	10.8	75.6	25.5
2014-15	23.2	69.7	11.9	17.1	110.0	11.9
2015-16	39.6	71.1	25.7	14.9	125.6	25.7
2016-17	34.6	53.6	29.4	17.3	105.5	29.4
2017-18	31.5	67.5	27.7	22.5	121.5	27.7
2018-19	31.7	64.2	25.2	25.8	124.5	25.6

Source: Department of Commerce

Table 2. Share of different edible oils in the import 2017-18 (% of total)

Oil	% share of total edible oils
Crude Palm Oil	42.01
Soybean Oil	22.01
Sunflower Oil	14.40
RBD Palmolein	19.01
Rapeseed oil	1.94
Palm kernel oil	0.57

Oil palm as a potential option

The Technology Mission on Oilseeds and Pulses (TMOP) was launched in 1986, which was the first comprehensive intervention aiming self-sufficiency in edible oils production through the spread of technology and provision of market support. This Mission met with early success, but, increasing demand for edible oils necessitated the imports in large quantities leading to a substantial drain on foreign exchange. Edible oil imports increased from around 15% of total edible oils consumption in 1995-96 to nearly 53% in 2009-10. The need for addressing this deficit motivated a systematic study of the oilseeds economy in order to formulate appropriate strategies to bridge the demand supply gap.

Among different oil producing crops, oil palm (*Elaeis guineensis* Jacq) with its potential to produce highest oil yield, is the right option to plug the gap between demand and supply of edible oil in India. It is known to be the highest edible oil yielding perennial crop and produces 4 to 6 tonnes of crude palm oil ha^{-1} yr^{-1} and 0.4 to 0.6 tonnes of palm kernel oil ha^{-1} yr^{-1} from 4th to 30th year of its productive life span. Palm oil is an important source of vegetable oil and also used in a wide range of products. The theorctical yield potential is reported at 18.5 tonnes of oil per hectare per annum in oil palm which is the highest ever in all the oil producing crops. It is a perennial monoecious, monocot tree crop (of family Arecaceae). Because of longer life cycle, release of GHGs is minimal in oil palm, as land preparation, ploughing and cleaning is not required every year. Carbon sequestration potential is high and it is in the range of 29.3 t ha^{-1} $year^{-1}$ of CO_2. In comparison with other oil producing crops such as soyabean, sunflower and rapeseed, oil palm has the highest land productivity. If other oil yielding crops are replaced by oil palm, the land requirement comes down to 6-10 times.

Therefore, oil palm is the best suitable option to minimize the import substitution for filling the widening gap between domestic demand and production of vegetable oils in the country. Accordingly, it was introduced as one of the crops under Technology Mission on Oil Seeds in 1986. As there were problems to grow this crop under forest land due to restrictions in the form of forest laws to convert forest land for agriculture, its expansion faced a setback during nineties. However concerted efforts from Department of Agriculture, Co-operation & Farmers' Welfare (DAC&FW), Ministry of Agriculture, Government of India could survive the crop through various developmental schemes to promote its cultivation. Among those, DRDA (District Rural Development Agency) plantations (1988) demonstration blocks of 1000 ha in Andhra Pradesh, Karnataka and Maharashtra states (1989-90) and introduction of OPDP (Oil Palm Development Programme in 1990-91 are significant.

In 1991-92, Technology Mission on Oilseeds & Pulses (TMOP) was taken up in the potential states. A comprehensive Centrally Sponsored Scheme, Oil Palm Development Programme (OPDP) was taken up during Eighth and Ninth Plans. During the Tenth and Eleventh Plans, Government of India had provided support for oil palm cultivation under Centrally Sponsored Integrated Scheme of Oilseeds Pulses, Oil Palm and Maize (ISOPOM). To boost oil palm cultivation, Government of India had implemented a Special Programme on Oil Palm Area Expansion (OPAE) under RKVY during 2011 to March, 2014. During the XII Five Year Plan a new National Mission on Oilseeds and Oil Palm (NMOOP) has been launched under which Mini Mission-II (MM-II) is dedicated to oil palm area expansion and productivity increases.

History of oil palm in India

Oil palm has its centre of origin in West African rain forests under humid tropical conditions. As per the available records, oil was extracted from its fruits in the year 1824 and recognised it as edible. British rulers introduced this crop into Malaysia in 1911 because of the existing climatic situation similar to Africa. Because of its high yielding nature, many nations took it forward and currently 43 countries in the world are cultivating it. In India, it was introduced for ornamental purpose during 1848 in the National Botanical Gardens, Kolkata. Subsequently during 1948 to 1954, Maharashtra Association for Cultivation of Sciences (MACS), Pune, introduced duras from Nigeria and planted on forest land and bunds. Later, the State Department of Agriculture, Kerala proposed using palm oil for edible purpose and a team of experts was deputed to Nigeria which suggested its introduction into Kerala as the climatic conditions are similar to its native country. Thus, some seedlings were imported from Nigeria by Government of Kerala and planted in 40 ha area at Thodupuzha near Kottayam in 1960. This was the first attempt of systematic cultivation of oil palm in the country. Later, during 1972-1984 its commercial cultivation started in Kerala and Little Andaman under rainfed conditions, which mimic Malaysia in temperature and rainfall. As a joint venture of Kerala state government and Union government of India, the first commercial scale plantation was taken up in 1971 under Oil Palm India Limited, Kottayam at Kulathupuzha, Bharatipuram and Chithara areas of Kollam district in an area of nearly 3705 ha. Later, in Athirapally near Thrissur, Plantation Corporation of

Kerala Limited (a Government of Kerala undertaking) planted oil palm in 705 ha in its Kalady Group of Estates lying all along the banks of Chalakudy river. To develop and manage the inaccessible forests in Little and North Andaman by growing agricultural and horticultural crops including red oil palm, Andaman and Nicobar islands Forest and Plantation Development Corporation Limited was established at Port Blair, as per the recommendations of National Commission on Agriculture, 1972. During 1979-81, the commission imported oil palm seed sprouts from Malaysia, IRHO and PNG and planted in 1593 ha out of proposed 2400 ha. All these plantations were under rainfed conditions.

Concept of irrigated oil palm

With the identification of 0.79 million ha of potential area in eleven states, the concept of irrigated oil palm came into existence only in 1986. Realizing that commercial production of oil palm would help in achieving self-sufficiency in edible oils, the Department of Biotechnology (DBT) was entrusted with the oil palm cultivation by initiating Oil Palm Demonstration Project (OPDP) during 1988-89 with technical expertise from Costa Rica in three states *viz.*, Andhra Pradesh, Karnataka and Maharashtra on farmers' fields (3164 ha) to demonstrate its suitability as an irrigated crop. The results indicated that it is possible to grow this crop under irrigated conditions in these states. Thus, during 1987-88, commercial cultivation of oilpalm was initiated by a few interested farmers in West Godavari district of Andhra Pradesh in large areas under DRDA project. The first planting was done in Sitaramapuram village of Pedavegi mandal and it is the foundation for modern oil palm industry of India using underground water for irrigation.

Identification of potential areas for oil palm cultivation

Various committees have been formed at different time periods to identify suitable areas for oil palm production. The first committee was appointed by DAC during 1986 under the chairmanship of Dr. K. L. Chadha. After surveying the coastal areas of India, the committee has identified an area of 0.79 m ha in 11 states. Accordingly, Govt. of India initiated the work plan with the help of Technology Mission on Oilseeds and Pulses (TMOP). TMOP under Ministry of Agriculture has started a project namely Oil Palm Development Programme (OPDP).This offered subsidy to the farmers on seedlings as well as cultivation practices for the juvenile period of first three years. By the end of eighth five-year plan, nearly 40,000 ha area was brought under this crop. In the year 2006, Govt. of India appointed Dr. K.L. Chadha to review the progress of oil palm expansion and further identify the potential areas. Thus, an area of 1.036 m ha was recommended as suitable for oil palm cultivation. Again in 2012, one more committee has been appointed under the chairmanship of Dr. P. Rethinam, which identified 1.93 m ha of area as suitable for oil palm cultivation in the country. Most recently in 2020, a committee appointed under the chairmanship of Dr BMC Reddy, identified suitable areas to the extent of 2.79 million ha in 22 states through scientific mapping using remote sensing and GIS techniques.

Novel research initiatives

Success of oil palm cultivation by giving irrigation is unique in the entire world as it is generally grown as rainfed crop in other South East Asian countries. During early stages of introducing oil palm in the country, work on mutation variability through gamma ray irradiation and tissue culture propagation was initiated at Bhabha Atomic Research Centre (BARC), Mumbai. During 1970-90, research was mainly confined to rainfed oil palm at CPCRI Research Centre, Palode where breeding and pest management aspects were given priority. Seed garden using the mother palms from Thodupuzha seed garden was established at this centre which has later become a big source for indigenous seed sprouts. Since growing of oil palm by giving irrigation is a new venture in the world, systematic research on oil palm started with the establishment of National Research Centre for Oil Palm, Pedavegi by ICAR in February 1995. The NRCOP has been elevated as Directorate of Oil Palm Research in 2009 and as Indian Institute of Oil Palm Research in 2014.

With concerted efforts since its inception, ICAR-IIOPR, developed following innovations for the development of oil palm in the country.

Varietal development

Oil palm with longer life cycle, narrow genetic base, monoecious nature poses problems for varietal development. Through conventional breeding, three tenera hybrids with desirable characters (high FFB yield, high oil yield and high sex ratio) have been identified for release at CVRC in 2020 and one (Godavari Swarna) among them has been recommended for release. The characteristics of these hybrids are;

Godavari Swarna (NRCOP-4): With high fresh fruit bunch yield (26.87 t/ha/yr) and high mesocarp oil yield (5.71 t/ha/year). Suitable for coastal ecosystem with moderate humidity and temperature under assured micro-irrigation system recommended for cultivation in Andhra Pradesh.

Godavari Ratna (NRCOP-2): With a fresh fruit bunch yield of 22.44 t/ha/yr average bunch weight of 18.31kg, high sex ratio (0.70), high mesocarp oil yield (5.36 t/ha/year) and high oil/bunch ratio (24.1), Suitable for high rainfall and assured irrigated conditions of coastal region of Maharashtra and Goa States.

Godavari Gold (NRCOP-17): It has special traits such as high fresh fruit bunch yield (27.23 t/ha/yr), more bunches/palm (11.74), high sex ratio (0.72), mesocarp oil yield (5.79t/ha/year) and oil/bunch ratio (21.28); suitable for coastal Tamil Nadu region under assured irrigated conditions.

Molecular approaches for rapid breeding

The rapid advances in DNA sequencing technology have made whole-genome sequencing (WGS) both technically and inexpensively feasible. Singh et al. (2013) published the first draft genome sequence of *E. guineensis* (African oil palm),using a combination of Roche/454 GS FLXTitanium (Roche/454) and Sanger bacterial artificial chromosome end sequencing. The availability of large genomic data in the open domain leads to pioneering approaches in identification of new genes for speed breeding approaches in developing high yielding oil palm cultivars. The following are the different approaches for rapid breeding of several traits with the genomic technologies to resolve the issue of high yielding cultivars in a short time.

Maintaining *pisifera* and *dura* mother palm blocks

The most cultivated high-yielding oil palm variety, the thin-shelled *tenera* having oil:bunch of more than 20% is produced when the thick-shelled *dura* (O/B approximately 17%) crosses with the shell-less *pisifera*. The *pisifera*, which is mainly female sterile, is used as the pollen parent. The oil palm breeding programme aims at improvement of mother palms (*Dura* improvement) and pollen parent (*Pisifera* improvement) independently and trying their combining ability in order to use the particular mother palms and *pisiferas* in hybrid seed production.Assortment of dura and pisifera palms at early stage helps in raising the separate dura and pisifera mother palm blocks which significantly enhances the rapid selection process to produce the high yielding tenera palms. It is also useful in certification of high yielding seed production by selective breeding of DxP crosses, chiefly on plantation that lack strict quality control or where natural pollination with small quantities of tenera or dura pollen can happen. Babu et al (2017) identified SNP based cleaved amplified polymorphic site markers which differentiates the dura, pisifera and tenera fruit forms at near the beginning stage (Fig. 2). Using this molecular marker technology, pisifera blocks are being planted for their effectual deployment in hybridization programmes.

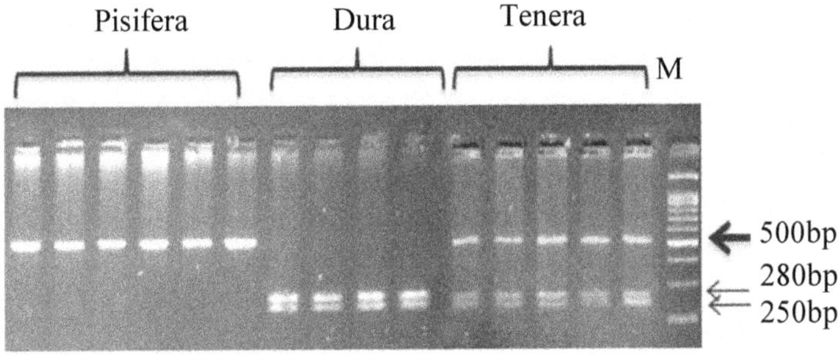

Fig. 2. Identification and differentiation of dura, pisifera and tenera genotypes using the CAPS marker

High yielding cultivars

In oil palm, varieties with high fresh fruit bunch (FFB) yield and good mesocarp content are in great demand. In conventional breeding it requires enormous selection intensity to reach a valuable rate of advancement in segregating generations. But through rapid breeding techniques involving genomics and high end phenomics, it could be simplified. Identification of genes/ QTLs linked to yield traits will assist in vetting a large number of germplasm for advantageous traits through genomic selection methods which is being widely used in several crops like rice, where as it is in initial stages in plantation crops like oil palm. Bai et al (2017) constructed high-density linkage map with 1357 SNPs and 123 SSR markers. DNA markers linked with oil content in these QTLs were identified. Teh et al. (2016) studied (GWAS) using 200K SNPs for oil-to-dry-mesocarp content (O/DM) on *tenera* palms that were selected based on the short-range linkage disequilibrium distance, which is inherent with long breeding cycles and heterogeneous breeding populations. Babu et al (2019a, 2020) used 4031 fully informative SNPs which resulted in identification of 40 highly significant loci, where additional genetic loci were found to be associated with oil to bunch (OB) ratio, followed by average bunch weight (ABW). Five QTLs were found to be linked for leaf area, bunch index and total dry weight. The nucleotide sequences of linked genetic loci for OB ratio were found to be similar to mitogen activated protein kinase-5 (MAPK-5) which is an early flowering protein. A highly dense SNP arrays also used by Kwong et al. (2016) for identifying genes linked to shell to fruit (SF) ratio and also studied the genomic selection. However few reports are available on using GWAS for QTL mapping using natural population consisted of diverse germplasm (Babu et al., 2017; Bhagya et al., 2020).

Dwarf oil palm cultivars

Oil palm trees attain a height of 40 meters, after 20-25 years, and make it difficult to harvest bunches. Dwarf palms with less annual height increment provide the solution for this problem. Molecular technologies aid in developing dwarf palms in less time. Babu et al. (2019b) identified five significant quantitative trait loci (QTLs) on chromosome 6, 7, and 9 through association mapping. These QTLs were found to be similar to oil palm B3 domain containing protein (LOC 105043719) transcript variant X3, mRNA. The B3 domain containing proteins consisted of families like auxin response factors (ARFs) and abscisic acid insensitive 3 (ABI3). Azni et al. (2014) identified 98 clones differentially expresses which found to be related to Aux/IAA, GA and brassino steroids related to height increment in oil palm.

Biotic and abiotic stress management

The impact of drought on oil palm has been extensively documented (Cochard et al., 2005; Carr 2011) and also in numerous other reports. Indeed, water deficit affects oil palm growth, sex ratio and abortion of female inflorescences, thereby lower yields. The major characteristics of the drought tolerant genotypes are higher Total Root Length (TRL), Total Root Surface Area (TRSA) and Potential Root Water Extraction Ratio (PRER) than susceptible genotypes based on half- distances between roots and the distance of water migration from soil to root (Nodichao et al., 2011). Rapid breeding for cold, drought and salinity tolerance with high yield are important for oil palm improvement in view of expansion of oil palm in less favourable climatic conditions. Identifying and validating genes related to abiotic stress responses in oil palm will help in producing elite cultivars suitable to harsh climatic conditions. Recently (Xiao et al., 2017), the *WRKY* genes were identified in oil palm through genome-wide screening and analyzed their expression analysis under various abiotic stress conditions. Zhou et al (2020) identified a set of 159 MYB genes in oil palm via genome-wide screening. The 159 identified oil palm MYB genes belong to two subfamilies (R2R3-MYB and 3R-MYB type), and were divided into 25 subgroups based on phylogenetic analysis.

The hemi biotrophic white rot fungi belonging to the *Ganoderma* genus causes Basal stem rot (BSR) in oil palm. With the availability of oil palm genome sequence and Ganoderma genome sequence, it is easier to researchers to identify all chitinases in oil palm genome. The mechanism of Ganoderma infection is elaborately discussed in oil palm by Sahebi et al. (2015).

Improved oil quality parameters

The main carotenes present in crude palm oil are ß-carotene (56%) and α-carotene (35%). Both ß-carotene and α -carotene are provitamin A. It is also a rich source of appreciable amounts of diglycerides (58%) and free fatty acids, which can have a substantial effect on its physical properties. Morcillo et al (2013) identified the lipase gene in oil palm and found that it co-segregates with the low-/high-lipase trait. They also identified a SNP based marker to rapidly identify potent elite genitors and introgress the trait into major cultivars which is a major asset to the breeders. In order to facilitate rapid breeding for healthy fatty acid content in oil palm, genome-wide association analysis (GWAS) was used to identify and validate SNP markers and associated candidate genes with several fatty acid components in a highly diverse germplasm of 200 oil palm individuals. The results identified key genetic markers for identifying low palmitic acid containing genotypes. The accumulation of oleic acid, influenced by key enzymes like namely stearoyl ACP desaturases (SAD) and acyl-acyl carrier protein (ACP) thio esterases (FAT) were found to play key role in large gene families with complex expression profiles across a wide range of tissue types and developmental stages. Zheng et al (2019) found 22 kn-miRNAs and 14 nov-miRNAs in FA metabolism pathways. Among them, eg-miR156c, eg-miR397, eg-miR444b and nov-miR129 regulated FA synthesis in plastids and the transport of FA-ACP from plastids to the endoplasmic reticulum by targeting *acetyl-CoA carboxylase 1* (*ACC1*), *long-chain acyl-CoA synthetase 9* (*LACS9*), *LACS4* and *enoyl-ACP reductase* (*ENR*), respectively. The mapping results showed that the key candidate genes in *Elaeis* related to oleic acid C18:1, using intra-gene SNPs, supported several QTLs underlying notably FATA and SAD enzymes.

Germplasm enhancement

Germplasm exchange and strengthening of oil palm gene bank

International collaborative research project on Oil palm germplasm exchange for R&D between India and Malaysia was taken up by IIOPR, which involved exchange of genetic resources of Sierra Leone and Senegal from MPOB, Malaysia with dura of Guinea Bissau and Zambia available in India. Sierra Leone and Senegal possess lower FFA compared to current material planted in Kluang. The screened accessions were imported to India through ICAR-NBPGR.The exploration surveys resulted in collection of 166 germplasm. The germplasm collected was evaluated, characterized and are maintained in the field gene bank of ICAR-IIOPR. Recently accessions from Mannargudi in Tamil Nadu; Belgaum and Marsein Karnataka were collected and bunch analysis was carried out. Three accessions from Zaire *viz.,* KLPA - D108 (dura with big fruit size); KLPA - D13 (dwarf dura with thin shell) and KLPA- P1(dwarf pisifera, 21 cm annual height increment, with compact canopy were collected from Oil Palm India Ltd., estate at Kulathupuzha, Kerala. Twenty palms belonging to African germplasm and interspecific hybrids were characterized and conserved. Registration of 63 indigenously collected oil palm germplasm was completed and accession numbers were received from ICAR-NBPGR, New Delhi as IC-0610000 to IC0610051-planted at ICAR-IIOPR, Pedavegi and IC-0599983 to IC-0599993–planted at ICAR-IIOPR Research Centre, Palode.

Establishment of seed gardens

Planting material (mother palms) was developed and supplied for establishment of four new seed gardens to augment indigenous hybrid seed production and meet planting material requirement. The new seed gardens were established at Taraka and Kabini in Karnataka and at Morampudi and Gopannapalem in Andhra Pradesh. DxD and T x T crosses were effected for the proposed seed garden in Mizoram and the sprouts were supplied.

Novel method for hybridization in oil palm

To aid in hybridization process of oil palm, a novel method was standardized for controlled pollination through pollinating weevil without climbing the tree. The pollinating weevils (*Elaeidobius kamerunicus*) were cleaned to make them devoid of pollen and sent through a controlled device with desired pollen to effect pollination in a controlled environment. Number of pollinating weevils, their pre-treatment, pollen quantity etc were standardized. The size of the FFB formed through this method is on par with the FFB formed through existing pollination method. This technique is labour effective and reduces the drudgery. A patent "A process and kit for insect facilitated controlled pollination in oil palm" has been filed (Appl. No. 201641044994) (Ravichandran et al., 2019).

Crop management strategies

Selection of suitable site

Through GIS and Remote Sensing techniques, digital suitability maps were developed for oil palm cultivation both under irrigated and rainfed conditions. Depending upon rainfall pattern, irrigated and rainfed locations were delineated and then by imposing suitability criteria for irrigated and rainfed areas separately digital maps have been developed. For irrigated conditions, ground water level rainfall, soil pH, soil salinity and soil texture are the five most important parameters considered. Similarly, in case of rainfed conditions, rainfall, minimum temperature slope, soil depth and length of continuous dry period are the parameters taken into account. Using multicriteria overlay analysis technique in GIS, separate maps have been generated for rainfed and irrigated conditions and the areas have been identified under highly suitable, moderately suitable, marginally suitable and not suitable categories.

Oil Palm Site Suitability Evaluator (OPSSE)

A unique computer aided Decision Support System (DSS) which facilitates an easy, precise and scientific way of evaluating the biophysical suitability of a given piece of land for cultivating oil palm has been developed by ICAR-IIOPR. OPSSE evaluates the land suitability both under irrigated and rainfed conditions depending upon the rainfall situation of the locality. This DSS considers climatic and soil parameters along with water resources for evaluating suitability of an individual site. This tool is highly useful for policy makers and all the stakeholders of oil palm research and development including researchers, oil palm processors state department of agriculture and horticulture officials, farmers and extension persons in taking appropriate decisions with reference to site selection for oil palm cultivation. In addition to this, an android mobile based application, Oil Palm Site Suitability Evaluator (OPSSE) also has been developed for site suitability evaluation and it is available in Google Play Store for free download (Manorama et al., 2020a).

Oil palm nursery management: Integrated use of bio-fertilizers with 25% of recommended dose of chemical fertilizers is recommended for commercial application in oil palm nurseries in India. POME sludge@10% in potting mixture can be an ideal substitute for chemical fertilizers and the same is recommended for commercial oil palm nurseries in India. NPK @ 30:38:25g/

seedling/year was found to be the most promising nutrient dose in improving the growth and vigour of oil palm seedlings and hence recommended for commercial application (Ramachandrudu, 2019a).

Scientific water management

Drip system of micro irrigation at 70% deficit of crop water requirement proved more economical for achieving higher water use efficiency in oil palm. Palms need continuous irrigation excepting rainy days with four drippers in basin area (Rao et al., 2019). Root distribution studies indicated that large volume of effective roots exists within 2 metre radius from the base of the trunk (Suresh et al., 2003). Three static android mobile apps on Water Requirement for oil palm have been developed for Andhra Pradesh, Tamil Nadu and Karnataka. With the help of this app, oil palm farmers can know the amount of water to be given to their plantations. The water requirement (per palm per day) can be calculated for a particular day or week or month, depending on their requirement (Suresh et al, 2019: Suresh et al, 2020b; Suresh et al., 2020c)).

Site specific nutrient management

Soil application of 1200:600:1200 g of NPK per palm per year is the recommended dose and it needs to be split into four equal parts to apply at quarterly intervals. Soil and leaf analysis based nutrient recommendations are to be followed for better nutrient use efficiency (Manorama et al., 2019). Diagnosis and Recommendation Integrated System (DRIS) norms and optimum leaf nutrients concentration of oil palm plantations in states of Karnataka, Gujarat, Goa, Mizoram, Tamil Nadu and Andhra Pradesh have been established for routine diagnostic and advisory purpose for balanced utilization of fertilizer. Assessment of soil properties and leaf nutrient concentration in oil palm plantations is required for effective nutrient management and higher crop yield. In West Godavari district of Andhra Pradesh, the leaf nutrients requirement order was B > Mg > K > N > P. The optimum concentration of leaf nutrients were 1.57-2.63% for N, 0.08-0.16% for P, 0.48-0.88% for K, 0.25-0.71% for Mg and 22.6-60.2 mg kg-1 for B. Information about soil nutrients status and nutrient requirement order and optimum leaf nutrient ranges can be used for effective management of nutrients in the OPP of study region (Behera et al , 2019). In Cauvery delta zone of Tamil Nadu, the order of requirement of leaf nutrients was found to be K > P > N > Mg > B. Optimum leaf nutrient range for N, P, K, Mg and B were estimated as 1.61 to 2.11%, 0.10 to 0.12%, 0.33 to 0.81%, 0.23 to 0.73% and 30.6 to 54.8 mg kg-1, respectively, which could be used for guiding balanced application of fertilizers. On the whole, 15, 31, 2, 8 and 10% of leaf samples had below optimum concentrations of N, P, K, Mg and B, respectively (Behera et al., 2018). Research studies on fertigation with NPK@600:300:600g/palm/year at monthly intervals coupled with irrigation based on Potential Evapo-transpiration (PET) is recommended for adult oil palm plantation (Reddy et al., 2002). Adoption of fertigation also reduced application cost, and increased nutrient use efficiency by two times. An android based mobile application to indicate nutrient requirement both under soil application and fertigation has been developed based on critical leaf nutrient ranges and nutrient sufficiency ranges estimated in different soils of oil palm cultivation. By feeding the soil and leaf analysis data of a particular location, the advisory for nutrient application can be obtained from this application both for soil application as well as for fertigation. The app is available in Google play store, (Manorama et al., 2020b).

Oil palm based farming/cropping systems

Standardized and perfected different oil palm based cropping systems (oil palm-cocoa, oil palm-red ginger, oil palm-heliconia, oil palm-bush pepper, oil palm-banana, oil palm-ornamental crops) in mature plantations. The cost benefit ratio for the above cropping systems varied from 1:2.38 to 1:2.86 and the net returns ranged from 1457 to 1771 US$ per ha.

Fig. 3. Oil palm + Heliconia intercrop system

Fig. 4. Oil palm + cattle farming system model

Oil palm based integrated farming system with fodder crops, dairy and back yard poultry has been found to be the most suitable and profitable system with a cost benefit ratio of 1:3.28. The technology has been spread to an area of 2900 ha in oil palm growing States of India. Hybrid Napier grass varieties DHN-6 (36.15t/ha) and KKM (34.18t/ha) and guinea grass varieties BG-1(38.38t/ha), DGG-1(28.67t/ha) and Co3(32.25t/ha) have been found as sustable shade tolerant and high fodder yielding varieties in mature oil palm plantations. (Ramachandrudu, 2019b).

Phenological growth stages of adult oil palm hybrids

The different phenological growth stages of adult oil palm hybrids grown in India under irrigated conditions were coded using the *Biologische Bundesantalt, Bundessortenamt and ChemischeIndustrie* General Scale. The duration between unfolding of 70 per cent spear leaf to bunch maturity in the hybrids ranged from

447.9 to 485.2 days and the duration from anthesis to maturity ranged from 145.8 to 153.7 days. Wide variation in the degree days from spear leaf development to bunch maturity were also observed among the hybrids ranging from 6320.2 to 6937.3. Among the hybrids, united plantations hybrid possessed the shortest duration and lowest degree days from spear development to bunch maturity. The study would help in estimating thermal time required for completion of different phenological stages, which could be used for yield forecasting models and studying climatic analogues in oil palm (Suresh et al., 2020a).

Mechanization in oil palm

Improved sickles and motorized harvester for harvesting oil palm

Two sickles DOPR-1 and DOPR-5 manufactured with stainless steel were designed and developed to harvest bunches from oil palm plantations. Two models of oil palm harvesting tools *viz.,* back pack mounted and trolley mounted were developed and tested in oil palm plantations. Based on the initial trials, it was observed that using the backpack model the operator could harvest comfortably up to a height of 5-6m and it has better adoptability and flexibility (Singh and Prasad, 2019).

Height adjustable hydraulic lift platform for harvesting in tall palms

Height adjustable hydraulic lift platform was designed and fabricated for harvesting bunches from tall palms. A person with average skill can reach up to a height of 4.54m, and with 20ft (6.1m) pole attached to sickle, he can reach a height up to 36 ft to harvest bunches (Singh and Prasad, 2019).

Oil palm trunk chipping bucket

An oil palm trunk chipping bucket was designed and fabricated and a trial was conducted in ICAR-IIOPR, Pedavegi. The experimental results showed that for complete fragmentation of 6-7m palm 5.47min in was required including felling time of 4.53 sec. The average number of strokes required was 35. The average palm girth at base, middle and crown was 3.0, 2.19 and 2.39 m respectively, so the base region required more than one stroke for fragmentation (Singh and Prasad, 2019).

Oil palm ablation tool

Ablation (removal of inflorescence during juvenile phase) tool developed by IIOPR eases the task of ablation, which is necessarily to be carried out in all juvenile oil palm plantations (1-3 years age oil palm plantation) for proper growth. This tool will have less drudgery, no damage to leaves, safe and manoeuvrable during intercrop period. On an average in one hour a person can ablate 125 inflorescences (Singh and Prasad, 2019).

ICT applications in oil palm

Information Communication Technologies (ICTs) are widely used to disseminate the agricultural technologies to reach the stakeholders in India. Among the ICTs mobile message services (text/Voice), mobile applications (Apps), Decision Support Systems (DSS) provide knowledge base to the stakeholders to adopt recommended / good agricultural practices, so as to obtain highest yields in respective crops. Accordingly, four static mobile apps viz., oil palm cultivation practices, nutrient management, pest management and disease management have been developed by ICAR-IIOPR in three languages (English, Hindi and Telugu). One interactive mobile app on 'oil palm crop doctor' was brought out in English. These apps are developed for android operating systems, placed on Google play store for free download. Oil palm technologies are being disseminated through oil palm kisan mobile message services by sending SMS to stakeholders.

ICAR-IIOPR developed short video films on management practices in oil palm cultivation *viz.,* irrigation, fertilizer, mulching, green manure and cover crops, intercropping, nutrient management, pest management and disease management. These films are placed on YouTube for free download. The ICTs enabled services i.e. mobile applications and decision support systems are targeted to disseminate technologies on oil palm to stakeholders across the country from a single authentic source of technology generation. This would help in adoption of recommended practices to achieve highest production and pave way for achieving self-sufficiency in vegetable oil production in the country (Mary Rani and Prasad, 2019).

Short video film title	Link - youtube
Oil palm Irrigation Management	https://www.youtube.com/watch?v=SUHUhvQcdDQ
Oil palm Fertilizer Management	https://www.youtube.com/watch?v=X1iJ8k6D42M
Oil palm Pest Management	https://www.youtube.com/watch?v=c1navCCwX8I
Oil palm Nutrient deficiencies	https://www.youtube.com/watch?v=oZS2xoPKlPc
Oil palm Mulching	https://www.youtube.com/watch?v=dTqwCt2GqmA
Oil palm Intercrops	https://www.youtube.com/watch?v=i-xl2ntSo90
Disease Management in Oil Palm	https://www.youtube.com/watch?v=41d7asABW-4
Green manure and cover crops	https://www.youtube.com/watch?v=_HtkDCNLHkQ

Policy support

To promote oil palm in India, government is providing good support to growers as well as processors in the form of subsidies during juvenile period, changes in import duty of crude palm oil etc. through provisions in oil palm act as well as depending on the needs of the stakeholders from time to time. However, few more interventions are necessary to further strengthen oil palm development programme with greater success.

Assured market price

Oil palm with an economic life span of 25-30 years is a perennial crop. In India, generally it is grown as a small holders' crop. Market volatility affects oil palm production greatly because farmers do not show interest to take up new and recommended technologies due to uncertainty in FFB prices which are dependent on international crude palm oil prices. This is the biggest hindrance for oil palm expansion programme even. To encourage farmers towards oil palm cultivation and to make them follow recommended package, price stabilization is very much required. This can be achieved by creating a Price stabilization fund or similar mechanism. Through that a minimum support price needs to be assured for the farmers by linking the cost of production of FFB.

Establishment of Oil Palm Board

Oil palm development in the country needs a continuous monitoring because of its perennial nature and longer time involvement in seed production to harvesting of bunches. A separate body, in the form of Oil Palm Development Board with dedicated staff at National level may help to look after the oil palm development in the country. The board can act as an advisory body to government while safeguarding the interests of all the stakeholders of oil palm.

Processing facilities and logistics

Oil palm fresh fruit bunches are perishable in nature and for better oil quality they need to be processed within 24 to 48 hours after harvest. Therefore, harvested produce should reach the processing unit at the earliest. To ensure this, processing factories are to be established timely in the areas where oil palm is being taken up for the first time. Further, road infrastructure needs to be improved for facilitating transport of produce from oil palm gardens and to transport inputs to oil palm fields.

Planting material

In oil palm planting material is produced through seed garden concept where dura mother palms and pisifera male parents are grown to produce tenera seeds. To take up oil palm expansion at a faster pace, availability of planting material is to be ensured by establishing sufficient number of seed gardens.

Crop insurance and institutional credit

To counter balance the impacts of vagaries of weather, an insurance scheme has to be designed and adopted. Provision should also be made for institutional credit from NABARD, commercial banks and cooperative societies for oil palm farmers for undertaking different farm operations relating to oil palm timely.

Conclusion

Growing oil palm entirely under irrigated conditions is a unique situation in India. Developing hybrids with better drought tolerance needs to be focused on. Further, latest developments in information technology and availability of sensors would pave the way for more precise water and nutrient management strategies. Although, oil palm cultivation is encouraged to meet edible oil demands of the country, from growers' perspective it is net benefit that matters ultimately. For achieving higher productivity from unit area of land, oil palm needs to be grown as a component of cropping / farming system. Location specific, viable cropping / farming system models have to be developed for this purpose. A study by Paterson (2017) on climate change predictions impacts revealed that suitable area for oil palm cultivation is going to turn unsuitable in most parts of the world. In 2050, the unsuitable climate increases by 6% at global level and highly suitable climate decreases by 22%. Therefore, suitable areas may shift to farther latitudes and higher altitudes. Accordingly, land suitability analysis needs to be carried out at regular intervals to identify these locations and thereby plan to shift its cultivation to suitable areas. Policy support from governments also is very much essential in this crop to ensure price stability and to provide logistic support for achieving success in oil palm.

References

Babu, B.K., R.K. Mathur, G. Ravichandran, P. Anitha and M.V.B . Venu. 2020. Genome wide association study (GWAS) and identification of candidate genes for yield and oil yield related traits in oil palm (*Eleaeis guineensis*) using SNPs by genotyping-based sequencing. Genomics, https://doi.org/10.1016/j.ygeno.2019.06.018

Babu, B.K., R.K. Mathur, G. Ravichandran, P. Anitha and M.V.B. Venu. 2019b. Genome-wide association study (GWAS) for stem height increment in oil palm (*Elaeis guineensis*) germplasm using SNP markers. Tree Genetics and Genomes. 15: 40. https://doi.org/10.1007/s11295-019-1349-2.

Babu, B.K., R.K. Mathur, G. Ravichandran, P. Anitha and M.V.B.Venu. 2019a. Genome-wide association study for leaf area, rachis length and total dry weight in oil palm (*Eleaeis guineensis*) using genotyping by sequencing. PLoS ONE. 14(8): e0220626. https://doi.org/10.1371/journal.pone.0220626.

Babu, B.K., R.K. Mathur, P.N. Kumar, D. Ramajayam, G. Ravichandran and M.V.B.Venu. 2017. Development, identification and validation of CAPS marker for SHELL trait which governs dura, pisifera and tenera fruit forms in oil palm (*Elaeis guineensis* Jacq.). PLoS One. 12(2): e0171933. https://doi.org/10.1371/journal.pone.0171933

Bai, B., L. Wang, M. Lee, Y. Zhang, Y.A. Rahmadsyah, B.Q. Ye, Y.W. Zi, C.H. Lim, A. Suwanto, N.H. Chua and G.H. Yue. 2017. Genome-wide identification of markers for selecting higher oil content in oil palm. BMC Plant Biology. 17: 93.

Behera, S.K., A.K. Shukla, K. Suresh and R.K. Mathur. 2018. Evaluation of Nutritional Status and Yield Limiting Nutrients in Oil Palm Plantations of Cauvery Delta Zone of Tamil Nadu. Journal of the Indian Society of Soil Science. 66(1) DOI:10.5958/0974-0228.2018.00010.5

Behera, S.K., A.K. Shukla, K. Suresh and R.K. Mathur. 2019. Estimation of soil properties and leaf nutrients status of oil palm plantations in an intensively cultivated region of India. Current Science. 117 (3): 497-502 doi: 10.18520/cs/v117/i3/497-502

Bhagya, H.P., B. Kalyana Babu, P.M. Gangadharappa, et al. 2020. Identification of QTLs in oil palm (*Elaeis guineensis* Jacq.) using SSR markers through association mapping. Journal of Genetics. 99, 19 (2020). https://doi.org/10.1007/s12041-020-1180-4

Carr, K.V. 2011. The water relations and irrigation requirements of oil palm. A Review. Experimental Ariculture. 47 (4), 629–652 doi:10.1017/S0014479711000494

Cochard, B., P. Amblard and T. Durand-Gasselin. 2005. Oil palm genetic improvement and sustainable development. Oil Seeds and Fats, Crops and Lipids. 12: 141- 147.

Corley, R.H.V. 2009. How much palm oil do we need? Environmental Science & Policy 12, 134–139

https://play.google.com/store/apps/details?id=com.cdac.CultivationPractices – Mobile app on oil palm cultivation

https://play.google.com/store/apps/details?id=com.cdac.diseasesmanagement – Mobile app on oil palm disease

https://play.google.com/store/apps/details?id=com.cdac.OilPalmPest_Eng – Mobile app on oil palm pests

https://play.google.com/store/apps/details?id=oilpalm.cropdoctor.iiopr.iioprdemo – Mobile app Oil palm Crop Doctor

Kwong, Q.B., C.K. Teh, A.L.Ong, H.Y. Heng, H.L.Lee, M. Mohamed, J.Z. Low, S. Apparow, F.T. Chew, S.Mayes, H. Kulaveerasingam, M.Tammi and D.R. 2916. Appleton. Development and Validation of a High-Density SNP Genotyping Array for African Oil Palm. Molecular Plant. 2016 Aug 1;9(8):1132-1141. doi: 10.1016/j.molp.2016.04.010. Epub 2016 Apr 22. PMID: 27112659.

Manorama, K., K. Suresh and B.N. Rao. 2019. Advances in oil palm nutrient management – Deficiency symptoms and advisory. In: Recent advances in oil palm production and special emphasis on emergence of new pest and its management. (Eds) K.Manorama and K Suresh. ISBN : 81-87561-58-0. Model Training Course manual. Pp: 82-91.

Manorama, K., K. Suresh, S.K. Behera, K.L. Mary Rani and R.K. Mathur. 2020a. https://play.google.com/store/apps/details?id=com.iiopr.opsse – Mobile app on oil palm site suitability

Manorama, K., S.K. Behera and K. Suresh. 2020b. https://play.google.com/store/apps/details?id=com.iiopr.fertapp – Mobile app on oil palm nutrient advisor

Mary Rani, K.L. and M .V. Prasad. 2019. Advancements in Oil Palm ICTs. Oil palm Crop Doctor and mobile apps. In: Recent advances in oil palm production and special emphasis on emergence of new pest and its management. (Eds) K. Manorama and K Suresh. ISBN: 81-87561-58-0. Model Training Course manual. Pp: 136-141.

Morcillo, F., D. Cros, N. Billotte, G.F.Ngando-Ebongue, H. Domonhédo, M. Pizot, T.Cuéllar, S. Espéout, R. Dhouib, F. Bourgis, S. Claverol, T.J. Tranbarger, B. Nouy and V. Arondel. 2013. Improving palm oil quality through identification and mapping of the lipase gene causing oil deterioration. Nature Communications. 4:2160. doi: 10.1038/ncomms3160.

NOM 2020. Challenges of quality seed production and supply in oilseed crops. Paper presented in National Oil Seed Meet -cum-"Research-Industry-Farmer Interface" during 23-25 September, 2020. Organized by Indian Institute of Soyabean Research, Indore, M.P. on virtual mode.

Paterson, Russell. (2017). Climate change affecting oil palm agronomy, and oil palm cultivation increasing climate change, require amelioration. Ecology and Evolution. 8. 10.1002/ece3.3610.

Ramachandrudu, K. 2019a. Scientific approaches in production of quality seedlings. In: Recent advances in oil palm production and special emphasis on emergence of new pest and its management. (Eds) K. Manorama and K. Suresh. ISBN : 81-87561-58-0. Model Training Course manual. Pp: 28-33

Ramachandrudu, K. 2019b. Oil palm based systems approach for higher net income. In: Recent advances in oil palm production and special emphasis on emergence of new pest and its management. (Eds) K. Manorama and K. Suresh. ISBN : 81-87561-58-0. Model Training Course manual. Pp: 120-126.

Rao, B N., K.Suresh, K.Ramachandrudu and .Manorama. 2019. Recent advances in irrigation management. In: Recent advances in oil palm production and special emphasis on emergence of new pest and its management. (Eds) K.Manorama and K Suresh. ISBN : 81-87561-58-0. Model Training Course manual. Pp: 41-50.

Ravichandran, G., G. Somasundaram, P. Anitha and H.P. Bhagya. 2019. Recent advances in oil palm seed production, seed processing and sprout production. In: Recent advances in oil palm production and special emphasis on emergence of new pest and its management. (Eds) K. Manorama and K. Suresh. ISBN: 81-87561-58-0. Model Training Course manual. Pp: 10-16.

Reddy, V.M., K. Suresh, K.N. Sarma, K.M. Sivasankara Kumar, A. Bhanusri, and P. Ramakrishna. 2002. Root biomass distribution in relation to its moisture pattern under different irrigation methods in oil palm. Proceedings PLACROSYM_XV, Edited by K.Sreedharan, P.K. Vinod Kumar, Jayarama and Basavaraj M. Chulaki. P. 344-347.

Sahebi, M., M. Mohamed Hanafi, Mui-Yun Wong, A. S. Idris, Parisa Azizi, Mohammad Faseleh Jahromi, Parisa Shokryazdan, Rambod Abiri and Hasmah Mohidin. 2015. Towards immunity of oil palm against Ganoderma fungus infection. Acta Physiologiae Plantarum, 37:195

Sapey, E., K.Adusei-Fosu, D.Agyei-Dwarko and G. Okyere-Boateng. 2012. Collection of oil palm (Elaeis guineensis Jacq.) germplasm in the northern regions of Ghana. Asian J. Agric. Sci. 4 (5), 325–328.

Singh, R., M.O. -Abdullah and E. Low. 2013. Oil palm genome sequence reveals divergence of interfertile species in Old and New worlds. Nature. 500, 335–339.

Singh, V.T. and M.V. Prasad. 2019. Recent advances in mechanization of oil palm cultivation. In: Recent advances in oil palm production and special emphasis on emergence of new pest and its management. (Eds) K. Manorama and K. Suresh. ISBN : 81-87561-58-0. Model Training Course manual. Pp: 142-156.

Suresh, K., K. Manorama and K. Lushaank. 2020b. https://play.google.com/store/apps/details?id=iiopr.waterrequirement.tamilnadu – Mobile app on oil palm water requirement for Tamil Nadu

Suresh, K., K. Manorama and K. Lushaank. 2020c. https://play.google.com/store/apps/details?id=iiopr.waterrequirement.karnataka – Mobile app on oil palm water requirement for Karnataka

Suresh, K., K. Manorama, K. Lushaank and C.S. Krishna. 2019. https://play.google.com/store/apps/details?id=iiopr.waterrequirement.andhrapradesh – Mobile app on oil palm water requirement for Andhra Pradesh.

Suresh, K., Behera, S.K., Manorama, K. and Mathur, R.K. 2020. Phenological stages and degree days of oil palm crosses grown under irrigation in tropical conditions. Annals of Applied Biology. 1–8. DOI: 10.1111/aab.12641

Suresh, K., S.K. Behera, K. Manorama1 and R. K. Mathur. 2020a. Phenological stages and degree days of oil palm crosses grown under irrigation in tropical condition. Annals of applied biology. 1-8.

Suresh, K., V.M. Reddy, K.N. Sarma, A. Bhanusri, and K.M. Sivasankara Kumar. 2003. Quantification of root biomass in oil palm grown under basin irrigation. Int. J. Oil Palm .2 (3&4): 39-41.

Teh, C.K., A.L. Ong, S.K. Apparow, F.T. Chew, S. Mayes, M. Mohamed, A. David and K. Harikrishna. 2016. Genome-wide association study identifies three key loci for high mesocarp oil content in perennial crop oil palm. Scientific Reports. 6, 19075. https://doi.org/10.1038/srep19075

Xiao, Y., L. Zhou, X. Lei, H. Cao, Y. Wang and Y. Dou. 2017. Genome-wide identification of WRKY genes and their expression profiles under different abiotic stresses in Elaeis guineensis. PLoS One 12 (12), e0189224.

Zheng, X., K. Liu, X. Wang, R. Zhang and X. Li. 2019. MicroRNA-192 acts as a tumor suppressor in colon cancer and simvastatin activates miR-192 to inhibit cancer cell growth. Molecular Medicine Reports. 19:1753–60. 10.3892/mmr.2019.9808

Zhou, L., Y. Rajesh, J. Longfeiand C. Hongxing. 2020. Genome-wide identification and expression analysis of MYB gene family in oil palm (Elaeis guineensis Jacq.) under abiotic stress conditions. Environmental and Experimental Botany. 180: 1042-1045.

Theme 3: Innovations and Strategies for Abiotic Stress Management

7

Natural Resource Management for Building a Self-Reliant India

Debashis Mandal, Nishita Giri, Gopal Kumar and Raman Jeet Singh

ICAR-Indian Institute of Soil & Water Conservation, Dehradun, Uttarakhand

Introduction

In the post-world trade organization (WTO) era, the fantastic imagination of "one world for self-reliance", has been shattered by the Corona induced calamity and the expansionists approach of a few rouge nations. Though the global welfare and peace has always been part of Indian traditional philosophy, approach and the concept of self-reliance, the practical boundary of self-reliance in its present context, starts from the country level itself. The idea of self-reliance at multiple levels (individual, family, society, and nation) has driven human civilization for most of its efforts, inventions and actions across the world.

A very high dependency of a large number of population on the natural resources in the developing world calls for proper management of natural resources for self-reliance. Since the early days of civilization, natural resources in the form of land, soil, water, biodiversity and climate have remained the very basis of supporting and sustaining life of human beings, plants and animals on the earth. However, in recent times intensive use and over-exploitation of pristine resources have robbed them of their legendry resilience. To effectively tackle the complex problems and to fulfil basic needs of livelihood, food security, poverty, employment, equity and environmental services, efficient and judicious utilization of natural resources for enhanced and sustained productivity is a matter of serious concern for policy makers, planners, scientists, conservationists and environmentalists.

Issues and Challenges

Global population (7.2 billion) is expected to be around 8 billion by 2030 and 9.2 billion by 2050 (GOS, 2011). Most of these increases will happen in sub-Saharan Africa and Asia including India. India's population is likely to reach 1.5 billion by 2030 and 1.8 billion by 2050 (Swaminathan, 2009). Therefore, India's cereal production has to be doubled by 2050 in order to meet the demand. Similarly, the demand of feed for livestock and poultry and biofuel production will also increase. As cereal products are increasingly used as a feed for livestock, which is estimated to be 45–50% by 2050 if meat consumption increases (FAO, 2003; 2006), finding alternative feed sources are essential for increasing the availability of cereals for human consumption. This poses a big challenge, since most of the easily available feed sources have already been fully exploited, although some alternatives still exist.

Without innovations and critical interventions, the global food production system will continue to degrade the environment, the biodiversity and eventually the world's carrying capacity. The widespread problems associated with the on-going intensification are topsoil erosion, depletion of soil fertility, soil acidification and salination. In addition to other forms of degradation, the rates of excessive water extraction for irrigation which in many places exceeds the rate of replenishment, and heavy dependence on fossil fuel energy for synthesis of fertilisers and pesticides caused severe damages to the ecosystem (GOS, 2011). For example, the production of nitrogen fertilisers is highly energy intensive: the roughly fivefold increase in fertiliser price between 2005 and 2008 was strongly influenced by the soaring oil price during this period.

The Dilemma

The real dilemma is that the existing food system depends extensively on non-renewable resources and consumes at rates far exceeding their replenishment. On the other hand, all such practices release greenhouse gases, nitrates and other contaminants into the environment. Unless the footprint of food system on the environment is reduced, the capacity of the earth to produce food for all, will be compromised with grave implications for food security in the near future. Sustainability need for all sectors of the food system, from production to consumption, and in education, governance and research must be addressed properly. One of the daunting tasks is to fulfil the demand of the society by 2050 without degrading the natural resources (Foley et al., 2011). Converting these challenges to opportunities is the key for self-reliance what India is aiming at.

Pressures on natural resources –competing demands

Globally, there will be a competing demand for land for different developmental activities by 2050 (UNEP, 2008). It is estimated that possible loss in land due to non-farm activities will be about 2-8% of cropland. Land degradation may result in abandonment of 1.5 to 2.5% of cropland. Additionally, cropland will lose productivity due to water scarcity, climate change and land degradation (Table 1).

Table 1. Causative factors of land pressures and possible production loss by 2050.

Cause of possible loss of cropland, UNEP 2008	
Causal factors	**Possible change by 2050**
Biofuel demand	-2% to -8%
Demand for others	
Non-food crops	-2.5% to -3.5%
Urban build-up	-2.5% to - 4.5%
Land degradation	-1.5% to -2.5%
Possible impact in yields	
Land degradation	-1% to -8%
Water scarcity	-1.7% to -12%
Climate change	-1.0% to -1.5%
Invasive species	-2.0% to -6.0%

Pressures on land resources

Several critical resources, on which food production relies, will come in stress in future. Conversely, growth in food system will itself exacerbate these pressures. According to an UN report, India will be the most populated country in the world and expected to add 273 million people between 2019 and 2050, exerting huge pressure on natural resources, particularly on soil and water to meet the increasing demand of food.

The rapidly increasing human population is making the man-to-land ratio more critical over the years. In India, per capita availability of land has fallen drastically from 0.91 ha in 1951 to about 0.32 ha in 2001 against world average of 2.19 ha. The availability of per capita net sown area has similarly reduced from 0.33 ha in 1951 to 0.14 ha in 2001 and is expected to decline further to 0.09 ha by 2050. According to an estimate, minimum economic holding size of 2 ha of unirrigated land and 1 ha of irrigated land has been suggested in India for sustaining a family of 5 or 6 persons (NAAS, 2007). Similarly, the minimum land area of 0.5 ha is required for the production of a diverse diet similar to that of United States and Europe against the present level of per capita availability of 0.32 ha in India (2001) (Pimental et al., 1999). The decline in quality of land is of greater concern for the quality of life. Although the quality of nearly 16% of total land area including cropland, rangeland and forests is improving, the ISRIC has estimated that out of the 11.5 billion ha of vegetated land on earth, about 24% has undergone human-induced soil degradation, particularly through erosion (GOS, 2011). Therefore, it is high time to examine the carrying capacity of the land under cultivation (Singh, 2019)

Even though, in principle globally a substantial additional land could be used for food production, in practice land will come under growing pressure for other uses and loss due to urbanisation, desertification, salinisation and sea level rise. In fact, the possibility of area expansion in India for cultivation is very limited, and the sustainable intensification appears to be the only feasible option towards realizing self-reliance. To produce 310 million tonnes (Mt) of food grains and 190 Mt of fibres, edible and non-edible oils by 2050 from about 150 mha is in high side, may be it is 140 mha. of net sown area is a big challenge for Indian agriculture. Further, it would require an average productivity of 3.3 t ha^{-1}. Therefore, good agricultural practices (GAPs) and proven technologies for different regions should be promoted to bridge the yield gaps. The land degradation due to unscientific developmental activities and management of land resources further aggravate the situation. The best available data indicates that about 25% of the world's land area currently suffers from soil degradation (Mandal et al., 2020). The extent and severity of soil degradation are high, particularly in the developing countries of the tropics and sub-tropics. Therefore, building self-reliancy requires a new economic model and agricultural production system that should be based on planetary boundaries with the ultimate goal to make sustainable human well-being and not merely a growth of resource spending (Rockstorm et al., 2009). Much of the current economic growth strategy is based on the false assumption that natural resources, such as land and water are inexhaustible, which they are not.

Land demand

The 'business as usual' trend analysis and demand projections of different kinds of phytomass by 2050 have been estimated for India (AFC, 2002; 2003). While the requirement from agricultural crops is estimated as 500 Mt, demands for other phytomass such as fodder, fuel wood and timber was 1980 Mt. This makes a total of 2.5 billion tonnes for a projected 1.7 billion population. The equivalent demand of land under various primary production systems has been 319 Mha which is 56 Mha in excess of a total 263 Mha land that can theoretically produce phytomass in India. Besides, there will be additional demand of another 26 Mha for settlements and 17 Mha for roads, railways, industries, mining, irrigation, and power projects. All of these together exceed India's total geographical area.

The minimum per capita land required for meeting the basic necessities of life even with the adoption of recommended agricultural practices, could be about 0.07 ha (Burngh et al., 1976). By this standard, a significant share of the world's population will be prone to land scarcity in the coming decade. The number of people dependent on less than 0.07 ha per capita is projected between 1.4 and 4.7 billion in 2050, depending on population growth rates (Engelman and Le Roy, 1995). These projections are based on the assumption that total arable land will remain the same over next three decades. Yet the area available for agriculture is progressively declining because of soil degradation and conversion to other land uses, including urbanization. The extent and severity of soil degradation are high particularly in the developing countries of the tropics and sub-tropics (Mandal et al., 2020).

Globally, pressure on the land is increasing due to population growth, conversion of agricultural land to other uses, and land degradation. The world's population grew from 0.6 billion in 1700 to 6.0 billion in 2000. Over the same period, the world's average per capita arable land decreased from 0.5 to 0.25 ha, and in many densely populated countries, it declined to 0.1 ha (or even less) (Engelman and Le Roy, 1995)

Intensification

In the whole history of human existence, almost 99% time was spent on hunting-gathering and fishing for obtaining food material. over the past 7000 yrs, it has changed remarkably. Increase in food production has a direct link with expansion of land, and there has been nearly 6-fold increase in the land area under agriculture over the past 300 years (Klein, 2001). Over half of the land expansion occurred in tropical rainforest by converting forests into land for cultivation or raising stock mainly for commodity crops such as soya bean and oil palm (Green et al., 2005).

A major global population, estimating to 1.3 billion, are engaged in producing about US $1.3 trillion worth goods at the farm gate every year. Globally, agribusiness produced US $420 billion in 1950 while farmers received a third of it. Researchers estimated that by 2028, global market for agricultural production would be US $10 trillion while farmers would receive only 10% of it (Clay, 2004).

Pressures on water resources

Agriculture by now consumes 70% of global 'blue water', mainly withdrawals from rivers and aquifers (GOS, 2011). Demand could rise by over 30% by 2030, while global water demand could increase by 35–60% between 2000 and 2025, and double by 2050 owing to pressures from industry, domestic use and the need to maintain environmental flows. In some arid regions of the world, several major non-renewable fossil aquifers are increasingly being depleted and cannot be replenished, for example in the Punjab state of India, and in countries of Egypt, Libya and Australia. Estimates suggest that exported foods account for 16-26% of the total water used for food production worldwide, suggesting significant potential for more efficient use of water via trade, although there is a risk of wealthy countries exploiting water reserves in low-income countries (GOS, 2011). One of the reasons for the poor use of rainwater in India is due to its ill-distributed and uneven nature, mostly occurring over a few months (June to September). Moreover, lack of appropriate water harvesting structures allows much of water lost as surface runoff. Prolonged dry periods are prevalent as a result of uneven and short-duration rains even in very high rainfall areas like north eastern Himalayas (NEH) where growing more than one crop becomes difficult if water is not retained or made available through water harvesting. Only 8% of the country receives very high or assured rainfall of above 2000 mm, and another

20% receives high rainfall of 1150 to 2000 mm. The rest of the country (72%), is in the low, dry, or medium rainfall range of less than 1150 mm, with 30% area with rainfall below 750 mm. Consequently, the vast 52% agricultural area is under monocropping with less fertilizer use and non-adoption of improved technologies (Singh, 2019). Thus, supplemental irrigation is must for crop production in these vast areas. For sustainable use of groundwater, adequate rainwater recharge through percolation tanks, check dams, restoration of dilapidated ponds, tanks and lakes are pre-requisite.

Over-exploitation of ground water

Ground water contributes to more than 60% of the irrigation demand in India, and thus it plays an important role in ensuring food security. However, over-exploitation of groundwater has become a crucial in most of the regions except eastern part of the country. Recent report revealed that out of the 6607 assessed blocks, 11% are in semi-critical (697), 3% in critical (217) and 16.2% in over-exploited (1071) category (Singh, 2019). By 2025, an estimated 60% of India's groundwater blocks will be in critical conditions. The 'business as usual' trend revealed that nearly 54% of the groundwater blocks in Gujarat, Haryana, Maharashtra, Punjab, Rajasthan and Tamil Nadu are likely to fall in these categories.

Pressures on biodiversity

Globally, land use land cover change (LULCC) has caused unprecedented changes in biodiversity mainly due to complex responses to several human-induced influences in the global environment. The degree of this change is so big (Sala et al., 2000) that in some important biomes, the ecosystem processes have been greatly disturbed with irreversible changes and the society's use of natural resources beyond the regeneration capacity. Biodiversity change is now considered one of important global changes (Sala et al., 2000). Deforestation, monoculture, wetland encroachment and rapid urbanizations altered the structure and compositions of species, and in addition it created pollutions, contamination and GHG emissions.

Increasing pressure on critical natural resources for survival and the misconception of its renewability has created immense damage to the ecosystem. Managing natural resources under competing and often contrasting demand is a challenge but inevitable for sustainable self-reliance. The challenges for bigger for population dense resource poor country like India.

Innovative management strategies towards self-reliance

Food production is the most basic need for human survival, and resource management has been centred around it through many decades. By repeatedly growing the same or combination of crops in an area depletes soil nutrients and can lead to build up of plant-specific pathogen infestations. External application of high doses of unbalanced fertilizers has also disrupted the nutrient cycling in many agro-ecosystems in India. In contrast, the resources and elemental cycles are more recycled in natural systems. Low factor productivity of fertilizers, high cost of phosphatic and potash fertilizers and their low efficiencies are well documented in Indian agriculture. Excess use of fertilizer, water, livestock intensification has known negative consequences and the knowledge gap needs to be bridged with innovative approach to achieve sustainability (Table 2)

Table 2. Consequences of land intensification and existing knowledge gaps

Biofuel demand -2% to -8%	Negative Consequences/ externalities	Knowledge gap/Innovation Required
Cropping Intensification (Continuing monoculture)	Poor soil quality and resilience, depletion of nutrients, soil erosion and breaking of soil aggregates	Understanding values of Ecosystem services Regeneration of soil functions and Critical limits of soil parameters as early warning systems
	Soil health, pesticide residue/soil pollution, toxicity	Assessment and understanding of Biological resilience Recovery mechanism Ecological intensification identifying Nature based solutions
Nutrient Intensification (Over fertilization)	Soil acidification, Water pollution, NO_2 emission and nitrate accumulation	Rate-reducing versus rate-balancing Increase in use efficiency, total- and partial factor productivity, biological N-fixation, Bio fertilizers, bio-inhibition Nano formulation
Irrigation	Water scarcity, Methane emission	Trade-offs between C and water Alternate wetting and drying, Direct seeded rice, promoting micro-irrigation, Water balance and recharge linked use
	Secondary Salinization, Water scarcity	Competition for water Water use efficiency, water management, water productivity, use of waste water improving conveyance infrastructure
Livestock Intensification	Soil degradation, Water storage, C loss	Forage versus feed crops, carrying capacity, alternative feeds, urea molasses block, improved livestock breed
	Waste, Water pollution, residue of veterinary medicine and antibiotics	Recycle and reuse of waste, promoting circular economy, Targeted use of agro-chemicals, alternatives to chemicals Integration with aquaculture
Forest Clearance, wetland, Drainage	Biodiversity, Natural wealth, C loss, Soil degradation	Agro-benefit versus natural value/ Biodiversity loss GHGs mitigation and adaptation emission/wetland reclamation Restoration though restocking and promoting natural succession

Sustainable soil management (SSM) for self-reliance

According to Inter Governmental Technical Panel on Soils (ITPS), soil management is sustainable, if the supporting, provisioning, regulating and cultural services provided by soil are maintained or enhanced without significantly impairing the soil functions that enable those services or biodiversity. The balance between supporting and provisioning services for plant production, regulating services for water quality and availability and for atmosphere GHGs composition are of specific concerns. Supporting services include primary production, nutrient cycling and soil formation, while provisioning services comprise of supply of food, fibre, fuel, timber and water, raw earth materials, habitat and genetic resources. Sustainable Land Management is a knowledge-based procedure that helps to integrate land, water, biodiversity and environmental management (including input and output externalities) to meet rising food and fibre demands while sustaining ecosystem services and livelihoods. Improper land management can lead to land degradation and a significant reduction in the productive and service functions (World Bank 2006). An innovative strategy for sustainable land management towards achieving self-reliance is presented in Fig 1. The 'revolutionary' changes are required to re-design and re-align the whole production system by incorporating the concept of ecological intensification. Action is needed simultaneously on many fronts where more sustainable food production is achieved through investment in existing technologies, reversing land degradation, innovation in improving use efficiencies of resources, minimising waste; improving the political and economic governance of the food system to increase productivity and sustainability and social infrastructure (a framework is presented below). According to some researchers, global food availability can be increased as high as 100-180%, which will meet the projected demands while lowering GHG emissions, biodiversity losses and other environmental problems if all these strategies are taken up through a holistic approach. To achieve sustainable development goals (SDGs) and land degradation neutrality (LDN), synergy should be created among scientific know-how, practical do-how and farmers 'participation.

Sustainable intensification

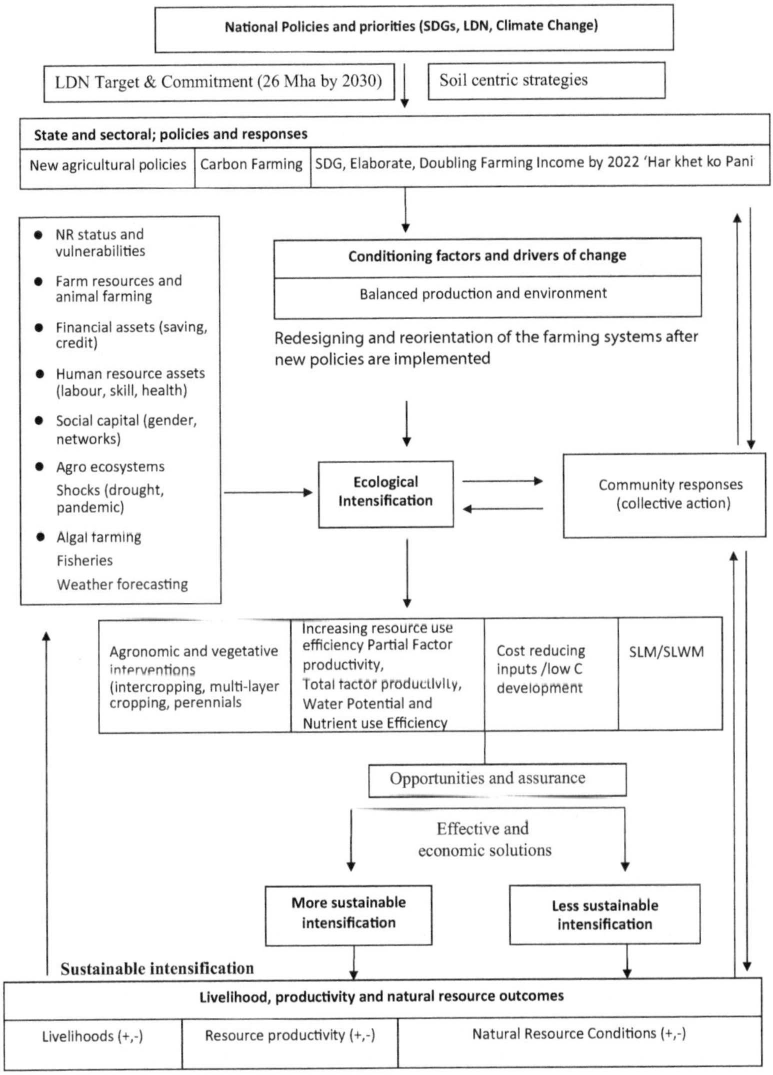

Fig. 1. NRM and policy linkages and factors involved in farm innovation

Sustainable intensification (SI) requires a system that is resilient, persistence and benevolence in the features of changing climates. The basic mechanism is to maintain the balance between intensification and biodiversity with a long-term objective to increase global food production without harming the ecosystem (Pretty, 2008).

Agricultural change often exists as an option between unsustainable intensification and extensive systems with less negative impacts. In SI system, we aim at producing more food and services on a sustainable basis with minimal use of additional land. Therefore, intensive agriculture is a complete knowledge-intensive approach where the most appropriate methods and techniques are imposed on a piece of land with a focus on improving the input-use efficiency and closing the yield gaps (Sapkota et al., 2017; Wollenberg et al., 2016). Researchers in South Asia and Africa have reported that only 40% of the potential crop production has been attained so far, particularly in rainfed areas. Accordingly, some promising strategies have been identified to produce more crop and income per unit of land through integrating mixed crop-livestock, crop-aquaculture and landscaping farming (Pretty et al., 2018; Pradhan et al., 2015). The intensity of use of non-renewable inputs must must be decreased in the long-term. This is particularly true for nitrogenous fertilizers that need to be produced in future by using renewable energy and hydrogen. Finding ways of reducing the processes of denitrification will also impact positively on GHGs emissions and the sustainability of agricultural systems.

The *Pradhan Mantri Krishi Sinchai Yojana* (PMKSY) is an exclusive programme for holistic management of all three categories of water resources - rainwater, surface water and ground water at the district level. This provides convergence of other similar programmes on conservation and management of water resources. Thus, the gaps identified under Strategic Research and Extension Plan (SREGP) of the district will be covered in the preparation of District Irrigation Plan (DIP), which will be the compendium of all the existing and proposed water resources network system in the district for effective demand-supply management.

Re-alignment of cropping systems and crop diversification in different production systems should be advocated with due consideration of water productivity of crops and by proper crop planning including both high and low water intensive crops in rotation. However, this can be enforced by suitable regulatory measures for an optimum coverage of the selected crops to increase the land use efficiency, and using smart technologies for water saving. It is a proven fact that water use efficiency can be increased by 20-25% through precision levelling followed the adoption of scientifically designed check basins or border strips and land configuration such as, ridge-and-furrow or raised-and-sunken beds. In other areas and selected crops, use of micro-irrigation system should be promoted in a big way which has potentiality for maximizing the water use efficiency (70% or more). Field experiments have shown that even rice cultivation is possible with drip system with multiple benefits (Singh, 2019). An integrated approach for nutrient and water management with balanced application of fertilizers in combination with irrigation has potential to enhance both water- and nutrient use efficiency.

Cultivation of short- and long-duration crops together for higher productivity and economic returns is most prevalent in dryland areas. However, in this high input farming and industrial crops era, this practice has minimum role to play. Pigeonpea is an ideal intercrop under diverse soil situations the rainfed regions. As an intercrop along with sorghum at 1:1 ratio, it is more productive and profitable. Apart from biological insurance against crop failure, roots of pigeonpea are able to break the hardpan in sub-surface layer of black (clayey) soils. Intercropping also gives a fairly good protection to mild sloping land against erosion and helps in runoff conservation by 13% (Somasundaram et al., 2014). It is also observed that approximately 1200-1300 kg of leaf litters ha from pigeonpea alone is added into soil every season. This in turn helps in enriching nutrient content and adds soil carbon during its decomposition (Somasundaram et al., 2014). Soil nourishing crops like green gram, black gram, pigeonpea and soybean followed by nutrient depleting crops like sorghum, maize, safflower and wheat offer an opportunity to manage the dryland soils efficiently.

Soil nutrient management

Nutrient management is one of the most practiced techniques for food production. However, the imbalanced and excess use of chemical fertilizers has affected the land and environment adversely. The balanced and integrated uses of nutrients have proven benefits, and the policy of fine-tuning including innovative marketing strategies must follow for its large-scale adoption. Biofertilizers and urea may be bundled while selling the chemical fertilizers. Purchase of biofertilizer may be made compulsory for every bag of urea to promote use of organic nutrients and reduce the use of chemical fertilizers. Drip fertilization technique need to be promoted in which fertilizer is combined with irrigation water in the drip system which conserves >40% nutrient and cut down water use by 50% (The Economic Times, 2020). Task force constituted by Govt. of India also suggested formation of crop based nutrient requirements to curb excess use of chemicals and fertilizers. Advisories should be issued; urea should be brought under nutrient-based subsidy (NBS) policy for better subsidy management. The Task Force also recommended in creating infrastructure in the villages for soil testing lab, agri clinics and agri-business centres for promotion of soil test based application of fertilizers. We must establish leaf analysis advisory laboratory for setting standard of fruits and other perennial crops. A clear picture of what the crop is experiencing from weather and disease can be seen from a leaf analysis (The Economic Times, 2020). The present use of fertilizer is far from balance (Table 3), which causes not only poor efficiency but also soil and environmental degradation (Sanyal and Chatterjee, 2007). Policy fine-tuning, technological advancement for fertilizer preparation, and application, promoting formulations including coated and mixed fertilizers, and site specific soil test based nutrient management are the viable option for improving the farm income.

In order to ensure correction of hidden deficiencies of micronutrients, Govt. of India has approved zincated urea and boronated super phosphate under Fertilizer Control Order (FCO) besides use of straight fertilizers of zinc sulphate, iron sulphate, manganese sulphate, copper sulphate, borax, and sodium molybdenum. *Fertilizer microdosing* is another technique developed by the International Crop Research Institute for Semi-Arid Tropics (ICRISAT). It is a practice of application of micronutrients to the crop with appropriate quantity and at right time. This method helps thousands of farmers in western and southern Africa to get their crops to mature faster and overcome the drought effects (Somasundaram et al., 2014).

Table 3. Imbalanced nutrient (NPK) usage

Sates	Nutrient (NPK) usage (kg/ha)	NPK ratio (Ideal ratio: 4:2:1)	Proposed strategy for promoting balanced use
Punjab	210.1	19.9: 5.9: 1	• Promoting mixed fertilizers with various combinations and customizing subsidy as per formulation. • Promoting/incentivising integrated source of nutrients –bundling of bio fertilizers with chemical fertilizers • Soil test based application of nutrients • Encourage corporate farming with high degree of resource sensitivity • Awareness of benefits of balanced fertilization, and the adversity of imbalanced use
Haryana	166.7	29.6: 8.8: 1	
Uttar Pradesh	140.4	12.1: 4.1: 1	
Andhra Pradesh	203.6	4.5: 2: 1	
Orissa	57.3	4: 1.5: 1	
Maharashtra	84.5	3.2: 1.8: 1	
West Bengal	127.5	23: 1.3: 1	

1. Sensors to detect nutrient deficiencies and early warnings of pests and diseases for precisely targeted control.
2. Use of aerial and global positioning technology for precise application of fertilizers- How much where and when it is needed in a field or forest.
3. Just in time irrigation, fertilization that maximize water and nutrient use efficiency.
4. New technologies with novel formulation of fertilizer materials and soil amendments based on nanotechnology and plant growth regulators (PGRs).
5. Smart fertilizers might be developed to release nutrients upon demand by the plant in response to specific molecular signals (Meeting et al., 2001).

Management of problem soils for self-reliance

Soil acidification has nearly doubled over the past 30 years, to which N-fertilizer is one of the factors which has contributed about 60% (Guo et al., 2010). Much of the acidification stems from the high inputs of ammonium-based N-fertilizers like urea, its uptake and removal of base cations rather than from acid rain. Impacts of acidification include suppression of rhizobial activity and reduced availability of micronutrients, in addition to direct impact of acidification on crop growth.

Restoration of degraded land

The area available for agriculture is progressively declining because of soil degradation and conversion to other land uses, including urbanization. The continuing degradation of land and soil is a severe threat to the provision of ecosystem services and economic development in India and globally. A recent national database on land degradation in India shows that 120.7 Mha or 36.7% of the total arable and non-arable land surface of the country suffers from various forms of degradation (NAAS, 2010) with water erosion being the chief contributor affecting 83 Mha (68.4%) area. Water erosion is the major threat to soil and water quality since it results in loss of organic C, nutrient imbalance compaction, decline in soil biodiversity, and contamination with heavy metals and pesticides. Annual soil loss rate in our country is about 15.35 t ha^{-1} yr^{1} (Sharda and Ojasvi, 2016), with an annual loss of 5.37 to 8.4 Mt of nutrients, reduction in crop productivity, occurrence of floods/droughts, reduction in reservoirs capacity (1% to 2% annually), and loss of biodiversity. Loss of crop productivity, being one of many negative impacts of soil erosion by water, has serious consequences for country's food, livelihood and environmental security. Major rainfed crops in India suffer from an annual production loss of 13.4 Mt due to water erosion which amounts to a loss of Rs. 205.32 billion in monetary terms (Sharda and Dogra, 2013). Apart from economic loss, erosion transports nutrients, contaminate water bodies, pollutes ground water and depletes soil fertility. Release of extra CO_2 into the atmosphere by organic matter dislodgement followed by decomposition has serious implications on climate change. Erosion also enfeebles soil's position in providing a secure habitat for both micro- and meso-biodiversity.

Soil erosion is inevitable in any production system and soil type with varying degrees. But with the advent of agriculture, the acceleration of soil erosion on mismanaged land area increased manifold. Of late, the rate of soil erosion reaches a point where it exceeds the soil formation rate and rates of soil loss tolerance limit which are recipes for disaster. This calls for a clear understanding of soil loss tolerance for the formulation of appropriate soil conservation strategies. In ideal situation, the erosion level should be contained within the permissible limit specified for a given location which ranges between 2.5 to 12.5 t ha^{-1} yr^{-1}. Once this threshold limit is crossed, the inherent fertility of the land begins to fall.

In India, the projected increase in population will occur, where the soil resources are most scarce and severely degraded. Therefore, to feed the projected global population of 9.2 billion by 2050, and to protect this scarce resource especially in ecologically fragile regions, the targeted second Green Revolution in the country must be focussed on the sustainable management of soil and water resources.

Important technological options

Based on the research experience and performance of soil conservation measures, site specific management strategies have been developed (Table 4). For reclamation of priority class 1, Sloping Agricultural Land Technology (SALT) is most suitable especially for shifting cultivation areas. Basically, the SALT method involves planting field and perennial crops in bands of 3-5 m width between double rows of N-fixing shrubs and trees planted along the contours. Similarly, Conservation Bench Terrace (CBT) system can be successfully adopted in mildly sloping agricultural lands under arid, semi-arid and sub-humid climates for conserving *in-situ* rainwater, reducing runoff and soil loss and enhancing crop productivity. Reliable and proven soil conservation technologies include ridge-planting, no-till cultivation, crop rotation, strip cropping, contour planting, and cover crops. Although specificity varies, all conservation practices reduce the erosion. Any conservation measures may be used singly or in combination with other practices. Medium-term experimental evidences at lower Himalayan region revealed that soil erosion of 1 t caused a reduction of 15 kg maize yield amounting to a loss of nearly 37.5 kg of total above ground biomass. The energy content in this biomass is calculated to be 6,55,500 kJ. Thus, any soil conservation measure or a combination of practices (minimum till + mulch; contour cultivation+ green manure) which reduces soil erosion by 1 t, has the potentiality to save 6,55,500 kJ of energy equivalent to 15 kg fuel oil (Sharda et al., 2018).

Table 4. Management strategies for different land use under various priority classes

Priority Class	Land Use		
	Forest	Shifting cultivation/ Open grazing	Traditional Agriculture
1	Diversion drain and Gully plugging	Afforestation, SALT	CBT, Stone terracing
2	Gully Plugging, Check dams, and Staggered trenches	Gully plugging, Check dams, Staggered trenches and SALT	CBT, Stone terracing
3	Check dams, Bio-Engineering, and Contour trenches	Increasing the gap of cultivation	Contour vegetative barrier, Strip cropping and Hedge-row cultivation
4	Maintenance of forest density and Trenching	Cultivation once in >5-6 years	Contour bund cultivation across the slope, Cover crops and Vegetative barrier

Priority Class	Land Use		
	Forest	Shifting cultivation/ Open grazing	Traditional Agriculture
5	Maintenance of forest density and Trenching	Cultivation once in >5-6 years	Mulch tillage, Soil surface management, Application of SOM, Strip cropping, Cover crops

Salt *(Sloping Agricultural Land Technology) is an integrated land management technique where different combinations like silvi-pastoral, agri-horti-pastoral, agriculture and aquaculture including wet cultivation is advocated as per land capability classes in sloppy lands.*

CBT (Conservation Bench Terrace) *system is suitable for 1-3% sloping lands. In the CBT system, land is divided along the slope in the ratio of 2:1. The lower 1/3rd area is levelled for collecting runoff from the upper 2/3rd area, which is left in its natural slope.*

Landscape approaches

An agro-ecosystem is a mixed production system where plants (crops, horticulture, grasses, and forages) and animals (livestock, poultry and fisheries) interact within a biogeochemical human-induced environment (Altieri et al., 2002). Because of the growing population pressures, it is not always possible to protect the entire landscape. However, high erosion risk areas with higher slopes should be converted to forest and low erosion risk areas may be transformed with agroforestry or terraced agriculture for better water management and biodiversity conservation. An Integrated Natural Sequence Farming method has been developed in north eastern Himalayas where the top 1.5 ha area was under protected forest with about 800 m^3 capacity of water storage in the middle with provision of well planed silt retention ditches and runoff water course. In this system, lower parts are used for paddy-cum-fish culture on about 1.9 ha areas. Splitted bamboo channels are used for water conveyance. This type of nature-based solutions has a lot of scope (Sharda, 2006).

Although agro-ecosystems are typically managed in isolation within a region, hill agro-ecosystem has numerous consequences in the neighbouring systems.

The following points are very important in sequence farming:

1. Recycling of biomass and balancing of nutrients flow and availability.
2. Securing favourable soil conditions for plant growth through enhanced SOC and soil biotic activities.
3. Minimizing losses of solar radiation, air, water and nutrients (soil surface management, micro-climate management and water harvesting).

4. Enhancing biodiversity (SALT for hilly areas, sequential farming; agroforestry, hedge row cultivation).
5. Enhancing beneficial biological interaction and synergisms among agro-biodiversity components.

Ground water management-watershed/community mode

The over-exploitation of ground water and continuation of high water-demanding crops in the production system has created an alarming situation and sustainability concern. Therefore, greater emphasis must be given on water harvesting and artificial recharge of groundwater. Although, overall groundwater development in the country is only 62%, in some of the states like Delhi, Haryana, Punjab Rajasthan, it has already crossed the 100%. There are several technologies and government initiatives for groundwater recharge, but the technology adoption is far from desired. The main reason of passive response is due to inadequate knowledge, where beneficiaries could not harvest the benefits of recharge. The community adoption in participatory watershed mode is the key to the success. The recharge linked withdrawal policy, incentivising judicious use of water groundwater recharge through various means including direct well recharge, are some of the options for groundwater amelioration.

Considering the slow progress of groundwater development in the eastern India and north-east, the Govt. of India has rightly decided to focus on this region under the programme aptly titled "Bringing Green Revolution to Eastern India" (BGREI).

Waste water management

Increasing urbanization and industrialization generate huge volume of waste water. These waste water would be of varying qualities depending upon the source, contamination level and quality of treatment before discharge. Considering the present and projected water availability, waste water needs to be considered as an reusable resource. In India, problems associated with wastewater reuse arise from lack of its proper treatment. On the whole, only 60% of industrial effluent and 26% of domestic sewage undergo treatment at present. Considering the size and vast population of the country and various sectoral demands, India has huge opportunity to reuse the waste water.

At the community/farm level, there are many water management technologies and practices that could increase the productivity and resilience of agroecosystems while enhancing other ecosystem services, such as rain water harvesting, soil management to facilitate rainfall infiltration, and conserve nutrients, drip irrigation, cultivation of water conserving resilient multipurpose tree crop mixtures, maintaining year round soil cover as well as hedges, tree rows and other vegetation corridors. All these practices often involve higher risk with lower returns and therefore, farmers need incentive schemes in order to adopt these schemes, such as payment for ecosystem services (FAO, 2007). Some problem specific strategies and technological interventions are presented in Table 5.

Table 5: Technological options for sustainable soil and water management

	Problems and issues	Proven technologies	Strategies and approaches
1.	Soil erosion and degradation	Nature based vegetative and agronomic measures such as residue recycling, mulching, cover cropping in conjunction with contour farming, vegetative strips and hedge-row cultivation.	• Minimize soil loss to permissible limit. • Adopt "Grain to Human & Residue to Soil" rule. • *In-situ* moisture conservation.
2.	Drought stress	Water harvesting and recycling, mulch farming, improving irrigation efficiency using drip and ridge-furrow methods especially in clayey soils.	• Farming technique for water conservation. • Providing technical support in installing farm ponds and efficient irrigation systems
3.	Nutrient depletion and low soil fertility	Nutrient cycling, manuring, biological nitrogen fixation, biosolids, judicious use of fertilizers, zeolites as slow release fertilizers, nanoenhanced materials and biochar-based amendments and development of promising formulations.	• Providing clean cooking fuel to the rural households so that animal manure and crop residues can be used as soil amendments. • Making fertilizers available to farming community through developing local sources of fertilizer.
4.	None or slow adoption of proven technologies	Involving farmers in the decision process, participatory approach	• Improving land tenure, and addressing gender and social equity. • Micro finance for purchasing inputs.
5.	Lack of resources for adopting recommended management practices	Enhancing farm income, growing high value crops, trading soil C credit	• Paying farmers for ecosystem/ environmental services

Modified from Lal, 2009

Conclusion

The lack of sustainable management and human-induced activities on ecosystems are causing environmental damage significantly through the process of soil erosion, nutrient depletion, nitrogen pollution and drying up of rivers and lakes. At the same time, there are increased competitions for, and scarcity of, resources and inputs into various primary production systems. Of these, land degradation and water scarcity are the most pressing, with significant effects on regional productivity likely to occur by 2050. Competition for land has also emerged as a significant factor in many countries including India. Addressing food security in India by 2050 will require new knowledge, technology, and much needed fund to carry out the basic and applied research. There is evidence of a slowdown in productivity gains today along with reduction in R&D investment in many countries including India over the last two decades.

Feeding the burgeoning population while simultaneously sustaining or improving limited natural resources is a major challenge before us today. It is possible to feed everyone without massive and irreversible damage to our ecosystems-damage that would ultimately endanger both water and food security in future. Seventy five percent of world's poor live in rural areas and most are involved in farming. In the 21st century, agriculture remains fundamental to economic growth, poverty alleviation and environmental sustainability. A good land management is essential for sustaining the productivity of agriculture, forestry, fisheries and hydrology as it affects a range of ecosystem services on which the sustainability of agriculture depends.

Advanced knowledge on the hydrological cycle, improved agricultural practices and new tools for mitigating the impacts of chemical pollutants and managing wastewater now offer a set of knowledge-rich solutions to reduce environmental impact. When combined with stakeholder-centred institutional approaches to resource management, the scope for effecting positive change across the key land and water systems that furnish the global food supply is very high. In landscape-based approaches, conservation of forests and wetlands will be particularly important as they play a crucial role as natural regulators of the hydrological cycle and elemental flow. Addressing systems at risk will require land and water management institutions to intervene in the way they engage with stakeholders and deploy solutions.

Conservation is not something that can be controlled exclusively by legislation. It is mostly a way of philosophy and a way of existence. It is as original as honesty and prudence, and it must be achieved in much the same way. The embedded behaviour of civilized people must be changed. We could see opportunities to increase not just "crop per drop" but also food and other ecosystem services per drop.

References

AFC. 2002. Land Resource Management Policy -Background Report for MoRD sponsored study. AFC, NRO, New Delhi.

AFC. 2003. Policy Frame work for Land Resource Management-Final Report for MoRD. AFC NRO, New Delhi.

Atieri, M.A. 2002. Agroecology: the science of natural resource management for poor farmers in marginal environments. Agriculture, Ecosystems and Environment 93:1-24.

Buringh, P., H.D.J. Van Heemst and G.J. Staring. 1976. "Comutations of the absolute maximum food production of the world, 'Wageningen, The Netherlands: Agricultural University.

Clay, J.W. 2004. World Agriculture and the Environment: A Commodity By Commodity Guide to Impact and Practices. Island Press. Washington. 585PP.

De Fraiture, C. and D. Wichelns. 2010. Satisfying future water demands for agriculture. Agric. Water Management. 97(4):502–11.

De Fraiture, C., D. Wichelns, J. Rockström, E. Kemp-Benedict, N. Eriyagama, L.J. Gordon, et al. Looking ahead to 2050: scenarios of alternative investment approaches. Water for food, water for life: a comprehensive assessment of water management in agriculture. 2007. London, UK: Earthscan; 91–145

Dhillion, B.S. and A. Agrawal. 2006. Biodiversity. In: Environment and agriculture. Chadha, KL. and Swaminathan, M.S. (eds). Pp.31-50. Malhotra Publishing House, New Delhi. India.

Engelman, R. and P. Le Roy. 1995. Conserving land: Population and sustainable food production, Washington, D.C: Population Action International.

FAO. 2003. World agriculture: towards 2015/2030. FAO, Rome. Available online at: ftp://ftp. fao.org/docrep/fao/004/y3557e/y3557e.pdf.

FAO. 2006. World Agriculture, towards 2030/2050. FAO, Rome. Available online at: http:// www.fao.org/es/ESD/AT2050web.pdf.

Foley, J. A., N. Ramankutty, K.A. Brauman, E.S. Cassidy, J.S. Gerber, M. Johnston, N.D. Mueller, C. O'Connell, D.K. Ray, P.C. West, C. Balzer, E.M. Bennett, S.R. Carpenter, J. Hill, C. Monfreda, S. Polasky, J. Rockström, J. Sheehan, S. Siebert, D. Tilman and P.M. David Zaks. 2011. Solutions for a cultivated planet. Nature. 478(7369): 337–342.

GOS. 2011. Foresight. The Future of Food and Farming. Final Project Report. The Government Office for Science, London.

Green, R.E., S.J. Cornell, J.P.W. Scharlemann and A. Balmford. 2005. Farming and the fate of wild nature. Science. 307:550–555.

Guo, J.H., Y. Zhang, J.L. Shen, W.X. Han, W.F. Zhang, P. Christie, K.W.T. Gowding, P.M. Vlyousek and F.S. Zhang. 2010. Significant acidification in major Chinese croplands. Science. 327:1008-1010.

Klein, G.K. 2001. Estimating global land use change over the past 300 years: the HYDE database. Global Biogeochemical Cycles. 15: 417–434.

Lal, R. 2009. Soils and World food security. Soil and Tillage Research. 102: 1–4.

Mandal, D., N. Giri, P. Srivastava. 2020. The magnitude of erosion-induced carbon (C) flux and C-sequestration potential of eroded lands in India. European Journal of Soil Sciences. 71: 151–168. https://doi.org/10.1111/ejss.12886.

Metting, F.B. et al. 2001. Science needs and new technology for increasing soil carbon sequestration. Climate Change. 51: 11-34.

NAAS. 2009. State of Indian Agriculture. National Academy of Agricultural Sciences, New Delhi.

NAAS. 2010. Degraded and Wastelands of India – Status of Spatial Distribution. National Academy of Agricultural Sciences, New Delhi.

Nellemann, C., M. MacDvette, T. Manders, B. Eickhout, B. Svihus, A.G. Prins, B.P. Kaltenborn, (Eds). February 2009. The environmental food crisis-The environment's role in averting future food crises: A UNEP rapid response assessment. UNEP, GRID-Arendal, www. grida.no. ISBN: 978-82-7701-054-0.

Pimental, D., O. Bailey, P. Kim. 1999. Will limits of the Earth's resources control human numbers? Environment, Development, and Sustainability. 1(1):19-39.

Pradhan, P., G. Fischer, H. Van Velthuizen, D. E. Reusser and J. P. Kropp. 2015. Closing yield gaps: How 21 sustainable can we be? PLoS One, 10, 1–18, doi:10.1371/journal.pone.0129487.22

Pretty, J., 2008. Agricultural Sustainability: Concept, Principles and evidence. Philosophical Transactions of the Royal Society of London, Series B. 363 (1491): 447-466.

Rockström, J et al. 2009. A safe operating space for humanity. Nature. 461: 472-475.

Rosegrant, M.W. and X.M. Cai. 2002. Global water demand and supply projections part - 2. Results and prospects to 2025. Water International. 27(2): 170-182.

Sala, O.E., F.S. Chapin, J. J. Armesto, et al. 2000. Global biodiversity scenarios for the year 2100. Science. 287 (5459): 1770-1774.

Sanyal, S.K. and S. Chatterjee. 2007. Efficient use of soil, water and plant nutrients for food and environmental security. Indian Journal of Fertilizers. Vol. 3(9), pp71-78, 81-88 and 132.

Sapkota, T. B., V. Shankar, M. Rai, M. L. Jat, C. M. Stirling, L. K. Singh, H. S. Jat and M. S. Grewal. 2017. Reducing global warming potential through sustainable intensification of Basmati rice-wheat systems in India. Sustainability. 9, doi:10.3390/su9061044.

Sharda, V.N. and P.R. Ojasvi. 2016. A revised soil erosion budget for India: Role of reservoir sedimentation and land-use protection measures. Earth Surface Processes and Landforms. 41: 2007-2023.

Sharda, V.N., D. Mandal and P. Dogra. 2019. Assessment of cost of soil erosion and energy saving value of soil conservation measures in India. Indian Journal of Soil Conservation, 47(1):1-6

Singh, P. 2019. Feeding 1.7 billion. Presidential address delivered on 5th June, 2019. National Academy of Agricultural Sciences, New Delhi.

Somasundaram , J., B.L. Lakaria, R. Saha, N.K, Sinha, P. Jha, R.S. Chaudhary, R.K. Singh, D. Mandal, K.M. Hati, K. Ramesh, C.M. Vassanda, A.K. Biswas, P. Dey, S. Reddy and A. Subba Rao. 2014. Management of stressed soils of dryland agriculture in semi-arid tropics –A review. Indian Journal of Soil Conservation, 42(2): 178-187.

Swaminathan, M.S. 2009. The media and the farm sector. The Hindu. 11 November, 2009.

The Economic Times. 2020. Biofertilizer and urea may be bundled. 12th August, 2020.

Wollenberg, E., Meryl Richards Pete Smith Petr Havlík Michael Obersteiner Francesco N. Tubiello Martin Herold Pierre Gerber Sarah Carter Andrew Reisinger Detlef P. van Vuuren Amy Dickie Henry Neufeldt Björn O. Sander Reiner Wassmann Rolf Sommer James E. Amonette Alessandra Falcucci Mario Herrero Carolyn Opio Rosa Maria Roman-Cuesta Elke Stehfest Henk Westhoek Ivan Ortiz-Monasterio Tek Sapkota Mariana C. Rufino Philip K. Thornton Louis Verchot Paul C. West Jean-François Soussana Tobias Baedeker Marc Sadler Sonja Vermeulen Bruce M. Campbell, 2016. Reducing emissions from agriculture to meet the 2 °C target. Global Change Biology. 22:1–6. doi:10.1111/gcb.13340.

World Bank, 2006. Sustainable Land Management: Challenges, Opportunities and Trade off. Washigton DC. ISBN: ISBN-13:978-08213-6597-7

Yang et al. 2003. A water resources threshold and its implications for food security. Environmental Science and Technology 37, 14, 3048–3054.

8

Innovations in Water Management in Agriculture

K.G. Mandal, Dibakar Ghosh, Amod K. Thakur and R.K. Mohanty

ICAR-Indian Institute of Water Management, Bhubaneswar, Odisha

Introduction

"--- initiation of exploitative agriculture without a proper understanding of the various consequences of every one of the changes introduced into traditional agriculture and without first building up a proper scientific and training base to sustain it, may lead us into an era of agricultural disaster in the long run, rather than to an era of agricultural prosperity."

— Dr. M.S. Swaminathan

The introduction of high yielding varieties, additional irrigation facilities, a great input flow through fertilizers and pesticides, farm mechanization, credit facilities and other rural infrastructure facilities ushered in the green revolution. The food grains area under irrigation as a percentage to its total sown area has been increased from 20.9 to 51.2%, rice irrigated area from 36.5 to 58.3%, wheat from 43.1 to 93.4%, pulses from 9.4 to 18.6%, oilseeds from 3.7 to 28.6% during the period from 1965-66 to 2013-14 (DES, 2015). Despite the net sown area being stagnant around 140.80 M ha, gross cropped area has been increased significantly from 155.28 M ha in 1965-66 to 195.25 M ha in 2011-12 due to increase in net irrigated area from 26.34 to 65.26 M ha (Fig. 1), gross irrigated area from 30.90 to 91.53 M ha during the period. There has been a sharp decline in average size of operational land holdings. The scope for expansion of the area available for cultivation has become limited. The average size of holdings for all operational classes (small & marginal, medium and large) have declined over the years and for all classes put together it has come down to 1.16 ha in 2010-11 from 2.28 ha in 1970-71. Therefore, it is the urgent need to use the natural resource like water efficiently and in a sustainable manner.

Water management challenges

India accounts for only about 2.4% of the world's geographical area and 4% of world's renewable water resources, but the country has to support about 18% of the world's human population and 15% of livestock (CWC, 2010). Agriculture is an important sector of the Indian economy, accounting for 14% of the nation's GDP, about 11% of its exports, about half of the population still relies on agriculture as its principal source of income. About 10 million farmers are being added every five years. Hence, the country has the twin challenges of meeting its water needs along with sustaining pace of development.

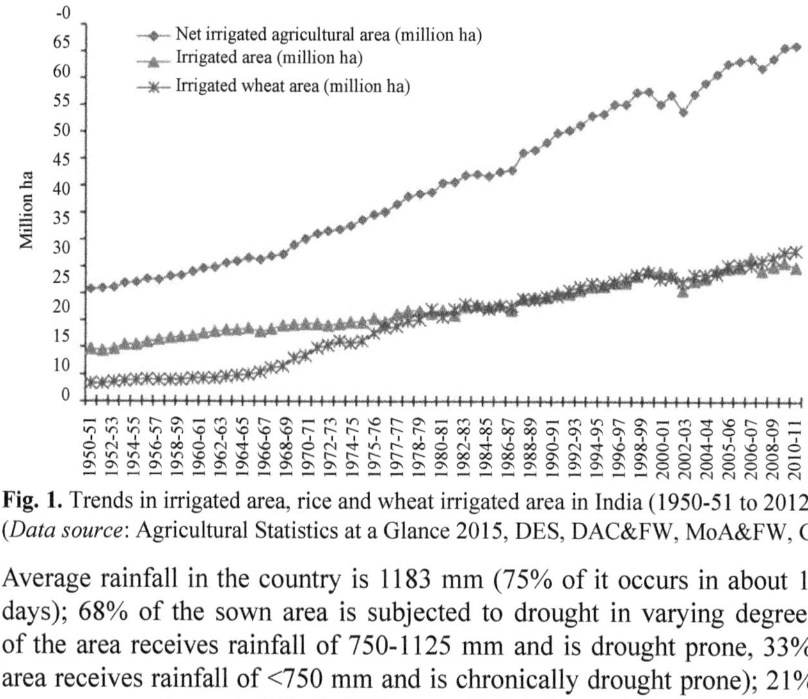

Fig. 1. Trends in irrigated area, rice and wheat irrigated area in India (1950-51 to 2012-13) (*Data source*: Agricultural Statistics at a Glance 2015, DES, DAC&FW, MoA&FW, GoI)

Average rainfall in the country is 1183 mm (75% of it occurs in about 100-120 days); 68% of the sown area is subjected to drought in varying degrees (35% of the area receives rainfall of 750-1125 mm and is drought prone, 33% of the area receives rainfall of <750 mm and is chronically drought prone); 21% of the sown area receives rainfall of <750 mm which is located in peninsular India and Rajasthan. Drought prone area lies in arid region (19.6%), semi-arid (37%) and sub-humid (21%) areas that occupy 77.6% of its total land area of 329 million ha. Rainfall is erratic (~ 4 out of 10 years).

India receives on an average annual precipitation of about 4000 BCM, which is its basic water resource. Out of this, after considering the evaporation, only about 1869 BCM is average annual natural flow through rivers and aquifers. Due to topographical constraints and spatiotemporal variations in resources, it has been estimated that only about 1123 BCM (690 BCM from surface water and 433 BCM from groundwater) is utilizable, which is just 28% of the water derived from precipitation (4000 BCM). There are further limits on utilizable quantities of water owing to uneven distribution over time and space, which may increase substantially due to climate changes, leading to water crisis and possible water related disasters, i.e., floods, increased erosion and increased frequency of droughts, etc. Agriculture, the largest user of water for producing food, is being seriously affected by the growing competition with industry, energy, domestic and other sectors for the available limited water resources.

Per capita availability of water is steadily declining, from 5177 m^3 in 1951 to 1820 m^3 in 2001, 1588 m^3 per year in 2010 due to increase in population, rapid industrialization, urbanization, cropping intensity and declining groundwater table; and it is expected to decline to 1341 and 1140 m^3 by 2025 and 2050, respectively (Fig. 2). The problems are likely to aggravate in future if suitable measures are not adopted.

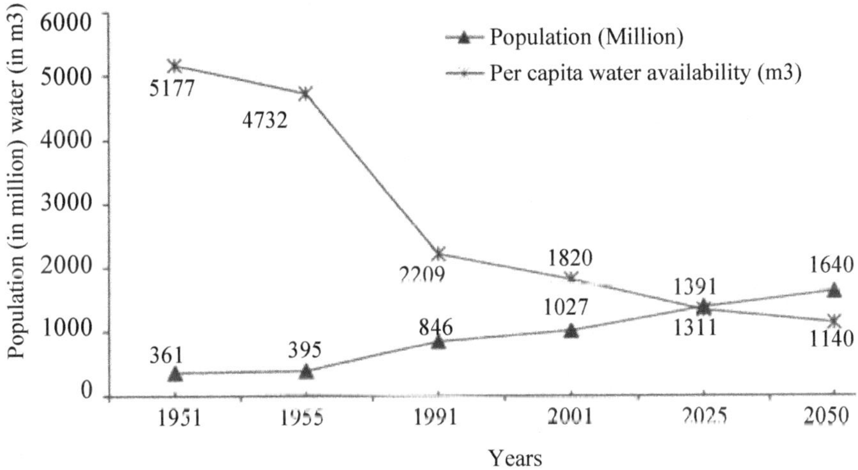

Fig. 2. Trends and projection of population and per capita availability of water per year in India (Data Source: MoWR, GoI)

Temporal variability of source-wise irrigated area

Canal irrigated area increased to 16.76 M ha in the year 1983-84 from 10.96 M ha in 1965-66, then it remained almost constant with the average of 16.63 M ha over years from 1988-89 to 2009-10 (Fig. 3). There was a decreasing trend in tank irrigated area, from 4.26 M ha in 1965-66 to 1.64 M ha in 2009-10. Tubewell irrigated area steadily increased to 28.948 M ha (2009-10), with an average of 27.04 M ha during 2002-03 to 2009-10. The area under other wells increased to the maximum of 12.61 M ha (1998-99), then decreased and remained almost constant with an average of 10.12 M ha (2003-04 to 2009-10). Hence, major irrigation source in India is through tubewells and other wells (60%) followed by canal irrigation (26%). Tank irrigation is confined to few pockets accounting for less than 5%. Although, there has been a significant achievement in water resources development, still there is a wide gap between Irrigation Potential Created (IPC) and Irrigation Potential Utilized (IPU), as well as Ultimate Irrigation Potential (UIP) and IPU. Hence, it is a great challenge to bridge the gap through use of potential technologies as well as evolving of innovative technologies.

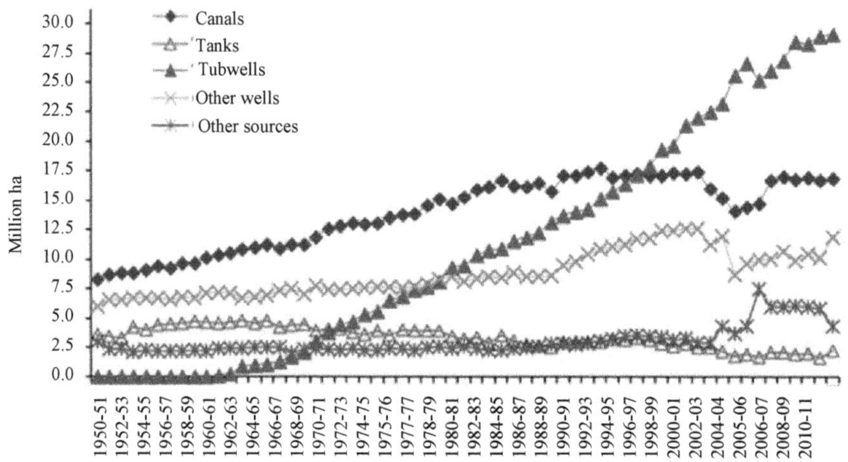

Fig. 3. Trends in net irrigated area from different sources area in India (1950-51 to 2010-11) (*Data Source*: MoWR, GoI)

Groundwater depletion

Excessive extraction of groundwater has occurred in northern Indian states viz. Rajasthan, Punjab and Haryana and a reliable estimate showed that groundwater was depleted at a rate of 54 ± 9 km^3 per year between April 2002 and June 2008; the depletion was equivalent to a net loss of 109 km^3 of water during the same period. With subsidized energy, farmers have few incentives to limit pumping. With rising energy prices groundwater pumping may become unaffordable, leaving smallholders dry and hungry, especially if subsidies are discontinued. Even the depletion of groundwater aquifer and reduced stream flows affect drinking water supplies, health and rural livelihoods. Regional variability exits in groundwater development: Punjab, Haryana, Rajasthan, parts of Gujarat, Tamil Nadu and Karnataka are over exploited; Maharashtra, Uttar Pradesh, Madhya Pradesh have groundwater development (GWD) more than 50%; eastern Indian states viz. Assam, Bihar, Chhattisgarh, Odisha and West Bengal have GWD less (22-43%) than that of country average (61%).

Poor irrigation efficiency

Irrigation efficiency is low. It is ~25-35% in most irrigation systems, 40-45% in a few exceptional cases; therefore, expansion of surface irrigation has become a major challenge with the concomitant reducing of conveyance and application losses and enhancing irrigation efficiency is a big challenge for irrigated systems. One of the main reasons for the limited expansion of irrigation facilities is the predominant use of the flood method of irrigation, using a large quantity of water per unit of land. It is a well-established fact that the flood method of irrigation leads to the inefficient use of water owing to enormous losses due to evaporation and distribution. The water use efficiency under the flood method of irrigation is only about 30–45% in India (Sivanappan, 1994; Rosegrant & Meinzen-Dick, 1996; Rosegrant, 1997). *Pradhan Mantri Krishi Sinchayee Yojana* (PMKSY), which has been initiated, is a sustainable approach to water management in the country as it has strong components viz. accelerated irrigation benefit programme, '*Har Khet ko Pani*', watershed and enhancing water use efficiency i.e. '*Per Drop More Crop*'.

Variable rate-precision irrigation

Irrigation in a variable rate, may be abbreviated as variable rate irrigation (VRI), leads to site-specific management of water. VRI encompasses a combination of innovative precision agriculture tools that enables a controlled irrigation systems to optimize irrigation application. VRI technology allows growers to easily apply varying rates of irrigation water based on individual management zones within fields. Management zones are field areas possessing homogenous features for landscape and soil properties. These features lead to similar crop yield potential and input-use efficiency for seed, nutrients and water (Doerge, 1998) and can be created via processes described by Zhang et al. (2002). Control systems on center pivots allow the right amount of water to be applied to these management zones within the field. Precision irrigation, defined as site-specific irrigation management that relies on the variable application of water, emerges as a potential solution to increase the productivity, and reduce the environmental impact, of irrigated agriculture. Precision irrigation reduces both water and energy use, while improving crop yield and quality. Substantial variations in soil properties and water availability exist across most fields. In these cases, the ability to apply site-specific irrigation management to match spatially and temporally variable conditions can increase application efficiencies, reduce environmental impacts, and even improve yields. The development of a distributed in-field sensor-based site-specific irrigation system offers the potential to increase yield and quality while saving water, but the seamless integration of sensor fusion, irrigation control, data interface, software design, and communication can be challenging.

Kim et al. (2008) have developed a conceptual system layout of distributed in-field wireless sensor network (Fig. 4). The system consists of five in-field sensing stations distributed across the field, an irrigation control station, and a base station.

The in-field sensing stations monitor the field conditions of soil moisture, soil temperature, and air temperature, whereas a nearby weather station monitors micrometeorological information on the field, i.e., air temperature, relative humidity, precipitation, wind speed, wind direction, and solar radiation. All in-field sensory data are wirelessly transmitted to the base station. The base station processes the in-field sensory data through a user-friendly decision making program and sends control commands to the irrigation control station. The irrigation control station updates and sends georeferenced locations of the machine from a differential GPS mounted at the cart to the base station for real-time monitoring and control of the irrigation system. Based on sprinkler head GPS locations, the base station feeds control signals back to the irrigation control station to site-specifically operate individual sprinkler to apply a specified depth of water.

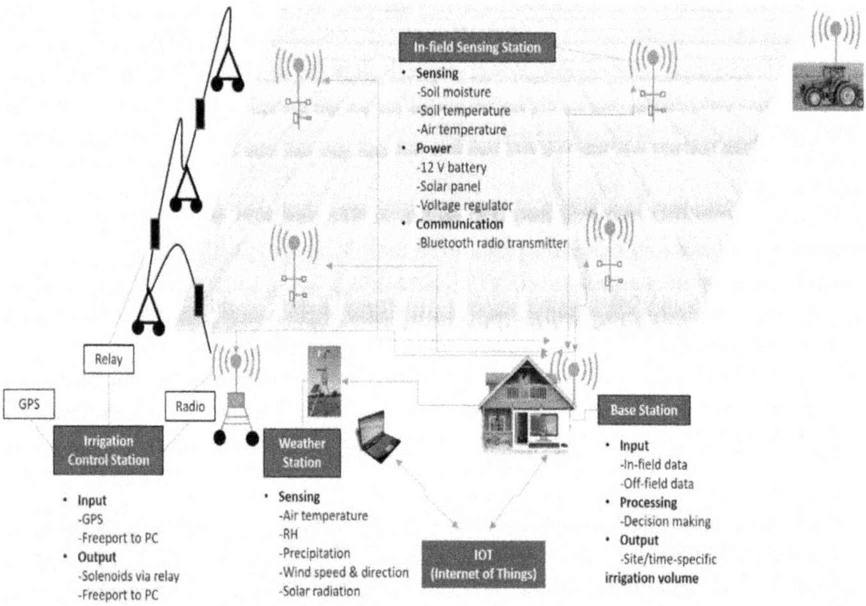

Fig. 4. Schematic diagram of a model showing its different components of sensor based variable rate irrigation in a field; Adapted from Kim et al. (2008)

The system components of the in-field sensing stations and weather station contained three main parts: data logging, wireless data communication, and power management (Fig. 1). A data logger measured field sensors and was self-powered by a solar panel. The sensory data were transmitted via a Bluetooth radio transmitter. The spatial variability of agricultural fields has been widely addressed in precision agriculture (Irmak et al., 2002; Ahmad et al., 1999) however, optimizing field configurations for site-specific management in each field remains a difficult task. The understanding about spatial variation of actual field is most important so that a minimum number of in-field sensor systems could be placed with optimal impact for characterizing the scope of the field information. In this case, the optimal distribution of the in-field sensing stations was determined on the basis of the spatial soil variability (Kim at al., 2005). Soil properties such as a water-holding capacity can have a major impact on crop yield (Farahani and Buchleiter, 2004). Apparent soil electrical conductivity (EC) was used to map the field for its variability, primarily as an indicator of water-holding capacity as well as salinity. EC mapping has been widely used as one way to characterize variability of agricultural fields (Farahani and Buchleiter, 2004; Drummond et al., 2000; Jabro et al., 2006). The EC is a measure of the amount of salt in soil, which is directly related to the water-holding capacity, and other soil properties such as the percentage of sand, clay, and organic matter.

Key role by soil moisture sensors

Appropriate sensors should be chosen for in-field soil moisture, soil temperature and other sensors. The accuracy of sensors need to be ascertained prior to use for irrigation. Soil moisture sensors that use the time domain reflectometry (TDR) function in a somewhat similar way to frequency domain reflectometry (FDR) probes, but the mechanics behind the measurement system are different. TDR sensors propagate a pulse down a line into the soil, which is terminated at the end by a probe with wave guides. TDR systems measure the water content of the soil by measuring how long it takes the pulse to come back. It requires a device to generate the electronic pulse and need to be carefully calibrated in order to precisely measure the amount of time it takes for the pulse to propagate down the line and back again. They are also sensitive to the saline content of salt and relatively expensive compared to some measurement methods. However, TDR devices do respond quickly to varying soil moisture.

Frequency domain reflectometry (FDR) sensors use an oscillator to propagate an electromagnetic signal through a metal tine or other wave guide, but with this method, the difference between the output wave and the return wave frequency is measured to determine soil moisture. FDR probes are considered accurate but must be calibrated for the type of soil they will be buried in. They offer a faster response time compared to TDR probes and can be connected to a standard data logger to collect readings. Soil moisture probes i.e. Stevens hydra probe, those use the coaxial impedance dielectric reflectometry method employ an oscillator to generate an electromagnetic signal that is propagated through the unit (usually by metal tines or other wave guide) and into the soil. Part of this signal will be reflected back to the unit by the soil, and the sensor will measure the amplitude of this reflected signal and the incident signal in volts. The ratio of these raw voltages is used in a mathematical numerical solution to Maxwell's equations to first calculate the impedance, then both real and imaginary dielectric permittivity which in turn is used to accurately estimate soil water content.

Gypsum blocks use two electrodes placed into a small block of gypsum to measure soil water tension. Wires connected to the electrodes are connected to either a portable hand-held reader or a data logger. The amount of water in the soil is determined by the electrical resistance between the two electrodes within the gypsum block. More water present in the soil will reduce the resistance, while less water will increase it. While gypsum blocks can be relatively inexpensive and easy to install compared to other types of soil sensors, they have to be replaced periodically as the gypsum disintegrates. Gypsum blocks are also more sensitive to having readings throwing off by soil with high salinity (salt content). When using a soil moisture sensor, it's important to know what type of soil you're dealing with in order to ensure the most accurate results possible. The following video will help you to understand what type of soil you have.

Canopy temperature-based variable irrigation rate

Crop canopy temperature is an index of moisture potential in plants, which can be linked to irrigation need. There are two methods for detecting water stress in crops- one is based on soil water content, and the other is based on crop temperature (Ihuoma and Madramootoo, 2017). Crop physiological changes, e.g., stomatal conductance (Gs) and leaf water potential, and biophysical changes, e.g., leaf and canopy structure, have been widely used to monitor crop water status (Gerhards et al., 2018). However, on-site measurements of soil water content and crop characteristics are time-consuming, laborious, and costly and cannot represent the spatial variability of crop water status (Li et al., 2010). In recent years, unmanned aerial vehicles (UAVs) i.e. drones have become an advanced field phenotyping platform to provide data with high spatio-temporal resolution. A schematic diagram shows the model (Fig. 5). These vehicles have boosted the use of near-earth aerial imagery to monitor crop water status (Zhang et al., 2019). For example, the water status of a cotton crop was evaluated using UAV thermal imagery with high spatio-temporal resolution (1-day revisits and 0.01-m resolution) in Yangling, Shaanxi, China (Bian et al., 2019). Crop water status is often monitored using UAV thermal remote sensing technology (Martínez et al., 2016), because canopy temperature (Tc) is one of the most important physiological parameters related to transpiration, leaf water potential, and Gs. When the water supply is adequate, rising environmental temperature results in increased Gs and a higher transpiration rate of crops to cool the leaves, resulting in insignificant changes in Tc (Tanner, 1963). However, under drought conditions, the leaf Gs and transpiration rate may decrease. Consequently, the Tc may increase because of the reduction in the cooling effect of transpiration (Tanner, 1963; Gates, 1968).

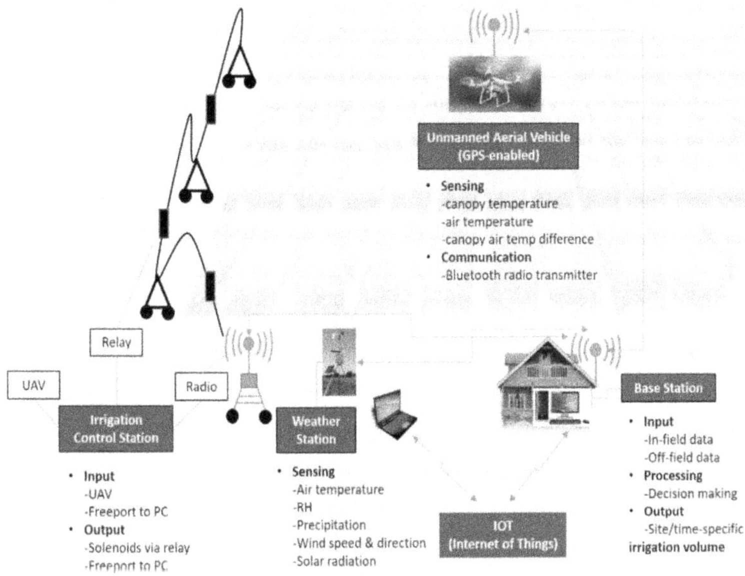

Fig. 5. Schematic diagram of a model showing canopy temperature-based sensors and variable rate irrigation components linked with internet and automatic irrigation; conceptually similar with soil moisture sensor-based irrigation. Adapted from Kim et al. (2008)

The use of UAV thermal remote sensing technology to monitor crop water status involves three important steps: temperature calibration, Tc extraction, and establishment of a Tc-based crop water stress indicator (Gerhards et al., 2018). To calibrate UAV thermal imagery, a linear regression model is often established between the measurements obtained using a handheld infrared thermometer and those obtained from UAV thermal images (Sagan et al., 2019). For example, Yang et al. (2018) collected ground-truth temperature data for maize canopy and white-black boards to calibrate UAV thermal images. The coefficient of determination (R^2) between the ground-truth temperatures and those estimated from UAV thermal images was 0.99 in the range of 25–55°C. It is important to establish specific linear regression models according to particular environmental conditions and measurement locations (Torres-Rua, 2017).

Crop water need-how much water to apply

The crop should be irrigated to fulfil its evapotranspiration (ET) demand at different growth stages. Once there is depletion of soil moisture and threshold canopy temperature shown in sensor-based IoT, crop should be irrigated through automatic irrigation i.e. irrigation control system, through micro-irrigation devices. The indicative values of crop water need have been presented in Table 1 for standard duration of different crops.

Table 1. Indicative values of the total growing period and seasonal crop water needs

Crop	Total growing period (days)	Crop water need (mm/total growing period)	Crop	Total growing period (days)	Crop water need (mm/total growing period)
Rice	90-150	450-700	Sugarbeet	160-230	550-750
Wheat/ Barley/ Oats	120-150	450-650	Citrus	240-365	900-1200
Maize grain	125-180	500-800	Banana	300-365	1200-2200
Sorghum	120-130	450-650	Alfalfa	100-365	800-1600
Millet	105-140	450-650	Tomato	135-180	400-800
Pea	90-100	350-500	Cabbage	120-140	350-500
Soybean	135-150	450-700	Onion green	70-95	350-550
Groundnut	130-140	500-700	Onion dry	150-210	650-900
Sunflower	125-130	600-1000	Bean green	75-90	300-500
Cotton	180-195	700-1300	Bean dry	95-110	450-650
Sugarcane	270-365	1500-2500	Melon	120-160	400-600
Potato	105-145	500-700	Pepper	120-210	600-900

Source: FAO (1986)

Precision-irrigation and crop quality

The high productivity with better quality of the produce can be achieved with precision irrigation practice. The application of irrigation water with the right rate at the right time may enhance the quality of the grape as well as wine quality. Opazo et al. (2010) found that precision irrigation in grapevines could be achieved using physiologically based irrigation scheduling methods. The midday stem water potential thresholds, imposed from post-setting may improve the wine quality by producing smaller berry with higher skin to pulp ratio, having greater amount of soluble solids and anthocyanin content in grapes. The problem of high pH for making quality wine may be minimized with applying limited irrigation in grape (Freeman and Kliewer, 1986). Suitability of variable rate in a pivot irrigation system for cotton production was evaluated by Booker et al. (2006) and found that the fiber strength improved with a decrease in water application rate.

Precision Irrigation-Enabled Integrated Farming System Model

In an integrated farming system model, efficient water management includes water resource conservation in farm-ponds, conservation of soil moisture, storage of excess canal water delivered in auxiliary ponds under canal commands, and conjunctive use of surface and groundwater using precision irrigation methods (Fig. 6). Advanced methods may be adopted under on-farm conditions using remote and proximal sensing such as wireless soil moisture sensors, crop temperature sensors using UAV-based thermal imaging systems, electromagnetic induction techniques and other non-invasive equipment.

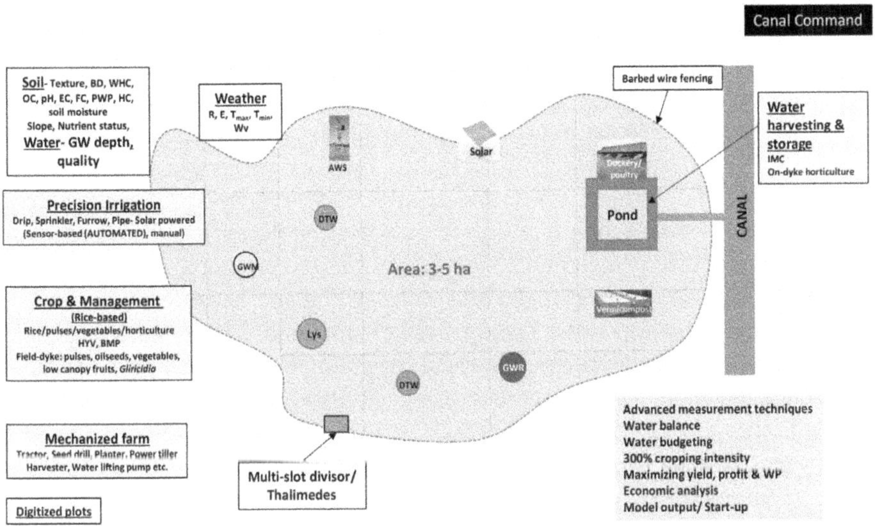

Fig. 6. Schematic model of precision irrigation-enabled pond-based integrated farming system

For a land-hydrological unit of an on-farm patch, the system components include construction of a water storage pond linked to canal irrigated area, deep tube wells, groundwater monitoring well & recharge shaft, installation of mini-weather station, lysimeter, runoff inflow and outflow measuring device, solar power systems, livestock/ poultry shed, vermicompost unit etc. Then there is need of installation of sensor-based micro-irrigation systems (drip and/ or sprinklers), pipe irrigation, furrow or other improved water-saving method with measuring devices connected to water storage pond and deep tube-wells, solar-operated pumps for irrigation. The operations would be use of farm mechanization with tractor, power tiller, planter, seed drill, rotovator, mini-paddy harvester etc; and the need for plantation of horticultural plants on pond-dyke and green manuring plants/ low canopy fruit plants on field-dykes. Variable rate irrigation should be adopted for the crops grown under integrated farming area.

These would enable developing sustainable and site-specific integrated farming systems with many-fold increase in overall water productivity. In eastern India, there is ample scope of developing integrated farming systems (crop-fish-livestock/ poultry) on a wider scale and in the enterprise development mode. But there is lack of such initiative by the farmers and/ or rural youths and farm women. Therefore it is the need of the hour to develop appropriate water management models with the active participation of farmers.

Improving on-farm water productivity

Water productivity is an important indicator of efficient water management. It is the ratio of farm produce to the volume of water used for farm production. A case study on increasing on-farm water productivity is described hereunder. This is the outcome of the efforts by a team of researchers from ICAR-Indian Institute of Water Management, Bhubaneswar. It was the participatory on-farm water management using auxiliary water storage ponds, dug-wells and development of integrated farming systems in a canal command viz. Kuanria medium irrigation project (KIP) in Daspalla block of Nayagarh district, Odisha; the site belongs to Agro-Climatic Zone VII (ACZ VII) and Agro-Eco Sub-region 12.2. The area is one of the most favourable eco-systems for agriculture, yet the agricultural production was much lower than its potential. Rice-fallow was the prevalent practice in most of the cultivated lands even under this canal command because of inadequate and unreliable irrigation water supply by the command. There was a large gap between demand and supply through canal irrigation. Hence, attempts were made to augment on-farm water resources using auxiliary water storage ponds and dug-wells through participatory approach. It was aimed at storage of rainwater, runoff water, and excess canal water delivered untimely during rainy season, and development of pond-based integrated farming systems involving fish culture, on-dyke horticulture and irrigated cropping. Conjunctive use of water from water harvesting pond and dug-wells facilitated irrigation to post-rain season crops in the command.

A view of the constructed pond and on-dyke banana cultivation

A view of community farmers harvesting fish from the constructed pond

Efforts were made to develop eight auxiliary water storage ponds (1630 m^3 capacity each) and four dug-wells (~160 m^3 each) plus previously existing four dug-wells in a participatory mode at 8 different sites of head, mid- and tail ends of the Kuanria command. Excess canal water, rainfall, runoff and seepage water stored in ponds, each pond-cum-well system has 3-4 ha of command area. Pond-based integrated farming systems (IFS) were developed as model IFS for the command with short-duration fish culture, with fish productivity of 1.32-2.96 t ha^{-1} and fish water productivity of 3.52-15.66 Rs m^{-3}, on-dyke horticulture (papaya, banana vegetables) and irrigated cropping in the command. Conjunctive use of water facilitated alternate cropping systems viz. rice + (fish in pond)-sweet corn, rice + (fish in pond)-sunflower, rice + (fish in pond)-green gram/ black gram, rice + (fish in pond) +pigeon pea (on dyke)-green gram, and rice + (fish in pond)-vegetables/ onion/ sesame with economic benefit ranged from Rs 1,37,210 to 1,74,012 per ha per year.

This agronomic-cum-engineering approach of increasing water productivity has been highly beneficial especially to the tail-end farmers of the canal command where, very often, canal water supply is inadequate. Auxiliary tank-cum-well system has brought additional area of about 30-35 ha under irrigation; which otherwise would have remained fallow during post-rainy season. The additional water (pond and well water) could be utilized for life saving irrigation to crops during dry spells and for post-rainy season crops. The intervention has tremendous impact on improving land and water productivity, increasing cropping intensity to the level of 230% from mere 150%, and increased livelihood of about 65 farm families through multi-enterprise systems viz. fish culture, horticulture and increased crop production. Water productivity has been improved significantly (20.5-53.46 Rs m^{-3}) due to such intervention on 8 different sites.

Conclusion

Precision irrigation should be adopted to management water efficiently. It must include variable rate irrigation based on soil moisture sensors and canopy temperature variations. Of course there is a need for standardization of soil moisture sensors, use of solar pump sets and android mobile app based operation of precision water application viz. drip or sprinkler irrigation. If the systems are set as per the site-specific conditions, then there will be enhanced irrigation efficiency from current level of about 30-40%, reduced over-exploitation of groundwater and enhanced net irrigated area. Participatory approach is required. Variable rate technology may result in varying degrees of beneficial and non-beneficial outcomes for each field (Pokhrel et al., 2018) because of the type and amount of variability, and the return on investment depends on a number of factors including crop type, useful life of the equipment, cost of water, and the cost to pump the water (Marek et al., 2001; Almas et al., 2003, O'Shaughnessy et al., 2019). The disadvantages viz. high initial cost of the systems and infrastructure development initial lack of technical aspects and operational know-hows should be overcome. This advanced irrigation approach and integrated farming will lead to efficient irrigation management and sustainable farming, and there will be 'more food more income with less water' and agricultural prosperity.

References

Ahmad, I.S., J.F. Reid, N. Noguchi and A.C. Hansen. 1999. Nitrogen sensing for precision agriculture using chlorophyll maps. Presented at the ASAE Annual International Meeting Toronto, ON, Canada, Jul. 18-22.

Almas, L.K., S.H. Amosson, T.H. Marek and W.A. Colette. 2003. Economic feasibility of precision irrigation in the Northern Texas High Plains. Presented in Southern Agricultural Economics Association Annual Meeting, Mobile, Alabama.

Bian, J., Z. Zhang, J. Chen, H. Chen, C. Cui, X. Li, S. Chen, and Q. Fu. 2019. Simplified evaluation of cotton water stress using high resolution unmanned aerial vehicle thermal imagery. Remote Sensing. 11: 267.

Booker, J.D., J. Bordovsky, R.J. Lascano and E. Segarra. 2006. Variable rate irrigation on cotton lint yield and fiber quality. Beltwide Cotton Conferences, San Antonio, Texas - January 3 – 6.

CWC, Water and Related Statistics 2010. Water Resources Information System Directorate, Information System Organization, Water Planning & Projects Wing, Central Water Commission, New Delhi.

DES, Agricultural Statistics at a Glance 2015. Directorate of Economics and Statistics, Department of Agriculture, Cooperation & Farmers Welfare, Ministry of Agriculture & Farmers Welfare, Government of India.

Doerge, T. 1998. Defining management zones for precision farming. Crop Insights.Vol. 8, No. 21. Pioneer Hi-Bred, Johnston, Iowa.

Drummond, P. E., C. D. Christy and E. D. Lund. 2000. Using an automated penetrometer and soil EC probe to characterize the rooting zone. in Proc. 5th Int. Conf. Precision Agriculture Bloomington, MN, Jul. 16–19, 2000.

FAO, 1986. Crop water needs (Chapter 3), In:Brouwer, C. and Heibloem, M. Irrigation water management, Training Manual No. 3, Food and Agricultural Organization (FAO). (http://www.fao.org/3/s2022e/s2022e02.htm)

Farahani, H. J. and G. W. Buchleiter. 2004. Temporal stability of soil electrical conductivity in irrigated sandy fields in Colorado. Transactions of the ASAE. 47: 79–90.

Freeman, B.M. and W.M. Kliewer. 1986. Effect of Irrigation, Crop Level and Potassium Fertilization on Carignane Vines. II. Grape and Wine Quality. American Journal of Enology Viticulture. 34: 197-207.

Gates, D. M. 1968. Transpiration and leaf temperature. Annual Review of Plant Physiology. 19: 211–238.

Gerhards, M., M. Schlerf, U. Rascher, T. Udelhoven, R. Juszczak, G. Alberti, F. Miglietta, and Y. Inoue. 2018. Analysis of airborne optical and thermal imagery for detection of water stress symptoms. Remote Sensing. 10: 1139.

Ihuoma, S. O. and C. A. Madramootoo. 2017. Recent advances in crop water stress detection. Computers and Electronics in Agriculture. 141: 267–275.

Irmak, A., J. W. Jones, W. D. Batchelor and J. O. Paz. 2002. Linking multiple layers of information for attribution of causes of spatial yield variability in soybean. Transactions of the ASAE. 45: 839–849.

Jabro, J. D., R. G. Evans, Y. Kim, W. B. Stevens and W.M. Iversen. 2006. Characterization of spatial variability of soil electrical conductivity and cone index using coulter and penetrometer-type sensors. Soil Science. 171: 627–637.

Kim, Y., R. G. Evans and J. D. Jabro. 2005. Optimal site-specific configuration for wireless in-field sensor-based irrigation, presented at the 26th Annual Irrigation Association International. Irrigation Show, Phoenix, AZ, Nov. 6–8.

Kim, Y., R.G. Evans and W.M. Iversen. 2008. Remote Sensing and Control of an Irrigation System Using a Distributed Wireless Sensor Network. IEEE Transactions on Instrumentation and Measurement. 57: 1379-1387.

Li, L., D. C. Nielsen, Q. Yu, L. Ma and L. R. Ahuja. 2010. Evaluating the crop water stress index and its correlation with latent heat and CO2 fluxes over winter wheat and maize in the North China plain. Agricultural Water Management. 97: 1146–1155.

Marek, T., L. Almas, S. Amosson and E. Cox. 2001. The feasibility of variable rate irrigation with center pivot systems in the northern Texas High Plains. ASABE Paper No. 011117.

Martínez, J., G. Egea, J. Agüera and M. Pérez-Ruiz. 2016. A cost-effective canopy temperature measurement system for precision agriculture: a case study on sugar beet. Precision Agriculture. 18: 95–110.

O'Shaughnessy, S. A., S. R. Evett, P.D. Colaizzi, M.A. Andrade, T.H. Marek, D.M. Heeren, F.R. Lamm and J.L. LaRue. 2019. Identifying advantages and disadvantages of variable rate irrigation: an updated review. Applied Engineering in Agriculture 35: 837-852.

Opazo, C.A., S.O. Farias and S. Fuentes. 2010. Effects of grapevine (*Vitis vinifera* L.) water status on water consumption, vegetative growth and grape quality: An irrigation scheduling application to achieve regulated deficit irrigation. Agricultural Water Management. 97: 956-964.

Pokhrel, B. K., K. P. Paudel and E. Segarra. 2018. Factors affecting the choice, intensity, and allocation of irrigation technologies by U.S. cotton farmers. Water. 10: 706.

Rosegrant, M. W. 1997. Water Resources in the Twenty-First Century: Challenges and Implications for Action. Food, Agriculture, and the Environment Discussion Paper 20, International Food Policy Research Institute, Washington, D.C.

Rosegrant, M. W. and R.S. Meinzen-Dick. 1996. Water resources in the Asia-Pacific region: managing scarcity. Asian-Pacific Economic Literature. 10: 32–53.

Sagan, V., M. Maimaitijiang, P. Sidike, K. Eblimit, K. Peterson, S. Hartling, F. Esposito, K. Khanal, M. Newcomb, D. Pauli and R. Ward. 2019. UAV-based high resolution thermal imaging for vegetation monitoring, and plant phenotyping using ICI 8640 P, FLIR Vue Pro R 640, and thermoMap cameras. Remote Sensing. 11: 330.

Sivanappan, R. K. 1994. Prospects of micro-irrigation in India. Irrigation and Drainage System 8: 49–58.

Tanner, C.B. 1963. Plant temperatures. Agronomy Journal. 55: 210–211.

Torres-Rua, A. 2017. Vicarious calibration of sUASmicrobolometer temperature imagery for estimation of radiometric land surface temperature. Sensors (Basel). 17: 1499.

Yang, W., C. Li, H. Yang, G. Yang, H. Feng, L. Han, Q. Niu and D. Han. 2018. Monitoring of canopy temperature of maize based on UAV thermal infrared imagery and digital imagery. Nongye Gongcheng Xuebao/ Transactions of the Chinese Society of Agricultural Engineering 34 (17), 68–75.

Zhang, L., H. Zhang, Y. Niu and W. Han. 2019. Mapping maize water stress based on UAV multispectral remote sensing. Remote Sensing. 11: 605.

Zhang, N., M. Wang and N. Wang. 2002. Precision agriculture-a worldwide overview. Computer and Electronics in Agriculture. 36:113-132.

9

Innovations of Controlling Desertification and in Renewable Energy

Priyabrata Santra, Mahesh Kumar, N.R. Panwar, R.K. Goyal, H.M. Meena R.N. Kumawat, Akath Singh and B.L. Manjunatha

ICAR-Central Arid Zone Research Institute, Jodhpur, Rajasthan

Introduction

The arid regions in India occupy 38.7 million ha (Mha), comprising 31.7 and 7 Mha under hot and cold arid region, respectively. The major part (90%) of the hot arid region lies in Northwest (NW) India and the rest in geographically isolated pockets in South India. The NW arid region extends from 22°30" to 32°50" N and 68°50" to 75°45"E , bounded by Aravalli hills in the east, Thar desert in the west the irrigated Indus plain in the north and the alluvial plains of the Sabarmati river in the south. The region mostly lies in western Rajasthan covering 12 districts (196150 km^2, 68.7%) followed six districts in northwest Gujarat (62180 km^2 21.8%), five districts in southwest Haryana in (12840 km^2 4.5%), and six districts in southwest Punjab in (14510 km^2, 5.0%) (Fig.1). The region is characterized by low rainfall (100 to 400 mm yr^{-1}), that is also erratic and highly unpredictable (coefficient of variation, CV = 30 to 70%); high evaporation (1600 to 2000 mm yr^{-1}); temperature extremes (-5.7° C to 50°C); frequent droughts (once in 2.5-5.0 years); strong winds (20-50 km h^{-1}) during summer; and short crop growing period (8 to 15 weeks) (Moharana et al., 2016). Despite the common characteristics of aridity and extremes of temperatures, there are enormous spatial variations in terms of rainfall pattern, physiography, soils, amount of available surface and ground water, and extent of vegetation cover (Joshi, 2012). Accordingly, the NW hot arid region has been classified in 4 sub-regions, 11 zones and 34 sub zones (Faroda et al., 1999).

Despite these intimidating conditions, the region supports a large human and livestock population and biodiversity. However, the ever-increasing human and livestock population and developmental activities exert enormous biophysical pressure on the slender natural resource base of the region. A major challenge is desertification due to wind erosion/deposition, water erosion, water logging and salinity. Degradation of permanent pastures has led to loss of fodder resources constraining livestock husbandry, the backbone of the economy of the region. Of late, industrial effluents and mining are also contributing to desertification.

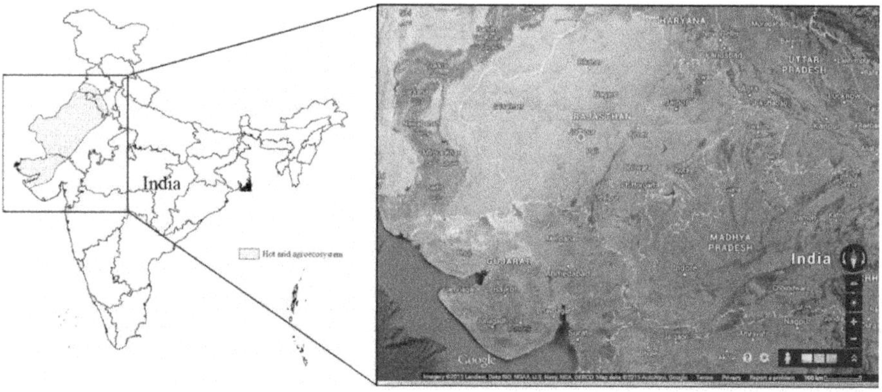

Fig. 1. Hot arid ecosystem of India

Farmers in western Rajasthan have a relatively larger operational land holdings (3.50 ha) compared to rest of the country (1.15 ha). Majority of the arable land is rainfed and only 20 per cent of it is under irrigation. The sources of irrigation are tube wells and some parts are benefitted by Indira Gandhi and Narmada canal water. Few villages in certain parts of Jalore, Pali, Barmer and Jodhpur districts are irrigated by open wells fed by tributaries of Luni river. However, recent research conducted by CAZRI revealed that 45% farm households in Jodhpur district have their land holdings under irrigation, either full or partial. Farmers generally have their fragmented holdings at two or three different locations and tube well was available in a minimum of one of the locations for these farmers.

Innovation for developing self-reliance in arid production system

Tackling water scarcity through rainwater harvesting

Agriculture and several of the other economic activities in arid areas depend on rain as the principal source of water, which augments soil moisture, groundwater and surface flows. This region is devoid of any well-defined perennial river system. Under such circumstances every drop of water becomes very precious. Rainwater harvesting, its conservation and efficient utilization can solve problem of water scarcity to a great extent. Rainwater harvesting in small ponds (*nadis*) underground tanks (*tankas*), *khadins* (Low lying areas) etc. are age-old traditional practices in the arid zone of Rajasthan. These structures vary in design, shape and size.

Rainwater harvesting in *nadi* and ponds

People of rural arid areas live in scattered settlements called *dhani's* distributed over sand dunes, inter-dunal plains and undulating landforms. Under such conditions, it is inconceivable that organized water supply will be feasible to fully meet the demand of thirsty land, human and livestock. Under such circumstances the system of rainwater harvesting in *nadi* and farm ponds are viable propositions for a group of farm families or community. *Nadi* is a dugout pond used for storing runoff water from adjoining natural catchment during rainy season. Farm Pond is an improved version of *nadi* with treated catchment and surplusing arrangement for removal of excess water (Fig. 2).

Fig. 2. Improved *nadi* with LDPE lining, boundary wall, inlet and outlet

Rainwater harvesting for safer and cleaner drinking water

Good quality potable water is a global issue, particularly in developing world because 80 per cent of the diseases arise from poor-quality drinking water. The problem of poor-quality groundwater used for drinking is more acute in Rajasthan. Concentration of fluoride in groundwater in this region ranges from 0.4 to 90 mg l^{-1} leading to various diseases like dental fluorosis, skeletal fluorosis and, non-skeletal manifestation etc. Appropriate harvesting of rainwater from roof top and its utilization can alleviate problem of fluoride to great extent. Surface rainwater can also be harvested in *tanka* using artificially prepared catchment. Traditional *tankas,* constructed with lime plaster, typically have a life span of three to four years. They suffer from seepage and evaporation losses and, in the absence of proper silt traps and pollutant-free inlets, the quality of conserved water deteriorates over time, making it unsafe for drinking. Also, in many situations, degradation of the catchment area means that it does not yield the quantity of water required to continuously replenish the structure. Planting of suitable tree species around the periphery of the catchment area of a *tanka* is recommended to improve the local environment.

Rainwater harvesting for supplemental/life-saving irrigation

Complete drought or long dry spells within a season are very common in hot arid regions of India. Harvested rainwater can be used to provide supplemental/life saving irrigation particularly to trees and crops. The system of rainwater harvesting by way of farm ponds, and subsequently its recycling for life saving irrigation provide effective check against dry spells and drought for economic yields (Goyal et al., 1995, 1997; Goyal and Sharma, 2000; Narain and Goyal, 2005).

Rainwater harvesting in *khadin* for crop production

Khadin is a unique practice of water harvesting, moisture conservation and utilization. It is suitable for the deep soil surrounded by natural rock outcrops in a catchment area. Collecting water in a *khadin* aids the continuous recharge of groundwater aquifers. Replenishment of aquifers means that subsurface water can be extracted through bore wells dug downstream from the *khadin.*

Rainwater harvesting for groundwater recharge

Percolation tanks, pondage in stock tanks with infiltration galleries, sand filled dam, anicuts across the stream, sub-surface barriers etc. are used for groundwater recharge (Ojasvi et al. 1996; Goyal, 1999). Adoption of conservation measures like anicuts, loose stone check dams, brush wood check dam etc. in watershed area has resulted in recharge of groundwater @ 0.33 to 0.75 m yr^{1} (Bhati et al., 1997; Narain et al., 2006; Goyal et al., 2007). Sub-surface barrier constructed across ephemeral streams traps subsurface flow to recharge the groundwater aquifer (Khan, 1998). Construction of two subsurface barriers each of 10 m length within 300 m from water supply well was enough to store runoff water for use a village having a population of 500.

Agro-meteorological advisory service and its impact in arid crop production

Weather plays pivotal role in agricultural production system, and is more crucial in hot arid areas where multiple challenges exist. Any abrupt changes in weather affect the crop growth environment, which has adverse impact on yield. Effective agrometeorological advisory services help farmers to take decisions for tackling both biotic and abiotic stresses. For example, proper selection of cultivars, timely sowing and harvesting, need-based application of fertilizers pesticides and insecticides, judicious application of irrigation water, can be managed with availability of weather-based agro-advisories to farmers. Weather forecast in short and medium ranges greatly contribute towards making short-term adjustments in daily agricultural operations which minimize losses resulting from adverse weather conditions, and improve the quantity and quality of agricultural produce (Rathore and Maini, 2008, Dupdal et al., 2020). India Meteorological Department

(IMD) has been providing weather forecasts (i) short-range (1-3 days), (ii) medium-range (4-10 days) and (iii) long-range (>10 days to the whole season). The medium-range weather forecast is especially of great value to farmers. Therefore, agroadvisory bulletins are prepared based on the medium range weather forecasts. Availability of such regular weather forecasts was earlier limited to meteorological departments, agricultural universities, research institutes etc. However, with advancement of digital technology, such agroadvisory services are getting into easy access to farmers; e.g. weather forecasts (e.g. temperature and rainfall) are available nowadays in smartphones.

The National Centre for Medium Range Weather Forecasting (NCMRWF) under the Ministry of Earth Sciences (MoES), Govt. of India in collaboration with IMD, Indian Council of Agricultural Research (ICAR) and State Agricultural Universities had been providing Agro-meteorological Advisory Services (AAS) to the farming community at the scale of agroclimatic zones based on location-specific medium-range weather forecast (MRWF). Since 2007, the entire framework of AAS, developed and successfully demonstrated by NCMRWF, has been relocated at IMD under MoES for extending the service (in operational mode) at the district level in each of 127 agroclimatic zones. It is now called the integrated AAS of MoES (Maini and Rathore, 2011).

Soil and nutrient management

Soil fertility under arid conditions, is constrained by environmental extremes of hot and cold temperatures, as well as by low water availability. With some exceptions, these soils are inherently low fertile. Nitrogen is generally deficient and availability of phosphorus is low to medium, while potassium is in adequately available (Mahesh Kumar et al., 2019; Praveen Kumar et al., 2009). Soils have low water-holding capacity, high pH, low organic carbon content (0.1 to 0.5%); are shallow, stony, and at times have other specific limitations. It is unlikely that Indian arid lands can be competitive for grain production with rest of India since arid areas have low production potential given their water and soil constraints. Further, given the poor access to markets and inadequate infrastructure in the region, low external-input technologies can be more impactful for sustainable production and livelihood improvement in the arid region. A few important low-input technologies and their impacts have been discussed below.

Efforts to improve nutrient utilization efficiency have emphasized on improved delivery of nutrients to the root zone in synchronization with crop demand, through band placement, fertigation, split fertilizer applications, etc. As N is universally deficient in arid soils, major emphasis is given to N fertilizers. To increase the access to N fertilizer by the crop, a variety of additives have been developed that inhibit urease activity, nitrification and denitrification. These approaches have been extremely successful in terms of increasing nutrient use efficiency and maximizing yields; however, their utilization on mass scale on small farms has met with limited success. This can be addressed by following approaches.

(a) Application of optimum quantity

A major step for achieving higher fertilizer use efficiency is to apply the right amount of fertilizer. But long-term estimations of N requirement of crops under arid rainfed arid systems are difficult as the yield levels vary from zero to three times of the average. Uncertainties in soil N supply can be reduced with site-specific N management approaches and choosing yield goals based upon (i) highest yield within past 5 years under proper crop management (ii) yield goal set at 1.5 times of long-term average, and (iii) yield goal based on soil capabilities. This approach is very different from most extension services in India which provide a single, standard fertilizer recommendation for an entire district or region.

(b) Internal nutrient circulation

The approach comprises of practices like recycling of crop residues either through direct application or as manure, or growing legumes, often referred as the integrated nutrient management. The practice of using inorganic and organic sources of nutrients in a proper proportion not only reduces the requirement of inorganic fertilizers, but also improves physical conditions of soil, especially soil water retention and its availability. Biological properties of soil are also improved considerably and fertilizer use efficiency is increased.

(c) Cereal-legume crop rotation

Cereal-legume rotations, intercropping of cereals and legumes are important practices for restoring soil fertility on large land holdings. Cultivation of green-gram in rotation with pearl millet supplied with 20 kg N ha^{-1} was found to provide similar yield as with application of 40 kg fertilizer N ha^{-1} (Singh and Singh, 1977). The amount of N returned from legume rotations depends on whether the legume is harvested for seed, used for forage, or incorporated as a green manure. Pearl millet in rotation with cluster bean has also been reported to yield higher than its continuous cultivation (Singh et al., 1985). Rotation of pearl millet with green gram or cluster bean was better than its rotation with moth bean. The beneficial effect of legumes to pearl millet also depended on the number of seasons of their cultivation before growing pearl millet.

Nitrogen-fixing trees: Leguminous trees can survive with arid soils' low levels of N due to their N-fixing capacity. These trees can fix 43-581 kg N ha^{-1} $year^{-1}$, compared to about 15-210 kg N ha^{-1} $year^{-1}$ from grain legumes (Dakora and Keya 1997). Nitrogen-fixing trees such as *Acacia* sp and *Prosopis* sp are some of the best sources for N enrichment in arid regions. In addition, most of these species are sources of highly nutritious fodder, fuel, food, charcoal, gums, fibre and timber.

(d) Conservation agriculture

Conservation agriculture aims to conserve and improve the natural resource base while using the resources available for agricultural production more efficiently. It avoids or minimizes soil tillage, maintains a permanent soil cover of crops and/or residues, and utilizes efficient crop rotations. It has been successful in many parts of the world, but has been least successful to date in the arid areas where production of organic matter is too low for permanent soil cover with crop residues because of water shortages. But it has been found that even small amounts of crop residues can reduce wind erosion considerably and increase soil water storage. Since significant quantities of soil and nutrients are lost by wind erosion where the soil remains bare for most part of the year, even small savings are worth pursuing. No dramatic increases in production and soil fertility can be expected by adopting conservation agriculture in arid areas in the short term, and farmers are unlikely to commit themselves to conservation agriculture without adequate incentives and policy support.

Solar energy applications in agriculture

Solar PV pumping system

For sustainable production from agricultural farms in arid region of India, irrigating crops at right stages is a crucial factor. Often, the lack of power supply to operate agricultural pumps does not guarantee the application of irrigation at the right stage. Solar photovoltaic (PV) pumping based irrigation could be a potential solution to irrigate the crops. It provides the self-reliance to farmers on energy access, and will also help greatly to achieve the set goal of 'more crop per drop' mission under the Pradhan Mantri Krishi Sinchayee Yojana (PMKSY) as well as to meet the national solar mission targets (details below). Solar PV pumping system generally consists of PV modules, mounting structure, pump unit (AC/DC) and tracking system. Sizing of PV module depends on the capacity of PV pump which is governed by the pumping head and the area to be irrigated. If the suction head is about 4-5 m, which is applicable in case of a surface water reservoir, 1 HP capacity pump is sufficient requiring 900-1400 peak watt (W_p) PV modules. However, to draw from deeper aquifer, PV module size needs to be increased accordingly. Further, if the designed irrigation network requires higher delivery head than the pressure, the pump can generate as per its capacity and available solar irradiance, the pump will not able to provide desired level of irrigation. Therefore, sizing of the PV pumping system is very crucial for applying right amount of water in agricultural field.

Considering the solar energy potential in India and the increasing energy requirement in agriculture, Government of India has announced the scheme '*Kisan Urja Surkasha evam Utthaan Mahabhiyan*' (KUSUM) in July 2019. The scheme includes three major components: (i) renewable energy-based power production in farmers' fields by installation of solar power plants of capacities ranging from 500 kW_p to 2 MW_p, (ii) installation of 17 lakh solar irrigation pumps of 7.5 HP capacity, and (iii) solarisation of existing grid-tied pumps for irrigation. While the first component of the scheme deals with PV energy generation and agricultural production together through agrivoltaic system, the last two components deal with solar PV pumps based irrigation in the agricultural field. Before the KUSUM scheme, different State Governments have been engaged in installation of solar PV pumping system in farmers' field by adopting different schemes and supports. Although Rajasthan state is quite ahead in installation of solar PV pumping systems in the country after installation of about 40,000 solar pumps in the state by the end of 2019, and there is still lot of scope to achieve the final target of national solar mission.

Agrivoltaic system

Food and energy are two basic requirements for human civilization. Therefore, competition for land may arise in future for agricultural use and PV based electricity generation. There is possibility that solar PV based electricity production will be preferred over agriculture on land resources. It is because of higher conversion efficiency of sunlight to electricity through photovoltaic process (~12-15%) than of sunlight to food through photosynthesis process (3-4%). Such preferability of PV based electricity generation on land surface will be more desirable in those areas where solar irradiation is available in plenty however, land productivity potential is low. Since food is the basic need for human civilization, utilization of land for food production cannot be ignored and, therefore, it is thought of producing both on a single land unit through agrivoltaic system (Fig. 3). Here crops are cultivated in inter space areas between and below PV arrays. Another important feature of the arrangement is rain water harvesting from top of PV module to a water storage tank, which is recycled for cleaning of PV module and also for supplemental irrigation to crops.

Fig. 3. Agrivoltaic system installed at ICAR-Central Arid Zone Research Institute, Jodhpur

The agrivoltaic system may bring huge opportunity to dryland farmers since rainfed crop production is risky due to uncertainty and scarcity of rainfall. Life cycle cost analysis of the system shows simple payback period of 5.87 years, discounted payback period of 10.40 years and internal rate of return of 16%. The concept of agrivoltaic system may be implemented in farmers' fields in arid and semi-arid regions of the country where land productivity is inherently low and solar energy is abundant. Renewable Energy Service Company (RESCO) model can be formulated for installation of agrivoltaic system in farmers' fields for food production and PV based electricity generation from a single land unit. Recently, Govt. of India has launched a scheme called KUSUM (*KishanUrja Suraksha Utthaan Mahaaabhiyan*) in which a component particularly deals with installation of solar PV systems in farmers field. To make the agrivoltaic system as a potential option to increase farmers' income in fragile arid and semi-arid ecosystems, it is necessary to provide grid network nearby farmers' fields to create selling opportunity of PV generated electricity from the system.

Use of digital science in soil management

Hyperspectral soil signatures and digital sensors have added the strength of DSM approach in recent times and improved the accuracy of digital soil products further. With the advancement of digital technologies, specifically in machine learning algorithms there is possibility to develop mathematical relationship between complex soil properties and several environmental and climatic variables (Murase, 2000, Banerjee et al., 2018, Jha et al., 2019). Machine learning tools are appropriate to analyze and model large soil database. Different machine learning tools are now available to identify the trend in large legacy soil database consisting data on several inter dependent soil properties. Few common machine learning tools are support vector machine (SVM), random forest regression (RF), artificial neural network (ANN), fuzzy logic (FL), decision tree (DT), deep learning (DL) etc. Artificial intelligence (AI) has been gaining popularity in recent times. Another recent development in computer science is 'internet of things' (IOT), which is a network of several communicating devices. These communicating devices may be computer, notepad, smartphone, Wi-Fi router, laptop etc. In the IOT system, chain of inputs and outputs has been made to take a final decision and each operation is fully based on automation.

Rapidly measurable soil spectral signatures may help to develop algorithms for estimation of soil properties (Shepherd and Walsh, 2002; Brown et al., 2006; Ben Dor et al., 2009; Viscara Rossel et al., 2016; Katuwal et al., 2017). Further, the developed algorithms may be translated to remote sensing platforms, low flying aircrafts, and even to drones to prepare map of soil properties in a quick time. A quick scan of geocoded soil samples can even estimate soil properties using the spectra-based algorithms; however, large soil spectral library representing different soil types in India is required. Even to attract the youth of the country in agriculture, such digital application may be appropriate. However, to apply digital technology in highly scattered land units of India following different land management practices is a challenging task. Soil health card mission of the country has been launched with the enthusiastic aim to recommend fertilizers and other input resources in the farmer's fields based on the soil test reports. With the advancement of digital technologies as mentioned above, the aim is achievable, provided soil information systems in the country are developed and made it available in digital platform. Even, sensor-based techniques need to be developed for in-situ characterization of soil in the field and communicating the output to IOT platform to take a decision on nutrient and water application in soil.

Android app for soil management

Soil information system in android platform may provide an optimistic solution for soil management by farmers. Such applications not only help in taking quick decisions but also make the farmers self reliant. Few examples of android application for soil management can be given here (Fig. 4). The android app, GypCal was developed by ICAR-CSSRI, Regional Station, Lucknow aims to provide decisions on sodic soil reclamation. The Fertilizer Calculator app, developed by ICAR-CCARI, Goa helps to calculate the amount of fertilizer to be applied in field based on the selection of crops by farmers. Both these android apps are available in Google plat store. The CAZRI soil moisture calculator helps to estimate two critical soil moisture contents at field capacity (FC) and permanent wilting point (PWP) based on the pedotransfer functions. It uses sand, silt, clay and organic carbon content as input. Therefore, these android applications may help farmers to decide on soil and nutrient management using smartphone and thus may make the farmers self-reliant.

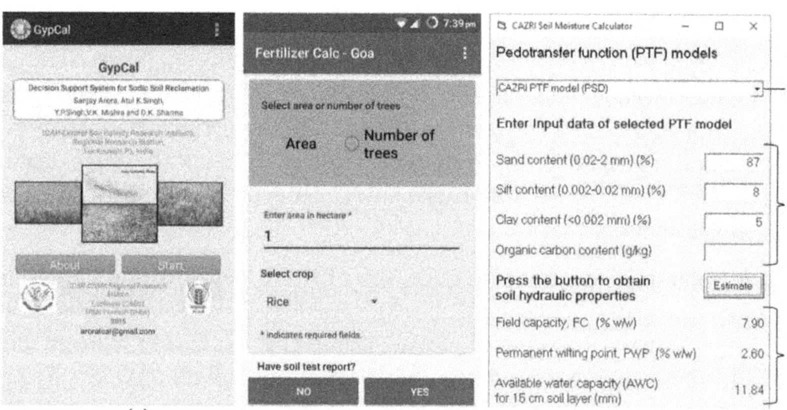

(a) (b) (c)

Fig. 4. Android applications for soil management; **(a)** GypCal developed by CSSRI, Regional Research Station, Lucknow, **(b)** Fertilizer Calc-Goa developed by CCARI, Goa and **(c)** CAZRI soil moisture calculator developed by CAZRI, Jodhpur

In-situ compost preparation

Soils of the hot arid region do not receive adequate nutrient replenishment and consequently soil productivity is relatively low despite of having high potential of improvement. Depletion of nutrients from agricultural fields is one main concern of declining soil productivity. Continuous cultivation with improper nutrient replenishment has led to depletion in soil phosphorus and potassium stock by 17.2% and 9%, respectively during last 27 years (Singh et al., 2007). Though improved crop varieties and chemical fertilizers have helped to increase in yields during last 15-20 years in the region, but improvement in soil fertility was not recorded. At the same time, with modernization of crop production system, soils are more intensively cultivated than before without any restoration measures for improving soil fertility. Looking in to these scenarios, it is very important to use compost as an organic source of nutrients, which can be produced from farm wastes at very low cost. One big advantage of compost preparation from farm waste is the efficient utilization of farm resources e.g. pre-monsoon grass vegetation, cow dung and farm labours, which are easily available. Dominant pre-monsoon vegetation commonly available in current fallow, permanent fallow, arable lands and permanent pasture lands in hot arid regions of Rajasthan includes *Crotolaria burhia* (*Saniya*), *Aerva javenica* (*Bui*), *Leptodenia pyrotechnica* (*Khimp*), and *Teprosia purpurea* (*Dhamasha*). Conventionally these native vegetations are removed from fields before cultivation of *kharif* crops. However, the above-ground biomass (15 mt) of these under shrub vegetations has great scope for in-situ preparation of compost in the hot arid region of India. Thus, preparation and use of compost can be a feasible solution to the tackle the problem of poor fertility of the soils and also will make the small and marginal farmers self-reliant in terms of supplying nutrient resources in the their fields.

Pit method for composting is better than the others for arid regions, where soil moisture is limiting. Pits keep the available moisture for a longer time to facilitate the decomposition process. Size of the compost mainly depends on farmers' need and/or capacity to prepare the compost. The width and length of a compost pit is not limited except its depth, which should not be more than 1.5 m. This is because it is difficult to control the temperature in a deep pit; it may be too hot and therefore, can easily loose moisture, and ends the decomposition process. Generally, the pit is filled with all kinds of green plant material of *saniya, bui, khimp* and *dhamasa* mixed with cow dung slurry in alternate 6-8 layers (6-8″ thick). A layer of rock phosphate is spread over each layer. Locally available weeds and grasses, etc. can also be used for filling the pits. The pit is filled up to the 1-2 feet above the surface and plastered with cow dung and soil. *Trichoderma harzianum* and *Bacillus firmus* can be applied at the rate of 50 g per layer depending on their availability. The moisture content during the entire process is maintained at 50-60% of water holding capacity. Turning of the compost pit at the interval of 30-35 days to provide aeration is essentially required. After 4-5 turnings, the compost should be ready for the field application. The pH and EC of compost vary from 7.2-7.9 and 0.025-0.350 dS m^{-1}, respectively. The compost thus prepared contains 15.5-19.7% C, 0.71-0.96% total N, 0.28-0.74% total P and 0.25-0.45% K, respectively. The C:N ratio of the prepared compost varies from 20-22.

Innovative institutional mechanisms

The country has witnessed several institutional innovations in the form of successful cooperatives in dairy and agriculture sector, Self-Help Groups in rural livelihoods and women empowerment. Recent addition in the list is Farmers Producer Organizations (FPOs). The Govt. of India enacted the Producer Companies Act, by incorporating a new part IXA in the Indian Companies Act 1956, through an amendment in 2002. The legislation enabled incorporation of cooperatives as companies and conversion of existing cooperatives into companies, while ensuring the unique elements of cooperative business with a regulatory framework similar to that of private companies (PC). The PC is formed with the equity contribution by the members and limited to them. The day-to-day operation is expected to be managed by the professionals, hired from outside, under the Board of Directors elected/selected by the general body of the PC for a specific tenure. Since farmers or the producers are the equity holders of the company, a PC as an organization provides an appropriate framework for owning the company by the producers themselves. The basic purpose of the PC is to make groups of small farmers or producers for backward linkage for inputs like seeds, fertilizers, credit, insurance, knowledge and extension services, and forward linkages such as collective marketing, processing and market-led agriculture production (Prabhakar et al., 2012).

FPOs provide following benefits to its member farmers: (i) provides collective bargaining power for small farmers/producers, (ii) training and consultancy services to farmers on good agricultural practices, (iii) opportunity for branding of the produce, (iv) ownership over FPO and empowering them to exercise their rights, and (v) equity to their production. There is less government control over PCs whereas the cooperative institutions are state controlled.

Major problems faced by the FPOs are related to financial and management issues. Some of the strategies to mobilize the capital are: (i) to get dealership with public and private companies to procure and supply inputs to its members and non-members; (ii) to utilize the existing government schemes to avail loans against pledging warehouse receipts; (iii) to develop effective linkages with ICAR institutes, SAUs and development departments to promote entrepreneurship and commercialization among its members while enhancing visibility of their work.

FPOs have to adopt following strategies for their success: (i) developing a brand value of the local crops, commodities and products, (ii) developing value added products, (iii) developing niche markets for its products at the national and international levels, and (iv) use of latest technologies in processing, value addition, packaging and marketing of products.

These FPOs would make farming community in arid regions of western Rajasthan more self-reliant by providing collective bargaining powder and reducing dependency on middlemen who enjoyed a great share of consumers' rupee at the cost of farmers. The FPOs can aggregate the produce of member farmers and decide on whom to sell, where to sell, when to sell and in what price to sell. The farmers would be greatly benefitted by effective delivery of backward and forward linkages, remunerative prices for their produce and enhanced social prestige associated with the identity of the FPOs and the brand name of the products.

Rejuvenation of community water sources and government intervention

Village ponds in hot arid region are the lifeline of people since it supports the need of drinking water and for other domestic use. Therefore, scientific interventions for monitoring and management of ponds are essential. Management of catchment area of the '*nadi*' or '*talab*', which is locally known as '*agore*' is an important factor to maintain the cleanliness and hygiene of the pond ecosystem. The community generally deputes a person from the village for proper care and management of the '*agore*'. If the villagers does not follow the rules as laid by the community for sustainable management of '*agore*', he/she is charged a penalty. Important rules are (i) open defecation is strictly prohibited in the catchment area, (ii) cutting of tree branches within the catchment area is not allowed, (iii) rearing of fish in stored rain water is also not permissible, etc.

Stored rainwater in '*nadi*' or '*talab*' is generally used for drinking purpose either by human being or by livestock. The general procedure to use the water is to carry the stored water from the '*talab*' to respective household. For carrying the water to household of the village, generally tractor mounted water tanks (~5000 L capacity) are used. No cost is levied by the village community as governed the local Panchayats for using the '*talab*' water, except for the carrying cost, which is to be paid to tractor owner by the villager. For pumping in the water from '*talab*' to water tank, suitable arrangements in the shaft of the tractor engine is made. In each household, water is stored in an underground water storage structure. Further, for use of water as drinking purpose by the people, water is kept in earthen pot, locally known as '*matka*' for 1-2 days for sedimentation of suspended solids. Chrome alum (chromium potassium sulfate) or white stone, locally known as '*fitkari*' is used for quick sedimentation of suspended solids in water. However, rejuvenation of village ponds need to be strengthened through scientific interventions and community participation for increasing the storage capacity and to maintain the cleanliness of the water. A successful scientific interventions have been recently carried out in village pond at Agoali village (pond area is 5.53 ha) and village pond at Baorli village (pond area is 3.33 ha) near Jodhpur, in which de-siltation was carried out to increase the pond storage capacity by 20-25%. During summer season, soil profile at the bottom of the pond was evaluated and about 0.4-0.5 m deposited sediment layer was removed. It is to be noted here that removal of soil material below the deposited sediment layer may lead to high percolation of stored water and thus reduce the storage capacity of pond. Satellite based algorithms was also developed to monitor the water spread area in the pond, which may help in monitoring and budgeting of water resources for multipurpose use. Villagers from Agolai actively participated in de-silting the village ponds through formation of "Agolai talab samiti" and contributing funds to carry out the excavation works. For maintaining cleanliness and hygiene of ponds, aquatic weeds from the ponds were removed and '*panghats*' were constructed at the boundary of the pond. Quality of the pond water was also improved significantly after intervention. Such scientific interventions may be multiplied to other village ponds of the region with the help of local Govt functionaries considering the recurrent drought situation in the context of climate change.

Blending with indigenous technologies

Farming community in arid regions have also evolved various Indigenous Traditional Knowledge (ITKs) pertaining to agriculture, livestock, soil and water conservation, water harvesting, prediction of weather, storage, processing and value addition aimed at solving problems while achieving a fine balance of efficiency and sustainability. Few examples are provided below:

Traditional water harvesting system of *Maldharis* in Banni grassland

Banni region has average annual rainfall of 317 mm with wide variability (CV 65%) and experience 2-3 drought or low rainfall years in every five years. Hence availability of drinking water during drought years becomes critical for survival of humans and livestock, and sustainability of pastoralist system (Manjunatha et.al. 2019a, Manjunatha et.al. 2019b). In response to this recurring challenge of drought, the *Maldharis* have developed their own traditional water harvesting system to conserve the rain water and keep it sweet despite the high salinity of land. They dig shallow wells in the land called *virda*. In a virda, the sand is removed up to a depth of one meter till the muddy water shows up (Fig. 5). This muddy water is removed till fresh water comes up. Wood/sticks are arranged in a crisscross manner against the inside walls of the well (*virda*) and grasses are placed in between the wall of well and sticks. Woods help *virda* to stay intact while grasses act as natural filters and help in purifying the water. Of late, the number of *virda*s and their use has declined on account of government's efforts to connect these villages with drinking water. However, certain old and abandoned *virdas* have been modified using modern-day technologies. Cement rings (concrete pre-cast well rings) of approximately 5′ diameter and 2-3′ height are used to construct wells in place of *virdas*. A gap of half to one inch is maintained between each ring (especially lower rings) by placing stones and gravels to allow seepage of water into the well from adjoining areas.

However, there are many old and abandoned *virdas* which need to be revived along with construction of new *virdas* to supplement the modern drinking facilities under the MNREGA. These *virdas* located at appropriate distances on the migratory route of pastoralists could act as boon for the livestock especially during drought years and summer months. These *virdas* could enhance the income of the pastoralists by making them self-reliant in water, one of the scarcest resources in the arid region.

Fig. 5. Virdas: Traditional water harvesting system in Banni grasslands of Gujarat

Sustainable green fodder production in arid region

The agriculture in the hot arid zone of India (western Rajasthan) was chiefly the subsistence farming and pasture land dominates as a land use system. Pasture lands provide support to the huge livestock population of the region. Rearing of animals play pivotal role in the economy of the people, and the animal husbandry sector provides round-the-year employment opportunity with more than 50% of total house hold income in arid region as against the national average of 22.5% (Anonymous, 2016, Manjunatha et al., 2029b). In arid regions of Rajasthan, common lands provide 66% of fodder requirement, whereas, crop residues contributed 22% in the total fodder requirements. The remaining demand of fodder is met out from the purchased fodder brought from the neighbouring states (Misra et al., 2015). The importance of these common resources can also be gauged from the fact that about 72% farmers graze their livestock on them extensively (Kumar and Singh, 2008). Traditionally, a close link was existed among crop, livestock and common grazing resources in the arid Rajasthan. However, with increase in population of both animals and humans, there is a proportional increase in competition for food to feed that is bringing agriculture and livestock production into direct conflict and led to deficit supply of feed and fodder to the livestock. The total annual availability of feed and fodder in arid Rajasthan is reported as 38.46 MT against a requirement of 58.39 MT (Kumawat and Misra, 2020). Similarly, the availability of green fodder and concentrate is estimated to the extent of 40% and 74% in arid Rajasthan. The deficit supply of feed and fodder is considered to be one of the major constraints in harnessing the productivity of milk and other animal products from the elite breeds of this part of Rajasthan.

Here, two case studies on green fodder cultivation in arid region are discussed. In the first case (CAZRI, 2019), rainwater was harvested and fodder crops were irrigated using the solar PV pumping system. The area of fodder crops under the system was 0.10 ha and hybrid napier (*Pennisetum perpureum* × *Pennisetum americarnum*) was taken as base crop to supply round-the-year green fodder. Hybrid napier was planted at 3 m × 1 m spacing in rows and the inter-row spacing area was used to grow cowpea (*Vigna unguiculata*) during *kharif* and lucerne (*Medicago sativa*) during *rabi seasons*. An average of 217.95 t ha^{-1} green fodder and 34.94 t ha^{-1} dry fodder was recorded from the system. In the second case, electricity operated pump was used to grow and irrigate conventional fodder crops (P. Santra; Personal Communication). Area under fodder crops was 0.38 ha and multi-cut fodder crops such as sorghum (summer and *kharif*), oat (*rabi*) and lucerne (*rabi*) were grown on 0.12, 0.12 and 0.14 ha, respectively. During *kharif* season, fodder sorghum was grown as rainfed crop. The system produced 295 t ha^{-1} green fodder in which contribution of sorghum, oat and lucerne was 41, 25 and 34%, respectively. Hence, it can be inferred that growing of green fodder even on a small area of land is much more lucrative than other agricultural enterprises in arid parts of Rajasthan.

Economic security through fruit production

Agricultural production system in arid region is highly unstable and often manifested through malnutrition and health hazards of dwelling people. Hence in traditional farming systems, inclusion of perennial components especially fruit crops is an important strategy for economical and sustainable production, which helps in developing self-reliance among the farmers. Since, per unit returns from horticultural crops are always higher, their inclusion in farming system improves net return of farmers and imparts resilience to aberrant weather. Recent introduction of tissue cultured date palm and pomegranate in the region proved worthy with net returns of 3 to 5 lakhs ha^{-1}. Besides, adopted species like *ber* (*Ziziphus* sp), *lasora* (*Cordia myxa*), guava (*Psidium guajava*), *karonda* (*Carissa carandas*), *khejri* (*Prosopis cineraria*), etc. may be profitable ventures under different production systems with improved marketing and processing facilities.

Moreover, most of the perennial horticultural crops are of multipurpose use meeting the requirement of fresh fruits, and of vegetables, fuel, fodder and fencing materials. Growing of pearl millet, moth bean, cluster bean, green gram etc. in the interspaces of fruit trees showed potential advantages in production, stability of output, resilience to climate change etc. There is substantial shift taking place towards horticulture, agri-horticulture and horti-pasture systems involving fruit trees with annual crops, perennial fodders, short duration fruit crops and vegetables. Few successful examples of horti-based cropping systems in arid region with high economic return are stated below.

Ber based agri-horti system

The common species of *ber* in arid region are *Ziziphus nummularia* (*jharberi*), *Z. rotundifolia* (*bordi*) and *Z. mauritiana (jujube)*. *Ber* plants utilize rainwater most efficiently as their growth and flowering coincides with monsoon rain. Growing of green gram, moth bean, cluster bean, and pearl millet between the trees of ber is a prevalent traditional practice in the arid region. If adequately spaced, the alleys of *ber* are most suited for growing green gram without any negative effect on yield of either of the crops. This agri-horti system has shown resilience to adverse weather conditions and was also remunerative. However, growing of *rabi* crops like wheat or mustard was not profitable in *ber* orchards.

Ber based horti-pasture system

Besides the economical part (fruit), leaves of *ber* are utilized as fodder for desert animals like sheep, goat, cattle etc. The palatability of air-dried leaves (locally named *pala*) for small ruminants is high. Trees such as *jharber* or *bordi* are commonly grown in rural areas and are an important top feed species. Even in drought years when crop fails, the top feed of this species comes to the rescue of the farmers. *Z. rotundifolia* is traditionally protected in pasture lands for additional availability of top feed. ICAR-CAZRI, Jodhpur has developed horti-pasture systems of *Z. mauritiana* and *Cenchrus ciliaris* (*dhaman* grass) that provides fruit with high TSS content, top feed for small ruminants and grass for cattle. It provides about 1.2-1.5 t of dry grass fodder along with 15-25 kg dry leaves of *ber*.

Research needs for developing self-reliance of farmers in arid region

- Identification of suitable cropping sequence with high water and nutrient use efficiency should be the major research focus in arid region of India in order to make the agricultural production system self-sustainable and self-reliant. For example, cluster bean-mustard cropping sequence has been found more resource efficient than cluster bean-wheat in Bikaner although the productivity is higher in the later. Similarly, resource efficient cropping sequence for other production environments of arid region need to be identified through research efforts.

- Use of locally available resources as reflective mulches for in-situ soil moisture conservation and providing nutrients to plants on long-term basis should be another research area. For example, gypsum, rock phosphate, feldspar etc. are abundant in pocketed areas of hot arid region and these materials can be used as reflective mulches since they are white in appearance. Although the stubbles and residue mulches are conventionally recommended as mulching materials, the application faces the challenge of termite infestation in the field. These are also low in availability after satisfying the feed demand of livestock.
- Water harvesting in farm ponds and use for irrigation purpose can always be promoted for sustainable use of water resources in the arid region. However, high evaporative demand of atmosphere often makes it impossible to fully utilize the harvested water during rainy season to irrigate successive *rabi* and pre-*kharif* crops. Therefore, research focus needs to be concentrated to develop economically feasible technology for minimization of water loss from farm ponds. Few farmers in the region use anti-evaporation chemicals and floating covers to control evaporation loss of water from farm ponds, although the high wind speed makes it difficult to maintain uniform coverage over the water surface.
- Housing facility for small ruminants in desert areas protects the animals from heat stress and thus increases the milk yield as well as improves the physiological growth performances. Scattered research efforts were made to develop thatched housing facility for goats; however, there is a need for efforts to develop modern housing facility with sensor-based monitoring of inside environment.
- Knowledge of soil properties helps in judicious application of both water and nutrient. However, measurement of soil parameters at multiple locations in the field through field sampling followed by conventional laboratory analysis involves time, which farmers cannot afford. Therefore, research efforts are required to develop sensor-based technologies to measure soil properties in the field in real-time so that farmers may take decisions on soil management quickly.
- Protected cultivation of vegetables and high value crops in arid region is the futuristic option to avoid harsh climatic condition. However, such protected cultivation system requires large amount of both energy and water to maintain congenial environment inside the structure for obtaining optimum crop yields. Research efforts should be directed to design and develop PV-driven greenhouse with water harvesting from the top of the protected structure to meet energy and water demand of the system.

Policy reforms needed

- Utilization of natural resources specially the water and plant resources is major concern in the fragile hot arid ecosystem. Policy guidelines are urgently required to conserve the fragile ecosystem for its sustainable and eco-friendly utilization.
- Pasture lands have been common property resources in the hot arid regions since long and meet the fodder requirements of livestock but degradation of these pasture lands is the major issue. At present, hot arid region is devoid of community pasture and is having very low per capita availability of fodder. Therefore, a strong policy intervention is necessary to regenerate and revive community pasture at panchayat level which leads to sufficient fodder availability to livestock as well as conserve the ecosystem at regional level.
- Urban and industrial zones have huge quantity of grey water (household wastewater from bathroom, kitchen and laundry). Policy interventions are urgently required to utilize this water after proper treatment for fuel and fodder production in nearby areas. This helps in effective management of soil, water and air pollution and increases availability of fuel and fodder at local level and keep environment safe.
- Land degradation is a major issue in hot arid region of India and wind erosion is the major land degradation process. Several wind erosion technologies were developed in past, however, their applications in the field is limited. In the context of land degradation neutrality (LDN) target of sustainable development goals (SDGs), there is a strong need for policy intervention to identify the degradation-prone areas and to implement technologies to combat it.
- Agrivoltaic system is found as a potential option to increase farm income. It will not only provide energy security but also improves the productivity of fragile land in hot arid environment. Suitable policy intervention is essential for large scale adoption of the system in farmers' field. Although the KUSUM scheme includes the target of installing large capacity solar PV systems in farmers' field, however, more focus needs to be given on small and marginal farmers. Even, the policy interventions include installation of agrivoltaic system through community participation in panchayat level.
- Solar PV pumping system has been found economical to irrigate crops in arid region of India. Presently, medium sized solar PV pumping systems (3 and 5 HP) have been distributed to farmers of the region through subsidy scheme. However, there is need for intervention to promote small sized solar PV pumping system (1 HP), which will be highly beneficial to small and marginal farmers. Moreover, such small sized solar PV pumping

system will not exploit the groundwater resources, rather it will use surface water resources most efficiently. Moreover, the PV modules may support the electricity needs of farm as well as home during off-time.

Conclusion

Enhancing the component diversity (crop, cultivar, tree, grass, flora, livestock) at farm level, in agreement with the site-specific bio-physical and socio-economic conditions, is prime-requisite to improve self-reliance of farming community and ensuring sustainable development in the arid region. For promoting implementation of different recommended technologies to a desired extent, there is need of creating sufficient infrastructural facilities (storage, transportation, and marketing), strengthening extension activities, promotion of co-ordination among different stakeholders, and ensuring enabling policy and institutional support for implementation of policies.

References

Anonymous. 2016. Dairying in Rajasthan - A Statistical Profile 2016.http://nddb.org/sites/default/files /NDDB_Rajasthan_21-9-16_Final.pdf.

Banerjee, G., U. Sarkar, S. Das, and I. Ghosh. 2018. Artificial Intelligence in Agriculture: A Literature Survey. International Journal of Scientific Research in Computer Science Applications and Management Studies. 7: 1-6.

BenDor, E., S. Chabrillat, J.A.M. Dematte, G.R. Taylor, J. Hill, M.L. Whiting, S. Sommer. 2009. Using imaging spectroscopy to study soil properties. Remote Sensing Environment 113: 38-55.

Bhati, T.K., R.K. Goyal and H.S. Daulay. 1997. Development of Dryland Agriculture on Watershed basis in Hot Arid Tropics of India - A Case Study. Annals of Arid Zone. 36: 115-121.

Brown, D.J., K.D. Shepherd, M.G. Walsh, M. Dewayne Mays and T.G. Reinsch. 2006. Global soil characterization with VNIR diffuse reflectance spectroscopy. Geoderma. 132: 273–290.

CAZRI 2019. Annual Report 2019. ICAR-Central Arid Zone Research Institute, Jodhpur, India. 216 p.

Dakora, F.D. and S.O. Keya. 1997. Contribution of legume nitrogen fixation to sustainable agriculture in Sub-Saharan Africa. Soil Biology and Biochemistry. 29: 809-817.

Dupdal, Ravi., Manjunatha, B.L., Rajkumar Dhakar and Patil, S.L. 2020. Perception and Economic Impact of Agromet Advisory Services: A Case Study of Thrissur AICRPAM Centre of Kerala State. Indian Journal of Extension Education 56(3), P.10-16.

Faroda, A.S., D.C. Joshi and B. Ram. 1999. Agro-ecological zones of north-western hot arid region of India. Annals of Arid Zone. 38: 1-8.

Goyal, R.K. 1999. Localized Ground Water Recharge System. Proceedings of brainstorming session on Groundwater modeling (Ed. K.D.Sharma) held at CAZRI Jodhpur pp. 170-173.

Goyal, R.K. and A.K. Sharma. 2000. Farm Pond: A well of wealth for the dryland dwellers. Intensive Agriculture. Vol. XXXVIII: 12-14 & 26.

Goyal, R.K., P.R. Ojasvi and J.P. Gupta.1997. Rainwater Management for Sustainable Production in Indian Arid Zone. 8th International Conference on Rainwater Catchment (Eds. B. Aminipouri and J. Ghoddoussi), Tehran, pp. 1060-1063.

Goyal, R.K., P.R. Ojasvi and T.K. Bhati. 1995. Economic Evaluation of Water Harvesting Pond under Arid Conditions. Indian Journal of Soil Conservation. 23: 74-76.

Goyal, R.K., T.K.Bhati and P.R. Ojasvi. 2007. Performance evaluation of some soil and water conservation measures in hot arid zone of India. Indian Journal of Soil Conservation.35: 58-63.

Jha, K., A. Doshi, P. Patel and M. Shah. 2019. A comprehensive review on automation in agriculture using artificial intelligence. Artificial Intelligence in Agriculture. 2: 1-12.

Joshi, D.C. 2012. Land use planning for sustainable agriculture in arid agro-ecosystem: Issues and strategies. Agropedology 24: 262-282

Katuwal, S., C. Hermansen, M. Knadel, P. Moldrup, M.H. Greve and L.W. de Jonge. 2018. Combining X-ray computed tomography and visible near-infrared spectroscopy for prediction of soil structural properties. Vadose Zone Journal. 17: 1-13.

Khan, M.A. 1998. Rainwater Management. Fifty Years of Arid Zone Research (Eds. A.S.Faroda and Manjeet Singh), Published by Central arid Zone Research Institute, Jodhpur. pp. 167-174.

Kumar, A. and Dhiraj K. Singh. 2008. Livestock production systems in India: An appraisal across agroecological regions. Indian Journal of Agricultural Economics. 63: 1- 4.

Mahesh Kumar, A. Kar, P. Raina, S.K., Singh, P.C. Moharana and J.C. Chauhan. 2019. Spatial variability of available nutrients in the Shekhawati region of Thar Desert. Journal of the Indian Society of Soil Science. 67: 21-33.

Maini, P. and L.S. Rathore. 2011.Economic impact assessment of the agrometeorological advisory service of India. Current Science. 101: 1296-1310.

Manjunatha, B. L., Shamsudheen, M., Sureshkumar, M., Pratibha Tewari and Yadav, O. P. 2019b. Occupational Structure and determinants of household income of pastoralists in Banni grasslands of Gujarat. Indian Journal of Animal Sciences 89(4): 453-458.

Manjunatha, B. L., Sureshkumar, M., Pratibha Tewari, Devi Dayal and Yadav, O. P. 2019a. Livestock population dynamics in Banni grasslands of Gujarat. Indian Journal of Animal Sciences 89(3): 319-323.

Misra, A.K., R.N. Kumawat, and M.M. Roy. 2015. Community grazing lands and livestock production in arid and semi-arid regions of India: status and way forward. In: Eco-Responsive Feeding and Nutrition: Linking Livestock and livelihood" organised by Animal Nutrition Association and Assam Agricultural University, Guwahati, India. pp. 28-43.

Moharana, P.C., P. Santra, D.V. Singh, S. Kumar, R.K. Goyal, D. Machiwal and O.P. Yadav. 2016. ICAR-Central Arid Zone Research Institute, Jodhpur: Erosion Processes and Desertification in the Thar Desert of India. Proceedings of Indian National Science Academy. 82: 1117-1140.

Murase, H. 2000. Artificial intelligence in agriculture. Computers and Electronics in Agriculture. 29 (1/2): 1-178.

Narain, P. and R.K. Goyal. 2005. Rainwater Harvesting for Increasing Water Productivity in Arid zones. Journal of Water Management. 13: 132-136.

Narain, P., R.K. Goyal and M.A. Khan. 2006. Watershed management in hot arid zone of India. Soil and Water Conservation Today. 1: 2-5.

Ojasvi, P.R., R.K.Goyal and J.P. Gupta. 1996. Hydrological Aspects of a localized Ground Water Recharge System. Proceedings of the symposium on New Strategies of Water Resource Management for 21st Century. J.N.V.U. Jodhpur, Dec. 20-21, p. 36-45.

Prabhakar, Itigi., Manjunatha, B.L., Nithyashree, M.L. and Dipika Hajong. 2012. Farmers Producer Company - An Innovative Farmers' Institution. Environment & Ecology 30(2): 427-430.

Praveen-Kumar, J.C. Tarafdar, D.K. Painuli, P. Raina, M.P. Singh, R.K. Beniwal, M.L. Soni, Mahesh Kumar, P. Santra, and M. Shamsuddin. 2009. Variability in arid soil characteristics. In: Trends in Arid Zone Research in India (Eds A. Kar, B.K. Garg, M.P. Singh and S. Kathju), ICAR-Central Arid Zone Research Institute, Jodhpur, pp. 78-112.

Kumawat, R.N. and A.K. Misra. 2020. An estimation of feed and fodder availability in the arid Rajasthan. Indian Journal of Animal Sciences 90: 249-253.

Rathore, L.S. and Maini, P. 2008. Economic Impact Assessment of Agro-Meteorological Advisory Service of NCMRWF, Report no. NMRF/PR/01/2008, 104p, Published by NCMRWF, Ministry of Earth Sciences, Government of India..

Shepherd, K.D. and M.G. Walsh. 2002. Development of reflectance spectral libraries for characterization of soil properties. Soil Science Society of America Journal. 66: 988–998.

Singh, K.C. and R.P. Singh. 1977. Intercropping of annual grain legumes with sunflower. Indian Journal of Agricultural Sciences. 47: 563-567.

Singh, S.D., Bhandari, R.C. and R.K. Aggarwal. 1985. Long term effects of phasing phosphate fertilizer on soil fertility and yield of pearl millet in rotation with grain legumes. Indian Journal of Agricultural Sciences. 55: 274-278.

Singh, S.K., Mahesh-Kumar, Sharma, B.K. and J.C. Tarafdar. 2007. Depletion of organic carbon, phosphorus and potassium under pearl millet based cropping system in the arid region of India. Arid Land Research and Management 21: 119-131.

Viscarra Rossel, R.A., T. Behrens, E. Ben Dor, D.J. Brown, J.A.M. Dematte, K.D. Shepherd, Z. Shi, B. Stenberg, A. Stevens, V. Adamchuk, H. Aichi, B.G. Barthes, H.M. Bartholomeus, A.D. Bayer, M. Bernoux, K. Bottcher, L. Brodsky, C.W. Du, A. Chappel, Y. Fouad, V. Genot, C. Gomez, S. Grunwald, A. Gubler, C. Guerrero, C.B. Hedley, M. Knadel, H.J.M. Morras, M. Nocita, L. Ramirez-Lopez, P. Roudier, E.M. Rufasto Campos, P. Sanborn, V.M. Sellitto, K.A. Sudduth, B.G. Rawlins, C. Walter, L.A. Winowiecki, S.Y. Hong, W. and W. Ji. 2016. A global spectral library to characterize the world's soil. Earth- Science Reviews. 155: 198-230.

10

Innovations on Climate Smart Agriculture and Structural Reforms for Self-Reliant India

Gouranga Kar and Ritesh Saha

ICAR-Central Research Institute for Jute & Allied Fibres, Barrackpore, Kolkata West Bengal

Introduction

Climate change effect is global and one of the most pressing issues of the 21st century. It has now emerged as a significant global environmental challenge affecting crop, animal, fish, egg and milk production, water availability, human health, and terrestrial and aquatic ecosystems. Evidences of changes in climate, particularly to the spatial and temporal variability of precipitation and temperature are many and as a result, uncertainty, risks and farmers distress have increased manifold over the current years. India, a developing nation and primarily an agriculturally based country will be highly affected due to greater dependence on agriculture for livelihood of larger proportion of population (Johnson and Hutton, 2014). Anthropogenic forces like rise in population, economic activities, consumption patterns, social behaviour, energy resource utilization and land-use changes in terms of deforestation, use of fossil fuels, industrialization, and race towards development, high concentration of green house gases (GHG) has been released resulting in global warming, a major factor leading to climate change. This heat of climate change has been experienced globally and has affected all forms of life on earth in many ways (Campbell-Lendrum et al, 2015).

As per the fifth assessment report of IPCC (2014), proportional sharing in GHGs emission of various sectors are energy production (34%), and agriculture, forestry and land-use (AFOLU) (24%), mainly contributed by biomass burning and resource intensive agronomic activities (Houghton et al., 2012; Tubiello et al., 2015). Global food security has been threatened by climate change more so in developing nations through changes in soil properties, and is one of the most important challenges in the 21st century to supply food for the increasing population.

Since climate change is inevitable, we must consider how best we can adapt it, so as to minimize its harmful consequences on the society. For subsistence farmers in India, adaptation in response to changes in climate patterns can be the difference between survival and calamity. There is a need to bring innovations on climate smart agriculture and also necessary agricultural reforms for emerging of a self-reliant India.

Climate change: Present Scenario

Global emissions of CO_2 (about 410 ppm now) have increased by almost 50% since 1990 with emissions increasing more quickly between 2000 and 2010 than in each of the three previous decades (Varallyay, 2010). The global mean temperature has increased by 0.74˚C during the past 100 years and by the end of 21st century, it is expected to rise by 1.8–4˚C. The fifth assessment report of IPCC projected an increase in temperature by 1.5 to 4.3 ˚C and increase in rainfall by 6-14% in India for all Representative Concentration Pathway (RCP) scenarios by 2080, compared with baseline 1961-1990 (Singh, 2017). Simulation studies showed mean warming in India is likely to be in the range 1.7-2 ˚C by 2030s, 1.9-2.4 ˚C by 2060s and 1.9-4.8 ˚C by 2080s relative to pre-industrial period (IPCC 2014). Using MARKSIM data generator model, the climate change projections for Indo-Gangetic region was made (Kar et al., 2019a) which indicated that maximum temperature will increase by 2.9 to 4.1 °C during 2050-2090 over baseline 2000-2018 temperature, while minimum temperatures are projected to increase in the range of 2.7 – 3.5°C by 2050 to 2080 over baseline 2000-2018 temperature under RCP 4.5 scenario.

The best-estimate linear trend in global surface temperature from 1906 to 2005 is a warming of 0. 74°C (likely range 0.56 to 0.92°C). Globally averaged temperatures in 2017 were 0.90 °C warmer than the 1951 to 1980 mean, and according to NASA 2016 was the third consecutive year in which temperatures were more than 1 °C above late nineteenth-century levels.

Issues and Challenges

Climate change affects crop productivity and quality primarily due to changes in growing environment and indirectly due to shifts in the geography and incidence of agricultural pests and diseases, impacts on soil fertility and biological activities. Studies indicated that the direct impact of climate change and variability is most significant on *kharif* crops due to increase in weather variability like decrease in rainy days and rainfall intensity, increase in duration and intensity of drought and floods, changes in atmospheric humidity, diurnal asymmetry of temperature, and pest incidence and virulence. The increased temperature also accelerates the depletion of soil moisture particularly under rainfed situations resulting in moisture stress and decline in crop yields.

Each one-degree celsius rise in mean temperature is estimated to reduce average global yields of wheat by 6%, rice yields by 3.2%, and maize yields by 7.4%. When averaged over all of the world's oceans, absolute sea level has risen at an average rate of 0.06 inches per year from 1880 to 2013. Since 1993, however average sea level has risen at a rate of 0.11 to 0.14 inches per year—roughly twice as fast as the long-term trend.

Observation records and climate projections provide ample evidence that fresh water resources are vulnerable and have the potential to be strongly impacted by climate change which can be linked to changes in the large-scale hydrological cycle such as: increasing atmospheric water vapour content; changing precipitation patterns, intensity and extremes; reduced snow cover and widespread melting of ice; and changes in soil moisture and runoff. Increased precipitation intensity and variability are projected to increase the risks of flooding and drought in many areas. All climatic parameters as well as potential evapotranspiration ET_0 are influenced by climate change. The spatial distribution of mean monthly ET_0 values under climate change RCP 4.5 scenarios was computed and its effects on water footprints of rice in Indo-Gangetic region were estimated (Fig. 1). At the end of the century, 8.5 to 14.3% water footprints of rice and wheat were projected to increase in different districts of Indo-Gangetic region. (Kar et al. 2019a).

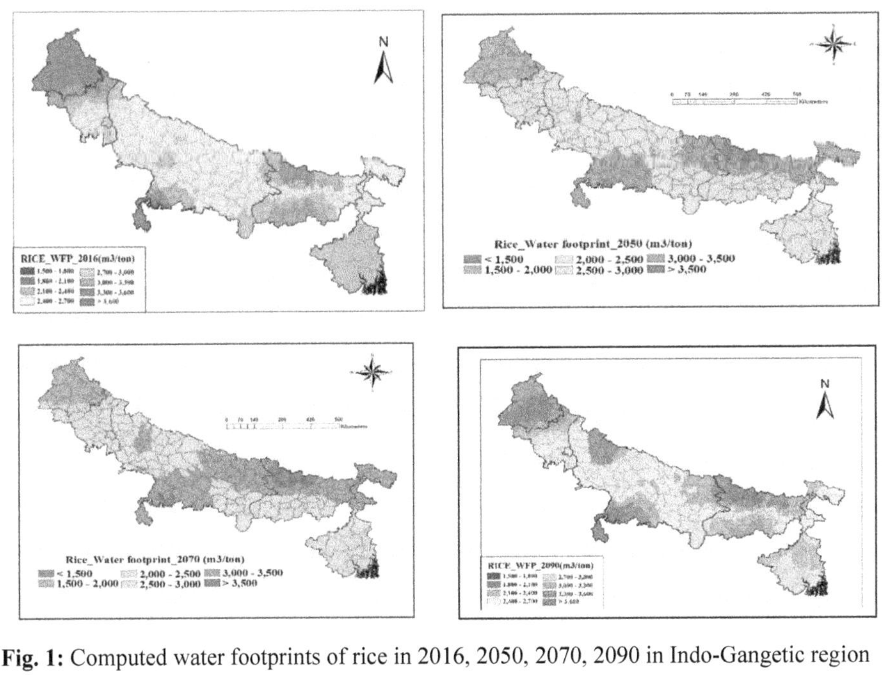

Fig. 1: Computed water footprints of rice in 2016, 2050, 2070, 2090 in Indo-Gangetic region

Thus, there are needs to change the agricultural practices and look for mitigation solutions which can be done by reducing the greenhouse gas emission and carbon sequestration. If the greenhouse gas emission is not controlled by strategic and sustainable approach then there are greater risks of fatal temperature rise which affects agriculture production. Further, agriculture sector contributing $1/4^{th}$ of the global GHGs emission is the major point of attraction for the climate change mitigation (IPCC, 2014). It is still possible, using a wide array of technological measures and changes in behaviour, to limit the increase in global mean temperature to 2 °C above pre-industrial levels. Affordable, scalable solutions are now available to enable countries to leapfrog to cleaner, more resilient economies. The pace of change is quickening as more people are turning to renewable energy and a range of other measures that will reduce emissions and increase adaptation efforts. But climate change is a global challenge that does not respect national borders. It is an issue that requires solutions that need to be co-ordinated at the international level and it requires international co-operation to help developing countries move toward a low-carbon economy.

Sustainable Development Goals (SDGs) and the commitment for reduction of GHG emissions

The SDGs are the blueprint to achieve a better and sustainable future for all. They address to mitigate the global challenges like poverty, inequality, climate change environmental degradation, peace and justice. To combat climate change Sustainable Development goals are:

- Strengthening resilience and adaptive capacity to climate-related hazards and natural disasters in all countries
- Integrating climate change measures into national policies, strategies and planning
- Improving education, awareness-raising and human and institutional capacity on climate change mitigation, adaptation, impact reduction and early warning
- Implementing the commitment of developed countries to the United Nations Framework Convention on Climate Change (UNFCCC) to mobilize jointly $100 billion annually by 2020 from all sources for addressing the mitigation strategies under Green Climate in developing countries.
- Promoting mechanisms for raising capacity for effective climate change-related planning and management in least developed countries and small island developing States, including focusing on women, youth and local and marginalized communities

In the Paris Agreement on 12 December 2015, parties to the UNFCCC agreed to work to limit global temperature rise to well below 2 °C above pre-industrial levels or even, to strive for 1.5 °C (Rhodes, 2016). Implementation of the Paris Agreement is essential for the achievement of the SDGs, and provides a roadmap for climate actions that will reduce emissions and build climate resilience. The Agreement entered into force on 4 November 2016. Additionally, the agreement aims to increase the ability of countries to deal with the impacts of climate change and make finance flows consistent with a low- GHGs emissions and climate-resilient pathway.

The Government of India has also adopted a National Action Plan on Climate Change in 2008 to address developmental activities and climate change together of which the National Mission for Green India is one of eight missions of the action plan. These national schemes are complemented by a host of specific programmes on solar energy, enhanced energy efficiency, sustainable habitats water, sustaining the Himalayan ecosystem, and to encourage strategic knowledge for climate change.

Adaptation and mitigation strategies

Agriculture is one of the most vulnerable sectors to climate change due to temperature rise (IPCC, 2007; Eze et al., 2013). The associated impacts of increased temperatures, altered pattern of rainfall and high frequency of cyclone, drought and floods are affecting crop yields and increase risks in agricultural productivity in different parts of the world (IPCC, 2007), especially in developing countries (Kurukulasuriya and Mendelsohn, 2008). A global analysis has projected that it is likely to affect crop production by 10–40 % in India by 2080–2100 (IPCC, 2007). Failure of farmers to adapt to such changes would likely have significant negative effects. India projects a decline in agricultural production due to climate change at varying magnitudes.

However, the agricultural sector holds significant climate change mitigation potential through reductions of GHGs emissions, and enhancement of carbon sequestration. In addition, it also has significant role to adapt climate change. Adaptation alone is not enough to offset the effects of climate change, and thus still need to be supplemented by concerted mitigation efforts (Vuren et al., 2009). Agricultural activities are relatively affordable form of mitigation option, for which many technical options are already readily available (FAO, 2011).

Adaptation and mitigation are often viewed as two closely related terms which vary in temporal and spatial scales, the former aiming to adjust to the actual or expected climate anomalies and the associated impacts and the latter aiming at reducing the GHGs emissions. Adaptation techniques can also contribute towards mitigation in a significant way making them a highly meaningful investment. Adaptation is achieved by enhancing adaptive capacity, or by minimising exposure and sensitivity to the harmful effects of climate change. The activities performed in mitigation will be evident after a long time on a large scale because of long residence time of GHGs in the atmosphere, whereas the effects of adaptation measures are effective immediately or in the near future on a local or regional scale (Vuren et al., 2009).

Technologies which aim at conserving resources and to efficiently utilise them such as zero tillage or reduced tillage, legume-based crop rotation, mulch/cover cropping, adjustment in sowing time, selection of resilient crop varieties and cropping pattern, manuring and balanced fertilization, efficient water harvesting and irrigation techniques can be effective adaptation strategies. Mitigation of climate change consists of strategies or technologies in agriculture and allied fields which can contribute to minimisation of the GHGs emissions and increase carbon sequestration of soils through changes in land-use management and enhancing input-use efficiency. Terrestrial carbon sequestration is one of the important measures for climate change mitigation to be focused.

Climate change adaptation promising technologies for India

Climate change adaptation is a continuous process requiring crop and location -specific response. Adaptation should enable agricultural systems to be more resilient to the consequences of climate change (Paul et al., 2009). Farming systems and farmers will differ enormously in their capacities to respond to climate change. Differentiated adaptation strategies and enhanced climate risk management support to agriculture and farming households are critical to counter the impacts of climate change (IPCC, 2007).

Climate ready crops

Development of new crop varieties with higher yield potential and resistance to multiple stresses will be the key to maintain yield stability. Improvement in germplasm for heatstress tolerance and oxidative stress tolerance should be targets of breeding programme. Improvement in water- and nitrogen use efficiencies assume more relevance in the climate change scenarios as water resources for agriculture are likely to dwindle in future and it is important to improve the root efficiency for mining the water and absorption of nutrients.

Conservation agriculture

The global empirical evidence shows that transformation of agricultural production systems based on conservation agriculture principles is already occurring and gathering momentum globally as a new paradigm. The conservation agriculture systems are currently practiced globally on about 157 million ha which corresponds to about 11% of global field cropland (Kassam et al., 2015). Recent estimates revealed that conservation agriculture technologies that include laser-assisted precision land leveling, zero/reduced tillage and direct drilling in the residues are being practiced over nearly 5 million ha of South Asia (Hafeez-ur-Rehman et al., 2015). Crop rotations especially with legumes and use of organic fertilizers have beneficial effect on soil microbial populations. Carbohydrates and proteins are profuse in crop-residues and act as excellent substrates for the growth of soil microbes. Easy and fast decomposable carbon is more effective for the growth of soil microbes.

Zero tillage/ Reduced tillage

Usually, tilling of the soil is done to prepare a good seedbed for easy sowing of crops and to manage weeds in the crop field. However, excessive tillage leads to the destruction of soil structure and stimulates the germination of weed seeds by bringing them to the soil surface (Das, 2008). Zero-tillage (ZT) can allow farmers to sow wheat sooner after rice harvest, which will prevent terminal heat stress in wheat. Field results have shown that conservation agriculture are increasingly being adopted by farmers in the rice-wheat belt of the Indo-Gangetic Plains because of several advantages of labour saving, water saving, early planting of wheat, mitigation of GHGs emission and adaptation to climate change (Pathak et al., 2013). In this technology, a crop, say wheat, is sown directly in undisturbed soil after the harvest of the previous crop, say rice, using a specially designed machine zero till-ferti-seed drill. This technology reduces the number of tractor operations required for field preparation and crop sowing and thus saves fuel, energy, cost and time. The sowing of zero tilled wheat after rice can be advanced by 7-10 days as compared to traditional methods and also resulted in a 2% higher wheat grain yield and 17% more net returns as compared to conventional tillage (Tripathi et al., 2013). Under continuous zero tillage, the annual weeds like *Phalaris minor* get suppressed as the deeply buried weed seeds do not get a chance to come up to soil surface. Zero till sown wheat witnesses less termite attack (Sharma et al., 2004) and Karnal bunt incidences (Sharma et al., 2007) as compared to conventionally tilled wheat.

Crop diversification

Crop diversification with low water requiring crops like maize, pigeonpea, groundnut, blackgram, cowpea in rainfed rice upland by substituting rice not only improves productivity and farmers income but also helps to develop climate resilient agriculture. Kar et al. (2003) developed drought mitigation strategies for enhancing productivity of rainfed upland though crop diversification, soil amelioration and soil-water conservation in Odisha. The technology of crop diversification with in-situ soil moisture conservation was developed through: (i) complete substitution of rice through sole cropping (maize, pigeonpea, groundnut, blackgram, cowpea), (ii) complete substitution of rice through double cropping (maize-horsegram/sesamum), (iii) partial substitution of rice through rice based intercropping (rice+pigeonpea), (iv) complete substitution of rice through legume based intercropping (groundnut + pigeonpea, maize + cowpea), (v) vegetable crops: cucumber, sweet potato, okra. Direct gain of Rs. 50,000-80,000/- per hectare per annum (with B:C ratio of 3.0) from the technology was obtained over existing upland rice (<Rs. 10,000 per ha). Alternative crops like maize, groundnut, blakgram, cowpea, pigeonpea not only improved agricultural productivity and income by 30-40%, but also facilitated to develop climate resilient agriculture (Kar et al., 2004a)

Pressurized irrigation system

The water use efficiency under conventional flood method of irrigation, which is predominantly practiced in Indian agriculture, is very low due to substantial conveyance and distribution losses. Recognizing the limited irrigation water potential available and increasing demand for water from different sectors, a number of demand management strategies and programmes have been introduced to save water and increase the existing water use efficiency in Indian agriculture. One such method introduced relatively last few decades in Indian agriculture are pressurized irrigation systems, which includes both drip and sprinkler method of irrigation. Since, these methods need water under pressure, they are classified as pressurized irrigation systems. Pressurized irrigation system is a well-established efficient method for saving water and increasing water use efficiency as compared to the conventional surface method of irrigation (Table 1).

Table 1. Irrigation efficiencies in pressurized irrigation system

Factors	Sprinkler irrigation system	Drip irrigation system	Surface irrigation system
Overall irrigation efficiency	50-60%	80-90%	30-35%
Application efficiency	70-80%	90%	60-70%
Water saving	30%	60-70%	NA

Micro irrigation is another vital option for high irrigation efficiency (> 90%). However, majority of the area covered under micro irrigation is in horticultural crops (55.40%) and only 7% area is under field crops (Alam and Kumar, 2001). Yield improvement due to drip-irrigation in sugarcane has been reported by 35-50%, in cotton by 5-10%, in castor by 15-42%, in groundnut by 20-66% and in potato 20-26.4% (Bhaskar et al., 2005). Sidhu et al., (2019) showed saving of irrigation water of 48–53% and 42–53% in rice and wheat respectively, under combination of sub-surface drip fertigation (SSDF) and CA compared to flood irrigation system and thus reduced the water footprint significantly. Dug well based drip-fertigation in hard rock areas facilitated to enhance water and nutrient use efficiency and to develop climate resilient agriculture (Kar et al., 2019b). Dug well based drip fertigation system for hard rock areas was developed and drip-fertigation schedule was standardized for vegetable crops like chilli, potato, bitter gourd, and ladies' finger (Fig. 2). The drip-fertigation technology has potential to enhance yield by 30-35%, to increase water use efficiency by 35-40% and nutrient use efficiency by 25-35%.

Fig. 2. Dug well based drip-fertigation system installed at Dhenkanal, Odisha

Residual soil moisture management using different tilling/seeding methods

Technology of utilizing carry-over residual soil moisture with different tillage/seeding methods *viz.* relay cropping/farmers' practice, reduced tillage (only two ploughing), conventional tillage and zero tillage could develop climate resilient agriculture. Study revealed that utilizing residual soil moisture, highest grain yields of 580, 630, 605 and 525 kg ha^{-1} were obtained from lathyrus (*Lathyrus sativus L.*), blackgram (*Vigna radiata roxb.*), pea (*Pisum sativum L. wilczek*) and green gram (*Vigna mungo L.*), respectively with reduced tillage treatment, higher than conventional tillage, zero tillage and relay cropping (Kar and Kumar, 2006). Water can be saved through mulching and supplemental irrigation which was adopted in potato, blackgram, gram, pea, lathyrus and safflower for climate change adaptation. Efficiency and crop productivity between mulched and non-mulched plots were derived on blackgram, gram, pea, lathyrus, safflower and potato. Study revealed that with straw mulching, 52-62%, 54-63%, 62%, 53-56%, 58-59% and 34-39% water use efficiency was enhanced in blackgram, gram, pea, lathyrus, safflower and potato, respectively. About 75-100 mm soil moisture was saved in mulched plots than that of non-mulched plots (Kar and Kumar, 2007).

Conjunctive use of water for enhancing water productivity

Conjunctive use of multiple water resources should be practiced to mitigate the effect of the shortage in canal water supplies; increase the dependability of existing water supplies; alleviate the problems of high water table and salinity resulting from introduction of canal irrigation; to facilitate the use of high salinity ground waters which, otherwise, cannot be used without appropriate dilution. Strengthening of knowledge based on geology and aquifer characteristics, hydrology of surface and groundwater facilities is required to develop appropriate conjunctive use system.

Ground water recharge to develop climate smart agriculture

Ground water played a major role in the success of green revolution and contributes to 60% of the total irrigated area of the country. Over exploitation of groundwater has reached critical level in Haryana, Punjab, Rajasthan and Tamil Nadu. These over-exploited regions could lose its production potential in a few decades if current pattern of groundwater extraction and pollution, soil salinization and rice wheat cropping system persists. Necessary steps must be taken immediately for groundwater recharge in these areas. Rainwater harvesting has considerable potential for recharging depleted groundwater aquifers meeting irrigation demand and to develop climate smart agriculture.

Improving water use efficiency of canal command for climate smart agriculture

Considerable amount of water is lost in canal command through evaporation seepage and percolation. In no irrigation project in India, the total losses in the canal distribution system and in the field has been less than 50% of the head discharge. A review of 90 irrigation projects of the world indicated generally low irrigation efficiencies, with only 20-40% of water diverted from the reservoir being effectively used by the crop, while in India, the irrigation efficiency is around 10-20%. The loss of irrigation water in unlined canal distribution system in north India is given in Table 2. To increase the overall irrigation efficiency of 40-50% for a project, improved irrigation and drainage practices should be adopted based on scientific land water management principles. The method to reduce conveyance losses are by lining the bed and sides of the canal, eradication of weeds, reduction of wastage from escapes and tail ends. The use of plastic films as lining material also has tremendous scope in India.

In any irrigation project, the total loss in the canal distribution system and field has not been less than 50% of head discharge. Studies conducted in Nagarjuna Sagar Project have revealed that only 40-60% of water released from the reservoir reaches the field and only 20-40% is used by the crops. The main component of water losses in an irrigation system are (a) water losses in storage- 10-20% (b) water losses in conveyance systems 25-40%, (c) water losses in operation 10-30% and (d) water losses in application 45-70%.

Table 2. Losses of irrigation water in unlined canal distribution systems and in the field

Source of loss	% of supplies at canal head		
	Seepage	Evaporation	Total
Main canals and branches	13.6	3.4	17.0
Distributaries (10% of supply at distributor)	6.4	1.6	8.0
Field water courses (27% of supply at outlet head)	16.0	4.0	20.0
Losses from field during application (30% of supply reaching the head)	13.2	3.3	16.5
Total	49.2	12.3	61.5

Climate smart water management in rice

With rising temperature and changes in precipitation patterns due to climate change, it is likely to have serious implications on water resources availability and thereby affecting agriculture. Rice cultivation requires huge amount of water, so climate smart water management practices need to be adopted. Direct-seeded rice (DSR), aerobic rice cultivation, alternate wetting and drying, deficit irrigation, partial root drying, intermittent irrigation and micro-irrigation management are well proven technologies for enhancing water use efficiency and to cope with challenges of climate change and variability. Similarly, aerobic rice system and DSR save 20 to 50% irrigation water in rice cultivation.

Precision land levelling

Declining irrigation water availability, sustainable crop production in increasing food demand necessitates quick adoption of modern scientific technologies for efficient water management. Land levelling at farmers' field is an important process which enables efficient water utilization through elimination of unnecessary land depression and elevated contours. Laser levelling is laser-guided precision levelling technique used for achieving very fine levelling of land. Precision land levelling must be treated as a precursor technology for improving crop yields, enhancing input-use efficiency and ensuring long term sustainability of the resources base in intensively cultivated areas.

Agroforestry

Agroforestry systems includes horizontal crop diversification, crop rotation systems, border planting, hedgerow intercropping, live fences and a mixed strata agroforestry with different land management practices. It has immense opportunities in climate changing scenario by producing assets for farmers along with promotion of sustainable production which in turn improves agro ecosystem diversity and climate resilience. Agroforestry systems buffer farmers against climate variability, and reduce atmospheric loads of greenhouse gases. Apart from the social and economic benefits, the agroforestry systems also helps in sustainable environment (Leakey, 2012) by recycling carbon and oxygen, fuel, construction material, and agro-industrial material. Further, a recent study showed that the replacing conventional feed ingredients with unconventional resources might reduce the carbon footprint of the diet and total global warming potential contributed by the ruminant species.

Kar et al., (2019c) developed agroforestry based water harvesting system for climate change adaptation in waterlogged areas (Fig. 3). The rainwater harvesting system was constructed by taking 25-30% of the total field of a farmer, rest of the field was utilized for intensive cultivation with harvested water. The pond was dug in inverted trapezoidal shape with the side slope of 1:1. The height of the bund was determined by the flooding depth. Study revealed that maximum flooding depth of 2.5 m occurred in saucer shaped flood prone coastal areas. Therefore, keeping a free board of 0.5m, maximum 3 m bund height was recommended for the water harvesting pond in flood prone coastal areas. The depth of pond ranged from 2-3 m depending upon the water requirement. The dyke of the pond was utilized to grow forestry species like *Acacia auriculiformis*, Casuarina etc. which has great demand in the local market. The harvested water inside the structure was utilized for fish rearing and also to grow second crops (rice, vegetables) surrounding the structure by providing supplemental irrigation from harvested water.

Fig. 3. Agroforestry based water harvesting system for climate change adaptation in waterlogged areas.

Integrated farming system

Per capita availability of land, water and other natural resources is already low and declining further. Agricultural diversification through integrated arming which has several components like on-dyke horticulture, poultry, duckery, fisheries, apiculture etc. along with diversified crop production or value-added products has proved to be an effective means in mitigating drought and eliminating poverty as it provides high income, regular employment, and balanced and quality food, even with less water. As the agricultural diversification comprises of several farm enterprises in addition to diversified crops, it provides opportunity to farmers to have a 'basket of complementary options, for reducing the weather risks involved in single commodity based agriculture and thus helps to develop climate resilient agriculture system.

Keeping in view the erratic distribution of rainfall, declined per capita availability of water resources, high cost of cultivation and labour and dryness of rivers, ponds/canals, integrated farming system is the best option to develop climate resilient agriculture and for doubling of farmers income. Changing circumstances necessitate the need to take a different approach to management practices. The integrated farming system does not necessarily mean integration by location; it essentially implies integration of resources such as manure, wastewater, crop residue, and animal waste with livestock feed. Benefits of integration of agricultural resources are synergistic (Gupta and Rai, 2012) and involve lesser external input with less negative environmental impact compared to the intensive system (Tipraqsa et al. 2007).

An *in-situ* jute retting based self-reliant integrated farming system has been developed in ICAR-CRIJAF farm (Fig. 4). In addition to retting of jute and mesta, the harvested water can be utilized in multiple ways 1. Dyke based horticulture (papaya, banana, seasonal vegetables (Profit of about Rs. 10,000- 12,000/- per tank); 2. Rearing of air breathing fish like telapia, magur and singhi, 50-60 kg; 3. Apiaries (Profit from honey around Rs. 7000/- per tank) and also honey bees will help in pollination; 4. Mushroom cultivation and vermicomposting; 5. Approximately 50 nos. ducks can be reared in the pond which result in additional income of Rs. 5000 /-; 6. After retting, the water can be utilized for providing supplement irrigation to crops in jute based cropping sequence resulting additional income of Rs 4000/acre. Thus, by loosing jute of Rs. 1000 to 1200 after construction of *in-situ* retting tank in the field, farmers can earn about Rs. 30,000/- in that area from multiple farming with saving of transport cost of Rs. 4000-5000/- This technology will also be helpful in reducing negative impacts from extreme weather events like drought, cyclone, flood etc.

Fig. 4. View of jute retting based self-reliant integrated farming system at ICAR-CRIJAF farm

Contingency crop planning and effective watershed management

As a case study, contingency crop planning and effective watershed management were adopted to develop climate smart agricultural practices in two watersheds of Odisha (Kar et al., 2004b). Contingency crop planning and effective watershed management (land levelling, summer ploughing, check dam, farm pond, pipe conveyance system, diversion channel, tapping of spring water, loose boulder structure, conjunctive use of groundwater and rainwater. Multiple use of water etc. facilitated to develop a drought resilient agriculture. Recommended technology provides an opportunity for increasing and stabilizing farm income by mitigating drought and economic gain obtained from the technology: Rs. 50,000-80,000/- per hectare per annum (with B:C ratio of 2.5 to 3.0) over existing upland rice even in rainfall deficit year (<Rs. 10,000/-) (Kar and Kumar, 2006).

Improved pest management

Changes in temperature and variability in rainfall would affect incidence of pests and disease and virulence of major crops since climate change will potentially affect the pest/weed-host relationship by affecting the pest/weed population, the host population and the pest/weed-host interactions. Thus, cultivars resistance to pests and diseases should be developed; integrated pest management should be adopted with more emphasis on biological control and changes in cultural practices, pest forecasting should be done using recent tools such as simulation modelling, and alternative production techniques should be developed that are resistant to infestations and other risks. Management of pests and diseases with use of resistant varieties and breeds; alternative natural pesticides; bacterial and viral pesticides; pheromones for disrupting pest reproduction, etc. could be adopted for sustainability of agricultural production process. Bioagents have a crucial role in pest management, hence practices to promote natural enemies like release of predators and parasites; improving the habitat for natural enemies; crop rotation and multiple cropping should be integrated in pest management practices. Reduction in use of pesticides will also help in reducing carbon emissions.

Better weather forecasting and crop insurance schemes

Weather forecasting and forewarning systems play pivotal role in minimizing the risks associated with climate change. In this context, information and communication technologies (ICT) can be of immense help in the dissemination of weather forecast and weather based agro-advisories. Crop insurance schemes help farmers in tackling the risk of crop failure due to climatic variability and extreme events. Weather Based Crop Insurance Scheme (WBCIS) reduces the distress of the insured farmers against the monetary loss resulting from adverse climatic conditions whereas yield-based insurance compensate farmers against crop yield loss. The recently introduced Pradhan Mantri Fasal Bima Yojana is a comprehensive insurance scheme for minimising the risk due to climatic vagaries, loss due to pests, etc. Such schemes need to be popularised among the farming community so as to enhance the coverage of farmers. Thus, there is an urgent need for development of sustainable insurance system and farmers need to be trained for availing such opportunities.

Mitigation based management practices

India has the potential to cut its annual GHGs emissions from agriculture and livestock which is equal to 85.5 mega tonnes of CO_2 equivalent per year. Climate change mitigation generally involves reduction in anthropogenic emissions of greenhouse gases. Examples of climate change mitigation include switching to renewable and nuclear energy, which is low carbon energy sources, reforestation to increase the capacity of carbon sinks to remove greater amounts of CO_2 from the atmosphere.

Conservation agriculture (CA)

Carbon sequestration can contribute to mitigating climate change through biomass production and conservation practices. In effect, improving agricultural management practices through CA enriches soil C content (Liu et al., 2011). Conservation agriculture has a role to sequester C and to increase soil organic carbon (SOC) via sequestration. Levels of SOC rise when there is more organic matter inputs coupled with lower rates of organic matter decomposition. This occurs whenever there is a diversified crop rotation, increased crop productivity, enhanced return of crop residues, use of organic and green manure, use of cover crops, agroforestry, improved irrigation, mulching, reduced soil tillage, reduced bare fallow periods and erosion control. Under CA, it could be achieved with minimal cost by efficient use of fertiliser, adoption of zero-tillage, direct seeding of rice, improved cultivations such as direct seeding of the corn, the wheat and maize furrow irrigated raised bed (FIRB) system; no-tillage; precise land levelling.

Precision water management and alternative planting methods in rice

This approach includes use of drip and sprinkler systems; precision agriculture sensor-based reliable nutrient management (Srinivasa Rao et al., 2016). By shifting water management from the current practice to intermittent flooding in all the irrigated rice-growing areas of the country, the CH_4 flux from irrigated rice fields could be reduced by 40% (Bhatia et al., 2012). It was also stated that by modifying water management from continuous to alternate flooding, or applying urea alone rather than urea coupled with farmyard manure would reduce global warming potential by 15 and 29%, respectively. Similarly, DSR practice decreases CH_4 emissions drastically thereby reducing the global warming potential by 35-75% over the conventional puddled transplanted rice. Such options capture the atmospheric C and store in the soil, and are important not only for global warming mitigation but also to contribute towards improving the fertility status of soils (Srinivasa Rao et al., 2017). Moreover, the GHGs emissions can be abridged by the substitution of fossil fuels by the agricultural feedstocks (e.g. crop residues, dung, and dedicated energy crops) for energy produce.

Alternate wetting and drying (AWD) is a water management technique in rice that was developed in the 1970s to address the problem of increasing water scarcity in agriculture (Bouman and Tuong, 2001). On-farm experiments on AWD shows 20–70% reduction in methane emissions along with 25-40% saving of water as compared to conventional flooded rice cultivation (Richards and Sander, 2014). Therefore, AWD serves the twin purpose of mitigating GHGs emissions by reducing methane and adapting to water stress.

Climate smart nutrient management

Recent IPCC and FAO reports have projected a 12-14% share of the total world's agricultural greenhouse gasses due to use of fertilizers (680-725 Tg CO_2-equivalent, in the year 2010-11). By enhancing the FUE by 5-10%, it could increase returns, reduce emissions of N_2O and significantly reduce production costs, thus giving farmers greater profits (Majumdar et al., 2000). Leaf N concentration is linked with leaf chlorophyll content and therefore, leaf greenness can be used as the basis of in-season nitrogen management. Chlorophyll meter (SPAD), leaf colour chart (LCC), or optical sensors (e.g., Green Seeker) are currently being use in farmer's fields. Similarly, Nutrient Expert and Crop Manager, decision support systems are progressively being used to enable the application of improved nutrient-management practices in farmers' fields (Pampolino et al., 2012). According to Bhatia et al. (2012) application of 120 kg N ha^{-1} as per leaf colour chart 4 (LCC4) in rice reduced emission of nitrous oxide and methane by 11% and 16 % respectively, which in turn reduced the GWP by 1297 kg CO_2 ha^{-1} than the conventional urea application.

Use of jute as potential carbon sequester

Jute has been considered as a potential carbon sequester since the carbon footprint of the jute fibre has been estimated as 566 kg CO_2 t^{-1} of jute (Singh et al., 2018a). Study also reveals that the most significant impact is carbon sequestration by green jute plants during the growth stages since, 0.97–2.8 t ha^{-1} of the left over above- and below- ground biomass of jute (leaves, stubbles and roots) is added annually to the soils under jute cultivation (Singh et al., 2018b). Approximately 4.88–5.30 t of CO_2 gets sequestered per hectare of raw jute fibre production which is much higher than many tree species (Rajagopal and Sanyal, 2012; Singh et al., 2018b). As reported by the International Jute Study Group (IJSG, 2013), 1 ha of jute plants consumes about 15 MT CO_2 and liberates 11 MT of O_2 in only 120 days. It can be estimated that on 0.75 mha area, India may reduce about approximately 12 MT of CO_2 from atmosphere every year which can be valued at approximately US$15 crores (CRIJAF, 2015).

Use of Biochar

The biochars can be produced by pyrolysis from the jute wooden stick material at low temperatures (~350–600°C) in an environment with little or no oxygen. Thus, biochar could be considered as the geo-engineering solution to control climate change by means of carbon sequestration and minimizing the emissions of the GHGs while supplying energy and improving the productivity of the cultivated crops. Use of biochar in agriculture can help in sequestering 62–66% of CO_2 emissions. Farmers of jute growing areas can be trained to prepare biochars by pyrolysis method in their jute farm by using locally available infrastructure at a low cost.

Use of digital science for mitigating climate change

Advance information and communication technologies (ICTs) offer great potential to address the climate change concerns by issuing forewarning, advisory, and decisions regarding change mitigation and adaptation. The ICTs can be tools for monitoring climate change through greater awareness and education. This can have a positive effect on fostering an environmental responsibility mind-set amongst stakeholders, including the ICT sector, policy makers, citizens and academia. This is extremely important as everyone needs to come together to take responsibility for this very complex problem, and to take concerted and timely action. Technological advancements, and the proliferation of frontier technologies (such as artificial intelligence, the Internet of Things, 5G, digital twins, robotics, etc.), in particular are contributing to the growing number of data centres and solutions.

ICTs can also facilitate the adoption of other innovative technologies such as digital twins to combat climate change. A digital twin is the virtual representation of a physical object or system across its life cycle. It uses real-time data and other sources to enable learning, reasoning and dynamic recalibration for improved decision making. Within the context of climate change and response, digital twins are an attractive proposition, particularly for urban areas that are rapidly growing in population, size and energy consumption. They are also attractive for those urban areas that, as a consequence, need to ensure the efficient, agile and responsive management and maintenance of all their systems. Another emerging example is 5G technology, which is the latest generation of cellular and wireless connectivity that is a key focus of International Telecommunication Union (ITU) current work agenda. The 5G technology is expected to offer faster speeds and greater coverage, in addition to longer battery life for devices, larger data transfer capabilities and more reliability for quick climate data transfer and communication.

There is potential for AI to be a tool in the effort to decouple economic growth from rising C emissions. In the not-too-distant future, many consumer purchases will be automated with algorithms that carry potential to influence consumer behaviour related to GHG emissions. However, there are still certain limitations as regards the C footprint of AI and machine learning (ML). These are explored later in this chapter. Renewable energy technologies remain very relevant in the discussion around climate change. ICTs can facilitate greater adoption – through the integration of automated sensors, data capture, performance measurement or other mechanisms – of renewable energy technologies that enable the creation of electricity, heat and fuel from renewable sources, such as solar, wind, hydro, wave and tidal power, heat-exchange/geothermal and bioenergy. According to the IPCC moving towards the use of renewable energy is vital, as it can contribute to social and economic development, energy access, secure energy supply, climate change mitigation, and the reduction of negative environmental and health impacts. This may, in fact, be realizable given that clean energy sources such as solar and wind power are quickly becoming more affordable than the conventional sources.

Some examples of ICT-use to address climate change include: remote sensing for monitoring of natural disasters; improved communications to help deal with natural disasters more effectively; and satellite and surface-based remote sensors for long-term environmental observations. The GIS-based decision rules and multi-criteria analysis can be adopted for assessing potential and prospects of existing soil hydro-physical properties, land use, slope etc. to develop action plan on alternative climate resilient land use/cropping system.

Public and private sector engagement

The concept of 'private sector engagement' is graining increased traction in the climate change mitigation and adaptation space. The rationale for bringing the private sector on board is two-fold. One is the businesses with track record of innovation, speed of delivery and financial resources, the private sector can support governments facing constrained public budgets and the rising costs of managing climate change to achieve global climate resilience. One study estimates that the costs of adaptation to climate change in developing countries lie between $70 billion and $100 billion per year for the period 2010–2050 (UNEP, 2016). The chronic and acute effects of a changing climate are already affecting people in South Asia. It is estimated that, by 2050, climate change could lead to a 50% drop in the region's gross domestic product (GDP) (Ahmed and Suphachalasai, 2014). Over 2015-2016, global public and private climate finance flows reached an average of $463 billion, according to the Global Landscape of Climate Finance by the Climate Policy Initiative (CPI), out of which 4.8% was allocated to adaptation (Fayolle et al., 2019). The other rationale of bringing private sector is that the private sector itself is gradually understanding the physical risks and opportunities arising from a changing climate, and there is nascent awareness of the measures it can implement to maintain the profitability of its activities. Although there is currently no definitive characterisation of what a resilient business is, climate-related disclosures on physical impacts and opportunities provide a lens for private sector actors to understand their adaptation needs. Hence, engaging the private sector in considering such risks and opportunities and in investing in adaptation-related measures also constitutes a key element in avoiding climate induced economic and financial shocks. Policy-makers and the international development community are therefore increasingly interested in how to leverage private sector finance for adaptation. Some public sector organizations like 'SKYMET' is providing short and medium weather forecasts which can be used for making agro-advisory and contingency crop planning by considering short term weather variability.

Role of FPO and NGO in developing climate smart agriculture

Collectivization of producers, especially small and marginal farmers, into Farmers' Producer organisations (FPO) has emerged as one of the most effective pathways to address the many challenges of agriculture but most importantly, improved access to investments on climate smart technology and inputs and markets. With the continuous challenges like low yield, lack of credit, market asymmetries, etc. and the absence of an effective organisational structure, farmers, especially the small and marginal farmers are more vulnerable to climate change. These constraints bring out the need for local collectivisation of farmers to tackle the problems of climate change. Many other countries which have introduced such structural changes are successful in agriculture production under challenging situations too.

Collectivisation helps in introducing economies of scale, bringing down the input costs, better bargaining power and magnifies the voices of farmers as collective voice which facilitates to take necessary measures collectively for climate change mitigation and adaptation. The most common form of collectivisation that Indian farmers have adopted over the years are cooperative societies, federations of self-help groups, joint liability groups, farmer clubs, common interest groups. These structures are always been inward oriented, i.e. focusing on the requirements of farmers and acting as a facilitating organisation to them while completely ignoring the outward orientation, i.e. the market needs, suppliers and end consumer demands and changing accordingly. The major focus of the producer companies is promoting and strengthening the governance system for production and market stabilisation, creating access to credit for production, introducing climate resilient technology for production and collective marketing of produces under climate change scenario.

NGOs can play a catalytic role in building up community awareness of climate change and its likely impact on their lives, livelihoods and habitats. They can help build up their capacities to undertake the needed adaptive actions to reduce vulnerability, mitigate risks and build resilience. In rural areas, for instance, they can help communities map and identify their local biodiversity resources with a view to their conservation and sustainable management in order to meet food, medicinal and livelihood needs, especially during times of disasters and crises (droughts, crop failures, epidemics). They can also assist in the establishment of ownership rights over resources that have economic value and familiarise community members with the legal, regulatory and commercial regimes and systems that affect and determine the extent of their access and claims to these resources. Similarly, developing early warning systems and contingency plans, conducting drills, organising communities to mobilise resources during disasters and linking them up with local resource and service providers goes a long way towards mitigation risks and building a climate smart agriculture.

Community participation

Countries' adaptation plans are largely state driven and top-down in approach while climate change is locally experienced and can only be effectively addressed by engaging local groups and institutions. All adaptation is local, and that local agencies, especially civil society organisations (CSOs), being closest to the problem, are best suited to creating adaptive capacities within communities.

Enlisting the active cooperation of local groups, communities, local institutions and stakeholders, building up their capacities and empowering them as active participants in decision making processes are the most important pre-conditions for efficient and effective adaptation measures. This is because the effects of climate change are experienced locally by communities, local institutions and stakeholders and they are best suited to addressing them. Social capital, in the form of local institutions, be they public, civic, private, formal or informal, play an important role in shaping the adaptive capacity of local communities. A particular sub-set of these, namely, CSO, by their very nature and function can make a powerful and unique contribution to this task as they can draw upon a vast base of purpose and cause-driven local constituencies and stakeholders. Examples of CSOs include local self-government bodies such as elected municipal councils, urban local bodies, village councils, intermediate governing bodies, various kinds of social and economic associations and NGOs. All adaptation is local and these bodies being closest to the sources of emissions and the impacts of climate change are best placed to do something about it.

The CSOs and communities have largely played limited roles in the formulation of these policies and strategies. While the active participation of enlightened and empowered citizens, communities and other stakeholders is recognised as necessary for successful implementation of the plans, hardly any thought has been given to establishing a framework to engage meaningfully with these stakeholders and incentivize their active involvement on an on-going basis. Technology interventions designed to address specific aspects of climate change can only be effective and sustainable, if they are suited to the local context, meet the perceived needs of the community, and are within its management capacity. Unless a community has been involved in the choice, design and deployment of the technology, they will not feel a sense of ownership and the very purpose of the climate change intervention will be defeated.

Skill development for adopting climate smart agriculture

Most vocational training programmes focus on mechanical/technical skills rather than 'soft' or 'green' skills. Green skills contribute to preserving or restoring environmental quality for sustainable future and include jobs that protect ecosystems and biodiversity, reduce energy and minimize waste and pollution. In line with the 'Skill India' (National Skill Development Mission), Ministry of Environment, Forest & Climate Change (MoEF&CC), Govt. of India has taken up an initiative for skill development in the environment and forest sector to enable India's youth to get gainful employment and/or self-employment. This programme is called the 'Green Skill Development Programme (GSDP)' which endeavours to develop green skilled workers having technical knowledge and commitment to sustainable development, which will help in the attainment of the Nationally Determined Contributions (NDCs), Sustainable Development Goals (SDGs), National Biodiversity Targets (NBTs), as well as Waste Management Rules (2016), which will facilitate to develop a clean and green India by bringing climate smart technologies.

Skilling, reskilling and upskilling covers not just technical skills but, core/soft skills (such as environmental awareness, analytical skills, teamwork, innovation, communications, leadership, negotiation abilities, and management and entrepreneurship skills), which can offer great help to the community to implement climate smart practices successfully.

Women empowerment

Empowering women and girls in developing countries ranked second among 76 solutions for curbing global warming to 2 °C, according to a new report by the climate research organization. Girls' education and family planning would facilitate to reduce greenhouse emission. Gender equality in health and education hasn't received much public attention as a climate solution, even though it's been on the global sustainability radar for years. In 2015, the United Nations included "achieve gender equality and empower all women and girls" in its 17 Sustainable Development Goals (United Nations, 2015). In both rural and urban areas, slower population growth relieves stress on ecosystems, allowing resources to recover from overuse without compromising local food access. With these additional resources, women also earn money, achieve career goals and face fewer health issues. Rural women see higher crop yields, providing better nutrition and financial stability for their families. Education of women and girls help them to combat climate disasters. Because of traditional gender roles, legal inequality and financial barriers, women face unique risks as climate change intensifies natural disasters and raises food and water prices. In agriculture, nutrition-smart and climate-smart agriculture emphasize the knowledge and legal rights of female farm owners to empower them, improve family health and protect the future.

Entrepreneurship

It is well documented that entrepreneurs play an important role in contributing to a country's growth and prosperity. By engaging in innovation activities, entrepreneurs help to bring new and improved technologies into broad usage, facilitating economic growth and disrupting the stagnation and hegemony of existing technologies. Because dealing with climate change will require far-reaching economic and social transformations, entrepreneurs have the potential to play a significant role in developing climate technologies that can help communities to reduce GHGs emissions and adapt to climate change. With the right encouragement, guidance and support, entrepreneurs can channel their efforts effectively into developing climate-friendly solutions. There are many ways in which entrepreneurs can make significant contributions to the global response to climate change. For instance, an entrepreneur may develop a new climate technology or adapt an existing one to the needs of a community. Another entrepreneur might develop a new business model that revolutionizes the use of an existing climate technology. The deployment of pay-as-you-go solar panels in Indian villages could be an example of such model. Alternatively, an entrepreneur may develop a service that enables the use of an existing technology in a climate friendly-way.

Likewise, there could be many different types of entrepreneurs and entrepreneurial activities that can contribute to achieving the purpose and goals of the Paris Agreement and the SDGs. However, entrepreneurs are confronted with many challenges to become successful. Insufficient encouragement to undertake their role, limited incentives to work on climate action, and a lack of support (including access to finance) are three major challenges that exacerbate the risks faced by entrepreneurs. By encouraging, guiding and supporting entrepreneurship governments can provide local talent – young and old, women and men, and the poor and marginalized – with opportunities to find employment, build livelihoods and contribute to the global response to climate change.

Policy initiatives and Government schemes for climate change mitigation and adaptation

Due to the efforts of National Agricultural Research System in the country, several climate resilient practices have been developed, particularly under the scheme "National Innovation for Climate Resilient Agriculture (NICRA). These resilient practices have been demonstrated in farmers' fields and their efficacy has been proven. There is a need to integrate these resilient practices into the developmental programmes and policies, so that these practices can reach a large number of farmers in the country and thus, can impart resilience. As the impacts of climate change are being observed in various parts of the country and the impacts of rainfall variability and extreme events are widespread and realized in several parts of every year, there is an urgent need to integrate these resilient practices into the policy paradigm of the State and Central Governments. Thus, there is an immediate need to streamline the resilient practices into the ongoing developmental programmes. Further there is a need to sensitize the policy establishment about the impending impacts of climate change and variability, the likely impacts of extreme events in the years to come and the technological options available to minimize their impact, and the need for their integration to on-going developmental programmes.

Indian government's first push towards a concrete climate policy came in 2008 with the establishment of Prime Minister's Council on Climate Change with the "purpose of formulating and coordinating a national action plan for assessment adaptation and mitigation of climate change".The Council came out with National Action Plan on Climate Change (Govt of India, 2008). It identified eight missions: *First*, National Solar Mission to promote the use of solar energy. *Second*, National Mission for Enhanced Energy Efficiency of domestic, commercial and industrial sectors *Third*, National Mission for Sustainable Habitat - to encourage "sustainable urban planning with the help of policy, infrastructural and research interventions in sectors such as buildings, waste management, water resources and transportation". It covers aspects like optimisation of energy demands by updating the designs for new and large commercial buildings; to improve urban planning; and recycling of material and urban waste management. It also includes research and development (R&D) programmes focusing on waste water use sewage utilisation, bio-chemical conversion, and recycling options. The initiative includes several flagship programmes of Government of India like Atal Mission for Rejuvenation and Urban Transformation (AMRUT), Smart City initiative Heritage City Development and Augmentation Yojana (HRIDAY), Swachh Bharat and National Urban Livelihoods Mission (NULM), etc. *Fourth*, National Water Mission – it aims to provide sustainable water supply throughout India by conserving water, minimising waste and ensuring equitable distribution of water resources. Within this, multiple initiatives like establishment of new Hydrological Observation Stations, groundwater monitoring wells, India-WRIS WebGIS etc.

have been taken. *Fifth,* National Mission for Strategic Knowledge on Climate Change – the idea is to create a knowledge system that supports climate change action in India. The key initiative taken is the Global Technology Watch Group (GTWG) which aims to keep track of state-of–the-art technologies emerging globally in eight sectors - coal, renewable energy, agriculture, water, sustainable habitat, manufacturing, energy efficiency and forestry. The mission aims to make the suitable technology commercially available by 2030. *Sixth,* National Mission for Sustainable Agriculture – the aim is to improve sustainability, productivity remuneration and climate resilience of agriculture in India. The four elements of the mission include - Rain-fed Area Development; On-Farm Water Management Soil Health Management; and Climate Change and Sustainable Agriculture Monitoring, Modelling and Networking. These are aligned with several other initiatives like MGNREGA, Integrated Watershed Management Programme (IWMP), Accelerated Irrigation Benefit Programme (AIBP), Rashtriya Krishi Vikas Yojana (RKVY), National Food Security Mission (NFSM), Pradhan Mantri Fasal Bima Yojna, etc. *Seventh,* National Mission for Green India to protect, enhance and restore forests. The aim is to increase forest cover on 2.5 mha and to further improve the quality of forest cover on another 2.5 mha and *Eighth* National Mission for Sustaining the Himalayan Ecosystem – six task forces have been established looking at various aspects of Himalayan ecosystems, like natural and geological wealth, forest resources, glaciers etc. so as to understand impact of climate change in the Himalayas.

Apart from this, Govt. of India announced two major initiatives to combat climate change. *First,* Jal Jeevan Mission aimed at providing functional tap-water to every household in the country by 2024 and establishment of a separate Jal Shakti ministry. *Second,* to end use of single-use plastic. On the international forum, India and France launched the International Solar Alliance in 2015 in which almost 80 countries have joined the efforts. The idea is to provide a "dedicated platform for cooperation among solar resource rich countries where the global community including bilateral and multilateral organisations, industry, and other stakeholders can make a positive contribution to assist and help achieve the common goals of increasing the use of solar energy in meeting energy needs of prospective International Solar Alliance (ISA) member countries in a safe, convenient affordable, equitable and sustainable manner".

Under the "Clean Energy and Climate Change", European Union (EU) supports development of India's off-shore wind development plans – the Facilitating Offshore Wind in India (FOWIND) project is case in point which is being facilitated by a consortium led by Global Wind Energy Council. Another important area of cooperation is the assistance provided by EU in India's Solar Park Programmes. During 2016-19, the number of solar parks in India has increased from one park with capacity of 345 MW to 12 parks of 6500 MW with additional 10 parks under various stages of development. Moreover, in 2018, the European Investment Bank signed a package of €800 million towards clean energy investments in India, of which €640 million were earmarked for various solar projects, making India the leading recipient of financing in solar energy outside EU. The EU technical assistance is aimed for the management and implementation of the solar parks, thereby contributing in mitigation of climate change impact.

India and EU are very well positioned to take up the leadership role to drive forward the climate agenda. The adoption of Agenda for Action 2020 during the 2016 India-EU Summit paved the way for intensification of cooperation on climate change issues. Both India and EU are working towards achieving the Intended Nationally Determined Contributions (INDCs) they submitted before the Paris climate agreement. They are also partnering with each other on energy efficiency where cooperation is focused on smart grids, off-shore wind projects, solar parks, energy research and innovation. The EU is an active partner in India's flagship programmes of 'Clean India and Clean Ganga'.

In order to avoid the worst impact of climate change, it is required for rapid cuts in global GHG emissions; switch from non-renewable resources to renewable resources like solar, wind, etc. and development of new technologies to push forward energy efficiency initiatives. India and EU have not only found consensus in their agendas but also share similar concerns on climate change. The cooperation on climate change presents India and EU as equal partners with joint responsibility. Therefore, further opportunities need to be explored in research and innovation, technology-transfer, energy efficiency, clean energy transition to mitigate the climate change impact.

Conclusion

Impact of climate change on agricultural productivity is multidimensional as several factors regulate the crop yield and these factors in turn are under the influence of climate change. Such climatic fluctuations could adversely affect agricultural sustainability resulting in unforeseen situational shortages. Under such situation, innovation on climate smart agriculture and structural reforms are very much needed to develop the self-reliant India ('*Atma Nirvar Bharat*'). If the temperature and rainfall continues to fluctuates like this, large scale agricultural research for developing climate tolerant varieties of crop and advancement in agricultural practices to adapt to the changing climate may probably prevent temperature-induced yield losses. A wider refinement in the existing technologies and concerted efforts for research and their adoption by farmers are needed for adaptation and mitigation. The most important and challenging task in tackling the threats to climate change is the need for effective and adequate mitigation measures worldwide by which reduction in the emission of greenhouse gases can be achieved to overcome the various threats associated with it. The mitigation of GHGs particularly CO2 emissions from agriculture sector could be attained by increasing carbon sequestration in the soil in the form of soil organic matter. CA technologies like zero or minimum tillage, surface organic residue retention, legume-based crop rotations have large potential to augment the use efficiency of natural resources like soil, water, nutrient, carbon, energy, and labour. Besides, other improved cultivation practices like precise land levelling; precision water management approaches including drip and sprinkler systems; precision nutrient management through LCC, SPAD can rationalise fertiliser application and minimise excessive use of resources. These approaches can advance the sustainability of agroecosystem by preserving the natural resource base with higher input use efficiency and also mitigating GHGs emissions. Stringent government plans and actions regarding the use of eco-friendly and organic techniques with the efficient use of the natural resources like water, land, will alleviate the threat to a certain extent.

References

Ahmed, M. and Suphachalasai, S. (2014) Assessing the Costs of Climate Change and Adaptation in South Asia. Mandaluyong City: ADB

Alam, A. and Kumar (2001) Micro irrigation system past, present and future. In: Proceeding of International Conference on Micro and Sprinkler Irrigation System (Eds Singh, et al. 2001), 8-10 February, 2000, Jalgaon, India, PP. 1-17.

Bhaskar, K.S., Rao, M.R.K., Mendhe, P.N. and Suryavanshi, M.R. (2005) Micro Irrigation Management. In: Cotton. CICR Technical Bulletin No 3:16.

Bhatia, A., Agarwal, P.K., Jain, N. and Pathak, H. (2012) Greenhousegas emission from rice and wheat growing areas in India: Spatial analysis and upscaling. Greenhouse Gas Science and Technology. 2:115-125

Bouman, B.A.M.,and Tuong, T.P., (2001) Field water management to save water andincrease its productivity in irrigated rice. Agricultural Water Management. 49(1), 11–30

Campbell-Lendrum D, Manga L, Bagayoko M, Sommerfeld J. (2015) Climate change and vector-borne diseases: what are the implications for public health research and policy? Philosophical Transactions. Royal Society. B 370: 20130552.

CRIJAF (2015) Vision 2050. ICAR-Central Research Institute for Jute and Allied Fibres Barrackpore, West Bengal.

Das, T.K. (2008) Weed Science: Basics and Applications. Jain Brothers.

Eze, S.O., Afolabi and Bola Abdul Hamid (2013) Effect of ambient temperature and rainfall on rice production in the Gambia: A case study of the Central River Region. Agriculture Science Developments, 2(9): 79-83.

FAO (2011) Strengthening Capacity for Climate Change Adaptation in the Agriculture Sector in Ethiopia. Proceedings from National Workshop held in Nazreth, Ethiopia.

Fayolle, V., Fouvet, C., Soundarajan, V., Nath, V., Acharya, S., Gupta, N., and Petrarulo, L. (2019) Engaging the private sector in financing adaptation to climate change: learning from practice. Learning paper. Oxford Policy Management.

Govt. of India (2008) National action plan on climate change. Technical document. pp.1-52.

Gupta, V and Rai, P. (2012) Integrated crop-livestock farming systems: A strategy for resource conservation and environmental sustainability. Environmental Science. Corpus ID: 55658681.

Hafeez-ur Rehman, Ahmad-Nawaz, A., Wakeel, Abdul, Saharawat, Y.S. and Farooq, M. (2015) Conservation Agriculture in South Asia. In: Conservation Agriculture. Farooq, M., Siddique, K.H.M. (Eds). Springer International Publishing, Switzerland, pp. 249–283.

Houghton, R. A.,House, J. I.,Pongratz, Julia,Werf, G. R. van der, DeFries, R. S.,Hansen, M. C.,Le Quéré, C.,Ramankutty, N. (2012) Carbon emissions from land use and land-cover change. Biogeosciences, 9(12): 5125-5142.

IJSG (2013) Carbon credit of jute and sustainable environment, International Jute Study Group (IJSG). Jute Matter 1(7): 1–2

IPCC (2007) Climate Change- Impacts, Adaptation and Vulnerability. in: Parry, M.L., Canziani, O.F., Palutikof, J.P., van der Linden, P.J., Hanson, C.E.(eds.) Cambridge University Press, Cambridge, UK, pp. 976.

IPCC (2014) Climate Change 2014: Impacts, Adaptation, and Vulnerability. Part B: Regional Aspects. Contribution of Working Group II to the Fifth Assessment Report of the Intergovernmental Panel on Climate Change [Barros, V.R., C.B. Field, D.J. Dokken, M.D. Mastrandrea, K.J. Mach, T.E. Bilir, M. Chatterjee, K.L. Ebi, Y.O. Estrada, R.C. Genova, B. Girma, E.S. Kissel, A.N. Levy, S. MacCracken, P.R. Mastrandrea, and L.L.White (eds.)]. Cambridge University Press, Cambridge, United Kingdom and New York, NY, USA, pp. 688.

Johnson, F.A. and Hutton, C. W. (2014) Dependence on agriculture and ecosystem services for livelihood in Northeast India and Bhutan: vulnerability to climate change in the Tropical River Basins of the Upper Brahmaputra. Climatic Change. 127: 107-121.

Kar, G. and Ashwani Kumar, (2006) Contingency Crop planning and effective Watershed Management to mitigate drought in eastern India. Published by Water Technology Center for Eastern Region, Bhubaneswar, p 8.

Kar, G. and Ashwani Kumar. (2007) Effects of irrigations and straw mulch on water use on tuber yield of potato in eastern India. Agricultural Water Management (Elsevier), 94: 109-116

Kar, G., Ananad, P.S.B., Raychoudhuri, M., Panda, D.K., A. Kumar, Ambast, S.K. (2019b) Dugwell based drip-fertigation system for hard rock areas – a success story, ICAR-IWM News Letter, 20 (1): 1-2. Published by Indian Institute of Water Management, Bhubaneswar.

Kar, G., Ananad, P.S.B., Raychoudhuri, M., Panda, D.K., Sahoo, H.N., Nayak, S.R., Roja Dora, K., Patra, P.K., Sahoo, N., Ashwani Kumar, Ambast, S.K. (2019c) Agro-forestry based water harvesting system in waterlogged areas for climate change adaptation- a success story. Published by Indian Institute of Water Management, Bhubaneswar, p 4.

Kar, G., Mohany, S., Anand, P.S.B., Panda, D.K., Raviraj, A., Rank, R.D., Singh, P.K. (2019a) National Innovations for Climate Resilient Agriculture, Annual Report, 2018-19: p 14-18. Published by Indian Institute of Water Management, Bhubaneswar.

Kar, G., Singh, R . and Verma, H. N. (2003) Crop diversification in rainfed rice area of eastern India to increase productivity and rainwater use efficiency. Research Bulletin 13/2003. Published by Water Technology Centre for Eastern Region, Bhubaneswar-23, Orissa, India, 51p.

Kar, G., Singh, R. and Verma, H. N. (2004a) Alternative cropping strategies for assured and efficient crop production in upland rainfed rice areas of eastern India based on rainfall analysis Agricultural Water Management (Elsevier), 67(1): 47-62.

Kar, G., Singh, R. and Verma, H. N. (2004b). Spatial variability of soil hydro-physical properties using GIS for sustainable crop planning of watershed of eastern India and it s testing in portion of rainfed rice area. Australian Journal of Soil Research, 42(4): 369-379.

Kassam, A., Friedrich, T., Derpsch, R. and Kienzle, J. (2015) Overview of the worldwide spread ofconservation agriculture. Field Actions Science Reports, 8. Available at: http://factsreports. revues.org/3966.

Kautsky, N., Berg, H., Folke, C., Larsson, J. &Troell, M. (1997) Ecological footprint for assessment of resource use and development limitations in shrimp and tilapia aquaculture. Aquaculture Research, 28: 753-766.

Kurukulasuriya, P and Mendelsohn, R. (2008) Crop switching as a strategy for adapting to climate change. African Journal of Agricultural and Resource Economics, 02(1): 105-126.

Leakey, R.R.B., Weber, J.C., Page, T., Cornelius, J.P., Akinnifesi, F.K., Roshetko, J.M., Tchoundjeu, Z., Jamnadass, R. (2012) Tree domestication in agroforestry: progress in the second decade (2003–2012). In: Nair PKR, Garrity D (Eds) Agroforestry – the future of global land use.Springer, Dordrecht, pp 145–173, 541 p

Liu, X., Duan, L., Mo, J., Du, E., Shen, J., Lu, X., Zhang, Y., Zhou, X., He, C., and Zhang, F. (2011) Nitrogen deposition and its ecological impact in China: an overview. Environmental Pollution 159:2251–2264

Majumdar, D., Rastogi, M., Kumar, S., Pathak, H., Jain, M. and Kumar, U. (2000) Nitrous oxide emissions from an alluvial soil with different nitrogenous fertilizers and nitrogen levels. Journal of the Indian Society of Soil Science 48(4):732-741.

Pampolino, M. F., Witt, C., Pasuquin, J. M., Johnston, A. and Fisher, M.J. (2012) Development approach and evaluation of the Nutrient Expert software for nutrient management in cereal crops.Computers and Electronics in Agriculture. 88:103-110.

Pathak, H., Sankhyan, S., Dubey, D. S., Bhatia, A., and Jain, N. (2013) Dry direct-seeding of rice for mitigating greenhouse gas emission: Field experimentation and simulation. Paddy and Water Environment, 11(1): 593–601. https://doi.org/10.1007/s10333-012-0352-0

Paul, H., Ernsting, A., Semino, S., Gura, S. and Lorch, A. (2009) Agriculture and Climate Change: Real Problems, False Solutions. Report published for the 15th Conference of the Parties of UNFCCC in Copenhagen, December.

Rajagopal, K. and Sanyal, T. (2012) Sustainable Infrastructure Development Including Limited Life Geosynthetics. Bangkok: Proceedings of Geosynthetics Asia.

Rhodes, C.J. (2016) The 2015 Paris Climate Change Conference: COP21. Science Progress. 99(1): 97-104.

Richards, M., and Sander B. O. (2014) Alternate wetting and drying in irrigated rice: implementation guidance for policy makers and investors. CGIAR Research Program on Climate Change, Agriculture and Food Security, Montpellier

Sharma, A. K., Babu, K. S., Sharma, R. K. and Kamlesh K. (2007) Effect of tillage practices on Tilletia indica disease of wheat in rice-wheat rotation of the Indo-Gangetic Plains. Crop Protection 26: 818-821.

Sharma, A. K., Srinivas Babu, K., Chhokar, R. S. and Sharma, R. K. (2004) Effects of tillage on termite, weed incidence and productivity of spring wheat in rice-wheat cropping system of North-Western Indian plains. Crop Protection 20(11):1049-1054.

Sidhu, H.S., Jat, M.L., Singh, Y., Sidhu, R.K., Gupta, N., Singh, P., Singh, P., Jat, H.S. and Gerard, B. (2019) Sub-surface drip fertigation with conservation agriculture in a rice-wheat system: A breakthrough for addressing water and nitrogen use efficiency. Agricultural Water Management, 216: 273-283.

Singh, A.K, Behera, M.S., Barman D, et al. (2018a) Modelling carbon sequestration under jute based agro-ecosystem. Annual Report 2017-18, Kolkata: ICAR-Central Research Institute for Jute and Allied Fibres

Singh, A.K., Mukesh K. and Mitra, S. (2018b) Carbon footprint and energy use in jute and allied fibre production. The Indian Journal of Agriculture Science 88(8): 1305–1311.

Singh, Rinku and Singh, G.S. (2017) Traditional agriculture: a climate-smart approach for sustainable food production. Energy, Ecology and Environment. 2: 296–316

Singh, V.K., Dwivedi, B.S., Singh, S.K., Majumdar, K., Jat, M.L., Mishra, R.P. and Rani, M. (2015) Soil physical properties, yield trends and economics after five years of conservation agriculture based rice-maize system in north-western India". Soil and Tillage Research 155:133-148

Srinivasa Rao, C., Kundu, S. and Srinivas, K., (2017) Carbon Sequestration: Semiarid Regions of India. In Encyclopedia of Soil Science (pp. 299-306). CRC Press

Srinivasarao, Ch., Rani, Y.S., Veni, V.G., Sharma, K.L., Sankar, G.M., Prasad, J.V.N.S., Prasad, Y.G. and Sahrawat, K.L. (2016). Assessing village-level carbon balance due to greenhouse gas mitigation interventions using EX-ACT model. International Journal of Environmental Science and Technology 13(1):97-112.

Tipraqsa, P., Craswell, E.T., Noble, E.D. and Schimdt-Vogt, D.(2007) Resource integration for ultiple benefits: Multifunctionality of integrated farming systems in Northeast Thailand. Agricultural Systems. 94(3):694-703.

Tripathi, S.C., Gill, S.C. and Meena, R.P. (2013) Intercropping of canola and mustard with wheat for higher profitability. Journal of Wheat Research 8(1):30–33.

Tubiello, F.N., Salvatore, M., Ferrara, A.F., House, J., Federici, S., Rossi, M., Biancalani, R., Golec, R.D.C., Jacobs, H., Flammini, A., Prosperi, P., Galindo, P.C., Schmidhuber, J., Sanchez, M.J.S., Srivastava, N., and Smith, P. (2015) The Contribution of Agriculture, Forestry and other Land Use activities to Global Warming, 1990–2012. Global Change Biology. Doi.org/10.1111/gcb.12865.

UNEP (United Nations Environment Programme) (2016) The Adaptation Finance Gap Report. Nairobi: UNEP.

United Nations (2015) Transforming our world: The Agenda. Goal 5. Achieve gender equality and empower all women and girls. United Nations.

Varallyay, G. (2010) The impact of climate change on soils and on their water management. Agronomy Research. 8 (Spl. Issue II):385-396.

Vuren, D.P.V., Ochola, W.O., Riha, S., Gampietro, M., and Ginze, H. (2009) Outlook on agricultural change and its drivers. In: McIntyre B.D., Herren H.R., Wakhungu J., Watson R.T. (Eds.). Agriculture at a Crossroads, Island Press, Washington, DC.

11

Carbon Neutral Roadmap for Green India

Shrikant Badole[1], *Dibakar Roy*[2], *Love Kumar Singh*[3], *Ashim Datta*[4]

[1]*Ramakrishna Mission Vivekananda Educational and Research Institute Morabadi, Ranchi, Jharkhand*
[2]*ICAR-Directorate of Weed Research, Jabalpur, Madhya Pradesh*
[3]*BISA, Ludhiana, Punjab*
[4]*ICAR-Central Soil Salinity Research Institute, Karnal, Haryana*

Introduction

Today, according to Global Assessment of Land Degradation and Improvement (GLADA), nearly one forth area of our world land has been degraded (Arora 2019). The United Nations Convention to Combat Desertification (UNCCD) is leading the issues of soil or land degradation to restore valuable natural resources for achieving a land degradation neutral world by 2030 (Cowie et al., 2018; IPBES, 2018; Bhattacharya, 2020). Land degradation neutrality (LDN) can be achieved primarily through adoption of sustainable land management practices. Soil organic carbon (SOC) plays a multifunctional role and is sensitive to land management practices and a key indicator of soil quality, soil health and LDN. Soil is a major reservoir of terrestrial C in the biosphere and plays an important role in global C cycle. Therefore, a little change in SOC pool can significantly influence the atmospheric CO_2 concentration. In general, 1 Pg of SOC stock is equivalent to about 0.47 ppm of CO_2 in the atmosphere, and vice versa (Lal, 2018). Besides this it maintains overall soil sustainability and enhances food security.

Achieving carbon neutrality in Indian soils is a major challenge. Because, productive land goes out of cultivation and marginal or degraded land comes under cultivation due to urbanization, industrialization and various development activities. Also, it is very difficult to increase SOC because climatic condition prevails with a high rate of decomposition. According to an UN report, India will be the most populated country in the world and expected to add 273 million people between 2019 and 2050 which will exert huge pressure on natural resources, particularly soil and water to fulfil the increasing food demands of such ever-growing population. The extensive use of natural resources along with intensive cultivation will enhance greenhouse gases (GHGs) emission, slow down agricultural productivity, enhance land degradation, and decline factor productivity of major production systems. Due to change in distribution of rainfall and concurrent rise in temperature owing to climate change, the fertile top soil becomes more prone to degradation.

In India, most of the intensive production systems are showing yield plateau and crop response to inputs is declining (Srinivasarao, 2018). Also, more and more inputs are required for a unit production of food which enhances emission of GHGs like CO_2 and N_2O. Most of the soil types in India have multiple production constrains however, depletion of SOC and subsequent soil health degradation is the major concern (Srinivasarao et al., 2014). Therefore, sustainable management of natural resources is required such as adoption of innovative C neutral technologies like balanced fertilization with organic and inorganics, enhancement of soil C through crop residues and waste management, farming C through innovative practices such as conservation agriculture (CA) and system of rice intensification (SRI), soil C '4 per mille' (4 per 1000) initiative, forestry/agroforestry system, climate smart agriculture (CSA), reclamation of salt affected soils and C credits/ trading to maintain or enhance SOC stocks.

Climate change impact on agriculture sector

Climate change and food insecurity are the two major concerns of the present world. Since 19th century, the average temperature of the earth atmosphere has increased by 0.9°C due to GHG emissions through various anthropogenic activities. As per the projection, it may rise by 1.5°C or even more up to 2050, which in turn may increase events like cloudburst, flood, drought, heatwave etc. throughout the world. According to estimates by AON's Weather, Climate and Catastrophe Insight 2019, direct economic losses and damage from natural disasters were USD 232 billion across the world in 2019 (AON, 2020). A ~95% of this loss came from weather events such as cyclones, floods and droughts that are directly related to climate change and affects the agricultural food production. World population is expected to reach to 9.7 billion by 2050, resulting in immense pressure on agricultural land to meet the growing demand of food already impacted by climate change. Agriculture is essentially reliable on the weather, and a little variation in rainfall and temperature greatly influence the food production. According to Food and Agricultural Organization (FAO) estimates (FAO, 2016), if the current rate of GHGs emission prevails up to the year 2100 then it will have profound effect on the major food grain production; yields will decrease by 20–45% in maize, 5–50% in wheat and 20–30% in rice. If the same trend of climate change continues in future, this will substantially decrease the food grain production and thereby will enhance the food demand and prices at uncontrolled level. It will become a daunting task to fulfil the rising demand of growing population.

As per the report of European Academies' Science Advisory Council (EASAC) (Arora, 2019), there have been increased in sporadic incidences of extreme events including flood by 50% in the last 10 years due to climate change. Recent heavy floods in Kerala, Maharashtra and West Bengal in the year 2018, 2019 and 2020, respectively are the glaring examples. These floods remove top fertile soil and nutrients, resulting in decline in crop productivity for years, unless and until proper management strategies are invoked. Rise in sea level due to climate change enhances soil salinity and also affect crop productivity in coastal areas. All these climate change activities directly or indirectly affect the agricultural crop productivity.

Innovations for making agriculture carbon neutral

a) Balanced fertilization with organic and inorganics

In India, farmers follow both balanced and imbalanced fertilization to their crops. Soil C sequestration gets influenced indirectly due to effects of these practices on crop performance. It has been observed that proportions of rhizodeposition C in below-ground C inputs through roots, rhizodeposition etc. averaged 54 to 63% in cereals (Hirte et al. 2018). In general, an apparent increase in SOC stock in soils upon balanced fertilization through organic and inorganic sources was reported by many researchers (Fig. 1). This was due to greater addition of organic C through plant residues and root biomass C in soils under balanced fertilization as evidenced by higher crop yields under balanced fertilizer treatments (Table 1) (Mandal et al., 2007). However, such build-up of SOC varied with ecology, soil type and climate of the region. Crops capable of producing higher residues and root biomass C should be included in the cropping systems for having a net C sequestration in the soil.

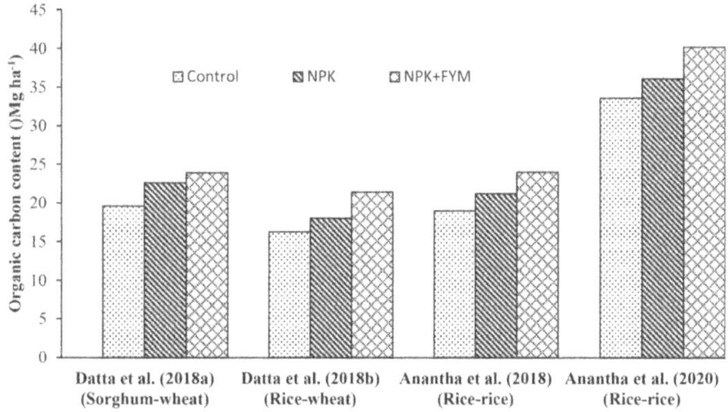

Fig. 1. Influence of balanced and imbalanced fertilization on organic C content (Mg ha^{-1}) in soils.

Table 1. Amount of crop residue C inputs (Mg C $ha^{-1}y^{-1}$) in soils across fertilization treatments and cropping systems (Mandal et al., 2007)

Cropping system	Fertilization treatment			
	Control	NPK	NPK+ FYM#/ Compost	Mean
Rice–mustard–sesame	1.88	2.76	3.75	2.80
Rice–wheat–fallow	1.82	3.33	3.97	3.04
Rice–fallow–berseem	2.45	3.17	4.16	3.26
Rice–wheat–jute	2.58	5.08	6.17	4.61
Rice–fallow–rice*	2.58	3.56	4.30	3.48
Mean	2.26	3.58	4.47	3.44

*Compost was used; # Farmyard manure

Table 2. Effects of balanced fertilization (NPK and NPK+FYM or compost) on C build up in soils under different cropping systems (Mandal et al., 2007)

Cropping system	C build-up over unfertilized control (%)		Rate of C build-up (Mg C ha^{-1} $soil^{-1}$ yr^{-1})	
	NPK	NPK+FYM#	NPK	NPK+FYM/ Compost
Rice–mustard–sesame	51.8	55.7	1.91	2.05
Rice–wheat–fallow	16.8	23.4	0.27	0.37
Rice–fallow–berseem	9.3	24.7	0.13	0.36
Rice–wheat–jute	14.9	32.3	0.11	0.25
Rice–fallow–rice*	33.5	54.8	0.28	0.45
Mean	25.2	38.2	0.54	0.70

*Compost was used; # Farmyard manure

Mandal et al. (2007) also found that there was disproportionately higher rate of SOC build-up under few cropping systems compared to their C inputs level. As for example, in spite of similar residue C inputs in R–W–F (3.33 Mg ha^{-1} yr^{-1}) and R–F–B (3.17 Mg ha^{-1} yr^{-1}) system, the annual C build-up rate in the former was more than the double (0.27 Mg ha^{-1} yr^{-1}) over the later (0.13 Mg ha^{-1} yr^{-1}) (Table 2). The decomposition of organic matter might have been accelerated due to higher N content in crop residues of berseem (2.6%) and jute (1.8%) with lower lipids and lignins in R–F–B and R–W–J systems. Therefore, not only the quantity but also the quality of the crop residue is important in regulating the SOC build up in soil. Rice and wheat crop residues with low N content are likely to be more efficient in C sequestration than the residues of crops like berseem and jute, with higher N content.

b) Soil C content enhancement through organic amendments, crop residues and waste management

Incorporation of crop residues or organic amendments is a well-known strategy for enhancing organic C in soils (Pathak et al., 2011, Ghosh et al., 2012). However, long term addition of organic amendments with increasing amount for achieving higher C sequestration in soil had shown mixed results (Fig. 2). In India, only 10-15% increase in SOC content was reported after application of 5-10 t FYM or crop residues for many years as evidenced in long-term experiments, thereby indicating loss of C from soil, and not actually contributing to cumulative build-up of SOC over years (Mandal, 2011).

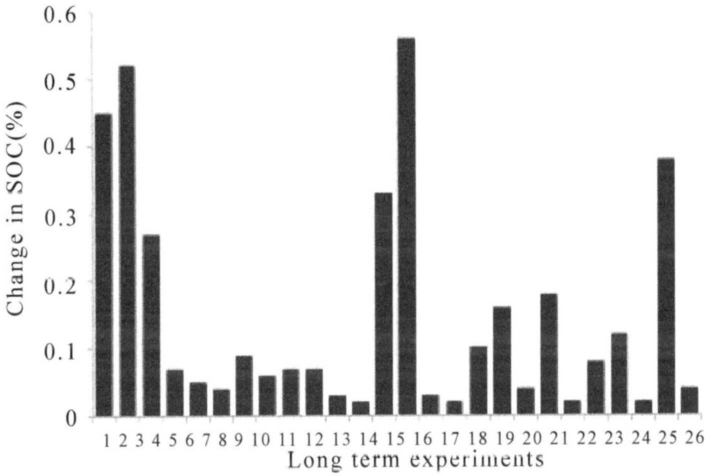

Fig. 2. Change in soil organic C in the NPK+FYM treatment compared to the NPK treatment in 26 long-term experiments throughout the country (Pathak et al., 2011)

Nearly 20.8 and 23.3 million tonnes (MT) of rice and wheat crop residues respectively are produced in a year in Punjab state only (Yadvinder-Singh et al. 2004). Most of the wheat residue is used as animal fodder, whereas majority of the rice residue is burnt in the field resulting in loss of C and N and GHGs emission to the environment. Therefore, to reduce N immobilization and improve soil fertility efficient residue management practices should be developed. Conservation agriculture utilizing the principles of zero tillage and residue management with suitable crop rotation is now gaining momentum in India.

As the population is increasing, huge amount of city wastes including municipal solid wastes are produced every day in cities and towns of the country. On an average, the quantity available could be more than 70 MT yr^{-1} and is growing at a steady annual rate of 1-1.33% (Mandal, 2011). These wastes can serve as a rich source of C and plant nutrients, but these also contain heavy metals, organic pollutants and pathogenic organisms. Proper mechanism must be followed for use of wastes particularly in process of segregation of organic materials from masses and subsequent composting of the same. It will have great benefits including a rich and cheap source of C and nutrients, and also as a pollution abatement measure.

All these organic amendments addition will cause at least a small enrichment of SOC in soils under tropical and subtropical climate in spite of rapid loss of SOC under these climates. In the context of soil functioning, significant improvement in soil physical, chemical and biological properties happen even with a small increase in SOC content (Gupta Choudhury, 2010).

c) Farming through innovative practices

Conservation agriculture (CA)

Conservation agriculture (CA) is the amalgamation of minimum soil disturbance mainly zero tillage (ZT) or reduced tillage (RT) with retention of crop residues and diversified cropping systems specially inclusion of legumes. Latest estimates suggest 157 Mha land is under CA globally which are mostly dominated in USA Brazil, Canada, Chile, Paraguay, Australia and a few other western countries (FAO, 2014). Total area under CA in India is still <1% of total cultivated area and mainly concentrated in Indo Gangetic Plains (IGP) of north western India (Haryana, Punjab and Uttar Pradesh) primarily as ZT-wheat which enable farmers to address timely showing of wheat after rice (basmati) harvest and to avoid *Phalaris minor* in wheat. However, CA is gaining importance among the farmers and increasing its acreage in various parts of India in other cropping systems as well (Mandal et al., 2020).

The focus on research and development of new technologies in green revolution period was primarily on higher production and productivity of a few cereal crops to ensure food security and profitability (Abrol, 2005). However, negative impact of green revolution was visible in late 90s with yield stagnation of cereal crops, lowering of ground water table, depletion of soil organic matter, deterioration of soil physical health, emergence of micro and macro nutrient deficiency and degradation of overall soil quality. All these factors triggered adoption of CA to address these above-mentioned issues. CA has been accepted as an updated crop production technology for better soil physical health, better nutrient recycling, enhanced microbial activity and biological diversity along with enhanced soil carbon sequestration vis-à-vis lowering GHGs emission. Soil C sequestration is the prime need for restoring soil fertility as C plays key role in maintaining soil quality. In India, tropical and subtropical climate coupled with conventional tillage (CT) and practices like residue burning decelerate SOC built up. CA practices specially ZT, residue retention enhances soil C content in the soil under long term perspectives.

Soil carbon accumulation under CA was clearly observed in top soil layer (0-5 cm) irrespective of soil and climate (Das et al., 2013; Bhattacharyya et al., 2015; Veloso et al., 2020) (Table 3). When cotton-wheat was grown under CA, ZT with bed planting (ZT-B) and ZT with flat planting (ZT-F) recorded 28 and 26% higher SOC stock compared to conventional tillage with bed planting (5.5 Mg ha^{-1}) at 0-5 cm layer in north India (Das et al., 2013). Inclusion of summer moong bean in between ZT-wheat and direct seeded rice (DSR) resulted in soil C build up in north-western IGP in 0-5 cm soil layer (Bhattacharyya et al., 2015). In Eastern India, after 7 years of CA practices, Samal et al. (2017) concluded that adoption of full CA practices like DSR-ZT wheat-ZT moong bean along with full residue retention registered highest SOC stock of 48 Mg C ha^{-1}, which was 15% higher over farmer's practice in 0-30 cm soil layer.

Yadav et al. (2017) reported that double rice cropping system with reduced tillage (RT), integrated nutrient management practice and partial residue retention (30%) recorded highest C accumulation (1.3 Mg C ha^{-1}) and highest sequestration rate (428 kg ha^{-1} yr^{-1}) over other treatments in Eastern Himalayan region. In rice-mustard cropping system, higher SOC sequestration, SOC pools (29.9 vs. 29.1 Mg ha^{-1} and 29.7–29.8 vs. 29.0 Mg ha^{-1}), carbon sequestration rate (450 vs. 265 kg ha^{-1} yr^{-1} and 391–428 vs. 221 kg ha^{-1} yr^{-1}) and carbon retention efficiency (7.7 vs. 4.6% and 6.6–6.9% vs. 4.7%) were observed in ZT with residue retention and mulched treatments over conventional tillage with residue incorporation and no mulched treatments were reported by Yadav et al. (2019). Again, growing of cow pea in pre-rainy season as cover crop doubled the carbon sequestration rate (478 kg C ha^{-1} yr^{-1}).

Table 3. SOC build up in soil under CA over CT system in India

References	Soil depth (cm)	Conventional tillage (CT)	Zero tillage (ZT) or reduced tillage (RT)
		SOC (Mg ha^{-1})	
Das et al. (2013)	0-5	5.7	6.9
Bhattacharyya et al. (2015)	0-5	5.6	6.0
Samal et al. (2017)	0-30	41.6	43.6
Yadav et al. (2019)	0-30	29.2	29.9
		SOC (g kg^{-1})	
Jat et al. (2018)	0-15	5.4	7.6

In Northwest India, enrichment of SOC content under CA is more pronounced with maize-based (7.7 g kg^{-1}) compared to rice-based (7.5 g kg^{-1}) cropping system over farmer's practice (4.5 g kg^{-1}) in 0-15 cm layer after 4 yrs of cropping cycle (Jat et al., 2018). Further, Jat et al. (2019) reported that adoption of ZT and full residue retention for 4 yrs recorded maximum improvement of SOC stock (70%) under ZT maize-ZT wheat- ZT moong bean treatment compared to farmer's practice (16.2

Mg C ha^{-1}) and ZT DSR-ZT wheat-ZT moong bean at 0-15 cm soil depth. Similarly, soil C pools were also higher under the above full CA treatments over the farmer's practice. Adoption of permanent bed (PB) and flat bed ZT (ZT-flat) practices in maize based cropping systems for 6 yrs improved C content and SOC stock over flat bed CT (CT-flat) at 0-5, 5-15 and 15-30 cm depths respectively in a sandy loam soil (Parihar et al., 2018). Parihar et al. (2020) calculated C sequestration and soil quality under maize-wheat-mung bean cropping system with varying tillage and nutrient management options. Permanent bed along with site-specific nutrient management (PB-SSNM) recorded highest C content (44.1%) over unfertilized-CT plot. Similarly, maximum SOC stock, carbon input (3.41 Mg ha^{-1} $year^{1}$) and C sequestration rate (1.15 Mg ha^{-1} yr^{1}) were also obtained in PB-SSNM treatment. In central India, higher proportion of large size water stable aggregates, aggregate associated C and higher proportions of non-labile C (C_{NL}) under NT over CT were reported by Somasundaram et al. (2018) in 0-5 and 5-15 cm layers after four cropping cycles. After 6 years of cropping cycles, Dey et al. (2020) reported that double ZT + residue retention in rice-wheat system sequestered ~2 Mg ha^{-1} total SOC in the 0–15 cm soil layer, where significant losses were registered in CT.

System of Rice Intensification (SRI)

Now-a-days, system of rice intensification (SRI) is advocated to economize the water use. The SRI is alternate rice growing technology in which 10-14 days old seedlings are transplanted with wide spacings (25 cm × 25 cm) and water level is maintained at near field capacity. Basically, SRI cuts down water requirement for growing rice crop required under continuous submergence (Satyanarayana et al., 2007; Chowdhury et al., 2014) and also enhances productivity over conventional continuous submerged rice (Ceesay et al., 2006). Rice plants under SRI have larger root volumes and root mass, tall and strong tillers with better photosynthetic ability (Thakur et al., 2010; Mishra and Salokhe, 2010; Gopalakrishnan et al., 2013). Alternate wetting and drying facilitates higher microbial activity under SRI condition compared to conventional submerged rice (Anas et al., 2011). Soil microorganisms present in rice rhizosphere are very efficient in sequestering atmospheric C into microbial biomass. Proliferation of root system, a greater number of tillers, enhanced microbial growth and higher nutrient availability under SRI increases yield of rice by 20- 50% (FAO, 2016). This increase is likely to be associated with better root proliferation and higher rhizodeposition in root zone. It was observed that rhizodeposition of C by below-ground plant parts specially roots accounts 54-63% of total below ground C input for cereals (Hirte et al., 2018). Higher biomass production under SRI accompanied with higher deposition of organic C under root zone in SRI. The carbon released by roots present in stable form, contributed to passive carbon a pool which is desired for carbon sequestration. Alternate wetting and drying and further cono weeding operation create favourable condition (inclusion of aerobic microorganisms) for humification process or enhanced degradation of non-humic substances in presence of air creates more humic and fulvic acid compounds under SRI than conventional submerged rice (Rajkishore, 2013). All these factors ensure higher soil C sequestration under SRI over conventionally grown continuous submerged rice. Furthermore, alternate wetting and drying check methane (CH_4) production and vis-à-vis emission by keeping redox potential of soil in mostly positive values. Jayadeva et al. (2009) estimated that CH_4 emission in SRI was 40% lower compared to submerged rice ecology. Later Rajkishore et al. (2013) reported that conventional submerged rice produced more CH_4 (44.6 and 55.5 kg ha^{-1}) than SRI (31.8 and 37.7 kg ha^{-1}) in summer and *kharif* seasons, respectively. Elsewhere lower CH_4 emission under SRI was also reported by various researchers (Suryavanshi et al., 2013; Dill et al., 2013; Thakur et al., 2017).

Further, alternate wetting and drying may increase nitrous oxide (N_2O) emission under SRI due to formation of NO_3^- in presence of air and subsequent reduction of NO_3^- under wetting period. Jain et al. (2014) observed on an average 22.5% higher N_2O and 61% lower CH_4 emission in SRI over transplanted puddled rice (TPR) in Indo-Gangetic plains of India. Recently Oo et al. (2018) observed a net increase in N_2O emission and a 41% decrease in CH_4 emission under SRI condition in south India. When total green house gas emission expressed in terms of global warming potential (GWP) it was observed that SRI had 28% lower GWP than TPR was observed in India (Jain et al., 2014). Similarly, reduction in net GHG emission in terms of GWP was also observed by Gathorne-Hardy et al., (2013) in south eastern India. Reduction in net total GHGs emission in SRI over TPR was also reported by many researchers across the globe (Kim et al., 2012; Choi et al., 2013; Chu et al., 2015). These reports indicate that adoption of SRI reduces net total GHGs emission from SRI system compared to TPR due to changes in water regime, however more studies are required to realize potential of SRI for SOC sequestration. SRI is gradually gaining popularity in north eastern regions of India such as in Tripura particularly in rice-wheat and other rice-based cropping systems mostly due to benefits such as higher input use efficiency.

d) Soil interventions: Soil carbon 4 per 1000 or 4 per mille initiative

The "4 per 1000 or 4 per mille" initiative was launched at the 21st conference (COP21) of the United Nations Framework Convention on Climate Change in Paris (November 11, 2015) whose main aim was to reduce the impact of climate change and limit the global warming to below 2°C. This is popularly known as "Paris Agreement on Climate Change". An ambition to launch '4 per 1000' initiative is to provide compensatory option for the global emission of GHGs through various anthropogenic sources by enhancing SOC stocks at 0.4% per year. This was accepted by stakeholders from different countries, with the commitment to prepare a volunteer action plan for implementing management practices that increase SOC stock in agricultural soils (Chambers et al., 2016; Lal et al., 2016; Minasny et al., 2017).

However, the value 4 per mille cannot be applicable all over the world as C storage potential varies with soil type, climate, vegetation, etc. Also, 4 per mille target is not easily achievable, because scarcity of scientific data, the finite capacity of SOC sink, resource poor small and marginal farmers, financial commitments and implementation as well as high spatial variability of SOC from field to regional, continental and global extent (Kerry et al., 2012; Scharlemann et al., 2014; Lal, 2016; Minasny et al., 2017). Studies on C sequestration suggested that agricultural soils have potential to sequester C in between 0.7 to 1.2 Gt C yr^{-1} with the adoption of improve management practices (Le Quere et al., 2015) *viz.*, CA, incorporation of legume in cereal-cereal rotations, integrated nutrient management etc. Many of the world agricultural soils under degraded condition have SOC below the critical limits. Restoring these soils by adopting best management practices not only improves SOC and offsets the GHGs emission but also improve soil quality and productivity which ultimately helps in achieving carbon neutrality and global food security.

Indian soils are inherently low in SOC (3.2 g kg^{-1}) (Rai et al., 2009). But it is difficult to increase the SOC content, because high temperature increases the rate of decomposition. Practices of burning or removal of crop residue also discourage SOC build up. Therefore, efforts have to make to maintain SOC stock by adding critical C inputs (Mandal, 2011). The estimated SOC stock for Indian soils is only 20-25 Gt for the top 1.0 m with an annual C emission of 566 MT (CIDAC database) (Bhattacharya et al., 2008; Banger et al., 2015; Sreenivas et al., 2016). Thus, to achieve C neutrality in Indian soils, the required C sequestration rate would be around 2.3-2.8% which is much higher that the global requirement of 0.4% (Minasny et al., 2017).

Different management techniques enhance SOC sequestration rate: organic amendments (~0.5 t C ha^{-1} yr^{-1}), residual incorporation (~0.35 t C ha^{-1} yr^{-1}), no or reduced till (~0.3 t C ha^{-1} yr^{-1}) and crop rotation (~0.2 t C ha^{-1} yr^{-1}) but it varied with location, soil types and climatic condition (Minasny et al., 2017). It can be possible to achieve 1-3% in soils with low SOC content by adopting the best management practices. Research studies from long-term experiments suggested that balanced fertilization along with organic manure and crop residues incorporation increased SOC by 10-20% (Table 4; Minasny et al., 2017). With estimated C sequestration rates of about 0.024 to 0.089 Gt $year^{-1}$ in rice-based cropping systems (Lal, 2004; Mandal et al., 2007, 2008; Pathak et al., 2011; Banger et al., 2015) and around 25 Gt of SOC stock, it may be possible to loss SOC around 1-4 per mille through C emission and contribute to global warming. Therefore, it is necessary to maintain the SOC level by adding critical C inputs (0.31 to 5.16 t C ha^{-1} yr^{-1}) through organic sources depending upon the availability under different agroecological zones based on the aridity index of the location (Mandal, 2011). Therefore, in order to achieve carbon neutrality in agricultural land, adoption of best management practices such as promotion of pulses, incentives for use of crop residues (reduce or zero tillage, compost), green manuring, improved nutrient management and incorporation of legumes in rice-based cropping systems are necessary for improving SOC stock in Indian soils.

In India, according to ICAR, 2010, about 121 Mha of land is under degradation (both natural and anthropogenic), of which 73 Mha is affected by water erosion and 6.7 Mha land area is affected by salinity/alkalinity; these requires immediate measures to achieve land degradation neutrality (Maji et al., 2010; Bhattacharyya et al., 2020). Because, it caused an annual economic loss of Rs 3.17 lakh crores ($46.90 billion) in 2014-15, which is equivalent to 2.5% of the country's gross domestic product (GDP) in the same year (Ministry of Environment, Forest and Climate Change, 2018). The highest mean SOC density in soils was observed under plantation crops (253 t C ha^{-1}) followed by forest (139.9 t C ha^{-1}) and crop land (58.5-67.4 t C ha^{-1}) (Sreenivas et al., 2016). Therefore, promotion of plantation programme for the restoration of degraded land, will not only help to achieve land degradation neutrality by improving SOC stock, but also improve soil quality, its stability and enhance food security.

Table 4. Effect of management practices on carbon sequestration

Management practices	Depth (cm) observed	Carbon sequestration rates t C ha^{-1} yr^{-1}	Average C stockt C ha^{-1}	Period of Observation (yrs)	References
Rice-Rice with NPK	0–20	0.23	31.3	36	Mandal et al. (2008)
Rice-Rice with NPK + compost	0–20	0.41	31.3	36	Mandal et al. (2008)
Rice-Wheat with NPK	0–60	0.66	34.4	19	Majumder et al. (2008)
Rice-Wheat with NPK + Farm yard manure (FYM)	0–60	0.99	34.4	19	Majumder et al. (2008)
Rice-Wheat with NPK + Paddy straw	0–60	0.89	34.4	19	Majumder et al. (2008)
Rice-Wheat with NPK + Green manuring	0–60	0.82	34.4	19	Majumder et al. (2008)
Inorganic fertilizer	0–15	0.16	13.3	6–32	Pathak et al. (2011)
Inorganic fertilizer + FYM	0–15	0.33	13.3	6–32	Pathak et al. (2011)

Forestry/agroforestry

Forests play a key role in adaptation and mitigation of climate change, as they act as both sink and source of carbon. According to Fifth Assessment Report of the Intergovernmental Panel on Climate Change (IPCC), the agriculture, forestry and other land use change activities are responsible for 24% global GHGs emission, whereas vegetation and soil account for about 30% C sequestration (IPCC, 2014). Forest vegetation and soils contain about 1240 Pg of C (Dixon et al., 1994). Also, they have high potential of C sequestration; it is about 0.4 Pg C yr^{-1} for forest soils and 1-3 Pg C yr^{-1} for forest biomes (Lal 2005). Forests currently contain 39% of all C stored in soils (Eliasch, 2008). The amount of C in forest soils varies among regions and forest types.

Forests play dual role as sink and source of C which offers cheaper option than other sectors for mitigation of climate change by removing C from the atmosphere and storing it in biomass and soil (Table 5). Beside this, it provides food, wood, fuel, shelter and income to billions of people. Considering the potential, it is necessary to increase forest and tree cover through afforestation, reforestation, forest management and forest products contribute to enhanced livelihoods and a lower C footprint.

Table 5. Key forest mitigation options for climate change

Activity	Mitigation action	Examples	Carbon benefits
Enhancing carbon stocks (i.e. sinks)			
Afforestation and reforestation	Increasing the forest area	Planting new forests	Increases atmospheric CO_2 removal from long-term forest stocks and tree plantations Augments the availability of HWPs, including for bioenergy use
Forest management	Increasing the carbon density/ stock in forests	Optimizing the use of appropriate fertilizers and irrigation Restoring degraded forests and soils Optimizing harvest systems and implementing reduced-impact harvesting Managing impacts of forest disturbances under a changing climate	Maximizes CO_2 removal and storage Enables long-term production of HWPs for bioenergy or other uses
Green building and furnishing	Replacing high-energy products with industrial wood products	Using wood and forest-based materials for green buildings, infrastructure and furniture	Increase carbon storage potential off-site through the use of harvested wood products
Reducing GHG emissions (i.e. sources)			
Reduced deforestation and degradation (REDD)	Maintaining forest area and site- and landscape-scale carbon stocks/ density	Preventing forest loss Suppressing forest disturbance via better management of fires, pests and diseases Improving management of soil carbon	Helps avoid GHG emissions and maintain CO_2 removal and optimal forest stock

Activity	Mitigation action	Examples	Carbon benefits
Use of wood energy	Improving practices related to bioenergy	Using sustainably produced wood products and residues for energy Promoting improved cook stoves	Avoids CO_2 emissions through substitution of fossil-fuel-based energy Avoids emissions from microbial decomposition of biomass
Green building and furnishing	Improving use and uptake of wood-based products	Using climate-smart products following a cascading principle, including increased use of wood in construction	Avoids CO_2 emissions through substitution of fossil-fuel-based products by green buildings and bio- based materials and chemicals Avoids emissions from microbial decomposition of biomass

Source: Nabuurs et al., 2007

Agroforestry systems also accounts for significant terrestrial C storage and offers both adaptation and mitigation option for global climate change. Agroforestry practices include alley cropping, home gardens, improved fallows, multipurpose trees on farms and rangelands, agri-silviculture, silvi-pasture and agri-horticulture systems, shaded perennial crop systems, shelterbelts, windbreaks and *taungya* systems etc. Agroforestry systems have provided range of economic, environmental and socioeconomic benefits, besides C sequestration (Srinivasarao, 2018).

At the national level, a number of national initiatives have been undertaken to achieve land degradation neutrality by 2030, which includes soil and water conservation programmes like Drought Prone Area Programme, Wasteland Development Program, Watershed Development Program, Joint Forest Management, Environment Policy, National Action Plan for Climate Change (NAPCC), Green India Mission, MGNREGA, CAMPA, etc. The National Action Plan for Climate Change, has an ambitious plan to create an additional carbon sink of 2.5 to 3 billion tonnes of CO_2 equivalent by 2030 through additional forest and tree cover and enhancing investment in vulnerable sectors to climate change particularly in agriculture, water resources, Himalayan and coastal regions, health and disaster management. It is also committed to restore an additional 5 Mha of degraded land by 2030, raising the land to be restored in India to 26 Mha through improving the condition of affected ecosystems, combating desertification/land degradation, promoting sustainable land management and contributing towards land degradation neutrality.

e) Climate-smart agriculture

Climate-smart agriculture (CSA) is agriculture that sustainably increases productivity, enhances adaptive capacity, and reduces GHGs emissions where possible. CSA is an approach to increase the technical knowhow, policy and investment on environment to get sustainable agricultural growth for food protection under climate change. At the local level, it shields farmers from the adverse effects of climate change, improves farm yields and household incomes, build stronger and more resilient farming communities. At the national level, it helps deliver food security and development goals, while reducing emissions.

A number of strategies can be adopted to achieve the objectives of climate smart agriculture. For example, use of integrated renewable energy technologies for farming like windmills, solar panels, pyrolysis units and bioenergy-operated water pumps are very crucial for energy-smart food systems. Resource conserving technologies (RCTs) like ZT permits the farmers to plant wheat soon after rice or cotton harvest to avoid warm weather detrimental for grain filling. The emergence of newly developed varieties which are tolerant to heat, drought and salinity is also a better strategy of CSA (Amin et al 2015).

Why CSA is needed?

World's population will rise by a third by 2050 and many of these will be residing in developing nations, and also more people will live in cities. If current trends in revenue and consumption growth persist, FAO predicts that agricultural production will need to rise by 60% by 2050 to meet projected demands.

Therefore, if it is to feed a growing global population and provide the foundation for economic development and poverty reduction, agriculture must reinvent itself. Climate change will make this task more difficult under a business-as-usual scenario, due to adverse impacts on agriculture, requiring spiraling adaptation and related costs (FAO, 2013). Adaptation to climate change and lower emission intensities per harvest will be needed to meet food security and agricultural growth goals. This conversion must take place without the destruction of the basis of natural resources. As a result of the increased occurrence of severe incidents and the growing unpredictability of weather patterns, climate change often impacts agriculture and food security. This would lead to a reduction in demand and a drop in incomes in developing countries. Improving food security while contributing to climate change mitigation and protection of the natural resource base and critical ecosystem resources needs the transition to more sustainable agricultural production systems that use inputs more effectively, have less uncertainty and greater yield reliability and are more resilient to threats, shocks and long-term instability of the climate. More sustainable and robust agriculture needs a significant change in the way land, water, soil nutrients, and genetic resources are controlled to allow more effective use of these resources. It needs major improvements in national and local government, laws, regulations, and financial structures to allow this transition. This transition would also require improving access to markets for farmers by reducing emissions of greenhouse gasses per unit land and/or agricultural. These changes will contribute greatly to reducing climate change and growing carbon sinks.

CSA is not a set of specific agricultural system or method which can be applied uniformly. It is a strategy that involves site-specific evaluation to determine the necessary technology and practices for agricultural development.

Climate Smart Village (CSV) Approach

The CSV is an agricultural research for development (AR4D) approach to assess technical and structural alternatives for addressing climate change in agriculture through participatory approaches (Agarwal et al. 2018).

The strategy of the CSV approach is to:

1. Understand the efficacy of a number of CSA alternatives (practices, technology, facilities, programs and policies) not only in enhancing output and rising revenue, but also in building climate resilience, increasing resilient capability and lowering GHGs emissions wherever possible;

2. Develop (no regrets) solutions in anticipation of future climate change impacts.
3. Understand the socioeconomic, gender, and biophysical constraints and enablers for adoption.
4. Test and define suitable route for adoption, finance possibilities, structural structures, and frameworks for scaling out or scaling up while maintaining consistency with local and national awareness, organizations, and growth plans.

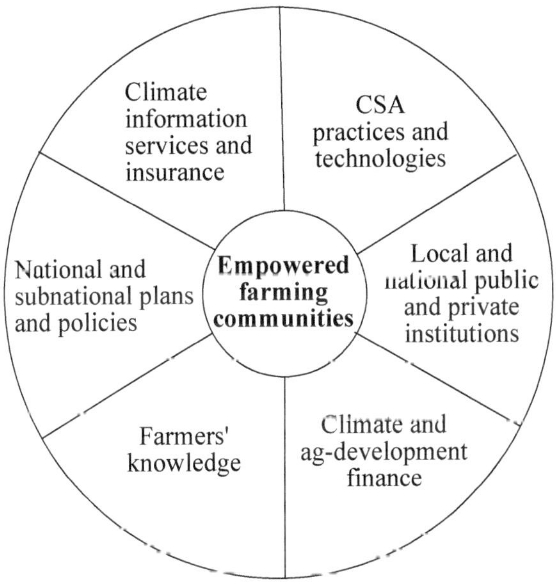

Fig. 5. Climate Smart Village Interventions
Source: Progressing towards climate resilient agriculture Top ten success stories from CCAFS in South Asia. Arun Khatri-Chhetri on 14 September 2017.

Table 6: Adaptation and mitigation options in agriculture#

Technology	Adaptation/Mitigation Potential
Water-Smart: Rainwater harvesting, Micro irrigation; Laser land levelling, Bed based furrow, Drainage management, Cover crops	Interventions to improve water use efficiency, e.g. minimisation of surface runoff and water loss from crop fields, effective control of irrigation and drainage.
Energy-Smart: Zero tillage/ Minimum tillage, Biofuels, Crop residue management, Direct seeded rice.	Interventions to improve energy use efficiency, e.g. reduce fossil fuel use and chemical fertilizers in agriculture.

Nutrient-Smart: Site-specific integrated nutrient management, Green manuring, Leaf colour chart, Intercropping with legumes.	Interventions to improve nutrient use efficiency, e.g. optimum application of soil nutrients matching with crop requirement, improvement in soil quality and nutrient supply
Carbon-Smart: Agro-forestry, Concentrate feeding of livestock to reduce methane production, Fodder management, Integrated pest management.	Interventions to reduce GHGs emissions, e.g. promote C sequestration, reduce nutrient losses, and reduce use of chemical fertilizers
Weather-Smart: ICT based weather forecasts and crop agro-advisory, Weather index based crop insurance, Climate analogues.	Interventions to provide services related to income security and weather advice to farmers, e.g. climate information-based value added agro-advisory to the farmers, crop-specific insurance to compensate income loss due to the vagaries of weather.
Knowledge-Smart: Contingent Crop Planning, Improved Crop Varieties, Seed and Fodder Banks, Farmer-to-Farmer Learning, Land Use Change.	Use of a combination of science and local knowledge, i.e. climatic risk management plan to cope with major weather-related contingencies: drought, flood, heat/cold stresses during the crop season, crop varieties that are tolerant to drought, flood and heat/cold stresses

Source: World Agriculture, Article # 1712, July 2017. www.world-agriculture.net (Arun et al., 2017)

CSA for Green India

Only about 35% of total agricultural land in India is irrigated and two thirds of cultivated land is entirely dependent on rainfall. As such, the agricultural production system in the country is more vulnerable to damage caused by severe climatic conditions, resulting in increased water stress leading to insufficient water sources for irrigation. In various agro-ecological areas of India, fluctuations in average temperatures, shifts in rainfall patterns, increased occurrence of extreme weather events such as prolonged droughts and floods and changes in agricultural seasons have already been observed. Long drought spells during *kharif* and increased temperatures and unseasonal heavy rains during the *rabi* season have caused serious distress to the farming communities in different states in recent years.

India has already begun the development and integration of policies on climate change adaptation in different sectors. For example, the National Missions for Sustainable Agriculture (NMSA) and the National Mission on Strategic Awareness for Climate Change seek to promote climate change adaptation in agriculture through the promotion of smart climate practices and technologies across the world. State-level climate change adaptation plans in India concentrate on resolving present and future climate change threats and taking steps to minimize related risks and vulnerabilities.

Climate change agriculture and food security (CCAFS) and CGIAR centers jointly involve national and sub-national stakeholders, including policy makers, NGOs and civil society groups, research organizations, farmers' groups, play a key role in India in the implementation of climate-smart agriculture in the Indo-Gangetic Plains of Northern India and dryland agriculture systems in the southern and western parts of India.

In the Indian state of Haryana, farmers have adopted climate-smart activities such as laser-land levelling and alternative wetting and drying of rice, minimizing water consumption, enhancing soil health and offering economic incentives. Farmers have also issued agro-advisories on their cell phones, with feedback from met offices, scientists, input dealers and farmers, allowing them to make timely decisions.

Due to the performance of the CSVs, the State Government of Haryana has initiated a program to launch an additional 500 CSVs in the state's rice-wheat systems. The program will be initiated by the Department of Agriculture, Government of Haryana, through integrated investment. The planning workshop with the most relevant stakeholders has already set out the selection of schemes that will be used in the villages. CSVs are catching on across the country, with new proposals for developing CSVs across five other Indian states, on over 237,000 ha of land. These states are particularly vulnerable to climate change, and the International Food Policy Research institute (IFPRI) has developed and submitted pilot projects worth USD 140 million as part of the CCAFS program. (CCAFS, 2015)

In eastern Indian state Bihar's agriculture production is affected by drought and floods while Bihar has fertile soils and ample ground water. For this cause, the Bihar State Government has prioritized the strengthening of agricultural resilience in its policy agenda, seeking measures to diversify production and enhance water management. For several years, CCAFS, CIMMYT and other partners have already tested agricultural practices that improve food security, build resilience to climate stresses, and reduce emissions in CSVs in Bihar. Consequently, the government of Bihar decided to implement CSA and CSV in all 38 districts of Bihar. Ever since, numerous events have been organized by CIMMYT, the Borlaug Initiative for South Asia (BISA) and other partners to assist the government of Bihar with their policy agenda 'the Krishi road map'. These events highlight the benefits of practices such as direct-seeding, site-specific nutrient management, and weather forecasting, and have already attracted hundreds of farmers and representatives of NGOs, companies and the government. In total, the policies inspired on the CSVs are expected to help hundreds of thousands small-holder farmers in Bihar improve their nutrition and livelihoods while coping with climate risks. (*CCAFS outcome 2016*. https://ccafs.cgiar.org/research/results/scaling-climate-smart-villages across-38-districts-bihar#.X6FZzYgzbIU).

The burning of crop residues in northwest India has created a crisis situation and a public health emergency. Currently, less than 15% of total rice residue is being utilized through on-farm recycling, for electricity generation, or other sustainable means of disposal, before the next crop requires labor, time, and capital. To advise and recommend proposals to the Indian Government, CCAFS the International Maize and Wheat Improvement Center (CIMMYT), the CGIAR Wheat Research Program (Wheat) and collaborators have produced a variety of policy briefs. In addition, policy dialogues, roundtable dialogs and high-level policy discussions helped bring the issue to the attention of policy makers. This work has motivated the Indian Government to prioritize strategies for managing crop residues and has now developed a large-scale routing investment of USD 170 million (INR 1150 crores). Investment would increase the popularity of "Happy Seeder" technology, one of the core areas of CCAFS research in CSVs. With this technology, wheat is sown through tractor using rice crop residues as mulch. This investment would support 2 million farmers, covering nearly 4 million hectares (*CCAFS case outcome 2017*, https://ccafs.cgiar.org/research/results/curbing-crop-residue-burning-improve-public-health-and-reduce-emissions#.X6FZyogzbIU). Therefore, all these efforts would help to sustain the crop productivity at the same time make the system resilient towards climate vagaries and play an important role in achieving green India.

f) Reclamation of salt affected soils

Globally, soil salinity and alkalinity prohibit significant cultivable area for reaching their actual productivity potential. The situation is further complicated by climate change. These degraded salt-affected lands need to be reclaimed and brought under productive cultivation to add to the country's food production. Salt affected soils hamper plant growth through higher osmotic stress and thereby causing impediments to successful crop production and sustainability of the production system. In India, salt affected soils comprise of nearly 6.73 Mha area, covering sodic soils of 3.78 Mha and saline soils of 2.95 Mha posing serious threats to food production of the country. This huge salt affected area results losses of about 230 billion INR every year which is expected to rise with further increase in salt-affected area to 16.2 Mha by 2050 (CSSRI, 2015). The ICAR-Central Soil Salinity Research Institute has been working since decades in developing technologies for reclamation of salt affected soils, and has been successful in reclaiming more than 2.0 Mha. Most prominent technologies are gypsum technology for reclamation of sodic soils, subsurface drainage for saline soils with lower ground water table, alternate land use, salt tolerant crop varieties of rice, wheat and mustard varieties. These technologies have helped in greening a vast tract of barren unproductive land in India and thereby enhanced SOC level and ultimately will help in achieving land degradation neutral India by 2030.

Biosaline agriculture is another promising technology developed at CSSRI. Here salt tolerant tree or plant species are successfully grown under highly saline soil or saline ground water. Halophytes such as *Salicornia, Atriplex* and *Salvadora* species are successfully grown in otherwise unproductive saline soils. Another promising technology is "Biodrainage" which involves growing of salt-tolerant trees for lowering of the ground water table in waterlogged areas (Sharma and Singh, 2019). The main criteria for selection of trees in biodrainage are salt tolerance and higher transpiration rate. *Eucalyptus tereticornis* is the most preferred tree species used in biodrainage. ICAR-CSSRI also developed a technology for the sustainable disposal of wastewater in tree plantations (Sharma and Singh, 2019). Recently Chandel et al. (2020) found that alternate application of saline and fresh water enhanced soil organic carbon and nutrient availability in seed spices crops. All these technologies will tremendously help to bring unproductive salt affected barren land with poor quality ground water in productive cultivation and reach carbon neutrality.

Soil organic matter turnover is significantly influenced both by salinity and sodicity and thereby plays an important role in GHGs emission mitigation. Many studies had shown that cultivation of appropriate crops, shrubs, and trees viz., acacia [*A. nilotica* (L.)], shisham [*D. sissoo*], mesquite [*P. juliflora*], and arjuna [*T. Arjuna*] in sodic and saline-sodic soils increased soil C through biomass production (Garg, 1998; Kaur et al., 2002; Datta et al., 2015a,b). Garg (1998) reported SOC build up in an alkali soil under shisham and mesquite in upper 0.6 m soil depth thereby facilitating decomposition of leaf litter and root decay leading to accumulation of humus. Maximum C sequestration rate (0.2–0.8 Mg C ha^{-1} yr^{-1}) was observed under *Prosopis juliflora* after 20 years of planting in an alkali soil (Singh and Gill, 2000).

Guava plantation recorded the highest SOC storage (133 Mg C ha^{-1}) and passive pool C (76 Mg C ha^{-1}) at 2.0 m soil depth in a reclaimed alkali soil in Northwest India (Datta et al., 2015a). Rehabilitation of salt affected soils serve many ecosystem functions like atmospheric CO_2 sequestration also contributes to livelihood security of famers. Recently C dynamics in salt affected soils was reviewed by Datta et al. (2019) and they emphasized the tremendous potential of the salt affected soils to sequester C which improved the quality of those degraded soils besides serving many ecosystem functions.

g) Carbon credits/trading

It is the outcome of achieving carbon neutrality. In Clean Development Mechanism (CDM) under Kyoto protocol, after offsetting the emission, net soil C gain with reference to a particular baseline can be traded. For the poor farming community of India, it is affirmed to be a win-win proposition for earning credits by providing a better niche for living. In tropical and subtropical countries like India, hungry soils for storing C and abundant sun light for higher photosynthetic activity, a great potential lies in such C trading. However, in Kyoto protocol agricultural systems has not been covered with such trading of C credits due to some uncertainties and inherent weaknesses in precisely accounting the amount of C sequestered in soils following emission offsetting practices. Lal (2007) emphasized the main barriers in trading of soil C credits. These are (i) upscaling the point data to large scale such as landscape, farm, watershed or a region comprising 100,000–200,000 ha, (ii) evaluation of soil C pool changes (in Mg ha^{-1} y^{-1}) with reference to a baseline for cultivated land unit comprising a large farm unit, (iii) verification of the permanence of sequestered C in soil and not re-emitted due to changes in land use or management practices. (iv) there is an urgent need of estimation of net C sequestration by considering hidden C costs of inputs such as fertilizers, pesticides, tillage operations, irrigation etc (v) determining worthiness of the soil C, considering all ecosystem services and looking beyond the short-term economics, is important for soil C commodification (Mandal, 2011). These are the major challenges which act as obstacles in earning C credits and thereby limiting the role of agricultural practices in reducing GHGs emissions.

For developing a broad and legitimate trading market, a robust, transparent, and accurate, yet inexpensive method should be developed for verifying the net amount of C sequestered. But accurate SOC measurement is always expensive and time-taking. SOC determination through wet oxidation and titration results in toxic dichromate waste and sometimes incomplete oxidation can lead to underestimation of C stocks. Accurate measurement can be done by elemental (C) analyzer but the equipment is costly and small amount of sample used in the analysis questions the representativity of the analysis. Separate measurement of inorganic carbon, if present must also be done. Loss-on-ignition method is not costly but has less accuracy; however, it is capable of measuring large number of samples, thereby increasing the representativity of the samples. Recently, infrared spectroscopy (IRS) is being used by the researchers as it is fast, inexpensive, non-destructive, with little or no pre-treatment of the samples. However, care should be taken for calibrations with traditional methods while using IRS.

Impact of Government Initiatives to Achieve Carbon Neutrality

Pradhan Mantri Krishi Sinchayee Yojana (PMKSY)

Government of India has so far initiated so many schemes for the benefit of farming and farmers. One of them is the PMKSY. The major objectives of PMKSY are increasing investments in the field of irrigation for expanding area under cultivation of assured irrigation at the field level; improving on-farm water use efficiency by efficient utilization of water and reduced the wastage of water; enhancing the area under precision agriculture and adoption of high-tech/ improved irrigation technologies; promotion of other water saving technologies such as drip (More crop per drop), recharging of ground water through adoption of different water conservation technologies and recharge of tube well and reusing municipal waste water after treated for peri-urban agriculture; and encouraging private investment in the field of precision irrigation system. The scheme will aid to conserve natural resources and bring more area under cultivation as more efficient utilization of water that will enhance greenery, biomass and SOC storage which ultimately help to achieve carbon neutrality.

National Food Security Mission (NFSM)

This centrally sponsored scheme for improved agricultural extension services, technology transfer and decentralized planning was launched in 2007-08 for sustainable increase in the crop production of rice, wheat and pulses. From 2016-17, NFSM is being implemented in 638 districts of 29 states; NFSM-Rice in 194 districts of 25 states, NFSM-Wheat in 126 districts of 11 states, NFSM-Pulses n 638 districts of 29 states and NFSM-Coarse cereals in 269 districts of 27 states. The interventions covered under NFSM are cluster demonstrations on improved package of practices, demonstrations on cropping systems, seed distribution of HYVs, farm machineries/resources conservation machineries/tools, efficient water application tools, plant protection measures, nutrients management/ soil ameliorants, cropping system based training of farmers etc. that not only help to achieve food security but also enhance soil health and SOC storage.

Paramparagat Krishi Vikas Yojana (PKVY)

The PKVY is a sub-component of Soil Health Management (SHM) scheme under National Mission of Sustainable Agriculture (NMSA) with the aim to develop sustainable models of organic farming through the mixture of traditional wisdom with modern science and technology to conserve natural resources, to ensure long term soil fertility build up, and to help in climate change adaptation and mitigation. It primarily focuses on increasing soil fertility and production of healthy food through organic practices without the use of agro-chemicals. This scheme will help substantially in promotion of integrated and sustainable farming systems through natural resource conservation, and thereby reduce the cost of production through sustainable integrated organic farming. Ultimately PKVY will promote eco-friendly low cost farmer friendly technologies for protecting the environment and produce chemical free and nutritious food for human consumption. All these practices will pave the pathway of green India through achieving carbon neutrality.

Research needed

1. It is essential to calculate critical and saturation level of SOC for estimation of SOC saturation deficit under different agro-climatic conditions and soil types to identify key spots for enhancing C sequestration.
2. Genotypes and cultivars which produce greater shoot and root biomass and help to allocate more C in deeper soil layers should be targeted.
3. Carbon sequestration varies with soil type, climate, vegetation, crop production and management factors, thus, it is important to identify most important site-specific key factors governing C sequestration.
4. Zero tillage has well known technology for achieving carbon neutrality but further research is needed to test its potential under different climate and soil types through long term experiments.
5. Biochar have potential to offset ~50% C emission from fossil fuel, at the same time to enhance soil fertility, crop yield and SOC stock. Research to be carried out to assess the impact of biochar prepared from different crop residues, which are abundant with Indian field crops.
6. It is necessary to develop agroecosystem-based soil, water and land management strategies to address the "Soil health problems" for agricultural development because production constrains and soil properties vary with ecosystem.

Policy reforms

1. Provision of financial incentive or subsidies to the farmers for adopting different SLM practices which enhance carbon sequestration.
2. It is necessary to implement intensive plantation programme for the restoration of degraded and deforested land which will not only help to achieve land degradation neutrality by improving SOC stock, but also improve soil quality, its stability and enhance food security.
3. Need to create a large-scale awareness programme against crop residue burning and promotion of alternate crop residue management practices such as in-situ crop residue management (direct incorporation into soils and mulching), composting, production of biochar and uses of crop residue in industrial applications for sustainable agriculture.
4. Urgent need for promotion and adaptation of different soil and water conservation technologies in hilly regions to reduce the soil loss through erosion and enhance carbon sequestration.
5. Need to implement integrated nutrient management and integrated farming system technologies for efficient utilization of farm resources and inputs along with soil health card scheme.

6. Promotion of resource conservation technologies such as zero tillage, use of crop residues, cover crops, green leaf manure, FYM and compost, crop rotation, crop diversification, site specific nutrient management, municipal waste management, rain water harvesting, contour farming, strip farming, precision farming, etc. are needed for sensitive ecosystems on priority basis.
7. Need to establish "Green Research Fund" at national level for the development of advance land management technologies and strengthening research on adoption, mitigation and impact assessment of climate change.
8. Need to take multilevel initiative for build-up of soil organic matter through promotion of organic farming by involving government bodies, extension staff, NGOs, organic fertilizer producers and farmers groups.
9. Need to establish climate smart villages for climate change adaptation and mitigation by implementing various climate resilient carbon positive technologies for managing soil resources.
10. Development of a system to measure change in soil organic carbon content after adaptation of different management practices by farmers and provide carbon credits to the farmers that can be sold in the market.
11. Since balanced fertilization is most important in this regard soil testing services of the country is required to be strengthened and more effective to ensure such fertilization.
12. Considering the enormous problem of deterioration of soil health throughout the country, serious attempts need be made urgently to use every source of C so that there is a C build up in soil. But at present we don't have any national soil C management strategies.
13. Capacity development of the farmers for CSA activities should be encouraged. Availability of farm machineries for CA should be enhanced through establishing custom hiring centres and FPOs in villages.
14. Need to undertake people centered watershed development programmes on priority basis in most vulnerable soil erosion prone areas for restoration of degraded land.
15. Need to convert barren, degraded, non-arable waste land and areas adjacent to road and railway line with planting of herbs, shrubs and trees under "Green India" programme with PPP mode.
16. Provision of incentive to prepare a stream of "National Green Corps (NGC)" by involving students of schools and colleges, like NCC/NSS for "Green India" movement and also the students may be incentivized for increasing the participation in this movement.

17. Need to protect degradation of our valuable natural resources by strict enforcement of regulatory measures for safe disposal of solid and liquid wastes by public-private investment support for low-cost, low-tech, user-friendly methods.
18. Need to promote land use planning based crops, cropping systems, cropping sequences and the application of related production technologies (soil conservation measures, balanced fertilization and best available irrigation management) to the respective potential area/zone for improving the soil quality, raise the productivity and profitability and arrest the land degradation.

Conclusion

In the 21st Century, to produce sufficient food for the increasing population of the country without any harmful effect on natural ecosystem, proper implementation of the above-mentioned innovations and practices are very much essential. For this purpose, awareness generation among the farmers is necessary. The National Agricultural Research System (NARES) with vast network in the country is well prepared and very much capable to achieve the carbon neutrality by successful implementation of those policies in the years to come. These will also help the country to cope up with the future climate change scenarios.

References

Abrol, I.P., Gupta, R.K. and Malik, R.K. 2005. Conservation agriculture- status and prospects. Centre for Advancement of Sustainable Agriculture, New Delhi, pp 1–242

Anantha, K.C., Majumder, S.P, Badole, S, Padhan, D, Datta, A., Mandal, B. and Sreenivas, C. H. 2020. Pools of organic carbon in soils under a long-term rice–rice system with different organic amendments in hot, sub-humid India. Carbon Management, 11(4): 331-339.

Anantha, K.C., Majumder. S.P, Padhan. D, Badole. S, Datta. A, Mandal. B. and Reddy. G.K. 2018. Carbon Dynamics, Potential and Cost of c sequestration in double rice. Journal of Soil Science and Plant Nutrition, 18 (2): 418-434.

Anas, I., Rupela. O.P, Thiyagarajan. T.M and Uphoff. N. 2011. A review of studies on SRI effects on beneficial organisms in rice soil rhizospheres. Paddy and Water, 9: 53-64.

AON. 2020. Weather, Climate & Catastrophe Insight 2019 Annual Report. Chicago, Illinois.

Arora, N.K. 2019. Impact of climate change on agriculture production and its sustainable solutions. Environmental Sustainability, 2: 95–96.

Arun, K.C. and Pramod, K.A. 2017. Adapting Agriculture to Changing Climate in South Asia. World Agriculture, Article # 1712, July 2017. www.world-agriculture.net.

Banger, K., Tian. H, Tao. B, Lu. C, Ren. W. and Yang. J. 2015. Magnitude, spatiotemporal patterns, and controls for soil organic carbon stocks in India during 1901–2010. Soil Science Society of America Journal, 79: 864–875.

Bhattacharyya, R., Das. T.K, Sudhishri. S, Dudwal. B, Sharma. A.R, Bhatia. A. and Singh. G. 2015. Conservation agriculture effects on soil organic carbon accumulation and crop productivity under a rice–wheat cropping system in the western Indo-Gangetic Plains. European Journal of Agronomy, 70: 11-21.

Bhattacharyya, T., Pal. D.K, Chandran. P, Ray. S.K, Mandal. C. and Telpande. B. 2008. Soil carbon storage capacity as a tool to prioritize areas for carbon sequestration. Current Science, 95: 482–484.

Bhattacharyya, T. 2020. Land degradation neutrality and land use options. Indian Journal of Fertilizers, 16: 830-841.

CCAFS, 2015. Scaling up Climate-Smart Villages in India. Available at: https://cgspace.cgiar.org/bitstream/handle/10568/67902/04outcomecase.pdf?sequence=3

Ceesay, M., Reid. W.S, Fernandes. E.C.M. and Uphoff. N. 2006. The effects of repeated soil wetting and drying on lowland rice yield with System of Rice Intensification (SRI) methods. International Journal of Agricultural Sustainability, 4: 5-14.

Chambers, A., Lal. R. and Paustian. K. 2016. Soil carbon sequestration potential of US croplandsand grasslands: implementing the 4 per thousand initiative. Journal of Soil and Water Conservation, 71: 68A–74A.

Chandel, S., Datta, A., Yadav, R.K., Dheri, G.S. 2020. Does saline water irrigation influence soil carbon pools and nutrient distribution in soil under seed spices? Journal of Soil Science and Plant Nutrition. (In Press)

Choi, J., Kim. G, Park. W, Shin. M, Choi. Y, Lee. S, Kim. S. and Yun. D. 2014. Effect of SRI water management on water quality and greenhouse gas emissions in Korea. Irrigation and drainage, 63: 263-270.Chu. G., Wang. Z, Zhang. H, Liu. L, Yang. J. and Zhang. J. 2015. Alternate wetting and moderate drying increases rice yield and reduces methane emission in paddy field with wheat straw residue incorporation. Food and Energy Security, 4(3): 238-254.

Cowie, A.L., Orr. B.J, Sanchez Castillo. V.M, Chasek. P, Crossman. N.D, Erlewein. A, Louwagie. G, Maron. M, Metternicht. G.I, Minelli. S, Tengberg. A.E, Walter. S. and Welton. S. 2018. Land in balance: The scientific conceptual framework for Land Degradation Neutrality. Environment Science Policy, 79: 25–35.

CSSRI Vision 2050. ICAR-Central Soil Salinity Research Institute, Karnal, Haryana.

Das, T.K., Bhattacharyya. R, Sharma. A.R, Das. S, Saad. A.A. and Pathak. H. 2013. Impacts of conservation agriculture on total soil organic carbon retention potential under an irrigated agro-ecosystem of the western Indo-Gangetic Plains. European Journal of Agronomy, 51: 34-42.

Datta, A., Basak, N, Chaudhari, S.K. and Sharma, D.K. 2015b. Effect of horticultural land uses on soil properties and organic carbon distribution in a reclaimed sodic soil. Journal of the Indian Society of Soil Science, 63 (3): 294-303.

Datta, A., Basak. N, Chaudhari. S.K. and Sharma. D.K. 2015a. Soil properties and organic carbon distribution under different land uses in reclaimed sodic soils of North-West India. Geoderma Regional, 4: 134–146.

Datta, A., Mandal. B, Badole. S, Chaitanya. A.K, Majumder. S.P, Padhan. D, Basak. N, Barman. A, Kundu. R. and Narkhede. W. 2018a. Interrelationship of biomass yield, carbon input aggregation, carbon pools and its sequestration in Vertisols under long-term sorghum-wheat cropping system in semi-arid tropics. Soil Tillage Research, 184:164-175.

Datta, A., Mandal. B, Basak. N, Badole. S, Chaitanya. K, Majumder. S.P, Thakur. N.P, Kumar. P. and Kachroo. D. 2018b. Soil carbon pools under long-term rice-wheat cropping system in Inceptisols of Indian Himalayas. Archives of Agronomy and Soil Science, 4(9): 1315-1320.

Datta, A., Setia. R, Barman. A, Guo. Y. and Basak. N. 2019. Carbon Dynamics in Salt affected soils. In: Research Developments in Saline Agriculture. J.C. Dagar, R.K. Yadav & P.C. Sharma, (eds.) pp. 369-389. Springer Singapore. ISBN 978-981-13-5832-6

Dey, A., Dwivedi. B.S, Bhattacharyya. R, Datta. S.P, Meena. M.C. Jat, R.K. Gupta, R.K. Jat M.L. Singh, V.K. Das, D. and Singh. R.G. 2020. Effect of conservation agriculture on soil organic and inorganic carbon sequestration and lability: A study from a rice–wheat cropping system on a calcareous soil of the eastern Indo-Gangetic Plains. Soil Use and Management, 36(3): 429-438.

Dill. J., Deichert. G. and Le Thi Nguyet Thu, 2013. Promoting the system of rice intensification: Lessons learned from Trà Vinh Province, Viet Nam. Hanoi.

Dixon, R.K., Brown. S, Houghton. R.A, Solomon, A.M, Trexler, M.C. and Wisniewski. J. 1994. Carbon pools and fluxes of global forest ecosystems. Science, 263:185-190.

Eliasch, J. 2008. Climate change: Financing global forests – the Eliasch Review. London Earthscan/James & James.

FAO. 2006. Global Forest Resources Assessment 2005 – Progress towards sustainable forest management.FAO Forestry Paper No. 147, Rome.

FAO. 2014. CA Adoption Worldwide, FAO-CA website. http://www.fao.org/ag/ca/6c.html

FAO. 2016. Save and grow: Maize, rice and wheat– A guide to sustainable crop production. UN Food and Agriculture Organization, Rome. p. 44–47.

FAO. 2016. The state of food and agriculture, climate change, agriculture and food security. Food and Agriculture Organization of the United Nations, Rome.

Garg, V.K. 1998. Interaction of tree crops with a sodic soil environment: potential for rehabilitation of degraded environments. Land Degradation and Development, 9: 81–93.

Gathorne-Hardy, A., 2013. A life cycle assessment (LCA) of greenhouse gas emissions from SRI and flooded rice production in SE India.

Ghosh, S., Wilson. B.R, Ghosal. S.K, Senapati. N. and Mandal. B. 2012. Organic amendments influence soil quality and carbon sequestrationin the Indo-gangetic plains of India. Agriculture, Ecosystems and Environment, 156: 134-141.

Gitz, V. 2013. Usage des terres et politiques climatiquesglobales: la physique, l'économie et les politiques del'usage des puits de carbone pour luttercontre le changementclimatique. Presses Académiques Francophones.Saarbrücken, Germany.

Gopalakrishnan S, Kumar. R.M, Humayun. P, Srinivas. V, Ratna. Kumari. B, Vijayabharathi. R, Singh. A, Surekha. K, Padmavathi. C, Somashekar. N, Raghuveer. Rao. P, Latha. P.C, Subba. Rao. L.V, Babu. V.R, Viraktamath. B.C, Vinod. Goud. V, Loganandhan. N, Gujja. B. and Rupela. O. 2013. Assessment of different methods of rice (Oryza sativa. L) cultivation affecting growth parameters, soil chemical, biological and microbiological properties, water saving and grain yield in rice–rice system. Paddy and Water Environment, 12: 79-87.

Hirte, J., Leifeld. J, Abiven. S., Oberholzer. H.R. and Mayer. J. 2018. Below ground carbon inputs to soil via root biomass and rhizodeposition of field-grown maize and wheat at harvest are independent of net primary productivity. Agriculture Ecosystem Environment, 265: 556-566.

ICAR. 2010. "Degraded and Wastelands of India – Status and Spatial Distribution", Indian Council of Agricultural Research, New Delhi.

IPBES. 2018. Summary for policymakers of the assessment report on land degradation and restoration of the Intergovernmental Science- Policy Platform on Biodiversity and Ecosystem Services, Secretariat of the Intergovernmental Science-Policy Platform on Biodiversity and Ecosystem Services. Bonn, Germany.

IPCC. 2000. Land use, land-use change, and forestry. A special report of the Intergovernmental Panel on Climate Change (IPCC), ed. Watson, R.T.,I.R. Noble, B. Bolin, N.H. Ravindranath, D.J. Verardo and D.J. Dokken. Cambridge, UK, Cambridge University Press.

IPCC. 2007b. Climate Change 2007: mitigation, B. Metz, O.R. Davidson, P.R. Bosch, R. Dave & L.A. Meyer, eds. Contribution of Working Group III to the Fourth Assessment Report of the IPCC. Cambridge, United Kingdom and New York, USA, Cambridge University Press.

IPCC. 2014. Summary for policymakers. *In* Edenhofer, O.,R. Pichs-Madruga, Y. Sokona, E. Farahani, S. Kadner, K. Seyboth, A. Adler, I. Baum, S. Brunner, P. Eickemeier, B. Kriemann, J. Savolainen, S. Schlömer, C. von Stechow, T. Zwickeland J.C. Minx, eds. Climate change 2014: Mitigation of climate change. Contribution of Working Group III to the Fifth Assessment Report of the Intergovernmental Panel on Climate Change. Cambridge, United Kingdom & New York, USA, Cambridge University Press.

Jain. N., Dubey. R, Dubey. D.S, Singh. J, Khanna. M, Pathak. H. and Bhatia. A. 2014. Mitigation of greenhouse gas emission with system of rice intensification in the Indo-Gangetic Plains. Paddy and Water Environment, 12: 355-363.

Jat, H.S., Datta. A, Choudhary. M, Sharma. P.C, Yadav. A.K, Choudhary. V, Gathala. M.K, Jat. M.L. and McDonald. A., 2019. Climate Smart Agriculture practices improve soil organic carbon pools, biological properties and crop productivity in cereal-based systems of North-West India. Catena, 181: 104059.

Jat, H.S., Datta. A, Sharma. P.C, Kumar. V, Yadav. A.K, Choudhary. M, Choudhary. V, Gathala. M.K, Sharma. D.K, Jat. M.L. and Yaduvanshi. N.P.S. 2018. Assessing soil properties and nutrient availability under conservation agriculture practices in a reclaimed sodic soil in cereal-based systems of North-West India. Archives of Agronomy and Soil Science, 64(4): 531-545.

Jayadeva, H. M., Prabhakara. Setty. T. K, Gowda. R. C, Devendra. R, Mallikarjun. G. B. and Bandi, A. G. 2009. Methane emission as influenced by different crop establishment techniques and organic manures. Agricultural Science Digest, 29: 241-245.

Kaur, B., Gupta. S.R. and Singh. G. 2002. Bioamelioration of a sodic soil by silvopastoralsystems in northwestern India. Agroforestry Systems, 54:13–20

Kerry, R., Goovaerts. P, Rawlins. B.G. and Marchant. B.P. 2012. Disaggregation of legacy soildata using area to point kriging for mapping soil organic carbon at the regionalscale. Geoderma. 170: 347–358.

Kim. G.Y., Lee. S.B, Lee. J.S, Choi. E.J. and Ryu. J.H. 2012. Mitigation of greenhouse gases by water management of SRI (System of Rice Intensification) in rice paddy fields. Korean Journal of Soil Science and Fertilizer, 45: 1173-1178.

Ladha, J.K., Khind. C.S, Khera. T.S. and Bueno. C.S. 2004. Effects of residue decomposition on productivity and soil fertility in rice–wheat rotation. Soil Science Society of America Journal, 68(3): 854-864.

Lal, R. 2004. Soil carbon sequestration impacts on global climate change and food security. Science. 304:1623–1627.

Lal, R. 2005. Forest soils and carbon sequestration. Forest Ecology and Management, 220: 242-258.

Lal, R. 2007. Farming Carbon. Soil and Tillage Research, 96: 1-5.

Lal, R. 2016. Beyond COP 21: Potential and Challenges of the "4 per Thousand" initiative. Journal of Soil and Water Conservation. 71: 20A–25A.

Lal, R. 2018. Digging Deeper: A Holistic Perspective of Factors Affecting SOC Sequestration. Global Change Biology, 28: doi:10.1111/gcb.14054.

Le Quéré, C., Moriarty. R, Andrew. R.M, Peters. G.P, Ciais. P, Friedlingstein. P, Jones. S.D, Sitch. S, Tans. P, Arneth. A, Boden. T.A, Bopp. L, Bozec. Y, Canadell. J.G, Chini. L.P, Chevallier. F, Cosca. C.E, Harris. I, Hoppema. M, Houghton. R.A, House. J.I, Jain. A.K, Johannessen. T, Kato. E, Keeling. R.F, Kitidis. V, Klein Goldewijk. K, Koven.

C, Landa. C.S, Landschützer. P, Lenton. A, Lima. I.D, Marland. G, Mathis. J.T, Metzl. N, Nojiri. Y., Olsen. A, Ono. T, Peng. S, Peters. W, Pfeil. B, Poulter. B, Raupach. B, Regnier. P, Rödenbeck. C, Saito. S, Salisbury. J.E, Schuster. U, Schwinger. J, Séférian. R, Segschneider. J, Steinhoff. T, Stocker. B.D, Sutton. A.J, Takahashi. T, Tilbrook. B, Van Der Werf. G.R, Viovy. N, Wang. Y.P, Wanninkhof. R, Wiltshire. A. and Zeng. N. 2015. Global carbon budget 2014. Earth System Science Data, 7: 47–85.

Maji, A.K., Obi Reddy. G.P. and Sarkar. D. 2010. Degraded and Wastelands of India Status and Spatial Distribution. Indian Council of Agricultural Research, New Delhi.

Majumder, B., Mandal. B, Bandyopadhyay. P.K, Gangopadhyay. A, Mani. P.K, Kundu. A.L. and Mazumdar. D. 2008. Organic amendments influence soil organic carbon pools and rice–wheat productivity. Soil Science Society of America Journal, 72: 775–785.

Mandal, B. 2011. Soil Organic Carbon Research in India-A way forward. Journal of the Indian Society of Soil Science, 59: S9-S22

Mandal, B., B.Majumder, P.K.Bandyopadhyay, G.C. Hazra, A.Gangopadhyay, R.N. Samantaray, A.K.Mishra, J. Chaudhury, M.N.Saha, and S.Kundu. 2007. The potentialof cropping systems and soil amendments for carbon sequestration in soils underlong-term experiments in subtropical India. Global Change Biology, 13: 357–369.

Mandal, B., Majumder. B, Adhya. T.K, Bandyopadhyay. P.K, Gangopadhyay. A, Sarkar. D, Kundu. M.C, Choudhury. S.G, Hazra. G.C, Kundu. S, Samantaray. R.N. and Misra. A.K. 2008. Potential of double-cropped rice ecology to conserve organic carbon under subtropicalclimate. Global Change Biology, 14: 2139–2151.

Mandal, B., Majumder. B, Bandyopadhyay. P.K, Hazra. G.C, Gangopadhyay. A, Samantaray. R.N, Mishra. A.K, Chaudhury. J, Saha. M.N. and Kundu. S. 2007. The potential of cropping systems and soil amendments for carbon sequestration in soils under long-term experiments in subtropical India. Global Change Biology, 13: 357–369.

Mandal, U.K., Sharma. K.L, Burman. D, Mandal. S. and Maji. B. 2020. Conservation Agriculture and C Sequestration in Tropical Regions. In Carbon Management in Tropical and Sub-Tropical Terrestrial Systems (pp. 175-199). Springer, Singapore.

Minasny, B., Malone. B.P, McBratney. A.B, Angers. D.A, Arrouays. D, Chambers. A, Chaplot. V, Chen. Z.S, Cheng. K, Das. B.S, Field. D.J, Gimona. A, Hedley. C.B, Hong. S.Y, Mandal. B, Marchant. B.P, Martin. M, McConkey. B.G, Mulder. V.L, O'Rourke. S, Richer-de-Forges. A.C, Odeh. I, Padarian. J, Paustian. K, Pan. G, Poggio. L, Savin, I, Stolbovoy, V, Stockmann. U, Sulaeman. Y, Tsui. C.C, Vågen. T.G, Van Wesemael. B. and Winowiecki. L. 2017. Soil carbon 4 per mille. Geoderma. 292: 59–86.

Ministry of Environment, Forest and Climate Change. 2018. Economics of Desertification, Land Degradation and Drought in India, Vol I: Macroeconomic assessment of the costs of land degradation in India. New Delhi.

Mishra. A. and Salokhe. V.M. 2010. The effects of planting pattern and water regime on root morphology, physiology and grain yield of rice. Journal of Agronomy and Crop Science, 196: 368-378.

MoEF (Ministry of Environment & Forests, India). 2011. National Mission for a Green India (GIM) (available at www.envfor.nic.in/major-initiatives/national-mission-green-india-gim).

Nabuurs, G.J., Masera. O, Andrasko. K, Benitez-Ponce. P, Boer. R, Dutschke. M, Elsiddig. E, Ford-Robertson. J, Frumhoff. P, Karjalainen. T, Krankina. O, Kurz. W.A, Matsumoto. M, Oyhantcabal. W, Ravindranath. N.H, Sanz Sanchez, M.J. and Zhang. X. 2007. Forestry. In Metz, B.,O.R. Davidson, P.R. Bosch, R. Dave and L.A. Meyer, eds. Climate change 2007: Mitigation. Contribution of Working Group III to the Fourth Assessment Report of the

Intergovernmental Panel on Climate Change, pp. 541–584. Cambridge, UK & New York, USA, Cambridge University Press.

Oo, A.Z., Sudo. S, Inubushi. K, Mano. M, Yamamoto. A, Ono. K, Osawa, T, Hayashida. S, Patra, P.K, Terao.Y, Elaykumar. P, Vanitha. K, Umamageswari. C, Jothimani. P. and Ravi. V. 2018. Methane and nitrous oxide emissions from conventional and modified rice cultivation systems in South India. Agriculture, Ecosystems and Environment, 252: 148-158.

Parihar, C.M., Jat. S.L, Singh. A.K, Datta. A, Parihar. M.D, Varghese. E, Bandyopadhyay. K.K, Nayak. H.S, Kuri. B.R. and Jat. M.L. 2018. Changes in carbon pools and biological activities of a sandy loam soil under medium-term conservation agriculture and diversified cropping systems. European Journal of Soil Science, 69(5): 902-912.

Parihar, C.M., Singh. A.K, Jat. S.L, Dey. A, Nayak. H.S, Mandal. B.N, Saharawat. Y.S, Jat. M.L. and Yadav. O.P. 2020. Soil quality and carbon sequestration under conservation agriculture with balanced nutrition in intensive cereal-based system. Soil and Tillage Research, 202: 104653.

Pathak, H., K. Byjesh, B. Chakrabarti. and P.K. Aggarwal. 2011. Potential and cost of carbon sequestrationin Indian agriculture: estimates from long-term field experiments. Field Crop Research, 120: 102–111.

Rai, M., S.S. Acharya, S.M. Virmani. And P.K. Aggarwal (Eds.). 2009. State of Indian Agriculture.NAAS Publ., New Delhi.

Rajkishore, S.K. 2013a. Carbon sequestration and greenhouse gas emission studies in SRI and conventional systems of rice cultivation. Ph.D Dissertation, Tamil Nadu Agricultural University, India.

Rajkishore, S.K., Doraisamy. P, Subramanian. K.S. and Maheswari. M. 2013b. Methane emission pattern in SRI and Conventional systems of rice cultivation. Taiwan Journal water conservancy, 64(4): 126-134.

Samal, S.K., Rao. K.K, Poonia. S.P, Kumar, R., Mishra, J.S., Prakash, V., Mondal, S., Dwivedi, S.K., Bhatt, B.P., Naik, S.K. and Choubey, A.K. 2017. Evaluation of long-term conservation agriculture and crop intensification in rice-wheat rotation of Indo-Gangetic Plains of South Asia: Carbon dynamics and productivity. European Journal of Agronomy, 90: 198-208.

Satyanarayana, A., Thiyagarajan. T. M. and Uphoff, N. 2007. Opportunities for water saving with higher yield from the system of rice intensification. Irrigation Science, 25: 99-115.

Scharlemann, J.P., E.V. Tanner, R. Hiederer. And V. Kapos. 2014. Global soil carbon: understanding and managing the largest terrestrial carbon pool. Cardiology Management, 5: 81–91.

Sharma, P.C. and Singh, A. 2019. Reviving the Productivity of Salt-affected Lands: Technological Options, Constraints and Research Needs. In: Research Developments in Saline Agriculture. J.C. Dagar, R.K. Yadav & P.C. Sharma, (eds.) pp. 591-628. Springer Singapore. ISBN 978-981-13-5832-6

Shrivastava, K.S. 2014. Rs. 13,000 crore to be spent on green India mission in next 5 years. Down to Earth, 24 February (available at www.downtoearth.org.in/news/rs-13000-crore-to-be- spent-on-green-india-mission-in-next-5-years-43600).

Singh, G.B. and Gill, H.S. 1990. Raising trees in alkali soils. Wasteland News, **6**:15–18

Somasundaram, J., Chaudhary. R.S, Awanish Kumar. D, Biswas. A.K, Sinha. N.K, Mohanty. M, Hati. K.M, Jha. P, Sankar. M, Patra. A.K. and Dalal, R. 2018. Effect of contrasting tillage and cropping systems on soil aggregation, carbon pools and aggregate-associated carbon in rainfed Vertisols. European Journal of Soil Science, 69(5): 879-891.

Sreenivas, K., Dadhwal. V.K, Kumar. S, Harsha. G.S, Mitran. T, Sujatha. G, Suresh. G.J.R, Fyzee. M.A. and Ravisankar. T. 2016. Digital mapping of soil organic and inorganic carbonstatus in India. Geoderma. 269: 160–173.

Srinivasarao, C. 2018. Soil carbon sequestration: research, technology and policy implementation needs in India. Journal of the Indian Society of Soil Science. 66: S26-S43.

Srinivasarao, C., Lal. R, Kundu. S, Babu. M.B.B.P, B. Venkateswarlu. and Singh. A.K. 2014. Soil carbon sequestration in rainfed production system in the semiarid tropics of India. Science of the Total Environment, 487: 587-603.

Suryavanshi. P., Singh. Y.V, Prasanna. R, Bhatia. A, and Shivay. Y.S. 2013. Pattern of methane emission and water productivity under different methods of rice crop establishment. Paddy Water Environment, 11: 321–329.

Thakur. A.K. and Uphoff, N.T. 2017. How the System of Rice Intensification Can Contribute to Climate-Smart Agriculture. Agronomy Journal, 109: 1163-1182.

Thakur. A.K., Uphoff. N. and Antony. E. 2010. An assessment of physiological effects of System of Rice Intensification (SRI) compared with recommended practices in India. Experimental Agriculture, 46: 77-98.

Veloso, M.G., Angers. D.A, Chantigny. M.H. and Bayer. C. 2020. Carbon accumulation and aggregation are mediated by fungi in a subtropical soil under conservation agriculture. Geoderma, 363: 114159.

Yadav, G.S., Das. A, Lal. R, Babu. S, Datta. M, Meena. R.S, Patil, S.B. and Singh. R. 2019. Impact of no-till and mulching on soil carbon sequestration under rice (*Oryza sativa* L.)-rapeseed (*Brassica campestris* L. var. rapeseed) cropping system in hilly agro-ecosystem of the Eastern Himalayas, India. Agriculture, Ecosystems and Environment, 275: 81-92.

Yadav, G.S., Lal. R, Meena. R.S, Babu. S, Das. A, Bhowmik. S.N, Datta. M, Layak. J. and Saha. P. 2019. Conservation tillage and nutrient management effects on productivity and soil carbon sequestration under double cropping of rice in north eastern region of India. Ecological Indicators, 105: 303-315.

12

Innovative Solutions to Reduce Greenhouse Gas Emissions

P. Bhattacharyya[1], S. Neogi[2], S.R. Padhy[1], P.K. Dash[1]

[1]Crop Production Division, ICAR-National Rice Research Institute Cuttack, Odisha
[2]Ashoka Trust for Research in Ecology and the Environment (ATREE) Bangalore, Karnataka

Introduction

Greenhouse gases (GHGs) absorb long-wave radiation emitted from the Sun and the Earth's surface, and radiate it back to Earth's atmosphere, creating the greenhouse effect. Gases like carbon dioxide (CO_2), methane (CH_4), nitrous oxide (N_2O) occur in traces but play crucial roles in controlling the temperature of Earth's atmosphere. The natural as well as anthropogenic activities have major impacts on the GHGs emissions that include continuously increasing concentration of CO_2, CH_4 and N_2O in the atmosphere. The heat-trapping properties of these above mentioned GHGs are well established. Different GHGs have different warming potential (radiative forcing) on the global climate system due to their different radiative properties and lifetimes in the atmosphere. Changes in the atmospheric concentrations of GHGs modify the energy equilibrium of the climate system which leads to subsequent climate change. They affect the absorption, scattering and emission of radiation within the atmosphere and at the earth's surface. Globally agriculture is estimated to account for 10%–20% of anthropogenic GHG emissions (Smith and Olesen 2010). Share of contribution of these gases towards global warming are different such as CH_4 20-25%, CO_2 40-50% and N_2O 5-10%; with present pace of increment every year 0.41, 0.42 and 0.25%, respectively IPCC, 2019.

At present, the concentrations of CO_2, CH_4 and N_2O gases in the atmosphere reached to 409 ppm, 1862 ppb and 329 ppb with net increases of approximately 143%, 254% and 122%, respectively, since the pre-industrial era (NOAA-ERSL 2019). Although emission from agriculture reduced from 6.9 to 5.1 Gt CO_2-eq (a metric measure to convert emissions from different GHGs to the equivalent amount of CO_2 for comparison) between 1990 and 2017, it is still contributing to one-fifth of the total CO_2-eq emissions globally (FAOSTAT, 2020). Thus contribution of agriculture to global warming and climate change continues. Consequences include warming the atmosphere and ocean, rise in mean sea level, melting of glaciers, severe heat waves, weather aberrations including frequent floods and droughts, and impacts on biodiversity and human health. The CO_2 is the most important GHG. It is a component of the global C-cycle, but burning of fossil fuels, deforestation and other emissions through various human activities add to its concentration in the atmosphere. As a result, nearly 1.7 ppmv (parts per million by volume) of CO_2 is added each year, globally (IPCC, 2007). The CO_2 in the atmosphere is utilised by plants for photosynthesis and is released through plant respiration, and organic matter decomposition. Fundamental activities contributing to atmospheric N_2O emission are coal burning, biomass and fertilizer use in agriculture, inappropriate waste disposal methods, industries, degradation of tropical forest and other land use changes. Total N_2O emission is recorded at $15.3 \pm 6.7 \times 10^{12}$ g N yr^{-1}. It is presently expanding at the pace of 0.7 ppbv (parts per billion by volume) $year^{-1}$ (IPCC, 2007, 2013). Livestock, flooded rice fields wetlands, and burning of biomass add to the atmospheric CH_4. In continuously flooded rice field, roots of rice plants trap CH_4 from rhizosphere and transport it through the plant vascular system to the atmosphere. Soil contributes about 20% to the total emission of CO_2 through soil and root respiration, 12% of CH_4 and 60% of anthropogenic N_2O emissions (IPCC, 2007). Therefore, soil is the largest source of CO_2 fluxes to the atmosphere. However, soil also acts as a sink of if effectively managed (Bhattacharyya et al., 2014).

Global rice production faces two contrasting challenges: providing food for rapidly growing population while also reducing the GHGs emissions. The South East Asia, where 90% of world's paddy is produced, contributes 10-20% of the total anthropogenic CH_4 emission and 40–60% of its agricultural emissions (UNFCC, 2019). Flooded rice cultivation is the primary source of atmospheric CH_4 which has 28 times more global warming potential than the CO_2 over a 100-year period. Globally the CH_4 emission rate from paddy fields was estimated to be 10% (21-30 Tg yr^{-1}) of total global CH_4 flux (Bhattacharyya et al., 2019). The N_2O emission from agriculture was approximately 4.1 Tg N-N_2O yr^{-1} (Bhattacharyya et al., 2019), which could be attributed mostly to N_2O originated from soil through synthetic fertilizer application, and to a smaller extent from crop residues and organic manures (Davidson, 2009).

It is apparent that methodologies to decrease the emission must concentrate on reducing their production, enhancing their utilization and decreasing their transport through plants. Agriculture has contributed to global climate change and received feedbacks from the changes. There have been considerable impacts on crops, soils, livestock and pests. Although these effects have been documented largely, several gaps remain in our knowledge on what progressions these changes will have for rural livelihoods, especially among the most vulnerable people.

Concept and approaches of Green India

The National Mission for a Green India (Green India Mission) is one of the eight national missions of India's action plans to address the challenges of climate change (MoEF, 2010). It was launched in 2014 in order to protect, restore and enhance India's depleting forest cover and responding to climate change through a combination of adaptation and mitigation measures. In addition to the improvement of the quality 'green' (forest) cover and its ecosystems services, the 'Mission' also aims at augmenting forest and tree cover on marginal agricultural lands/fallows and other non-forest areas under agroforestry/social forestry (3 m ha) (Venkataramani, 2015). The areas of working on the mission revolve around the climate adaptation/mitigation, and food, water, environmental and livelihood security of tribal and forest dwellers (MoEF&CC, 2015). Hence, through the lenses of climate-smart agriculture (CSA)/climate-smart farming (CSF), a climate resilient and sustainable agriculture system may be achieved for the nation in line with 'Green India' approach.

Principles of mitigation of GHGs emissions

To execute effective mitigation options for decreasing GHGs emissions from agriculture, one should initially know the mechanism of production, transformation, emission as well as consumption of important GHGs, which are essential in agriculture. Net GHG emission is the balance between total production and the consumption. So, a detailed understanding of the mechanisms and factors responsible for GHGs production and emission processes is essentially the first step to chalk out the mitigation strategies for different crop production systems grown under varied agroecologies and management practices.

Mechanisms and principles of CO_2 production

Two significant influencing C parity of a terrestrial ecosystem are photosynthesis of vegetation and soil respiration. Net ecosystem exchange (NEE) of CO_2 between the biosphere and the atmosphere lies in the harmony between fluxes associated with photosynthetic assimilation by the vegetation (Gross primary production, GPP) and respiratory effluxes (ecosystem respiration, RE) from autotrophs (roots) and heterotrophs (microbial and soil fauna). Soil respiration is the important pathway of C efflux from ecosystem. This includes root and microbial respiration, and a mass turnover of organic matter (OM) to release of CO_2. Agricultural practices have impact on soil pH, temperature, moisture, aeration, and the C:N ratio of substrates, which influence the CO_2 fluxes. Soil microbes and their activities are affected, thereby modifying the rate of decomposition of plant-derived C to form (SOM) and subsequently CO_2.

Mechanisms and principles of N_2O production

Alternate wetting and drying (AWD) cycles make the soil a unique system wherein both aerobic and anaerobic nitrogen (N) metabolisms take place. Physicochemical properties of the rhizospheric soil affects N transformation and thereby emission of N_2O. Nitrous oxide released from the soil to the atmosphere is a by-product of both nitrification and denitrification, which are essentially microbial-mediated processes. Nitrification is the main source of N_2O under aerobic condition, whereas denitrification [reduction of nitrate (NO_3-) to NO, N_2O and N_2], dominates under flooded or anaerobic condition. The production of N_2O depends also on the presence of water-soluble and readily metabolized OM available in the rhizosphere. In general, addition of organic materials in soils containing NO_3^- increases N_2O production and subsequent emission if condition favoured. Salinity also influences the N_2O emission through its effect on both nitrification and denitrification processes.

Mechanisms and factors in production and emission of CH_4

The CH_4 production and its consumption are biologically controlled in submerged/ anaerobic rice paddy system. Methane is released to the atmosphere by processes like ebullition, diffusion across water-air interface and transport through aerenchyma. In undisturbed paddy fields, up to 90% of CH_4 is emitted through aerenchyma. Flooded paddy soils are conducive to the production and emission of CH_4 due to the presence of methanogenic bacteria that utilize readily decomposable OM under anaerobic soil condition. Both CH_4 production and emission from flooded rice soils are strongly influenced by a number of soil processes including changes in soil redox status and pH, substrate dynamics and nutrient availability, and textural makeup. Methane escapes to the atmosphere predominantly through aerenchyma of rice plants, which allows the vertical gas transfer including the passage from root to culm (jointed stem of rice, hollow at internodes, and solid at nodes). Nutrient supply affects development of aerenchyma as well as root exudation and thus the budget of CH_4.

Net emission of CH_4 from a production system is a balance between its production (methanogenesis) and consumption (methanotrophy). The net emission would be positive or negative, depends on the relative magnitudes of these processes. Methanogens (anaerobic microbes associated with highly anaerobic condition like flooded rice soils) use simple C compounds (in particular $H_2 + CO_2$ and acetate) for methanogenic fermentation. About two thirds of CH_4 produced in nature comes through the reduction of the methyl group of acetate and the rest from reduction of CO_2 with electrons from H_2 or formate. Very less amount of $CH_4$is produced by the oxidative and reductive dismutation of methanol or methylamines associated with marine sediments.

Adaptation cum mitigation options for CH_4, CO_2 and N_2O emissions

Agricultural systems have major impacts on GHGs budget. Crops use atmospheric CO_2 in photosynthesis and emit CO_2 through the process of respiration. Lowland flooded fields (rice) are important source of CH_4, while upland fields (all other crops including upland non-flooded rice crop) enriched with nitrogenous fertilizers emits mostly N_2O and CO_2. The source and sink capacities of agriculture system depends on the management practices. The net emissions of CO_2, CH_4 and N_2O from agricultural fields to atmosphere are regulated by various factors *viz.* their production, transport processes and utilization through crop varieties, soil types, fertilizer management and other agricultural practices etc. In rice cultivation, a trade-off exists between N_2O and CH_4 emissions relying on the field water management. A balanced approach is, therefore, required involving mitigation options which favours yield sustainability and economic benefit of farmers, and is simultaneously eco-accommodating.

Different adaptation cum mitigation options for reducing CH_4, CO_2 and N_2O emissions from rice fields are briefly given below:

(i) ***Judicious water management:*** Water management practices in flooded (rice) soil should not allow the redox potential Eh to drop below -200 millivolts (mV), so that CH_4 production is reduced drastically. Alternate wetting and drying in the rice field between maximum tillering to panicle initiation period maintains the soil Eh, and therefore this could be one of the effective practices in mitigating CH_4 emissions from rice fields.

(ii) ***Introduction of rice cultivars with low CH_4 emitting potential:*** Rice varieties should be developed/introduced that have lower CH_4 emission potential. This plays an important role in reducing the amount of CH_4 emission from lowland ecologies.

(iii) ***Balanced use of fertilizers, manure and amendments:*** In *rainfed* lowland ecology, deep placement of urea super granules decreased both CH_4 and N_2O emissions compared to urea or prilled urea broadcasting. Sulphate has an inhibitory effect on methanogens and therefore, application of major nutrients through sulphate containing N fertilizers (i.e. ammonium sulphate instead of urea) reduces CH_4 emissions from soils. Saha et al. (2017) reported that use of ammonium sulphate reduced 6.8 mg $m^{-2} d^{-1}$ CH_4 as compared to other treatments in a sandy clay loam paddy soil. The supply of potassium (K) through K_2SO_4 also decreased CH_4 emission. Adequate K fertilization is crucial in maintaining oxygenated zone around rice roots, inducing iron oxidation, and maintaining favourable pH in the rhizosphere. The addition of K could induce higher oxidizing conditions in the rhizosphere, affecting adversely the methanogenic environment, and inhibiting CH_4 production (Babu et al., 2006; Zhang et al., 2019) and its subsequent release into the atmosphere through rice plants. For example, K applied as muriate of potash significantly reduced CH_4 emission from a submerged Inceptisol by 49% compared to control (no K fertilizer used) planted to rice. Cumulative seasonal emission was 63.81 kg CH_4 ha^{-1} where 30 kg K ha^{-1} was applied compared to control (125.34 kg CH_4 ha^{-1}). Potassium prevented the drop in redox potential and reduced the contents of active reducing substances and Fe^{2+} content in soil along with inhibited the methanogenesis and stimulated methanotrophic bacterial activities (Babu et al., 2006).

Soil amendment like phosphogypsum, which is rich in sulphate could reduce CH_4 emissions (Ali et al., 2012; Bhattacharyya et al., 2020). Biochar enhances C-sequestration in the soil in the long-run, which is a 'negative feedback of gaseous C emission and it also improves the physical properties of soil. Basic slag, a by-product of steel industry having 20-35% of silicate and free iron and manganese oxides act as electron acceptors (compete with CO_2 and acetate), which reduce CH_4 production in the soil (Ali et al., 2008, 2015).

(iv) ***Real time N management**:* Real time N management through leaf colour chart (LCC), Soil Plant Analysis Development (SPAD) meter, other sensor-based applications reduce N_2O emission from agriculture. In this connection, drone-based monitoring through sensors could be the future.

(v) ***Nitrification inhibitor**:* Organic neem-based nitrification inhibitors (like Nimin, an alcohol extract of neem plant) have been shown to inhibit N_2O emission from flooded rice soils. The use of botanical as such or as formulation of chemical inhibitors (like Dicynamide) has huge potential to mitigate N_2O emission. It was evident that application of dicynamide @ 20 kg ha^{-1} reduced total N_2O emissions by 65% as compared to application of cow urine at a rate of 700 kg N ha^{-1} (Treweek et al., 2016). These need to be promoted through pilot projects or policy framework.

(vi) ***Innovative cropping practices:*** Direct seeded rice (DSR), with paired-row-rice-*dhanicha* system, introduction of short-duration pulses in rice-rice or rice-fallow system with efficient micro-irrigation practices are a few important climate resilient technologies to mitigate GHGs emissions and climate change adaptation.

Innovations on mitigating GHGs emissions in agriculture

Technological interventions

Ministry of Agriculture & Farmers Welfare, Government of India and Indian Council of Agricultural Research (ICAR) have been promoting various innovative technologies and policy framework for ensuring the food security of the country (GoI, 2018). Convergence of agriculture, digital and engineering technologies with modern biotechnological tools is reshaping Indian agriculture to accommodate in changing times. Sustainable increase in production and productivity with enhanced inputs and resource use efficiency is necessary. India's estimated population would touch 1.7 billion by 2050 (DAC&FW, 2018; GoI, 2019). Ensuring food and nutritional security for such a large population would be a great challenge. On top of that, the land, water and biodiversity resources are under stress. This can only be accomplished through the use of innovative technologies in agriculture. Efforts are needed to generate, transfer and adoption of green-technologies in the coming 2-3 decades. Encouraging smart farming with greater technological support to reduce post-harvest losses, strengthening the value-chain, and providing markets to increase farmers' income could be the vital interventions (WGMI, 2019). Introduction and adoption of cutting-edge technologies like biosensors, genomics, biotechnology, nano-technology and alternative energy sources in agriculture will be of great values. Biotechnology, encompassing wide range of technologies such as cloning, marker-assisted selection, transgenics, gene editing and veritable 'omics' (an umbrella of technologies dealing with cellular molecules) should be used judiciously for augmenting tolerance and resilience of plant materials to biotic and abiotic stresses, improving nutritional quality of the produce, enhancing physiological and nutrient use efficiencies of crops. Recent developments in genome editing have far reaching implications for future agricultural and food systems (Ganguli et al., 2017; Singh, 2019). Central and State Governments are engaged in developing science-based policies and institutions to harness the best of science, technology and product quickly reach the farmers and other users along the value chain.

Science and technology are the drivers of change in farm operations and output. New technologies to help enhancing productivity of land and water are needed for overcoming the prevailing technology fatigue in agriculture sector. There is a need for synergistic blending (amalgamation) of frontier technologies like information and communication, biotechnology and nanotechnology for explaining opportunities for achieving an ever-green revolution capable of improving productivity in perpetuity without ecological harm. Alternative site-specific eco-friendly cropping systems with higher and stable profit should be identified for diverse agroecological regions of India. In this regard, the integrated farming systems (IFS), internalizing synergies of different components for enhanced resource utilization, income and livelihood generation and minimizing environmental loading that need to be developed and validated in different agroecologies. The eco-friendly site-specific IFS would provide more diversified food basket and better food and nutritional security to the people living therein.

The specific GHGs mitigation technologies that could be up-scaled and taken into policy levels:

(i) The DSR in medium and upland with aerobic rice varieties and AWD water management.

(ii) Conservation agriculture (CA) in belt having light textured soil with Block wise custom hiring centres for CA equipment, weather and climate advisory services.

(iii) Double transplanting in rice, with mid-season drainage and LCC based N management in lowland ecologies.

(iv) Large scale promotion of rainwater harvesting (RWH) both structure for RWH and recycling with drainage. The watershed approach with incentives could be useful.

(v) Community nursery, seed and fodder bank should be promoted. In Indian perspective adaptation measure having mitigation potential should be promoted first.

(vi) We should give more focus on demand-side mitigation than supply-side. However, building of series of food-storage structure and prevention of wastage of crop-harvest could reduce GHGs emission up to 30%. In this aspect policy intervention and large-scale Government and Non-Governmental finance is necessary.

Climate-smart agriculture and climate-smart villages

The Landscape Approach which aims to optimize the production and resource use in combination at a larger scale (landscape) without affecting the environment is the key to climate-smart agriculture (CSA). The World Bank defines CSA as an integrated approach for managing cropland, livestock, forests and fisheries – components of a landscape and all are connected to address the challenges of food security and climate change (World Bank, 2019). It seeks to achieve three major outcomes namely, (a) increased and sustained productivity and income, (b) building resilience (adaptation), and (c) reduced GHGs emissions (mitigation) (FAO, 2013; 2020). In the arena of CSA, the 'Climate Smart Villages (CSVs)'are sites where farmers, researchers, local government and private sectors come together to understand and execute the CSA practices best suited for a particular location (CCAFS-CIMMYT, 2014). A bag of interventions to increase farmers' incomes through higher productivity have been identified, while building their resilience to extreme and variable climatic events. The interventions aim to reduce GHGs emissions and thus ensure that resources are used sustainably. The concept of CSVs must be introduced in large scale for a successful 'Green India Mission'.

The CSVs have successfully bridged global knowledge with local action to help farmers to sustainably produce more food, while curbing GHGs emissions and increasing resilience to climate change. In India, farmers in Haryana have started implementing climate-smart practices like laser land levelling and alternate AWD in rice for reducing water use, improving soil health and bringing economic benefits. The laser land levelling technology in rice under rice-wheat systems of north-west India reduced irrigation time by 47-69 h ha^{-1} season^{-1} and enhanced the yield by 7% compared with conventional fields. In wheat, irrigation time was reduced by 10-12 h ha^{-1} season^{-1} and yield enhanced by 7-9 %. Farmers gain an additional INR 10,045 ha-1 year-1 (approximately) through increased yields and reduced electricity used in laser levelled fields compared to conventional one when the electricity tariff was considered. Therefore, if around 50% of crop-area under the rice-wheat system in Punjab and Haryana were laser levelled, this could provide an additional production of 0.7 million tonnes of rice and 0.9 million tonnes of wheat, amounting to INR 26,950 million, annually (Aryal et al., 2015). In those CSVs, the farmers have also been receiving agro-advisories on their mobile phones, with inputs from meteorological departments, research-scientists and agri-input dealers that allow them to make timely decisions (Taneja et al. 2014; Aryal et al., 2015). Moving towards CSA through the CSV project is one way to insulate farmers from the impact of climate change.

Projects run by The Consultative Group on International Agricultural Research (CGIAR) Research Program on Climate Change, Agriculture and Food Security (CCAFS) and The International Maize and Wheat Improvement Centre (CIMMYT) have witnessed farmers implementing CSA practices viz. laser land levelling zero-tillage, residue management, direct dry-seeded rice, AWD of rice, precision nutrient management decision-support tools (Nutrient Expert) and sensors (Green Seeker), agroforestry, crop diversification and climate information services among others. When used in combination, resource saving practices could save water energy, make soil healthy and economically viable too.

The ICAR-NICRA project initiative: A successful case study

The ICAR-NICRA project offers opportunity to work with climate change and GHGs emission mitigation technologies at farmers' fields. Under that exhaustive project on-farm participatory demonstrations for climate resilience technologies are being implemented through Krishi Vigyan Kendras (KVKs) (36 in number) and seven core institutes of ICAR at village clusters in 121 climatically vulnerable districts across the country. Studies on developing a framework for climate resilient villages (CRVs) and quantifying the resilience was done along with huge capacity building works. A number of climate resilient practices were taken up in 1572 demonstrations covering 1360 ha (2014-15). The resilient technologies primarily focused were rainwater harvesting and recycling in rainfed areas; improved planting methods (918 demonstrations); zero till cultivation in *rabi* crops (594 demonstrations); raised and sunken bed planting in N-E region for cultivation of vegetable and legumes. The contingency crop plans for delayed planting (after mid-July) with suitable crop (sesame, castor, black gram, pigeon pea, sunflower, foxtail millet, pearl millet, horse gram, cluster bean, toria and basmati rice) and soil moisture practices were implemented. Apart from that short duration and stress tolerant cultivars and resilient intercropping systems were demonstrated to

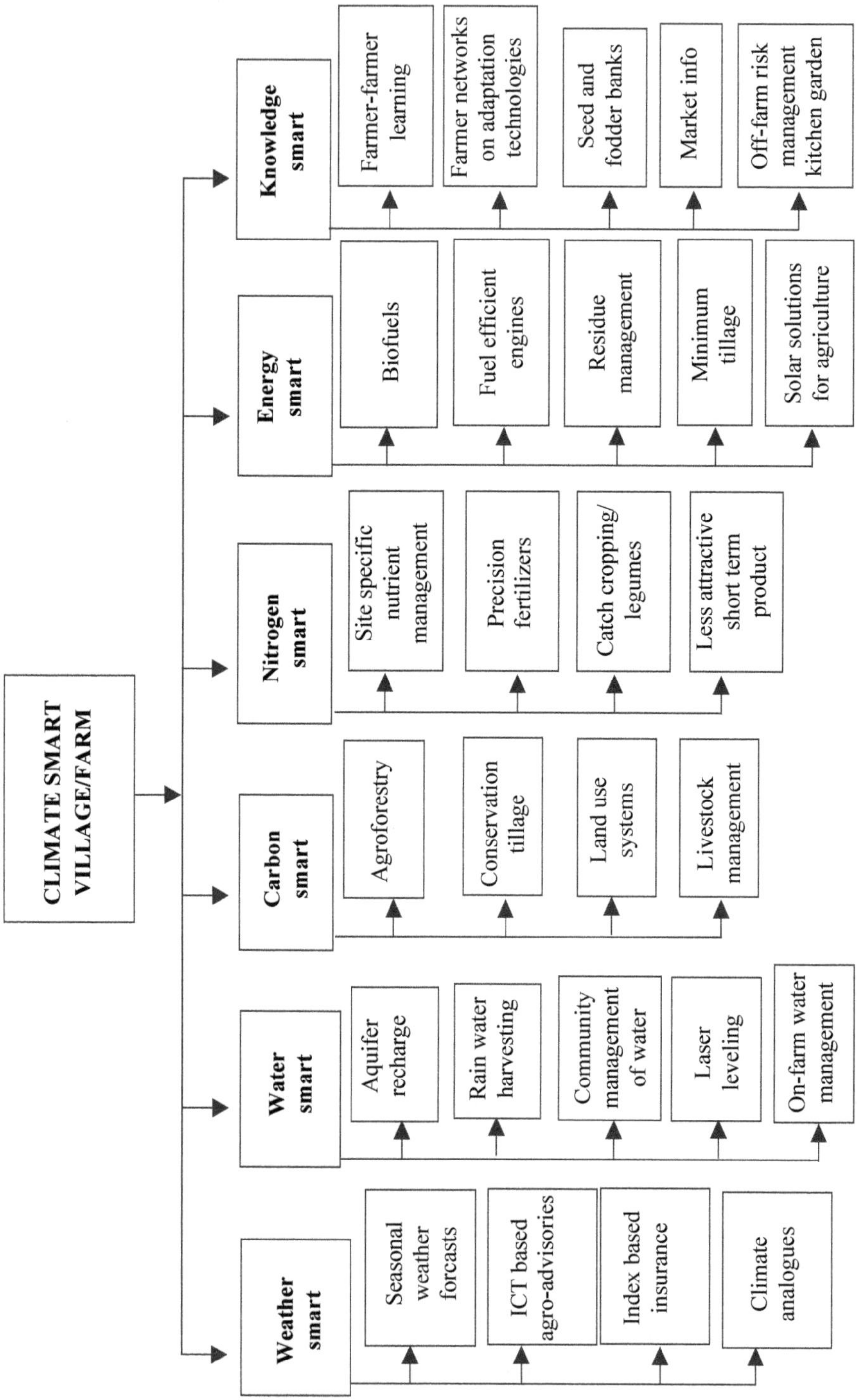

Fig. 1. Varying approaches adopted in a climate-smart village (*Source*: CCAFS-CIMMYT, 2014.)

stabilizing productivity in vulnerable districts (3000 demonstration in sole crop systems). 'Village Level Climate Risk Management Committees (VCRMC)' managed the 'Custom Hiring Centres' for farm machineries. At policy level the 27 proven climate resilient practices and technologies were identified for up-scaling in ICAR-DAC interface workshop on NICRA-National Mission on Sustainable Agriculture (NMSA).

Forecasting and agro-advisory services by using digital science and Apps

Rural India is fast progressing towards digitalization and technology usage. 'Digital India' programme was launched in 2015 'with a vision to transform India into a digitally empowered society and knowledge economy' (https://digitalindia.gov.in). The aim of this programme is to promote digital literacy and digital infrastructure for the rural and farming community. 'Digital Agriculture' powered with user-friendly Apps for the farming community (AgriApp, Iffco Kishan App, Kisan Yojna etc.) are excellent path to guide and educate the farmers. Through the interactive Apps, particularly in local language, farmers could solve their farming problems related to fertilization, sowing, harvesting, pest (diseases and insects) control conveniently on real-time basis. These tools should be promoted in 'Green India initiative' (GoI, 2019; Ayaz et al., 2019).

The Govt. of India has been promoting 'farm-to-fork' movement over decades to make the farming more profitable, and enhancing self-resilience to the farming communities. Major focus was to build up a large-start-up-ecosystem at each district, block and village by 2024 (MANAGE, 2017). In that direction, launching of 'Pradhan Mantri Kisan Samman Nidhi (PM-KISAN)' scheme in 2019 by the Govt. of India was a big step. It is to be noted that the Govt. of India's 'Kisan Suvidha App' and 'Pusa Krishi Mobile App' have large downloads since their launch in 2016. The 'Kisan Suvidha App' disseminates information on market prices, seeds, pesticides, weather, agricultural machinery, fertilizers and crop insurance. 'Pusa Krishi Mobile App' developed by the 'ICAR-Indian Agricultural Research Institute', provides information both in English and Hindi, on the latest technologies to tackle farming problems (DAC&FW, 2018; Singh, 2019). Another mobile application App, 'Meghdoot', a joint initiative of India Meteorological Department (IMD), Indian Institute of Tropical Meteorology (IITM) and ICAR, launched in 2019, disseminates information on weather and weather-based advisories to crop and livestock in ten local languages. 'Meghdoot' was initially introduced for 150 districts; presently it covers >600 districts. It is integrated with 'Facebook' and 'WhatsApp' also for further coverage and is also available on app store and Google Play Store. The plan is to extend it further on state and country scales with phases. Meghdoot provides maps, images and pictures for better understanding by farmers (Ganguly et al., 2017; Gopalakrishnan and Subramanian, 2020). This App should be utilized to its full extent in the 'Green

India' mission. A few other user-friendly Apps include 'IFFCO Kisan Agriculture', 'RML Farmer- Krishi Mitr', 'Kheti-Badi', 'Agri App', 'Agri-Market' and 'Crop Insurance' (MANAGE, 2017; DAC&FW, 2018). It is required to extend the Apps in local languages as much as possible.

It becomes increasingly clear that the 'Digital-Agriculture' is going to play the key role in shaping the Indian agriculture in near future and contributing in 'Green India'. Few points need considerations:

(i) Real-time weather forecasting is essential (using Drone or ground-based sensors) particularly for growing seasons (*kharif/rabi*). An up-gradation and value-added advisory should be provided with weather forecasting (Figures 2, 3).

(ii) Decision support system (DSS) (climate/weather-forecasting based) on GIS-Remote sensing platform need to be strengthened at the district level initially and then at block level. The DSS must include fertilizer use schedule, irrigation and pest management.

(iii) Mapping of vulnerable disaster-prone-blocks and biotic-stress-hotspots in GIS-Remote sensing platform need to be promoted. Interlinking of automatic weather stations with data analysis and dissemination throughout the country is the prerequisite for 'Green India' and sustaining food security in the country.

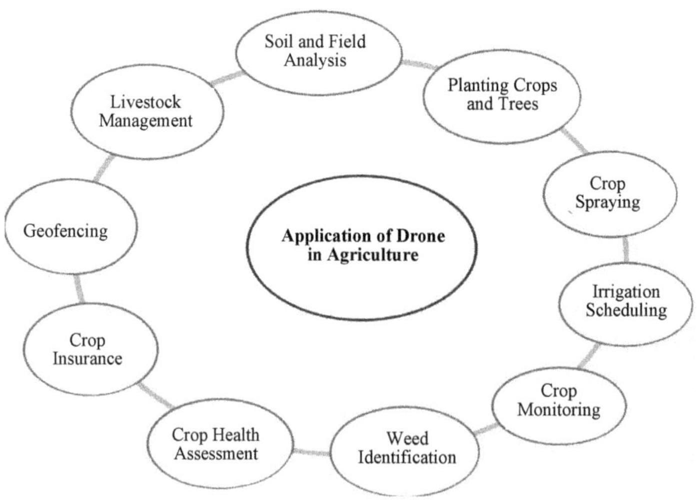

Fig.2. Applications of Drones in agriculture (*Source*: Adapted from Rani et al., 2019).

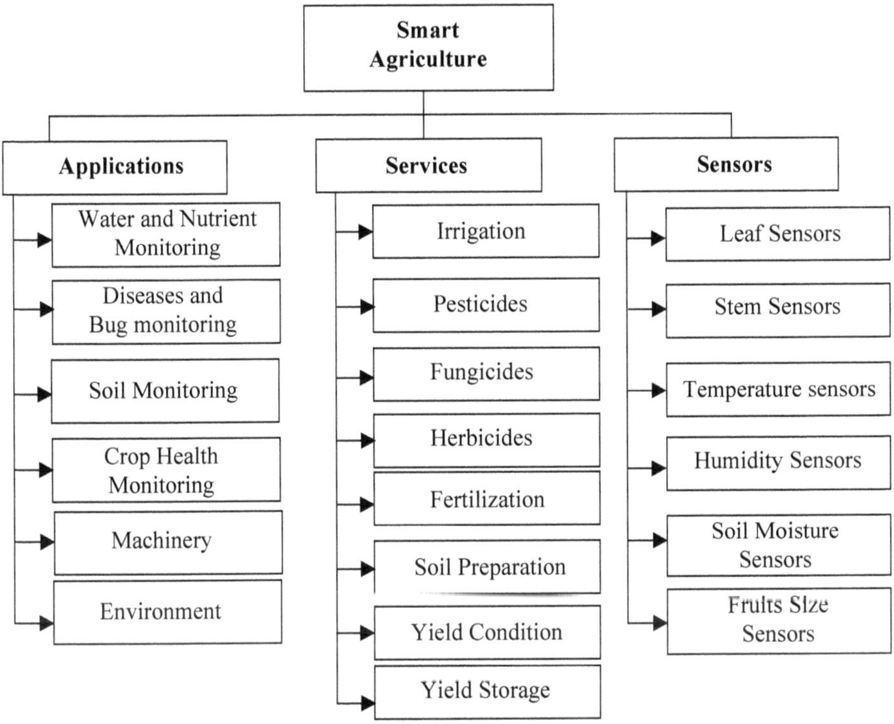

Fig. 3. Applications, services and sensors for smart agriculture (*Source*: adapted from Ayaz et al., 2019)

Institutional innovation through private-public-NGO partnership crop insurance, contract farming and custom hiring

Indian agriculture has also been facing several climatic and anthropogenic challenges with unseen opportunities. Challenges in terms of land degradation and climate change also provides opportunities to redefine the production system and post-harvest strategies (GoI, 2019). At the recent past, huge potentiality has been seen on application of biotechnology and information and communication technology (ICT) in agriculture for achieving higher efficiencies (Singh, 2019). Institution innovation, specifically refers to transparent 'Public-Private-Partnerships' with minimum transaction costs, co-ordinating and enforcing relations in production system both in the sector of goods and services (GoI, 2018) (Figure 4). A policy approach for promoting economic and social development with the flexibility of mid-term correction must be facilitated both in private and public sector with long-term perspective. These should be in synergy with the mutual gains of both the partners with transparent credit sharing and competitiveness. In public benefits angle, positive environmental, social and economic benefits for livelihoods of beneficiaries also should be looked into. The process of public-private-partnership has been found to increase in recent past (Ganguly et al., 2017). A major step in this direction is the recently formalized comprehensive guidelines for 'Intellectual Property Management and Commercialization of Technologies', enabling stronger partnerships. Discussions on these partnerships might play greater roles in having insights on varied pertinent aspects *viz.* seed quality improvement, on-farm testing of new technologies, commercialization of technologies, soil testing laboratories, value-chain in processed food, smaller machinery, custom-hiring services, contract farming, crop insurance, use of fertilizers and micronutrients and extension systems etc. (OECD, 2018a,b). The attempts covering the interface indicating various dimensions of the partnerships hold immense possibilities. Linking of extension education and research for transfer of technology through IT and approach is the key of 'Green India Mission'. It should include dissemination evaluation, monitoring and mid-term correction for keeping face with ever changing climate scenario.

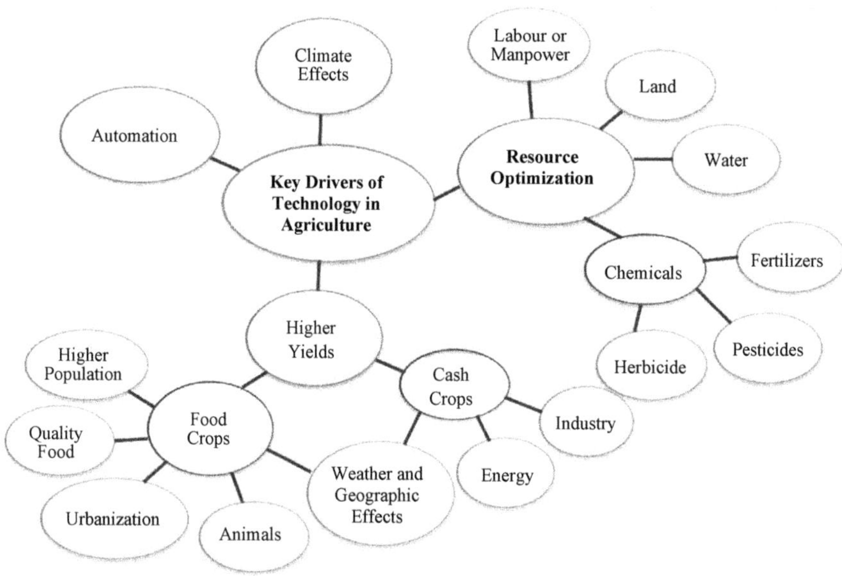

Fig. 4. Agriculture industry-key drivers (*Source*: adapted from Ayaz et al., 2019)

Policy innovation for Green India

The climate action plan/ policy must be incorporated into core National policies.

The climate action and implementation of GHGs mitigation technologies is mostly long term in nature and fetch environmental intangible benefit rather than immediate short-term economic benefit, so financing is must. Here, top-down approach for financing is required. But execution and implementation should follow approach with proper transparent regulation. Here, Ministry of Agriculture and Farmers' Welfare could play a potential role along with state government and NGOs (Ganguly et al., 2017; OECD, 2018a).

Several agricultural mitigation options are having both co-benefits and trade-offs. Therefore, following issues in terms of mitigation might need attention in future course: (a) wide scale deployment of CSF and efficient use of agricultural resources; and (b) judiciously growing more food on the land available for agriculture meanwhile increasing adaptation, resiliency and mitigation in agriculture.

Climate change threatens food security gains. Food security, mitigation and adaptation goals are in priority and going to be more focussed in future climate change scenario, however these are reinforcing each other. Gains one activity generate co-benefits for another. Therefore, CSA approach could be followed with national and international funding with site specific action plan. The CSA should be a part of policy goals.

The "cooperative approaches" and 'carbon credit compliance' could be two handy approaches for 'Green India Mission' (Kritee et al., 2014, 2015, 2018, 2019; IPCC, 2019).

Research Innovation

An advance agronomy involving economic benefits, soil health maintenance and efficient energy use should be the research priority to sustain food security in view of shrinking resource base. Development and adaptation of improved package of practices for sustainable increase in crop yield through technology-weaned (e.g. agriculture with minimum of fossil fuels, crop growth tracking, nutrition-labelling, partnerships between stakeholders-growers, suppliers, and retailer buyers) are crucial (Thea et al., 2017). Use of sustainable IT-based sensors and communication technologies is necessary for effective utilization of farm land. Research project should integrate IT technologies with crop-agronomy and natural resource management. Use of existing IT may not be enough, and the IT-based interventions should be customized at the local level based on farmers' demand and supply chain perspectives. A variety of equipment employed for monitoring and obtaining real-time information on crops need to be explored. In this regard the range of sensors and equipment *viz.* unmanned aerial vehicles (UAVs), robots communication devices, and other heavy machinery must be promoted (Tzounis et al., 2017).

The emergence of farm technologies in India, in conjunction with robust ICT framework, is still evolving and holds tremendous aspirational promises and overwhelming potential to positively (progressively) impact agricultural performance and enhance farmers' income. The adoption of modern technologies in agriculture might help in significantly enhancing risk and damage assessments, resulting into revolutionizing the way meant for responding to disasters affecting the farm productivity and livelihoods of vulnerable farming community and country's food security. Supporting research and development endeavours in agriculture can effectively contribute to creating knowledge and preparing for the future. Technology-driven agriculture might be instrumental in developing new mechanism and institutions to ease access and affordability of technology adoption and popularization amongst farmers.

Training, capacity and skill building of youth in agriculture and start-ups

Skill development for rural youth is an important scheme of the Govt. of India (Bhattacharyya and Mukherjee, 2019). The Ministry of Agriculture and Farmers Welfare, Govt. of India, in line with the National Policy on Skill Development and Entrepreneurship, has taken steps for promoting the skill development. The Skill Training of Rural Youth (STRY) programme under National Mission on Agricultural Extension and Technology is aimed to rural youth for imparting knowledge and hands-on training on agri-based areas and allied schemes to promote employment in village areas and empowering the youth. The National Institute of Agriculture Extension Management (MANAGE) and Ministry of Agriculture and Farmers Welfare coordinate these activities at the national level through State Agriculture Agencies. The training activities are primarily implemented through State Agricultural Management & Extension Training Institutes (SAMETIs) at the state level and coordinated through Agriculture Technology Management Agency (ATMA) at the district level. The training to rural youths is imparted through KVKs, Nehru Yuva Kendras (NYKs), Farmer Training Centers (FTCs) and other identified training institutions at state/district level (Singh, 2019).

Skill development of farmers is a must to enhance productivity and profitability in Indian agriculture sector. It is the need of the hour for agriculture to become more attractive and enterprising. Skills need to be essentially developed amongst the farmers on varied aspects of farming practices in order to gradual replacing and phasing out time (energy) and cost intensive (consuming) methods by more efficient, scientific, environmentally as well as economically viable practices. Skilled agricultural work force might convert weaknesses and threats lying in agriculture into strengths and opportunities. It is anticipated that through proper skill development program Indian agriculture might be transformed towards a more productive, remunerative and sustainable one.

Conclusion

It is pertinent from the afore-mentioned discussion that the climate smart agriculture through lens of climate resilience approach under Green India Mission might potentially aim at better accounting for GHGs emissions being generated from Indian agriculture to understand how best these might be managed while improving farm productivity and farmers' livelihoods, considering the national commitments under the Paris Agreement. Farmers need to essentially adopt CSF because of direct economic benefits that help promote the well-being of their families and communities, meanwhile improving resource use efficiency leading towards climate mitigation and agroecosystem sustainability.

Acknowledgment

Authors acknowledge the support of ICAR-NRRI, ICAR-National Fellow Project; NICRA

References

Ali, M., Farouque, M., Haque, M., ul Kabir, A.J., 2012. Influence of soil amendments on mitigating methane emissions and sustaining rice productivity in paddy soil ecosystems of Bangladesh. J. Environ. Sci. Nat. Res. 5: 179-185.

Ali, M.A., Kim, P., Inubushi, K., 2015. Mitigating yield-scaled greenhouse gas emissions through combined application of soil amendments: a comparative study between temperate and subtropical rice paddy soils. Sci. Total Environ. 529: 140-148.

Ali, M.A., Oh, J.H., Kim, P.J., 2008. Evaluation of silicate iron slag amendment on reducing methane emission from flood water rice farming. Agric. Ecosyst. Environ.128: 21-26.

Aryal, JP, Bhatia M, Jat ML, Sidhu HS. 2015. Impacts of Laser Land Leveling in Rice-Wheat Systems of the North-western Indo-Gangetic Plains of India. Food Security 7: 3.

Ayaz, Md., Ammad-uddin, Md., Sharif, Z., Mansour, A., Aggounc, E.H.M., 2019. Internet-of-Things (IoT) based Smart Agriculture: Towards Making the Fields Talk. IEEE Access. DOI 10.1109/ACCESS.2019.2932609.

Babu, Y.J., Nayak, D.R. and Adhya, T.K., 2006. Potassium application reduces methane emission from a flooded field planted to rice. Biology and fertility of soils, 42(6), pp.532-541.

Bhattacharyya, P., Dash, P.K., Padhy, S.R., Roy, K.S., Neogi, S. and Nayak, A.K., 2020. Soil Carbon Dynamics in Different Land-Use and Management Systems in Tropical Coastal Regions of India. In Carbon Management in Tropical and Sub-Tropical Terrestrial Systems (pp. 89-102). Springer, Singapore.

Bhattacharyya, P., Dash, P.K., Swain, C.K., Padhy, S.R., Roy, K.S., Neogi, S., Berliner, J., Adak, T., Pokhare, S.S., Baig, M.J. and Mohapatra, T., 2019. Mechanism of plant mediated methane emission in tropical lowland rice. Science of The Total Environment, 651: 84-92.

Bhattacharyya, Pratap, Neogi, Suvadip, Singha Roy, Koushik, Dash, Pradeep Kumar, Nayak, A.K. and Mohapatra, T., 2014. Tropical low land rice ecosystem is a net carbon sink. Agriculture, Ecosystems & Environment. 189. 127–135. 10.1016/j.agee.2014.03.013.

Bhattacharyya, S. and Mukherjee, A., 2019. Importance of Skill Development in Indian Agriculture. In: ICT and Social Media for Skill Development in Agriculture. (Editors) Om Prakash, Anirban Mukherjee and Pratibha Joshi, Today & Tomorrow's Printers and Publishers, New Delhi - 110 002, India, pp. 47-62.

CCAFS-CIMMYT, 2014. Climate-Smart Villages in Haryana, India. Copenhagen, Denmark: CGIAR Research Program on Climate Change, Agriculture and Food Security (CCAFS). Available at: https://hdl.handle.net/10568/42346.

DAC&FW, 2018. A farmer friendly handbook for schemes and programmes 2018-19. Department of Agriculture, Cooperation & Farmers Welfare (DAC&FW), Ministry of Agriculture and Farmers Welfare, Government of India, New Delhi, pp. 106.

Davidson, E.A., 2009. The contribution of manure and fertilizer nitrogen to atmospheric nitrous oxide since 1860. Nature Geoscience, 2(9), pp.659-662.

FAO, 2013. Climate-smart agriculture sourcebook. CCAFS: CGIAR Research Program on Climate Change, Agriculture and Food Security. Available at: https://ccafs.cgiar.org/publications/climate-smart-agriculture-sourcebook#.WaQfGT5JbIU

FAO, 2020. Food and Agriculture Organization of the United Nations. http://www.fao.org/climate-smart-agriculture/overview/en/

FAOSTAT, 2020. http://www.fao.org/economic/ess/environment/data/emission-shares/en/

Ganguly, K., Gulati, A., Braun, V.J. 2017. Innovations spearheading the next transformations in India's agriculture. ZEF Working Paper Series, ISSN 1864-6638. Center for Development Research, University of Bonn, Editors: Christian Borgemeister, Joachim von Braun, Manfred Denich, Till Stellmacher and Eva Youkhana, Working Paper 158, pp. 62.

Gopalakrishnan, T. and Subramanian, S., 2020. Agrometeorological Advisory Services in India: An Assessment, Centre for Science and Environment, New Delhi, pp. 60.

Government of India, 2018. Report of the Committee on Doubling Farmers' Income (Chaired by Ashok Dalwai). 14 Volumes, Department of Agriculture, Cooperation and Farmers' Welfare, Ministry of Agriculture & Farmers' Welfare, 201 pp.

Government of India, 2019. Report on Policies and Action Plan for a Secure and Sustainable Agriculture. Agriculture (Chaired by R.S. Paroda). Submitted to The Princiapal Scientific Adviser to the Government of India, 198 pp.

https://digitalindia.gov.in

IPCC, 2007: Climate Change 2007: Synthesis Report. Contribution of Working Groups I, II and III to the Fourth Assessment Report of the Intergovernmental Panel on Climate Change [Core Writing Team, Pachauri, R.K and Reisinger, A. (eds.)]. IPCC, Geneva, Switzerland, 104 pp.

IPCC, 2013. Summary for policymakers. Climate Change 2013: The Physical Science Basis. Contribution of Working Group I to the Fifth Assessment Report of the Intergovernmental Panel on Climate Change [Stocker, T.F., D. Qin, G.-K. Plattner, M. Tignor, S.K. Allen, J. Boschung, A. Nauels, Y. Xia, V. Bex and P.M. Midgley (eds.)]. Cambridge University Press, Cambridge, United Kingdom and New York, NY, USA.

IPCC, 2019. Climate Change and Land: an IPCC special report on climate change, desertification, land degradation, sustainable land management, food security, and greenhouse gas fluxes in terrestrial ecosystems [P.R. Shukla, J. Skea, E. Calvo Buendia, V. Masson-Delmotte, H.- O. Pörtner, D. C. Roberts, P. Zhai, R. Slade, S. Connors, R. van Diemen, M. Ferrat, E. Haughey, S. Luz, S. Neogi, M. Pathak, J. Petzold, J. Portugal Pereira, P. Vyas, E. Huntley, K. Kissick, M. Belkacemi, J. Malley, (eds.)]. Available at: https://www.ipcc.ch/srccl/.

Kritee, K., Nair, D., Tiwari, R., Rudek, J., Ahuja, R., Adhya, T. and Dava, O., 2015. Groundnut cultivation in semi-arid peninsular India for yield scaled nitrous oxide emission reduction. Nutrient Cycling in Agroecosystems, 103(1), 115-129.

Kritee, K., Nair, D., Zavala-Araiza, D., Proville, J., Rudek, J., Adhya, T. K. and Hamburg, S. P., 2018. High nitrous oxide fluxes from rice indicate the need to manage water for both long- and short-term climate impacts. Proceedings of the National Academy of Sciences, 115(39), 9720–9725.

Kritee, K., Nair, D., Zavala-Araiza, D., Reddy, M., Proville, J. and Ahuja, R., 2019. Climate smart farming in India: A pathway to poverty alleviation, food security, and climate adaptation and mitigation. An online report with greenhouse gas flux data from rice and non-rice cropping systems from four agro-ecological regions in India. Published by Environmental Defense Fund, New York, NY.

Kritee, K., Tiwari, R., Nair, D., Loecke, T.D., Adhya, T.K., Rudek, J., Ahuja, R. and Hamburg, S., 2014. Poster: Greenhouse gas emissions from rice, peanut and millet farms in peninsular India: Effects of water and nitrogen management. EDF: http://www.edf.org/sites/default/files/agu_poster-final.pdf.

MANAGE, 2017. Mobile Apps Empowering Farmers. Extension Digest, 1(2), 1-35.

MoEF&CC, 2015. Guidelines for convergence of MNREGS with GIM of Ministry of Environment, Forest and Climate Change (MoEF&CC).

MoEF, 2010. The National Mission for a Green India. Ministry of Environment and Forests (MoEF), Govt. of India.

NOAA-ERSL, 2019. http://www.esrl.noaa.gov/gmd/.

NOAA-ESRL, 2012. Available at. http://www.esrl.noaa.gov/gmd/ccgg/trends.

OECD, 2018a. Review of Agricultural Policies in India. OECD Publishing, Paris. pp. 286.

OECD/ICRIER, 2018b. Agricultural Policies in India, OECD Food and Agricultural Reviews, OECD Publishing, Paris. https://doi.org/10.1787/9789264302334-en.

Rani, A., Chaudhary, A., Sinha, N.K, Mohanty, M., Chaudhary, R.S. 2019. Drone: The Green Technology for Future Agriculture. Soil Health: Technological Interventions. Harit Dhara, 2(1), 1-4.

Saha, G., Kar, B. and Karmakar, S., 2017. Methane and nitrous oxide emission from Kharif rice field as influenced by nutrients and moisture regimes in new alluvial agroclimatic region of West Bengal. Curr Sci, 112(5), p.989.

Singh, R.B., 2019. Agricultural Transformation-A Roadmap to New India. National Academy of Agricultural Sciences, New Delhi, pp 415.

Smith, P. and Olesen, J.E., 2010. Synergies between the mitigation of, and adaptation to, climate change in agriculture. The Journal of Agricultural Science, 148(5), pp.543-552.

Taneja, G., Pal, B.D., Joshi, P.K., Aggarwal, P.K. and Tyagi, N.K., 2019. Farmers' Preferences for Climate-Smart Agriculture—An Assessment in the Indo-Gangetic Plain. In Climate Smart Agriculture in South Asia (pp. 91-111). Springer, Singapore.

Thea K., Martin C., Jeffrey M., Gerhard E, Dimitrios Z, Edward M. and Jeremy P., 2017. Food safety for food security: Relationship between global megatrends and developments in food safety. Trends in Food Science & Technology. 68, 160-175.

Treweek, G., Di, H.J., Cameron, K.C. and Podolyan, A., 2016. Effectiveness of the nitrification inhibitor dicyandiamide and biochar to reduce nitrous oxide emissions. New Zealand Journal of Agricultural Research, 59(2), pp.165-173. DOI: 10.1080/00288233.2016.1161651

Tzounis A, Katsoulas N, Bartzanas T, Kittas C., 2017. Internet of things in agriculture, recent advances and future challenges. Biosystems Engineering, 164, 31-48.

UNFCC, 2019. United Nations Framework Convention on Climate Change. http://di.unfccc.int/ghg_profile_non_annex1

Venkataramani, V., 2015. National Mission for a Green India. Centre for Development Finance, IFMR LEAD. Available at: http://ifmrlead.org/napcc-progress-and-evaluation/.

WGMI, 2019. Mitigation Instruments for Achieving India's Climate and Development Goals: A White Paper by the Working Group on Mitigation Instruments. New Delhi: Council on Energy, Environment and Water (CEEW) and Environmental Defense Fund (EDF).

World Bank, 2019. The World Bank Group. https://www.worldbank.org/en/topic/climate-smart-agriculture/

Zhang, G., Sun, B., Zhao, H., Wang, X., Zheng, C., Xiong, K., Ouyang, Z., Lu, F. and Yuan, Y., 2019. Estimation of greenhouse gas mitigation potential through optimized application of synthetic N, P and K fertilizer to major cereal crops: A case study from China. Journal of Cleaner Production, 237, p.117650.

13

Disaster Management: New Strategies

Joydeep Mukherjee and K.K. Bandyopadhyay

Division of Agricultural Physics, ICAR-Indian Agricultural Research Institute New Delhi

Introduction

Disaster may be defined as a sudden event that causes damage to life, property and destroy the economic, social and cultural life of people. It could be of natural or anthropogenic origin. Disasters impact the poor and the most vulnerable section of the community. Over the past 30 years, more than 2.5 million people and nearly $4 trillion have been lost to natural disasters in India; globally the losses jumped from $50 billion a year in the 1980s to $200 billion in the last decade (NIDM, 2019).According to a World Bank report, nearly 75% of the losses are attributable to extreme weather events and climate change threatening to push an additional 100 million people into the extreme poverty by 2030 (Hallegatte et al., 2016). The Bank's Unbreakable report finds that natural disasters had larger and long-lasting impacts on the poverty. Population growth and rapid urbanization are major driving factors for an increase in the disaster risks.

The United Nations estimates that more than two-thirds of the world's population will live in cities by 2050 (NIDM, 2019). This could put 1.3 billion people and $158 trillion assets at risk from river and coastal floods alone. According to the Bank's Investing in Urban Resilience report (2016), by 2030, without significant investment into making cities more resilient, natural disasters may cost cities worldwide $314 billion each year.

Mainstreaming disaster risk management into development planning can reverse the current trend of rising disaster impact. Furthermore, when countries rebuild stronger, faster and more inclusively after disasters, they can reduce the impact on people's livelihoods and well-being by as much as 31%, potentially cutting global average losses (NIDM, 2019). Decisive act can save the lives and assets, however, many developing countries lack the tools, expertise, and instruments to factor the potential impacts of disasters into investment decisions.

Following enactment of the Disaster Management Act, 2005, the National Disaster Management Authority (NDMA) of India was constituted act on 27th September,

2006 with Prime Minister as Chairperson and nine other members (one such member as Vice-Chairperson). Under the Act, the National Executive Committee has to prepare a National Plan, and coordinate and monitor the implementation of the National Policy. Similarly, under this act the State Authority in each state of India is responsible for laying down policies and plans for disaster management in the respective state. A ten-point agenda towards disaster risk reduction has been developed with the most important points as: (i) Work involving risk coverage for all-from poor households to micro, small and medium enterprises to multi-national corporations to nation states; (ii) Leverage technology to enhance the efficiency of disaster risk management and (iii) Build on local capacity and initiatives.

It has been stressed that appropriate capacity building among the local people is the key to face a disaster event. Working with local people in a participatory mode, empowering self-help groups involving local youth can aid in disaster management. The local indigenous technical knowledge (ITK) may be explored and documented in forewarning of disaster. This will reduce the load on Govt. machinery and will also lead to self-reliant India with capacity to deal with any kind of disaster, and minimize the loss and make way to normalcy soon after the disaster.

Types of Disasters

Disaster is the exposure of a group of people to a hazard, leading to a serious disruption of the functioning of a society and causing human, material, economic or environmental losses which exceed the ability of the affected community or society to cope. A disaster results from a combination of hazards and vulnerability that exceeds the capacity of a society to reduce the potential negative consequences of risk. It is often described as a result of the combination of: the exposure to a hazard; the conditions of vulnerability that are present; and insufficient capacity or measures to reduce or cope with the potential negative consequences. The consequences of disaster may include loss of life, injury, disease and other negative effects on human physical, mental and social well-being, together with damage to property, destruction of assets, loss of services, social and economic disruption and environmental degradation.

Thus, a disaster combines two elements: the hazard, and the vulnerability of affected people. A disaster occurs when a hazard exposes the vulnerability of individuals and communities in such a way that their lives are directly threatened or sufficient damage has occurred to community's economic and social structure which makes difficult for their ability to survive. Therefore, the types of hazards and vulnerabilities aggregate to form different disaster categories.

A. Hazards

Hazards are defined "as dangerous conditions or events, that threat or have the potential for causing injury to life or damage to property or the environment." A hazard may be any event that threatens life and property and disrupts normal life. Hazards can be both natural and man-made (Roy and Pandey, 2016). It is a process phenomenon or activity that may cause loss of life, injury or other health impacts property damage, social and economic disruption or environmental degradation.

Hazards may be natural, anthropogenic or socio-natural in origin. Natural hazards are predominantly associated with natural process and phenomenon, and are caused by physical forces working on or within the earth's surface. Examples include earthquakes, volcanoes, floods, cyclones, droughts etc. Because natural hazards are created by physical forces, it is not possible to stop them. However, by proper planning and policymaking, the risk to these hazards can be reduced drastically.

Anthropogenic Hazards (Man-made/Unnatural Hazards): These hazards are caused by human activities either through sheer negligence or for vindictive purposes. Examples include fire, accidents, terrorism, etc. Through raising awareness, better education, and better urban and rural management, these hazards can be tackled and removed.

Hazards may be single, sequential or combined in their origin and effects. Each hazard is characterized by its location, intensity or magnitude, frequency and probability. Biological hazards are also defined by their infectiousness or toxicity or other characteristics of the pathogen such as dose-response, incubation period case fatality rate and estimation of the pathogen for transmission.

Multi-hazard analysis involves (1) the selection of multiple major hazards that the country faces, and (2) the specific contexts where hazardous events may occur simultaneously, cascading or cumulatively over time, and taking into account the potential interrelated effects.

B. Vulnerability

Vulnerability is a relative concept. It refers to the inability of an individual, group, society, or a nation to deal with disasters effectively. It is observed that when a disaster strikes a region, the poor, women, children, and old are the most-affected. These are the groups that are referred to as 'vulnerable'. Many economic, social, and political factors affect the capacity of people to deal with disasters. In the past, vulnerability has been limited to physical susceptibility; however, over the time, this concept became comprehensive to include susceptibility, exposure, coping and adaptive capacity, social inequalities, and physical, institutional, and economic weaknesses. Anderson and Woodrow (1990) categorized vulnerability into three areas namely:

- Physical/ Material vulnerability: For example, poor people who have limited physical/ material resources usually suffer more from disaster than rich people do. Poor people often live on marginal lands, have minimum or no personal savings or insurance, and are mostly in poor health condition. These make them more vulnerable to disaster and will have harder time surviving and recovering from a calamity than people who are better off economically.
- Social/Organizational vulnerability: People who have been 'marginalized' in social, economic or political terms have been found vulnerable, while groups which are well organized and members are in high commitment to each other suffer less when catastrophe strikes. For example, deep divisions can lead to conflict and war. Conflict on resources due to poverty can also lead to violence. A second area of vulnerability then, is in the social/ organizational and economic realm.
- Attitude and Motivational vulnerability - People who lack confidence in their ability to affect change and feel defeated by events they cannot control, are hit harder by disasters. This is in contrast to those who have a sense of their ability to bring about the changes they desire (NIDM, 2017).

Typologies of Disasters

List of natural hazards as identified by NDMA has been summarized in the following table 1 (NIDM, 2019)

Table 1. List of Natural Hazards

Family	Primary hazard	Secondary hazard	Affected States/ Regions in India
Geo-Physical	Earthquake	Landslide Fire Liquefaction Surface Displacement Tsunami	58.6% of the landmass of India is prone to earthquakes of moderate to very high intensity
	Volcano	Surface Displacement Ash fall, Lahar Pyroclastic flow Landslide Fire Tsunami	Barren island in the Andaman is the only confirmed active volcano.
Hydrophysical	Flood Landslide Wave action	Avalanche Coastal Erosion Coastal flood Debries flow, Mud flow Rock fall Flash flood Flood Hydrological wave ction	Over 40 million hectares/12% of land is prone to floods and river erosion

Family	Primary hazard	Secondary hazard	Affected States/ Regions in India
Meteorological	Hazards caused by short-lived, micro to meso scale extreme weather and atmospheric conditions that may vary from minutes to days	Cyclone, Storm, Surge, Tornado, Convective storm, Extratropical storm, Wind, Cold wave, Extreme temperature, Fog, Frost, Freeze, Hail, Heat wave, Lightening, Heavy rain, Sand storm, Snow, Ice, Winter storm, Blizzard, Cloud burst etc.	Out of 7,516 km long coastline of India, nearly 5,700 km is prone to cyclones and tsunamis.
Climatological		Drought, Extreme hot/cold conditions, Forest/Wildfire, Fires, Glacial lake outburst, Subsidence	5,11,300 km^2 of the landmass of India is drought prone. About 50% of the forests in India are prone to wildfires. All the states bordering the Himalayas are prone to Glacial lake outburst
Biological		Epidemics: Viral, Bacterial, Parasitic, Fungal, Prion infections, Insect infestations, Animal stampedes	Almost the entire country is prone to various epidemics. There are cases of Nipah virus, cases of dengue, chikungunya, cases of Zika virus, Avian influenza, Japanese encephalitis, cases of Swine flu and Novel Corona Virus has been reported across the country.

Natural Hazards

Hazards include (as mentioned in the Sendai Framework for Disaster Risk Reduction 2015-2030) biological, environmental, geological, hydro-meteorological and technological processes and phenomena.

- **Biological hazards** are of organic origin or conveyed by biological vectors, including pathogenic microorganisms, toxins and bioactive substances. Examples are bacteria, viruses or parasites, as well as venomous wildlife and insects, poisonous plants and mosquitoes carrying disease-causing agents.
- **Environmental hazards** include chemical, natural and biological hazards. They can be created by environmental degradation or physical or chemical pollution in the air, water and soil. However, many of the processes and phenomena that fall into this category may be termed drivers of hazard and risk rather than hazards in themselves, such as soil degradation, deforestation, loss of biodiversity, salinization and sea-level rise etc..

- **Geological or geophysical hazards** originate from internal earth processes. Examples are earthquakes, volcanic activity and emissions, and related geophysical processes such as mass movements, landslides, rockslides, surface collapses and debris or mud flows. Hydro-meteorological factors are the important contributors to some of these processes. For example, Tsunamis are difficult to categorize. Although they are triggered by undersea earthquakes and other geological events, they essentially become an oceanic process that is manifested as a coastal water-related hazard.
- **Hydro-meteorological hazards** are of atmospheric, hydrological or oceanographic origin. Examples are tropical cyclones (also known as typhoons and hurricanes); floods, including flash floods; drought; heat waves and cold spells; and coastal storm surges. Hydro-meteorological conditions may also be a factor in other hazards such as landslides, wild land fires, locust, plagues, epidemics and in the transport and dispersal of toxic substances and volcanic eruption material.

Anthropogenic/ Human Induced Hazards

According to the NDMP, increasing levels of urbanization and industrialization within identified high risk zones has increased the vulnerability of human population to disasters. The key hazards leading to a possible disaster that can be attributed to human activity are: chemical, biological, radiological and nuclear (CBRN) hazards; and terrorism, violence, conflict and fire hazards. Other mentions are accidents, poor planning and construction, over-population or overcrowding, environmentally insensitive practices and so on.

Technological hazards originate from technological or industrial conditions, dangerous procedures, infrastructure failures or specific human activities. Examples include industrial pollution, nuclear radiation, toxic wastes, dam failures, transport accidents, factory explosions, fires and chemical spills. Technological hazards also may arise directly as a result of the impacts of a natural hazard event.

Human induced hazards may be identified using the following approaches

- Records with local emergency responders, such as fire departments, police departments, local hospitals and medical facilities and records within the site of previous events, caused due to human intervention
- Local area maps highlighting nearby industries, nuclear plants, possible sources of contamination, sources of pollution, sources of ignition and so on
- Socio-economic mapping and other demographic information of the precinct to identify any potential sources of conflict among different interest groups or potentially volatile situations (NDMA, National Disaster Management Guidelines for Cultural Heritage Sites and Precincts September, 2017).

Scales of disaster

The scale of the disaster is an important factor to assess the degree of preparedness and vulnerability. The disaster situations can be categorized into three 'levels': L1, L2, and L3 (NIDM, 2007). Level zero (L0) is the period of normalcy, which should be utilized for monitoring and preparatory activities, and involves planning and training and resource base assessment.

- Level-1 (L1): This is the first level of disaster that can be managed within the capabilities and resources at the district level. Both the state and central authorities however remain prepared to provide assistance if needed.
- Level-2 (L2): This signifies the disaster situations that require assistance and active mobilization of resources at the state level and deployment of state level agencies for disaster management. The central agencies must remain vigilant for immediate deployment if required by the state.
- Level-3 (L3): This corresponds to a nearly catastrophic situation or a very large-scale disaster that overwhelms the state and district authorities.

Hazard Profile of India

The unique geo-climatic and socio-economic conditions made India vulnerable to natural as well as manmade disasters. Most of the Indian states are vulnerable to cyclones, droughts, floods, earthquakes and forest fires. More than 50 per cent of Indian land mass is earthquake prone in varying degrees, more than 12 per cent area are prone to floods, while long coastlines are highly cyclone and tsunami prone. Even 68 per cent of cultivable lands are drought prone (NIDM, 2011). Both natural and human induced, including adverse geo-climatic conditions, population growth, environmental degradation, topographic features, industrialization, urbanization, non-scientific development practices etc. are the main factors behind the vulnerability of the country towards disasters. The desert of western India receives very low rainfall, therefore are prone to drought and famine, while the coastlines of India have depressions, cyclone and storms (Fig 1). The Himalayan regions are affected by flash floods, earthquakes and landslides; plain lands of India are also prone to floods. For this, major river systems of India originated from Himalayan glaciers bring lot of sediments which causes siltation of river beds and causes floods frequently particularly in monsoon season in UP and Bihar. Western states namely Gujarat and Rajasthan are affected due to occurrence of droughts. Situation further get worse in monsoon failure year, thus covered whole India. The disturbed pressure conditions over oceans, forms cyclones in coastal regions. Coastal regions are also vulnerable to tsunami disaster due to tectonic movements that occur below the ocean floor. Indian long coastlines are getting on an average five to six tropical cyclones annually. Projections from different GCMs indicate that extreme events i.e. heat waves, cold waves, more floods, more droughts, more intense cyclones and flash floods will increase. Extreme rainfall has substantially increased over large areas, particularly over the west coast and west central India. The poor section of the society is the worst affected in disasters. The situation further aggravates due to the compulsion of the poor to exploit environmental resources for their survival, increasing the risk and exposure of the society to disasters, in particular those triggered by flood, drought and landslides. Poverty also compels the poor to migrate and live at physically more vulnerable locations, often on unsafe land and in unsafe shelters. These inhabitations of the poor at such locations are either due to the fact that there is no other land available at reasonable cost or it is close to the employment opportunities. The inhabitations of the poor people on marginal land are prone to all types of disasters. The type of construction of these houses further deteriorates the condition. These dwellings made up of low cost material without giving much consideration to technical aspect are easy targets of various hazards.

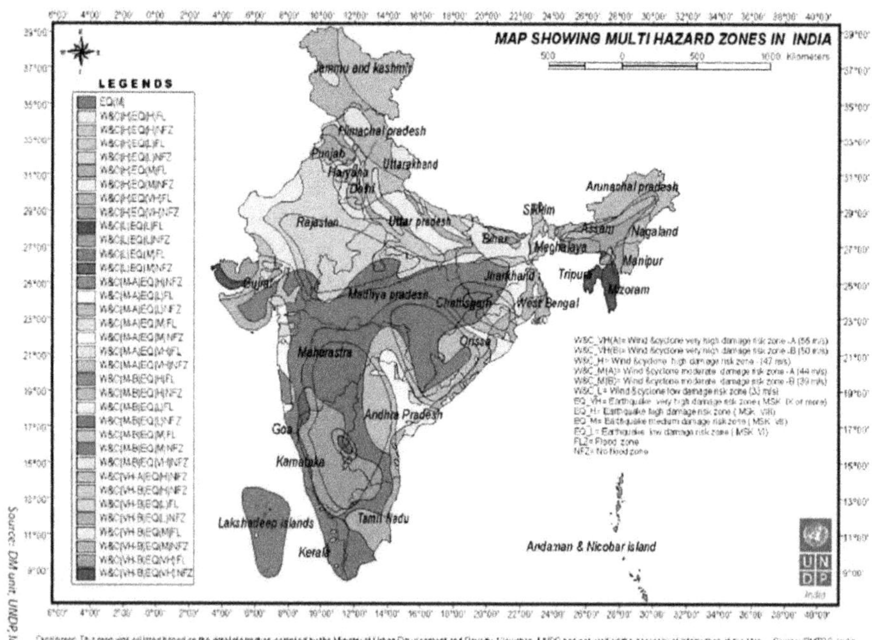

Fig. 1: Multi Hazard Map of India (*Source*: DM Unit, UNDP, India)

Cause and effect of disasters

Over the years the occurrence and severity of extreme weather events have increased and thus make India as one of the most vulnerable country in the world. India is an agriculture dominant country and more than 70% of the population is depending on agriculture directly or indirectly. So the impacts of these extreme events are tremendous on human population and other living beings. State of

Odisha alone faced 49 years of floods, 30 years of droughts and 11 years of severe cyclones. Extreme weather events like floods and droughts occurrence in India is associated with extent and nature of summer monsoon. Summer monsoon rainfall fluctuations over whole India are the most important factor for floods or droughts. Both floods and droughts affect large areas of India. One-sixth area of the country is drought-prone. Out of 40 million hectares of the flood prone area in the country, on an average, floods affect an area of around 7.5 million hectares per year.

Box 1.2: Examples of extreme weather events

Primary weather Events (climate-driven)

- Cold wave, fog, snow storms and avalanches
- Hailstorm, thunderstorm and dust storms
- Extreme temperature
- Tropical cyclone and tidal wave
- Floods, heavy rain
- Droughts (hydrological, meteorological and agricultural etc.)

Secondary weather Events (may be climate-driven)

- Incidence of epidemics or diseases
- Urban and rural water shortage
- Crop plantation failure or harvest failure
- Malnutrition or under nutrition and hunger
- Landslides, saline water intrusion and mudflows

(i) Drought

Insufficient rainfall is the primary cause of any drought and the distribution and intensity of this deficiency of rainfall and in relation to existing reserves. A prolonged period of relatively deficit rainfall leading to drought, which is a widely recognized climate anomaly. Drought can be devastating as water sources dry up, crops dried up, animals die and malnutrition and ill-health become widespread. The environmental effects of drought, including salinization of soil and groundwater table decline, increasing pollution of freshwater and regional extinction of animal species.

(ii) Flood

India is one of the most flood prone countries in the world (Fig. 2). Flood results havoc and brings miseries to the rural people. It also causes outbreak of epidemic of water borne diseases, huge scarcity arises in flooded areas.

Fig. 2. Flood hazards map of India. (*Source*: NDMA, 2011)

Flood has severe impact on agricultural production even stagnant water over larger areas for long time hampers production *rabi* crops. The principal reasons for flood lie in the very nature of natural ecological systems in this country, namely, the monsoon, the highly silted river systems and the steep and highly erodible mountains, particularly those of the Himalayan ranges. The average rainfall in India is 1150 mm with significant variation across the country. The annual rainfall in the western coast and Western Ghats, Khasi hills and over north eastern parts amounts to more than 2500 mm. Heavy rainfall in monsoon season causes flood and mostly it is associated with cyclonic depressions, active and break monsoon conditions. Flash flood is also occurring due to high intensity rainfall and less permeable soils. Area having hardpan below the ground levels are more prone to flood as water fails to infiltrate down.

(iii) Cyclones

India is having more than 7500 km of coastline which is frequently facing one of the most natural disaster cyclone. Nearly 10 per cent of world's tropical cyclone. More than 70 per cent area of these coastlines is in ten states (Gujarat, Maharashtra, Goa, Karnataka, Kerala, TamilNadu, Puducherry, Andhra Pradesh, Orissa and West Bengal) of India. Both the islands (Andaman, Nicobar and Lakshadweep) are very prone to cyclones. At least five to six tropical cyclone is formed in Bay of Bengal and Arabian Sea which affects this area severely. Tropical cyclone associated with high speed winds, storm surges, heavy rainfall and floods enhance the risk of loss and damages of the inhabitants of coastlines. Storm surge is very pronounced as a wide and shallow bays in northern part of Bay of Bengal is exposed. Five to six cyclones in every year is very common. The ratio of occurrence of cyclone in Bay of Bengal and Arabian sea is approximately 4:1. The Odisha and West Bengal receives the maximum cyclone events. The highest annual number of storms, severe storms occurs in the Odisha - West Bengal coast during south west monsoon season.

(iv) Heat Wave

Heat wave occurs due to extreme positive departure of daily maximum temperature from normal during summer. Northwestern parts of India experienced rising maximum temperature in pre monsoon time may continue up to June month. During April, 2002 country was experienced abnormally high temperature. The maximum temperature at Gannavaram (Vijayawada) 49°C (WMO 2003) was recorded on 10th May, 2002. The cause of decreased Diurnal temperature range is mainly due to urbanization which leads to human discomfort and mortality. Increase of minimum temperature in summer is mainly responsible for high nocturnal respiration of plants during a heat wave epoch.

(v) Cold Wave and Fog

North India experienced cold wave during winter months when minimum temperature falls very less than normal in association with dry cold winds from higher latitudes. The northern parts of India, especially the hilly regions and the adjoining plains, are influenced by transient disturbances in the mid latitude westerlies which often have weak frontal characteristics. Hilly regions and its adjoining area of north India, are affected by disturbances in mid latitudes westerlies which sometimes have a weak frontal characteristic, known as western disturbances. Generally, the areas above 20°N is affected by cold waves but sometimes southern states like Karnataka and Maharashtra is also affected with large amplitude of troughs. Due to poor air quality in urban areas people are suffering from discomfort and deaths in recent years. States like Uttar Pradesh and Bihar recorded highest causalities in cold waves due to lack of proper shelters and poor infrastructural development to cope with extreme events to the farmers and labourers.

(vi) Thunderstorm, Hailstorm and Dust Storm

Temperature starts rising after ending of spring seasons initially in southern states which leads to development of thunder storms and gusty weather which sometimes happen to be very hazardous.

Another extreme events dust storm and hail storm (due to development of Cumulonimbus clouds) affects the northern, north eastern, north western and central states of India, but southern parts of the country are generally free from these events. The occurrence of hailstorm events is highest in North eastern states followed by Uttarakhand, Jharkhand and Maharashtra. Thunderstorms also occur in Delhi, Kolkata, Jaipur and Ahmedabad.

Disaster Management

Disaster management can be defined as a "collective term encompassing all aspects of planning for and responding to disasters, including both pre- and post-disaster activities" (CERO, 2004). It may refer to the management of both the risks and consequences of disaster. Amarasinghe et al. (2006), defined disaster management as "an applied science, which seeks, by the systematic observation and analysis of disasters to improve the measures to prevent, respond and recover from the effects and consequences of a disaster".

Therefore disaster management can be identified as a cycle of inter related activities. According to Warfield (2004), the Disaster management cycle illustrates the ongoing process by which governments, businesses, and civil society plan for and reduce the impact of disasters, react during and immediately following a disaster, and take steps to recover after a disaster has occurred. Therefore it is important to take appropriate actions at all points in the disaster management cycle and it will lead to greater preparedness, better warnings, reduced vulnerability or prevention of disasters (Adams and Wisner, 2003). According to Kawata (2001), disaster management can be divided into four parts Emergency response and relief, Recovery and reconstruction, Mitigation and Preparedness, which are mainly, categorized under two main headings - Crisis Management, and Risk Management (Fig. 3). Disaster management cycle is an open ended process or an ongoing attempt to limit the effects of the disasters (Amarasinghe et al, 2006). The individual phases are often overlapped but it is crucial that they operate as a closed loop because a major aim of hazard management is to learn from experience and feedback (Smith, 2001). Further, Smith (2001) has classified disaster management under two major headings Pre Disaster Protection and Post Disaster Recovery including sub elements of Risk Assessment, Mitigation, Preparedness, Relief, Rehabilitation and Reconstruction.

Kelly (1996) defined Disaster Management as the range of activities designed to maintain control over disaster and emergency situations and to provide a framework for helping those who are at risk to avoid or recover from the impact of the disaster. This includes preparedness before disaster, response and recovery i.e. rebuilding and supporting society. Disaster management aims to reduce or avoid the potential losses from hazards, assure prompt and appropriate assistance to victims of disaster, and achieve rapid and effective recovery. The disaster management cycle illustrates the ongoing process by which governments, businesses, and civil society plan for and reduce the impact of disasters, react during and immediately following a disaster, and take steps to recover after a disaster has occurred. Appropriate actions at all points in the cycle lead to greater preparedness, better warnings, reduced vulnerability or the prevention of disasters during the next iteration of the cycle. Some of actions cut across the emergency response and recovery stages viz., coordination and ongoing assistance) while other activities are unique to each stage (e.g. early warning and evacuation during emergency response, and reconstruction and economic and social recovery as part of recovery). This cycle works best for relatively sudden-onset of disasters, such as floods, earthquakes, bushfires, tsunamis, cyclones etc., but is less reflective of slow-onset disasters, such as drought, where there is no obviously recognizable single event which triggers the movement into the Emergency Response stage.

Fig. 3. Disaster management cycle (Kawata, 2001)

The three key stages of activities that are taken up within disaster risk management are as follows

- Before a disaster (pre-disaster). Pre-disaster activities are those which are taken to reduce human and property losses caused by a potential hazard. For example, carrying out awareness campaigns, strengthening the existing weak structures, preparation of the disaster management plans at household and community level, etc. Such risk reduction measures taken under this stage are termed as mitigation and preparedness activities.
- During a disaster (disaster occurrence). These include initiatives taken to ensure that the needs and provisions of victims are met and suffering is minimized. Activities taken under this stage are called as emergency response activities.
- After a disaster (post-disaster). There are initiatives taken in response to a disaster with a purpose to achieve early recovery and rehabilitation of affected communities, immediately after a disaster strike. These are called as response and recovery activities.

The mitigation and preparedness phases occur as disaster management improvements are made in anticipation of a disaster event. Developmental considerations play a key role in contributing to the mitigation and preparation of a community to effectively confront a disaster. As a disaster occurs, disaster management actors, in particular humanitarian organizations, become involved in the immediate response and long-term recovery phases. The four disaster management phases *i. e.* mitigation, preparedness, response and recovery, do not always, or even generally, occur in isolation or in this precise order. Often phases of the cycle overlap and the length of each phase greatly depends on the severity of the disaster.

- Mitigation - Minimizing the effects of disaster. Examples: building codes and zoning; vulnerability analyses; public education.
- Preparedness - Planning how to respond. Examples: preparedness plans; emergency exercises/training; warning systems.
- Response - Efforts to minimize the hazards created by a disaster. Examples: search and rescue; emergency relief.
- Recovery - Returning the community to normal. Examples: temporary housing; grants; medical care.

To analyze the scope of disaster management in the revised context, the cycle of the phenomenon should be studied. Disasters are as old as human history but the dramatic increase and the damage caused by them in the recent past have become a cause of national and international concern.

Institutional framework

Disaster management in India has evolved from an activity-based reactive setup to a proactive institutionalized structure; from single faculty domain to a multi-stakeholder setup; and from a relief-based approach to a 'multi-dimensional pro-active holistic approach for reducing risk. The beginnings of an institutional structure for disaster management can be traced to the British period following the series of disasters such as famines of 1900, 1905, 1907 & 1943, and the Bihar-Nepal earthquake of 1937. Over the past century, the disaster management in India has undergone substantive changes in its composition, nature and policy. During the British administration, relief departments were set up for emergencies during disasters. Such an activity-based setup with a reactive approach was functional only in the post-disaster scenarios. The policy was relief-oriented and activities included designing the relief codes and initializing food for work programmes. Post-Independence, the task for managing disasters continued to rest with the Relief Commissioners in each state, who functioned under the Central Relief Commissioner, with their role limited to delegation of relief material and money in the affected areas. Every five-year plan addressed flood disasters under "Irrigation, Command Area Development and Flood Control". Until this stage, the disaster management structure was activity-based, functioning under the Relief Departments. The institutional structure for disaster management in India is in a state of transition. The new setup, following the implementation of the Act, is evolving; while the previous structure also continues. Thus, the two structures co-exist at present. The National Disaster Management Authority has been established at the centre, and the SDMA at state and district authorities at district level are gradually being formalized. In addition to this, the National Crisis

Management Committee, part of the earlier setup, also functions at the Centre. The nodal ministries, as identified for different disaster types of function under the overall guidance of the Ministry of Home Affairs (nodal ministry for disaster management). This makes the stakeholders interact at different levels within the disaster management framework. Within this transitional and evolving setup, two distinct features of the institutional structure for disaster management may be noticed. Firstly, the structure is hierarchical and functions at four levels – centre, state, district and local. In both the setups – one that existed prior to the implementation of the Act, and other that is being formalized post-implementation of the Act, there have existed institutionalized structures at the centre, state, district and local levels. Each preceding level guides the activities and decision making at the next level in hierarchy. Secondly, it is a multi-stakeholder setup, i.e., the structure draws involvement of various relevant ministries, government departments and administrative bodies.

National Disaster Management Authority (NDMA)

The National Disaster Management Authority (NDMA) was initially constituted on May 30, 2005 under the Chairmanship of Prime Minister vide an executive order. The NDMA was constituted by the Section-3(1) of the Act on 27th September, 2006 with Prime minister of India as its Chairman with nine other members among which one member is Vice Chairperson.

Prevention and Mitigation

Prevention and mitigation contribute to lasting improvement in safety and should be integrated in the disaster management. The Government of India has adopted mitigation and prevention as essential components of their development strategy. Accordingly, the Tenth Five Year Plan document has a detailed chapter on disaster management. The plan emphasizes the fact that development cannot be sustainable without mitigation being built into the developmental process. Mainstreaming DRR involves incorporating disaster risk reduction into development policy and practice. It means radically expanding and enhancing disaster risk reduction so that it becomes normal practice, fully institutionalized within an agency's relief and development agenda. Mainstreaming has three purposes:

(a) To make certain that all the development programmes and projects that originate from or are funded by an agency, are designed with evident consideration for potential disaster risks and to resist hazard impact,

(b) To make certain that all the development programmes and projects that originate from or are funded by an agency, do not inadvertently increase vulnerability to disaster in all sectors: social, physical, economic and environment,

(c) To make certain that all the disaster relief and rehabilitation programmes and projects that originate from or are funded by an agency are designed to contribute to developmental aims and to reduce future disaster risk.

Preparedness and Response

Facing disaster by way of mitigation, prevention, preparedness, response, evacuation, relief and rehabilitation has been part of the administrative ethos. India has a long history of rendering relief in an organized fashion in times of drought and famine. The states have antiquated relief code which deals with the general principles of administration of relief. It starts with the responsibility of the Government for combating distress, defining scope of object of such measures etc. India, with a vast agrarian economy in the past had focused on distress relief mainly related to agricultural activities. Preparedness included collection of statistical data on the rainfall, weather conditions, crop pattern activities relating to management of cattle etc. The relief work focused on departmental work and village work for generation of employment during drought. With the changing pattern of disaster and with the introduction of technology, material and new financial terms into disaster management, several modifications have been incorporated in the administrative measures for relief work The scope of disaster has since changed and so the response thereto. With the constitution of successive Finance Commission under the provision of the Constitution, the measures for relief and the scope of disaster have enlarged. An attempt has been made in the succeeding paragraphs to capture the information relating to timely prediction/forecasting of different kind of disasters and development of Standard Operating Procedures (SOPs) for responding such disasters. Forecasting about climate change is prerequisite for taking preparedness measure to respond to the disaster is the most important element of disaster management. The Ministry of Environment & Forest (MoEF), Ministry of Earth Sciences (MoES), Ministry of Science & Technology (MST), Ministry of Agriculture (MoA), Ministry of Water Resources (MWR), Ministry of Human Resource Development (MHRD), Ministry of Non-conventional Energy (MNES), Defence Research & Development Organization (DRDO), Ministry of Defence (MoD), Ministry of Health and Family Welfare (MoHFW), Indian Space Research Organization (ISRO) and Indian Meteorological Department (IMD) promote and undertake climate and climate change related research in the country. India Meteorological Department's tradition of monitoring weather and climate spans more than 135 years giving it a sound and useful dataset to fall back upon for environmental assessment There are 25 types of atmospheric monitoring networks that are operated and coordinated by the IMD. This includes meteorological, climatologically, environment, air pollution and other specialized observation of atmospheric trace constituents. It maintains 559 surface meteorological observatories, about 35 radio-stations and 64 pilot balloon stations for monitoring the upper atmosphere. Specialized observations are made for agro meteorological purposes at 219 stations and radiation parameters are monitored at 45 stations. There are about 70 observatories that monitor current weather conditions for aviation. The IMD collects meteorological data over oceans by an establishment of cooperation fleet of Voluntary Observing Ships (VOF) comprising merchant ships of Indian registry, some foreign merchant vessels and a few ships of the Indian Navy. These ships, while sailing on the high seas, function as floating observatories. Records of observations are passed on to the IMD for analysis and archival.

Disaster Risk Reduction

Disaster risk reduction (DRR) includes the systematic development and application of policies, strategies and practices to avoid (prevention) or limit (mitigation and preparedness) the adverse effects of hazards (ISDR, 2010). The impact of disasters can be reduced or prevented with the proper adoption of the disaster risk reduction strategies. The World Conference on Disaster Reduction, held from 18 to 22 January 2005 in Kobe, Hyogo, Japan, has adopted a Framework for Action 2005-2015: Building the Resilience of Nations and Communities to Disasters. Hyogo Framework was adopted by 168 governments at the World Conference for Disaster Reduction in January 2005. It seeks to achieve a substantial reduction of disaster losses in lives and in the social, economic and environmental assets. It identifies five priorities for action:

- Ensure that disaster risk reduction is a national and a local priority with a strong institutional basis for implementation.
- Identify, assess and monitor disaster risks and enhance early warning.
- Use knowledge, innovation and education to build a culture of safety and resilience at all levels.
- Reduce the underlying risk factors.
- Strengthen disaster preparedness for effective response at all levels.

The disaster risk reduction measures can be categorized in several ways and one such method of categorizing disaster risk reduction is policy and planning measures, physical preventative measures, physical coping and/or adaptive measures and community capacity building measures (DFID, 2005). Policy and planning measures are implemented at the national or regional level and help to integrate DRR into the policy framework. Physical preventive measures are designed to reduce the vulnerability and exposure of infrastructure to natural hazards. Physical coping and adaptive measures includes coping and adaptive infrastructure which are capable to withstand in case of a disaster. Community capacity building measures are designed and implemented at the community level, particularly the strengthening of communities to better respond and cope to a disaster event through training and capacity building.

The Prime Minister's 10-point Agenda on Disaster Risk Reduction (DRR)

Following are the 10-point agenda envisaged by the Honourable Prime Minister for the Disaster Risk Reduction (DRR).

- All development sectors must imbibe the principles of disaster risk management.
- Work towards risk coverage for all-starting from poor households to SMEs to multi-national corporations to nation states.
- Encourage greater involvement and leadership of women in disaster risk management.
- Invest in risk mapping globally. For mapping risks related to hazards like earthquakes we have accepted standards and parameters.
- Leverage technology to enhance the efficiency of our disaster risk management efforts.
- Develop a network of universities to work on disaster issues.
- Utilize the opportunities provided by social media and mobile technologies.
- Build on local capacity and initiative.
- Opportunity to learn from a disaster must not be wasted. After every disaster there are papers on lessons that are rarely applied.
- Bring about greater cohesion in international response to disasters.

Conclusion

Disaster is the exposure of a group of people to a hazard, leading to a serious disruption of the functioning of a society and causing human, material, economic or environmental losses which exceed the ability of the affected community or society to cope. Disaster management includes sum total of all activities, programs and measures which can be taken up before, during and after a disaster with the purpose of avoiding, reducing the impact or recovering from its losses. This chapter deals with types of disaster, causes and effects of disaster, hazard profile of India, institutional framework, policy and guidelines to meet the disaster, prevention and mitigation strategies and preparedness and response to mitigate the disaster. Prime Minister of India, Chairman of the National Disaster Management Authority has put forth a ten point agenda for Disaster Risk Reduction. One of the important points among them is to build on local capacity and initiative to meet the disaster. So capacity building of local people is very much essential to minimize the ill effects of disaster and for speedy recovery after the disaster is over. This will help in confidence building among the local people and also make them self-reliant to manage any disaster. This will also lead to building resilience among local people, meeting the Sustainable Development Goals, and improving the well-being of the people.

References

Adams, J. and B. Wisner. 2003. Environmental health in emergencies and disasters: a practical guide [online]. World Health Organization. Available from: http://www.who.int/water_ sanitation_health/hygiene/emergencies/emerge ncies 2002/en/ [Accessed 19 May 2006].

Amarasinghe, M.K.D.W., H.T.K. Pathmasiri and M.H.A. Sisirakumara. 2006. The Public Sector Role in a Disaster Situation: An Empirical Analysis based on Tsunami Disaster on 26th December 2004. ICBM – 2006, University of Sri Jayewardenepura, Sri Lanka, p.372-379.

Anderson, M.B. and P.J. Woodrow. 1990. Reducing vulnerability to drought and famine: Developmental approaches to relief. Disasters. 15(1): 43-54.

Central Emergency Relief Organization. 2004. Disaster Management for Students, Managing Disasters [online]. Barbados, Central Emergency Relief Organization. Available from:http://cero.gov.bb/pages/ students.html [Accessed 13 May 2006].

DFID. 2005. "Natural disasters and disaster risk reduction measures", A Desk Review ofCosts and Benefits – Draft Final Report, DFID, London.

Disaster Reduction in United Nations. 2011. Roles mandates areas of work of Key United Nations entities (ISDR).

Disastrous weather Events annual reports, IMD, 1968 -77 Extreme Weather Events over India in the last 100 years; De, U.S., R.K. Dube and G.S.J. Prakasa Rao,. IND. GEOPHYS. UNION. .9(3):173-187, 2005.

Hallegatte, Stephane; Bangalore, Mook; Bonzanigo, Laura; Fay, Marianne; Kane, Tamaro; Narloch, Ulf; Rozenberg, Julie; Treguer, David; Vogt-Schilb, Adrien. 2016. Shock Waves : Managing the Impacts of Climate Change on Poverty. Climate Change and Development;. Washington, DC: World Bank. © World Bank. https://openknowledge.worldbank.org/ handle/10986/22787 License: CC BY 3.0 IGO."

ISDR. 2010. Strategy outline for the 2010-2011 ISDR world disaster reduction campaign onbuilding resilient cities, addressing urban risk [online]. International Strategy for Disaster Reduction– ISDR. Available from: www.unisdr.org/.../campaigns/ campaign2010-2011/.../230_CampaignStrategy.doc [Accessed 28 Feb 2010].

Kawata, Y. 2001. Disaster mitigation due to next Nankai earthquake tsunamis occurring in around 2035 [online]. Kyoto, Kyoto University. Available from: http://www.pmel.noaa. gov/its2001/Separate_Papers/1-08_Kawata.pdf [Accessed 25 February 2006].

Kelly, C. 1996. "Limitations to the use of military resources for foreign disaster assistance". Disaster Prevention and Management. 5(1):22-29.

Mall R.K., R. Kumar and R. Bhatla. Climate change and Disasters in India. 2010. Journal of South Asia Disaster Studies, SAARC Disaster Management Centre, New Delhi.Crisis Management from Despair to Hope, Third Report- Second Administrative Reforms Commission, September 2006, New Delhi.

NIDM. 2011. Ministry of Home Affairs, Government of India. Page: 12-22.

NIDM. 2017. Basics of Disaster Management, National Institute of Disaster Management Available at: https://nidm.gov.in/PDF/trgreports/2017/August/07-11_NIDM.pdf

NIDM. 2019. Handbook on disaster management for nodal officers (Compilation). National Institute of Disaster Management.

Report of Working Group on Disaster Management for the Eleventh Five year Plan (2007-2012), Department of Planning Commission, December 2006, New Delhi.

Smith, K. 2001. Environmental Hazards. 3rd ed. London and New York: Routledge.

The Report of the High Powered Committee on Disaster Management in India, National Centre for Disaster Management, Indian Institute of Public Administration.

Warfield, C. 2004. The Disaster Management Cycle [online]. Kobe, The Global Development Research Center. Available from: http://www.gdrc.org/uem/disasters/1-dm_cycle.html [Accessed 15 April 2006].

World Bank Group. 2016. Investing in Urban Resilience: Protecting and Promoting Development in a Changing World. World Bank, Washington, DC. © World Bank. https://openknowledge.worldbank.org/handle/10986/25219 License: CC BY 3.0 IGO

Theme 4: Innovations and Strategies for Biotic Stress Management

14

New Innovations and Approaches for Biotic Stress Management of Crops

Kiran Kumar[1], J. Sridhar[4], V.K. Choudhary[2], H.K. Singh[3], B. Parameshwari[5] K.M. Senthil Kumar[6], Bhimeshwari Sahu[4], Narasimham Dokka[4], P.N.Sivalingam[4]

[1]*ICAR-Central Citrus Research Institute, Nagpur, Maharashtra*
[2]*ICAR-Directorate of Weed Research, Jabalpur, Madhya Pradesh*
[3]*Department of Plant Pathology, Indira Gandhi Krishi Vishwavidyalaya Raipur, Chhattishgarh*
[4]*ICAR-National Institute of Biotic Stress Management, Raipur, Chhattishgarh*
[5]*ICAR-Sugarcane Breeding Institute, Regional Centre, Karnal, Haryana*
[6]*ICAR-Central Tuber Crops Research Institute, Thiruvananthapuram, Kerala*

Introduction

Agriculture is the backbone of India as 50% of its work is engaged in agricultural activities enabling India self-sufficiency in food production. However, agriculture sector is facing several challenges such as reduction in arable land due to urbanization and industrialization, decrease in farm size due to family division, climate change, yield loss due to abiotic and biotic stresses like pest, disease and weed, proliferation of non-genuine/illegal agro-chemicals etc (Murali-Baskaran et al., 2019). FAO estimates that a total annual loss of 20 to 40 percent of global crop production due to pests. Each year, plant diseases cost the global economy around $220 billion, and invasive insects around US$70 billion. While in India, yield loss to crops is estimated as 26, 20, 6 and 8% by insects, plant pathogens, rodents and others, respectively with an annual loss of Rs. 2,25,000 crores annually. Traditionally, ITKs have been practiced by farmers to control pests based on three categories viz., (a) cultural practices, physical, mechanical methods and use of botanicals. Later, IPM has been implemented in various crops to curb such losses caused by biotic stresses. Although cultural, mechanical and biological control strategies are in practice, insecticide application has been the most widely used tactics to reduce such losses to achieve quick control and particularly in case of outbreaks/upsurges. In India 76% of the pesticide used is insecticide, as against 44% globally (Mathur, 1999). First generation insecticides such as sulphur, arsenic, calcium arsenate, copper acetoarsenite, hydrogen cyanide, lead were introduced during 1950's which are simple, persistent, usually very toxic chemicals used in large scale which harm almost everything they come into contact with. Later second generation insecticide molecules like DDT, chlorinated hydrocarbon, organophosphates, carbamates, and pyrethroids have been introduced with few advantages over first generation molecules. Subsequently, insect growth regulators (IGRs) that mimic insect growth hormones preventing natural young insect moulting which are less harmful to mammals have been introduced. Then, during past decades, newer insecticide molecules have been introduced which are target specific, relatively safe to non-target organisms and low residual activity. In terms of pesticide production and consumption, India is the fourth largest user of pesticides in the world (FAOSTAT, 2017). India is one of the lowest consumer i.e. 0.5 kg/ha (0.06 mt) of pesticides in the world as compared to China (13.06/ha), Japan (11.85 kg/ha), Brazil (4.57 kg/ha) etc. The United Nations has declared 2020 as the International Year of Plant Health (IYPH). The main aim is to raise global awareness on how protecting plant health can help end hunger, reduce poverty, protect the environment, and boost economic development. FAO estimates that up to 40% of food crops are lost due to plant pests and diseases annually. This leaves millions of people without enough food to eat and seriously damages agriculture - the primary source of income for rural poor communities. .

Plant parasitic nematodes (PPNs) cause 12.6% ($215.77 billion) economic damage to food crops at global level (Abd-Elgawad and Askary, 2015). Root knot (*Meloidogyne* spp.) and cyst nematodes (*Heterodera* and *Globodera* spp.) are economically important species. Various methods have been employed for the management of PPNs, which include chemical pesticides, crop rotation, and development of resistant cultivars and using of biological control agents. Use of chemical pesticides has been most effective for the management of PPNs. However, majority of them have been withdrawn due to their environmental toxicity and health effects (Davies and Curtis, 2011; Kumar et al., 2019). Crop rotation, a cultural method is inadequate to control PPNs (Shojaei et al. 2019). Nematode resistant genes are present in some plant cultivars. But they are limited to specific pathotypes and overtime, this strategy lead to the replacement by another species.

The emerging of new approaches *viz*., nanotechnology, molecular mechanisms (RNAi, CRISPR/Cas9) and identification of highly virulent biological control agents and other new approaches has resolved many concerns prevailed with the earlier strategies for pest management. Weed usually requires external efforts to be excluded from the cropping and eventually reduce the economy, deteriorate the environment, human health and amenity. Weeds alone account for 37% of total annual crop yield loss in India (Yaduraju, 2012). That is why weed management has become an integral part of the crop husbandry. Conventionally, mechanical, cultural, chemical and biological method of weed management is adopted for keeping or minimizing the weed population and their growth below the level of economic injury to the crop. Weed management are thus essential to the production of high-yielding and high-quality crops.

Pesticide issues

Newer molecules such neonicotenoids, avermectins, milbimycins, microbial disruptors of insect midgutmembranes, nereistoxin, neuromodulators entomopathogens, novel biological pesticides have been introduced to control insect pests. However, pesticide regulation mechanism in India needs to be strengthened to check manufacturing, sale and distribution of spurious and substandard pesticides, encouraging environmental friendly molecules of botanical biological origin, quality assurance etc. Due to enormous use of spurious pesticides by uneducated small and marginal farmers of country, pest problems remain high. Non availability of branded pesticides and microbial insecticides on credit basis and lack of awareness among large rural farmers are the reasons for use of spurious pesticides. The size of the spurious pesticide market in India is as high as 30% by volume and 25% by value and spurious chemicals market is growing at 20% per year leading to a loss of about 10.6 million tonnes in food grain production, particularly in 2015 (FICCI, 2015).States like Uttar Pradesh, Madhya Pradesh, Andhra Pradesh, Maharashtra, West Bengal, Haryana, Maharashtra, and Karnataka are some of the worst affected. Moreover, non-judicious uses of synthetic pesticides lead to insecticide resistance, resurgence, and residue. Few such problems are resistance in polyphagous pests like *Helicoverpa armigera, Spodoptera litura, Pectinophora gossypiella, Plutella xylostella*; resurgence of sucking pests like *Bemisia tabaci*. Indian exports of food products such as mangoes, table grapes, okra, peanuts, chillies, are facing rejections and even bans in markets such as the US, Vietnam, EU, Saudi Arabia, Japan and Bhutan due to issues such as presence of higher than approved levels of chemical residues, and pest infestation (Mukherjee et al., 2017). While importing planting material and other agricultural commodities, India also bearing risk of introducing exotic pests and diseases along with imported material also rises. Therefore, plant biosecurity seeks to prevent, minimise and control the introduction and spread of these pests in India in the course of international trade.

In this chapter, forecasting/forewarning techniques, innovative agronomic practices, precision pest management technologies, nanotechnology based formulations, innovative biological pest suppression, novel biotechnological interventions, strengthening biosafety and biosecurity issues, reforms in the pesticide management and future prospects in India are discussed.

Innovative agronomic practices

Employing release of egg parasitoids, *Trichogramma* has been successfully demonstrated worldwide and advocated to be an alternative to chemical insecticides in many crops including rice. Organic mulches apart from bringing advantages for many abiotic stresses also act as an Innovative pest management strategy. In sweet potato, mulches of dry cypress, eucalyptus and lucerne reduced movement of sweet potato weevils, *Cylas formicularis* from a release point to reach sweet potato plants. Similarly, fruitflies infesting wide variety of economically important fruit crops can be controlled by a low-cost fruitfly trap consists of extract along with crushed leaves (20 gm) of *Ocimum sanctum* (holy basil) placed inside coconut shell and shell filled with 100 ml water and poisoned the solution by adding 0.5 gm carbofuran 3G and 0.5 gm citric acid (4 traps/tree) and fruit flies feeding on *Ocimum* will die. Increase research on crop rotations to consider the inclusion of new major crops, including industrial crops, companion crops and crops which are likely to be displaced from other regions due to climate change.

Monitoring, diagnosis, forecasting and forewarning

Timely monitoring of the notorious insect pests is very essential as they can cause countrywide devastating crop losses in a short span of time. Invasive and transboundary insect pests such as locusts, fall armyworm, fruitflies, bollworms, whiteflies etc are of great economic importance in agriculture and horticultural ecosystems in India. Recent advancements in remote sensing, machine learning, artificial intelligence has enabled to monitor pests in inaccessible sites. Survey and surveillance are very crucial to record dynamics of key pests and also invasive pests on crops. They also play a pivotal role in preventing and containing potential and harmful biotic stresses from causing serious crop losses. Monitoring, providing early warning and acting promptly to control outbreaks of migratory pests are crucial to ensure timely action when an emergency arises. Recent advances in robotics, space research has enabled to monitor pest populations in remote and inaccessible areas also as in case desert locust. Machine-learning algorithm was also effective to identify desert locust breeding sites in East Africa using key climatic factors, temperature and rainfall as well as edaphic factors like sand and moisture contents along with greenness. The prediction model developed based on these factors could predict oviposition and breeding sites accurately in East African countries, Kenya, Uganda, Sudan, Saudi Arabia which would aid in preventing pandemics by destroying eggs and immature stages in small breeding sites. Such novel technological interventions can save crops from devastating losses caused by desert locust pandemics across the continents. Recently, innovative harmonic radar has been developed and utilised for tracking the movement of *Vespa velutina* a pest of honey bees and to identify its nests by fitting on insects (Maggiora et al., 2019). This technology can be very well adopted to other insect pests such as locusts, bees, butterflies, moths, bumble bees, dragonflies etc.

Precision pest management technology

Precision pest management technologies offer better opportunities for integrated control of key insect pests as they are spatially aggregated by reflectance-based pest hot spot monitoring as well as precision control systems, like distributors of natural enemies and pesticide spray (Iost Filho et al., 2019). Drone-based multispectral sensor have been successfully employed to detect crop pests such as sugarcane aphid *Melanaphis sacchari* Zehntner in sorghum (Stanton et al. 2017), Colorado potato beetle *Leptinotarsa decemlineata* Say damage in potato (Hunt and Rondon 2017), whitefly induced stress in soybean (Iost Filho 2019), stress induced by two-spotted spider mite in strawberry, spatial and temporal distribution of the *P. solenopsis* mealybug (Prabhakar et al., 2013). These technologies are low cost, may cover larger areas, have higher monitoring frequencies and time saving.

A web enabled forecasting models for monitoring rice insect pests viz., yellow stem borer (*Scirpophaga incertulas* Walker; Pyralidae; Lepidoptera), leaf folder (*Cnaphalocrocis medinalis* Guenee; Pyralidae; Lepidoptera), brown plant hopper (*Nilaparvata lugens* Stal; Delphacidae; Hemiptera), white backed plant hopper (*Sogatella furcifera* Horváth; Delphacidae; Hemiptera) and gall midge (*Orseolia oryzae* Wood-Mason; Cecidomyiidae: Diptera) have been developed with an accuracy of 65-100% and can be employed to issue pest alerts for timely management of these key pests. Farmers can make use of this web based forewarning tool for timely control measures thereby by they can protect the rice crop (Vennila et al., 2016).

New innovations in biological pest suppression

Biological control is a potential sustainable alternative to synthetic pesticide use. Drones may be particularly useful tool for augmentative biological control, which relies on the large-scale release of natural enemies for immediate control of pests. Drone-released *Trichogramma* parasitoids are employed to combat the sugarcane borer in Brazil and also against various other lepidopteran pests in other crops. Other types of natural enemies can be drone-applied as well, such as green lacewing, Chrysoperla spp. and minute pirate bug to control aphids and thrips, and mealybug destroyer. Chemical ecology is one of the promising areas of pest control in the recent past. Bioefficacy of biological control agents can be enhanced by spraying insect/plant originated and also synthetic kairomonal compounds which was witnessed in rice. The parasitic rate of *Trichogramma japonicum* has been enhanced from 25% to a maximum of 95% on eggs of yellow stem borer a key pest of rice in India (Murali-Baskaran et al., 2019).

Genetically modified crops

Genetically modified (GM) crops are being widely accepted and successfully cultivated in many developed countries. GM crops offer great advantage by providing continuous inbuilt resistance which was witnessed in Bt cotton resistant to bollworms as they contributed significantly in reduction of chemical insecticide use and survival of non-target coexisting beneficial organisms. Greater success has been achieved through bollworm resistant *Bt* cotton not only in India but also in the world.

Nanopesticides

Nanopesticides or nano-plant protection offer a variety of benefits including increased effectiveness, durability, less environmental contamination through reduction in pesticide application rate, enhanced efficacy of controlled release of pesticide, low photo-degradation and safe handling that are being used in protecting crops against diseases and insects. Nano-encapsulation is currently the most promising technology for protection of host plants against insect pests. List of nanopesticides available in Indian market are presented in the table below:

Type of product	Nature of nanocontents	Manufacturer	Application
Herbicide	Nano-formulated	Tamil Nadu Agricultural University (India) and Technologico de Monterry (Mexico)	Attacks the seed coat of weeds, destroy soil seed banks and prevent weed germination
Nanopesticide	Nanoparticles	Comprising a Crop Protection Agent Nano pesticide for crop protection (10 and 150 nm)	BASF
Nanoemulsion	Nanoemulsion	Bayer Crop Science	Nanoscale droplets in the range of 10-400 nm in insect pest control
Agro Nanotechnology Corp	Nanopesticide	Nano-Gro	Mimics stress conditions, increasing crop activity and yield. It increased average crop yield by 20% and for some crops even more: sunflower by 50%;rice by 35% and cucum

Source: Bhan et al., 2018

Biodegradable nanoparticle carriers would be ideal for delivering nanopesticides for environmentally safe pest management. In this connection, a clay nanotube, halloysite have been developed as carriers of pesticides which will greatly reduce the required amount of pesticides. Despite of sincere efforts nanopesticide market is yet to be picked up well as research efforts on a) green chemistry with environmental sustainability principles in nanopesticide development to maximize their efficiency, b) process development for up scaling of nanopesticide for commercial level, c) environmental impact assessment of nanopesticide to determine toxicity level, d) improvement in regulation for nanomaterial application in agriculture and e) development of smart nanopesticide solution to pest problems (Chhipa, 2017). The potential application of engineered nano-materials in agriculture is also ascertained in disease and weed management. Inorganic nano particles, such as zinc oxide (ZnO), copper (Cu), calcium oxide (CaO), Magnesium oxide (MgO) and Ag nano particles play important roles in various area of crop protection including microbial activity and bacterial diseases. Nano-silica is hydrophobic in nature and can absorb into the cuticle layer of the insects upon contact, leading to ultimate death of insects. Zinc oxide nanoparticles have been shown to reduce the growth of *Penicillium expansum, Fusarium graminearum*, *Alternaria alternata*, *Rhizopus stolonifer* and *A. flavus* as well as against plant pathogenic bacteria viz., *Pseudomonas aeruginosa* (Servin et al., 2015). Studies conducted by Luksiene and co-workers in 2019 revealed that ZnO nano particles in the presence of sun-light activated by ultra violet and visible light can protect strawberry fruits from *Botrytiscinerea* infection more effectively than Fenhexamid fungicide. The potential of nano-materials encourages a new green revolution with reduced farming risks. However, there are still huge gaps in our knowledge of the uptake capacity, permissible limit and the eco-toxicity of different nano-materials. Hence, further research is urgently needed to unravel the behavior and fate of altered agriculture inputs and their interaction with bio-macromolecules present in living systems and environments.

Nanotechnology is an emerging and promising technique for nematode management due to its characteristics such as small particle size, controlled delivery of functional molecules and strong inhibitory or antimicrobial activities towards PPNs (Kalaiselvi et al., 2019). Metal based nanoparticles such as silver, gold, copper, titanium oxide and zinc oxide have been utilised for nematode management. Among these, silver nanoparticles (AgNP) have shown evidence of being a potentially effective nematicide (Roh et al., 2009) and its toxicity is associated with the induction of oxidative stress in the cells of targeted nematodes (Lim et al., 2012). Other metal based nanaoparticles have been reported effec tive against PPNs. For example, application of PEGylated CuFe NPs were highly effective against *M. incognita* and *M. javanica* (EC50 value= 0.03 μg ai g/soil) Gkanatsiou et al. (2019) which further resulted in increased shoot weight and root weight in tomato.

Although, several protocols have been followed, the inclusion of toxic chemical substance during synthesis of nanoparticles cannot be avoided (Hardman, 2006). Hence, the development of ecologically safe green synthesis methods using natural bioresources such as plants and microorganisms could act as an eco-friendly alternative to chemical and physical approaches.

Biotechnological approaches

Exploiting the RNA interference (RNAi) gene mechanism to silence essential genes in pests is one of the novel and innovative approach since the year 2000. Much work has been done which proved well in the laboratory studies for the management of whitefly, aphid, fruit flies, brown plant hopper, lepidopteran larvae, coleopteran grubs, hymenopterans etc (Christiaens et al., 2020). However, Bayer Crop Science for the first time in the world had successfully developed RNAi based transgenic corn which expresses a hairpin dsRNA targeting the snf7 gene resistant to Western corn rootworm, *Diabrotica virgifera* and this new RNAi construct is also stacked with two *Bacillus thuringiensis* Cry proteins (Cry3Bb1 and Cry34/35Ab) named as SmartStax Pro, in an effort to delay the evolution of resistance (Head et al., 2017). It is only one translational research of RNAi/dsRNA that has reached the field, the same can extrapolated to other economically important insect pests. Sincere attempts were made to develop dsRNA based insecticides for controlling insect pests but not much success.

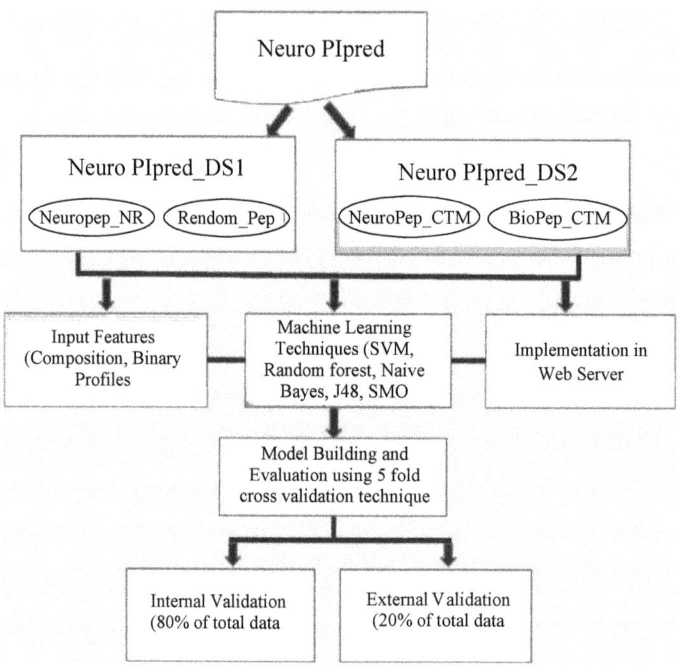

NeuroPIpred, a web based tool to predict, design, scan neuropeptides to identify novel target site for the management of insect pests. Insect neuropeptides are responsible for vital physiological functions of insect such as moulting, homeostasis, diuresis, signaling processes, pheromone synthesis and muscle activities. Agarwal et al. (2019) has developed this tool using machine learning technique by employing more than 50 neuropeptide families and over 400 different insect species which will help only in discriminating insect neuropeptides with non-neuropeptides but also new target sites for efficient pest control.

RNAi-Based Biofungicides

Chemical based fungicides are mainly used for management of fungal diseases, including the grey mold diseases (*Botrytis cinerea*). The application of these fungicides impairs the crops metabolic & physiological mechanisms viz., photosynthesis, reproductive organ development, CO_2 assimilation and retardation of plant growth etc. Further extensive use of fungicides generates long term residues in the environment. Hence, new eco-friendly approaches viz., spray induced gene silencing is a new RNAi-based approach for silencing targeted genes in plant pathogens using exogenous applications. The exogenous application of double stranded RNA (dsRNA) and siRNA has been reported to reduce grey mold infection in strawberry and tomato fruits as well as detached leaves of oilseed rape (Islam and Sherif, 2020). Spray induced gene silencing could be efficiently used for the pre & post-harvest management of *Botrytis cincrea* in major horticultural crops such as tomato, strawberry, and grapes etc. However the target genes for RNAi based fungicides are still limited, hence the detailed transcriptomics analyses are needed to identify specific genes for biotrophic and necrotrophic phases of grey mold infections. Moreover, the uptake mechanisms of exogenous dsRNA by plants or fungal cells remain inexplicable. The clear understanding of the roles of membrane-bound proteins and receptors of plant and fungal cells may improve our understanding of the underlying small RNA uptake mechanisms. The stability of naked dsRNA molecules under field conditions is another major concern that may limit the application of SIGS-based disease management strategies. However, the utilization of nanoparticles and other stabilizers for enhanced stability and sustained release of the RNAi-biofungicides could overcome these limitations. The development of cost-effective and scalable approaches for the production of RNAi-biofungicides is probably the major challenge facing the practical agricultural utilization of this technology.

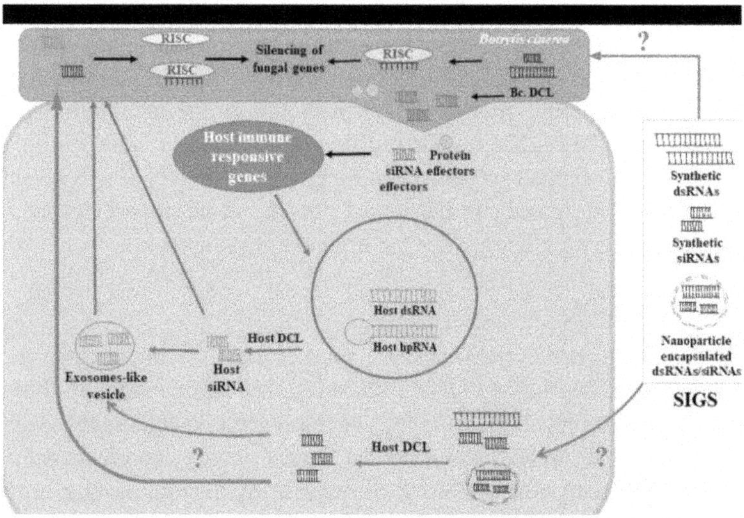

Fig. 1. Spray induced gene silencing (SIGS) by using exogenous applications of dsRNA or siRNA alleviates gray mold infections by targeting essential genes of *Botrytis cinerea* virulence and pathogenicity. hpRNA: hairpin RNA; RISC: RNA induced silencing complexes

(*Source*: Islam M.T., and Sherif S. M., 2020)

Gene editing

Genome engineering or genome editing has emerged as a new breeding technique with the ability to edit the genomes of plants, animals, microbes, and human beings. Gene editing, using CRISPR-Cas9, has formed the foundation for novel and innovative control strategies for insect vectors of human diseases. Similar tool is being largely investigated for control of agricultural pests across the world. The scientific community has unraveled the utility of various editing tools from endonucleases to CRISPR/Cas in several aspects related to plant growth, development, and mitigation of stresses including designing insect pest resistance in plants. In the case of Drosophila, a mutation in Or83b gene disrupted the selection of the egg-laying site (host) and impaired olfactory detection capacity and knock out of the Orco (olfactory receptor coreceptor) gene in *Spodoptera litura* through CRISPR/Cas9 demonstrated distraction in the mating partner selection and loss of identity toward host plants leaving them anosmia. CRISPR/ Cas9-based knock out of odorant receptor 16 (OR16) in *H. armigera* made males unable to receive pheromone signals from mature females, thus resulting in mating with immature females that subsequently led to dumping of sterile eggs. Therefore, knock out of OR16 receptor using CRISPR/Cas 9 tool in lepidopteran pests can therefore be a novel and effective strategy to regulate mating time for pest management in agricultural crops.

Prime editing, a unique protein fusion technology has been developed to improve homology directed repair (HDR) provides advantages over traditional CRISPR-Cas9-based gene editing as off-target instances are less. This innovative and novel technique, used in case of in rice and wheat protoplasts, has demonstrated the success of plant prime editors in point mutations, insertions, and deletion (Tyagi et al., 2020). To harness the benefits of this technology, there is a need to develop extensive phenotyping and enrichment of the resistance gene pool that needs to be addressed assessment of the available germplasm including wild relatives of specific crops for insect response and identification of stress responsive genes using multi omics approaches.

Biosafety and Biosecurity

Plant Biosecurity is of paramount importance to any country especially India to safeguard food-security, sustainability of agricultural/horticultural production and also in protecting livelihood of people. The recent concern in the national agriculture scenario is to manage many types of invasive alien pests (insects / mites / nematodes / diseases etc.) species that cross over from across the globe. The following examples are illustrative of the paramount alert that the farming community has to keep the onerous task of sustaining well-planned national crop production and productivity. Some recent examples of success stories of managing such invasions in various states are (i) Sugarcane mealy bugs, (ii) Cotton mealy bugs, (iii) Papaya mealy bugs, (iv) Cotton leaf curl virus, (v) Rugose Spiralling Whitefly in Palms, (vi) Fall Army worm (*Spodoptera frugiperda*), (vii) unusual invasion of locusts in new areas etc. Therefore, there is a need to protect the entry establishment & spread of these exotic & invasive plant pests in India. Step to combat invasion of exotic pests are (i) pre border -off shore inspection, certification and treatments, Pest Risk Analysis & quarantine regulations (ii) border/ entry point - application of quarantine regulations, inspection, sampling, testing and treatment (iii) post border- regulations for growing in post entry quarantine facilities, pest surveillance, exotic pest monitoring, containment and/ or eradication of pests. Strict monitoring of entry of invasive/quarantine pests at potential international borders to prevent entry of potentially harmful pests, pathogens, microbes nematodes, infected planting material etc. Model developed at high altitudes of Himalayas for monitoring entry of wheat rust pathogens in India should be extrapolated to other potential pests. In India, the Destructive Insects and Pests Act 1914 need to be adopted for plant protection and quarantine regulatory measures. The import of plants and plant materials is regulated under the Plant Quarantine (Regulation for Import into India) Order 2003 issued under the act. The Plant Quarantine Order addresses gaps in the previous Plants, Fruits and Seeds Order regarding the import of germplasm, genetically modified organisms (GMOs), transgenics, biocontrol agents and similar specimens. The Directorate of Plant Protection, Quarantine and Storage under the Ministry of Agriculture carries out plant quarantine operations in India. The manufacture, import, use, research and release of GMOs are controlled and regulated by the Ministry of Environment Forests and Climate Change (MoEFCC) and the Department of Biotechnology of the Ministry of Science and Technology, pursuant to the 1989 rules notified under the Environment (Protection) Act 1986 and MoEFCC regulates the commercial release of GMOs through the Genetic Engineering Appraisal Committee. While Department of Biotechnology deals with all aspects of transgenics research. It reviews, permits and monitors experiments which use GMOs and recombinant DNA products. Recently, introduction of potato tubers from neighbouring countries has been stopped due to high risk of introducing of plant viruses.

Remote sensing usage in biotic stress management

Progress within the remote sensing domain along with advances in computer science and sensor technologies, have served as basis for the development of tools and methods for the assessment of plant diseases (Prabhakar, et al., 2012). Major technological advancements in monitoring of plant stresses depend on the capacity of remote sensed observations providing spatial, temporal and non-invasive field based information related to the plant condition and further diagnostics of plant disease stress at early stages which potentially represents one of the major contributions of remote sensing to agriculture. Many important diseases of wheat, peanut, rice, Apple, Sugarcane and mustard etc. were detected by remote sensing technologies (Prabhakar, et al., 2012). Details are given below:

Remote sensing based detection of plant diseases

S.No.	Crop	Disease	Platform	Spectral resolution	Optimum bands (in nm)/ indices/ technique used
1.	Wheat	Leaf rust, yellow rust	Space borne	Multispectral (Landsat)	NDVI
2.	Peanut	Leaf spot	Ground based	Multispectral	800 nm
3.	Apple	Scald	Ground based	Hyperspectral	550 700 nm; 551 BRI
4.	Rice	Panicle blast	Ground based &Airborne	Hyperspectral	485,675 nm; (R470/ R570);(R520/R675); (R570/R675);(R550/ R970);(R725/R900)
5.	Alfalfa	Leaf spot	Ground based	Multispectral	810 nm
6.	Sugarcane	Orange rust	Spaceborne	Hyperspectral (Hyperion)	DWSI

S.No.	Crop	Disease	Platform	Spectral resolution	Optimum bands (in nm)/ indices/ technique used
7.	Rubber	Corynespora	Spaceborne	Multispectral (IRS1C)	NDVI
8.	Tomato	Late blight	Ground based Airborne	Hyperspectral Multispectral (ADAR)	5-index feature vector method
9.	Rice	Sheath blight	Airborne	Multispectral (ADAR)	RI, SDI
10.	Mustard	Alternaria	Space borne	Hyperspectral (Hyperion)	DWSI
11.	Wheat	Powdery mildewand leaf rust	Airborne and Space borne	Hyperspectral (HyMap) Multispectral (QuickBird);	SMA, MTMF, NDVI
12.	Rice	Bacterial leaf blight	Ground based	Hyperspectral	943 and1039 nm MLR
13.	Tomato	Bacterial leaf spot	Ground based	Hyperspectral	Partial least squares (PLS) regression stepwise multiple linear regression

(*Source*: Prabhakar, et al., 2012 & Rojas F.A., 2018)

Though application of air-borne remote sensing for biotic-stress has been in vogue in several developed countries, it is yet to find wide usage in many of the developing countries. One of the main reasons could be the high cost involved, availability of suitable sensors, small farm holdings and diverse cropping systems. Limited availability of fine spatial resolution, near real-time data has slowed the application of satellite remote sensing in the past, but now with the launch of new generation satellites this might not be a limiting factor. The narrow bands in the hyper-spectral sensors are able to measure the characteristic absorption peaks of plant pigments and other related parameters more precisely and thereby provide better information related to plant health. But availability of hyperspectral data from satellite platforms is still in its infancy. Nevertheless, airborne and space borne remote sensing can provide spatial variability of biotic stress and a synoptic view of the large area in a non-destructive and non- invasive way.

Artificial intelligence in biotic stress management

Remote sensing technology provides the capability to acquired huge amount of data, complex analyses were explored to find hidden regularities in the growing volumes of data that could contribute to the establishment of suitable solutions to the crop management community. Many authors have been explored different machine learning approaches to provide new insights into the complex host-pathogen systems using hyperspectral information. For instance, Rahman, et al. (2017) compared different machine learning algorithms like support vector machines (SVM), random forest (RF) and multilayer perceptron (MLP) to classify healthy and non-healthy plant leaves of sorghum, citrus and cabbage. They concluded that the use of color-based features, together with the use of a non-parametric classifier such as the random forest, provided a good combination of procedures to distinguish between healthy and infected plant leaves. Studies conducted by Selvaraj, et al. (2019) showed that the DCNN AI was a robust and simple deployable strategies for digital banana diseases and pests detection. A pre trained disease- recognition models were able to carry out deep transfer learning to produce a networks that can make clear predictions. This important high success rate makes the model a useful early pest diagnostic tool and this study can be further extended to develop a completely automated mobile app to help millions of farmers in developing countries. Another study conducted by Selvaraj, et al. (2020) revealed that using combination of vegetation-indices along with principal-component analysis shown promising option to map banana under mixed African landscapes. The high resolution sensors: UAVs, PS and WV-2, used in this study were found to be more accurate to map banana than medium resolution satellite images (S2). Accurate banana mapping using open-source medium resolution satellites (S2) under mixed complex system is still challenging, but integrating informations from different (medium and high resolution) sensors is way forward. Through UAV-based RGB imagery systems, we were able to detect banana and their major diseases with higher accuracy and fewer errors.

Biopesticides

Biopesticides as a green technology have emerged as an alternative to the chemical pesticides offering solutions to the concerns of pathogen and pest resistance/ resurgence, detrimental effect of synthetics on public health, soil microorganisms, ecosystem and environment as well as increasing cost of pesticides. There is a galaxy of antagonistic microorganisms which are used in biocontrol of plant diseases and pests as mentioned as:

Potential Bioagents to mitigate biotic stress

Bacteria	Fungi	Viruses
Disease management *Bacillus subtilis B. cereus B. lichenformis Pseudomonas fluorescens P. putida P. aeruginosa P. alcaligens P. aureofaciens P. cepacia Agrobacterium radiobacter Erwinia amylovora (Harpin Protein) Burkholderia cepacia Serratia marcesans* GPS 5	**Disease management** *Trichoderma viride, Trichoderma virens, Trichoderma harzianum, Trichoderma hamatum, Trichoderma koningii Chaetomium globosum, Penicillium oxalicum Penicillium islandicum, Aspergillus niger Ampelomyces quisqualis, Candida oleophila Myrothecium verrucaria, Fusarium oxysporum* (non pathogenic), *Coniotryrium minitans Gigaspora margarita Gigaspora gilmori Glomus mosseae G fasciculatum G. etunicatus G. macrocarpum G. intraradices G. manihotis G. margarita G. gigantean G. epigaeus G. deserticola Acaulospora laevis Acaulospora marrowae Sclerocystis* sp. *Piriformospora indica Phelbia gigantean*	**Disease management** Mycoviruses Bacteriophages Cross protecting mild strains of plant viruses
Nematode Management *Pasteuria penetrans Pseudomonas oryzihabitans Pseudomonas fluorescens Bacillus firmus B. thuringiensis B. subtilis*	**Nematode Management** *Drechslerella anchonia Drechslerella dactyloides Drechslerella brochopaga Dactylellina candida Dreschmeria coniospora Arthrobotrys oligospora Arthrobotrys dactyloides Coprinus comatus Stropharia rugosoannulata Pleurotus* spp. *Pythium caudatum Harposporium leptospira Catenaria anguillulae Catenaria auxiliaries Hirsutella rhossiliensis Verticillium chlamydosporium Monacrosporium ellipsoporum Paecilomyces lilacinus Pachonia chlamydosporia Aspergillus niger*	**Nematode Management**
Insect Pest Management *Bacillus thuringienis* var. *sphrarericus* *Bacillus thuringienis* var. *galleriae* *Bacillus thuringienis* var. *israelensis* *Bacillus thuringienis* var. *kurstaki*	**Insect Pest Management** *Metarhizium anisopilae* *Beauveria bassiana* *Verticillium lecanii* *Paecilomyces lilacinus* *Nomurea rileyi* *Photorhabdus luminescences*	**Insect Pest Management** Nuclear Polyhedrosis Virus NPV of *Helicoverpa armigera* NPV of *Spodoptera litura* Granulosis Viruses (GV)

Despite having a huge potential to manage devastating diseases and insect pests biopesticides are lagging behind in terms of usage, accessibility to the farmers and lack of registered products. They are still not popular and need sincere efforts in order to take off at a large scale in India. There is enormous market potential for microbial biopesticides and a number of initiatives at national and state level are underway to promote them as alternatives to chemical pesticides. Biopesticides are target-specific affecting only target pests and their close relatives. They do not destroy friendly insects, birds and mammals. Few registered biopesticide formulations are given below:

Biopesticide formulations registered under Central Insecticide Board (CIB) Act 3 and 3 (B) in India

S.No	Biopesticide	Formulation registered	Parameters tested by CIL
1.	*Ampelomyces quisqualis*	2.00%WP	CFU, Antagonistic capability pH, Suspensibility, Pathogenic contaminants
2.	*Bacillus subtilis*	2.00%AS	Heat resistant Spore count (CFU)
3.	*Bacillus thuringienis* var. *spharericus*	1.3%FC	pH, Moisture, Suspensibility
4.	*Bacillus thuringienis* var. *galleriae*	1.3%FC	
5.	*Bacillus thuringienis* var. *israelensis*	Liquid and WP formulations 5%AS, 12%AS, 5.0%WP	
6.	*Bacillus thuringienis* var. *kurstaki*	5%WP, 2.5% AS, 0.5%WP, 3.5% AS	
7.	*Beauveria bassiana*	1.15%WP, 1.0%WP, 1.15%SC, 10%SC, 5.0%WP	CFU count, pH, Moisture content Suspensibility, Pathogenic contaminants
8.	*Metarhizium anisopilae*	1.15%WP, 1.0%WP, 1.5% Liquid	
9.	NPV of *Helicoverpa armigera*	0.43%AS, 2.0%AS, 0.5%AS	POB Count, Pathogenic contaminants, pH, Moisture
10.	NPV of *Spodoptera litura*	0.5%AS	content, Suspensibility,
11.	*Pseudomonas fluorescens*	0.5% WP, 1.75%WP, 1%WP, 2.0%AS, 1.5%LF	Viable cell count, Antagonistic capability, pH, Moisture content Suspensibility, Pathogenic contaminants
12.	*Trichoderma viridae*	1%WP, 0.5%WP, 5.0%WP	CFU, Antagonistic capability pH, Suspensibility, Pathogenic
13.	*Trichoderma harzianum*	0.5%WP, 2.0%WP, 5.0%WP	contaminants

S.No	Biopesticide	Formulation registered	Parameters tested by CIL
14.	*Verticillium lecanii*	1.15%WP, 1.5% Liquid	CFU count, pH, Moisture content Suspensibility, Pathogenic contaminants
15.	*Verticillium chlamydosporium*	1.00%WP	

(*Source*: DPPQS, DAC Ministry of Agriculture and Farmers Welfare, Government of India)

The cost of registration of biopesticides in India is very high. Also there is a long-drawn-out process involved in registering a biopesticide as per the laid norms. Major task for bio-pesticide formulations is the verification of microorganism for extensive bio-safety and allergy through laid protocols/tests before registration. Initial research on development of bio-pesticides by many universities and research institutions are unable to take off as commercial formulations due to extensive battery of safety and allergy tests requiring high additional costs. Their efficacies are also dampened by humidity, temperature fluctuations and exposure to ultraviolet radiation. Its effectiveness in field conditions is severely reduced by any other microbial contamination. Selection of highly virulent strains requires continuous selection stable, location and host specific strain.

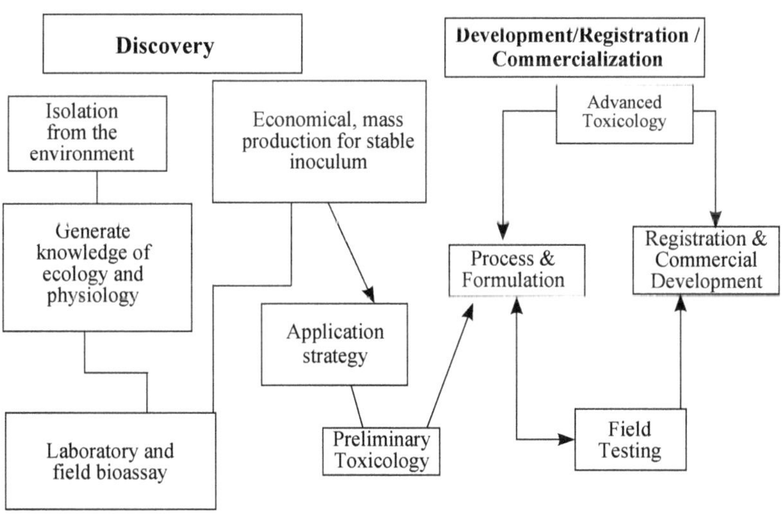

Fig. 1. Steps involved in development and commercialization of microbial product

Percentage distribution of different biopesticides

There have also been many mycoinsecticides as well as mycoacaricides developed worldwide. Among them the products that are commercial worldwide are based on *Metarhizium anisopliae* (33.9%), *Beauveria bassiana* (33.9%), *B. brongniartii* (4.1%) and *Isaria fumosorosea* (5.8%). They have been either registered or undergoing registration (deFaria and Wraight, 2007; Sandhu et al., 2012). Primarily focus of recent research in entomopathogens is basically hinged on the hyphomycetes belonging to genera *Beauveria, Metarhizium, Paecilomyces* and *Verticillium*,. They have a broader host range and are highly specific. The usage of entomopathgenic fungi is lagging far behind in India.

Cross Protection

Plants systemically infected with one strain of a virus, will not develop additional symptoms when inoculated with a second strain of the same virus. This phenomenon forms the basis of cross-protection tests. Commercially the technique was used in tomato industry to protect crops against severe strain of TMV by developing mild strains by mutagens in the Netherland and U.K. by growers. The yield was increased upto 15%. The technique has also been used in citrus in Florida, Brazil and some other countries.

Disease forecasting systems

Forecasting systems may aim on two aspects. Infection forecasting which prediction of initial appearance of disease is. It is generally based on amount of primary inoculums and prevailing environmental conditions. In disease forecasting, the emphasis is on further development of disease. Amount of secondary inoculums and number of infection cycles are important. Fundamental models involve research and experimentation for establishing fruitful relationships, identifying critical factor(s) and relating them with appearance (time and amount) of disease. First disease forecasting system in plant pathology was developed for vine downy mildew (*Plasmopara viticola*) in 1913 (Muller's incubation calendar, Germany). For prediction of plant diseases the criteria used are related with survival of inoculums, production and dispersal of primary and secondary inoculums and role of vector population. Some of the classical forecasting models are:

Apple scab forecasting

Apple scab (*Venturia inaequalis*) is one of the most important diseases of quality apples. Plants are exposed to attack by the fungal spores for very long period (bud breaking to fruit maturity). Apple – *Venturia inaequalis* system was a fit case for development of forecasting system to guide the growers for first (Prophylactic spray) and further need based sprays to reduce the pesticide usage. In epidemic years apple losses have been as high as 70%. In Himachal Pradesh, when the apple scab appeared in epidemic form in late 1970's onward, 5 Apple Scab Monitoring and Research Laboratories were established at Mashobra, Kotkhai, Thanedhar and Sarahan in Shimla, and Bajaura in Kullu district. Apple scab infection periods were predicted by using equipments, like de Wit 7 day leaf wetness recorder, Reuter Stokes apple scab predictor and SWG Biomat at different locations. When the conditions congenial for the development of apple scab were recorded, this information was given to the farmers by blowing sirens, radio broadcasts and personal contacts. Of all the systems, Reuter Stokes predictor was found to be the most efficient in predicting the infection periods at Kotkhai (Sharma and Gupta, 1995). Other Automated or electronic apple scab predictors are available. Recently Drones are also explored as a new tool to combat the apple scourge.

Forecasting late blight of potato

Van Everdingen (1926) analyzed the combined effect of several weather conditions on the development of *Phytophthora infestans* and evolved four rules, popularly known as Dutch rules, on which the appearance of blight was observed to depend. Krause, et al. (1975) devised for USA a complete forecasting system known as BLITECAST. The output of BLITECAT was in the form of recommendations for the grower, tailor made to his conditions. Other computer based expert system for late blight post BLITECAST were TOMCAST, BLITECAST and SIMCAST. NEGFRY is a computer based programme for control of potato late blight. It is based on two existing models: (a) model for forecasting risk of primary attack (first spray) and (b) model for timing of subsequent fungicide applications.

Control of nematodes

Biological control acts as an eco-friendly alternative to harmful chemicals in the management of plant parasitic nematodes. Currently, several fungi and bacteria have been widely used as biocontrol agents of PPNs.

The majority of research work has focused using various strains of *Fusarium oxysporum* which reduced the infection and reproduction of root knot nematode, burrowing nematode, lesion nematode and spiral nematode (Waweru, et al., 2014). While, another endophytic fungi, *Trichoderma* spp. parasitizes nematode eggs and juveniles by secretion of chitinolytic enzymes, serine proteases and trypsin like proteases (Szabó, et al., 2012).

Members of bacterial genera such as *Pasteuria*, *Bacillus* and *Pseudomonas* have shown great potential as biocontrol agents of PPNs. They infect the nematodes by either parasitizing (production of toxins, antibiotics, enzymes) or interfering with nematode-host recognition. Recently, plant growth promoting rhizobacteria (PGPR) were reported to be effective against PPNs besides enhancing the plant growth (Tabatabaei, et al., 2017).

A number of new nematicides have recently become available in different parts of the world. These chemicals provide better option for growers in managing nematode problems in the field. However, fluensulfone showed enhanced fruit yield compared to other nematicides. Application of fluensulfone (above 24 ml/100 m) through drip irrigation system resulted substantial reduction in *Meloidogyne* sp. population and increased fruit yield on tomato and cucumber under greenhouse conditions (Giannakou and Panopoulou, 2019). However, seed germination and root growth were affected with the recommended dose of fluensulfone under laboratory conditions.

Approaches for weed management

Advancements in weed control technology have had a huge impact on agricultural productivity. Robust weed control technology will successfully control weeds in spite of variability in the field conditions. Adaptable weed control technology has the capacity to change its strategy in the context of evolving weed populations, genetics, and climatic conditions. Weed control methods are improving over time as technologies and research become available. New approaches for weed control can assess the effectiveness of different weed control and the best weed management approach. Specialized weed management activities are required for weed control in specific circumstances including identification of new populations of weeds, weed control on verges and post-fire weed management. The emerging trends in technology and innovation that offer hope of providing sustainable weed management into the future.

1. **Infra-weeder weed control equipment:** The new Infra-weeder weed control equipment uses infrared heat to kill undesired vegetation. Infra-weeder equipment uses a propane- fuelled ceramic heating element that develops temperatures up to 1000 degrees Celsius, which applies infrared radiation to weeds. Because of the high heat produced by this machine, it eliminates wind borne weed seeds, bacteria, and moulds on the ground surface.
2. **Thermal weed control**: These include: fire, directed flaming, hot water, steam, microwave, infrared, ultraviolet radiation, electrocution and freezing. Heating results in the coagulation of proteins and bursting of protoplasm due to expansion, which kills the tissue.
3. **Microwaves and radiations:** It is based on the high energy of microwaves, which can kill weeds very efficiently. This method is highly targeted and there is no fear of non-targeted damage. This technology is effective against many weeds, but the energy required is very high, which increases its cost. However, its efficiency and energy budget may be decreased by flux configuration and through induction of thermal runaway in weed plants, making it comparable with other weed-control tools in terms of cost (Brodie, et al., 2011).
4. **Hot Water, Steam, and Hot Air:** Heat is the main source of energy can be used to kill weeds through hot water application. Hot water treatment for weed control has been evaluated in many countries with a great deal of success (Rask and Kristofferson, 2007). Hot water application proved

effective against most of the annual and a large number of perennial weeds. Its effectiveness is greater under dense weed population because of increased penetration ability (Hansson and Ascard, 2002). Use of steam instead of hot water has been observed as a more effective, quick, and sustainable method, especially in cases where weed control is on relatively hard surfaces.

5. **Pneumatic weed control**: An implement has been developed that injects compressed air into the soil to loosen and uproot small weeds on either side of the crop row. It has been used successfully in carrot, maize and sugar beet. These machines works best in dry soils.
6. **Lasers:** Lasers has been shown to inhibit the growth of water hyacinth (*Eichornia crassipes*). The treatment did not generally kill the weed but treated plants were smaller, propagated fewer daughter plants and covered less water surface than the untreated. More recently, the possibility of using a CO_2 laser as a device for cutting down weeds has been demonstrated. Lasers transfer high energy to plant tissues and raise the water temperature at the cellular level, resulting in cell death.
7. **Robotic selective application:** The ecoRobotix spraying robot is autonomously guided through crop rows using GPS and machine vision sensors to follow crop rows. This robot detects weed plants, and then uses two spray nozzles on delta robot arms to position two nozzles over weed plants and selectively apply herbicide directly to detected weed plants (Fennimore and Cutulle, 2019).
8. **Flame weeding:** Flame weeding uses the heat generated from one or more propane burners to kill weeds. Intense heat sears the leaves of the weeds, causing the cell sap to expand, damaging cell walls (Singh, 2014). This causes leaves to wilt and prevents water from moving from the roots to the leaves. In a short period of time, the plant withers and dies.
9. **Remote Sensing:** Remote sensing is a modern technology used in agriculture to ensure the precision management of inputs as well as to frame out weed presence (Thorp and Tian, 2004). In agricultural field, weeds are present in patches. Thus, remote sensing is a good option to minimize the herbicide load and also production cost by enhancing the herbicide application efficiency. It is largely based on differential spectral reflectance of weeds and other vegetation, like crops and spectral resolution of the instrument in use. These two factors govern the efficacy of remote sensing in mapping weeds. The precision and accuracy of weed management largely depends on sharpness of the pixels, quality of the picture, which is crucial for subsequent mapping.
10. **Robotics:** In recent past, with the advancement in robotics research and their use in agriculture, it is now possible to use them weeds control

(Young et al., 2014). Robotic weed management is a four-step process, involving guidance, identification, precision robotic removal, and mapping of weed species. The feasibility of a robotic weed control system depends upon machine vision analyses, robotic efficiency/suitability, variable rate application technology, decision support system, and strength of weed-sensing tools, directly or indirectly (Slaughter et al., 2008).

Innovations in pesticide management in India

New pesticide management bill to make amendments in light of the today's context is very important. In order to encompass all gamut of agro-chemicals management in agriculture and public health that are expected to be in use in for farming, agriculture commodity storage, animal health / veterinary purpose, public health and any related fields, this comprehensive draft Pesticide Management Bill (PMB), 2018 has been presented for public view seeking relevant comments. All commodities have to be compliant to the notified Good Agriculture Practices (GAP). Pesticides of biological origin and developed from plants and useful microorganisms such as fungi and bacteria are also to be included in the Schedule for being registered with the following features; Bill should replace the Insecticides Act, 1968 and should have strict regulation on sale of spurious, substandard and misbranded pesticides and should give State Governments more power to deal with the issue and take action against culprits. Bill provides for paying of compensation to the affected farmers or users under the provisions of Consumer Protection Act, 1986. Recently, the Indian Government in its Gazette notification proposed blanket ban on 27 pesticides. These pesticides are registered for protecting against a wide array of pests and diseases in 74 crops and does not have suitable alternatives. The government may have contingent plan to combat pest problems before banning 27 pesticides so that farmers can save their crops. Moreover, PMB has been drafted to replace age old Insecticide Act 1968 with many amendments such as a) reorienting the regulatory mechanism, b) establishing environment for growth of pesticide industry, c) strengthening pesticide research and innovation for development etc.

Policy reforms

The experts of plant protection in India have made a roadmap including policy issues to the Government of India.

- There is an urgent need for national Policy on Agrochemicals with an emphasis on safe use of pesticides.
- There is an urgent need for fast track transparent time bound on-line registration system by revamping the current registration process based on the recommendations of duly constituted independent expert committee. Guidelines for registration of newer molecules shall also be harmonised and also re-registration of pesticides which will take a mandatory period of 10 years after registration.

- Complete check on sale and use spurious pesticides indeed is the need of the hour.
- High quality well equipped quality control accredited national pesticides testing laboratories across the country should be established with international standards.
- Bio-safety data for new molecules to be generated preferably through notified National Accreditation Board for Testing and Calibration Laboratories (NABL) only.
- Pesticides of biological origin, growth regulators whose toxicity levels are very low can be relaxed from stringent regulatory ambit.
- Strengthen and encourage manufacturing of new molecules under Make in India initiative.
- Capacity building and skill development in assessing unregistered pesticides in imported commodities need to be strengthened.
- Urgent need to move forward to innovations in biopesticides of biological origin, green chemistry pesticides etc.
- A centre of excellence (CoE) on agrochemicals with multifaceted wide spectrum and modern bioscreening facilities needs to be established urgently.
- Greater thrust needs to be given to develop low cost technologies for mass production and bulk availability of biocontrol agents and biopesticides.

Future challenges

- Establishment of national referral laboratories for quick identification/ diagnosis and containment in order to prevent the wide spread of pest. A well-defined and coordinated mechanism should be developed for safe guarding our country. Artificial Intelligence for monitoring, tracking breeding sites, migration of movement of locusts and aerial application of pesticides for site specific management of transboundary quarantine pests.
- Exploitation of advanced breeding techniques, including agricultural biotechnology, to introduce insect and disease resistance into major crop plants. Much success has been achieved for enhancing yield traits using such technologies but research is limited in breeding biotic stress resistant varieties against key pests (except Bt cotton).
- Establish Public-Private Partnerships to contribute to control biotic stresses.
- Modelling of crop pest epidemics and disease outbreaks for major pests across the crops should be developed.
- Development and implementation of a national crop protection strategy founded based on the principles of integrated management.

- Investment in training provision among researchers and users to improve use of new technologies and innovation.
- A nationwide programme on robust monitoring and advisory system for quarantine, invasive, transboundary biotic stresses need to be developed.
- Robust forewarning systems for timely management of invasive, quarantine pests are to be developed in tune with changing climate, Extrapolation model, INDO-BLIGHTCAST being employed for controlling late blight disease for other biotic stresses in India.
- Exploration of semiochemicals/pheromones or allelochemicals for the management of sucking pests is a big challenge ahead of scientific community. Utilization of such chemicals in trapping technology would be of great help to faming community in overcoming losses caused by vector and vector borne viral diseases.
- There is a need to put in lot of efforts on development of suitable consortia of microbes for the management of biotic/abiotic stresses and reduced cost of cultivation.
- The biocontrol agents need to be assessed for their compatibility with commonly used agrochemicals periodically.
- Induction of higher tolerance in strain / isolates of the bio – agents against need based pesticides for control of soil borne diseases.
- Compatibility studies of biopesticides with biofertilizers
- Fast-tracking of biopesticide registration is one block that needs to be addressed without bias and involvement off red tapism
- Early warning system and pest advisory needs to be developed and strengthened in view of changing climatic scenario.
- Climate risk management through developing DSS (Decision Support System) and improving agro-advisory network services
- Develop simulation models and forecasting system to predict the outbreak of diseases
- Extensive field studies under various conditions should be needed to understand the interaction between nanoparticles, microorganisms, soil and plants against PPNs.
- Strategies on utilisation of multiple biocontrol agents with complementary and synergistic modes of action and integration of biocontrol agents with other management methods such as chemicals, crop rotation and use of resistant cultivars could be highly effective against PPNs
- There is a need to develop strict regulatory mechanisms to maintain the quality and availability of the biocontrol agents at affordable cost to farmers.

- Weed management policy need to be formulated in urgent basis to control the severity, spread and threat to the environmentEmpower the community to manage weeds under the policy.
- Supporting and contributing to a regional approach to weed management and control.
- Human resource development in order to support the weed management programme at village/community level.

References

Bhan, S., L. Mohan and C.N. Srivastava. 2018. Nanopesticides: A recent novel ecofriendly approach in insect pest management. Journal of Entomological Research, 42:263-270.

Brodie, G., C. Ryan and C. Lancaster. 2011. Microwave technologies as part of an integrated weed management strategy: a review. International Journal of Agronomy. DOI: 10.1155/2012/636905.

Chhipa, H. 2017. Nanopesticide: current status and future possibilities. Agric Res Technol, 5:1-4.

Davies, K.G. and R.H.C. Curtis. 2011. Cuticle surface coat of plant-parasitic nematodes. Annual Review of Phytopathology. 49: 135-156.

de Faria, M.R. and S.P. Wraight. 2007. Myco insecticides and myco acaricides: a comprehensive list with worldwide coverage and international classification of formulation types. Biological control. 433:237-256.

Fennimore, S.A. and M. Cutulle. 2019. Robotic weeders can improve weed control options for specialty crops. Pest Management Science. doi:10.1002/ps.5337.

Giannakou, I.O. and S. Panopoulou. 2019. The use of fluensulfone for the control of root-knot nematodes in greenhouse cultivated crops: Efficacy and phytotoxicity effects. Cogent Food & Agriculture. 5:1643819.

Gkanatsiou, Ch., N. Ntalli, U. Menkissoglu-Spiroudi and C. Dendrinou-Samara. 2019. Essential metal-based nanoparticles (Copper/Iron NPs) as potent nematicidal agents against *Meloidogyne* spp. Journal of Nanotechnology Research. 1:44-58.

Hansson, D. and J. Ascard. 2002. Influence of developmental stage and time of assessment on hot water weed control. Weed Res 42:307–316.

Hardman, R. 2006. A toxicologic review of quantum dots: toxicity depends on physicochemical and environmental factors. Environmental Health Perspectives. 114:165-172.

Head G. P., M. W. Carroll, S. P. Evans, D. M. Rule, A. R. Willse, T. L. Clark, N.P. Storer, R.D. Flannagan, L.W. Samuel, and L.J. Meinke. 2017. Evaluation of SmartStax and SmartStax PRO maize against western corn rootworm and northern corn rootworm: efficacy and resistance management. Pest Management Science. 73:1883-1899.

Hunt, E. R., and S. I Rondon. 2017. Detection of potato beetle damage using remote sensing from small unmanned aircraft systems. Journal of Applied Remote Sensing. 11(2):026013.

Iost Filho, F.H., W.B. Heldens, Z. Kong, and E.S de Lange. 2020. Drones: innovative technology for use in precision pest management. Journal of Economic Entomology. 113(1):1-25.

Islam, M.T., and S.M. Sherif. 2020. RNAi-Based Biofungicides as a Promising Next-Generation Strategy for Controlling Devastating Gray Mold Diseases. Int. J. Mol. Sci. 21:2072. doi:10.3390/ijms21062072

Kalaiselvi, D., A. Mohankumar, G. Shanmugam, S. Nivitha and P. Sundararaj. 2019. Green synthesis of silver nanoparticles using latex extract of *Euphorbia tirucalli*: A novel approach for the management of root knot nematode, *Meloidogyne incognita*. Crop Protection. 117:108-114.

Krause, R.A., L.B. Massie and R.A. Hyre. 1975. Blitecast: a computerized forecast of potato late blight. Plant Disease Reporter. 59:95-98.

Kumar, K.K., J. Sridhar, R.K. Murali Baskaran, S. Senthil Nathan, P. Kaushal, S.K. Dara and S. Arthurs. 2019. Microbial biopesticides for insect pest management in India: Current status and future prospects. Journal of Invertebrate Pathology.165:74-81.

Lim, D., J.Y. Roh, H.J. Eom, J.W. Hyun and J. Choi. 2012. Oxidative stress-related PMK-1 P38 MAPK activation as a mechanism for toxicity of silver nanoparticles to reproduction in the nematode *Caenorhabditis elegans*. Environmental Toxicology and Chemistry. 31: 585-592.

Maggiora, R., M. Saccani, D. Milanesio, and M. Porporato.2019.An innovative harmonic radar to track flying insects: The case of vespa velutina. Scientific reports. 9:1-10.

Mathur S.C. 1999. Future of Indian pesticides industry in next millennium. Pesticide Information. 24:9-23.

Mukherjee, A, A. Kapoor and T.M. Goyal. 2017. Toxicity, the bane of our food exports, https://www.thehindubusinessline.com/opinion/toxicity-the-bane-of-our-food-exports/article9909816.ece.

Murali Baskaran, R.K., S.K. Jain, P.N. Sivalingam, P. Mooventhan, J. Sridhar, K.C. Sharma, P. Kaushal and J. Kumar. 2019. National Status of Biotic Stresses of Crops. ICAR-National Institute of Biotic Stress Management, Raipur, Chhattisgarh, India. pp174 .

Prabhakar, M., Y. G. Prasad, S. Vennila, M. Thirupathi, G. Sreedevi, G. R. Rao, and B. Venkateswarlu. 2013. Hyperspectral indices for assessing damage by the solenopsis mealybug (*Hemiptera: Pseudococcidae*) in cotton. Computers and Electronics in Agriculture. 97:61-70.

Prabhakar, M., Y.G. Prasad and M.N. Rao.2012. Remote Sensing of Biotic Stress in Crop Plants and Its Applications for Pest Management. In: Venkateswarlu B., Shanker A., Shanker C., Maheswari M. (eds) Crop Stress and its Management: Perspectives and Strategies. Springer, Dordrecht. https://doi.org/10.1007/978-94-007-2220-0_16

Rahman, H., N. Jabbar Ch, S. Manzoor, F. Najeed, M. Siddique, and R. Khan. 2017. A comparative analysis of machine learning approaches for plant disease identification. International Quartely Journal Of Biological Sciences, 4:120-126. Retrieved from www.als-journal.com/ ISSN 2310-5380/ August 2017.

Rask, A.M. and P. Kristoffersen. 2007.A review of non-chemical weed control on hard surfaces. Weed Res. 47:370–380.

Rehman, M., J. Liu, A. C. Johnson, T. E. Dada, and G. M Gurr. 2019. Organic mulches reduce crop attack by sweetpotato weevil (*Cylas formicarius*). Scientific reports. 9(1):1-9.

Roh, J.Y., S.J. Sim, J. Yi, K. Park, K.H. Chung, D.Y. Ryu and J. Choi. 2009. Ecotoxicity of silver nanoparticles on the soil nematode *Caenorhabditis elegans* using functional ecotoxicogenomics. Environmental Science and Technology. 43:3933-3940.

Rojas F.A. 2018. Exploring machine learning for disease assessment from high-resolution UAV imagery. M.Sc. theis. at Wageningen University and Research Centre, The Netherlands.

Selvaraj, M.G., A. Vergara, F. Montenegro H.A. Ruiz, N. Safari, D. Raymaekers, W. Ocimati, J. Ntamwira, L. Tits, A.B. Omondi G. Blomme. 2020. Detection of banana plants and their major diseases through aerial images and machine learning methods: A case study in DR Congo and Republic of Benin. ISPRS Journal of Photogrammetry and Remote Sensing 169:110-124.

Selvaraj, M.G., A. Vergara, H. Ruiz, N. Safari, S. Elayabalan, W. Ocimati and G. Blomme, 2019. AI-powered banana diseases and pest detection. Plant Methods. 15:92. https:// doi.org/10.1186/s13007.

Servin, A., W. Elmer, A. Mukherjee, R. De la Torre-Roche, H. Hamdi, J.C. White, P. Bindraban and C.A. Dimkpa 2015.review of the use of engineered nanomaterials to suppress plant disease and enhance crop yield. J. Nanopart. Res. 17:92.

Sharma, H.D. and A.D. Gupta. 1995. The objectives of waste management in India: a futures inquiry. Technological Forecasting and Social Change, 48:285-309.

Singh, R. 2014. Weed management in major kharif and rabi crops. National Training on Advances in Weed Management. pp. 31-40.

Slaughter, D.C., D.K. Giles and D. Downey. 2008. Autonomous robotic weed control systems: a review. Comp Elec Agric. 61:63–78.

Stanton, C., M.J. Starek, N. Elliott, M. Brewer, M.M. Maeda, and T. Chu. 2017. Unmanned aircraft system-derived crop height and normalized difference vegetation index metrics for sorghum yield and aphid stress assessment. Journal of Applied Remote Sensing. 11(2):026035.

Szabo, M., K. Csepregi, M. Galber, F. Viranyi and C. Fekete. 2012. Control plant-parasitic nematodes with *Trichoderma* species and nematode-trapping fungi: The role of chi18-5 and chi18-12 genes in nematode egg-parasitism. Biological Control. 63:121-128.

Tabatabaei, F.S. and A. Saeedizadeh. 2017. Rhizobacteria cooperative effect against *Meloidogyne javanica* in rhizosphere of legume seedlings. Hellenic Plant Protection Journal. 10:25-34.

Thorp, K.R. and L.F. Tian. 2004. A review on remote sensing of weeds in agriculture. Prec Agric. 5:477–508.

Tyagi, S., K. Kesiraju, M. Saakre, M. Rathinam, V. Raman, D. Pattanayak and R. Sreevathsa. 2020. Genome Editing for Resistance to Insect Pests: An Emerging Tool for Crop Improvement. ACS omega.5:20674-20683.

Van Everdingen, E. 1926. Het verband tusschen de weersgesteldheid en de aardappelziekte (*Phytophthora infestans*).Tijdschrift over plantenziekten, 32: 129-139.

Vennila, S., J. Singh, P. Wahi, M. Bagri, D.K. Das and M. Srinivasa Rao. 2016. Web enabled weather based prediction for insect pests of rice. ICAR-National Research Centre for Integrated Pest Management, New Delhi-110 012 India. 50p.

Waweru, B., L. Turoop, E. Kahangi, D. Coyne and T. Dubois. 2014. Non-pathogenic *Fusarium oxysporum* endophytes provide field control of nematodes, improving yield of banana *Musa* sp. Biological Control. 74:82-88.

Yaduraju, N.T. 2012. Weed management perspectives for India in the changing agriculture scenario in the country. Pak. J. Weed Sci. Res. 18:703-710.

Young, S.L., G.E. Meyer and W.E. Woldt. 2014. Future directions for automated weed management in precision agriculture. Pages 249–259 in Young SL, Pierce FJ, eds. Automation: The Future of Weed Control in Cropping Systems. Dordrecht, Springer.

15

Pest and Disease Monitoring and Surveillance for Ensuring Food Security

S.N. Mandal[1] *and Parimal Sinha*[2]

[1]*Department of Information Technology, Kalyani Government Engineering College, Kalyani, Nadia, West Bengal*
[2]*Division of Plant Pathology, ICAR-Indian Agricultural Research Institute New Delhi*

Introduction

Surging food demand and shrinking land resources to produce more from the limited natural resources like soil and water is poised to environmental degradation. Increased production must be achieved mainly by increasing crop productivity per unit area while sustaining the environment and enhanced input use efficiency. Agricultural production systems are increasingly under pressure to adapt decision-making process to be make more informed and efficient use of irrigation water, fertilizers and pesticides. Adverse effects of climate change are added challenges to agricultural productivity. Efficient crop management strategies to improve resource use efficiency and reduce crop losses are vital issues to be taken care for a sustainable sustainably enhanced production system. Genetic improvement approaches involving high throughput phenotyping and genomics-assisted breeding enhancing resource use efficiency, stress tolerance, yield has become imperative. Precise crop health monitoring is one of the keys to the modern production system. Obviously, disease forecasting and surveillance systems are in great demand to adopt precise management against the losses caused by diseases. Predicting disease risk at regional and global scales is of great interest for on-farm management and for assessment of agricultural policies and practices.

As a part of crop management system, disease forecasting systems or otherwise called disease forewarning systems are developed to help growers take economic decisions about disease management (Magarey et al. 2005). Early warning services for crop diseases are valuable when they provide timely forecasts that farmers can utilize accurate disease management decisions (Campbell and Madden 1990). Importance of disease forecasting being realized way back early 20th century and many disease models for many diseases for economically important crops are made (Krause and Massie, 1975, Savary and Willocquet, 2014). However, most of these models could not be utilized in disease management decisions as they are unable to meet the users need. Models to evaluate the impact of plant diseases on crop production under current and future climatic conditions are increasingly requested by different stakeholders (Garrett et al., 2006). A forecasting system requires to assess spatial and temporal variability of crop, soil and environmental variables at real-time to facilitate management decision with robust mapping systems is a challenge. Forecasting systems are very often complex and require expertise to operate. Additionally, input need, their accessibility and operations are the major hindrances in their use. On the other hand, users need real-time information easily accessible, clearly understandable and actionable. Access to forecast information is one of the major challenges to solve.

Current chapter mentions about the possibilities of disease forecasting and surveillance systems required for prioritization of research and strategic disease management decisions warranted by climate change urgencies.

Plant disease forecasting systems

Fundamental rule of disease forecasting is that diseases are caused by mostly pathogens whose growth or life cycles are influenced by weather conditions in spatiotemporal dimension. A cause-effect relationship between pathogen development (life-cycle) and favorable weather period is important in development of a forecasting system. They are developed from the data, obtained experimentally in the laboratory or field regarding the relationships of biological and environmental conditions governing host-pathogen interactions (Krause and Massie, 1976). Infection process leading to disease development is a complex interaction between host, pathogen, and their environment. Disease epidemics occur under conditions when crop growth stage and weather conditions favour faster infection process. Microclimate weather plays a significant role in disease occurrence as pathogen growth is favoured by certain weather conditions. Therefore, infection is preceded by the happenings of some specific weather spells generally characterized by temperature and leaf wetness or high relative humidity hours (Huber and Gillespie, 1992). Therefore, most of the diseases are seasonal in nature. Infection model for fungal diseases is the most critical component in disease forecasting systems. Infection process consists of germination, penetration, colonization, sporulation and spread are usually limited by duration of high RH or surface wetness greatly influenced by temperature and leaf wetness (Magarey et al. 2005, Madden et al., 2007). Therefore, most often these two factors are combined to construct an index for the occurrence of plant diseases (Wilks and Shen, 1991 Guyader et al., 2013). Traditionally, plant disease models have used leaf wetness duration (LWD) combined with temperature, to predict infection and colonization and then identify the risks of an epidemic. These types of models have been used with observed climate records to track the favorable periods, indicating tactics or strategies of control. Therefore, disease monitoring and forecasting depends on monitoring of favorable weather factors, specifically at the threshold level.

Temperature response on infection process is typically characterized by a unimodal curve and can be explained by three cardinal temperatures. To describe the effect of temperature (*T*),a temperature-dependent function f(T) is used. The function uses cardinal temperatures (minimum T_{min}, maximum T_{max} and optimum T_{opt} of the pathogen) for estimation of infection index (Yan and Hunt 1999, Magarey et al. 2005, Guyader et al. 2013, Viswanath et al. 2017):

$$f(T)= [(T_{max} - T)/(T_{max} - T_{opt})]*[(T - T_{min})/(T_{opt} - T_{min})](T_{opt} - T_{min})/(T_{max} - T_{opt}) \quad \text{if } T_{min} \leq T \leq T_{max}$$

and 0 otherwise.

Leaf wetness or high RH hour response is a sigmoid curve indicates infection starts with a minimum period of RH hours and increases with increase in RH hours and then reaches to saturation (Butler and Jadhav1991; Scherm and van Bruggen1993; de Vallavieille-Pope et al. 1995):

For modeling the combined effects of temperature (T) and high RH hours (D) on infection index as Y(D), Richards's function for RH-hours (D) with temperature-dependent parameters *m* and f(T) were used (Butler and Jadhav1991; Scherm and van Bruggen1993; de Vallavieille-Pope et al. 1995):

$$Y(D) = f(T)(1\text{-}\exp\{-[A(D\text{-}m)]\})) \text{ and } Y=0 \text{ for } f(T) < 0$$

A is the intrinsic rate of increase in infection index (Y); *m* is the minimum leaf wetness duration for initial infection response. The response surface for Y has been estimated for spot blotch (*Bipolaris sorokininana*) infection in wheat indicates the combination of temperature and RH hours to explain the infection process (Viani et al., 2017). Based on the response surface, infection or otherwise disease risk of various levels are estimated (Fig 1 and 2).

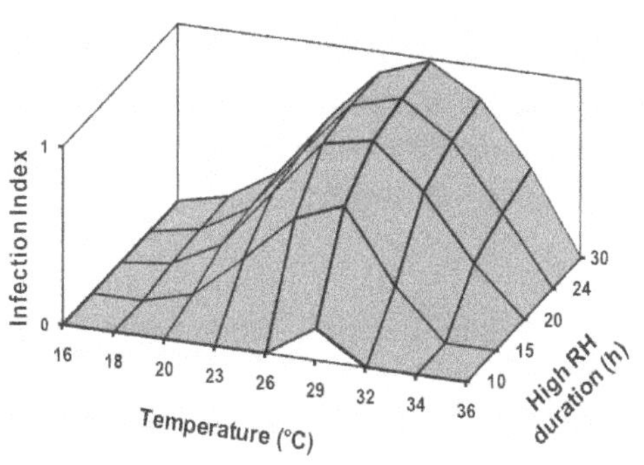

Fig 1. Spot blotch infection index in wheat in terms of temperature and high RH duration (h)

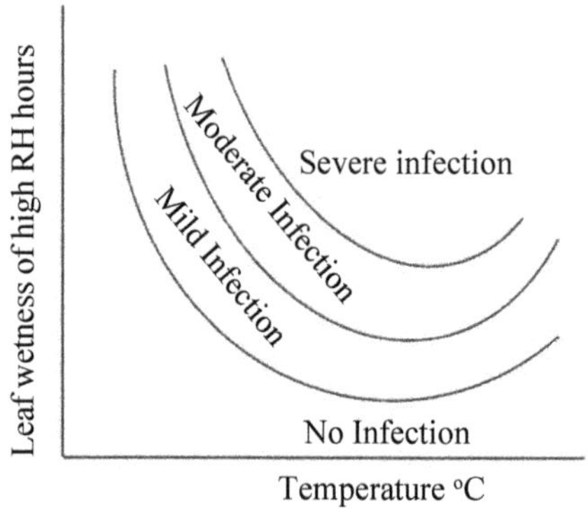

Fig 2. Spot blotch infection risk levels for various combination of temperature and leaf wetness high RH duration (h)

Although various types of relationships have been used to simulate disease risk or index but temperature and leaf wetness duration in non-linear fashion is the most appropriate (Scherm and van Bruggen 1993). Disease dynamics involves a complex interaction between a host, a pathogen, and their environment in spatiotemporal dimension. A forecasting systems requires to assess spatial and temporal variability of crop, soil and environmental variables at real-time.

A robust model for disease risk incorporating the influence of temperature and leaf wetness has been established for stem rust in wheat and leaf rust in coffee (Bebber et al., 2015).Survival analysis to assess time to infection or time to healthy has been incorporated in disease risk assessment. Probabilistic assessment of time to infection ($T<t$) or time to remain healthy ($T>t$) based on Cox Hazard function in addition, can help associate predictor variables especially temperature, age and time of plant exposure to favourable period for infection. Infection (germination and appressorium formation) is as a random process with probabilities varying through time (Jong et al., 1997). Time T to transition (i.e. from a spore to germination, or from germination to appressorium formation) is a random variable with probability density f (t) and cumulative distribution F(t) at time t after initiation. To model F(t), which can be interpreted as the fraction of a population transitioned by t, for germination and infection by pathogens. F(t) can be determined by the cumulative hazard function H(t), where the hazard function h(t) is the instantaneous risk of transformation given survival to t:

$$F(t)=1-e^{-H(t)}$$

Survival function S(t)=1-F(t) is commonly modelled using the Weibull distribution which allows the hazard to vary over time:

$$h(t)==(\gamma/\alpha)(t/\gamma)\gamma\text{-}1H(t) = \int_0^t h(s)ds = (t/a)\gamma - 1$$

where α is the scale parameter, γ the shape parameter and h(s) is a dummy function. The processes of spore germination and infection are affected by temperature, hence we multiply h(t) and thus H(t) by f(T), so that rates are greatest at the optimum temperature and decline to zero if outside the temperature range.

Integrated model-based forecasting frameworks have the potential to improve the timeliness, effectiveness, and foresight for controlling crop diseases. For forecasting stem rust in wheat turbulent atmospheric dispersion and deposition of *Pgt*-spores was simulated using Numerical Atmospheric-dispersion Modelling Environment (NAME), to simulate release, turbulent atmospheric transport, loss of viability during transport and deposition of *Pgt*-urediniospores from RDLs on susceptible and environmentally suitable wheat producing areas (Jones et al. 2007).

Current challenges – integrated forecasting systems

Forecasting information is in great demand and challenges for precise disease risk assessment to cater management decision and phenotyping of resistance. To make forecasting information to be useful integrated modeling framework keeping components of forecasting rules, algorithms, data on crop, soil and weather, potential inoculums using accessing tools made available to users is required (Fig 3). Assessment of disease is important firstly, for timely plant protection measures; secondly, making strategic decision to choose a crop and need of application for certain measures; like, soil treatment, planting a resistant variety etc. Recent advances in weather sensors, satellite-based technologies and speed in machine learning offer new opportunities for robust predictive models to process information the required output for the users. Web-based technologies are increasingly available to extract specific information required for management decisions available to the users. It requires robust integrated model-based forecasting frameworks to improve the timeliness, effectiveness, and foresight for controlling crop diseases, while minimizing economic costs and environmental impacts, and yield losses (Newlands, 2017).

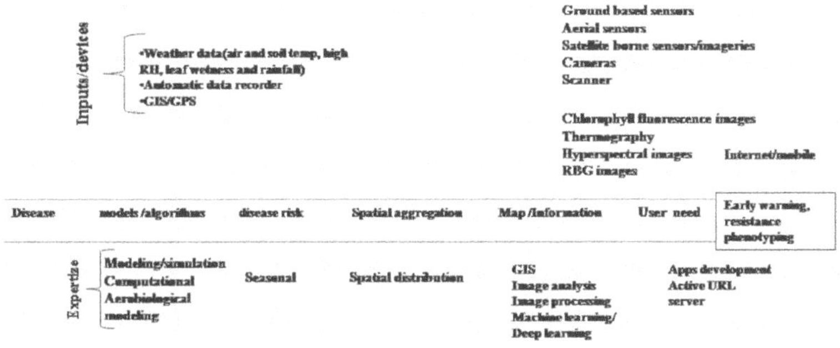

Fig 3. Integration of forecasting components for real-time crop advisories

Web-based forecasting systems and IoT based forecasting systems

Currently, recent advances in information technology, easy access to programming languages, global web connectivity gave ways to be explored further in the production and transfer of forecasting information. Rapid advances in web-based technologies are bringing major changes in the development and use of decision-support systems by producers and specialists in the management of plant diseases. Spatial aggregation pattern in infected units or initiation of foci is due to cluster or lump of inoculumn load and other heterogeneities in the field is a motivation to identify infection foci for control measures. Advancement in imaging techniques from remotely sensed data-multispectral, hyper-spectral and chlorophyll fluorescence help detecting foci of infection based on the changes in the reflectance.

Disease simulators implemented through web tools become more powerful and provide a large amount of information for the decision-makers and stakeholders in general. *Agroclimate* is a web-based forecasting system developed for prediction of Anthracnose and Botrytis fruit rot risk on strawberries (Pavan et al., 2011). It has climate information and decision support system to facilitate extension agents, producers, and natural resource managers with tools to aid their decision-making aimed at reducing risks associated with climate variability. Information available in such systems includes climate forecasts combined with risk management tools and information for selected crops, forestry, pasture, and livestock. The system was developed to allow easy expansion of the topic areas, number of commodities and risk management tools available for users.

An agricultural monitoring system can provide a service to store the environmental and soil information collected from a wireless sensor network (Khattab et al. 2019). The system uses Wireless Sensor Network (WSN) technology and is accessible through the Internet for precision agriculture applications such as epidemic disease control.

Forecasting based on pathogen detection and phenotyping of resistance

In addition to robust forecasting models crop health monitoring demands early pathogen detection or phenotyping resistance against diseases. Imaging techniques are useful for 'healthy and diseased plant classification' with an emphasis on classification accuracy, early detection of stress, and disease severity (Lowe et al. 2017). Hyperspectral imaging (Rumpf et al., 2010, Lowe et al., 2017; Franke et al., 2005; thermography and chlorophyll fluorescence (Mahlein et al., 2013) have potential for application in detection of pathogen infections.

Hyperspectral remote sensing based disease detection

Weather-based forecasting models mainly emphasize infection on plants. Plant infection in the field normally happens in patches which are called as infection foci. Therefore, disease risk prediction along with detection of infection foci can facilitate localized application of pesticides. Hyperspectral imaging uses high-fidelity colour reflectance information over a large range of the light spectrum (240-2500 nm). Identification of any subtle changes in plant growth and development are detectable with a fine resolution (Lowe et al., 2017). Remote sensing techniques are used to detect and map crop diseases for site-specific management of diseases (Yang 2020). Plant diseases are assessed based on temporal changes of spectral indices and one of the most popular and widespread indices is the normalized difference vegetation index (NDVI), which is used for measuring the general health status of crops (Table 1). It is calculated via a simple ratio of near-IR and visible light. NDVI has been used to detect stress in plants. Powdery mildew risk predicted by weather parameters (temperature, solar radiation, RH and rainfall) and remotely sensed red band was able to depict approximately spatial of powdery mildew in wheat and its temporal dynamics. Plant vigor has a certain association with diseases and used as important indicator of crop health status (Lowe et al., 2017). The triangular vegetation index (TVI) and soil adjusted vegetation index (SAVI) are proven to be suitable for representing plant vigor and its stress status. The plant water status is also associated with susceptibility of plants to disease. To capture this information from remote sensing data, the disease water stress index (DSWI) and the shortwave infrared water stress index (SIWSI) used as candidate model input variables. Both indices contain the NIR and SWIR bands and have a potential in detecting water stress of crops at a canopy level.

Table 1. Plant health status indicators derived from remote sensing data

Spectral indices	Formula	Reference
Normalized difference vegetation index (NDVI)	(RNIR-RRED)/(RNIR+RRED)	Bravo et al., (2003))
Red Edge NDVI	(R705-R750)/(R705+R750)	Zhang et al., (2005)
Triangular vegetation index (TVI)	0.5[120(RNIR-RG)-200(RR-RG)]	Broge and Leblanc, (2001)
Soil adjusted vegetation index (SAVI)	(1+L)(RNIR-RR)/(RNIR-RR+L)	Huete, (1988)
Disease water stress index (DSWI)	(RNIR+RG)/(RSWIR+RR)	Galvao et al.,(2005)
Leaf rust disease severity index (LRDSI)	$6.9 \times (R605/R455) - 1.2$	Ashourloo et al., (2014)

Airborne (Drone based and light aircraft) and satellite images (from Landsat, SPOT 6, QuickBird and WorldView -2) acquired during growing seasons are useful for disease detection although still a challenge at early stage. For detection of Phytophthora foot rot in citrus orchards and mapping of late blight in tomato fields airborne digital multispectral imagery was evaluated (Fletcher et al 2001 Zhang et al 2005). Airborne hyperspectral imagery was used for identification of stripe rust in wheat (Huang et al 2007), grapevine leaf roll virus (MacDonald et al., 2016), and yellow leaf curl on tomatoes (Lu et al., 2018). Aerial multispectral and hyperspectral imaging techniques were used for detecting citrus greasy spot (Du et al., 2004), cotton root rot (Yang et al., 2010), and huanglongbing or citrus greening (Kumar et al., 2012, Li et al., 2014). QuickBird satellite imagery was used for detecting powdery mildew and leaf rust in winter wheat, and high accuracies were achieved with severe infections at late growth stages (Franke and Menz 2007). QuickBird imagery was also used to map and identify basal stem rot in oil palms (Santoso et al., 2011). SPOT 6 satellite imagery was used for mapping powdery mildew in winter wheat in multiple regions (Yuan et al 2016). Unmanned aircraft systems have been evaluated for the detection of crop diseases such as citrus greening (Garcia-Luiz et al 2013), and root rot in alfalfa (Mattupalli et al., 2018).

Therefore, site-specific disease monitoring is possible complementing disease risk forecast with remote sensing data as substantial improvements in space-borne sensors in spatial and temporal resolution increases the chances to acquire images and predicting plant diseases.

Chlorophyll fluorescence imaging

Pathogen infection leads to destruction of the photosynthetic apparatus and finally degradation of chlorophylls (Mahlein et al., 2013). Hence, monitoring leaf photosynthetic activity can rapidly assess the increasing effects diseases on the vitality of plants. Monitoring leaf photosynthetic activity can be useful to assess early changes in photosynthetic properties. The potential maximum photochemical quantum yield of electron transport through photosystem II (Fv/Fm) has been well established as a rapid indicator of both maximum photochemical efficacy and intactness of the entire photosynthetic machinery. Fv/Fm readily declines in response to many abiotic and biotic stressors such as pathogen action. CFA parameter can be easily and rapidly determined by a simple measurement procedure as a straightforward method capable of objectively and rapidly distinguishing resistant and susceptible lines.

Expert systems –machine learning based detection and forecasting

Application of machine learning tools is in demand as plant disease management decision is beset with problem of classification or identification (disease diagnosis phenotyping resistance), optimization (precise quantification of infected tissue) and disease prediction. For crop management decision aids machine learning based expert systems are the potential tools to make an interface between complex forecasting systems and users. Availability of software in computer vision facilitates development of experts systems for plant pathological diagnosis and precise quantification and feature identification. Machine learning techniques have been used for leaf blast forecasting in combination with web server tools (Kaundal et al., 2006) and to predict seedlings infected by bakanae (Chung et al., 2017). An image based disease rating scale for phoma blight of potato and the scale has been segmented by K-means clustering and the percentage of affected area has been estimated (Roy et al., 2020). Applicability of deep convolutional neural networks (CNN) for image classification increases with the availability of popular architectures, namely AlexNet (Krizhevsky et al., 2012). Using CNN, infected and healthy leaves has been identified with 99.95% accuracy.

Future directions

Crop production requires precise information for efficient use of resources. For accurate and efficient disease management forecasting systems are yet to cater the need of the users. Integration of forecasting components require is expected to generate real-time user information. Future challenges remain to make integrated systems to reach out to the end users.

References

Ashourloo, D., M.R. Mobasheri and A. Huete.. 2014. Developing two spectral disease indices for detection of wheat leaf rust (*Puccinia triticina*). Remote Sens. 6:4723-40.

Bebber, D.P., A.D. Castillo, and S.J. Gurr. 2016. Modelling coffee leaf rust in Columbia with climate reanalysis data. Philos. Trans. R.Soc. B 371:20150458.doi: 10.1098/rstb.2015.0458.

Bravo, C., D. Moshou, J. West, A. McCartney and H. Ramon. 2003. Detailed spectral reflection information for early disease detection in wheat fields. Biosystems Engineering. 84(2):137e145.

Broge, N.H. and E. Leblanc. 2001. Comapring prediction power and stability of broadband and hyperspectral vegetation indices for estimation of green claf area index and canopy chlorophyll density. Remote Sens Environ. 76:156-172.

Butler, D.R. and D.R. Jadhav. 1991. Requirements of leaf wetness and temperature for infection of groundnut rust. Plant Pathol. 40:395-400.

Campbell, C.L., and L.V. Madden. 1990. Introduction to Plant Disease Epidemiology. John Wiley & Sons, New York.

de Vallavieille-Pope, C., L. Huber, M. Leconte and H. Goyeau. 1995. Comparative effects of temperature and interrupted wet periods on germination, penetration, and infection of *Puccinia recondita f. Sp. tritici* and P. striiformis on wheat seedlings. Phytopathology. 85: 409-415.

Du, Q., J.V. French, M. Skaria, C. Yang and J.H. Everitt. 2004. Citrus pest stress monitoring using airborne hyperspectral imagery. In: Proceedings of the International Geoscience and Remote Sensing Symposium; 2004 Sep 20-24; Anchorage, AK, USA. New York: IEEE.. p. 3981-4.

Fletcher, R.S., M. Skaria, D.E. Escobar and J.H. Everitt. 2001. Field spectra and airborne digital imagery for detecting Phytophthora foot rot infections in citrus trees.Hort Science. 36(1):94-7

Franke, J. and G. Menz. 2007. Multi-temporal wheat disease detection by multi-spectral remote sensing. Precis Agric.8(3):161–72.

Galvao, L.S., A.R. Formaggioo and D.A. Tisot. 2005. Discrimination of sugarcane varieties in Southern Brazil with EO-1 Hyperion data. Remote Sens Environ. 94:523-534.

Garcia-Ruiz, F., S. Sankaran, J.M. Maja,, W.S. Lee, J. Rasmussen and R. Ehsani. 2013. Comparison of two aerial imaging platforms for identification of huanglongbing-infected citrus trees. Comput Electron Agric. 91:106-15.

Garrett, K.A., S.P. Dendy, E.E. Frank, M.N. Rouse and S.E. Travers. 2006 Climate change effects on plant disease: genomes to ecosystems. Annu. Rev. Phytopathol. 44, 489–509. (doi:10.1146/annurev.phyto.44. 070505.143420)

Guyader. S., J. Crombez, M. Salles, F. Bussiere and T. Bajazet. 2013. Modelling the effects of temperature and leaf wetness on monocyclic infectionin a tropical fungal pathosystem. Eur. J. Plant. Pathol. 136:535-545

Huang, W., D.W. Lamb, Z. Niu, Y. Zhang, L. Liu and J. Wang. 2007. Identification of yellow rust in wheat using in-situ spectral reflectance measurements and airborne hyperspectral imaging. Precis Agric. 8(4–5):187-97.

Huber, L. and T.J. Gillespie.1992. Modelling leaf wetness in relation to plant disease epidemiology. Annu. Rev. Phytopathol. 30:553-577

Huete, A.R. 1988. A soil-adjusted vegetation index (SAVI). Remote Sens Environ 25:295-309.

Jones, A.R., D.J. Thomson, M. Hort and B. Devenish. 2007. In Air Pollution ModelingAnd Its Application XVII (eds Borrego, C. & Norman, A.-L.) Ch. 62 (Springer,New York, 2007).

Jong, E.J.D., A.B. Eskes, J.G.J. Hoogstraten and J.C. Zadoks. 1987. Temperature requirements for germination, germ tube growth and appressorium formation of urediospores of *Hemileia vastatrix*. Neth. J. Plant Pathol. 93:61-71.

Kaundal, R., A.S. Kapoor and P.S.R. Gajendra. 2006. Machine learning techniques in disease forecasting: a case study on rice blast prediction. BMC Bioinformatics. 7: 485.

Khattab, A., S.E.D. Habib, H. Ismail, S. Zayan, Y. Fahmy and M.M. Khairy . 2019. An IoT-based cognitive monitoring system for early plant disease forecast. Computers and Electronics in Agriculture. doi.org/10.1016/j.compag.2019.105028.

Krause, R.A. and L.B. Massie. 1975. Predictive systems: Modern approaches to disease control. Annu. Rev. Phytopathol. 13:31-47.

Krizhevsky, A., I. Sutskever and G.E. Hinton. 2012.Imagenet classification with deep convolutional neural networks,"in Advances in Neural Information Processing *Systems*, eds F. Pereira, C.J.C. Burges, L. Bottou and K.Q. Weinberger (Curran Associates, Inc.). 1097–1105.

Kumar, A., W.S. Lee, R.J. Ehsani L.G. Albrigo, C. Yang and R.L. Mangan. 2012. Citrus greening disease detection using aerial hyperspectral and multispectral imaging techniques. J. Appl. Remote Sens. 6(1):063542.

Li, H., W.S. Lee, K. Wang, R. Ehsani and C. Yang. 2014. 'Extended spectral angle mapping (ESAM)' for citrus greening disease detection using air borne hyperspectral imaging. Precis Agric. 15(2):162-83.

Lowe, A., N. Harrison and A.P. French. 2017. Hyperspectral image analysis techniques for the detection and classification of the early onset of plant disease and stress. Plant Methods. 13:80.. https://doi.org/10.1186/s13007-017-0233-z.

Lu, J., M. Zhou, M.Y. Gao and H. Jiang. 2018.. Using hyperspectral imaging to discriminate yellow leaf curl disease in tomato leaves. Precis Agric.19(3):379-94.

MacDonald, S.L., M. Staid, M. Staid and M.L. Cooper. 2016. Remote hyperspectral imaging of grapevine leafroll-associated virus 3 in cabernet sauvignon vineyards. Comput Electron Agric.130:109-17.

Madden, L.V., G. Hughes and F. Van den Bosch.2007. The study of plant disease epidemics. The American Phytopathological society, APS Press St. Paul, Minnesota, p 241.

Magarey, R.D., T.B. Sutton and C.L.Thayer. 2005. A simple generic infection model for foliar fungal pathogens. Phytopathology. 95:92-100.

Mahlein, A.K., T. Rumpf , P. Welke, H.W. Dehne, L. Plumer, U. Steiner, E.C. Oerke. 2013. Development of spectral indices for detecting and identifying plant diseases. Remote Sens Environ. 128:21-30.

Mattupalli, C., C.A. Moffet, K.N. Shah C.A.Young. 2018. Supervised classification of RGB aerial imagery to evaluate the impact of a root rot disease. Remote Sens .10(6):917.

Newlands, N.K. 2017. Model-Based Forecasting of Agricultural Crop Disease Risk at the Regional Scale, Integrating Airborne Inoculum, Environmental, and Satellite-Based Monitoring Data. Frontiers in Plant Science. doi: 10.3389/fenvs.2018.00063.

Pavan, W., Fraisse, C.W., and Peres, N.A. 2011. Development of a web-based disease forecasting system for strawberries. Computers and Electronics in Agriculture 75:169-175.

Roy, K., S. Dan, K, Mukherjee, S. Mustafi, P. Ghosh, S.N. Mandal, S. Dutta and A. Chakraborty. 2020. "Development of Image Based Disease Scale of Phoma Blight of Potato using K-means Clustering", Proceedings of Electronic Systems and Intelligent Computing.

Rumpf. T., A.K. Mahlein, U. Steiner, E.C. Oerke, H.W. Dehne and L. Plumer. 2010. Early detection and classification of plant diseases with Support Vector Machines based on hyperspectral reflectance. Computers and Electronics in Agriculture. 74: 91-99.

Santoso, H., T. Gunawan, R.H. Jatmiko, W. Darmosarkoro and B. Minasny. Mapping and identifying basal stem rot disease in oil palms in North Sumatra with Quick Bird imagery. Precis Agric. 12(2):233–48.

Scherm, H. and A.H.C. Van Bruggen. 1993. Response surface models for germination and infection of *Bremia lactucae*, the fungus causing downy mildew of lettuce. Ecol Model. 65:281-296.

Serge, S. and L. Willocquet. 2014. Simulation modeling in botanical epidemiology and crop loss analysis. Plant Health Instructor. pp.147. 10.1094/PHI-A-2014-0314-010. hal-02636693

Viani, A., P. Sinha, T. Sharma and L.M. Bhar.2017. A model for forecasting spot blotch disease in wheat. Australasian Plant Pathology. 46:601-609.

Viswanath, K., P.S. Sinha,, K.S. Naresh, T. Sharma, S. Saxena, S. Panjwani, H. Pathak and S.M.Shukla. 2017.Simulation of leaf blast infection in tropical rice agro-ecology under climate change scenario. Climatic Change. DOI 10.1007/s10584-017-1942-z.

Wilks, D.S. and K.W. Shen. 1991. Threshold relative humidity duration forecasts for plant disease prediction. J. Appl. Meteorol 30:463-477.

Yan, W. and L.A. Hunt. 1999. An equation for modeling the temperature response of plants using only the cardinal temperatures. Ann. Bot. 84:607-614. (doi:10.1006/anbo.1999.0955).

Yang, C. 2020. Remote Sensing and Precision Agriculture Technologies for Crop Disease Detection and Management with a Practical Application Example. Engineering 6:528-532.

Yang, C., C.J. Fernandez and J.H. Everitt. 2010. Comparison of airborne multispectral and hyperspectral imagery for mapping cotton root rot. Biosyst Eng.107 (2):131-9.

Yuan, L., R. Pu, J. Zhang, J. Wang and H. Yang.. 2016. Using high spatial resolution satellite imagery for mapping powdery mildew at a regional scale. Precis Agric.17(3):332–48.

Zhang, M., Z. Qin and X. Liu. 2005. Remote sensed spectral imagery to detect late blight in field tomatoes. Precis Agric.6(6):489–508.

16

Novel Approaches for Biotic Stress Management in the Emerging Production System

Jeer, M[1]*., Sahu, M.P.*[2] *and Choudhary, V.K.*[2]

[1]*ICAR National Institute of Biotic Stress Management, Raipur, Chhattisgarh*
[2]*ICAR-Directorate of Weed Research, Jabalpur, Madhya Pradesh*

Introduction

With growing concern about climate change, environmental sustainability, soil health and food safety, conservation agriculture, organic agriculture and protected cultivation are emerging production systems being popular in India. The changing farm practices in these production systems invite many biotic stresses and demand novel approaches for their sustainable management for profitable farming. The dynamics of biotic stresses vary from system to system; therefore they require system specific approaches for sustainable management. The different agronomic/ plant protection measures practiced in above mentioned emerging systems are listed in Table 1. Protected cultivation is meant for the high value crops are grown, the management of biotic stresses is the foremost important to minimize losses. Organic farming demands more preventive measures for sustainable control of biotic stresses rather than curative measures because of complete restriction on use of harmful chemical pesticides. Conservation agriculture demands more of eco-friendly measures for biotic stress management with minimal use of pesticides. In this chapter, we discussed novel approaches for biotic stress management in emerging production systems.

Table 1. Comparison of agronomic and plant protection practices in 3 emerging production systems

Agronomic/Plant protection practices	Conservation Agriculture	Organic Farming	Protected Cultivation
Deep ploughing	Not permitted	Permitted	Permitted
Crop rotation	Permitted	Permitted	Permitted
Permanent soil cover with crop residues	Highly recommended	Not always necessary	Not always compulsory
Use of chemical fertilizers	Permitted	Prohibited	Permitted
Use of herbicides	Permitted	Prohibited	Permitted
Use of agrochemicals	Permitted	Restricted	Permitted
Soil mulching with polythene	Not permitted	Restricted	Highly recommended

Novel approaches for conservation agriculture

Cover crops and their residues

Cover crops are those crops which are not only protect soil from erosion, build up soil fertility, increase biodiversity and improve water infiltration but also suppress insect pests, diseases, weeds and nematodes etc. Some of the crop pests controlled by growing cover crops are mentioned in Table 2. Surfaces covered with plant residues will provide favorable microhabitats for beneficial soil arthropods such as ground beetles, rove beetles, spiders, annelids which can effectively control pests (Altieri and Schmidt, 1985). Many studies have proved that fire ants could be potential insect predators under plant residues in various cropping systems (Morrill, 1977). Cover cropping of non-Bt cotton with crimson clover encouraged natural enemies of bollworm (*Helicoverpa armigera*) and reduced the pesticide inputs. Populations of the predator (*Geocoris punctipes*) built up on crimson clover in the early spring and moved onto cotton and subsequently preyed on bollworm pests in early season cotton. The crimson clover also provided nectar to predators, honeybees, and other insect pollinators (Reddy, 2017a). Straw mulch is responsible for reduction of the early-season Colorado potato beetle's (CPB's) ability to locate potato plants and for the suppression of its activity in potatoes due to the creation of a micro-environment that enhances the population of predators such as lacewings, lady beetles, and ground beetles (Reddy, 2017b).

Table 2. Some of the cover crops for pest control

Cover crops	Pests controlled
Oat (*Avena sativa*) The mulch is persistent and very effective at suppressing	Broadleaf weeds (Dube et al. 2012)
Rye grass (*Secale cereale*)	Has excellent weed suppression ability through allelopathy. It also breaks nematodes cycles (Clark, 2007)
Vetches (*Vicia* spp.)	Residues have fast decomposition but can provide early weed suppression. Host for soybean cyst nematode (*Heterodera glycines*) and root-knot nematode (Smith, 2006)
Barley (*Hordeum vulgare*)	The mulch suppresses weeds. However, barley is host to the root-knot nematode
Sorghum (*Sorghum bicolor*)	Residues of sorghum have allelopathic effects on weeds. However, it harbours root knot nematode
Buckwheat (*Fagopyrum esculentum*)	The flowers are attractive to bees, wasps and parasitic flies. It is allelopathic to many weeds but harbours root lesion nematode (*Pratylynchus penetrans*) (Marks and Townsend, 1973)
Canola (*Brassica napus*)	Its flowers are attractive to bees, wasps and parasitic flies. It is susceptible to *Sclerotinia sclerotiorum*
Cowpea (*Vigna unguiculata*)	Attracts beneficial insects such as wasps and honey bees. It can reduce root-knot nematodes (*Meloidogyne arenaria* and *M. incognita*) and soybean cyst nematode (*Heterodera glycines*) (Clark, 2007)

Adapted from Fanadzo et al. (2018)

Crop rotation

Insect pests, diseases, weeds and nematodes can be effectively managed by rotating the crops, which is one of the principles of conservation agriculture. Rotating with non-host crops effectively suppress the pest load and breaks the life cycle. Some of the insect pests, diseases and nematodes that can be controlled by crop rotation are given in Table 3. The rotation between susceptible crops should be three to seven years for effective management of pests (Linker et al. 2009) especially under conservation agriculture where crop residues are retained (Fanadzo et al., 2018).

Some common characteristics of crop rotation for managing insect pests includes

- Fairly narrow host range of herbivore,
- Restriction of pest within limited area,
- Herbivore's incapability to survive long periods without a crop,
- Herbivore needs to be sedentary in nature.

Maize rootworm (*Diabrotica virgifera*) eggs usually overwinter in maize fields which can hatch to larvae to feed on maize roots the following year, and rotating maize with different non-host crop such as soybeans denies food to such larvae, thereby breaking the life cycle (Spencer et al., 2009). The frequency of corn fields with severe rootworm damage which is resistant to Bt corn was significantly reduced by rotating with other crops like soybean and not planting same type of Bt corn year after year (Carrière et al., 2020). Many of the soil-borne diseases were effectively managed by crop rotation. The root rot and damping-off disease pathogens on onion were markedly reduced and yield increased when planted after *Daucus carota*. Club root of cabbage, *Plasmodiophora brassicae*, was effectively controlled by growing tomato, cucumber, snap bean, and buckwheat in rotation (Reddy, 2017). *Streptomyces scabies* which causes scab disease in potato can be effectively managed by sequential cropping of fallow-rice-potato, mungbean-rice-potato and sunhemp-rice-potato in a single cropping sequence over four year in northern India (Hunjan and Sabhikhi, 2020). Crop rotation of processing tomato with soybean, maize, sweet corn and bean crops was able to significantly reduce numbers of volunteer plants, which were the main source for bacterial spot pathogen, *Xanthomonas perforans* and significantly reduced its inoculum (Moura et al., 2020). The incidence of field bean pod borer (*Adisura atkinsoni*) can be reduced by crop rotation with cereals (sorghum, maize). In the same way, the weevil (*Cylas* spp.) damage on sweet potato can be minimized by crop rotation (rice – sweet potato – cowpea) (Reddy, 2017). No-tillage soil was naturally suppressive to soybean cyst nematode (SCN), but rotating soybean with corn in no-tillage soil significantly decreased the SCN population and increased relative abundance of fungivores nematodes (Neher et al., 2019).

Table 3. Insect pests, diseases and plant parasitic nematodes that can be managed by crop rotation

Insect-pests/crops	Root-knot nematodes/crops	Diseases/crops
Hypera postica (lucerne)	Bacterial blight (wheat, barley)	*Meloidogyne incognita* (velvet bean, rapeseed, common vetch, castor bean, French marigold)
Diabrotica virgifera (maize)	Bacterial wilt (alfalfa)	*M. javanica* (velvet bean, rapeseed, sesame, common vetch, French marigold)
Leptinotarsa decemlineata (potato, tomato)	Granville (bacterial) wilt (tobacco, potato, eggplant)	*M. arenaria* (velvet bean, sesame, common vetch, castor bean, French marigold)
Meromyza saltatrix (*Triticum* spp.)	Black Shank (tobacco)	*M. hapla* (French marigold)
Zonosemata electa (pepper)	Black Dot (potato) *Exserohilum turcicum* (maize)	
Anasa tristis (*Cucurbita* spp.)	*Ustilago maydis* (maize) *Sclerotium rolfsii* (peanuts)	

Insect-pests/crops	Root-knot nematodes/crops	Diseases/crops
Mayetiola destructor (*Triticum* spp.)	*Verticillium dahliae* (*Solanum tuberosum, Dianthus annus*)	
	Sclerotinia sclerotiorum (*Arachis hypogea, Solanum tuberosum, Glycine max*)	

Adapted from Reddy, (2017)

Mixed cropping

Mixed cropping systems usually tend to decrease pest density by increasing biological diversity because in this system the habitat is manipulated in one or the other way that is favourable for biocontrol agents (parasites, parasitoids and predators) and not favourable for insect pests, diseases and nematodes. It is one of the effective measures to combat soil borne pests and diseases in conservation agriculture systems where tillage is not allowed.

Intercropping

Intercropping is nothing but growing one or more crops in between main crops for sustainable crop production and income apart from increasing soil health. Intercrops may be a host or non-host for a particular pest and disease. In some cases the pest may be attracted to intercrop and that can be killed by other means, while in other cases the intercrops may act as refugee upon which the natural enemies flourish and keep feeding on insect pests on main crop. Enhanced ratio between natural enemies and herbivores was noticed in pear orchard intercropped with aromatic plants (Beizhou et al., 2011).

Some of the intercrops may have allelopathic effect against soil-borne diseases and nematode pests. The root exudates of intercrops effectively suppress the fungal spore germination mycelial growth and hatching of nematode juveniles and thereby protect main crop. This mechanism was well documented in many systems with soil-borne pathogens like *Fusarium* spp. (Hao et al., 2010), *Verticillium dahliae* (Fu et al., 2015), *Cylindrocladium parasiticum* (Gao et al., 2014). Some of the Asteraceae species are known to produce thiaphenes and thiarubrines which can suppress parasitic nematodes and when planted as intercrop they protect main crop by negatively affecting nematodes (Tsay et al., 2004). Similarly, *Solanum sisymbriifolium* produces high levels of egg-hatching chemicals and when planted as intercrop it safeguards main crop against susceptible nematode-host for potato cyst nematode, *Globodera* spp. (Dias et al., 2012). Likewise there are other example *viz.*, Tomato along with mustard as intercrop significantly affected root-knot nematode population (Abrar et al., 2020). Okra intercropped with soybean, red pepper and *Amaranthus* significantly reduced nematode infestation but okra intercropped with watermelon, bitter gourd and maize aggravated the nematode population (Singh and Hemalatha, 2011).

Push-pull technology

It involves growing trap crop that attracts insect pests and an intercrop which repels the insect pests, thereby main crop will be protected from herbivore damage. Khan et al. (2006) developed a push-pull model in corn for managing stem borers and weeds where in Desmodium (*Desmodium intortum*) for repelling the stem borers away (push) and Sudan grass (*Sorghum vulgare sudanense*) in the crop border which attracts female moths (pull) for egg laying. Maize crop with desmodium as push crop and napier grass/*Brachiaria* as pull crop significantly reduced incidence of stem borer, fall army worm, striga weed and ear rot in maize crop (Njeru et al., 2020). Ultraviolet (UV)-reflective mulch and foliar applications of kaolin as push and *Bidens alba* (L.) as pull significantly reduced flower thrips infestation in tomato (Julian et al., 2018).

Trap cropping

Trap crops are grown by the side or in the periphery of the main crop to attract, distract, repel or divert the insect pests so that pest load/damage on main crop may be reduced. Once the insect pests got attracted on trap crop, they can be killed by localized application of pesticides or by any means. Trap crops are very much suited for pest management in conservation agriculture systems where one of the objectives is low usage of pesticides, minimal disturbance or no tillage. Selection of trap crop based on its attractiveness to pests, canopy coverage, and spatial/temporal growth are critical for effective insect pest management. Dead-end trap cropping is the growing of trap crop which are highly attractive to insect pests on which they cannot survive ((Shelton and Nault, 2004). Sequential trap cropping is growing trap crops earlier or later than the main crop where insect pests are trapped and main crop got escaped from insect pest incidence and damage. In multiple trap cropping, one or more trap crops grown along with main crop to attract or divert insect pests. Diamond back moth is successfully managed by growing Indian mustard as trap crop in cabbage at different row ratios (Srinivasan and Krishna Moorthy, 1991).

Novel approaches for Organic Agriculture (OA) and Protected Cultivation

With growing concern about food safety and ecological sustainability, organic agriculture got much more importance in last two decades. Indiscriminate use of agro-chemicals has led to environmental pollution, residual problems in food produce and difficulty in managing the insect pests with multifold resistant to pesticides. In the era of WHO and policy making, world community demands safer food, healthy environment with sustainable food production. In this connection, organic farming shows great promise. At the same time management of insect pests and diseases becomes major issue in OA as the use of agro-chemicals is completely restricted.

Protected cultivation is another emerging production system for high valued vegetable and ornamental crops. Soil-borne pathogens and plant parasitic nematodes (PPN) apart from some sucking pests are the major bottleneck under protected cultivation systems and have become national problem due to their vast damaging nature. Some of the common pests and diseases under protected cultivation are listed in Table 4. In India, on an average, a national loss of Rs. 21,068.73 millions has been estimated due to plant parasitic nematodes (IIHR, 2015) and overall 60% yield loss has been reported due to PPN under protected cultivation. Some of the novel approaches for insect pests, diseases management in OA and protected cultivation are discussed below:

Table 4. Some of the common pests and diseases under protected environment in India

Insect pests		Host
Common name	Scientific name	
Aphids	*Aphis gossypii*	Capsicum
	Macrosiphoniella sanborni	Chrysanthemum
	Myzus escalonicus	Strawberry
	Myzus persicae	Capsicum, Gerbera
	Toxoptera aurantii	Orchid
Caterpillars	*Helicoverpa armigera*	Capsicum, tomato, carnation
	Spodoptera litura	Rose, tomato, capsicum, cucumber
Leaf-miners	*Liriomyza trifolii*	Tomato, cucumber, chrysanthemum, gerbera
Mites	*Aceria lycopersici*	Tomato
	Polyphagotarsonemus latus	Capsicum
	Stenotarsonemus fragariae	Strawberry
	Tetranychus cinnabarinus	Carnation
Thrips	*Scritothrips dorsalis*	Rose
	Thrips palmi	Gerbera
Whiteflies	*Bemisia tabaci*	Gerbera, capsicum
	Trialeurodes vaporariorum	Tomato, cucumber, capsicum, beans, gerbera
Powdery mildew	*Erysiphe cichoracearum* *Podosphaera xanthii* *Leveillula taurica*	Cucumber, lettuce, capsicum
Gray mould	*Botrytis lactucae*	Lettuce
	Botrytis cineria	Tomato
Fusarium wilt	*Fusarium oxysporum* f. sp. lycopersici	Tomato
	Fusarium oxysporum f. sp. *cucumerinum*	Cucumber
Fusarium crown and root rot	*Fusarium oxysporum f.* sp. *radicis lycopersici*	Tomato

Insect pests		Host
Common name	Scientific name	
Damping off/ root rot	*Pythium ultimum, Pythium aphanidermatum*	Cucumber, lettuce
	Phytophthora spp., *Rhizoctonia solani*	Capsicum
Downy mildew	*Pseudoperenospora cubensis*	Cucumber
	Bremia lactucae	Lettuce
Root knot nematode	*Meloidogyne incognita*	Tomato, Capsicum Cucumber

Adapted from Sabir and Singh, (2013)

Cultural practices

Some of the insect pests and diseases can be managed by modification of agronomic practices. Deep ploughing of field in summer exposes the hibernating and soil dwelling insects which can be picked by birds or killed by desiccation. Selection of resistant varieties is best practice to control insect pest and diseases. Resistant cultivars give first hand protection against biotic stresses. If resistant cultivars are not available, tolerant or moderately tolerant cultivars also provide required defenses when integrated with other management practices. It may be be noted that genetically modified crops (GMO) are not allowed in organic agriculture. Some of the insect pests and diseases can be well managed by sequential cropping. Rotating with non-host crops significantly reduce pest populations in alternate seasons. Trap crop also gives protection against insect pests by trapping and arresting movement of insect pests to main/cash crop. Trap crops may be dead-end trap crop, sequential trap crop, push-pull mechanism, perimeter trap crop etc. Adjusting the sowing/planting date and time can reduce pest infestation. Most susceptible stage of the crop to insect pests can be escaped by adjusting the planting time. Early sowing or late sowing can escape from incidence of insect pests and thereby protect the main crop. By increasing seed rate, the loss can be minimized, one such example is shoot fly in sorghum. Proper spacing between the rows of main crop significantly reduces the incidence of many biotic stresses including weeds. Proper water management can significantly reduce incidence of soil borne pathogens and nematode pests.

Physical methods

Using of traps like pheromone traps, light traps, sticky traps, bait traps etc are highly encouraged in organic agriculture. Pheromone traps can be used for monitoring, mass trapping and mate disruption. Most of the pheromone traps attract male insects thereby reducing the frequency of male insects for mating and further decline in population. Light traps with different light sources can be employed in fields for collection of light loving insects. One of the disadvantages of light trap is it collects both beneficial and harmful insects. Sticky traps with different colour combinations are useful in catching sucking pests like thrips, whiteflies, aphids, hoppers etc. Bait traps can be installed below ground for collecting soil dwelling insects like root grubs, *Spodoptera* spp etc.

Mechanical methods

These methods include hand collection of larvae from the field, beating the bush to dislodge the insect pests, digging trench around the field to arrest mass movement of insect pests as in case of hairy caterpillars etc.

Biological control methods

Use of biological control agents like parasites, parasitoids, predators and insect pathogens are highly encouraged in organic farming as these are eco-friendly and give sustainable control of insect pests. Some of the biological control agents used in organic farming is mentioned in Table 5.

Table 5. Potential biological control agents for pest management in organic crop production

Biocontrol agents	Group of insects	Crops
Bacteria *Bacillus thuringiensis*	Lepidopteran pests	Cotton, sunflower and vegetables
Entomopathogenic fungi *Metarhizium anisopliae* *Beauveria bassiana* *Verticilium lecanii*	Coleopteran grubs Lepidopteran and coleopteran Aphids and whiteflies	Coconut, cotton, and green house vegetables
Insect predators Lady bird beetles *Chrysoperla* spp	Aphids, whiteflies and mealy bugs	Fruits and vegetables
Insect parasitoids *Trichogramma* spp. *Chelonis blackburni*	Lepidopterans	Sugarcane and tomato
Entomopathogenic nematodes *Heterorhabditis bacteriophora* *Steinernema carpocapsae*	Coleopteran and Lepidopterans	Sugarcane and plantation crops
Entomopathogenic Viruses Nuclear polyhedrosis virus (NPV) Granulosis virus	*Helicoverpa* and other Lepidopterans	Pulses and vegetables

Adapted from Kumaranag, et al. (2013)

Cotesia plutellae is a larval parasitoid of diamond back moth in India and it causes 40-60 per cent parasitism under filed conditions (Krishna Moorty et al., 2006). Many biocontrol agents like *Trichoderma* spp., *Azospirillum lipoforum*, *Gliocladium* spp., *Psuedomonas fluorescens, Bacillus subtilis*, *Aspergillus niger*, *Azotobacter chroococcum*, etc. have been exploited for the management of major plant diseases. Antagonistic microbes like *Bacillus* spp., *Pseudomonas* spp., *Trichoderma viride* and *T. harzianum* against *Phytophthora capsici*, *T. viride* provided maximum inhibition (72.5%) of the pathogen (Shashidhara et al., 2008). Chowdappa et al. (2013) reported that *B. subtilis* OTPB1 and *T. harzianum* OTPB3 induced systemic resistance in tomato seedlings against early and late blight. Adhikari *et al.* (2013) isolated rhizobacteria from different vegetables including tomato and studied their antagonistic properties against *R. solani*. According to them *P. fluorescens* isolate PF8 and PF7 inhibited the mycelial growth of *R. solani* (72.05 and 68.25%) respectively in dual culture method.

Chemical control methods

The use of agro-chemicals for management of biotic stresses like insect pests, diseases, weeds and nematodes are completely restricted in organic farming. A list of chemicals/products which are permitted or restricted to use in OA are mentioned in Table 6.

Table 6: The list of products which can be used for pest and disease management in organic farming

List of products	Conditions for usage
I. Substances from plant and animal origin	
Azadirachta indica [neem preparations]	Permitted
Neem oil	Restricted*
Preparation of rotenone from *Derris elliptica, Lonchocarpus, Thephrosia* spp.	Restricted
Gelatine	Permitted
Propolis	Restricted
Plant based extracts (e.g. neem, garlic, pongamia, etc.)	Permitted
Preparation on basis of pyrethrins extracted from *Chrysanthemum cinerariaefolium*, containing possibly a synergist *Pyrethrum cinerafolium*	Restricted
Preparation from *Quassia amara*	Restricted
Release of parasite predators of insect pests	Restricted
Preparation from *Ryania* species	Restricted
Tobacco tea	Prohibited
Lecithin	Restricted

List of products	Conditions for usage
Casein	Permitted
Sea weeds, sea weed meal, sea weed extracts, sea salt and salty water	Restricted
Extract from mushroom (Shiitake fungus)	Permitted
Extract from Chlorella	Permitted
Fermented product from *Aspergillus*	Restricted
Natural acids (vinegar)	Restricted
II. Minerals	
Chloride of lime/soda	Restricted
Clay (e.g. bentonite, perlite, vermiculite, zeolite)	Permitted
Copper salts / inorganic salts (Bordeaux mix, copper hydroxide, copper oxychloride) used as a fungicide, maximum 8 kg per ha per year depending upon the crop and under the supervision of inspection and certification agency	Restricted
Mineral powders (stone meal, silicates)	Prohibited
Diatomaceous earth	Restricted
Light mineral oils	Restricted
Permanganate of potash	Restricted
Lime sulphur (calcium polysulphide)	Restricted
Silicates (sodium silicate, quartz)	Restricted
Sodium bicarbonate	Permitted
Sulphur (as a fungicide, acaricide, repellent)	Restricted
III. Microorganisms / Biocontrol agents	
Viral preparations (e.g., Granulosis viruses, Nuclear polyhydrosis, viruses etc.).	Permitted
Fungal preparations (e.g., *Trichoderma* species etc.)	Permitted
Bacterial preparations (e.g., *Bacillus* species etc.)	Permitted
Parasites, predators and sterilized insects	Permitted
IV. Others	
Carbon dioxide and nitrogen gas	Restricted
Soft soap (potassium soap)	Permitted
Ethyl alcohol	Prohibited
Homeopathic and Ayurvedic preparations	Permitted
Herbal and biodynamic preparations	Permitted
V. Traps	
Physical methods (e.g., chromatic traps, mechanical traps, light traps, sticky traps and pheromones)	Permitted
Mulches, nets	Permitted

*Restricted means that the conditions and the procedure for use shall be set by the certification programme

Adapted from NPOP, 6th Edition;2006.

Use of botanicals and microbial extracts

Pesticides from plant or microbial origin are highly encouraged in organic agriculture for effective and sustainable pest management. The crude extracts or commercial formulations of some plants like neem, pongamia, tobacco or garlic are extensively used for pest management in conventional and organic agriculture. Haldhar et al. (2014) tested different organic IPM modules against melon aphids, cucurbit fruit fly, hadda beetle and found that integration of growing resistant genotype (RM-50), spray of neem oil at 20 DAS, installation of pheromone trap (10/ hectare) at 42 DAS, spray of tumba fruit extract (TFE 5%) at 50 DAS and spray of spinosad 46 SC at 60 DAS was the best module with high B: C ratio (8.8:1).

Biofumigation

Biofumigation has emerged as an effective non-chemical alternative to manage nematode and soil borne pest and pathogens in recent years. Karavina et al. and co workers (2015) reported that mustard was the most effective brassica in controlling nematodes, while cabbage, radish and rape significantly reduced *M. javanica* population when compared to the untreated control. Hamaguchi et al. co workers (2019) reported that α terthienyl action was much higher after photoactivation, but was still effective against *C. elegans* Dauer larvae and *M. Incognita* second stage juveniles, even without photoactivation. Yan et al. co workers (2019) tested the efficacy of dimethyl disulfide (DMDS) and reported excellent field efficacy of 80%–94% on root-knot nematode applied at 10–100 g m^{-2} on tomato in China. Mao et al. (2014) and Yu et al. (2019) reported the significant effect of DMDS against nematode and fungal diseases in cucumber and tomato, respectively. Ren et al. (2018) tested efficacy of AITC (allyl isothiocyanates) against soil-borne pests in tomato and observed that AITC has high efficacy against root-knot nematodes (LC_{50} of 18.046 mg per kg), moderate efficacy against fungal pathogens (LC_{50} of 27.999 - 29.497 mg per kg), and weeds (LC_{50} of 17.300 - 47.660 mg per kg) and further reported that AITC used as a soil fumigant (30 - 50g per square meter) effectively controlled major bacterial and fungal pathogens, root-knot nematode and weeds. Ghosh et al. (2020) evaluated suspension concentrate formulation of allyl isothiocyanates and defatted mustard seed meal powder as pre-planting treatment of soil @ 500 ml per acre through drip irrigation against soil-borne pests and pathogens in capsicum and reported that an excellent suppression of the soil-borne fungal pathogens (70-80%) like *Sclerotium rolfsii*, *Fusarium* sp., bacterial pathogens (50-60%) like *Ralstonia* sp., *Xanthomonus* sp. and root knot nematodes (60-70%).

Novel approaches of weed management in emerging production systems

Presently, there are major transformation is taking place in respect of weed management. Newer weed management approaches for weed control that revolutionized the weed science sector and minimizes economic losses caused by weeds without impairing ecological and environmental health. It has been addressed by selective chemical application.

Spot application or selective spraying systems are the innovative approach where nozzles are located in such a way that they work with on and off nozzle basis. In this system, selective herbicides are applied in the area where weed exists but not in a area where weeds are not present. It is operated by robotic systems (Steward et al., 2002). However, in organic farming the chemical herbicides are completely prohibited.

Robotic selective application

Recently, selective herbicide spraying technology have been linked with smaller automatically guided vehicle technology attached with hardware and control technology more precise herbicide application. Example- Eco Robotics is based on automatically sensor based technology. It guide through crop rows using GPS and machine vision sensor and follow crop rows (EcoRobotics, 2019).

Source: EcoRobotics, 2019

Mechanical weeders

It is very useful approach for that sector where chemical weed control was not possible. Mechanical weeder is very useful equipment for organic farming because chemical application is totally restricted. It works on the basis of three basic physical techniques for weed control which include burying, cutting and uprooting the weeds (Ahmad, 2012).

Robocrop automated mechanical weeding system.
Source: Tillett and Hague Technology Ltd.

Allelopathic approach

The plant releases the chemicals, which suppressed the growth and development of undesirable plants. This can be potentially used in various crops. This may be either placed in soil surface or extract can be directly applied in to the weeds (Table 7). Use of eucalyptus leaf litter as a mulch material recorded highest weed control efficiency (62.0 and 51.8%) at 20 and 40 DAS respectively (Maheshwari and Kartik, 2019).

Table 7. Allelopathic effect of different crop mulches on weeds of various crops

S.No.	Crop residue	Crop	Control weeds	References
1.	Paddy straw	Chickpea	*Cichorium intybus, Medicago truncatula, Melilotus indica, Anagallis arvensis, Chenopodium album*	Sahu et al. (2020)
2.	Wheat straw	Chickpea	*Cichorium intybus, Medicago truncatula, Melilotus indica, Anagalis arvensis, Chenopodium album*	Sahu et al. (2020)
3.	Eucalyptus leaf litter	Maize	*Cyperus rotundus, Sorghum helpense, Digeria arvensis, Alternanthera pungens, Amaranthus hybrids, Echinochloa colona, Tribulus terrestris*	Mahmood et al. (2018)

S.No.	Crop residue	Crop	Control weeds	References
4	Wheat and Rye straw	Summer crops	*Portulaca oleratia, Amaranthus retroflexus, Echinocloa colona*	Boz (2003)
5	Neem leaves	Chickpea	*Chenopodium album, Cyperus rotundus, Anagallis arvensis*	Varshney and Kumar (2008)

Ecological intensification

Compatible crop intensification in system leads to suppression of weeds and favoures the crop for better establishment. It provides opportunity for better resource utilization and act as potential weed management tools. Competitive crops, smother the field and doesn't allow the weeds to establish eventually reduced the density and dry biomass accumulation of weeds (Choudhary and Choudhury 2018). It has been found that short duration pulses i.e. greengram, cowpea and soybean smothers weed effectively (Mubeen, 2013).

Thermal weed control

Heat is a potential source of energy to be used for weed control. Recently many experiments have been conducted and established that fire, directed flaming, hot water, steam, microwave, infrared, ultraviolet radiation; electrocution and freezing are the possible ways can be used for weed control. In these, heat coagulates protein and burst the protoplasm mainly due to explosion which kills weed tissue. Sometimes, exposure of very low temperature can also kill the weeds. For example aquatic weeds were control by using low air temperature by removing water from pond or lake (Maheshwari and Kartik, 2019).

Thermal weed control, Plant Protection and care

(*Source*: fruithill.com)

Pneumatic weed control

It compressed the air into the soil for loosening and uprooting of small weeds between the row and aside. This can be operated in row crops like maize, carrot and sugar beet. The operating depth of machine is about 15 cm with speed of 5-6 km/h^{-1} (Vale, 2003). It is useful in organic farming and protected cultivation systems

Pneumatic weed control tools, *Source*: OACC-2005

Mulching

Mulching reduce the weed problem by covering soil surface. It prevents weed seed germination or suppress the growth of emerging weeds (Choudhary et al., 2015; Choudhary 2016). Mulches generally cover the ground by organic or inorganic materials. These are spread uniformly on soil surface for the purpose of weed control, moisture conservation and reduce the loss of water by wind and water. This is very effective economically viable technology in conventional as well as organic agriculture (Maheshwari and Kartik, 2019).

Biological control of weeds

It is a complex of natural enemies or biological agent for weed suppression. Biological weed control is emerging tool for weed control in organic farming. Different types of bio-agents are being used in agriculture for weed control i.e. Phytophagous arthropods (insect and mites), plant pathogen (fungi, bacteria, viruses and nematodes), fish, birds and other animals like sheep and goat (Westwood et al. 2018).

Bio-herbicides

Bio-herbicides consist of phytotoxins, pathogens and other micro-organism used for biological weed control. It derived from micro-organism such as fungi, bacteria and protozoa or plant residue extracts or single compounds derived from other plant species. Successfully control of *Cuscuta* spp. (Dodder) by *Colletotrichum gleosporiodes* f.sp. *cuscutae* (Kremer, 2005) is one such example.

Development of herbicide resistant crops

Herbicide resistance is inherent ability of individual plants that survive after herbicide application but kill a normal population of same species. Herbicide resistance crop increase herbicide selectivity resulting in minimizing the yield loss by weeds. Some herbicide resistant crop namely tobacco and cotton resistant against 2, 4-D responsible gene mono-oxygenage and tobacco resistant against pendimethalin responsible gene carbonate hydroxynage (Datta and Pilli, 2019). Similarly, herbicide resistant crops like corn, soybean, cotton and canola has been developed, although in many countries it is not yet approved to grow in large scale.

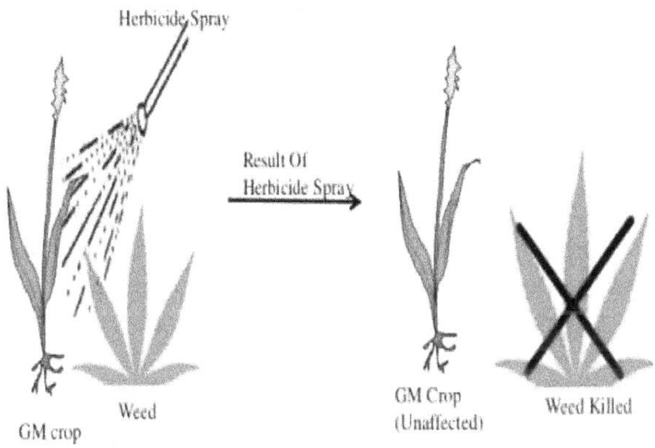

Development of herbicide resistant crop
Source- Datta and Plli , 2019.

Modeling of crop weed interference

It is a complex task for development of integrated weed management system for better understanding of crop-weed interference. Computer based models have been developed to collect the integrated information about interference, critical period and other parameters for better outcome prediction of crop weed interference (Ghersa and holt, 1995).

Integration of cultivation and herbicide banding

Integration of cultivation and herbicide banding is a complex phenomenon for reducing herbicide uses in row crop. In row crops primary flushes of weeds can be control by shallow manipulation of the soil. It requires two to four passes (Bhowmik and Inderjit, 2004). However, it is not applicable in organic farming systems.

Future strategies for weed management

New herbicide targets and Bio-pesticides

Future of chemical weed control totally depends on discovery of new herbicide molecules and mode of actions, but during last two decades no new mode of herbicides has been developed. Therefore, efforts are required in order to develop herbicides with new mode of action for broad-spectrum weed control. For example 5- enolpyruvilshikimate-3-phosphate synthase and acetolactate synthase herbicides are good inhibitor of sikimate and branch chain amino acid pathway (Westwood et al., 2018). But, these are no more effective in some of the weeds as they have evolved resistance. New research direction is required on this line.

RNA interference herbicides and genetic engineering

Potential of new technologies is the uses of RNA to hidden feature of weed genes by the process of RNA interference. It enhanced weed susceptibility to herbicides or death ability of weeds. As a spray, RNA develop more potential for weed management because sequences can be design to target a specific weed species or a group of related weed species by selectivity (Westwood et al., 2018).

Precision agriculture and robotics

Precision agriculture based on more precise use of agricultural inputs like-herbicide, fertilizers, water and other agro inputs. It is possible through combination with robot techniques.

The need for precision

In present agriculture, use of agro-chemicals is very high and thus increases the production cost and causing serious threat to soil, water and environment. There is urgent need to use the agro-chemicals site specific by using precision tools for maintaining sustainability and ecological balance. It considerably reduces the investment on herbicide, minimizes cost of weeding and also minimizes the crop injury (USDA- ERS 2015). It is very important approach for minimizing the residual effect of herbicide on soil, plant and water bodies due to uses of limited doses of herbicides under precision agriculture through advanced techniques of herbicide applications.

Co-robotics, the symbiosis among man, machine, and crop plants

Automatic weed controls have been demonstrated through advanced smart-machine technology. It manages the herbicide application on the basis of crop row spacing and weed intensity. However, the large proportions of weeds are automatically controlled and may also be used at commercial scale. However, their adoptability for many other weeds, crops and growing conditions has yet to be standardized (Fennimore et al., 2014).

Beginning of the robotic weeding era

Technologies automation and mechatronics are more effective and commercially viable for future weed control strategies. It is already used in industrialized countries for specific crops (King, 2017). Denmark developed an equipment Robovator intra-row cultivator. It has two flat blade tines per crop row that cut the weeds below the soil at 1 to 2 cm depth (Melander et al., 2015).

Harnessing biology for weed management

The main focus of weed scientists would be to develop crop varieties through crop breeding that compete with weeds. Three soybean cultivars 'Harosoy 63, Amsoy, and Corsoy were identified as a form of weed competitive cultivars. An alternative strategy that focus a research on molecular, physiological and morphological mechanism for inter-specific and intra-specific competition (Westwood et al., 2018).

Information Technology for Extension

a) Communication and Computer Power: Communication and computer power play an important role for future scenario for transfer of advance techniques for weed control. It is very useful for taking a decision how to control weeds and minimize yield loss by weeds. Currently, University and industry professionals use all form of communication resources to carry research findings for better understanding (Lubell et al., 2014).

b) Training the next generation of weed scientists: The key principles of weed science will be acquiring essential knowledge for students for the future. Continuous use of herbicides is not merely a routine, therefore training in the sector of herbicide chemistry, physiology and technology must be provided the importance of weed biology and ecology continue to grow by 2050. Similarly, recent scenario of weed development and competitive interaction gives new options for weed management.

Policy related issues

1. **Promoting ITKs**: Use of indigenous technologies like panchagavya etc for biotic stress management should be validated scientifically and subsidy like encouragements should be given to organic farmers for using these traditional technologies.
2. **Organic pesticides**: Pesticides from plant or animal origin should be tested scientifically and their registration with CIBRC for use in organic farming should be in fast track mode.
3. **Pesticides registration**: The climatic/edaphic factors in protected cultivation and conservation agriculture are altogether different from conventional

agriculture. Hence, the testing and registration of pesticides for these two production systems should be exclusively separate.

4. **Herbicide resistance and weed plasticity**: Herbicidal weed management came into existence due to more labour wages and lack of availability of labour during critical period of weed management. Continuous use of herbicides with similar mode of action, leads to herbicide resistant toweed. Some weed species show multiple herbicide resistance e.g., *Amaranthus* spp. In soybean, *Echinochloa* spp., *Phalaris minor* in cereal and cereal-based rotations like *Avena futua* and *Chloris truncate* have limited option for control through herbicide (Chauhan et al., 2017).
5. **Misconceptions about integrated weed management and neglected areas of research in weed science**: The execution of real integrated weed management programs demands efficient and diverse approaches, instead of just relying on herbicides e.g. sequential application, premix and tank mixture. At present, weed research is dealing just in the form of herbicide research and several findings released in this direction. Many experts argued that weed science is a science of herbicide instead of weed science (Harkar and Donovan, 2013).
6. **Herbicide related contamination**: Herbicidal weed control give better option for timely control of weeds reduced cultivation and production cost, increase profit per unit land but create many problems related to soil, plant and environment. Herbicide contaminates soil and water and also interfere enzymatic and microbial activity of soil, which is responsible to many reactions and transformations that regulate soil health. For example- sulphonylurea herbicides affect nitrification, nitrogen fixation and mineralization at recommended doses of herbicides applications (Rose et al., 2016).
7. **Lack of trained weed scientists in developing countries**: Except entomology and pathology, no or very few departments related to weed science in agriculture universities exists. Only few universities facilitate research related to weed control of fruits, orchards, ornamentals, aquatic, pastures and forest in developing countries. Faculties related to cropland weed management, vegetables, turf, ecology and statistics also not available in true sense. Different degree awarding institutes from various locations in a country have almost similar course curriculum, do not consider area specific variations in crops cropping sequence, weed species, management practices, input levels, socio-economic background and skilled farmer of weed management (Chauhan et al., 2017).
8. **Plant quarantine**: Main aim is to forest abolishment of statutory body is to control the entry of exotic weed species, disease and fungal spores by following the strict rules and regulation, but at present, this body is

fail to prevent the entry of exotic weeds and diseases in one country to another country due to poor checking systems, as a result many weeds like, *Parthenium* spp., *Eichhornia* spp. and *Lantana camara* cover most of the country in the world.

9. **Climate change**: Lack of precise information about the effect of climate change on agricultural pests particularly weeds are major issue and create big problem for agricultural management practices (Ramesh et al., 2017). Under natural condition weeds grow simultaneously with crops. Climate change affect directly and indirectly to the life cycle of weeds, crops, herbicidal efficacy etc. According to different studies it has been established that effect of CO_2 have considered weeds and crops species separately, elevated CO_2 increase quantification of crop-weed competition. Weed developed mechanism according to climate situation so that they are easily survived. Under changing climate development of new weed management approaches is a challenging task for weed scientists.

References

Abrar, S., A. Seid and M. Dejene. 2020. Integrated management of *Meloidogyne incognita* in tomato (*Solanum lycopersicum*) through botanical and intercropping. African Journal of Agricultural Research. 15:492-501.

Adhikari, A., S. Dutta, S. Nandi, I. Bhattacharya, M. Roy, G. De Sarkar and T. Mandal. 2013. Antagonistic potentiality of native rhizobacterial isolates against root rot disease of okra, incited by *Rhizoctonia solani*. African Journal of Agricultural Research. 8:405-412.

Ahmad, B. M. T. 2012. Development of an automated mechanical intra-row weeder for vegetable crops. Iowa State University Graduate Theses and Dissertations. 12278. Available at: lib. dr.iastate.edu/etd/12278

Altieri, M. A. and L. L. Schmidt. 1985. Cover crop manipulation in northern California orchards and vineyards: effects on arthropod communities. Biological Agriculture and Horticulture 3:1–24. https://doi.org/10. 1080/01448765.1985.9754453

Beizhou, S., Z. Jie, H. Jinghui, W. Hongying, K. Yun and Y. Yuncong. 2011. Temporal dynamics of the arthropod community in pear orchards intercropped with aromatic plants. Pest Management Science. 67:1107–1114.

Belfry, K. D., C. Trueman, R. J. Vyn, S. A. Loewen and L. L. Van Eerd. 2017. Winter cover crops on processing tomato yield, quality, pest pressure, nitrogen availability, and profit margins. PLoS ONE 12: e0180500. https://doi.org/10.1371/journal. pone.0180500

Bhowmik, P. C. and Inderjit. 2004. Weed Biology and Management. 363-373. Kluwer Academic Publishers.

Boz, O. 2003. Allelopathic effects of wheat and rye straw on some weeds and crops. Asian Journal of Plant Sciences. 2:772-778.

Carrière, Y., Z. Brown, S. Aglasan, P. Dutilleul, M. Carroll, G. Head, B. E. Tabashnik, P. S. Jørgensen and S. P. Carroll. 2020. Crop rotation mitigates impacts of corn rootworm resistance to transgenic Bt corn. PNAS. 117:18385-18392

Chauhan, B. S., A. Matloob, G. Mahajan, F. Aslam, S. K. Florentine and P. Jha. 2017. Emerging challenges and opportunities for education and research in weed science. Frontiers in Plant Science. 8:1537.

Choudhary, V. K. 2016. Response of land configuration and mulches on maize-french bean-toria cropping system. Agronomy Journal. 108:2147-2157. DOI: 10.2134/agronj 2016.02. 0088.

Choudhary, V. K. and B. U. Choudhury. 2018. A staggered maize-legume intercrop arrangement influences yield, weed smothering and nutrient balance in the Eastern Himalayan Region of India. Experimental Agriculture. 54:181-200.

Choudhary, V. K., A. Dixit, R. Bhagawati, A. K. Vishwakarma and Brajendra. 2015. Influence of locally available mulches on soil moisture content, root behaviour, weed dynamics and productivity of pea (*Pisum sativum* L.). Progressive Research- An International Journal. 10:1372-1375.

Chowdappa, P., S. P. Mohan Kumar, M. Jyothi Lakshmi and K. K. Upreti. 2013. Growth stimulation and induction of systemic resistance in tomato against early and late blight by *Bacillus subtilis* OTPB1 or *Trichoderma harzianum* OTPB3. Biological Control. 65:109-117.

Clark, A. 2007. Managing cover crops profitably, 3rd edn. Sustainable Agriculture Network. United Book Press, Inc., p 247. http://www.sare.org /publications/ covercrops/ covercrops. pdf. Accessed 24 Feb 2016

Datta, M. K. and K. Plli. 2019. Biotechnological Approach in Weed Management https://www. researchgate.net/publication/333981979

Dias, M. C., I. L. Conceic¸a˜o, I. Abrantes and M. J. Cunha. 2012. *Solanum sisymbriifolium*—A new approach for the management of plant-parasitic nematodes. European Journal of Plant Pathology. 133:171-179.

Dube, E., C. Chiduza, P. Muchaonyerwa, M. Fanadzo and T. S. Mthoko. 2012. Winter cover crops and fertilizer effects on the weed seed bank in a low-input maize-based conservation agriculture system. South African Journal of Plant and Soil. 29:195–197. https://doi.org/1 0.1080/02571862.2012.730637

EcoRobotix. 2019. Switch to smart weeding. Available at: www.ecorobotix.com (accessed on May 11, 2019).

Fanadzo, M., M. Dalicuba and E. Dube. 2018. Application of Conservation Agriculture Principles for the Management of Field Crops Pests S. Gaba et al. (eds.), Sustainable Agriculture Reviews 28, Ecology for Agriculture. 28. https://doi.org/10.1007/978-3-319 90309-5_4.

Fennimore, S. A., R. F. Smith, L. Tourte, M. LeStrange and J. S. Rachuy. 2014. Evaluation and economics of a rotating cultivator in bok choy, celery, lettuce, and radicchio. Weed Technology. 28:176–188.

Flint M.L. and P.A. Roberts. 1988. Using crop diversity to manage pest problems: some California examples. Am J Altern Agric 3:163–167.

Fu, X., Wu, X., Zhou, X., Liu, S., Shen, Y., and Wu, F. 2015. Companion cropping with potato onion enhances the disease resistance of tomato against Verticillium dahliae. Frontier in Plant Science. 6:726.

Gao, X., M. Wu, R. Xu, X. Wang, R. Pan, H. J. Kim and H. Liao. 2014. Root interactions in a maize/soybean intercropping system control soybean soil-borne disease, red crown rot. PLoS One. 9:e95031.

Ghersa, C. M. and J. S. Holt. 1995. Using phenology prediction in weed management: a review. Weed Research. 35:461-470.

Ghosh, S. K., T. Nataraj, N. Usha and Asha. 2020. Biofumigation - An effective tool in enhancing yield of capsicum by suppressing soil-borne pathogens and augmenting biopesticide under protected cultivation in India. Journal of Biological Control. 34:59-65. DOI:10.18311/ jbc/2020/23013

Haldhar, S. M., B. R. Choudhary, R. Bhargava and S. K. Sharma. 2014. Development of an organic integrated pest management (IPM) module against insect-pests of muskmelon in arid region of Rajasthan, India. Journal of Experimental Biology and Agricultural Sciences. 2:19-24.

Hamaguchi, T., K. Sato, C. S. L. Vicente and K. Hasegawa. 2019. Nematicidal actions of the marigold exudate α-terthienyl: oxidative stress-inducing compound penetrates nematode hypodermis. The Company of Biologists. 8:bio038646. doi:10.1242/bio.03864

Hao, W., L. Ren, W. Ran and Q. Shen. 2010. Allelopathic effects of root exudates from watermelon and rice plants on *Fusarium oxysporum* f. sp. *niveum*. Plant Soil. 336:485-497.

Harker, K. N. and J. T. O'Donovan. 2013. Recent weed control, weeds management, and integrated weed management. Weed Technology. 27:1–11. doi: 10.1614/WT-D-12-00109.1

Hunjan, M. S. and H. S. Sabhikhi. 2020. Designing a crop rotation strategy to manage *Streptomyces scabies* causing potato scab in north India. Journal of Phytopathology. https://doi.org/10.1111/jph.12911

IIHR, 2015. Nematode Management in Protected Cultivation. Technical bulletin No.48.

Julian, K. T., J. Funderburk, M. Srivastava, S. Olson and S. Adkins. 2018. Evaluation of a push-pull system for the management of *Frankliniella* species (Thysanoptera: Thripidae) in tomato. Insects. 9:187. https://doi.org/10.3390/insects9040187

Karavina, C. and R. Mandumbu. 2012. Biofumigation for crop protection: potential for adoption in Zimbabwe. J. Anim. Plant Sci. 14: 1996e2005.

Khan, Z. R., J. A. Pickett, L. J. Wadhams, A. Hassanali, C. A. O. Midega. 2006. Combined control of *Striga hermonthica* and stem borers by maize-*Desmodium* spp intercrops. Crop Protection. 25:989–995.

King, A. 2017. Technology: The future of agriculture. Nature. 544:S21–S23.

Kremer, R. J. 2005. The role of bioherbicides in weed management. Biopesticides International 1:127-141.

Krishna Moorthy, P. N., K. K. Pandey, K. K. Pandey and N. K. Krishna Kumar. 2006. Status and prospects of integrated pest management strategies in selected crops: Vegetables. In: Integrated Pest Management. Principles and Applications Volume 2: Princiles, Eds. Singh, Amerika., Sharma, O. P. and Garg, D. K. CBS Publishers and Distributors, New Delhi, pp. 340-92.

Kumaranag, K. M., S. C. Kedar and M. D. Patil. 2013. Insect Pest Management in Organic Farming Systems. Popular Kheti. 1:3.

Linker, H. M., D. B. Orr and M. E. Barbercheck. 2009. Insect Management on Organic Farms. North Carolina Cooperative Extension Service. https://cefs.ncsu.edu/w content/ uploads/ insectmgmtfinaljan09. pdf?x47549.

Lubell, M., M. Niles and M. Hoffman. 2014. Extension 3.0: managing agricultural knowledge systems in the network age. Society and Natural Resources 27:1089–1103.

Maheswari, M. U. and A. Karthik. 2019. Recent trends and techniques of weed management practices in organic farming. Acta Scientific Agriculture. 32:150-157.

Mao, L., Q. Wang, D. Yan, P. Liu, J. Shen, W. Fang, et al. 2016. Application of the combination of 1, 3-dichloropropene and dimethyl disulfide by soil injection or chemigation: effects against soil borne pests in cucumber in China. Journal of Integrative Agriculture. 5:145–52. https://doi.org/10.1016/S2095-3119 (15)61065-6

Marks, C. F. and J. L. Townshend. 1973. Multiplication of the root lesion nematode *Pratylynchus penetrans* under orchard cover crops. Canadian Journal of Plant Science 53:187–188. https://doi.org/10.4141/ cjps73-034.

Mehmood, T., S. U. Khan, A. Qayyum, A. R. Gurmani, W. Ahmed, M. Liaquat and A. Farid. 2018. Evaluation of organic and inorganic mulching as an integrated weed management strategy in maize under rainfed conditions. Planta daninha. 36. http://dx. doi.org/10.1590/s0100-83582018360100143

Melander, B., B. Lattanzi and E. Pannacci. 2015. Intelligent versus non-intelligent mechanical intra-row weed control in transplanted onion and cabbage. Crop Protection. 72:1–8.

Morrill, W. L. 1977. Red imported fire ant control with diazinon and chlorpyrifos drenches. J Georgia Entomol Soc. 12:96–100.

Moura, D. R., J. K. Yamada, L. C. Albuquerque and N. C. Pontes 2020. Crop rotation reduces the density of volunteer plants in processing tomato fields and the inoculum of bacterial spot. European Journal of Plant Pathology. 156:299–304.

Mubeen, M. A. 2013. Growth and yield performance of various spring planted sunflower (*Helianthus annuus* L.) hybrids under semi-arid conditions of Sargodha, Pakistan. Science International. 25:341-344.

Neher, D. A., T. Nishanthan, Z. J. Grabau and S. Y. Chen. 2019. Crop rotation and tillage affect nematode communities more than biocides in monoculture soybean. Applied Soil Ecology. 140:89-97.

Njeru, N. K., C. A. O. Midega, J. W. Muthomi, J. M. Wagacha and Z. R. Khan. 2020. Impact of push–pull cropping system on pest management and occurrence of ear rots and mycotoxin contamination of maize in western Kenya. Plant Pathology. https://doi.org/10.1111/ppa.13259

NPOP, 2015. National Programme on Organic Farming. India Organic. Sixth edition.

Ramesh, K., A. Matloob, F. Aslam, S. Florentine and B. S. Chauhan. 2017. Weeds in a changing climate: vulnerabilities, consequences, and implications for future weed management. Front. Plant Science. 8:95. doi: 10.3389/fpls.2017.00095.

Reddy, P. P. 2017. Trap cropping, pp. 133–147. *In* P. V. Reddy (ed.), Agro-ecological approaches to pest management for sustainable agriculture. Springer. https://link.springer.com/book/10.1007 %2F978-981-10-4325-3

Ren, Z., Y. Li, W. Fang, D. Yan, B. Huang, J. Zhu, X. Wang, X. Wang, Q. Wang, M. Guo and A. Cao. 2018. Evaluation of allyl isothiocyanate as a soil fumigant against soil-borne diseases in commercial tomato (*Lycopersicon esculentum* Mill.) production in China.

Rose, M. T., T. R. Cavagnaro, C. A. Scanlan, T. J. Rose, T. Vancov, S. Kimber, et al. 2016. Impact of herbicides on soil biology and function. Advances in Agronomy 136:133–220. doi: 10.1016/bs.agron.2015.11.005

Sabir, N. and B. Singh. 2013. Protected cultivation of vegetables in global arena: A review. Indian Journal of Agricultural Sciences. 83:123–35.

Sahu, M. P., M. L. Kewat, A. K. Jha, J. K. Sharma and S. Sondhia.2020. Weed dynamics as affected by practices and straw mulches in chickpea. International Journal of Chemical Studies. 8:1857-1859.

Shashidhara, S., M. S. Lokesh, S. Lingaraju and M. G. Palakshappa. 2008. In vitro evaluation of microbial antagonists, botanicals and fungicides against *Phytophthora capsici* Leon., the causal agent of foot rot of black pepper. Karnataka Journal of Agricultural Sciences. 21: 527-531.

Shelton, A. M. and B. A. Nault. 2004. Dead-end trap cropping: a technique to improve management of the diamondback moth, *Plutella xylostella* (Lepidoptera: Plutellidae). Crop Protection. 23:497–503. https://doi.org/10.1016/j.cropro.2003.10.005

Singh, M. K. and H. Pant. 2018. Influence of intercropping on root-gall nematode disease on okra (*Abelmoschus esculentus* L.). Journal of Natural Resource and Development. 13: 114-119.

Smith, B. 2006. The farming handbook. University of KwaZulu-Natal Press, South Africa, p 431.

Spencer, J. L., B. E. Hibbard, J. Moeser and D. W. Onstad. 2009. Behaviour and ecology of the Western corn rootworm (*Diabrotica virgifera* LeConte). Agricultural and Forest Meteorology 11:9–27. https://doi. org/10.1111/j.1461-9563.2008.00399.x

Srinivasan, K. and P. N. Krishna Moorthy. 1991. Indian mustard as a trap crop for management of major lepidopterous pests on cabbage. Tropical Pest Management 37:26–32. https://doi. org/ 10.1080/09670879109371532

Steward, B. L., L. F. Tian and L. Tang. 2002. Distance-based control system for machine vision-based selective spraying. Transactions of the ASAE. 45:1255–62.

Tsay, T. T., S. T. Wu and Y. Y. Lin. 2004. Evaluation of Asteraceae plants for control of Meloidogyne incognita. Journal of Nematology 36:36-41.

USDA-ERS (United State Department of Agriculture Economic Research Service). 2015. Organic Agriculture. http://www.ers.usda.gov/topics/naturalresources- environment / organic-agriculture.aspx. Accessed: July 28, 2016

Vale, 2003. Pneumatic weed control- A Review. Biological Weed Control. 34:354-358.

Varshney, J. G. and L. Kumar. 2008. Influence of covering materials on weed control and yield of chickpea. Indian Journal of Weed Science. 40:22-25.

Westwood, J. H., R. Charudattan, S. O. Duke, S. A. Fennimore, P. Marrone, D. C. Slaughter, C. Swanton and R. Zollinger. 2018. Weed Management in 2050: Perspectives on the Future of Weed Science. Weed Science 10.1017/wsc.2017.78.

Xu, Q. C., S. Fujiyama and H. L. Xu. 2011. Biological pest control by enhancing populations of natural enemies in organic farming systems. Journal of Food Agriculture and Environment 9:455–463.

Yan, D., A. Cao, Q. Wang, Y. Li, O. Canbin, M. GuM, et al. 2019. Dimethyl disulfide (DMDS) as an effective soil fumigant against nematodes in China. PLoS One. 14:e0224456. https:// doi.org/10.1371/journal.pone.0224456

Yu, J., C. J. Land, G. E. Vallad and N. S. Boyd. 2019. Tomato tolerance and pest control following fumigation with different ratios of dimethyl disulfide and chloropicrin. Pest Manage Science 75:1416 24. https://doi.org/ 10.1002/ps.5262 PMID: 30417562

Theme 5: Innovative Technologies for Agroecosystem Development

17

Innovations on Farming System Models

U.K. Behera

College of Agriculture, Central Agricultural University, Kyrdemkulai Meghalaya

Introduction and Background

Agriculture at present has the trend of using an industrialized production model, where crops and animals are produced in systems that are increasingly specialized, simplified and concentrated, to address growing food, feed and fibre demands. Within this model, external inputs (e.g. irrigation, synthetic fertilizers chemical pest control, feeds, and growth hormones) have been utilized to achieve production goals, often at the expense of environmental quality and key ecosystem services. Continuation of this model of agricultural production has resulted in loss of ecosystem services, increased ecosystem simplification and species extinction. Responses to future challenges in agriculture should include the development of novel systems that are highly productive, minimize damage to the environment and effectively utilize renewable resources. Achieving these multiple goals will be a momentous task, as it will require development and implementation of more complex, diverse and management-intensive production systems than currently employed. Furthermore, future agricultural production systems will need to be adaptable to respond to unforeseen environmental challenges. Integrated agricultural systems have been purported to possess attributes to address these challenges. Agro-chemicals, fuel-based mechanization and irrigation operations are the heart of industrial agriculture which is derived entirely from dwindling and ever more expensive fossil fuels. Intensification of agriculture by the use of high yielding crop varieties, fertilization, irrigations and pesticides are impact heavily on natural resources with serious health and environmental implications. Practice of monoculture in some parts of the country and the world has an adverse environmental impact due to lack of ecological regulation mechanisms and heavily dependent on external inputs.

In order to break the vicious cycle of external industrial inputs linked to industry middlemen and farmers and address the COVID-19 challenges, Govt of India has introduced an innovative vision and programme in the name of "AATMA NIRVAR BHARAT (Self reliant India)". In its agriculture perspective there is need to make farming systems self-sufficient and self-reliant by generating major inputs and resources at the farm level. To meet this vision, Integrated Farming System is a potential option, where as per the thumb rule 70% of the resources should be generated and recycled in the farming systems. In this chapter, information relating to farming systems and integrated farming systems for making farm self reliant and doubling the farmers income, while addressing the issue of climate change and enhancing resource use efficiency and rural bio-entrepreneurship is provided.

In other words, to keep pace with the food requirements of a large population there is an urgent need to accelerate all aspects of agricultural food production with due consideration to restoration and conservation of natural resources, which can only be achieved through sustainable resource management and the adoption of farmer-participatory holistic approaches (termed the farming systems approach). The farming systems approach is considered a resource management strategy to achieve economic and sustained productivity that meets the diverse requirements of the farm household whilst preserving the resource base and maintaining a high level of environmental quality (Lal and Miller, 1990).This multidisciplinary whole-farm approach is very effective in solving the problems of small and marginal farmers (Mahapatra and Behera, 2011). The approach aims at increasing income and employment from small-holdings by integrating various farm enterprises and recycling crop residues and by-products within the farm itself (Ahlawat et al., 2002; Rautaray et al., 2005).

The structure of Indian agriculture is undergoing transformation. A notable feature of changing farm structure is the dominance of smallholders whose number has increased over time and will continue to do so in future (Table 1). The traditional monoculture and disciplinary approach is unable to meet the growing and changing food demand and improve the livelihood of these smallholders on a sustainable basis (Mahapatra and Behera, 2011). Therefore, an integrated approach to farming is critical to sustain agricultural production, maintain farm incomes, safeguard the environment and respond to consumer concern about food quality issues (Singh et al., 1998; Yadav and Prasad, 1998). However, the potential contribution of IFS to the development of a more sustainable agriculture has largely been ignored (Paroda, 2014). Moreover, decline in per capita availability of land from 0.5 ha in 1950-51 to 0.15 ha in 2011 and a projected further drop to less than 0.1 ha by 2020 (Mahapara and Behera, 2011) point to the urgent need of developing strategies and agricultural technologies that enable adequate employment and income generation, with small and marginal farmers at the fore front. The problems encountered by these 2 groups of farmers are different than those having large holdings. These farms need multi-enterprise farming activities that are complementary and technically compromising in the interest of the productivity of the whole farming system.

Table 1. Number and Area of Operational Holdings by Size Group, 2010-11

Category of holdings	Number of holdings			Area			Average size of holdings		
	2000-01*	2005-06*	2010-11*	2000-01*	2005-06*	2010-11*	2000-01*	2005-06*	2010-11*
Marginal (less than 1 hectare)	75408 (62.9)	83694 (64.8)	92826 (67.1)	29814 (18.7)	32026 (20.2)	35908 (22.5)	0.40	0.38	0.39
Small (1-2 hectares)	22695 (18.9)	23930 (18.5)	24779 (17.9)	32139 (20.2)	33101 (20.9)	35244 (22.1)	1.42	1.38	1.42
Semi-medium (2-4 hectares)	14021 (11.7)	14127 (10.9)	13896 (10.0)	38193 (24.0)	37898 (23.9)	37705 (23.6)	2.72	2.68	2.71
Medium (4-10 hectares)	6577 (5.5)	6375 (4.9)	5875 (4.2)	38217 (24.0)	36583 (23.1)	33828 (21.2)	5.81	5.74	5.76
Large (10 hectares and above)	1230 (1.0)	1096 (0.8)	973 (0.7)	21072 (13.2)	18715 (11.8)	16907 (10.6)	17.12	17.08	17.38
All holdings	119931 (100.0)	129222 (100.0)	138348 (100.0)	159436 (100.0)	158323 (100.0)	159592 (100.0)	1.33	1.23	1.15

No. of holdings: ('000), Area operated: ('000 hectares), Average size: (hectares). The figures in parenthesis represent % of total. (*Source*: ASG, 2015.)

In IFS research, integration of landbased enterprises such as aquaculture, poultry duckery, apiary, field and horticultural crops within the biophysical and socio-economic environment of the farmers is important to make farming self reliant (Behera et al., 2004; Rautaray et al., 2005). Adoption of an individual farm enterprise in isolation cannot sustain the farm family, but the IFS approach holds the promise of addressing the issues of sustainable economic growth of Indian farming communities. Integrated farming systems benefit from synergisms among different enterprises, diversity of produce, and environmental soundness (Behera et al., 2010). For this reason, the IFS model has been suggested by several workers for developing small and marginal farms across the country (Behera and Mahapatra, 1999; Singh et al., 2006).

Concept of Farming Systems

"Farming System" is a complex inter-related matrix of soil, plants, animals implements, power, labour, capital and other inputs controlled in parts by farming families and influenced to varying degrees by political, economic, institutional and social forces that operate at many levels" (Mahapatra, 1992). The term "farming system" refers to a particular arrangement of farming enterprises that are managed in response to physical, biological and socio-economic environment and in accordance with farmer's goals, preferences and resources (Shaner et al 1982). "The household, its resources and the resource flows and interactions at the individual farm levels are together referred to as a farm system" (FAO, 2001) The Farming System, as a concept, takes into account the components of soil water, crops, livestock, labour, capital, energy and other resources with the farm family at the centre managing agricultural and related activities. The farm family functions within the limitations of its capability and resources, the socio-cultural setting, and the interaction of these components with the physical, biological and economic factors.

Farming system focuses on

- The interdependencies between components under the control of household and,
- How these components interact with the physical, biological and socio-economic factors, which is not under the control of household.
- Farm household is the basic unit of farming system and interdependent farming enterprises carried out on the farm.
- Farmers are subjected to many socio-economic, bio-physical, institutional administrative and technological constraints.
- The operator of the farming system is farmer or the farming family.

The primary inter-relationships at the farming system level are illustrated in Figure 1. This highly simplified model puts the farmer the decision maker, at the centre. Decisions are influenced by the priorities of the household, farmer's knowledge and experiences, and resource at his command. External factors - natural, economic and socio-cultural, also plays significant roles.

Integrated farming systems (IFS)

Two major features of an IFS are: (i) waste or by-product utilization in which the wastes or by-products of one subsystem become an input to a second subsystem; and (ii) improved space utilization in which the two subsystems essentially occupy part or all of the space required for an individual subsystem. Integrated farming has also defined as the biologically IFS which: (i) integrates natural resources and regulation mechanisms into farming activities to achieve maximum replacement of off-farm inputs; (ii) secures sustainable production of high quality food and other products through ecologically preferable technologies; (iii) sustains farm income (iv) eliminates or reduces sources of present environment pollution generated by agriculture; and (v) sustains the multiple function of agriculture (IOBC, 1983).

Thus, an IFS represents multiple crops (e.g. cereals, legumes, tree crops, vegetables) and multiple enterprises (e.g. livestock, apiary, aquaculture) on a single farm in an integrated manner (Behera et al., 2015). The IFS approach is holistic, multi-disciplinary, problem solving, location specific and farmer oriented (Singh et al., 1998). The basic aim of IFS is to derive a set of resource development and utilization practices, which leads to a substantial and sustained increase in agricultural production (Kumar and Jain, 2005). However, there exists a web of interactions among the components within farming systems and it becomes difficult to deal with such inter-linking complex systems. This is a likely reason for the slow and limited progress achieved in the field of farming systems research in Asian countries.

For the small and marginal farmers of India, IFS can play a vital role in enhancing their economic situation and livelihood (Devendra and Thomas, 2002a). Unlike specialized farming systems, IFS activity is focused around selected, interdependent, interrelated and often interlinking production systems based on several crops, animals and related subsidiary professions. An IFS involves the utilization of primary and secondary produce of one system as basic inputs to the other systems, thus making them mutually integrated as one whole unit. There is need for effective linkage and complementarities of various components to develop effective holistic farming systems (Singh et al., 2007).

Integrated agriculture: principles

Practising integrated agriculture (IA), we give justice by providing opportunity to others to co-exist with a natural harmony. This has been well recognised and valued by ICAR and Ministry of Agriculture in their demonstration of the same by showcasing the achievements in tableau of 2018 and 2019. In 2018, ICAR showcased its thrust and achievement in the form of Integrated Farming Systems for the development of vast small and marginal farmers of our country. This was more appropriately designed in the context of north eastern hill (NEH) region. NEH region is rich in its culture and traditions and bio-diversity. In this model, the rich culture was depicted at the background which is important that the existing farming systems of the region must not be tempered. Other principles on which IA is based are: maximum replacement of off-farm inputs by recycling and efficient utilization of farm wastes and by-products; exploring the natural mechanisms of a farming systems in the way of complementarities of co-existence of various organisms, enterprises, water harvesting and natural resource conservation.

Benefits of IFS

The advantages of IFS include pooling and sharing of resources/inputs, efficient use of family labour, conservation, preservation and utilization of farm biomass including non-conventional feed and fodder resources, effective use of manure/animal waste, regulation of soil fertility and health, income and employment generation for many people and increase economic resources. It improves space utilization and provides diversified products. IFS is a part of the strategy to ensure sustainable use of the natural resources for the benefit of present and future generations. The followings are the important benefits from implementation IFS in a region/country.

(i) **Productivity:** IFS provides an opportunity to increase economic yield per unit area per unit time by virtue of intensification of crop and allied enterprises (Manjunath and Itnal 2003a; Ravisankar et al., 2007; Rathore and Bhatt, 2008).

(ii) **Profitability:** Improves profitability by reducing production costs through recycling wastes and by-products of one enterprise as inputs to other enterprises (Maheswarappa et al., 1998; Manjunath and Itnal, 2003b Ravisankar et al., 2010).

(iii) **Sustainability:** Organic supplementation through effective utilization of by-products of linked components is practiced, thus providing an opportunity to sustain the potential of the production base for much longer periods. It gives emphasis to achieving agro-ecological equilibrium through reduced build-up of pests and diseases (Korkanthimath and Manjunath, 2009; Gill et al., 2010; Kumar et al., 2011).

(iv) **Balanced food:** Components of a varied nature are linked to produce different varieties of products and produce, which serve to provide a balanced diet for the farm family (Kumar et al., 2011, 2012, 2013).

(v) **Environmental safety:** In an IFS, waste materials are effectively recycled by linking appropriate enterprises and components, thus minimizing environment pollution. It is recognised that single enterprise based farming endangers ecological security. For example, burning rice residues is common practice in intensively rice-wheat cropped areas of India (e.g. the Punjab, Haryana, Western Uttar Pradesh), resulting in vast nutrient loss and increasing the concentration of GHG in the atmosphere (Kumar et al., 2013). Such situations could be avoided by agricultural diversification with the introduction of more enterprises (e.g. animal husbandry) on the farm. Rice straw can be used as animal feed and turned into manure for sustaining soil health. Also, as IFS takes into account effective resource use and nutrient recycling and makes the farming less dependent on external inputs, it helps minimize environmental pollution occurring due to heavy use of external inputs (Shukla et al., 2002).

(vi) **Resource recycling:** Effective recycling of waste materials and by-products (crop residues and livestock wastes) is practiced in IFS. Therefore, there is less reliance on outside inputs (e.g. fertilizers, agrochemicals, feeds, energy). This leads to a more stable production system (Kumar et al., 2013).

(vii) **Income year round:** IFS provides a flow of money for the farmers throughout the year by way of the sale of a variety of farm produce (e.g. milk, egg mushroom, vegetables, fruits, food grains) (Behera and Mahapatra, 1999 Maheswarappa et al., 2001; Kumar et al., 2013).

(viii) **Risk minimization:** As an IFS provides a stable and sustainable production system through diversified crops and enterprises, this helps in risk minimization and resilience to climate change (Ayyappan and Arunachalam, 2014). Single commodity based agriculture is always endangered by natural hazards such as floods, drought, and disease epidemics. During 1999-2000 in India, many cotton growers in Andhra Pradesh, Maharashtra and Karnataka committed suicide as their crops were heavily damaged by pests. Adoption of IFS would help farmers escape such situations and reduce the risk involved in crop failure (Shukla et al., 2002). As an IFS comprises several enterprises like livestock, fishery, poultry in addition to cropping, it provides farmers with a range of opportunity to reduce risk. About 63% of the gross cropped area of India where mono-cropping is prevalent is rain-fed. These areas have a higher risk of crop failure due to drought and other aberrant weather conditions. Diversification of farming with intercropping agroforestry, horticulture, livestock and plantation crops would help in minimizing risk and raising the income of farmers in these areas.

IFS for self-reliant farm, nutritional security and doubling farmers income

Integrated Farming System model for irrigated agro-ecosystem

In integrated agriculture systems a variety of produces are produced which helps in family nutrition at the farm and regional level, thus contributing nutritional security. Citing one example, in the integrated agriculture production systems at ICAR-IARI, New Delhi a variety of products such as cereals – wheat, maize; pulses pigeon pea, pea, greengram; oilseeds – mustard, groundnut; fibre – cotton; leafy vegetables – raddish, mustard, spinach; vegetables – brinjal, raddish, baby corn tomato, drumstick; legume vegetables – country bean; spices –ginger, coriander onion; fodder– berseem, corn (stover), baby corn (stover), sweet corn (stover) fruits – guava, kinnow, lemon, banana; milk, egg, meat, fish and biogas etc. were produced which helped in the better nutrition of farm family and livelihood of the farmers (Behera et al., 2018). Such an approach helped in self-reliance with respect to meeting the nutrition of the farm family.

In this model, the farm become self-reliant and self-sufficient with respect to meeting the various requirement of resources. The pond (1000 m^2) in the IFS model served as a water harvesting structure and around 30% (6.0 lakhs litres) requirement of water for pond was met by water harvesting during rainy season. Fish-duck integration helped in fertilising the pond partly since duck droppings go directly to the pond and similarly the requirement of feed of ducks reduced since, ducks could meet their food from the fish pond. Similarly, poultry droppings also directly fall in the pond and fertilised the pond water. This water when irrigated to crops resulted in higher productivity of the crop. Dairy unit of the IFS model generated around

20.0 t/year of cow dung, which was recycled through biogas plant for manuring the crops and fish pond, thus cutting down the fertilizer/nutrient requirement of the system. The biogas produced was sufficient for requirement of gas for cooking food for 4-5 members of a farm family. Similarly integration of apiary and crop, helped in pollination of crops and increased the yield. Nutritional garden and medicinal garden are also important components of the IFS, which could helped in livelihood improvement of the farmer. Overall, such an IFS approach can be considered as an ideal model for bringing self-reliance at the farming system level of vast small and marginal farming communities in our country.

Multi-enterprise agriculture has shown the potential to decreases cultivation costs through less dependence on external/industrial inputs by synergetic recycling of by-products and residues of various components within the system and also provide a regular source of income and employment (Behera and Mahapatra 1999). The IFS approach targeted at synergistic blending of crops, forestry, apiary horticulture, dairy, poultry, fishery, piggery, etc. ensures recycling of the leftover by-product of one enterprise as a input for other enterprises. Under conditions in the north-eastern hill regions, interventions involving crop-fish-pig (pig-based IFS) and crop-fish-duck helped in multiple use of water (MUW) and resulted in significant improvement in water productivity, employment generation, income and livelihood of the farmers over current practice (Behera et al., 2012; Das et al., 2013). The pig and duck-based MUW through diversified farming (crop fruit, livestock and fishery) enhanced system productivity by 352% and 190% respectively, and generated a net return of Rs. 28,250 and Rs. 20, 350 from an area of 1500 m^2, which were 284 and 176% higher than present farmer practice (without irrigation), respectively.

Integrated Farming System model for hill agro-ecosystem

Integrated Farming Systems (IFS) is considered as the master plan and potential approach for improving the livelihood and double the farmers' income in NEH region, where most of the farmers are small and marginal and there exist fragile agro-ecology. An IFS model was developed at College of Agriculture, Kyrdemkulai with the objectives: (i) To meet the food, fodder, energy and cash need of the small and marginal farmers and improve their livelihood, (ii) Generate a net income of about Rs. 1.25 to 1.50 lakhs/annum from 0.5 ha. of land area, (iii) To enhance the resource use efficiency and system productivity by recycling the farm by-products and wastes within the system, and (iv) Promote bio-diversity and sustainability through inclusion of variety of crops and enterprises. The details of model description is given below (Table 2) Figure 1 & 2 also given.

Table 2. Integrated Farming Systems Model for small and marginal farmers in NEH region (0.6 ha) (*Source*: College of Agriculture, Kyrdemkulai)

Module	Enterprise / Crop details	Area (m^2)
Module 1	**Crop enterprise** Field crops: (Rice, Maize, Black gram, Potato & Mustard)	1555
	Horticultural crops Fruits (Mango, Assam Lemon, Guava, Papaya, Aonla, Carambola, Jamun, Dragon fruit, Ber, Pineapple, Banana and Litchi	2455
	Vegetables (Brinjal, Bottleguard, Pea, Onion, Radish, Carrot, Countrybean, Greens: Sarsonsag (mustard leaf, laipatta), Spinach and Lettuce	548
Module 2	Fishery: Composite pisciculture with Catla, Rohu, Mrigal and Grass carp	700
Module 3	Dairy: 2 units Sahiwal cows for milk purposes	100
Module 4	Poultry: Layer birds (50 number): Gram Priya and Rainbow Rooster	44
Module 5	Low cost mushroom production unit: For Oyster mushroom production	26
Module 6	Agroforestry: Turmeric, Ginger, Drum Stick, Elephant apple, Tamarind, Jamun, Bamboo, Bay leaf, Cardamom, and Tree bean	550
Module 7	Apiary: 3 units of ISI mark bee hives with Apis cerana	12
Module 8	Composting: Compost making from crop residues, farm bio-wastes to produce 2.5 t compost round the year from 2 pits (3mx1m) (2 cycles)	10
	Total area	6000

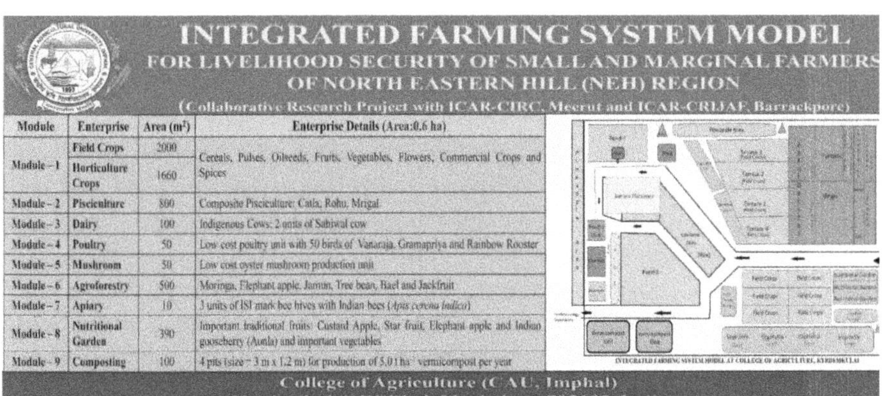

Fig. 1. Integrated Farming System Model (*Source*: College of Agriculture, Kyrdemkulai)

Fig. 2. Layout of Integrated Farming Systems Model for small and marginal farmers in NEH region (0.6 ha)
(*Source*: College of Agriculture, Kyrdemkulai)

Performance of the IFS model

The model developed at College of Agriculture, Kyrdemkulai demonstrated the farmers and rural youth how in a hilly ecology spring water can be harvested in the form of ponds and utilised for various purposes ensuring higher productivity of water through multiples uses of water. IFS model attracted the rural youth and farmers for adopting IFS as rural bio-entrepreneurship option and doubling the farmers' income. The IFS model served as an example for self-reliant movement in agriculture (*Aatma Nirvar Bharat*). In this model, a farmer could produce cereals, pulses, fruits, vegetables, flowers, spices, mushroom, milk, eggs and meat (from poultry unit). The model generated 10 t/year of cow dung and 5.0 t/year of vermi-compost by utilising the crop residues and forest litter. Around 1.0 t/year of poultry droppings were also collected from the low-cost poultry unit. These manures were recycled for crop production and manuring fish pond. The cow urine and cow dung and other local resources were utilised for preparation of liquid manure and bio-pesticides (Jivamruta, Panchagavya, Dashagavya etc.) which were utilised for manuring crops and controlling the pests. The system become highly self-reliant and avoided external inputs (fertilizer and pesticides). The overall success of the model was commendable with monthly income of Rs. 25000-Rs.30,000 from 0.6 ha area. In year a marginal farmers could generate an income of Rs. 3.0 to 3.6 lakh/ha, out of which Rs. 150,000 to Rs 180,000 was net income and Rs. 150,000 to Rs 180,000 was towards the family labour charges engaged in the system. This resulted in generating employment for the farm family and served as source of livelihood improvement and doubling the farmers' income.

Overall, integrated farming system has become a boon for the small and marginal farmers by providing additional income, employment, better family nutrition, better farm ecology and soil health. This is an important approach for the upliftment of small and marginal farmers in NEH region.

IFS for doubling farmers' income

Mono-cropping of rice-wheat system in Indo-Gangetic Plains (IGP) has resulted in natural resource degradation, decline in farm profitability, factor productivity and environment security. An Integrated Farming System (IFS) model involving land-based enterprises - crops, dairy, fishery, duckery, poultry, biogas plant and agro-forestry was developed in 1.0 hectare area at Indian Council Of Agricultural Research (ICAR)-Indian Agricultural Research Institute (IARI), New Delhi, India, during 2015-16 to 2017-18. The objective of the study was to find an alternative approach to replace rice-wheat system and double the farmers' income in a sustainable manner. The IFS model was found to be potential to improve the farmers livelihood by producing a variety of produce, recycling the resources in the system and doubling the farmers' income with Rs. 3,78,784 as against INR 68200 from conventional rice-wheat system (Shah et. al., 2015). The waste and by-products of one enterprise served as an input for the other, and reliance on off-farm inputs is reduced to a great extent aiding in strengthening environmental sustainability. Net returns obtained from all the components was Rs.3,78,784 with 4 times higher than conventional rice-wheat system (Shah et. al., 2015). Unlike conventional rice-wheat system, the IFS model in the present study generated an employment of 628 man days/ha that helps in engaging two labourers throughout the year (Table 3). This assured employment reduced migration of agricultural labourers to adjacent towns and also helps rural youth practicing farming as a rural bio-entrepreneurship option.

Table 3. Productivity (wheat equivalent yield) and profitability and employment generation of different components of integrated farming system (pooled data of 3 years)

Enterprise	Area (ha)	Productivity (WEY kg year^{-1})	Cost of cultivation (INR)	Net returns (INR)	Employment generation (man days)
Cropping system	0.625	10335	72,156	93,198	150
Dairy	3 cows	30758	3,30,482	1,61,638	365
Duckery	35 birds	3818	30,679	30,411	26
Fishery	0.1	5693	53,792	37,288	26
Poultry	50 birds	3316	24,272	28,778	26
Fruit production (kinnow + lemon)	0.05	1244	8,658	11,242	15

Enterprise	Area (ha)	Productivity (WEY kg year^{-1})	Cost of cultivation (INR)	Net returns (INR)	Employment generation (man days)
Agroforestry	0.012	285	1,331	3,229	3
Biogas	KVIC (2 m^3)	563	4,000	5,000	12
Country bean	Fencing area	625	2,000	8,000	5
			5,27,370	3,78,784	628
Rice – wheat	1	9860	78630	68200	148

(*Source:* Bhargavi and Behera, 2020.)

IFS for biodiversity conservation and ecosystem services

Integrated agriculture which is based on agro-biodiversity principles has the potentiality to reverse the trend. Agro-biodiversity can be used as a resilient crop production system due to its usefulness in self-sufficiency, erosion control, carbon sequestration, soil fertility build up, drought resistance, functional biodiversity and organic production systems. Agrobiodiversity and its link to ecosystem properties have cultural, intellectual, aesthetic and spiritual values that are important to society.

Such a diversity not only serves as strong pillar to sustainability, risk minimisation and climate resilience, but also provides ecosystem service for the farm family to maintain a healthy life by meeting around 70-80% of the farm family requirement for various products including modern form of renewable energy (biogas). The produces are of better quality due to no or minimal use of chemicals (Behera et al., 2018). Agricultural intensification through intensive use of fertilizers and mechanization reduces the biodiversity and this made the necessity for integration agriculture which depends less on these modern inputs. Specialization of farms as pure cropping or pure intensive livestock production systems is considered a mainstream path for modernization and development, which creates a negative impact on biodiversity and ecosystem services. For example, in Argentina, the cultivated area of crops increased 45% between 1990 and 2006, while use of fertilizers increased 400% (Gavier-Pizzaro et al., 2012). Consequences were loss of habitat heterogeneity, particularly loss of avian diversity and associated ecosystems services that benefit crop production.

Resource recycling in IFS

IFS is an important approach to enhance the resource use efficiency and system productivity by efficient utilization of the available resources and also enhance the income of the farmers by integrating different land-based enterprises, and reduce production cost by recycling farm by-products and wastes within the system. A close examination of resource recycling (Figure 3) indicated the interdependence of the different components of the total farming system to make the farmer self-sufficient in terms of ensuring the family members a balanced diet for leading healthy life and also making farm self-sufficient through recycling of by-products/ wastes. Infact, the IFS operates with the principle that there is no wastes in the system. The waste/wastes of one sub-system becomes the inputs of other sub-system. IFS has been a better approach to convert wastes in to wealth. The by-product of crops like wheat straw, maize stover, legume straw and other crop residues are served as feed for dairy animals. Litter from cowshed and weeds are utilized for compost making and valuable compost is utilized for production of vegetables, fruits and other crops thus cutting the fertilizer cost. The by-product of dairy, i.e. cow dung, forms a major raw material for biogas. Digested slurry of biogas forms a major part of feed of pisciculture for increasing plankton growth as well as supplying valuable manure to raise the productivity of crops. The by-product of poultry and duckery, i.e. droppings, form an important ingredient of pisciculture for increasing the plankton growth and also increases the fertility of land. Fruits, fence crops and moringa provided diversified nutritious food thus ensuring nutritional security of the farm family.

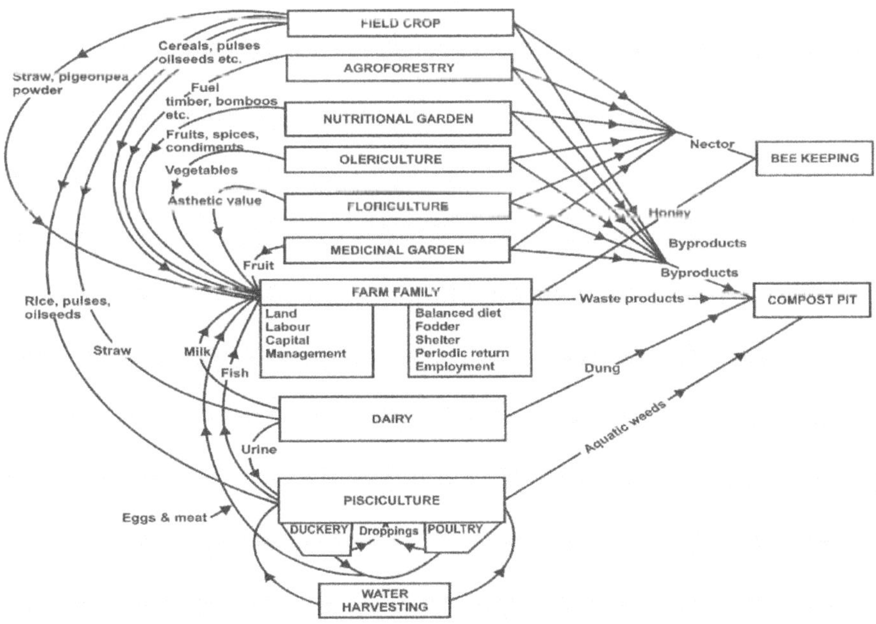

Fig. 3. Resource flow in IFS model at research farm, ICAR-IARI, New Delhi

Innovative Approaches for Promotion of IFS for self reliant India

(i) Extension strategy for 21st century agriculture

System mode of production incorporating crop, livestock, fish, horticulture and agro-forestry is a potential option for doubling farmer's income and self reliance in farming. In the national agriculture systems, usually the extension and development officers visit farmers' field and provide number of recommendations and advice to the farmers related to commodities, varieties, breeds and technologies in isolation. There is need to advice in holistic ways by projecting the whole farm income, expenditure, outcome and activities round the year. This is lacking in national agricultural research and extension system. Enterprise/activities combination in farming systems and optimization of scarce resources at the command of farmers must revolve round optimization - optimization of enterprise combination and optimization scare resource use at farm level fitting to the enterprise combination. These two aspects are weak in national agriculture systems in India and other Asian countries also. This is very important for doubling the farmers' income in our country. No agency in national agricultural system is equipped to do so. There is need to modify our education/training programme to develop scientific and skilled manpower in this area.

(ii) The developmental farm model (DFM)

To help bring agricultural growth to the rural areas of South Asia and in keeping with the sustainable development goal of reducing world poverty by 2030, we advocate another model namely the DFM (Behera et al, 2013; Behera and France 2014). This is a holistic farm model that assures adequate income generation and family employment opportunities for the farmer, particularly resource-poor small-holder farmers, thereby helping to reduce poverty and improve livelihood. It is based on a bottom-up approach and is applicable to all classes of farmer to meet their multi-objective needs. Whether they are subsistence or corporate farmers this model can help them achieve their desired objectives. The use of optimization methodology (e.g., linear programming and its variants) helps enhance the application of DFM in an efficient way.

The vision behind the DFM concept is to provide the farmers of South Asia and other developing regions with the opportunity to have their own model of development. This takes full account of the fact that all farms are not alike. Hence with the DFM approach, we visualize providing each and every farmer with a model of their farm for their development that has the following characteristics.

Thus, DFM serves as a powerful capsule for helping cure the multiple problems of agriculture at the farm level. The ingredients of the capsule might be the energy component, soil health improvement by developing a soil health card farm mechanization to reduce drudgery, value addition, etc. Farm activities can be planned simultaneously to produce renewable energy, generate income and minimise risk by introducing different enterprises, achieve food security and maintain biodiversity and prevent erosion by way of planting trees, shrubs, etc. This creates ecosystem awareness by encouraging diversity at the farm level and redesigning the farming system. Thus, by implementing the DFM approach, aspects/issues such as income generation, risk minimisation, climate change, ecosystem service and energy security can all be well addressed, which is difficult to achieve using conventional approaches.

(iii) Energy Self-sufficient Integrated Farming Systems (E-IFS)

The question arises as to how and what forms of renewable modern energy generating capacity can be added to the existing resource recycling capabilities of an IFS so that the whole system becomes self-sufficient or even surplus in energy (Behera et al., 2013). In the E-IFS concept, the objective is to integrate all direct and indirect sources of energy (which so far have not been fully explored) and to utilize resource recycling and resource conservation to reduce the energy embodied in inputs such as fertilizers, pesticides and irrigation water. This would lead towards a reduction in carbon emissions (i.e. towards carbon neutrality) and cleaner and greener farming.

An E-IFS would be one that uses little or no fossil fuel energy and even produces more energy than is required, so that it becomes another product off the farm. However, such an energy surplus needs to be achieved without competing with food crops for land and resources. Finally, the very nature of farming within an E-IFS, with its diversity of enterprise, can act as a risk reduction mechanism in the wake of climatic changes that may possibly lead to crop failure. The incorporation of modern energy sources within an IFS can provide an ideal farming system if some of the limitations mentioned previously are overcome. Next we briefly address some policy measures that could make such a system attractive to farmers under current Indian circumstances (Behera et al., 2015).

(iv) Farm Development Card

The system view is a useful conceptual device for all concerned with agriculture as it helps researchers understand the context of their research and in defining its content, thus contributing to its relevance. It helps them see their specialization in perspective and in relation to other forces and creates a better climate for cross-disciplinary work. It has been demonstrated that research programmes based on the traditional approach (i.e. commodity based) are not wholly sustainable, equitable and stable over a long period (Behera and France, 2016). Doubling the farm income by 2022 is a major challenge before National Agricultural Research System, which can be achieved through large-scale promotion of agri-entrepreneurship by small and marginal farmers, who constitute more than 85% of the farming community, and by the rural youth. For the sustainable development of small and marginal farmers in India there is need of system and holistic approach, which can better achieved with the help of farm development card (Behera and France, 2016).

Farm development card (FDC) is a business plan for the farmers, which is based on holistic concept of farm development – which considers overall situations of the farm including the knowledge and skill possessed by the farmer and activities needed to achieve the farmers goal within the physical, biophysical and socio-economic and institutional forces under which the farmers operate. Farm development card also accommodates soil health card. Hence, beyond soil health card farm development card can better address the farmers' problem and double the farmers' income. FDC includes the following vital components (Behera, 2018): (i) Ecological Development, (ii) Economic development, (iii) Energy security, (iv) Enhancing water productivity, (v) Soil health management (vi) Enhancing system productivity and sustainability. In addition, FDC considers the farmers resource availability, constraints and market opportunity (Figure 4). For development of FDC, the procedures - (i) Collecting preliminary information, (ii) Formulation of FDC, (iii) Revalidation of the plan to know the feasibility, (iv) Implementation of the Farm Development Card and (v) mid-term correction of the plan are followed.

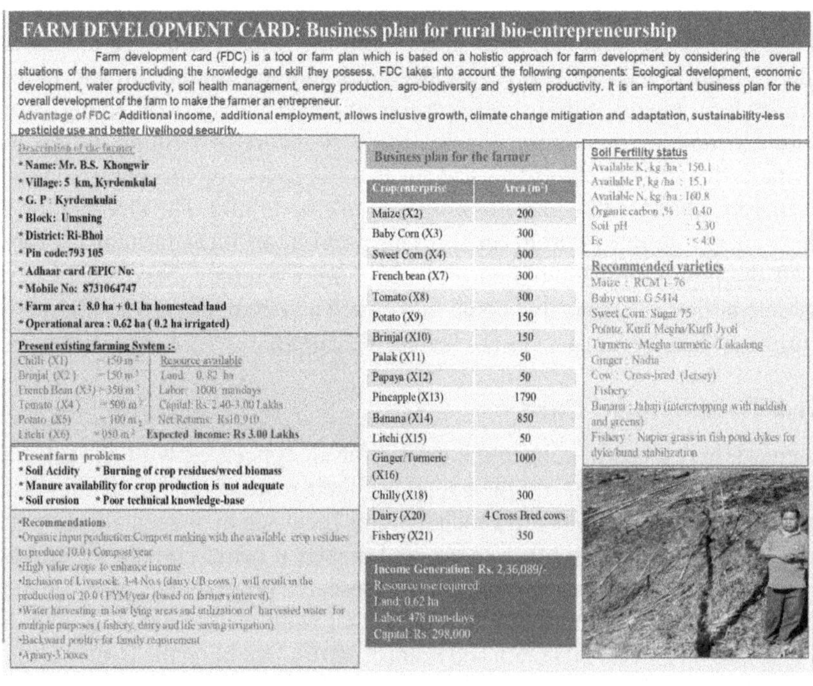
FARM DEVELOPMENT CARD: Business plan for rural bio-entrepreneurship

Farm development card (FDC) is a tool or farm plan which is based on a holistic approach for farm development by considering the overall situations of the farmers including the knowledge and skill they possess. FDC takes into account the following components: Ecological development, economic development, water productivity, soil health management, energy production, agro-biodiversity and system productivity. It is an important business plan for the overall development of the farm to make the farmer an entrepreneur.

Advantage of FDC: Additional income, additional employment, allows inclusive growth, climate change mitigation and adaptation, sustainability-less pesticide use and better livelihood security.

Description of the farmer

* Name: Mr. B.S. Khongwir
* Village: 5 km, Kyrdemkulai
* G. P: Kyrdemkulai
* Block: Umsning
* District: Ri-Bhoi
* Pin code:793 105
* Adhaar card /EPIC No:
* Mobile No: 8731064747
* Farm area : 8.0 ha + 0.1 ha homestead land
* Operational area : 0.62 ha (0.2 ha irrigated)

Present existing farming System :-

Chilli (X1) = 150 m²
Brinjal (X2) = 150 m²
French Bean (X3) = 350 m²
Tomato (X4) = 500 m²
Potato (X5) = 100 m²
Litchi (X6) = 050 m²

Resource available
Land: 0.82 ha
Labor: 1000 mandays
Capital: Rs. 2.40-3.00 Lakhs
Net Returns: Rs10,910
Expected income: Rs 3.00 Lakhs

Present farm problems
* Soil Acidity * Burning of crop residues/weed biomass
* Manure availability for crop production is not adequate
* Soil erosion * Poor technical knowledge-base

•Recommendations
•Organic input production:Compost making with the available crop residues to produce 10.0 t Compost/year
•High value crops to enhance income
•Inclusion of Livestock: 3-4 Nos (dairy CB cows) will result in the production of 20.0 t FYM/year (based on farmers interest).
•Water harvesting in low lying areas and utilization of harvested water for multiple purposes (fishery, dairy and life saving irrigation)
•Backward poultry for family requirement
•Apiary-3 boxes

Business plan for the farmer

Crop/enterprise	Area (m²)
Maize (X2)	200
Baby Corn (X3)	300
Sweet Corn (X4)	300
French bean (X7)	300
Tomato (X8)	300
Potato (X9)	150
Brinjal (X10)	150
Palak (X11)	50
Papaya (X12)	50
Pineapple (X13)	1790
Banana (X14)	850
Litchi (X15)	50
Ginger/Turmeric (X16)	1000
Chilly (X18)	300
Dairy (X20)	4 Cross Bred cows
Fishery (X21)	350

Income Generation: Rs. 2,36,089/-
Resource use required
Land: 0.62 ha
Labor: 478 man-days
Capital: Rs. 298,000

Soil Fertility status
Available K, kg /ha : 150.1
Available P, kg /ha : 15.1
Available N, kg /ha : 160.8
Organic carbon ,% : 0.40
Soil pH : 5.30
Ec : < 4.0

Recommended varieties
Maize : RCM 1-76
Baby corn: G 5414
Sweet Corn: Sugar 75
Potato: Kufri Megha/Kufri Jyoti
Turmeric: Megha turmeric /Lakadong
Ginger : Nadia
Cow : Cross-bred (Jersey)
Fishery:
Banana : Jahaji (intercropping with raddish and greens)
Fishery : Napier grass in fish pond dykes for dyke/bund stabilization

Fig. 4. Design of a farm development card for a farmer in Ri-bhoi district in Meghalaya with objective to double the farmer's income

(v) Prescriptive Agriculture

Many sick farms of the country need a proper recommendation for corrective measures in the form of a prescription. However, even after more than 6 decades of our efforts in agricultural development, scientists advice farmers mostly orally without handing over a prescription as a doctor does when a patient consults him. It is observed that our National agricultural systems are not properly equipped to do so. Providing a prescription to the farmers may build mutual confidence and strengthen partnership, which is very important from farmers development point view. Present adhoc and random approach with more emphasis on component agriculture for advising the farmers for their development needs to be changed to a system and holistic approach in a linked manner, which is possible by providing a prescription. While we provide a prescription, we must look farmers as an agri-preneur, but not only a beneficiary. We should have the plan to groom farmer as an agri-preneur. While providing a prescription or developing a prescription, we should try to customize the policy and interventions to meet the need of an agri-preneur considering his own concept and understanding.

(vi) Synergy Through Collaboration and Linkages

In India, for accelerating agricultural growth key lies with exploring synergy through collaboration and linkages. India is having a well developed network system of transfer of technology. Most of the agencies are working in parallel lines. Collaboration and proper linkage is needed to avoid duplicity and bring synergy in efforts. We have developed vast infrastructure in our country in comparison to other countries. We have around 740 Krishi Vigyan Kendras, 343 zonal research centres, around 100 farming system research centres (net work of AICRP on IFS), vast agricultural line departments and more than 29000 ATMA workers. All are mandated to improve farming systems and benefit farmers. It is seen that they work in parallel lines and hardly collaborate and converge. Now, time has come we should think how to get better output from the Govt. mechanism/from these vast infrastructure country has developed. This is possible through exploring synergy by linkages and collaborations. For this, above mentioned innovative approaches viz. DFM, FDC, Prescriptive agriculture play a key role. Besides, public-private partnership, involvement of NGOs, watershed-basis of participatory approaches and involvement of Farmers Producers Organisations are important for the large scale promotion.

Conclusion

"*Aatma Nirvar Bharat* (Self reliant India)" has been the focus of todays' agricultural development. This is important to break the vicious cycle of external (industrial) inputs. There is need to make farming systems self-sufficient and self-reliant by generating major inputs and resources at the farm level. Integrated Farming System is a potential option, where about 70% of the resources should be generated and recycled in the farming systems. The adoption of multiple farm enterprises in an integrated manner can ensure a higher income generation, enhance resource-use efficiency and recycling of farm by-products and wastes, thus helps in bringing self-reliant farming systems and sustainable livelihood of farmers. The adoption of IFS is the right approach in this direction and should be supported through institutional, extension, policy and marketing interventions in a system approach. System mode of production incorporating crop, livestock, fish, horticulture and agro-forestry is a potential option for doubling farmer's income. Integrated Farming Systems (IFS) is extremely important for the efficient management of available resources at the farm level to generate adequate income and employment for the rural poor. A shift from the existing discipline-oriented strategy to more holistic approaches would require perceptible attitudinal changes from the policy makers, scientists and developmental personnel. The key innovative approaches to promote IFS are DFM, E-IFS, FDC, prescriptive agriculture and synergy through collaboration and linkages.

References

Ahlawat, S P S, Mandal, A B and Pramanik, S C. 2002. Integrated crop-livestock fish farming system for sustained production – Bay island perspective. Proceedings of the national symposium on agriculture in changing global scenario (R.C. Gautam, B.N. Mishra, B.G. Shivkumar, R.K. Rai, S.K. Mishra and R.N. Garg, Eds.). Indian Society of Agricultural Sciences, New Delhi.

Ayyappan, S and Arunachalam, A. 2014. Crop diversification for social security of farmer in India, pp. 1-6. In Souvenir (D.S. Rana and A.K Singh, Eds.). Indian Society of Agronomy, New Delhi.

Behera U K, Jha K P, Mahapatra I C. 2004. Integrated management of available resources of the small and marginal farmers for generation of income and employment in eastern India. Crop Research 27(1): 83-89.

Behera U K, Rautray S K and Sahoo, P. K. 2010. Farming system-based composite pisciculture in harvested pond water under eastern Indian situations. Journal of Soil and Water Conservation 9: 143-147

Behera, U K and France, J. 2016. Integrated farming systems for the livelihood security of small and marginal farmers in India and other developing countries. Advances in Agronomy 138: 235-282.

Behera, U K and Mahapatra, I C. 1999. Income and employment generation of small and marginal farmers through integrated farming systems. Indian Journal of Agronomy. 44(3): 431-439.

Behera, U K and France, J. 2014. Developmental Farm Model for revolutionizing agricultural growth. ICAR News 20(3): 5-6.

Behera, U K, Babu, A, Kaechele, H and France, J. 2015. Energy self-sufficient sustainable integrated farming systems for livelihood security under a changing climate scenario in an Indian context: a case-study approach. CAB Reviews 10 (19): 11pp.

Behera, U K, Bhargavi, B, Meena, S L, Bhattacharyya, R, Singh, R, Singh, R K, Singh, V K, Gupta, N and Rana, K S. 2018. Research Bulletin on Integrated farming system model for livelihood security of small and marginal farmers. Division of Agronomy, Indian Agricultural Research Institute, New Delhi – 110 012.

Behera, U K, Choudhary, A K, Dass, A and Vyas, A K. 2013. Integrated Farming Systems for Livelihood Security of Small and Marginal Farmers. Division of Agronomy, Indian Agricultural Research Institute, New Delhi – 110 012, 327 pages.

Behera, U K.2018. Farm Development Card: A potential tool for attracting rural youth for agro-entrepreneurship and doubling farmers' income. Upadhyay, P.K., Rajanna, G.A. Sepat, S., Bana, R.S., Choudhary, A.K., Singh, T., Singh, R.K.,Rathore, S.S., Kumar, D. Nath, C.P., Shiva Dhar, Ghosh, P.K. and Yadav, R.L. 2018. Doubling farmers' income through agronomic interventions under changing scenario. In: Extended summaries of XXI Biennial National Symposium. Indian Society of Agronomy, IARI, New Delhi 110 012, p. 652

Behera, U.K., Panigrahi, P. and Sarangi, A. 2012. Multiple water use protocol in integrated farming systems for higher productivity. Water Resources Management 26: 2605-2623.

Bhargavi, B and Behera, U.K. 2020. Securing the livelihood of small and marginal farmers by diversifying farming Systems. Current Science 119 (5):854-859

Dalsgaard, J P T and Prein, M. 1999. Integrated smallholder agriculture-aquaculture in Asia: Optimising tropic flows. In Nutrient disequilibria in agro-ecosystems: Concepts and case studies (Smaling, E.M.A., Oenema, O., Fresco, L.O., Eds.), pp. 141-155. CABI Wallingford, UK.

Das, A., Choudhury, B.U., Ramakrishna, G.I., Jripathi, A.K., Singh, R.K., Ngachan, S.V. Patel, D.P., Layck, J., Munda, G.C. 2013. Multiple use of pond water for enhancing water productivity and livelihood of small and marginal farmers. Indian Journal of Hill Farming 26(1): 29-36.

Devendra, C. and Thomas, D. 2002a. Smallholder farming systems in Asia. Agricultural Systems 71: 17-25.

Devendra, C. and Thomas, D. 2002. Crop-animal interactions in mixed farming systems in Asia. Agricultural Systems 71: 27-40.

FAO, 2001. Farming Systems and Poverty: Improving Farmers' livelihoods in a changing World. Food and Agriculture organization of the United Nations, Rome pp 412.

Gavier-Pizzaro, G. I., Calamari, N.C., Thompson, J.J., Canavelli, SB., Solari, L.M., Decarre, J. Goijman, A.P., Suarez, R.P., Bernardos, J.N. and Zaccagnini, M.E., 2012. Intensification of row crop agriculture in te Pampas and Espinal of Argentina can reduce ecosystem service provision by changing avian density. Agric. Ecosyst. Environ. 154: 44-55.

Gill, M S, Singh, J P, Gargwar, K S. 2010. Integrated farming systems and agriculture sustainability. Indian Journal of Agronomy 54(20): 128-139.

IOBC 1983. International organization for biological and integrated control. Technical bulletin 16(1).

Korikanthimath, V S., Manjunath, B L. 2009. Integrated farming systems for sustainability in agriculture production. India Journal of Agronomy 54(2): 140-148.

Kumar, S. and Jain, D.K. 2005. Are linkages between crops and livestock important for the sustainability of the farming system? Asian Economic Review 47(1):90-101.

Kumar, S., Shivani and Singh, S S. 2013. Sustainable food and nutritional security through IFS. Indian Farming 63(7): 30-36.

Kumar, S., Singh S S., Shivani and Dey, A. 2011. Integrated faming system for eastern India. Indian Journal of Agronomy 56(4): 297-304.

Kumar, S., Singh, S.S., Meena, M.K., Shivani and Dey, A. 2012. Resources recycling and their management under an integrated faming system for lowlands of Bihar. Indian Journal of Agricultural Science 82(6): 504-510.

Lal R and Miller F P. 1990. Sustainable farming system for the tropics. Sustainable Agriculture: Issues and Prospective. R.P. Singh (Ed.) pp 69-89. Indian Society of Agronomy, IARI, New Delhi.

Mahapatra, I C and Behera, U K. 2011. Rice-based farming systems for livelihood improvement of Indian farmers. Indian Journal of Agronomy 56(1): 1-19.

Mahapatra, I.C. 1992. Farming systems research challenges and opportunities. Eastern Indian Farming System Research and Extension Newsletter 6(4): 3-10.

Maheswarappa, H P, Hegde, M R, Dhanpal, R, Biddappa, C C. 1998. Mixed farming in coconut garden: its impact on soil, physical, chemical properties, coconut nutrition and yield. Journal of Plantation Crops 26(2): 139-143.

Maheswarappa, H P, Hegde, M R, Dhanpal, R, Sairam, C V , Singh, TV. 2001. Impact of integrated mixed farming systems in coconut garden on coconut yield and economic analysis. Indian Journal of Agronomy 46(1): 56.63.

Manjunath, B L and Itnal, C J. 2003a. Integrated farming system in enhancing the productivity of marginal rice (oryza sativa) holdings in Goa. Indian Journal of Agronomy 48(1): 1-3.

Manjunath, B L and Itnal, C J. 2003b. Farming system options for small and marginal holdings in different topographics of Goa. Indian Journal of Agronomy 48(1): 4-7.

Paroda, R. 2014. Sustainable livelihoods through agriculture diversification and climate smart agriculture. In Souvenir (D.S. Rana and A.K. Singh, Eds.). Indian Society of Agronomy, New Delhi.

Rathore, S S , Bhatt, B P. 2008. Productivity improvement in jhum fields through an integrated farming system. Indian Journal of Agronomy 53(4): 167-171.

Rautaray, S K, Das, P C and Sinhababu, D P. 2005. Increasing farm income through a rice-fish based integrated farming system in rainfed lowlands of Assam. Indian Journal of Agricultural Sciences 75(2): 79-82.

Ravisankar, N, Pramanik, S C, Rai, R B, Nawaz, S, Biswas, T K, Bibi, N. 2007. Study of an integrated farming system in hilly upland areas of Bay Islands. Indian Journal of Agronomy 52(1): 7-10.

Ravisankar, N, Subrmani, T, Ambast, S K, Srivasyava, R C. 2010. Enhancing farm income in an island ecosystem. Indian Farming 60(6): 16-19.

Shah, M. S., Verma, N. and Vishwkarma, S.K.2015. Diversification of rice (Oryza sativa)–based cropping systems for higher production efficiency, water productivity and economic viability in Madhya Pradesh. Indian Journal of Agronomy 60(1), 25–30.

Shaner, W.W., Philipp, P.F. and Schmehl, W.R. 1982. Farming Systems Research and Development: Guide for Developing Countries. West View Press, Colorado, USA.

Shukla, A K, Shukla, N D, Singh, V K. 2002. Farming system approach for food security and sustained rural economy. Indian Journal of Fertilizers 47(11): 55-62.

Singh, D P, Singh, K P and Yadavika. 1998. Integrated farming systems – a key issue for research education – extension linkages to sustain food security and ecofriendly environment in 21st Century, pp. 474-486. In Proceedings of the First International Agronomy Congress: Agronomy, Environment and Food Security for the 21st Century (Singh, P., Prasad, R., Ahlawat, I.P.S., Eds.), pp. 474-486. Indian Society of Agronomy, New Delhi.

Singh, J.P., Gill, M.S. and Tripathi, D. 2007. Development of integrated farming system models for marginal and small farmers. In Extended summaries of 3rd national symposium on integrated farming system and its role towards livelihood improvement, pp. 51-57. ARS, Durgapura, Jaipur, 26-28 November 2007.

Singh, K, Bohra, J.S., Singh, Y. and Singh, J.P. 2006. Development of farming system models for the north-eastern plain zone of Uttar Pradesh. Indian Farming 56(2): 5-11.

Yadav R L and Prasad K. 1998. Farming systems research in India: needs and strategies. In Proceedings of first international agronomy congress: Agronomy, environment and food security for 21st century (P. Singh, R. Prasad and I.P.S. Ahlawat, Eds.), pp.333-341. Indian Society of Agronomy, New Delhi.

18

Innovative Farming Systems for Self-Reliance of Farmers in Eastern India

M.S. Malik[1], *P.K. Ghosh*[2], *Pinaky Rai, Nikita Kumari and Christina Minz*

[1]*Agroforestry Division, Birsa Agricultural University, Ranchi, Jharkhand*
[2]*ICAR-National Institute of Biotic Stress Management, Raipur, Chhattisgarh*

Introduction

In India the economy is based on rural and agrarian and decline with land holding area poses a serious challenge to the sustainability and profitability of farming. In view of the decline in per capita availability of land from 0.5 ha in 1950-51 to 0.15 ha by the turn of the century and a projected further decline to less than 0.1 ha by 2020, it is imperative to develop strategies and innovate new agricultural technologies that enable adequate employment and income generation, especially for small and marginal farmers who constitute more than 80% of the farming community. The crop and cropping system based perspective of research needs to make way for farming systems based research conducted in a holistic manner for the sound management of available resources by small farmers (Jha, 2003). Under the gradual shrinking of land holdings, it is necessary to integrate land based enterprises like fishery, poultry, duckery, apiary, field and horticultural crops, etc. within the bio-physical and socio-economic environment of the farmers to make farming more profitable and dependable (Behera et al., 2004). No single farm enterprise is likely to be able to sustain the small and marginal farmers without resorting to integrated farming systems (IFS) for the generation of adequate income and gainful employment year-round (Mahapatra, 1992; 1994). Farming systems approach, therefore, is a valuable approach to address the problems of sustainable economic growth for the farming communities in India.

Rapid population growth, urbanization and growth in income in developing countries like India, the demand for more food of animal origin thus accelerating the competition between crops and livestock. Recent decades have a witnessed rapid socio-economic transformation globally. Structural changes to economies have boosted per capita incomes, reduced poverty and enhanced food security almost everywhere. Despite these positive changes, 700 million people still live in extreme poverty, and nearly 815 million suffer from chronic hunger (FAO

2017, IFAD, UNICEF, WFP and WHO, 2017). Unless economic growth is made more inclusive, the first two Sustainable Development Goals (SDGs) i.e. to end poverty and achieve zero hunger by 2030 cannot be achieved. A system approach is therefore the need of the hour for satisfying demand of an ever-increasing population without disturbing the ecological balance. The IFS seems to be the solution to the continuous increase in demand for food production, stability of income and nutritional security particularly for the small and marginal farmers with limited available resources. It is not only a reliable way of obtaining a fairly high productivity with substantial input economy but also a concept of ecological soundness, leading to sustainable agriculture. Therefore, the Ministry of Agriculture and Farmers' Welfare, Government of India has given major emphasis on IFS while planning for doubling of farmers' income by 2022.

System Overview

"Farming System is a complex inter-related matrix of soil, plants, animals, implements, power, labor, capital and other inputs controlled in parts by farming families and influenced to varying degrees by political, economic, institutional and social forces that operate at many levels" (Mahapatra, 1992). The term "farming system" refers to a particular arrangement of farming enterprises that are managed in response to physical, biological and socio-economic environment and in accordance with farmer's goals, preferences and resources (Shaner et al 1982). FAO declares - The household, its resources and the resource flows and interactions at the individual farm levels are together referred to as a farm system" (FAO, 2001). The components and elements of the farming system are listed in Table 1

Table 1. Components (a) and (b) of farming system (FAO, 2001)

Component (a)

Agriculture	Mushroom cultivation	Seed Production	Sheep rearing
Horticulture	Sericulture	Vermiculture	Piggery
Forestry	Azolla Farming	Pigeon rearing	Rabbitory
Dairy	Kitchen gardening	Apiary	Value addition
Fish Farming	Fodder production	Poultry	-
Duck rearing	Nursery	Goat rearing	-

Component (b)

Watershed	Farm Ponds	Bio- pesticides	Bio- Fertilizers
Plant products as pesticides	Bio-gas	Solar energy	Compost making
Green manuring	Rain water harvesting	-	-

Multiple benefits of a farming system

Productivity: Farming system provides an opportunity to increase economic yield per unit area per unit time by virtue of intensification of crop and allied enterprises.

Profitability: The system as a whole provides an opportunity to make use of produce/waste materials of one enterprise as an input to another enterprise at low/ no cost. Thus by reducing the cost of production the profitability and benefit cost ratio works out to be high.

Potentiality: Soil health, a key factor for sustainability is getting deteriorated and polluted due to faulty agricultural management practices viz., excessive use of inorganic fertilizers, pesticides, herbicides, high intensity irrigation etc. In farming system, organic supplementation through effective use of manures and waste recycling is done, thus providing an opportunity to sustain potentiality of production base for much longer time.

Balanced food: In farming system, diverse enterprises are involved and they produce different sources of nutrition like proteins, carbohydrates, fats & minerals etc. form the same unit of land, which helps in solving the problem of malnutrition prevalent among the marginal and sub-marginal farming households.

Environmental safety: The very nature of farming system is to make use or conserve the by-product/waste product of one component as input in another component and use of bio-control measures for pest & disease control. These eco-friendly practices bring down the application of huge quantities of fertilizers, pesticides and herbicides, thus greatly reducing environmental pollution.

Income/cash flow round the year: In conventional single enterprise crop activity, the income is expected only at the time of disposal of economic produce. This happens after several months depending upon the duration of the crop. The IFS enables cash flow round the year by way of sale of produce from different enterprises viz., eggs from poultry, milk from dairy, fish from fisheries, silkworm cocoons from sericulture, honey from apiculture etc. This not only enhances the purchasing power of the farmer but also provides an opportunity for him to invest in improved technologies for enhanced production.

Saving energy: Availability of fossil fuels is declining at a rapid rate leading to a situation where in the whole world may suffer for want of fossil fuels by 2030. In farming system, recycling of organic wastes to generate energy from biogas plants can mitigate to a certain extent, this energy crisis.

Meeting fodder crises: In IFS every inch of land area is effectively utilized. Alley cropping or growing fodder legumes along the border or water courses, intensification of cropping including fodder legumes in cropping systems helps to produce the required fodder and greatly relieve the problem of non- availability of fodder to livestock component of the farming system.

Solving timber and fuel crises: The current production level of 20 million m^3 of fuel wood and 11 million m^3 of timber wood is no match for the demand estimated at 360 m^3 of fuel and 64.4 million m^3 of timber wood in the year 2000. Hence the current production needs to be stepped up several-fold. Afforestation programmes besides introduction of agroforestry component in farming system without detrimental effect on crop yield will greatly reduce deforestation, preserving our natural ecosystem.

Employment generation: Various farm enterprises viz., crop +livestock or any other allied enterprise in the farming system would increase labor requirement significantly and would help solve the problem of under employment. An IFS provides enough scope to employ family labor round the year.

Scope for establishment of agro- industries: When once the produce from different components in IFS is increased to a commercial level, there will be surplus for value addition in the region leading to establishment of agro-industries.

Enhancement in input use efficiency: An IFS provides good scope for resource utilization in different components leading to greater input use efficiency and benefit- cost ratio.

Major types of farming systems

The major types of farming systems can be broadly classified as

Integrated Farming System

Integrated farming system is based on the concept that there is no waste, and waste is only a misplaced resource which can become a valuable material for another product (Edwards et al. 1986). This approach is not only a reliable way of obtaining fairly high productivity with substantial input economy but also deriving maximum compatibility and replenishment of organic matter by way of effective recycling of organic residues/wastes obtained through integration of various land-based enterprises (Jayanti et al. 2003). The IFS combines livestock, aquaculture, agriculture and agro-industry in an expanded symbiotic or synergistic system, so that wastes of one process become inputs for other processes, with or without treatment to provide the means of production, such as energy, fertilizer, and feed for optimum productivity at minimum costs. The concepts associated with IFS are practiced by numerous farmers throughout the globe. A common characteristic of these systems is that they have a combination of crop and livestock enterprises and in some cases may include combinations of aquaculture and trees.

An example may be cited here. An economic analysis was carried out to identify the economically best farming system in Burakocha village of Ranchi district in Jharkhand (Khan and Sengupta 2018) (Table 2). People mostly depend on agriculture and livestock and lac cultivation is also prevalent. Survey identified the most popular farming system as paddy + tomato + black gram + Lac + goatery + poultry. However the most viable Farming system was Paddy + tomato + bottlegourd + Lac with a BC ratio of 3.38.

Table 2. Major farming systems and their economics (Khan and Sengupta 2018)

Sl. No	Types of Farming System	Total Investment (Rs)	Total Return (Rs)	Net Income (Rs)	B:C
Farming system 1	Paddy + Tomato + Bottle gourd + Lac + Goatery + Poultry	109748	346550	236802	3.15
Farming system 2	Paddy + Tomato + Lac + Goatery + Poultry	86744	282550	195806	3.25
Farming system 3	Paddy + Tomato + Cabbage +Lac + Goatry + Poultry	114688	382550	267862	3.33
Farming system 4	Paddy + Tomato + Goatery + Poultry	72119	208100	135981	2.88
Farming system 5	Paddy + Tomato + Bottle gourd + Lac	91073	308450	217377	3.38

Integrated Farming Systems modules for small and marginal farmers of Eastern region in lowland and midland irrigated ecosystems

IFS modules for small and marginal farmers of Eastern region in lowland and midland irrigated ecosystems of Bihar have been developed (Kumar et al. 2012). Details of the modules are given as:

A) Two-acre IFS module (for lowland situation) components: Crop + Livestock + Fishery Allied: Duckery / Vermicomposting / Bee keeping/ FYM

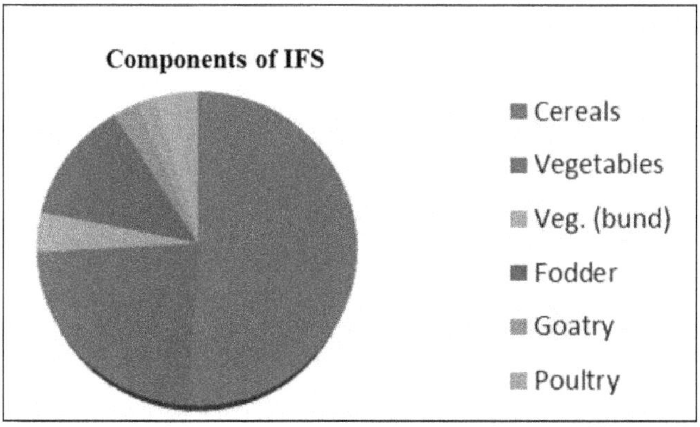

Fig. 1. Pie chart showing percentage of different components (Kumar et al. 2012)

Cereal crops: (50% area)

Kharif: Rice

Rabi: Wheat/Maize/Lentil/Mustard

Horticultural crops (Fruits +Vegetables): 12.5 % area

Vegetables: Cucurbits/Brinjal/Okra

(Kharif):-Brinjal/Boro/ Okra/ Bitter gourd / Cucumber etc. (Summer)

Fruits: Papaya (On pond's dike and field bunds); Banana (On pond's dike); Lemon (On pond's dike and Horticultural block); Guava (On pond's dike and Horticultural block)

All around the field bunds cucurbits or seasonal vegetables having lesser water requirement may be raised by making wire fences.

Fish + Duck integration (17.80% area)

Mix carp culture: Rohu (20% as column feeder), Catla (30% as surface feeder), Mrigal/common carp (50% as bottom feeder)

Duck: For 1000 m^2 water area, 40 numbers of ducks are sufficient *Khakhi Campbell* breed of duck is right choice for this area (Dual purpose) A thatched hut of 10 x 15' size is optimum for 40 ducks above the water or on the pond's dike.

Livestock (1.8% area)

A size of 3 adult cows + 3 calves is optimum for two acre land in respect of farm yard manure requirement for the fields and fodder requirement for the livestock. A thatched hut of 20' x 30' with sufficient paddock space is sufficient for above no. of animals. The cowshed should be connected with the pond with a drainage channel so that urine and water can move into the pond. A storage hut for storing of animal feed should be also made near the animal shed.

Fodder production (12.5% area)

For feeding of 3 cows and 3 calves, 1000 m^2 land is sufficient, if year round fodder production is carried out. In addition to green fodder, straw, leaves, stems of different cereals and vegetables can be also used as animal feed.

Kharif: M.P. Chari/Sudan grass/ Napier/Maize

Summer: Boro/Lobia/Maize/Sudan grass

Rabi: Berseem/Oat/Maize etc.

Spices: In the sheds or where light intensity is less like orchards, spaces between the huts etc. turmeric, ginger or guinea grass can be taken.

FYM/ vermi-composting pits: (1.4 % area)

Optimal size pits for preparation of FYM and vermi-compost should be made depending upon the land available near the cowshed. Thus it will ensure that required raw materials for making manures are available nearby for convenience and to avoid transportation charges.

Specialized Farming Systems

Specialization involves the intensification of the agricultural activity aimed at maximization of the production/unit area/unit time. This involves improvement of operational efficiency and speed of operation/execution at each step. The specialized farming system is thus focused on single cropping system or sequence of farming enterprises like animal breeding and dairy, so as to achieve the 'highest degree of precision management' with minimal diversion of resources/attention to diverse crops or enterprises, differing widely in managerial skills and involving extremely different implements and/or handling capabilities for the produce.

Conventional Subsistence Farming Systems

In this case, conventional farming systems refers to the farming systems usually followed with either single crop or multiple crops, and single or multiple enterprises that the farmers manage for the subsistence of their families. There is not much emphasis on resource recycling, optimization of resources within farming systems and urge to enhance the farm income. However such system has an important role in livelihood support of the farm family.

Indigenous farming systems

Indigenous technical knowledge (ITK) is the knowledge that people in a given community have developed over times, and also continue to develop. It is based on experience, often tested over long period of use, adapted to local culture and environment, dynamic and changing, and lays emphasis on minimizing risks rather than maximizing profits. The ITK covers a wide spectrum –soil water and nutrient management; pasture and fodder management; crop cultivation; plant protection; farm equipment, farm power, post-harvest preservation and management; agro-forestry; bio-diversity conservation and also exploitation; animal rearing and health care; animal products preservation and management; fisheries and fish preservation; and ethnic foods and homestead management. Thus, the ITK of a farmer has a great influence in managing the farm and farming system.

Different farming systems exist in India and other parts of the world. For example: shifting cultivation, is an indigenous and ancient form of farming system; the Taungya system, is an organized and scientifically managed form of shifting cultivation originated from Myanmar; Zabo is an indigenous farming system practiced in north eastern hill regions, particularly in Nagaland, which refers to combination of forest, agriculture, livestock and fisheries with well-founded soil and water conservation base.

Efficient and effective nutrient recycling in farming systems

Efficient nutrient recycling within the system is an integral part of farming system research. In an integrated system, crops and livestock interact to create a synergy, with recycling allowing the maximum use of available resources. Crop residues can be used for animal feed, while livestock and livestock by-products and its processing can enhance agricultural productivity by intensifying nutrients that improve soil fertility, reducing the use of chemical fertilizers. For agricultural use, animal excreta can be used for fertilizer, feed and fuel. Excreta can be dried, composted, or liquid-composted for the production of biogas and energy for household use (e. g. cooking, lightning) or for rural industries (e.g. powering mills and water pumps). Fuels in the form of biogas or dung cakes can replace charcoal and wood. It can be methane-fermented, directly combusted, or made into solid fuel. Furthermore, biomass production of feed is possible; the excreta is treated to be used as feed again (Moriya and Kitagawa 2007, Matsumoto and Matsuyama 1995). In a crop based IFS crop residues are recycled. It is pre-requisite to ensure the efficient recycling of resources particularly crop residues, because 80-90% of the micronutrients remain in the biomass.

An efficient nutrient recycling /input-output flow diagram under IFS is illustrated in Fig 2 (Kumar et al., 2011). Bhatt and Bujarbaruah (2005) reported that crop residue, weed biomass could be recycled for vermicomposting in an intensive IFS. On an average, 2.43 t of vermin-compost could be obtained from 7.02 t of biomass (dry weight basis). Removal from the nutrient pool includes primarily uptake by the trees and crops which becomes either locked up in the vegetative parts or exported through harvested produce. Nutrient removal through harvested produce is compensated by nutrient input through manures, fertilizers, recycled crop residues and tree nutrient cycling processes. The tree species by virtue of their deep roots intercept, absorb and recycle nutrients that would have been otherwise lost by leaching. A dynamic equilibrium can be expected with respect to organic matter and plant nutrients in the soil due to continuous addition of leaf litter, other plant residues and animal wastes and its continuous removal through decomposition (Varughese and Thomas 2009).

Energy budgeting in farming system

Farming system is a resource management strategy to avail maximum efficiency of a particular system. Integration of poultry and mushroom enterprises with rice-brinjal system required the highest energy input whereas rice cropping alone recorded the least. The energy output was maximum under rice-brinjal + mushroom + poultry system. The output of multi-rice-based enterprise was reasonably good varying from 1.9 to 10.5 lakh MJ/ha. It is thus evident that efficient utilization of scarce and costly resources is the need of the hour, and can be accrued by following the concept of IFS through supplementation of allied agro-enterprises (Korikanthimath and Manjunath 2009).

Behera et al. (2014) undertook a case study in a small farm (1.25 ha) in eastern India involving IFS (crop livestock-fishery-agroforestry) and presented a concept of energy self-sufficient IFS. The total energy requirement involving farming and household was 315 MJ and there was net deficit of 62 MJ (5259 KWH). The energy deficit can be met by exploring renewable energy production from biogas, solar panel and windmill. Its whole idea is to produce modern form of energy at the farm itself by linking various interdependent enterprises in order to bridge the energy deficit and future energy demands and offset emissions.

Sustainability of farming systems

The most important consideration is that the farming systems practiced in different forms must be sustainable. There are many ways to improve the sustainability of a given farming system, and these vary from region to region, However, there are some common sets of practices among farmers trying to take a more sustainable approach, in part through greater use of on-farm or local resources each contributing in some way to long- term profitability, environmental stewardship and rural quality of life.

Soil conservation: Many soil conservation methods, including contour bunding, graded bunding, vegetative barriers, strip cropping cover cropping, reduced tillage etc help prevent loss of soil due to wind and water erosion.

Crop diversity: Growing a greater variety of crops on a farm can help reduce risks from extremes in weather, market conditions or crop pests. Increased diversity of crops, trees and shrubs, can also contribute to soil conservation, wildlife habitat and increased populations of beneficial insects.

Nutrient management: A proper management of nitrogen and other nutrients can improve the soil and protect environment. Increased use of farm nutrient sources such as manure and leguminous cover crops, also reduces fertilizer consumption, and thereby minimized cost.

Integrated pest management (IPM): IPM is a sustainable approach to managing pests by combining biological, cultural, physical and chemical tools in way that minimizes economic, health and environmental risks.

Cover crops: Growing plant such as sun hemp, horse gram, *pillipesara* in the off-season after harvesting a grain or vegetable crop can provide several benefits, including weed suppression, erosion control, and improved soil nutrients and soil quality.

Rotational grazing: New management- intensive grazing systems take animals out barn into the pasture to provide high-quality forage and reduced feed cost.

Water quality and water conservation: Water conservation and protection have important part of agricultural stewardship. Many practices have been developed viz., deep ploughing, mulching, micro irrigation techniques etc, which contribute in protecting the quality of drinking and surface water.

Agroforestry: Trees and other woody perennials are often underutilized and covers a range of practices viz., agri-silviculture, silvi-pastoral, agri-silvi-horti-pastoral, horti/silvi-pastoral, alley cropping, tree farming, lay farm that help to conserve, soil and water.

Marketing: Farmers across the country are finding that improved marketing ways to enhance profitability, direct marketing of agricultural product from farmers to consumers are becoming more common, including through Rythu bazaar road side stands.

Economics of integrated farming system for small farmers

Manohar Lal in 2018 assessed the economics of different integrated condition of India (Table 3 and 4). Case studies at three different locations were undertaken. One of the case studies revealed higher net income generated during third year compared to first and second year due to proper recycling of farm resources through vermin-compost and FYM, horticulture components, floriculture and vegetables, according to seasonal demand which contributed to better returns. Among the components cotton+ vegetable cultivation+ dairy+ vermicomposting unit+ fodder cultivation on bunds was higher profitable. The study from farming systems in southern Rajasthan revealed that crop-livestock-horticulture model was more economical for small farmers compared to crop-livestock system suggesting the scope of improvement in total income of farmers through adoption of improved livestock technologies.

Table 3. Economics of IFS for small farmers (Manohar Lal, 2018)

Types of Model	Components	Area (ha)	Cost of cultivation (Rs/ha)	Gross return (Rs/ha)	Net return (Rs/ha)
1	Crops-Livestock	1.0	28292	61677	33385
	Crops		18952	45870	26918
	Livestock		9340	15807	6467
2	Crops-Livestock-Horticulture	1.47	28647	80808	52161
	Crops		12534	33958	21424
	Livestock		8506	18558	10052
	Horticulture		7607	28292	20685

Table 4. Economics of the farming system (complete one year cycle) (Manohar Lal, 2018)

Components	Gross income (Rs)	Net income (Rs)
Rice crop	13893	8398
Vegetable/Fruits	30240	21015
Goatry	50000	39180
Poultry	52500	32500
Beekeeping	16500	10800
Mushroom production	22000	16800
Total (Rs)	185133	128693

Selected IFS models for arid and semi-arid regions

Various suitable IFS for arid and semi-arid regions has been developed in farmers' field and at research station by integrating various compatible enterprises such as crops (field crops, horticultural crops), agroforestry (agri-silvi culture, agri-horticulture, agri-pastoral, silvi-pastoral, horti-pastoral), livestock (dairy pigs, poultry, small ruminants), fishery, mushroom, biogas production and tree plantation (Bhati 1997, Bhati et al., 2008, Singh et al., 1999, Jayanthi et al., 2001 Singh et al., 2007, Gill et al., 2009, Surve et al., 2014). Some of the suitable IFS models identified for arid and semi-arid regions of India are presented in Table 5. The agroforestry based integrated production system is suitable for improving productivity, employment opportunities, economic condition and nutritional security in arid regions (Chundawat, 1993; Pareek, 1999; Chadha, 2002; Harsh and Tewari, 2007).

Table 5. Suitable IFS models for arid and semi-arid regions of India

Region	IFS model	Reference
Semi-arid vertic Inceptisols of Tamil Nadu	Crop + poultry (20) + goat (4) + sheep (6) + dairy (1)	Solaiappan et al. (2007)
Tungbhadra project area of Karnataka	Crop (0.74 ha)+ horticulture (0.18 ha) + fodder crop (0.02 ha) + livestock (2 cow, 1 buffalo,14 goat)+ vermicompost units (4) + compost unit (1) + Azolla unit (1Crop (0.74 ha)+ horticulture (0.18 ha) + fodder crop (0.02 ha) + livestock (2 cow, 1 buffalo,14 goat)+ vermicompost units (4) + compost unit (1) + Azolla unit (1Crop (0.74 ha)+ horticulture (0.18 ha) + fodder crop (0.02 ha) + livestock (2 cow, 1 buffalo,14 goat)+ vermicompost units (4) + compost unit (1) + Azolla unit (1 Crop (0.74 ha)+ horticulture (0.18 ha) + fodder crop (0.02 ha) +	Basavanneppa and Gadi (2017)

Region	IFS model	Reference
	livestock (2 cow, 1 buffalo,14 goat)+ vermicompost units (4) + compost unit (1) + Azolla unit (1 Crop (0.74 ha)+ horticulture (0.18 ha) + fodder crop (0.02 ha) + livestock (2 cow, 1 buffalo,14 goat)+ vermicompost units (4) + compost unit (1) + Azolla unit (1)	
Dryland western zone of Tamil Nadu Dryland western zone of Tamil Nadu	Sorghum + cowpea (grain purpose), sorghum + cowpea (fodder) and *Conus glaucus* each in 0.33 ha intercropped with *Emblica officinalis* + goat (5+1)	Radhamani (2001)
Semi-arid Gujarat	Crops: Pearl millet – wheat (0.44 ha), mustard – pearl millet (0.22 ha), cotton- fodder sorghum (1 ha) + horticulture: papaya (0.04 ha) + dairy buffalo (6)	Patel et al. (2007)
Semi-arid irrigated Punjab	For 2 ha land: Crop (1.14ha) + dairy (0.22) + Fishery (0.56) + piggery (0.24)	Gill et al. (2009)
Semi-arid Haryana	Cropping + buffalo or cropping + crossbreed cattle	Singh et al (1993)
Arid Rajasthan	IFS for 7 ha land: Agroforestry (*Prosopis cineraria* + crop), agroforestry (*Hardwickia binata* + crop), agri-horti (*Zizyphus mauritiana* + crops), hortipasture (*Zizyphus mauritiana* + grass), silvipasture (*Colophospermum mopane* + grass) with six adult cattle units (4 cow, 8 bucks and 4 rams)	Tanwar et al. (2016)

The pomegranate + cluster bean/horsegram/mung bean/henna (Lal 2008), and pomegranate + pearl millet/ mung bean/isabgol/sorghum/cumin (Gupta et al. 2000) are suitable production systems for Pali and Jalore regions of Rajasthan, respectively. Samadia et al. (2004) proposed crop components for horticulture based production systems in the hot arid regions of Rajasthan (Table 6)

Table 6. Vegetable crop components for cropping systems in hot arid/regions (Samadia et al. 2004)

Rainfed/Irrigated	High storey crops	Medium storey crops	Ground crops			Micro-wind break, biodefense or field divider
			Vegetable	Agronomic	Grasses	
Rainfed areas (150-300 mm rainfall)	Khejri, ber	Ber, Khair	Mattera, Kachri, snapmelon	Guar, moth, bajra, til	*Cenchrus, Lasiurus*	Ker, phog, khimp, jharber
Rainfed areas (300-500 mm rainfall)	Ber, Lasora, Khejri	Sehjana, lasora	Mattera, kachari, snapmelon, tinda, brinjal, Indian bean, cluster bean, cowpea	Guar, moth, bajra, til	*Cenchrus, Dichanthium, Panicum*	Ker, phog, khimp, jharber
Irrigated areas	Date palm, ber, aonla	Lime, guava pomegranate	Cucurbits, chilli, tomato, brinjal, cole crops, pea, beans, onion, okra and leafy vegetable	Cumin, isabgol, groundnut, mustard		Lasora, sehjana, karonda

Empowerment of women through IFS

Women play significant role in household management including agricultural operations. This is especially true for hilly and tribal areas. There is a vast scope to improve the household profitability by judiciously utilizing family labour using innovative practices and ensuring multiple uses of household resources. This is possible by providing them women empowerment through location-specific trainings and critical need-based support. With the improvement in educational status in the years to come, the role of women in agriculture and management of household resources will be increasingly important. As such, feminization of agriculture in the long run is expected and developing women-centric farming system models will be a real challenge as men are migrating to rural non-farm sectors.

Strategic approaches to harness potential for development of agriculture

Keeping in view of the present situation and the target of doubling farmers' income, strategies, expected contribution and contributing factors that emerged can be summarized as-

Table 7. Areas of consideration for different components

Area to be considered	Components to be taken
Increase in Productivity	Crops, horticulture, fisheries & livestock
Agricultural Diversification	Cropping system, IFS, NWFP, Lac, Sericulture, Honey
Decreasing input cost	Per drop more crop, Farm mechanization
Utilizing available resources for maximum production	Soil acidity, Labor, Capital and Fallow land management
Utility creation & distribution	Skill & entrepreneurship development
Decreasing post harvest losses	Skill & entrepreneurship development, training

Strategies for increasing production of vegetables, flowers and post harvest produce

- Promotion of quality seed through Public Private Partnership (PPP)
- Application of integrated nutrient management (Source: Organic-25%, Inorganic-75%) and integrated pest management
- Promotion of cultivation of rainy season tomato and brinjal crops
- Increase of area under marigold and gladiolus cultivation
- Promotion of protected horticulture
- Post harvest and value addition of horticultural Crops

Strategy to introduce agroforestry activities in degraded land

- Plantation of multipurpose trees species (MPTS) species on bunds of farms and waste land
- Improvement of nutrient status of soil through agro forestry models
- Market support
- Production of fodder grass under silvi pasture system
- Capacity building of farmer for adoption of agro-forestry system

Strategy for improving dairy sector to yield higher milk

- Promotion of crossbred cattle- Supporting viable dairy units (2-5 lactating cows) with the convergence of Department of Animal Husbandry.
- Providing preventive and therapeutic healthcare service- Providing veterinary service to the remote areas through para-veterinary experts.
- Breeding Policy - Promoting exotic milch breeds for progressive farmers and good quality indigenous milch breeds for resource poor farmer .
- Commercial fodder production model- Fodder supply to the farmers on cost basis in cut and carry methods. Distribution of planting material/seed of improved varieties of drought tolerant perennial grasses/legumes species (Perennial Fodder trees like Subabul, Sesbania, Moringa).
- Promotion of Milk Federation Cooperative network at village level and milk plants for value chain.

Strategy for improving backyard poultry production

- Introduction of improved breeds of chicken for egg and meat *e.g.* Jharsim, Vanraja, Gramapriya.
- Improved management practices for backyard poultry such as regular vaccination and timely deworming.

Strategy for improving fisheries sector

- Ensuring availability of fish feed- Fish Feed Production at cottage level should be done.
- Fish production- Bringing more number of farm families under fisheries sector.
- Target oriented area expansion & coverage for aquaculture- Renovation of old ponds and digging of new ponds.
- Capacity building and skill development- Training of fish seed growers & fish farmers.

Strategy for farm mechanization

- Adoption of Improved Agricultural Implements and Machinery such as Paddy seeder, Zero till seed-cum-fertilizer drill, Cono-weeder, Vertical conveyor reaper, Paddle operated paddy thresher and Wheat thresher.
- Use of hand/ animal/ power operated Implements like Birsa Ridger, Rice transplanter, Paddy seeder, Dutch-hoe, Naveen sickle, Hand Maize sheller, Weeder , Rotavator, Power operated thresher.

Strategy for improving NWFPs- Lac Sector

- Formation of National Lac Development Board.
- Declaring Lac as agricultural produce.
- Establishment of Regional Stations and Quality Evaluation Laboratories.
- Introduction of Lac Integrated Farming System Models for assured livelihood in rain fed agriculture.

Way forward for self-reliant India

In the context of making India self-reliant, the task of achieving self-reliance and self-sustainability in agriculture is easier and cost-efficient compared with other sectors of the economy. The first set of measures rely on branding of local farm products to be sold globally, the branding helping to tide over price spirals and making farmers quality-conscious. A branded quality product has the potential to increase India's post-Covid global export share beyond 2%, and aid rural prosperity. The Mission Swadeshi 2.0 must include standardization and promotion of indigenous technology knowledge (ITK) in agriculture, where techniques are dependent on local resources in dealing with nutrition, disease and pests. These are the sort of rural innovations that need to be registered at a national-level innovation registry, which, in turn, can help in the filing of patents and get micro-venture capital support for enterprises.

ITK (Indigenous Technical Knowledge) businesses need to be honored with incentives like recognition at community, state and national levels. Their success beyond the local will depend on how far they resonate to be in sync with sustainability-related goals (SRGs). For instance, as the level of pesticide use per hectare in India is about 0.3 kg, one of the lowest in the world (China uses about 13 kg per hectare), this provides an immediate opportunity to brand and promote ITK.

Finally, recognizing allied activities in agriculture - dairy, animal husbandry, beekeeping, herbal cultivation and fisheries - in Govt of India *Atmanirbhar Bharat Abhiyan* (Self-reliant India Movement) is welcome. It is also essential to highlight the fact that their marketing channels have been mostly distorted and less cooperative-driven. Infrastructure supporting these allied activities must be encouraged as a standalone enterprise, especially among the tribal farming communities. It is the right time to establish an Agricultural Development Council on the lines of the GST Council so as to accelerate the pace of reforms to enhance land leasing private investment, agricultural R&D, etc. With more than 50% population directly or indirectly dependent on agriculture and allied industries, it is essential that big ticket reforms reach these people as we head into a period where we run a vibrant economy adapting to Covid-19 conditions.

References

Behera U.K., Babu Amjath, Kasechele Harald and France J. 2014. Energy self-sufficient sustainable integrated farming systems for livelihood security under changing climate scenario. (In) Proceedings of National Symposium on Agricultural Diversification for Sustainable livelihood and Environmental Security, held during 18-20 November 2014 at Ludhiana, Punjab, p 576.

Behera, U.K., Jha, K.P. and Mahapatra, I.C. .2004. Integrated management of available resources of the small and marginal farmers for generation of income and employment in eastern India. Crop Research 27(1): 83-89.

Bhati T.K., Tewari J C and Rathore S S. 2008. Productivity dynamics of integrated farming systems in western Rajasthan. (In) Diversification of Arid Farming Systems, pp 23-30. Pratap Narain, Singh M P, Kar A, Kathju S, Praveen-Kumar (Eds.). Arid Zone Research Association of India and Scientific Publishers (India), Jodhpur, India.

Bhati T K. 1997. Integrated farming systems for sustainable agriculture in drylands. (In): Sustainable Dryland Agriculture, pp 102–5. Central Arid Zone Research Institute, Jodhpur.

Bhatt B P and Bujarbaruah K M. 2005. Intensive integrated farming system: A sustainable approach of land use in eastern Himalayas. Technical Bulletin No, 46, ICAR Research Complex for NEH Region, Umian, Meghalaya, pp 1-43.

Chadha K L. 2002. Diversification to horticulture for food, nutrition and economic security. Indian Journal of Horticulture 59: 209–29.

Chundawat B S. 1993. Intercropping in orchards. (In) Advances in Horticulture, Vol 2 pp 763–767. Chadha K L and Pareek O P (Eds). Malhotra Publishing House, New Delhi.

Edwards P, Kaewpaitoon K, McCoy E W and Chantachaeng C. 1986. Pilot small scale crop/ livestock/ fish, integrated farm, AIT Research Report 184, pp. 131 Bangkok, Thailand, p 131.

FAO (2017).The State of Food and Agriculture- Laveraging food system for inclusive rural transformation. pp. 2.

FAO, 2001. Farming Systems and Poverty: Improving Farmers' livelihoods in a changing World. Food and Agriculture organization of the United Nations, Rome pp 412.

Gill M S, Singh J P and Gangwar K S. 2009. Integrated farming system and agriculture sustainability. Indian Journal of Agronomy 54: 128–39.

Gupta J P, Joshi D C and Singh G B. 2000. Management of arid ecosystem. (In) Natural Resource Management for Agricultural Production in India, pp 551-668. Yadava J S P and Singh G B (Eds).Indian Society of Soil Science, New Delhi.

Harsh L N and Tewari J C. 2007. Agroforestry system in arid region (Rajasthan) of India. (In): Agroforestry Systems and Practices, pp 175–90. Puri S and Panwar P (Eds).New India Publishing Agency, New Delhi.

Jayanthi C, Balusamy M, Chinnusamy C and Mythili S. 2003. Integrated nutrient supply system of linked components in lowland integrated farming system. Indian Journal of Agronomy 48: 241–6.

Jayanthi C, Rangasamy A, Mythili S, Balusamy M, Chinnusamy C and Sankaran N. 2001. Sustainable productivity and profitability of integrated farming systems in low land farms. (In) Extended Summaries of National Symposium on Farming System Research on New Millennium, PDCSR, Modipuram, pp 79–81.

Jha, D. 2003. An overview of farming systems research in India. Annals of Agricultural Research 24(4): 695-706.

Khan, A. & Sengupta, A. (2016). An Economic Analysis on Integrated Farming System Model Burakocha village in Angara block of Ranchi district, Jharkhand.Journal for Scientific Research & Development (IJRD). Vol.6 (Ed.): 474-476.

Korikanthimath V S and Manjunath B L. 2009. Intgrated farming system for sustainability in agricultural production. Indian Journal of Agronomy 54(2):140-8.

Kumar S, Joshi P K, Suresh Pal (Eds). National Centre for Agricultural Economics and Policy Research, ICAR, New Delhi.

Kumar Sanjeev, Singh S S, Shivani and Dey A. 2011. Integrated farming system for eastern India. Indian Journal of Agronomy 56(4): 297–304.

Lal G. 2008. Pomegranate based cropping systems for sustainable production in arid ecosystem. (In) Diversification of Arid Farming Systems, pp 55–60. Partap Narain, Singh M P , Kar A, Kathju S, Parveen-Kumar (Eds.), AZRAI and Scientific Publishers (India), Jodhpur, India.

Lal M et.al,2018. Different integrated farming system model for irrigated condition of India on the basis of economic assessment . International Journal of Chemical Studies 6(4):166-175.

Mahapatra, I.C. 1992. Farming systems research challenges and opportunities. Eastern Indian Farming System Research & Extension, Newsletter 6(4): 3-10.

Mahapatra, I.C. 1994. Farming system research – A key to sustainable agriculture. Fertilizer News, 39(11):13-25.

Matsumoto H and Matsuyama S.1995. Livestock Hygiene. Japan Agricultural Development and Extension Association, Japan.

Moriya K and Kitagawa M. 2007. Old and new near-future style agricultural production. In: New challenges in Livestock Production, pp 217-44. Imai H (Ed.). Kyoto University Press.

Pareek O P. 1999. Dryland Horticulture. (In) Fifty Years of Dryland Agricultural Research in India, pp 475–84.

Samadia D K, Vashishtha B B and Raturi G B. 2004. Prospects of vegetable production in the arid zone of India. (In) Impact of Vegetable Research in India, pp 101–14.

Shaner, W.W., Philipp, P.F. and Schmehl, W.R. 1982. Farming Systems Research and Development: Guide for Developing Countries. West View Press, Colorado, USA

Singh H P, Ramakrishana Y S, Sharma K L and Venkateswarlu B (Eds). Central Research Institute for Dryland Agriculture, Hyderabad.

Singh J P, Gill M S and Tripathi D. 2007. Development of integrated farming system models for marginal and small farmers. (In) Extended Summaries of 3rd National Symposium on Integrated Farming System and its role Towards Livelihood Improvement held at ARS, Durgapura, Jaipur, pp 51–53.

Singh R, Singh N, Phogat S B, Sharma U K, Singh R and Singh N. 1999. Income and employment potential of different farming system. Haryana Agricultural University Journal of Research 29 143–145.

Surve U S, Patil E N, Sindhi J B, Bodake P S and Kadlag A D. 2014. Evaluation of different integrated farming system under irrigated situation of Maharashtra. Indian Journal of Agronomy 59: 518–26.

Varughese K and Mathew T. 2009. Integrated farming system for sustainability in coastal Ecosystem. Indian Journal of Agronomy 54(2) 120-7.

19

Agroforestry for Sustainable Intensification in Agriculture Necessary Structural Reforms for Self-Reliancy

R. Kaushal[1], S.B. Chavan[2], A.K. Handa[3], R. Dhiman[4], D. Mandal[1]
Chhavi Sirohi[5], and O.P. Chaturvedi[6]

[1]ICAR-Indian Institute of Soil and Water Conservation, Dehradun, Uttarakhand
[2]ICAR-National Institute of Abiotic Stress Management, Baramati
Pune, Maharashtra
[3]ICAR-Central Agroforestry Research Institute, Gwalior Road
Jhansi, Uttar Pradesh
[4]Ex General Manager, WIMCO Seedlings (ITC-PSPD), Rudrapur, Uttarakhand
[5]Chaudhary Charan Singh Haryana Agricultural University
Hisar, Haryana
[6]Dr. Rajendra Prasad Central Agricultural University, PUSA
Samastipur, Bihar

Introduction

Expanding human and livestock population, fast-growing industrial requirements and urbanization are putting ever-increasing pressures on land resources, creating competition and conflicts which results in sub-optimal use of both land and its natural resources. With no additional land for horizontal expansion of agriculture, there is an urgent need to find viable technologies for enhancing productivity and resource conservation on a sustainable basis. In this context, Agroforestry offers huge potential not only to diversify and increase overall land productivity to meet the demands of fuel, fodder, timber, medicine and other non-woody forest products but also to enhance green cover, and reduce the pressure on forest (Chavan et al. 2016; Handa et al., 2016). Agroforestry offers an affordable alternative in place of expensive conventional conservation measures for providing insurance against risks caused due to weather aberrations, controlling erosion hazards and ensuring sustainable production of the land on a long-term basis. The main objective of agroforestry is to take the advantage of complementary relationships between trees, crops, and livestock in such a way so that sustainability, productivity stability, and profitability of the system is maintained. In India, agroforestry practices are important component in the various developmental programmes schemes like Flood Control/Management Programmes, Multipurpose River Valley Projects, Agriculture Development Programmes, Integrated Rural Development Programmes (IRDP), National Watershed Development Programme for Rainfed Areas (NWDPRA), Forestry Development Scheme, Drought Prone Area Development Programme (DPAP) and Desert Development Programme (DDP) from government sector and wood based industry from private sector. After 2000 a transitional shift in agroforestry observed to tackle the ill effects of intensive agriculture, climate change and nutritional security. Under changing climatic scenarios agroforestry rose as a silver bullet to offer not only single benefits but also a basket of multiple benefits. In the present chapter a journey of agroforestry research and developments, region-specific models, area under agroforestry, initiative of government on policy aspects and challenges are elaborated extensively.

Historical development of agroforestry

Agroforestry is not a new concept although it has been practiced in India for thousands of years. Since the 1970s, agroforestry has come in limelight for tackling various issues of natural resource degradation in India as well as world. The term 'agroforestry' was coined by a visionary study in the mid-1970s led by John Bene of International Development Research Centre (IDRC), Canada (Bene et al., 1977). Agroforestry was incorporated into national agricultural and forestry research agendas in 1983 with the launch of All India Coordinated Research Project (AICRP) on Agroforestry by ICAR with 20 centres (Handa et al., 2015). It was followed by the establishment of National Research Centre on Agroforestry in 1988 at Jhansi, U.P. which was further upgraded to Central Agroforestry Research Institute (CAFRI) in 2014. The AICRP is continuing in 37 Centres of AICRP on Agroforestry located at 26 State Agricultural Universities (SAUs), 10 ICAR institutions, and 1 ICFRE institute tackling the problems of different major agro-ecological zones of the country. Indian Council of Forestry Research and Education (ICFRE) also initiated agroforestry research in different parts of the country (ICFRE, 2011). Besides, many business corporations limited companies such as ITC, WIMCO, BAIF, West Coast Paper Mills Ltd., Hindustan Paper Mills Ltd. and other institutions initiated agroforestry research with emphasis on the production of improved planting material of the fast-growing species for catering to their demand for raw material (Dhyani et al., 2015).

Traditional agroforestry systems in India

Traditional agroforestry systems in India ranges from apparently simple forms of shifting cultivation to complex home-gardens, from systems involving sparse stands of trees on farmlands to high-density complex multi-storied homesteads of humid lowlands, from systems in which trees play a predominantly 'service' role to those in which they provide main commercial products. Most of these are anecdotal but in some enough research efforts have been carried out in recent times. The traditional systems/practices having diverse characteristics of agroforestry are listed in Table 1. Traditionally many well-proven and region specific trees, tree-based systems which being practiced from centuries for uplifting livelihood of a grief-stricken rural population. Such systems are evolved over a period of time and testified under difficult situation are widely known as indigenous tree-based systems or traditional agroforestry (Chavan et al., 2020).

Table 1. Traditional agroforestry systems prevalent in India

Systems Practices	Agro-ecological adaptations	Features
Shifting Cultivation	In Tropical forest areas (mainly in north eastern hilly states, odisha and Andhra Pradesh)	Cyclic slash and burn, growing of arable crops, fallow and back; Public and private lands; subsistence livelihoods
Taungya	Earlier in all regions but now has been discountinued in many states	Tree planting, intercropping upto 3 years, moving to next place; Govt. land and trees, intercrops for cultivators; Sharing of land resources between state and landless
Homegardens	Mainly humid tropics (Southern and north eastern states)	Transitory stage between tropical forest ecosystem and arable cropping; Plantation, fruit, spices, vegetable, fuel and fodder and MPT in backyard; Domestic needs and biodiversity
Large Cardamom and Alder Plantation	Mainly humid tropics (north eastern hilly states)	Cardamom supported by alder and other species; Tree harvesting at 15cm basal dia, Lopping after cardamom maturation and pollarding at 2-2.5 m. height for regeneration of younger seedlings and open canopy to regulate light; Soil fertility, shade, fodder, employment for labour and better income.
Plantation Crops and Fruit Orchard	All regions	Intercropped with cereals and other field crops; Short term fruits; Tubers and fodder crops; Enhanced resource use efficiency, Employment, livelihood support and income
Scattered Trees on Farm Lands	In all regions specially semi-arid and arid	Naturally occurring or planted timber, fruit, MPT and main field crops; Income, Wood supply and some fodder in critical period.
Shelterbelts and Windbreaks	In wind prone arid, alpine areas and coastal areas	Plantations against sea tides, Hot winds etc., Income; Environmental protection and domestic needs.

Systems Practices	Agro-ecological adaptations	Features
Boundary Plantations and Live Hedges	All regions	Trees for various uses on field crops and orchard boundaries; Income, Employment, Supply for domestic needs and industry (e.g. plywood), crop protection and wind break.
Woodlots for Soil Conservation	In hilly areas, along sea coast and ravine	Resource conservation, ground water recharge, microclimatic amelioration
Trees on Rangelands	All regions	Shrubs and bushes in community grazing lands; Resilience against climatic aberrations, Green biomass during lean months and higher production in organized silvi-pastures.
Plantation crops with pastures and animals	Mostly humid and sub-humid regions	Green biomass during lean months and higher production in organized silvi-pastures.
Industrial Plantations with Crops	Intensively cropped areas in north India	For fuel, pole and pulpwood; Income and domestic needs and Industry
Seasonal Forest Grazing	Semi-arid and mountainous ecosystem	Migratory grazing, reduce fire hazards, increase in soil fertility, livelihood support and income.
Aquaforestry	Low lying lands	Fish pond dikes and surface planted with suitable species (e.g. Coconut, Mangroves); Source of Nutrition to aquatic life, Breeding ground, Nutrient cycling; Shade for crustaceans, Employment and higher Income
Apiculture with Trees	All regions	Trees as a medium for apiculture; Employment and Income for landless and resource poor; Better pollination of other crops.

Agroforestry Research in India

To identify and document existing agroforestry systems in the country, a Diagnostic and Design (D&D) survey was conducted by different centres. The survey revealed agrisilviculture as the most prominent (AICRP) agroforestry system in seven agroclimatic regions, followed by agrihorticulture in six agroclimatic regions, and agrihortisilviculture and silvipastoral systems in two agroclimatic regions. The specialized systems such as *Jhum* (shifting cultivation) in Eastern Himalaya, energy plantations in Lower Gangetic plains, block plantations in Eastern plateau and hills and Central plateau and hills, homesteads in Middle Gangetic plains, homegardens in East coast plateau and hills, West coast plains and ghats and Islands, and shelterbelts in East coast plateau & hills and Western dry regions are practiced. Research was conducted on evaluation of 184 promising germplasms in different agroclimatic zone. On the basis of study, the suitable species for different agroclimatic zone are given in Table 2. Dhyani et al. (2009) reported twenty common agroforestry systems being practised in different agro-ecological regions of India presented (Table 2). ICAR-CAFRI and AICRPAF published a summary on "*Agroforestry technologies for different agroclimatic zones of Country*", whereas it consists of 84 grassroots levels farmers adopted technologies developed by 26 AICRPAF centers from five different agroclimatic regions of the country (Chaturvedi et al., 2016).

Table 2. Prominent agroforestry systems in different regions of India

Agro-climatic Zone	Agroforestry systems	Tree Component	Crop/Grass
Western Himalayas	Silvipasture (RF)	*Grewia optiva*	*Setaria* spp.
		Morus alba	*Setaria* spp.
	Agrihorticulture	*Malus pumila*	Millets, Wheat
	Agrihorticulture	*Prunus persica*	Maize, Soyabean
Eastern Himalayas	Agrisilviculture	*Anthocephalus cadamba*	Rice
	Agrihorticulture	*Alnus nepalensis*	Large Cardamom/Coffee
	Silvipasture	Bamboos, *Parkia roxburghii, Morus alba*	-
	Silvipasture	*Bauhinia variegate, Ficus, Morus alba*	Napier
Lower Gangetic Plains	Agrisilviculture (Irri)	*Eucalyptus, Albizia lebbeck*	Rice
	Agrihorticulture (Irri)	Mango/Banana/Litchi	Wheat, Paddy, Maize
	Silvipasture	*Morus alba, Albizia lebbeck*	*Dicanthium, Pennisetum*
Middle Gangetic Plains	Agrisilviculture (Irri)	*Populus deltoides*	Sugarcane-Wheat
	Agrisilviculture (Irri)	*Eucalyptus* spp.	Rice-Wheat
	Agrisilviculture	*Dalbergia sissoo*	Sesamum
	Agrihorticulture (Irri)	Mango/Citrus	Rice-Wheat
	Silvipasture	*Albizia lebbeck*	*Chrysopogan Dicanthium*
Trans Gangetic Plains	Agrihorticulture (Irri)	*Emblica officinalis*	Black gram/Green gram
	Agrisilviculture	*Azadirachta indica*	Black gram- Wheat/ Mustard
	Silvipasture	*Bauhinia variegata Albizia deltoides*	*Cenchrus, Pennisetum*
Upper Gangetic Plains	Agrisilviculture (Irri)	*Populus deltoids*	Wheat, Bajra fodder
	Agrisilviculture (Irri)	*Eucalyptus*	Rice-Wheat
	Silvipasture	*Bauhinia variegata Albizia lebbeck*	*Chrysopogon, Poa*

Agro-climatic Zone	Agroforestry systems	Tree Component	Crop/Grass
Eastern Plateau & Hills	Agrisilviculture	*Gmelina arborea*	Rice, Linseed
	Agrisilviculture	*Acacia nilotica*	Rice
	Silvipasture	*Acacia mangium, A. nilotica, Bamboos*	-
	Silvipasture	*Leucaena leucocephala*	*Chrysopogon, Pennisetum, Dicanthium*
Central Plateau & Hills	Agrihorticulture (Irri)	*Psidium gujava*	Bengal Gram/Groundnut
	Agrihorticulture (RF)	*Emblica officinalis*	Black gram/Green gram
	Agrisilviculture	*Acacia nilotica / Leucaena leucocephala /Azadirachta indica / Albizia lebbeck*	Soyabean, Black gram-Mustard/Wheat
	Silvipasture (RF- and degraded lands)	*Albizia amara, Leucaena leucocephala, Dichrostycus cinerea*	*Chrysopogon, Stylosanthes hamata S. scabra*
	TBOs (RF)	*Jatropha curcas*	-
Wastern Plateau & Hills	Agrihortisilviculture (Irri)	*Tectona grandis, Achrus zapota*	Rice, Maize
	Agrihorticulture	*Areca catechu*	Black pepper Cardamom
	Silviculture	*Prosopis juliflora, Ailanthus excelsa*	-
	Silvipasture	*Acacia mangium, Albiza amara*	*Cenchrus*
Southern Plateau & Hills	Agrisilviculture (RF)	*Eucalyptus, Casuarina equisetifolia, Ailanthus excelsa*	Cotton, Groundnut
	Agrisilviculture (Irri)	*Eucalyptus tereticornis, Melia dubia*	Chilli
	Silviculture (RF)	*Leucaena leucocephala, Acacia leucopholea*	-
		Eucalyptus	-
	Agrihorticulture	*Tamarindus indica*	Chilli
	TBOs	*Pongamia pinnata*	-
East Coast Plains & Hills	Agrisilviculture (RF)	*Ailanthus excelsa, Acacia leucophloea*	Cow pea
	Silviculture	*Casuarina equisetifolia, Leucaena leucocephala*	-
	TBOs	*Pongamia pinnata*	-
	Silvipasture	*Artocarpus* spp.	*Chrysopogon, Napier Cenchrus*

Agro-climatic Zone	Agroforestry systems	Tree Component	Crop/Grass
West Coast Plains & Hills	Agrisilviculture (RF)	*Acacia auriculiformis*	Black Pepper
	Agrihorticulture (RF)	*Artocarpus heterophyllus*	Black Pepper
	Agrisilviculture (RF)	*Acacia auriculiformis*	Rice
	Agrihorticulture	*Cocos nucifera/Areca catechu*	Rice
	Agrisilviculture	*Casuarina equisetiofolia*	Rice
	Silvipasture	*Hardwickia binnata Albizia lebbeck*	*Cenchrus*
Gujarat Coast plains & Hills	Agrisilviculture	*Azadirachta indica Ailanthus excelsa*	Cow pea, Green gram
	Silviculture	*Prosopis juliflora Acacia nilotica*	-
	Silvipasture	*Leucaena leucocephala*	*Cenchrus, Setaria*
Western Dry Region	Agrisilviculture	*Prosopis cineraria Tecomella undulata Acacia nilotica, Azadirachta indica*	Pearl millet
	TBOs	*Jatropha curcas*	-
	Silvipasture	*Albizia lebbeck Hardwickia binnata*	*Cenchrus*
All Islands	Agrihorticulture	*Cocos nucifera*	Rice
	Silvipasture	*Bauhinia* spp, *Erythrina indica, Leucaena leucocephala*	*Cenchrus, Pennisetum*

Irri: irrigated; RF; rainfed, TBOs: tree borne oil seed
(*Source*: Dhyani et al., 2015; Newaj et al., 2016)

Agroforestry technologies have been developed for rehabilitation of degraded lands affected by water erosion, mines affected areas, river affected areas etc. Agroforestry-based watershed interventions with erosion control structures such as half-moon terraces, contour bunds, grassed waterways, gully plugging bench terracing silvipasture system, bamboo plantations have been developed by ICAR-IISWC (previously CSWCRTI) for the rehabilitation of degraded lands in northwestern Himalayas (Kaushal et al. 2016, 2019, 2020; Chaturvedi et al. 2014, Sharda et al. 2012). *Chrysopogon fulvus* + *Aonla* has been identified as an economically-viable and eco-friendly system for rehabilitation of marginal lands in Shiwalik hills. In Doon valley, *Dalbergia sissoo, Acacia catechu, Albizia lebbek Eucalyptus* hybrid, *Leucaena leucocephala, Grewia optiva* based silvipasture system provided good returns when grown with *Chrysopogon fulvus* (gorda) *Panicum maximum* (guini grass), *Eulaliopsis binata* bhabar) and *Pennisetum purpureum* (napier grass). These silvipasture systems are capable of yielding mean annual production of 8-10 t ha^{-1} which comprised 4.5-5.0 t ha^{-1} of fodder from grasses, 1.5-2.0 t ha^{-1} of leaf fodder from tree loppings and 2.0-2.5 t ha^{-1} of fuelwood from the lopped branches. Benefit: Cost ratio of these ranged from 1.6-1.9 (Chaturvedi et al., 2014). The replacement of field crops and horticulture by silvipasture and agroforestry for, the productive and protective purposes has led to a reduction in runoff in watershed management of 48 to 99% and reduction in soil loss 81 to 98%. Silvipasture horti-pasture, agri-silvipasture and silvihortipasture land use systems have been developed for salt affected soils by ICAR- Central Soil Salinity Research Institute Karnal (CSSRI), Haryana. Among the tree species tested, *Acacia nilotica, Albizia procera, Leucaena leucocephala, Azadirachta indica* and *Eucalyptus* hybrids were found most suitable tree species for rehabilitation of sodic soils (Dagar 2012). Mesquite (*Prosopis juliflora*) and *Kallar* grass (*Leptochloa fusca*) silvi-pastoral practice has been identified most promising for firewood and forage production and also for soil amelioration. By using the agroforestry technologies developed at CSSRI, the state forest departments, non-governmental agencies (NGOs) National Wastelands Development Board (NWDB), and other developmental agencies have rehabilitated more than one million hectares (out of a total of 6.75 million hectares) of salt-affected soils, particularly the village community lands, government lands, absentee lands, areas along roadsides, canals and railway tracks (Tomar 1998; Dagar 2012). A compiled list of recommended tree species suitable for different problem soils are given in Table 3:

Table 3: Tree species suitable for problem soils

Type of problem	Recommended species
Sodic soils (pH 8.62-10)	*Acacia nilotica, Acacia auriculiformis, Casuarina equisetifolia, Prosopis cineraria, Prosopis juliflora, Tamarix articulate, Salvadora persica, Pithecellobium dulce, Pongamia pinnata, Sesbania sesban, Terminalia arjuna, Poplars, Eucalyptus tereticornis, Butea monosperma, Azadirachta indica*
Wetland	*Alnus trabeculosa, Alnus cremastogyne, Salix xuchonensis, Salix babylonica, Paulownia tomentosa, Taxodium distichum, Taxodium scandens and Morus alba*
	Bambusa arundinacea, Eucalyptus spp., *Corymbia tessellaris, Casuarina equisetifolia, Terminalia arjuna, Tamarix, Lucerne, Syzygium cumini, Barringtonia acutangula, Ficus glomareta and Psidium gujava,*
Sand dunes stabilization	*Prosopis cineraria, Prosopis juliflora, Acacia tortilis, Acacia nilotica Azadirachta indica, Acacia senegal, Ziziphus* spp., *Tecomella undulata Capparis deciduas Ailanthus excelsa and Parkinsonia aculeta*
Coastal Area	*Anacardium occidentale, Ailanthus malabarica, Casuarina equisetifolia Pongamia glabra, Terminalia catappa, Calophyllum inophyllum,* species *of Pandanus, Nypa fruticans, Salicornia bigelovii, Sesbania aculeate Salvadora persica, S. oleoides and Borassus flabellifer*

Type of problem	Recommended species
Ravine area	*Acacia catechu, Acacia nilotica, Dalbergia sissoo, Morus alba, Dendrocalamus strictus, Azadirachta indica, Albizia* spp and *Holoptelia integrifolia*
Acid soils	*Pinus roxburghii, P. wallichiana, Celtis australis, Pinus, Cassia, Acacia mearnsii, A. decurrens, A. dealbata, A. pycnantha, Manilkara hexandra Phoenix* spp. *A. auriculiformis, Pterocarpus marsupium, Shorea robusta Hardwickia binnata, Terminalia* spp., *Tectona grandis, Gmelina arborea Xylia xylocarpa, Bamboo* spp., *Alnus* spp

Source: Uthappa et al. (2015a)

Research work on tree selection and improvement conducted at different centres lead to development and registration of the elite germplasm. BAIF a research-based famous NGO introduced high yielding germplasm of *Leucaena leucocephala*, Hawaiin type for improving the degraded lands in 1970s. To meet the shortage of raw materials, WIMCO Seedlings Ltd. (UK) laid the foundations of the world-renowned success story of commercial-scale poplar plantations by farmer which was followed by development of high-yielding Eucalyptus clones by the Indian Tobacco Company (ITC) in Khammam, Andhra Pradesh. These models have ensured the backward integration of raw material with wood-based industries. Germplasm of Shisham was developed by NRCAF now CAFRI, Jhansi (Bundel-1 and Bundel-2) and GBPUAT, Pantnagar (PS 52); Poplar clones (L and PL series by Lalkuan nursery and PAU, Ludhiana, Kranti, Uday, Bahar, W-series from WIMCO, Pant Poplar 5 by GBPUAT, Pantnagar, teak clone (PDKV/AF/1) by College of Agriculture, Nagpur and Eucalyptus (Clone No. 7, 10, 413, 405, Hybrid clone 2070 by ITC, K series from KFRI, SRY-16 by MPKV, Rahuri, *Acacia nilotica* (KSH- 9, 33, 89, 104, PT-4, PT-14, provenances from Banaskantha, Akola, Adilabad, Shivpuri, Hastinapur), clonal and seed orchards of *Dalbergia, Acacia, Prosopis, Pongamia, Simarouba* and other species were also established. CAFRI also registered 220 accessions of Neem, 32 of Shisham, 63 of Babul, 16 of *Aogeissus* (Kardhai), 138 of Karanj and 284 of Jatropha with NBPGR. ICFRE and its institute have developed clones of *Casuarina equisetifolia* (7), *C. junghuhniana* (5), meliadubis (5), *Eucalyptus* (4). ICFRE also initiated provenance testing program in collaborationwith CSIRO, Australia and exchange new germplasm of important agroforestry species like *Melia dubia, Casuarina, Eucalyptus* 'mallees', *Grevillea robusta, Acacia mangium* and *Acacia melaxylon* from Australia and Teak from Thailand.

Private sector in promotion and development of agroforestry

Organized agroforestry which is also referred sometimes as industrial agroforestry or commercial agroforestry (Dhiman 2013) has developed in certain specific locations in India. Private sector especially certain wood based companies like safety match company -WIMCO and paper company- ITC PSPD played a key role in its development and promotion from their own resources. This kind of agroforestry is not evenly distributed and has significant presence in isolated locations around the clusters of wood based industry indicating that this activity during the last four decades is sustainable with synergic and symbiotic relationship between farmers and industry. If there is market for wood in accessible locations, farmers continue its practice and if not they usually shift to other production system as happens during crash in wood prices. With the success of these two case studies of poplar by WIMCO and Eucalypts by ITC, many other paper companies and a couple of panel industry have also started the promotion of agroforestry by supplying quality planting stock and guarantee support price on wood procurement.

Of late, the private sector has started contributions in the promotion of agroforestry through their Corporate Social Responsibility (CSR) and Corporate Environmental Responsibility (CER) commitments. CSR activities have been made binding w.e.f. April 2014 for profit making industrial establishments. According to the CSR rules issued by the Government, companies having a net worth of Rs. 500 Crores or having a maximum turnover of Rs. 1000 crore or those having a net profits of at-least Rs. 5 crores have to mandatory spend 2% of the average of last three years profits in CSR activities. Schedule VII of section 135 enlists number of activities on which CSR money could be spent. Agroforestry and plantation activities are covered under category No IV on Environment related activities. According to the Ministry of Corporate Affairs, GoI CSR expenditure of 460 listed companies, which have placed annual returns on CSR on their websites is estimated at Rs. 6337 crores during the year 2014-15 (http://www.mca.gov.in / Ministry/pdf/Rajyasabha_unstarred _que999.pdf). According to a survey, N100 listed companies in India have spent Rs. 8691 Crores during the year 2018-19 under CSR activities. The amount spent on Environment activities including agroforestry was Rs. 700 crores during 2018-19 (KPMG 2020).

ITC- A leader in green field projects among private sector, is pursuing its CSR program on agroforestry and farm forestry in 200 km^2 area around factories for maximum social, economical and environmental impact (https://www.itcportal.com.). Some other wood based industries especially paper and a couple of panel and plywood industries have also started spending their CSR share in promotion of plantation activities including agroforestry as back-end linkages of wood raw material production with their wood product manufacturing and improving their social and environmental impact. It is estimated that the CSR amounts of five leading listed panel industries to be around Rs. 10 crores/annum. Some of them like Green Ply and Century among some others are already spending some of their CSR allocations on production/procurement and distribution of seedlings to farmers at subsidized prices

Based on the recommendations of the parliamentary Standing Committee, Ministry of Environment, Forest and Climate Change (MoEF&CC), issued guidelines that banks should allocate a minimum of 25% of CSR for afforestation efforts. The proposal is to make a Corporate Social Responsibility Tree Planting Fund (CSR TPF) which could be spent on number of plantation activities including growing nurseries and medical plants. Some names of non wood-based industry including banking and financial institutions engaged in promoting tree plantations and also in distribution seedling to local communities under CSR projects by Indian Companies includes Reliance Industries, Mahindra and Mahindra, Cummins India Ltd., Karnataka Bank Ltd., Tata Sponge Iron, Power Finance Corporation Ltd., Bharat Heavy Electricals Ltd., Rallies India Ltd., Atul Ltd etc. WIMCO Seedlings is also supplying planting stock to some of the NGOs promoted by some companies for CSR activities.

MoEF&CC also issued guidelines for Corporate Environment Responsibility (CER) on new projects in May, 2018. As per these guidelines, each corporate seeking green clearance has to set aside up to 2% of its capital investment for CER. Companies seeking EC (environment clearance) to expand the capacity of projects (brownfield projects), would be required to earmark 0.125%–1% of the capital investment on CER. The slab for new projects (greenfield projects), is in the range of 0.25%–2% of capital investment.(https://economictimes.indiatimes.com/news/economy/policy/pay-2-of-capital-investment-for-green-clearance-environment-ministry-to-corporates/articleshow/ 64008830.cms?utm_source=contentofinterest&utm_medium=text&utm_campaign=cppst).

Agroforestry area in India

Agroforestry is practiced by 1.2 billion people (World Bank, 2004) on 10% area of total agricultural lands; this accounts for more than 1 billion ha (Zomer et al., 2016). Despite large number of researchers and agencies engaged to work out the accurate assessment of area under Agroforestry Systems, the statistics of the area under agroforestry is not very clear due to complexities, scatteredness unavailability of proper database involvement of several institutes and different methodologies. Dixon (1995) reported that worldwide, 585-1215 million hectare area is suitable for agroforestry, whereas Nair et al. (2009) estimated the agroforestry area of 823 m ha. Remote sensing data show that in 2010, 43% of all agricultural land globally had at least 10% tree cover which has increased by 2% over the previous ten years (Zomer et al., 2016). Chavan et al. (2015) reported the agroforestry area based on data from CAFRI, In India, Jhansi and Bhuvan LISS III, was 13.75 m ha. However, Forest Survey of India estimated the same as 11.54 m ha, which is 3.39% of the geographical area of the country (FSI 2013). Figure 1 shows the various agroforestry systems and areas under them in the India.

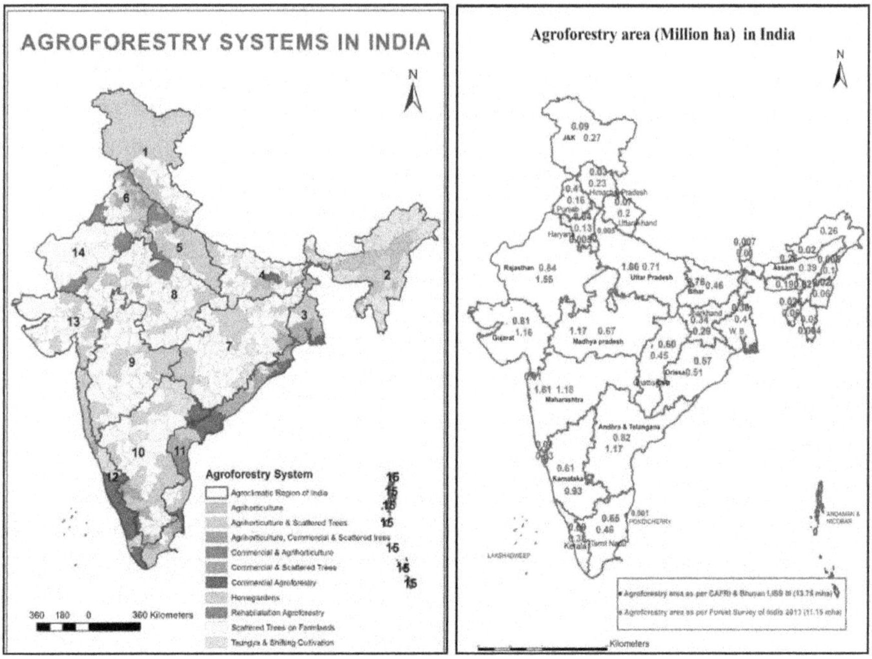

Fig. 1. Agroforestry systems (www.cafri.res.in) and area in India (Chavan et al., 2015)

Agroforestry for ecosystem services

Globally, the services provided by ecosystem are getting increasing policy attention. In October 2010, the Conference of Parties to the Convention on Biological Diversity held in Nagoya witnessed the acceptance of environmental goods part of the national accounts. Milder et al. (2010) estimated that by the year 2030, markets for biodiversity conservation could benefit 10–15 million low-income households in developing countries, carbon markets could benefit 25–50 million people, markets for watershed protection could benefit 80–100 million people, and markets for landscape beauty and recreation could benefit 5–8 million people. Agroforestry has the potential to enhance soil fertility, improves soil health, reduce soil erosion, improve water quality, conserve soil moisture, enhance biodiversity, increase aesthetics, and sequester carbon and provide a number of ecosystem services. However, in the agroforestry literature , evidences supporting these perceived benefits are lacking due to methodological difficulties in quantifying them. Globally some of the researches quantified ecosystem services in monetary terms (Alam et al., 2014). In India, though methodologies have been framed for ecosystem services at international level and Indian Institute of Forest Management (IIFM) at national level for pure forest or natural plantation, however, methodologies for quantification and payment of ecosystem services (PES) in agroforestry system is still in a neglected state. In an era of environmental consciousness and ecological sustainability, the role of agroforestry in providing ecosystem services, therefore, needs to be fully explored. As of now methodologies for carbon sequestration (regulating services) are framed clearly. For other services, the criteria, indicators, field methods for quantification and evaluating need to be framed in a scientific manner.

Agroforestry for mitigating climate change

Agroforestry plays an important role in adaptation and mitigation of climate change through improvement of microclimate, efficient utilization of natural resources, low carbon emissions, carbon assimilation and encouraging gender equity. Agroforestry has great potential for eco-restoration of degraded lands/forests and mitigating the accumulation of C as CO_2 in the atmosphere by maintaining and increasing the land-based carbon sinks (Verma et al., 2014; Goswami et al., 2014; Chavan et al., 2020). Agroforestry has been recognized as a component of climate-smart agriculture and is frequently mentioned for strong potential for climate change adaptation and mitigation (Newaj et al., 2015), which is extensively lightened under various international initiatives such as the Kyoto Protocol of 2001 (Nair et al., 2009) REDD+ mechanism (Minang et al., 2011), Sustainable Developmental Goals (United Nation Assembly 2015) and Paris Agreement (UNFCCC 2015). Ajit et al. (2017) also reported that existing agroforestry systems in India are estimated to mitigate 109.34 million tons CO_2 annually, which may offsets one-third (33 %) of the total GHG emissions from agriculture sector. Different extrapolated estimates of carbon storage in agroforestry ranges 233-279.3 million tonnes reported by FSI (2013), Zomer et al. (2016) and CAFRI-NICRA (2015) as shown in figure 2. By using IPCC methodology IPCC (2013) estimated that agroforestry contributes to mitigating 27±14 t CO_2 equivalents ha^{-1} y^{-1} at least for the first 14 years after establishment of agroforestry on agricultural land. Soils under agroforestry also oxidized (uptake) 1.6±1.0 kg CH_4 ha^{-1} y^{-1} and emitted 7.7± 3.3 kg N_2O ha^{-1} y^{-1}.

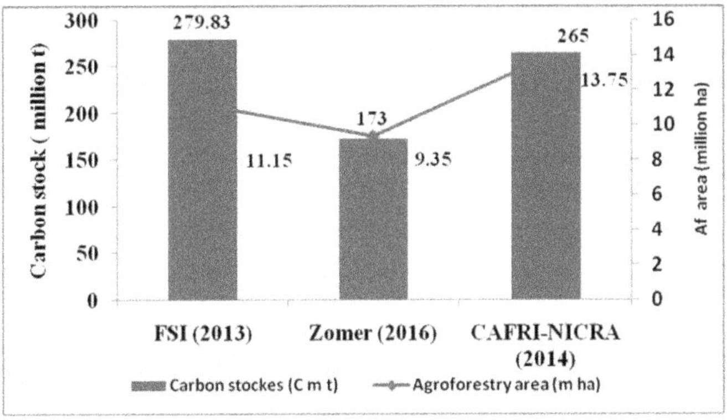

Fig. 2. Agroforestry area and total carbon sequestration in India

The objective of mitigating climate change through tree plantations can be achieved by encouraging the farmers to retain and maintain the existing trees on farm boundaries or farmlands and by motivating them to adopt tree-based farming of short rotation, on their farmlands. It is estimated that restoring 900 million hectares through agroforestry could stabilize global GHG emissions for 15-20 years. As per the International Panel on Climate Change (IPCC) it is estimated that currently there is 400 million ha area under agroforestry throughout world which captures about 0.72 Mg carbon per ha per year.

Agroforestry for doubling farmers' income

The dream of Hon'ble Prime Minister of India Sh. Narendra Modi is to double the farmer's income by 2022. The report of Doubling Farmers Income (DFI), mentioned that it has difficult to double the income by 2022 and it may be required till 2025 with proper target-oriented efforts. Chand (2017) also provided six majors steps of achieving or sources of doubling farmer's income in India. It includes improving in productivity, resource-use efficiency or saving in cost of production, increasing in cropping intensity, diversification towards high-value crops, shifting cultivators to farm no non-farm occupations and improving in terms of trade or farmers or real price received by the farmer, etc. Among these six majors steps, agroforestry has the capacity of qualifying under four parameters due to its complex nature, multiple production systems, resource conservation and deriving numerous environmental services (microclimate moderators to soil health improvement).

The economic analysis of agroforestry systems from different parts of the country suggests that agroforestry is more profitable to farmers than agriculture or forestry for a particular area of land. A comprehensive analysis (Table 4) indicated economic viability with IRR ranging from 25 to 68 and B:C ratio of 1.01 to 4.17 (Planning Commission, 2001 and Newaj et al., 2016)). Poplar, Eucalyptus in Indo-gangetic region; *Eucalyptus, Melia dubia* and *Leucaena* based agroforestry in Andhra Pradesh and other southern states are very remunerative. Poplar can give gross income of INR 0.6-0.8 millions ha^{-1} under block plantation with productivity of 20 m^3 ha^{-1} yr^{-1} while income from intercrop is additional (Dhiman 2012; Dhiman et al. 2013). Average productivity of 20 to 25 m^3 ha^{-1} yr^{-1} has also been reported for Eucalyptus under un-irrigated conditions. In Tamil Nadu, very high productivity of 41.54 cubic metre/ha/yr has been reported for *Melia dubia* for which assured income of INR 8500/- per tonne is available (Parthiban and Cinthia, 2020).

Table 4. Financial Analysis of prominent agroforestry systems in various Agro-climatic zones India

Agro-climatic Zone/ Region	Situation	Agroforestry system	Period (year)	Species	Crop/ Grass	Total discounted cost @15% (Rs/ha)	NPV (Rs/ ha)	B/C ratio	FIRR (%)	LEV (Rs/ha
Western Himalayan	Rainfed	Agrihorticulture	15	*Prunus persica*	Maize, Soyabean	57,144	12,039	1.21	30	13,726
Eastern Haimalayan	Irrigated	Agrisilviculture	8	*Anthocephalus cadamba*	Paddy	43,776	24,348	1.56	32	36,173
Lower Gangetic Plains	irrigated	Agrisilviculture	10	*Eucalyptus hybrid*	Paddy, Wheat	86,237	114,532	2.33	33	152,138
Lower Gangetic Plains	Rainfed	Agrisilviculture	15	*Acacia auriculiformis*	Jute	27,016	4,558	1.17	30	5,197
Middle Gangetic Plains	Rainfed	Agrisilviculture	20	*Dalbergia sissoo*	Seasamum	14,360	9,030	1.63	36	9,618
Middle Gangetic Plains	Non- arable land	Silvipasture	10	*Bambusa vulgaris*	Marvel grass	24,796	18,907	1.76	56	25,115
Upper Gangetic Plains	Irrigated	Agrisilviculture	7	*Populus deltoides*	Paddy, Wheat	82,292	190,241	3.31	69	304,843
Trans Gangetic Plains	Rainfed	Agrisilviculture	20	*Acacia nilotica*	Bajra, Guar	17,263	27,788	2.61	39	29,596
Eastern Plateau and Hill	Irrigated	Agrisilviculture	12	*Gmelina arborea*	Paddy, Linseed	69,140	50,857	1.74	34	62,548
Eastern Plateau and Hill	Non- arable land	Agrisilviculture	15	*Hardwickia binata*	Dinanath grass	60,062	68,035	2.13	48	77,568
Central Plateau and Hill	Rainfed	Agrihorticulture	8	*Emblica officinalis*	Groundnut, Gram	86,494	7,409	1.09	28	11,007
Western Plateau and Hill	Non- arable land	Silvipasture	16	*Albizia amara*		63,236	43,768	1.69	44	49,005
Southern Plateau and Hill	Irrigated	Agrisilviculture	15	*Ceiba pentendra*	Groundnut	109,840	46,746	1.43	29	53,296
Southern Plateau and Hill	Rainfed	Agrisilviculture	10	*Tamarindus indica*	Chilli	28,163	29,101	2.03	33	38,657

Agro-climatic Zone/ Region	Situation	Agroforestry system	Period (year)	Species	Crop/ Grass	Total discounted cost @15% (Rs/ha)	NPV (Rs/ ha)	B/C ratio	FIRR (%)	LEV (Rs/ha
West Coast and Ghat	Irrigated	Agrisilviculture	7	*Casurina equisetifolia*	Paddy	86,889	33,907	1.39	26	54,332
West Coast and Ghat	Irrigated	Agrihorticulture	8	*Cocos nucifera*	Elephant Foot, Ginger	64,387	17,650	1.27	33	26,222
West Coast and Ghat	Rainfed	Agrisilviculture	20	*Tectona grandis*	Sweet Potato	209,715	210,246	2.00	31	223,928
Gujarat Plains and Hill	Rainfed	Agrisilviculture	20	*Dalbergia sissoo*	Castor	7,080	8,037	2.14	31	8,560
Western Dry	Rainfed	Agrisilviculture	20	*Prosopis cineraria*	Bajra	35,665	54,409	2.53	37	57,950
Western Dry	Non- arable land	Silvipasture	10	*Ziziphus nummularia*	Cenchrusciliaris	37,131	19,810	1.53	34	26,315
The Islands	Rainfed	Agrisilviculture	6	*Cocos nucifera*	Paddy	37,452	6,325	1.17	32	11,143

Source: Planning Commission (2001) Report of the Task Force on Greening India for Livelihood Security and Sustainable Development, Government of India, New Delhi. Newaj et al. (2016)

Note: i) NPV: Net present value is determined by discounting all revenues and costs to the present; ii) B/C ratio: benefit/cost ratio is calculated by dividing the sum of discounted revenues by the sum of discounted costs; iii) FIRR: Financial internal rate of return is a discount rate at which an investment's NPV equal zero; iv) AEV: Annual equivalent value is an indicator that expresses NPV in annual equivalents distributed equally over the years of the lifespan of the investment; and v) LEV: Land expectation value is interpreted as the maximum amount of money a landowner can pay for the land and still earn the minimum acceptable rate of return of an agroforestry investments.

Government initiatives for promoting agroforestry

The National Forest Policy 1952 & 1988, National Agriculture Policy 2000, Task Force on Greening India 2001, National Bamboo Mission 2006, National Policy on Farmers 2007 and Green India Mission, recognized agroforestry as an effective program for efficient nutrient cycling, enhancement of organic matter in the soil for sustainable agriculture and for improving tree cover. The Government of India (GoI) has launched several schemes/projects like the National Bamboo Mission (NBM) under the Ministry of Agriculture, GoI, Rashtriya Krishi Vikas Yojana (RKVY), National Horticulture Mission (NHM), National Biofuel Policy, Task Force on Greening India 2001, National Policy on Farmers 2007 and Green India Mission, 2010 for promoting agroforestry in the country. The National Bank for Agriculture and Rural Development (NABARD) also provides financial and banking institutional support for agroforestry.

The National Agroforestry Policy of India is a comprehensive policy framework designed to improve agricultural livelihoods by maximizing agricultural productivity for mitigating climate change (NAP, 2014). The Department of Agriculture Cooperation & Farmers Welfare (DAC & FW) under the Ministry of Agriculture and Farmers Welfare (MOA & FW) is the nodal Ministry for implementing agroforestry policies and initiatives. Based on the recommendations of the National Agroforestry Policy (NAP), a dedicated Sub-Mission on Agroforestry (SMAF) was established since 2016-17 by the Government of India under the framework of the National Mission for Sustainable Agriculture (NMSA) intending to expand the tree coverage on farmland in complementary with agricultural crops (https://nmsa.dac.gov.in/pdfdoc/Agroforestory_Guidelines_English.pdf). The SMAF is focused to achieve increased tree cover to enhance carbon sequestration, enrichment of soil organic matter, availability of quality planting material, improved livelihood and productivity enhancement of crop and cropping systems, development of information system etc. Presently the scheme or mission is being implemented by more than 25 states and 2 Union territories.

The mission interventions have main objectives to expand coverage under Trees Plantation suited to local agro-climatic condition, growing trees on farmlands, Peripheral and Boundary Plantations and Promoting various Agroforestry practices/models suited to different agro-ecological regions and land use conditions that will enhance livelihood resources by creating additional income opportunities for farmers; Bio-diversity/Environmental protection and supplement to livestock farming; Help facilitate adaptation and mitigation efforts in climate change and Productive use of waste and degraded lands. Under these missions, five major interventions such as nursery development for quality planting material (QPM) plantation on farm boundaries as peripheral and boundary plantation, low-density plantation, high-density block plantation, capacity building and trainings and demonstration of agroforestry models are being promoted with funding patter of

60:40 between Centre and State Govt. for all States excepting NE & Hilly states, where it would be 90:10 and 100% in case of UTs & National Level Agencies. Under the Mission, multipurpose tree species with short, medium, and long term returns are encouraged, so that farmers may get additional income at regular intervals.

Based on Arun Kumar Bansal Committee, 2011, GoI has exempted important multipurpose tree species from all restrictions related to harvesting, transportation and marketing grown under agroforestry systems (Chavan et al. 2015; Handa et al. 2017). Based on report, table 5 provides compiled information on regulated and restricted tree species existing in different states. Adding sugar to this, Government of India cleared an ordinace amending Indian Forest Act, which omits bamboo grown in non-forest areas from the definition of trees (as mentioned in the section 2(7) of Indian Forest Act 1927 after 90 years) thereby exempting bamboo grown in non-forest areas from the requirement of felling/transit permit. Such moves required on more species to encourage growth of small and medium industries in the villages and smaller towns and reduce dependence on imports. In another significant move, the Corporate Social Responsibility (CSR) laws of India were modified in 2014, and agroforestry was included as a legitimate CSR activity. As of now, 27 states have ratified the recommendations of the NAP.

Table 5. List of exempted and restricted tree species from different states of India

State	Exempted (No Permission)	Restricted species (Permission required)
Andhra Pradesh	*Mangifera indica, Casuarina equisetifolia, Acacia nilotica, Syzygium cuminii, Psidium guajava, Azadirachta indica, Anacardium occidentale, Cocos nucifera, Ficus religiosa*, Eucalyptus and Subabul	*Pterocarpus santalinus, Tectona grandis* and *Santalum album*
Bihar	Poplar, Eucalyptus, *Anthocephalus cadamba, Mangifera indica, Bombax ceiba, Gmelina arborea*, Lichi, Phoenix and Bamboo except *Dendrocalamus strictus*.	*Tectona grandis*
Gujarat	Eucalyptus, *Casuarina equisetifolia* and *Prosopis juliflora*	*Tectona grandis, Mangifera indica Madhuca latifolia, Acacia catechu Dalbergia latifolia, Santalum album Madhuca indica, Gmelina arborea and Anogeissus pendula*
Jharkhand	*Acacia nilotica, subabul, Bamboo, Cane, Shorea robusta, Tectona grandis, Pterocarpus marsupium, Boswellia serrata, Terminalia elliptica*	*Mangifera indica, Madhuca indica Gmelina arborea, Dalbergia sissoo Artocarpus heterophyllus, Syzygium cumini*

State	Exempted (No Permission)	Restricted species (Permission required)
Karnataka	*Eucalyptus* Casuarina, Subabul, Rubber, Coconut, Arecanut, orange, Erythrina, Gliricidia, Sesbania, Silver oak.	*Dalbergia latifolia, Santalum album Acacia catechu*
Kerala	61 species permitted. Some of the important are *Pongamia pinnata, Hevea brazilensis, Ailanthus excelsa, Mangifera indica, Garcinia cambogia*, and *Terminala* spp.	*Santalum album, Tectona grandis Dalbergia latifolia, Xylia xylocarpa Hopea parviflora, Michelia chempaca Grewia tiliaefolia, Cedrus toon*
Maharashtra	*Acacia nilotica, Leucaena leucocephala*, Prosopis, Eucalyptus, Moringa, Phoenix, Sapota, *Acacia auriculiformis* and Poplar	*Terminalia chebula, Tectona grandis Madhuca latifolia, Tamarindus indica, Mangifera indica, Artocarpus integrifolia Acacia catechu, Santalum album, Pterocarpus marsupium, Adina cardifolia, Terminalia tomentosa Hardwckia binata Syzigium cumini Mangrove*
Madhya Pradesh	*Eucalyptus*, Casuarina, Subabul, Poplar, *Pithocellobium dulce*, Prosopis	*Acacia nilotica, Albizia lebbeck Azadirachta indica, Zizyphus mauritiana, Butea monosperma Syzygium cumini, Acacia leucophloea* Bamboo
Odisha	*Bambusa nutan, Bambusa vulgaris, Bambusa tulda, Samania saman, eucalyptus Hybrid, Acacia auriculiformis, Cassia siamea, Casuarina equisetifolia*, Silver oak	*Mangifera indica, Artocarpus heterophyllus, Schleichera oleosa Madhuca indica, Tamarindus indica Santalum album*
Punjab & Haryana	*Eucalyptus*, Poplar and *Melia composita*	*Acacia catechu, Emblica officinalis* and bamboo
Tamil Nadu	*Grevellia robusta, Eucalyptus* spp, *Casuarina equsetifolia, Leucaena leucocephala*	*Santalum album, Diospyros melanoxylon, Pterocarpus santalinus Dalbergia latifolia and Tectona grandis*
Uttar Pradesh	Poplars, *Eucalyptus*, Subabul., Casuarina, *Ailanthus* spp*; Gmelina arborea; Grevillea robusta, Morus alba, Anthocephalus cadamba, Melia composita, Acacia spp., Albizzia* spp,, *Borassus flabeliformis, Butea monosperma, Tamarindus indica, Grewia oppositifolia*	*Mangifera indica, Madhuca indica Ficus bengalensis, Azadirachta indica and Carpinus viminae*
West Bengal	No tree exempted	*Acacia catechu, Bombax ceiba Dalbergia sissoo, Diospyrous melanoxylon, Gmelina arborea Madhuca indica, Michelia champaca Shorea robusta, Swietenia mahogany* Teak and Mangroves

Adapted from Chavan et al. 2015

Price fluctuation of wood has been observed in past due to which farmers loose their interest in growing trees on their farm. For example, in last decade, in 2012-13 the poplar wood received highest prices but later years extreme reductions was reported due to increased supply and middlemen influence (Singh *et al.* 2017). To avoid such situation, the Government of India has announced MSP prices for ten of the minor forest products like tamarind, mahua seed, lac, sal leaves, chirongi, honey, gum karaya etc. which can help the farmers for getting proper returns from the tree plantations. In Tamil Nadu, the industrial agroforestry model has assured to famers that *Melia dubia* plantation could possible to get INR 8500/- per tonne as MSP and other tree species varies from 3000 to 8500/tonne for dendro-biomass (Parthiban and Cinthia, 2020). The contract farming experience of big guns of paper industries like ITC Bhadrachalam (Andhra Pradesh), BILT, Ballarsha (Maharashtra), TNPL, Karur (Tamil Nadu) and JK papers, Songarh (Gujarat) emerged successful model of industrial agroforestry where farmers are supported with quality planting stock and improved package of practices for assuring greater returns from the plantations. Such MSP is also required for other agroforestry species so that farmers can get assured returns for their produce.

Recently National Bamboo Mission has provided adequate support to farmers and micro scale industries towards value addition of bamboo products for enhancing the income of bamboo growers. In Tamil Nadu, a “Consortium of Industrial Agroforestry (CIAF)” to create a value chain in agroforestry has been established. The consortium has 208 members, including 41 farmers practicing agroforestry, 28 entrepreneurs, 3 financial institutions, 12 nurseries, 6 line departments and 29 small and medium–scale industries from Tamil Nadu, Kerala, Karnataka, Telangana, Andhra Pradesh, and Pondicherry. They also prioritized 30 tree species suitable for agroforestry and efforts are being taken to developing high yielding short-rotation clones/varieties (HYSR), institutions for Production Qulaity planting material, harvesting and marketing institutions, desgning of multifunctional agroforestry models (MFAM), development of price supportive system, value addition technologies, involving credit and insurance support and creating sound database. Till today, industrial species such as melia, eucalyptus, casuarina, teak, kadamba, ailanthus, acrocarpus and leucaena are being grown on an area of 75,000 ha area in Tamil Nadu through CIAF. The intiative of consortium have created 300 mandays/ha of employment and increased productivity of wood plantation from 10 m^3 ha^{-1} yr^{1} to 25 m^3 ha^{-1} yr^{1} (Handa et al. 2020 and Parthiban and Cinthia, 2020). In this chain wood based industries are are coming forward with assured marketing and minimum support price. MoU have been signed with tree growers to encourage them to adopt agroforestry. On similar lines, Ministry of Environment, Forests and Climate Change is also looking to encourage investments through budgetary allocations or public private partnership for increasing productivity of degraded forests

For overcoming natural disasters such as floods, cyclones, fire and various human causes which results in huge losses of agroforestry to the farmers, insurance schemes have been launched. These tree insurance schemes cushion the shock of crop losses by providing farmers with a minimum amount of protection and guard them against disastrous effect of losses due to natural hazards or low market prices (Uthappa et al. 2015b). At present, three insurance companies namely United India Insurance, Agriculture insurance company of India and oriental insurance company are providing the facility of tree insurance. The species alongwith risks covered, premium amount and sum assured varies with different companies is given in Table 6.

Table 6. Different Tree Insurance policies in India

S. No.	Trees covered	Premium	Sum assured	Policy Name	Insurance Company	Period of insurance
1	Casuarina, *Eucalyptus, Melia dubia* Ailanthus, Gmelina, Leucaena and Dalbergia sissoo	1.25 per cent of the input cost	Input cost	Agroforestry Plantation Insurance'	United India Insurance, Chennai,	Annual
2	Eucalytptus, Poplar, Subabul and Casuarina	Depends on species, risks, geographical location etc	Equivalent to input cost and extended upto 125-150% of input cost	Pulpwood tree insurance	Agriculture insurance company of India	Annual
3	Jatropha, Karanja, Neem, Mahua, *Callophyllum & Simarouba*	Depends on species, risks, geographical location etc	Equivalent to input cost and extended upto 125-150% of input cost	Bio-Fuel Tree /Plant Insurance	Agriculture insurance company of India	Annual
4	Rubber, *Eucalyptus*, Poplar, Teak, Arecanut, Cocoa,citrus, chickoo, pomgranate	1.25% of Sum Insured	Sum Insured shall be based on the cost of cultivation	Plantation / Horticulture Insurance	The Oriental Insurance Company Limited	Annual
5	Rubber	Depends on input cost and stage of crop	Based on input cost and yield deduction	Rubber plantation insurance	Agriculture insurance company of India	Immature: 7 years Mature :8 to 25 years

Source: Uthappa et al. (2015b)

Stratergies and challenges for promoting agroforestry

Significant progress has been made in the agroforestry sector in different agroclimatic region. Various agroforestry models have been developed for enhancing productivity and providing ecological and livelihood security. The development of quality planting material have resulted in significant improvement in productivity and profitability. The government of India has also taken many initiatives for promoting agroforestry in the country. However, the following issues still need to be addressed for promoting agroforestry on large scale:

- Despite the high economic potential and its huge utility agroforestry has not been able to find its desired place in the rural economy due to lack of awareness. There is a need of creating awareness on the commercial viability and profitability of different agroforestry species. Region specific agroforestry models for small, marginal and big farmers are needed to be developed.
- Long term impact of identified agroforestry systems needs to be assessed in terms of employment generation, livelihood support, conservation of land and water, quality of produce, increase in biomass productivity and micro-climate improvement.
- Methodologies for quantification and payment of ecosystem services (PES) in agroforestry system needs to be framed. This has become more important in Himalayn region where there is ban on green felling. Besides, link of consumers of environmental services with suppliers needs to be framed so that ultimate benefits of these PES can be passed to the stakeholders involved.
- Despite so many efforts, still no clear cut policy exists for placing the farm-grown trees in agriculture or forestry sector. As a result, trees are not covered under Farmers (Endowment and Protection) Agreement on Price Assurance and Farm Services Act, 2020 and the Farmers Produce Trade and Commerce (Promotion and Facilitation) Act 2020 .
- Simplified regulations urgently formulated to explore the full potential of agroforestry. As still many important trees like teak requires permission, which hinders farmers for adopting. However, the list varies with state as a result most of the farmers do not know the trees exempted in their states. Therefore, there is an urgent need to bringing uniformity in tree and land tenure laws across the country in interests of multiple stakeholders.
- District level Nursery accreditation committee should be formulated for approval. Recently many private nurseries and industries come-up with tree planting scheme and selling the seedling (low quality) with higher rates. For example tissue cultured teak sold at a rate of INR 100-150 per seedling with assurance to yield 20-25 cft wood in a span of 8-10 years. Similarly,

Mahagony, Melia, red sanders and sandalwood seedlings are sold at INR 80-200 with promise of buyback and higher returns. NAP 2014 and SMAF should provide guidelines and strict legal aspects against such scam.

- Statutory ceiling limits on agricultural land holdings for wastelands should be liberalized. Long term finance, fiscal incentives, and tax benefits should be provided to promote technology-based reforestation and farm forestry projects. For encouraging the growth of agroforestry sector in the country, import duties also need to be rationalized.
- There is a strong need for farmer-to-farmer extension and capacity building for farmers and extension staff in timber tree selection, management and marketing.
- Under SMAF mission, agroforestry models are implemented on farmers field by providing subsidy. The experience or ground realities tell that market development is a crucial part to flourish the agroforestry like Yamunangar, Rudrapur, Saharanpur, Khammam, Chandrapur and some parts of Tamil Nadu. Therefore there is an urgent need for the development of wood-based industrial hubs under "Skill India" or "Make in India" and "*Atmanirbhar Bharat*". Also there is a need to increase import duty and reduce Goods and Services Tax on wood and wood products.
- Farmers should get a facility of subsidy under a single-window program. At presently, farmers have to apply separately for planting, drip irrigation, and farm pond. Therefore single window for agroforestry subsidy will increase the adoption rate among farmers.
- Though insurance schemes and credit cover mechanisms have been provided the farmers are reluctant to adopt these schemes due to a lack of clarity. There is an urgent need to pass the information in a clear cut manner so that benefits can be passed on to the farmers
- All KVK'S may have agroforestry subject matter specialist or where agroforestry is prominent; Krishivaniki Vigyan Kendra (KVVK'S) can be established for smooth dissemination of successful agroforestry models (or) it would be better to strengthen the existing KVKs, Regional Agri Research Stations of ICAR, ICFRE and SAUs, R&D Extension Institutions operating under various bodies of the State Govts.
- There is an urgent need to revive traditional agroforestry systems like the Oraons of Rajasthan, Kangeyam system of TN, Home-gardens, and Cardamom Hill Reserves of Kerala.
- There is many community-based agroforestry (predominantly silvi-pastoral) systems that are breaking down as community rights are being extinguished as a result these community property resources (CPRs) are divided and distributed among individuals resulting into change in land use pattern and

community access. The issue therefore need to be addressed for the benfits of stakeholders.

- Though agroforestry contributed significantly towards enhancing tree cover, the main weakness had been the absence of a linkage between the producers (farmers) and the industry and a strong value addition component. Emphasis is required for product development and value addition including primary processing and treatment; micro, small & medium enterprises as well as high value products; markets, and skill development, thus ensuring a complete value chain for growth of the agroforestry sector to boost wood based industry which would also have a ripple effect on rural economy.
- While, there are many schemes dealing with tree planting / agroforestry, there is an absence of a dedicated and focused Laws, Acts, Rules and Regulations and lack of an institutional mechanism for coordination and convergence among the schemes / ministries to pursue agroforestry in as thematic manner.

Conclusion

Agroforestry research in past has lead to development of many agroforestry models/technologies which are highly profitable and sustainable. These models have enough potential to double the farmers income and address the various issues of ecological and livelihood security. Dissemination of these agroforestry models to farmers field however, is not very encouraging except for few commercial agroforestry models (Poplar, Eucalypts, Subabul, Casuarina) where quality planting material with assured market supply was available. Therefore there is need to develop strong linkage between research organizations and line departments. Agroforestry needs to be made an economically viable option through policy and institutional interventions such as R&D, development and provision of quality planting material and nursery accreditation, price support instruments and mechanisms, transport and marketing infrastructure in remote areas, monitoring, certification and review processes.There is need of an open policy in which the regulations are relaxed. States will have to work with the Centre in tandem to ensure implementation of the reforms.

References

Ajit, S.K. Dhyani, A.K. Handa, R. Newaj, S.B. Chavan, B. Alam, et al. 2017. Estimating carbon sequestration potential of existing agroforestry systems in India. Agroforestry Systems. 90(4):1101–1118. https ://doi.org/10.1007/s1045 7-016-9986-z.

Alam, M., A. Olivier, A. Paquette, J. Dupras, J.P. Reveret and C. Messier. 2014. A general framework for the quantification and valuation of ecosystem services of tree-based intercropping systems. Agroforestry Systems. 88:679-691. https://doi.org/10.1007/s10457-014-9681-x

Bene, J.G., H.W. Beail and A. Cote. 1977.Trees, food and people.IDRC, Ottava.

Chand, R. 2017. Doubling Farmers' Income: rationale, strategy, prospects and action plan. Niti Policy Paper no. 1/2017.

CAFRI. 2015. Vision 2050. Indian Council of Agricultural Research, New Delhi, 36p.

Chaturvedi, O.P., A.K., Sikka, A.K. Handa and C.K. Bajpai.2016. Agroforestry Technologies for different Agro-climatic zones of the Country. All India Coordinated Research project on Agroforestry, ICAR-Central Agroforestry Research Institute, Jhansi (UP).

Chaturvedi, O.P., R.. Kaushal, J.M.S. Tomar, A.K. Prandiyal and P. Panwar. 2014 Agroforestry for wasteland rehabilitation: Mined, ravine, and degraded watershed areas. In: Agroforestry systems in India: Livelihood security and ecosystem services, Advances in Agroforestry vol. 10 (eds. J.C. Dagar, A.K. Singh and A. Arunachalam). Springer, Dordrecht. pp 233-272.

Chavan, S.B., A. Keerthika, S.K. Dhyani, A.K. Handa, R. Newaj and K. Rajarajan. 2015. National Agroforestry Policy in India: a low hanging fruit. Current Science.108:1826-1834.

Chavan, S.B., A. Keerthika, S.S. Bhat, A.K. Handa, K. Rajarajan and S. Ahmad. 2020. Poplar (*Populus deltoides*) in Jammu and Kashmir, India: facts and fiction. Current Science. 119(6):910-911.

Chavan, S.B., A.R. Uthappa, K.B. Sridhar, A. Keerthika, A.K. Handa, R. Newaj, N. Kumar, D. Kumar and O.P. Chaturvedi. 2016. Trees for Life: Creating Sustainable Livelihood in Bundelkhand Region of Central India. Current Science, 11:994-1002.

Chavan, S.B., R. Newaj, R.H. Rizvi,. et al. (2020). Reduction of global warming potential vis-à-vis greenhouse gases through traditional agroforestry systems in Rajasthan, India. Environ Dev Sustain https://doi.org/10.1007/s10668-020-00788-w

Dagar, J.C. 2012. Utilization of degraded lands/ habitats and poor quality water for livelihood security and mitigating climate change. Indian Journal of Agroforetsry. 14(1):1-16.

Dhiman, R.C. 2012. Diagnosis of intercrops in poplar based agroforestry in India. Indian Forester. 138 (7): 600-609.

Dhiman, R.C. 2013. Status and impact of commercial agroforestry in India. Ind. J. Agrofor. 15:55–67.

Dhyani, S.K., K. Kareemulla, Ajit and A.K. Handa. 2009. Agroforestry potential and scope fordevelopment across agro-climatic zones in India. Indian J For. 32:181–190.

Dhyani, S.K., R. Newaj, B. Alam and I. Dev. 2015. Agroforestry: Present status and Way Forward. Biotech Books, New Delhi.

Dixon, R.K. 1995.. Agroforestry systems: sources of sinks of greenhouse gases? Agroforestry systems. 31(2): 99-116.

FSI, India State of Forest Report. 2013. Forest Survey of India, (Ministry of Environment &Forests), Dehradun, India.

Goswami, S, K.S. Verma, R. Kaushal. 2014. Biomass and carbon sequestration in different agroforestry systems of a Western Himalayan watershed. Biol Agric Hortic. 30(2):88-96.

Handa, A.K., O.P. Toky, S.K. Dhyani, S.B. Chavan. 2016. Innovative agroforestry for livelihood security in India. World Agriculture. 6(2):7-16.

Handa, A.K., S.B. Chavan, A.R. Uthappa and O.P. Chaturvedi. 2017. Agroforestry as a solution for India's timber deficit. In: Status and Recent Researches on Important Timber Trees o f India, C. Buvaneswaran, S. Senthilkumar, S. Saravanan, P. Kathirvel, S. Murugesan and R.S. Prashanth eds. ICFRE Dehra Dun, pp.110-127.

Handa, A.K., S.B. Chavan, V. Kumar, R. Vishnu, S. Suresh Ramanan, R.K. Tewari, A. Arunachalam, S. Bhaskar, S.K. Chaudhari and T. Mohapatra. 2020. Agroforestry for Income Enhancement, Climate Resilience and Ecosystem Services. Indian Council of Agricultural Research, New Delhi.

Handa, A.K., S.K. Dhyani, and Uma. 2015. Three Decades of Agroforestry Research in India: Retrospection for way forward. Agricultural Research Journal. 52:1-10.

Kaushal R., J. Jayaparkash, D. Mandal, A. Kumar, N.M. Alam, J.M.S.Tomar, H. Mehta, O.P. Chaturvedi. 2018.Canopy management practices in mulberry: Impact on fine and coarse roots. Agroforestry Systems. 93(2): 545-556.

Kaushal, R., Tewari, S., Banik, R.L., Thapliyal, S. D.. Singh, I., Reza, S., Durai, J. (2020). Root distribution and soil properties under 12-year old sympodial bamboo plantation in Central Himalayan Tarai Region, India. Agroforest. Syst. 94:917-932

Milder, J.C., S.J. Scherr, C. Bracer. 2010. Trends and future potential of payment for ecosystem services to alleviate rural poverty in developing countries. Ecology and Society. 15(2): 4.

Minang, P.A., F. Bernard, M.V. Noordwijk and E. Kahurani 2011. Agroforestry in REDD+: Opportunities and Challenges. Nairobi, Kenya: ASB Policy.

Nair, P.K.R., B.M. Kumar, and V.D. Nair. 2009. Agroforestry as a strategy for carbon sequestration. Journal of Plant Nutrition and Soil Science. 172(1):10-23.

National Agroforestry Policy, Department of Agriculture and Cooperation, Ministry of Agriculture, Government of India, New Delhi, 2014 pp 1-13.

Newaj, R., Chavan, S. B., and Prasad, R. 2015. Climate-smart agriculture with special reference to agroforestry. Indian J. of Agroforestry, 17(1), 96-108.

Newaj, R., S.B. Chavan, B. Alam and S.K. Dhyani. 2016. Biomass and carbon storage in trees grown under different agroforestry systems in semi arid region of central India. Indian Forester. 142 (7):642-648

Parthiban, K.T. and C. Cinthia Fernandaz. 2020. Contract plywood farming: A fillip to agroforestry promotion in India. Indian Journal of Agroforestry. 22 (2):1-6

Planning Commission. 2001. Report of the Task Force on Greening India for Livelihood Security and Sustainable Development. Planning Commission, GoI, New Delhi, 254 p.

Ram-Newaj, Chavan, S.B., Kumar, D., Uthappa, A.R., Singh, M. and Sridhar, K.B. 2016. Agroforestry research in India: relevance in livelihood and climate change. In: Agroforestry Rresearch Developments (eds. J.C. Dagar and J.C. Tewari). Nova Publishers, New York. pp. 327-359.

Sharda, V.N., A.K. Sikka, G.P. Juyal, 2012. Participatory Integrated Watershed Management – A Field Manual. CSWCR&TI, Dehradun: pp 366.

Singh, M., K.B. Sridhar, D. Kumar, A.R. Uthappa, R.P. Dwivedi, I. Dev and R.H. Rizvi. 2017. Doubling farmers' income through agroforestry in north-western India : A policy perspective. Indian Journal. of Agroforestry, 19: 90-95.

Tomar, O.S., R.K. Gupta, and J.C. Dagar. 1998. Afforestation techniques and evaluation of different tree species for waterlogged saline soils in semiarid tropics. Arid Soil Res. Rehab. 12(4):301-316.

UNFCCC. 2015 Adoption of the Paris Agreement.

Uthappa, A.R., S.B. Chavan and A.K. Handa. 2015b. Tree Insurance: A new initiative in India. Krishiseva. https://www.krishisewa.com/articles/miscellaneous/630-tree-insurance.html .

Uthappa, A. R., Chavan, S. B., Dhyani, S. K., Handa, A. K., and Newaj, R. 2015a. Trees for soil health and. Indian Farming, 65(3), 02-05.

Verma, A., R. Kaushal, N.M. Alam, H. Mehta, O.P. Chaturvedi, D. Mandal, J.M.S. Tomar, A.C. Rathore and C. Singh. 2014. Predictive models for biomass and carbon stocks estimation in *Grewia optiva* on degraded lands in western Himalaya. Agroforestry Sysytems. 88:895-905.

World Bank. 2004. Sustaining forests: a development strategy. ISBN 0-8213-5755-7 (The World Bank, Wash., D.C.,

Zomer, R.J., N. Henry, J. Xu, A. Antje, B. Deborah, T. Antonio, M.V. Noordwijk and W. Mingcheng. 2016. Global Tree Cover and Biomass Carbon on Agricultural Land: The contribution of agroforestry to global and national carbon budgets. Scientific Report, 6: 29987; doi: 10.1038/srep29987.

20

Innovations on Hill Agriculture Research and Development

Anup Das[1], Lopa Mudra Sahoo[1], Subhra Saikat Roy[2], Jayanta Layek[3] Raghavendra Singh[4], G.S. Yadav[5], Saurav Saha[6], M.A. Ansari[2], Vinay Singh[1] Subhash Babu[5], K.P. Mohapatra[7] and B.K. Kandpal[3]

[1]*ICAR-Research Complex for NEH Region, Tripura Centre, Lembucherra Tripura*
[2]*ICAR-Research Complex for NEH Region, Manipur Centre, Lamphelpat Manipur*
[3]*ICAR-Research Complex for NEH Region, Umiam, Meghalaya*
[4]*ICAR-Research Complex for NEH Region, Sikkim Centre, Tadong, Sikkim*
[5]*ICAR-Indian Agricultural Research Institute (IARI), New Delhi*
[6]*ICAR-Research Complex for NEH Region, Mizoram Centre, Kolasib, Mizoram*
[7]*ICAR-National Bureau of Plant Genetic Resources (NBPGR), New Delhi*

Introduction

Majority of the India's population (about 70%) depend directly and indirectly on agriculture for a living, and thus, the sector must attain self-reliance for their future survival. The North Eastern Region (NER) of India comprising the states of Arunachal Pradesh, Assam, Manipur, Meghalaya, Mizoram, Nagaland, Tripura and Sikkim covers a total area of 26.2 million hectare (7.97% of the India's total geographical area). Of this area, hill and mountain region accounts for about 70% area that excludes Assam plains. The region is characterized by unique geophysical socio-cultural and environmental features. Drought and floods are the two miseries for the farmers of North Eastern Region of India (NER). The poorest people are likely to be the hardest hit by impacts of climatic factors and its change because they rely heavily on climate-sensitive sectors such as rainfed agriculture and fisheries. They are also less capable to respond due to limited human, institutional and financial capacity and have very limited ability to cope with climate impacts and adaptation options in agriculture sector to a changing hazard burden (Ngachan and Das, 2018). The NER of India is endowed with enormous water resources potential that accounts for about 34% of the country's total water wealth. In spite of the abundant water resources, the ratio of percent irrigated area to net sown area varies from 5% in Assam to 28% in Arunachal Pradesh with an average of 10.6 % in NER. This is less than one-fourth of the national average of net irrigated area which currently stands at 43.2% (Sharma et al., 2010). The water scenario in the region is skewed and is characterized by plenty and scanty syndrome, i.e. excess water during rainy season and severe scarcity during post- and pre- monsoon season. There is wide variation of rainfall in space and time. It is projected that by 2021, an additional 15 million population will be added to the current 45 million in the region (Choudhury et al., 2012). Hence, adequate planning and management of these bountiful but most vital resources are required for overall sustainability of agriculture in NER.

The extent of land degradation in NER varies from state to state. Entire area under shifting cultivation is classified under lands prone to degradation. About 0.76 M ha is under shifting cultivation (NRSC, 2010). The percent degraded land in NER ranges from 28.2% in Assam to as high as 89.2% in Mizoram (Das et al., 2020). The major reasons of degradation can be clubbed together as over exploitation of forest for fuel, timber and fodder; soil acidity/Aluminium toxicity; deforestation; shifting cultivation/biomass burning; improper land use practices; mining activities; soil erosion/soil carbon loss; nutrient mining and population explosion including livestock (Sharma et al., 2006; Das et al., 2011a.). Of various degradation types, soil erosion, reduced infiltration, acidification, nutrient leaching, burning of vegetation, decline in vegetative cover and biodiversity are important in the context of NER (Sharma and Sharma, 2004).

Organic agriculture by default is eco-friendly and sustainable with recycling of crop biomass, use of organic manure and adoption of indigenous technical knowledge that prevails in the NEHR of India since ages. As on 31st March 2020, the total area under organic certification (registered under National Programme for Organic Production) is 3.67 million hectare with a total volume of export of 6.389 lakh tonnes and foreign monetary realization of around INR 4,686 crore (APEDA, 2020). Being bestowed with favourable endowments of soil, agro-climate, agro and forest biodiversity, flora and fauna, and no to low usage of chemicals in different states, the NER has tremendous potential to grow organic foods and make the region a 'global organic hub'. The region has potential to produce 46 mt of organic manure to sustain organic farming. In addition to this, the region has a vast resource of weed biomass, green manure and litter falls through forest species. The region also has about 0.76 m ha area under shifting cultivation where no inorganic input and tillage has ever been used.

The richness in agro-biodiversity of this mountainous region warrants the great concern for its sustainable management and utilization of plant genetic resources of agri-horticultural crops. The region accounts for about 43% of the flora in India. The ICAR-National Bureau of Plant Genetic Resources, Regional Station at Shillong collected/augmented about 23989 crop germplasm (till 2010) which will be of immense use for further improvement of crop plants for food security and livelihood. The region is hot spot of bio-diversity having many endemic species which accounts germplasm collections for more than 50 bamboo species, 6000 rice, 14 banana, 17 citrus, 600 orchids and many medicinal and aromatic plants. The aquatic bodies of the region harbours a rich diversity of ichthyo fauna, 274 fish species constitute about 34% of the total fresh water fish species of the country. There is a chronic shortage (about 67%) of green as well as dry fodder in organized sector in hills of north east India in particular. As results the NER is still deficit in milk, meat and eggs.

Status of food security

The NER requires a total of 8.74 mt of food grains for a population of about 49 million. During the last decade, food grain production of this region increased by 26.11% (from 5496.20 thousand tonnes in triennium (TE) 2008 to 8744 thousand tonnes in TE 2016-18 with a compounded growth rate of 3.3% over the last ten years (RCM, 2019). Increased production and productivity of food grains is possible because of introduction of high yielding varieties, availability and timely supply of of quality seed materials and use of fertilizer and manures etc. Among the NE states, Tripura achieved highest productivity of paddy (2.82 tonnes ha^{-1}) and cropping intensity (186%). Similarly, in organic farming, Sikkim is leading among all other states especially in spice crops. The data on food grain production and requirement indicate that there is a deficit and surplus existing in different NE states varying from surplus of 29% in Arunachal Pradesh to deficiency of 41% in Mizoram. In rice and maize production, the region is almost self-sufficient.

By 2050 the NER will need about 14 mt of food grain which is almost 60% higher than the 2012 production. About 32% food grain deficiency is projected by 2050 in NER with present level of technology. Simulation study revealed that despite changes in average temperature , elevated CO_2 level up to 550 ppm can substantially improve the rice productivity under farmers practice (without N fertilizer) while at higher level of CO_2 concentration (700 ppm), intensive application of fertilizer-N (at 60-90 kg N ha^{-1}) can have synergistic effect on rainfed rice productivity. Decrease in temperature is the most detrimental in reducing the rice productivity since cardinal temperature for optimum rice growth at mid-altitude (Meghalaya) is already at lower threshold level (<23°C). However, application of N-fertilizer at of 90 kg ha^{-1} could be possible immediate adaptation strategies under changing climate in the subtropical climate of mid-altitude Meghalaya (Choudhury et al. 2013). Fruit production increased with an annual compound growth rate (CGR) of 7.6%. In case of vegetables there is a need of market complexities in remote area requiring special attention. Though fish, egg, milk and meat production have increased substantially as compared to production level in 2018 the region is still deficient in these products (Table 1).

Table 1. Food grain production and requirement (2018 population) scenario for NE States

Commodity	Production in Triennium (2016 to 2018) in '000 tons	Requirement in '000 tons as per 2018 population	Deficit/ Surplus in '000 tons	Per cent (%) Deficit/ Surplus
Rice	7213	7923	43	1
Maize	496	-	-	-
Pulses	238	920	-682	-74
Food grains	8056	8744	-688	-8
Oilseeds	380	1830	-1450	-79
Vegetables	6053	5520	533	9
Fruits	4523	1473	3050	67
Milk	1398	5520	-4122	-75
Eggs (lakh nos.)	10655	90748	-80093	-88
Fish	426	655	-229	-35
Meat	230	553	-323	-58

Source: RCM, 2019; Requirement : Rice @ 157 kg per head per annum, Pulses @ 18.25 kg per head per annum, Oilseed @ 36.5 kg per head per annum, Fruit @ 43.8 kg per head per annum, Vegetables @ 109 kg per head per annum, Fish @13 kg, milk @200 g per person per day, 150 eggs per person per annum, Meat @11 kg per person per year

A specific long-term strategy is needed with commodity wise plans that factor in comparative advantages, available market, production, productivity, consumption, subsidies and incentives. India cannot be an economic superpower without becoming self-reliant in agriculture sector. Hills must have to play a significant role in achieving self-reliant agriculture in India emphasizing followings strategic plans.

a. Strategic plan for food security in NER

- **Seed production strategy:** Seed is one of the major limitations for the steady progress of Agriculture in NER. Quality seeds of high yielding varieties of different prominent agricultural and horticultural crops are at uttermost importance. As majority of the NE states are rich in agro-biodiversity; indiscriminate introduction and popularization of the high yielding mainland crop varieties cause rapid genetic erosion. Therefore, balanced introduction of high yielding open pollinated varieties should be executed with the identified selection of high yielding lines of different horticultural crop. Multi-location institutional advanced variety trial may be executed for the release of the identified potential lines as state varieties will maintain the net ecological balance of the subsistence agro-ecosystem in North East India. For example in Mizoram, ICAR Kolasib has released a local selection of anthocyanin rich purple bean named 'Zorin bean' for the long term conservation of the selected lines (termed as MZFB 48) with subsequent rise in average bean productivity of both *jhum* (*kharif*) and winter (*rabi*) agriculture. Similarly, ICAR Research Complex for NEH Region, Umiam released composite maize variety RCM 1-61 which is acid tolerant and perform well under diverse ecosystems of the NEH Region. Recently ICAR Sikkim Centre also released 9 varieties of different crops (maize, rice, *Rajmash* and Soybean) suitable for organic farming. Village level seed bank generation of open pollinated varieties or commercial cultivation of such selected potential lines through seed village programme implementation will reduce the dependence on hybrids (not readily available in rural markets of North East India) and increase in average agro-ecosystem productivity of north eastern Agriculture. Seed multiplication of cross pollinated crops require due attention for maintaining isolation distance (~400 m for maize).

- **Target based programme on nutri- cereals**: Potential millets for NER (including cold tolerant lines for Sikkim and Arunachal Pradesh) and major nutrient management constraints that mask the available genetic yield potential under on-farm conditions need to be investigated and required skill development programme should be implemented for popularizing the millets.
- **Integrated farming system (IFS) for small holder** : Integrated farming with different topo-sequential cropping involving agri-horti-silvi-pastoral system or Agro-pastoral based land use system was found to be the most economical with effective soil and water conservation measures in northeast India.

Achieving *Atmanirbhar Bharat* and building resilience in farming for small and marginal farmers, to cope up during pandemic, unforeseen conditions like lockdown, the IFS comes for the rescue. For providing adequate food and nourishing security for sizable populace, efficiency upgradation may give a fundamental solution. This includes the selection of logical and scientific agronomic practices and advance technologies which guarantee an increase of the profitable limit of customary agrarian frameworks. To satisfy the fundamental needs of house hold including food (grain, pulses, oilseeds, milk, fruit, honey, egg, meat etc.), feed, fodder fiber etc., IFS is seen as a feasible option in contrast to conventional farming systems for food and wholesome security especially for small and marginal farmers with balance the farm income. National Mission for the Sustaining Himalayan Ecosystem (NMSHE) Task Force-6 venture has been begun with new mediation to improve the livelihood through IFS at Charkalak Autonomous District Council (ADC) village at Jampuijala, Sepahijalla region. Agriculture especially rice based farming system is the major farming system of the villages in Jampuijala sub-district under Sepahijala district in Tripura, but the productivity of this system is low. About 85% of the population in this village is resource poor and their main employment is depend agriculture. Therefore, for enhancing the productivity to get nutritional security and income generation ICAR Tripura Centre, Lembucherra diversified the rice based system through integration of improved rice varieties like Gomati with improved methods of rice cultivation like system of rice intensification (SRI) and also to utilize the land near the households as kitchen garden. Apart from crop cultivation, fish cultivation, rearing of livestock like pigs and poultry can be secondary sources of income. For this, ICAR-NEH Tripura centre, conducted several training and demonstration on IFS to sensitize the farmers about the benefits they can derive by adopting this new technology to achieve *Atmanirbhar Bharat*. Farm as a unit is to be considered and planned for effective integration of the enterprises (Das et al., 2013).Economics of various IFS models adopted are presented in Table 2.

Table 2. Economic return of different Integrated Farming Systems at Charkalak ADC village Tripura (Das et al., 2013)

Integrated farming systems		Cost of cultivation (Rs)	Total benefit (Rs)	Net benefit (Rs)	B:C ratio
IFS-1	Rice- Fish- Kitchen garden	23,950	70,616	46,666	2.95
IFS-2	Rice- Pig- Kitchen garden	23,899	66,713	42,814.	2.79
IFS-3	Rice- Fish – Poultry	26,274	74,965	48,691	2.85
IFS-4	Rice – Vegetable	29,176	58,900	29,724	2.02
IFS-5	Rice - Pig - Fish - Kitchen garden	41,101	1,25,588.	84,487	3.06

- **Developing water harvesting technique and efficient water use**: Rain water harvesting (RWH) and its efficient use must be promoted to reduce pressure on ground water and increase irrigated areas. In north east India *jalkund* (Polylined, 30,000 L capacity) in hill tops and uplands (Saha et al., 2007.) and farm ponds in valleys and plains (Das et al., 2013) have been found a viable option for rain water conservation and creating multiple livelihood opportunities. Considering undulating topography and steep slopes, micro irrigation has maximum potential in the region. Thus, efficient methods like drip irrigation and sprinkler irrigation must be popularized for efficient water use and higher water productivity.
- **Soil fertility building strategy**: Customized fertilizers and micro nutrients will have to be given special emphasis. Efficient recycling of available resources, looking for alternative sources of phosphorus like phosphate rocks, promotion of biofertilizer etc. will substantially reduce important requirement of these two vital nutrient sources.
- **Promotion of high-tech automated farming**: Polyhouse cultivation, hydroponic and aeroponic system of crop production has emerged as commercially viable technology for maximizing the yield of high value crop per unit of area, time and input used and can play a vital role in generating employment and income in NER.

- **Promotion of specialized farming**: Poultry farming, commercial fish farming, ornamental fish, dairying, high tech nursery, agri-clinics has got very good opportunity in the NER. Location specific modules may be established and promoted through public-private-partnership (PPP) mode.
- **Promotion of maize as food-feed crop**: Maize is the second most important crop and occupying ~2.61 lakh hectare area with less than 2.0 t ha^{-1} productivity in North East states (Rakshit, 2019). Piggery and poultry farming is one of the major enterprises in farming system, where it is the most preferred meat among tribal farmers of NE region. It is also important to mention that most of the farmers facing problems of feed for their livestock. Since, maize is an essential ingredient for feeds of animal, piggery, poultry and dairy livestock farming, there is a need to address the farmers' requirement to and make the farmers to self-reliant in full fill the demand of food, feed and fodder. ICAR-NEH Region-ICAR-IIMR analysed the impact of introduced improved technologies to make the farmers self-reliant in fulfilling the requirement of food, feed and fodder as well as sustain their economic returns. It has been observed that farmers integrate the maize with their livestock (poultry/piggery/dairy) to full the requirement of not only food but also feed and fodder along with strengthening in value addition and market chain. The valley farmers integrate maize with dairy farming and hills farmers integrated with poultry and piggery farming. Earlier, farmers were not integrating livestock farming with their produce (main products and by products). Adoption of integrated maize based cropping system enhanced the maize productivity from 1.5 t ha^{-1} to 3.4 t ha^{-1} maize equivalent yield (MEY), net farm income by 132.6 to 266.8% and benefit cost ratio by 20.6 to 64.6% over without integration in foot hills area. From this technology, the beneficiary finds himself, self-reliant in meeting the needs of the family and feed for the livestock in integrated farming system. Use of QPM as poultry feed leads to early development of broilers, save energy and feed, and also the extra cost incurred on lysine and tryptophan fortification. The economics of some major individual integrated farming systems of Mizoram are presented in Table 3.

Table 3. Economic return of different Integrated Farming Systems at Kolasib, Mizoram

Integrated Farming Systems	Total cost of cultivation*	Gross return*	Net return*	Benefit cost ratio
Livestock – fish culture at ICAR research farm	23550	50340	26790	2.13
Paddy cum fish culture at Champhai	30150	104160	74010	3.45
Integrated Agri-Horti-Animals-Fish-Silvi Production system at Kawnpui	28717	86427	57710	3.01
Agri-Horti-Silvi-Pastoral system at Sethawn	23123	60790	37666	2.62

Integrated Farming Systems	Total cost of cultivation*	Gross return*	Net return*	Benefit cost ratio
Horti-Silvi-Pastoral system at Bairabi	11700	13252	1552	1.13
Grapes ginger (Horti-Horti system) at Champhai	24200	66944	42744	2.76

Note: *Expressed in INR ha^{-1} $year^{-1}$

- **Organic Farming:** The organic farming has emerged as an alternative system of farming which not only addresses the quality and sustainability concerns, but also ensures a profitable livelihood option. The state of Sikkimhas been declared completely organic after a decade of transition (conversion) period. Organic agriculture could contribute significantly to reduce the release of greenhouse gases (GHG) and to sequester carbon in soils and biomass towards assistance in building resilience and adaptation in farming systems. Thus, it will help in reducing the adverse effect of changing climate on agriculture. Realizing the potential of organic farming in the region, Ministry of Agriculture and Farmer Welfare, Govt. of India launched a central sector scheme entitled "Mission Organic Value Chain Development for North Eastern Region (MOVCDNER)" for development of certified organic production in a value chain mode to link growers with consumers. Among the states of NE India, Sikkim has achieved a remarkable distinction of converting its entire cultivable land (more than 75000 ha) under organic certification. The states like Meghalaya are also going to be organic in near future. Having huge biomass and high number of domestic animal, the region become self-sufficient in the desired production of quality manure for sustaining the soil fertility for organic production of crops. However, there is need of scientific package of practices for quality input preparation including vermi-compost, organic pesticides etc. to make the region self-reliant and sustainable crop production. Although majority areas remain organic by default, the farmers are not getting the premium price for their organic produce due to lack of certification. Therefore, there is urgent need of organic certification for the organic produce of the region for getting good remuneration and encouragement for farmers.

  Organic certification system is a quality assurance initiative, intended to assure quality, prevent fraud and promote commerce, based on set of standards and ethics. It is a process certification for producers of organic food and other organic plant products. Organic farming certification in India is governed by The National Programme for Organic Production (NPOP) under the Ministry of Commerce & Industry, GOI. Agriculture remains the major cornerstone for economic development of the NER of India by ensuring food security, alleviating poverty and conserving the vital natural resources of fragile ecosystem.

Any farmers or producer can choose a system for certification ether i) Third Party Certification System of Agricultural and Processed Foods Export Development Authority (APEDA) under Ministry of Commerce, Govt. of India or ii) Participatory Guarantee System (PGS) of National Centre on Organic Farming (NCOF), MoA & FW, GoI. While the Third part certification is pre-requisite for exporting any produce to other countries, PGS certification is only for domestic market. The PGS is an internationally applicable organic quality assurance system [like ISO 9000] implemented and controlled by the committed organic farmer-producers through active participation, along with the consumers, in the process based on verifiable trust. It is not an "inspection raj" certification system but, rather, one that is based on personal integrity and peer pressure. The "Local Group" of five or more organic farmers is the fulcrum of the self-regulatory support system of PGS. The quality assurance standards are harmonized by the PGS Organic Council, which permits the use of its PGS label on a product as a mark of quality. Before submission of application, it is important for the applicant or farmer to ensure that his/her farm conforms to the standard laid out by the National Programme for Organic Production (NPOP) for organic crop production. The guidelines for organic certification are laid down by NPOP. When organic planting materials are available they shall be used. However, the absence of certified organic planting materials (especially for remote area of NEHR), chemically untreated conventional materials is permissible to use. Recently Govt. of India through the Ministry of Development of North Eastern Region (DoNER) has initiated Mission Organic in NER with an outlay of 125 crores which will be implemented through State Government.

- **Promoting Secondary and speciality agriculture**: Scented/aromatic products/products with Geographical Information (GI) Tags, niche items etc. must be promoted for higher income through marketing in metros and international markets. Some of such items are cardamom, aromatic rice (*joha* of Assam, *kalikhasa* of Tripura, purple rice of Manipur etc), ginger turmeric, pineapple and many others.
- **Smart farming**: Strengthening weather monitoring through the establishment surface weather stations in each districts, promotion of contract farming community nursery, seed and fodder storage, district contingency planning developing soil quality index (SQI) maps and updating them on 5 year basis must be undertaken for climate resilient agriculture (Roy et al., 2018).
- **Soil management strategy**: The relationship between soil erosion, nutrient runoff losses, organic matter depletion, and beneficial effects of conservation and management practices occur simultaneously. Maintenance of optimum fertility level with application of regular doses of manure and fertilizers

adjusted to soil pH 5.5 to eliminate the aluminum toxicity is recommended. Multiple cropping, inter-cropping, relay cropping, inclusion of legumes in rotation, strip cropping etc. and adoption of integrated nutrient management to ensure better crop productivity, besides maintaining soil fertility.

Acid soils occupy nearly 85% of geographical area in the eastern Himalayas. Acidic soils below 5.5 pH occupy around 16.2 m ha. The productivity of cereals in such acid soil is hardly about 1.5 -2.0 t ha^{-1}. Furrow application of lime 250-500 kg ha^{-1} of high quality uniform grades /sizes at furrows every year can optimize the yields of crops in acid soils of NEH Region. Use of acid tolerant varieties and application of organic manure also improves productivity of such soils.

- **Prevention of loss of biodiversity**. Decline in biodiversity reduces productivity, environment quality, and human wellbeing. Indeed biodiversity plays a critical role in ecosystem productivity and ecological stability (Hautier et al., 2015), which must be enhanced. Thus, re-wilding of marginal/depleted hill lands would be an essential pre-requisite to reversing the degradation trends (Lal, 2015).
- **Land configuration for increasing cropping intensity**: In permanent raised and sunken beds, the raised area is used for cultivation of vegetables and other remunerative crops, whereas sunken area is used for double cropping of rice (Das et al., 2015). The land utilization is 100 per cent in these systems. For temporary system, after harvesting *kharif* rice, temporary raised beds are constructed to cultivate vegetables. About 1 lakh hectare of marshy land available in the NEH which could be benefited from such land configurations.
- **Integrated mountain development:** It is a process whereby optimum use of mountain resources can be sustained over several generations in the context of available technology. It includes preservation of gene pool, augmentation of the well-being of the local people, controlled and acceptable downstream effects. Major efforts are needed to diversify the mountain economy and living standard of people with emphasis on hill environmental protection and sustainable development. The validation of indigenous knowledge based on latest technical know-how by inter-generations wisdom of local inhabitants of the region through native means to suit their conditions.
- **Utilization of rice fallows:** In north east India about 2 million ha area is available as rice and maize fallow areas. At least 30% of this areas can be brought under oilseeds and pulses with adoption of adequate technologies like suitable varieties, crop and soil management practices. Even if 30% of this area (about 6 lakh ha) is brought under oilseeds, it will produce about 420,000 tonnes of oilseeds for the region. Thus, oilseed deficiency will reduce to about 50% from present level of 79%.

Rapeseed and mustard are economically important crops and constitute important component of diets for people of the region. In addition, mustard oilcake is important source of fish and cattle feed in the region. Intensifying rice fallows will not only improve the productivity of the land but also increase the income of the smallholder farmers. There is evidence that large scale adoption of conservation effective agro-ecological methods like best management practices will not only increase productivity but also reduce use of water resources, and reduce cost of cultivation for farmers. However, to bring the rice (*rabi*) fallow lands under oilseed production, a systematic research effort is required to develop a sustainable rapeseed-mustard production technology, since rice fallow lands have many associated constrained/factor like soil moisture stress, terminal drought and heat stress, poor soil fertility, low pH, lack of irrigation, very long window of rice harvesting (from October to December) and many more which affects crop establishment, causes low plant stand and leads to low crop productivity. The productivity of rapeseed-mustard can be substantially increased through development/standardizations of best combination of improved technologies especially a combination of high yield plant type, suitable crop management practices, conservation tillage practices, abiotic stresses management, balanced crop nutrition, managing soil health and many others.

- **Location specific farm mechanization strategy**: Mechanization in hills needs a different approach than those of plains. Light weight machineries, farm tools should be promoted through local skill and competence development. Special emphasis on artificial intelligence (AI) technology like use of drone for pest management, remote sensing tools for planning and disaster management may be given for enhancing efficiency and effectiveness.
- **Processing, value addition and packaging**: A large quantity of raw materials like arecanut, ginger, turmeric, cardamom, and other are exported to outside the region. If even 50% of these materials are processed and packed with adequate branding farmers income may enhance substantially. Potential for processing and value addition of some raw materials available in NER is highlighted in Table 4.

Table 4. Potential post-harvest processing, value addition and entrepreneurship development options:

Raw material available	Potential	Packaging/ post-harvest potential	Reference
Bird eye chilli (Zo chilli)	Bird eye chilli has the GI for Mizoram having distinct spiciness equivalent to King chilli	Preparation and packaging of chilli powder mas the exclusive potential to ensure high state revenue from the export chain to be developed inside and outside India, as the distinct product of jhum agro-ecosystem in southern Mizoram	Dutta et al. (2017)
Hatkhora (Citrus macroptera)	Antioxidant rich citrus species often reported for its high health benefits, available in the parts of Kolasib, Aizawl and Mamit district only.	Higher degree of natural preservation has been reported for the extracted juice. Therefore value addition with post-harvest processing to make some marketable product may flourish in near future.	Hazarika, (2002)
Fermented local soybean (Bekang)	The local small seeded soy based fermented food product is bad in taste but reported for its low fat/cholesterol content and high nutritive value with anti-aging properties.	Commercial value added product may be popularized for achieving good market opportunity based on the beneficial prospect of human consumption.	Tamang, (2015)
Popcorn (Puakzau)	Multi cob high yielding aromatic popcorn cultivated in eastern Mizoram (Serchhip and Champhai district)	ICAR-Kolasib developed a line of red colored (anthrocyanin rich) popcorn having distinct aroma. The popcorn line may be released as state variety followed by mass scale popularization of the final popcorn product for commercial utilization.	Singh et al. (2016)
Mango/ dragon fruit based post-harvest product	Jam/ prickle industry for the local Rangkui mangos (southern Mizoram i.e. Lawngthlai and Siaha) and dragon fruit (for northern Mizoram i.e. Kolasib, Aizawl, Mamit etc.)	Post-harvest technology should be calibrated for the subsequent value addition of these fruit crops for establishment of village level industry in different parts of Mizoram.	Mandal et al. (2018)

Raw material available	Potential	Packaging/ post-harvest potential	Reference
Ginger – Thinglaidrum	The local Thinglaidrum ginger is popularly cultivated in the secondary jhum fallows of steep hill slopes of Kolasib, Aizawl and Champhai district	May be used for making candy and other ginger based post-harvest formulations; commercial cultivation is already available for its high export potential values. However, post-harvest processing is still lacking for the indigenous ginger product.	Singh et al. (2018)
Fermented Fish (Ngari/ Sidal)	It is the single most popular traditional fermented fish product of Manipur which is eaten daily in almost every household. The preservation used is a traditional practice and has its own cultural identity. Also popular in other states of NER.	It can be used in many ways either as a compulsory side dish called "eromba" (prepared with seasonal vegetables, potato and chilli, tree beans etc.) or cooked with seasonal vegetables and dry fish in a stew locally called as "kangsoi" or "chamthong". It has a characteristic smell that might be repulsive to those new to it but when added in food preparations in required amounts, it increase appetites .	Jeyaram et al. (2009)

- **Strengthening feed and fodder sector**: There is ample opportunity for promoting food-feed crops like tuber crops, maize, soybean, green fodder especially perennial forage grasses and fodder trees in the region to meet feed and fodder requirement within the region while sustaining food security.
- **Promoting alternative source of income**: Each farmer must have at least one alternative source of income either through a shop, processing unit, custom hiring, weaving unit, processing unit or any other source of income.
- **Use of digital different apps in agriculture**: Information is power and will underpin future progress and prosperity. Efforts must be made to strengthen the informatics in agriculture by exploiting the cyberspace through development of new databases, linking state databases with national databases and adding value to information to facilitate decision making at various levels. The magic of E-extension and M-extension services has not been felt in North Eastern Region. Considering the topographical constraints and remoteness, these services must be given priority.Real time dissemination of agricultural information is essential for increasing farming resiliency of North East Indian states including Mizoram. Moreover, location specific weather information through Meghadūta mobile App, developed by India Meteorological Department (IMD), Indian Institute of Tropical Meteorology (IITM) and ICAR will be useful to promote climate proof agriculture in north east India. Moreover, for precision farming use of exact quantity of seeds and fertilizers 'Farm Calculators' app 'Saur Shakti ICAR' for utilization of solar power in agriculture, fruit management practices through 'Fruitcrops' app etc. has huge potential for future use in shifting agriculture of north eastern state of Mizoram.

b. Nutritional security in North East India through horticulture sector

Horticulture sector is the sunrise sector for North Eastern Region, because it plays a unique role for economy of the region by improving the income of the rural people. Cultivation of these crops is labour intensive and as such they generate lot of employment opportunities for the rural population. The region has diverse agro-climate due to varying altitude and has immense potential for the development of horticultural crops.

Due to COVID-19 pandemic, the horticulture sector in the region has been severely affected as most of the horticultural produces are perishable in nature. Considering the nutritional security, population pressure, need for market-driven crop production and employment generation; the self-reliance in horticultural sector should target self-reliant production (srP), self-reliant value addition (srV), self-reliant nutrition (srN) and self-reliant marketing (srM). All these approached should be implemented adequate R&D back up.

3-Tier Strategy

Tier –I: Development of Crop Pockets

- Improving system productivity and cropping intensity in the existing horticultural areas.
- Establishment of accredited nurseries with buy-back agreement.
- Identification of area/villages and/or formation/mobilization of growers' group/FPCs for development of crop pockets.
- Procurement and delivery of seeds/planting materials and other agri-inputs.
- Creation of water resources in the identified pockets under Pradhan Mantri Krishi Sinchayee Yojana (PMKSY)/MGNREGS/Command Area Development Programme.
- Promotion of agri-inputs delivery system through PPP mode.
- Cluster based skill training of farmers and nurserymen.

Tier-II: Transformation of Crop Pockets into Crop Clusters

- Consolidation of various crop pockets to form crop cluster.
- Integration of clusters with pre-processing, packaging and storage infrastructure.
- Establishment of custom hiring centre through PPP mode.
- Creation of water harvesting tank, small check dam, etc. under PMKSY/ MGNREGS.
- Bringing all farmers under KCC coverage / crop insurance plan (PMFBY).
- Linking all farmers to ICAR Agro-advisory services.
- Intra-state market linkage (wholesaler, trader, processor, hotelier, e-commerce etc.).

Tier-III: Transformation of Crop Clusters into Horti-Hub

- Expansion of crop cluster for few potential specialty crops.
- Linkage of crop cluster with cold supply chain or processing units.
- Increase the institutional credit flow to the farmers.
- Establishment of mechanisms for waste recycling and by-product utilization to convert waste in to wealth at commercial scale.
- Inter-state market linkage (including e-platform) supported by transport infrastructure for surplus production.

Establishment of agro-input delivery mechanism at block level through PPP mode is urgently required for reliable and timely availability of quality inputs at reasonable prices. Establishment of accredited horticultural nursery, improving productivity and cropping intensity in the existing horticultural area, creation of water resources for assured irrigation during rabi season, development of protected cultivation cluster, introduction of innovative farm mechanization for hilly terrains and women friendly farm tool, promotion of secondary horticulture based agripreneurship, increasing credit flow, establishment of market linkage, capacity building programme and use of information technology are the important action points for achieving self-reliance in horticulture sector. Farm gate post-harvest infrastructures have potential to reduce the spoilage by many folds and thus indirectly contribute towards increase in production.

c. Strategy for nutritional and income security from allied sectors

i. Livestock sector: Animal Husbandry is a very important sector in the NEH region as the hill farmers integrate crop farming with a large number of livestock which supports 13.38 m cattle, 0.51m buffalo, 0.37 m sheep, 5.4 m goat, 4.24 m pig, 69.22 m poultry, 0.39 m mithun and 0.029 m yak (20th Livestock Census - 2019, GOI). These systems benefit the farming community from the complementarities of crop-livestock system, provide insurance coverage during risk period and also meet their demand for livestock products as about 95% people are non-vegetarian in their dietary habit. Meat, milk and egg deficiency in this region is basically due to indigenous type of animal that constitutes bulk of the population. Huge gap in supply and demand of livestock products provides tremendous scope for the growth of the sector and may enhance the farmer's income significantly if constraints restricting the growth of the sector are recovered.

Strategic plan and policies for self-sufficiency in livestock sector

- Development and promotion of cross bred cattle by using exotic (HF & Jersey) and indigenous milch breed (Sahiwal) with indigenous local cattle of the region.
- Development of new high yielding, stress tolerant varieties/breeds of cattle, pig and poultry suitable for the region. Artificial Insemination technology in both pig and poultry must be propagated at village level State veterinary officials and the para-vets including subject matter specialists in KVK need to be so that quality pig / poultry production is ensured.
- The improved quality of the chicks and duckling is one of the major critical inputs for the poultry production. There are so many improved germplasms of chicken, (Vanaraja, Gramapriya and BND Cross), duck (Khaki Campbell) are developed specially for backyard poultry farming, need to be popularized which are better than native chicken and duck with respects to egg production, and growth rate.
- Hybrid Bajra-Napier, maize, multi-cut perennial grasses and leguminous fodders like cow pea, lucerne has been identified as a potential green fodder

for dairy animals and can be cultivated very easily in wastelands and farm fences. Community fodder bank may be developed to meet the fodder availability during lean period (November –April).

- Most of the upland is under single cropping which can effectively be used to cultivate maize, mustard and pulse crops through suitable in situ soil moisture practices. Maize grain, and by-products of mustard and pulse besides rice bran could be used as feed ingredient while maize and pulse straw could be good source of dry fodder.
- There is a need to develop cheaper balance ration using locally available alternate feed resources so the farmers can afford to buy the feed for feeding as a minimum supplement to the animals and birds. Research on enrichment of the feed through mechanical, chemical and biological means are envisioned to be taken up. Tuber (Tapioca, yam, cassava) cultivation should be done for pig feeding through convergence of fund under Mahatma Gandhi National Rural Employment Guarantee Act (MGNREGA).
- The region being a high rainfall and high humidity zone, livestock suffer from various parasitic diseases leading to poor production. The parasitic disease diagnostic kit developed by the institute is proposed to be mass produced through genetic engineering technique for assisting the field veterinarians in prompt diagnosis of parasitic diseases. The molecular disease diagnostic protocols developed and standardized by ICAR Research Complex, Barapani need to be used in a systematic way to facilitate faster diagnosis of important animal diseases so that preventive measures could be taken to save the loss from both the morbidity and mortality.
- The Indigenous Technical Knowledge (ITK) available in the region for treatment of animal diseases may be validated to establish their efficacy either in the present form or by way of blending the ITKs with modern technological know-how.
- Organized marketing of livestock and livestock products like milk, meat and eggs remains relatively insignificant despite efforts made in the past to develop and promote collective market mechanisms. This needs due attentions.
- Milk, meat, eggs and their products are highly perishable products. The shelf life of these products could be increased by using cold storage facilities.
- Training of private AI workers and provision of logistic support for all AI service at farmers' door step is essentially required for this region.
- Improve information dissemination mechanisms to reach wider population.

ii. Fishery sector: It is noteworthy that 95% of the population in the region relish fish in their diets. But, there is a large gap between demand and local production with net deficit of almost 50%. The culture technology including seed production, feed development and stocking density, etc. have not been standardized so far. In addition to difficult terrain, non- availability of quality fish seeds in time is identified as one of the major constraints in expanding aquaculture in the hilly region. Most of the capture fishery resources are under severe pressure from habitat alterations and over exploitation.

Strategic plan for self-reliance in fishery sector

Mapping of the fishery resources: The GIS inventory will help the policy maker to take better decision for a particular area or cluster. It will help policy maker to identify the performance of a particular cluster/ area and also to identify the area to be focused for fish production/ fisheries enhancement purpose.

Species prioritization for efficient breeding: Many indigenous fish including pabda, chital, magur, koi and minor carps require lesser feed, self-recruiting and fetch handsome price in local market. Thus, diversification to indigenous fish aquaculture has great potential to reduce cost of production and enhance net income besides conserving the biodiversity.

Recirculation Aquaculture System (RAS): It is a technology wherein water is recycled and reused after filtration and removal of suspended matter and metabolites. The method is used for high-density culture of various species of fish utilizing minimum land area and water. This method can be used for culture of high valued trout in Sikkim and Arunachal Pradesh and Mahaseer in Meghalaya.

Biofloc technology: Biofloc is the macro-aggregation of bacteria, algae, detritus and other decomposed components. The principle of this technology is nutrient recycling. Manipulation of Carbon: Nitrogen ratio in culture system helps in conversion of inorganic nitrogen to organic microbial protein which is good proteinaceous food for the fishes. Thus it helps in high density fish farming in limited area. It also reduces maintenance cost because it can be used as feed supplement for the cultivable fishes that provide an additional value by reducing Food conversion rate (Azimm and Little, 2008).

Integrated Multi Trophic Aquaculture (IMTA): The IMTA is the farming of different aquaculture species together in a way that allows one species waste to be utilized as feed by other. Farmers can combine fed aquaculture (Fish, Prawns) with inorganic extractive (Aquatic weed) and organic extractive (Shell fish) aquaculture to create balanced system for environment remediation (biomitigation), economic stability (improved output, lower cost, product diversification and risk reduction) and social stability (better management practices) (Barrington *et al.*, 2009). This model can be developed for augmenting the average productivity of open water bodies.

Organic aquaculture: Periphyton based farming: Periphyton refers to the complex of attached aquatic biota on submerged substrates comprising phytoplankton, zooplankton, benthic organism and detritus. A study conducted on Nile Tilapia reported higher production and specific growth rate of 5.61% for Nile Tilapia using this periphyton technology (Abwao *et al.*, 2014). This technology is cheap and is appropriate for fish farmers in the hilly states.

Improvement of stock: Use of quality seed in fish farm plays vital role for higher production and good return. Inbreeding & unplanned hybridization is a major problem in seed production in most of the hatcheries of Tripura.. To minimize inbreeding depression & brood stock deterioration, genetic improvement through breeding programs need to be undertaken with the involvement of hatchery owners, farmers & other stakeholders. Upgradation of broods stock is a continuous process to be done once in four year.

Feed development: The region is feed-grain deficit region and feed ingredients are mostly imported into the region from long distances which in turn increases the feed cost of production. Augmenting local aquaculture production utilizing local resources would be a positive effect on the local economy..

Health: The NER is one of the vulnerable places of the country to Transboundary aquatic animal diseases (TAADs) as this state is encircled by different countries, causing serious socio-economic and possibly public health consequences. At present EUS (Epizootic ulcerative syndrome) is the major disease concern in the region followed by parasitic infections. . Supplementing fish feed with 2-3% Aswagandha (*Withania somnifera*) root powder could significantly improve the immunity level of fish (DARE-ICAR Annual Report, 2016-17).

Breeding and Seed Production of Ornamental Fishes: There are some fascinating species inhabiting in the region, in which, some of these have already recognized as ornamental fishes in other part of the country. Captive breeding of indigenous ornamental fish and their propagation and marketing could be a low-input enterprise for livelihood security in future.

Conclusion

Concerted effort by extension agencies with technical support from research organizations, universities, and adequate policy support can make India self-reliant sooner than later in agriculture sector. Prioritization of commodities and development of strategy and time frame based action plan for implementation is the key for success. The recent success of self-sufficiency in pulses is a great example that adequate policy, technological support and outreach programme can bring the desired change in agricultural production. Even if 50% of rice fallow areas in NER are brought under oilseed cultivation, the full oilseed demand of the NER cannot be meet out. Thus, emphasis must be given in other areas like organic farming, cardamom, aromatic rice, spices, pineapple, beans, among many other items which has comparative advantages in the region and has large opportunity for enhancing farmers' income due to premium price. For farmers livelihood and bare necessity, IFS must be promoted which can provide them required nutrition and additional employment. However, for a dignified living emphasis should be on livestock, fruits, vegetables, fishery and others which have very good potential in the region and can even earn premium price by supplying to outside the region. New crops like dragon fruits and kiwi has shown very good potential in the region to earn high income. With a view to this now the region is identifying one commodity-one district concept to promote indigenous niche items for increasing their commercial value and trades. Secondary agriculture, specialty agriculture along with emphasis on farm mechanization will enhance farmer income and create additional employment opportunities. While enhancing income, employment and productivity, the natural resource base specially soil and water should be optimally used and improved through resilient farming practices. Adequate skill development programme and hand holding will be required to implement and achieve the vision of self-reliant north east with a focus on self-reliant India.

References

20th Livestock Census Report. 2019. Department of Animal Husbandry and Dairying, Animal Husbandry Statistics Division, Krishi Bhawan, New Delhi, (GOI).

Abwao, J.O., N.P. Boera, M.J. Munguti, P.S. Orina, E.O. and Ogello. 2014. The potential of periphyton based aquaculture for Nile Tilapia (*Oreochromis niloticus* L.) production, a review. International Journal of Fisheries and Aquatic Studies. 2(1): 147-15.

Amonette, J.E. and S. Joseph. 2009. Physical properties of biochar. In 'Biochar for environmental management.Science and Technology'. (Eds J Lehmann, S Joseph) pp. 33-53. (Earthscan: London).

Animal Resources Development Department of Tripura, Government of Tripura.

APEDA. 2020. Agricultural and Processed Food Products Export Development Authority. http://apeda.gov.in/apedawebsite/organic/Organic_Products.htm

Babu, S., R. Singh, R.K. Avasthe, G.S. Yadav, and D.J. Rajkhowa. 2016. Intensification of maize (*Zea mays*)–based cropping sequence in rainfed ecosystem of Sikkim Himalayas for improving system productivity, profitability, employment generation and energy-use efficiency under organic management condition. Indian Journal of Agricultural Sciences.86(6): 778-84.

Barrington, K., Chopin, T. and Robinson, S. 2009.Integrated multi-trophic aquaculture (IMTA) in marine temperature waters. In: Integrated mariculture: a global review (ed. Soto, D.). FAO Fisheries and Aquaculture Technical Paper, 529: 7-46.

Basic Animal Husbandry Statistics. 2019. GOI, Ministry of Fisheries, Animal Husbandry and Dairying. Department of Animal Husbandry and Dairying, Krishi Bhawan, New Delhi.

Bujarbaruah, K.M. 2004. Organic Farming: Opportunities and Challenges in North Eastern Region of India. Paper presented at International Conference on Organic Food, held at ICAR Research Complex for NEH Region, Umiam, Meghalaya, during 14–17 February. pp. 13–23.

Choudhury, B.U., A. Das, M. Kumar, T. Ramesh, K.P. Mohapatra, B.C. Verma and S.V. Ngachan. 2013. Impact of elevated carbon dioxide and temperature on rice productivity in the mid altitude Meghalaya, Northeast India - Results from a simulation study. Journal of Agrometeorology. 15 (Special Issue-I): 1-3.

Choudhury, B.U., A. Das, S.V. Ngachan, L.J. Bordoloi. and P. Chowdhury. 2012. Trend Analysis of Long Term Weather Variables in Mid Altitude Meghalaya, North-East India. Journal of Agricultural Physics. 12(1):12-22.

DARE ICAR Annual Report. 2016-2107. Immunostimulation of *Labeo rohita* against EUS through *Withania somnifera*. Pg 131.

Das, A, G.S. Yadav, J. Layek, R. Lal, R.S. Meena, S. Babu and P.K. Ghosh. 2020. Carbon Management in Diverse Land- Use Systems of Eastern Himalayan Subtropics Springer Nature Singapore Pte Ltd. 2020 123. P.K. Ghosh et al. (eds.), Carbon Management in Tropical and Sub-Tropical Terrestrial Systems, https://doi.org/10.1007/978-981-13-9628-1_8

Das, A., B.U. Choudhury, G.I. Ramkrushna, A.K., Tripathi, R.K. Singh, S.V. Ngachan, D.P. Patel, J. Layek, and G.C. Munda. 2013. Multiple use of pond water for enhancing water productivity and livelihood of small and marginal farmers. Indian Journal of Hill Farming.26(1):29-36.

Das, A., G.C. Munda, P.K. Ghosh, G.I. Ramkrushna, D.P. Patel, B.U. Choudhury, S. Mandal, S.V. Ngachan and J. Layek. 2011a. Managing degraded land and soil organic carbon for climate resilient agriculture in North East India. In: Soil Carbon sequestration for climate change mitigation and food security (Ed. Ch. Srinivasa Rao, B Venkateswarlu, K Srinivas, S Kundu and AK Singh). Central Research Institute for Dry land Agriculture, Hyderabad. PP.241-265.

Das, A., J. Layek, , G.I. Ramkrushna, D.P. Patel, B.U. Choudhury, S. Chowdhury and S.V. Ngachan. 2015. Raised and sunken bed land configuration for crop diversification and crop and water productivity enhancement in rice paddies of the north eastern region of India. Paddy & Water Environment.13:571-580.

Das, S.K., R.K. Avasthe, , M. Singh, and A. Yadav. 2018b. Soil health improvement using biochar application in Sikkim: a success story. Innovative Farming. 3(1). 48-50.

Dutta, S.K., S.B. Singh, Vanlalhmangaiha, A. Banerjee, R.S. Akoijam, T. Boopathi, S. Saha, and S. Roy. 2017. *Capsicum frutescens* L. Landraces of North-East India: From Phenotypic Diversity Perspective of Unexplored Collection. Proceedings of the National Academy of Sciences, India Section B: Biological Sciences. doi: 10.1007/s40011-017-0850-3.

GOI (DAHD). 2016-17. Department of Animal Husbandry, Dairying and Fisheries, Ministry of Agriculture and Farmers' Welfare, GoI.http://dahd.nic.in/.

Hautier, Y., D. Tilman, F. Isbell, E.W. Seabloom, E.T. Borer, and P.B. Reich. 2015. Anthropogenic environmental changes affect ecosystem stability via biodiversity. Science. 348 (6232):.336-340.

Hazarika, T.K. 2012. Citrus genetic diversity of north-east India, their distribution, eco-geography and eco-biology. Genetic Resources and Crop Evolution. 59: 1267–1280.

Jeyaram, K., T.A. Singh, W. Romi, A.R. Devi, W.M. Singh, H. Dayanidhi, N.R. Singh and J.P. Tamang. 2009. Traditional fermented foods of Manipur. Indian Journal of Traditional Knowledge. 8(1):115-121.

Lal, R. 2015. International year of soil and the sustainable development goals of the united Nation. In: Souvenir of sustaining hill agriculture in changing climate, 5-7 Dec, (2015), Agartala, Tripura.

Lal, R.2015. Restoring soil quality to mitigate soil degradation. Sustainability, 7(5):5875-5895.

Layek, J., A. Balusamy, S. Hazarika, S. Devi, R. Krishnappa, K.P. Mohapatra, S. Mandal, B.U. Choudhury, and B.K. Kandpal. 2020. Prospects of Bio-char for Sustainable Agriculture and Carbon Sequestration in North Eastern Hill Region of India. ICAR Research Complex for North Eastern Hill (NEH) Region, Umiam – 793 103, Meghalaya, India, pp.46.

Mandal, D., L. Sailo, T.K. Hazarika, and A.C. Shukla. 2018. Effect of edible coating on shelf life and quality of local mango cv. Rangkuai of Mizoram. Research on Crops 19 (3): 419-424.

Mandal, S., B.C. Verma, G.I. Ramkrushna, R.K. Singh, and D.J. Rajkhowa. 2015. Characterization of biochar obtained from weeds and its effect on soil properties of North Eastern Region. Journal of Environmental Biology. 36(2):499-505.

MoAFW. 2020. 2nd Advance Estimates of 2019-20 of Horticulture Crops. Ministry of Agriculture and Farmers' Welfare, Govt. of India.

MoHA. 2011. Size, Growth Rare and Distribution of Population. In: Provisional Population Total – India. Ministry of Home Affairs, Govt. of India. pp 36-58. Available at: https://censusindia.gov.in/2011-provresults/data_files/india/Final_PPT_2011_chapter3.pdf

Ngachan, S.V. and A. Das. 2018.Climate Proofing of Agriculture for Food Security. In: Conservation Agriculture for Advancing Food Security in Changing Climate, Editors : Das A, Mohapatra KP, Ngachan SV, Panwar AS, Rajkhowa DJ, Ramkrushna GI and Layek J. Today & Tomorrow's Printers and Publishers, New Delhi - 110 002, India, Vol. 1 (2018): 37-57

NIN. 2011. Dietary Guidelines for Indians – A Manual, 2nd Edition. National Institute of Nutrition, Indian Council of Medical Research, Hyderabad – 500 007, India.127 pages.

Patil, P.J. 2020.*Aatmanirbhar Bharat*: Self-reliance in agriculture first step towards economic power. Policy Circle, July, 12, 2020, https://www.policycircle.org/economy/aatmanirbhar-bharat-self-reliance-in-agriculture-first-step-towards-economic-power/

Rakshit, S. 2019. Status of maize research in India and potential in North East Region. (In) Souvenir cum compendium of seminar papers of National seminar on "promotion of maize in North East India- opportunities and strategies organized during 22-23 Oct, 2019 at ICAR Lembucherra, Tripura. PP: 130-132.

RCM 2019.24th Regional Committee Meeting no. III of ICAR, Agenda notes & status of Centre –State Co-ordination in Agricultural Research, Education and Extension in Zone III, 23-24 November, 201, Guwahati, Assam.

Roy, S.S., M.A. Ansari, S.K. Sharma, B. Sailo, D. C. Basudha, I.M. Singh, A. Das, D. Chakraborty A. Arunachalam. N. Prakash and S.V. Ngachan. 2018. Climate resilient agriculture in Manipur: status and strategies for sustainable development. Current Science.115 (7): 1342-1350

Saha, R., P.K. Ghosh, V.K.Mishra, and K.M. Bujarbaruah. 2007.Low-cost micro-rainwater harvesting technology (*Jalkund*) for new livelihood of rural hill farmers. Current Science.92(9): 1258–1265.

Santhosh, B., N.P. Singh, M. Datta, and K.R. Dhiman. 2006. Production technology for composite fish culture in Tripura, Publication No. 14., ICAR Research Complex for NEH Region, Tripura Centre, Lembucherra, West Tripura, pp. 16.

Sharma ,U.C. and V. Sharma. 2004. Implications of soil and nutrient transfer with runoff in NE region of India. In: Sediment transfer through the fluvial systems (Eds. V Golosov, V Belyaev and DE Walling), IAHS Publ. 288: 488-493.

Sharma, B.R., M.V. Riaz, D. Pant, D.L. Adhikary, B.P. Bhatt and H. Rahman. 2010. Water poverty in the northeastern hill region (India): potential alleviation through multiple –use water systems- cross learnings from Nepal Hills. New Delhi, India: International Water Management Institute (IWMI-NAIP Report 1). pp. 44. doi 3910/2009.200

Sharma, U.C., J.S. Samra and M.A. Prodhani. 2006. Land and environmental degradation. In: Soils and their management in north east India (Eds. Sharma UC, Datta, M and Samra, JS). ICAR Research Complex for NEH region, Umiam, Meghalaya, India, pp. 122-170.

Singh, A.R., S.B. Singh, S.K. Dutta, T. Boopathi, S. Saha, M. Thoithoi Devi and N.H. Singh 2016. Multi cob-bearing popcorn (Puakzo) maize: A unique landrace of Mizoram, North East, India. Current Science 110 (8):1392- 1393.

Singh, A.R., S.K. Dutta, T. Boopathi, S.B. Singh, L.S. Saha, V. Dayal, P. Anita, L.S. Singh. 2018. Integrated management of soft rot of ginger in North eastern hills of India. Indian Phytopathology, 71: 83–89. DOI: 10.1007/s42360-018-0001-7.

Singh, R. S., S. Babu, R.K. Avasthe, G.S. Yadav and D.J. Rajkhowa. 2015. Influence of tillage and organic nutrient management practices on productivity, profitability and energetic of vegetable pea (*Pisum sativum* L.) in rice – vegetable pea sequence under hilly ecosystems of North East India. Research on Crops.16(4) : 683-688.

Srinivasarao, Ch., B. Venkateswarlu, R. Lal, A.K. Singh, and S. Kundu. 2013. Sustainable Management of Soils of Dryland Ecosystems of India for Enhancing Agronomic Productivity and Sequestering Carbon. In Donald L. Sparks, editor: Advances in Agronomy, Vol. 121, Burlington: Academic Press. pp. 253-329.

Tamang, J.P. 2015. Naturally fermented ethnic soybean foods of India. Journal of Ethnic Foods. 2(1): 8-17.

21

Innovative Approaches for Agriculture in Coastal Fragile Ecosystem in India

Uttam Kumar Mandal, Subhasis Mandal, Tashi Dorjee Lama and Dhiman Burman

ICAR-Central Soil Salinity Research Institute, Regional Research Station Canning Town, West Bengal

Introduction

Coastal zone represents a broad transitional zone between land and sea which is strongly influenced by both the shoreline ecosystems and the near shore sub-littoral land-based ecosystems. The coastal areas of the country occupy about 10.8 million ha of land and are spread over the 7517 km long coastline along the Bay of Bengal in the East coast, Arabian sea in the West coast and two islands territories. The coastline of the country in its journey covers 9 states (West Bengal Odisha, Andhra Pradesh, Tamil Nadu, Kerala, Karnataka, Maharashtra, Gujarat Goa) 2 union territories (Puducherry and Daman and Diu) and 2 groups of islands (Andaman & Nicobar group in the Bay of Bengal and Lakshadeep and Minicoy group in the Arabian sea) in the country. The region has varied geomorphic and topographical features of mountains, valleys, coastal plains, riverine systems climatic conditions, soil conditions, water budgets and a wide range of cultivated crops. It also supports diverse vegetation ranging from rich tropical rain forests to coastal mangroves and highly rich in biodiversity. The entire coastal ecosystem is highly fragile and vulnerable to climate change and highly endangered by the sea level rise following global warming. Agriculture is the major occupation of the people in the rural areas of coastal regions of the country but it is highly complex risk-prone and entirely dependent on the vagaries of nature.

Since, the problems of coastal agriculture are widely variable there is no universally applicable remedial measure for self-reliant coastal agriculture and it is likely that the remedial measures or the technological interventions will be variable according to the problems associated with the concerned area. It is highly necessary that technological improvement aimed at enhancement of the productivity of coastal agriculture and the livelihood of the farming communities must be simple in use to suit the skill and resource poor farmers and, to be operative at small scale farming as the majority of the farmers are in marginal and small categories with very poor landholdings.

Coastal agricultural across the countries are the major contributor to food productions systems, particularly rice and fish. Coastal communities are largely dependent for their livelihood on fisheries, forestry, animal husbandry and tourism. Because of population pressure and unsustainable development, coastal areas are being exploited unscientifically that affects the natural biological wealth of the region. Major challenges to achieve the self-reliance goal could be, managing the natural resources of the small-holder farmers with gainful engagement, uses of critical resources (land and water) sustainably, reducing the farmers distress through either scientific input management or direct benefit transfer schemes, managing agricultural risks, managing environmental risk, enhancing marketable surplus through grassroots level institutional innovations (like farmers organization) and linking farm to markets.

Major issues and degradation in the coastal zone

The coastal zone has many resources like fertile land, fishing, mangrove forests, marine and terrestrial biodiversity, scenic beauty, marine resources, meeting point of flora and fauna of fresh and saline water; salt fields, and minerals: Quartzes and Zircon, Uranium etc.; and easy transportation and sailing facilities. These have facilitated developments of ports, industries and tourism. Deforestation, over-fishing and cutting of hills for unplanned construction, ship breaking industries and tourism have accelerated destruction of the ecosystem. Un-regulated removal of sand, gravel and pebble deposits from beaches and under water coastal slopes add to the ongoing threats to the coasts. More than 34 species of tropical rain forest plants, including *Podocarpus nerifolia* and *Enteda phaseoloids* are facing extinction from the coastal forests. Many animals such as cats, bear, porcupine wild boars, pythons and anteater are in the process of being wiped out from the coastal areas. Among the marine and coastal species, Redcrabs, jelly-fish, sharks and dolphins are becoming rare but these were the major species before 1980. Coastal resource depletion leads to frequent conflict between users and causes serious socio- economic and cultural problems, such as weakening of the social fabric, marginalization, unemployment and destruction of property by erosion.

Sundarbans is under the threat of severe erosion and drowning due to relative sea level rise and increasing cyclonic activity. Already, Lohachara Island and New Moore Island/South Talpatti Island have disappeared under the sea, and Ghoramara Island is half submerged.

Mangrove forests in coast provide critical ecosystem services, fulfil important socio-economic and environmental functions, and support coastal livelihoods. Their unique root systems create a great deal of physical roughness, thus capturing and storing vast quantities of sediment from upland and oceanic origin. While the total mangrove area did not change significantly over the last few centuries change detection analyses clearly show natural internal mangrove dynamics with small-scale land loss and gain by erosion and accretion of sediments within the tidal channels. Mangrove forests generally possess a high resilience to natural disturbances such as tropical storms and tsunamis. According to Islam (2014) recovery time in storm disturbed mangrove forests of the Sundarbans is around 25 years. However, following a sequence of recent severe disturbing events, with several cyclones in a row (1988, 1991, Sidr in 2007, Nardis in 2008 and Aila in 2009) and the Asian tsunami of 2004, there is concern that the mangroves' regeneration ability has been weakened. On the Bay of Bengal along the Sundarbans, the occurrences of cyclones increased by 26% between 1881 and 2001.

Another most serious anthropogenic impact on the mangrove forest are wastewater pollution from large cities and industry, and a reduction in freshwater supply owing to the construction of upstream embankments for diversion of irrigation water. The most prominent example is the Farakka Barrage, constructed in 1975, which has altered the regional hydrological balance, and has contributed a great deal to differences in freshwater distribution, and the increased salinization of the soils.

The areas along the coast in India are endowed with a variety of soils. Soils vary from alluvial to lateritic, coarse sand to clay, non-saline to highly saline, alkaline to highly acidic, well-drained to poorly drained, low to high organic matter content, deficiency as well as toxicity of some nutrient elements, etc (Velayutham et al., 1999). Depth of water table also shows wide variation. Areas in East and South-West receive high rainfall, whereas, most tracts in Gujarat receive very low rainfall. Regions like Sundarbans suffer from excess water in monsoon season with the problem of prolonged deep waterlogging land. The salt affected soils occupy extensive area spread over both East and West coast regions, and include saline, sodic, acid sulphate, marshy and waterlogged subgroups situated in the low lying areas (Yadav, 2004). Soil salinity hampers crop production in the coastal ecosystem to such an extent that term 'coastal saline soil' has become almost synonymous, though not correctly, with the entire coastal ecosystem. Impeded drainage, inundation, and ingress of seawater have led to the development of saline and alkali soils, rendering vast land as tracts of *Khar*, *Pokhali* and *Kole* lands unsuitable for economic cultivation of major crops. But the soil salinity in the coastal region is temporal. There is a gradual increase in salinity of soil after rainy season till the onset of next monsoon season due to upward capillary movement of saline ground water. During the rainy season most of the lands turns non-saline as the soil salinity is much reduced due to leaching and washing of salts through rainwater.

Certain location-specific problems are also encountered in the coastal soils *viz.*, Fe toxicity in the soils of Odisha, highly permeable sandy soils in parts of Gujarat, and highly leached low fertility lateritic soils of Maharashtra, Karnataka and Kerala. A detailed account of different coastal soils is shown in Table 1. Crop failures due to acidification and salinization are common in the acid sulphate and tidal marshy areas of Kerala, West Bengal and Andaman & Nicobar Islands. The acid sulphate soils with distinct characteristics of high acidity occur in the lowlying areas of Kerala, Sundarbans of West Bengal, and Andaman & Nicobar Islands and they usually are having toxic content of soluble Fe and Al. These soils mostly develop as a result of drainage of those soils that are rich in pyrites (FeS_2) which on oxidation produce sulphuric acid in the presence of excess SO_2^- ions, and are very poor in available P but rich in organic matter.

In coastal areas of Andhra Pradesh and West Bengal, many rice fields are being converted into brackish water fish farms due to high remuneration from aquaculture for more than a decade. Farmers draw brackish water from the sea through creeks and drain into the land and store millions of gallons of this salt laden seawater on the surface in big tanks. This activity is practiced within 10 km of the seacoast. The stored seawater is used for raising valuable prawns (Singh, 2009).

Table 1. Characteristics of soils and ARSRs of coastal area of India (State-wise)

State	AESR No.	AESR	Length of growing period (days)	Soil/Textures	Org. C (%)	Dominant Soils	Area (Km^2)
West Bengal	15.1	Bengal Basin north Bihar plains/hot moist sub humid	210-240	Silty clay loam to silty clay, deep loamy to clayey alluvium	0.2-0.5	Vertic Haplaquepts	14,152
	18.5	Gangetic delta, hot moist sub-humid to humid	240-270	Silty loam to clay loam, deep loamy to clayey coastal and deltaic alluvium	0.3-0.8	Typic Fluvaquents	
Odisha	18.4	Utkal plains and east Godavari delta, hot dry sub-humid	180-210	Sand to silty clay loam, deep loamy to clayey coastal and deltaic alluvium	0.1-0.2	Aertic Tropaquepts	7,900
	18.5	Gangetic delta, hot moist sub-humid to humid	240-270	Clay, deep loamy to clayey coastal and deltaic alluvium	Tr-0.2	Vertic Tropaquepts	
Andhra Pradesh	18.3	Andhra plains/hot dry sub-humid	150-180	Silty clay to clay, deep clayey coastal and deltaic alluvium	0.9-1.1	-	35500
	18.4	Utkal plains east Godavari delta, hot dry sub-humid	130-210	Loam to clay, deep loamy to clayey coastal and deltaic alluvium	0.9-1.2	Chromic Haplusterts	
Tamil Nadu	18.1	South Tamil Nadu plains/ hot dry semiarid	90-120	Sand and Sandy loam, deep loamy to clayey alkaline coastal and deltaic alluvium	0.1	Psammentic Paleustalfs	7424
	18.2	North Tamil Nadu plains/ hot moist semiarid	120-150	Clay, cracking deep clayey and cracking coastal and deltaic alluvium	0.6-1.0	Psammentic Paleustalfs	

State	AESR No.	AESR	Length of growing period (days)	Soil/Textures	Org. C (%)	Dominant Soils	Area (Km^2)
Kerala	19.3	Kerala coastal plains / hot humid to per humid	240-270	Clay to loam, deep, clayey to loamy, acidic, coastal alluvium	0.3-2.6	Typic Sulfaquents	7719
Karnataka	19.3	Karnataka coastal plains / hot humid to per- humid	240-270	Sand to sandy loam, deep, clayey to loamy, acidic, coastal alluvium	0.07-1.2	Aquic Ustifluvents	7424
Maharashtra	19.1	Konkan coast/ hot humid	210-240	Loam to clay, medium to deep loamy to clayey mixed red and black soils	0.3-0.6	Udic Paleustalfs	10000
	19.3	Konkan coast plains/ hot humid to per-humid	240-270	Clay to loam, deep, clayey to loamy, acidic, coastal alluvium	0.4-0.7	Udic Paleustalfs	
Goa	19.1	North Sahyadris and Konkan coast/ hot humid	210-240	Clay to silty clay loam, medium to deep mixed red and black soils	0.3-0.6	Fluventic Ustropepts	220
	19.3	Coastal plains/ hot humid to per-humid	240-270	Clay to loam, deep, clayey to loamy, acidic, coastal alluvium	0.2-0.6	Fluventic Ustropepts	

State	AESR No.	AESR	Length of growing period (days)	Soil/Textures	Org. C (%)	Dominant Soils	Area (Km^2)
Gujarat	2.2	Kachchh peninsula (Great Rann of Kachchh)/ hot hyper -arid	<60	Saline marsh of the Rann of Kutch, salts, light textured and high EC, preponderance of chloride and sulphate, low AWC	0.1	Typic Natrargids	17465
	2.4	South Kachchh and north Kathiawar peninsula/ hot arid	60-90	Loam, deep loamy saline and alkali soils, low AWC	0.3-0.5	Typic Calciorthids	
	5.3	Coastal Kathiawar peninsula /hot moist semi-arid	120-130	medium to deep black soils of deltaic and coastal semi-arid region,, Loam to clay, deep loamy coastal and deltaic alluvium	0.1-0.2	Fluventic Ustochrepts, Typic haplusterts	
Andaman & Nicobar	20.1	Coasts of Islands/ hot per-humid	>300	Loam to clay, shallow to deep loamy coastal alluvium			
Lakshadeep	20.2	Coastal Lakshadweep and group of Islands / hot humid	240-270	Sand, shallow to medium deep sandy and littoral soils	0.1-0.8	Typic Ustipsamments	26
Pondicherry and Karaikal	18.4	Coast Pondicherry and Karaikal / hot dry sub –humid	130-210	Loam to clay, deep loamy to clay coastal and deltaic alluvium		Typic Ustorthents	3

Source: Velayutham *et al.* (1999)

Almost all the small and marginal farmers and also the progressive and big farmers are engaged in this practice as a livelihood. It is reported that an area of nearly 2 lakh ha is under saline aquaculture in the coastal districts of Andhra Pradesh. Farmers can't revert back to rice cultivation because of severe salinity in the pond soils. The farmers are abandoning the prawn cultivation activity because of the contamination of pond waters with some fatal virus due to its mixing with drained canal/creek water. The farmers further revealed that the whole crop of prawns is destroyed within a few days of its infection with the virus. As such there is no treatment for the control of this virus once it infects the prawns. Continuous cultivation of prawns in the coastal belt for about a decade has now resulted in the serious problems of environmental degradation, deterioration in the soil and water quality and associated socio-economic concerns. Adjoining cultivated fields are also affected severely by salinization leading to a reduction in crop yields.

Approximately 10% of the world's population lives in coastal areas of less than 10 m elevation above sea level and nearly 40% of the world's population live within 100 km from the coast. A substantial number of people migrate to the coastal cities because they are regarded as "happening places". The result of this increased population is the urban growth of cities and the associated built-up areas, such as roads, houses and industries that ultimately cause an increase in the pollution of the soil, water and air. Dramatic population growth in coastal cities also results in a concentration of people and businesses that are producing waste that remains mostly untreated prior to its release into the water and soil. Excessive extraction of groundwater from geologically young sedimentary coastal cities can cause subsidence, which in turn can also enhance the relative rise in sea level. Population and economic growth alongside the coast (i.e., littoralisation) are changing the coastal morphodynamics, limiting the sustainability of coastal ecosystems and agriculture (Gopalakrishnan et al., 2019). Furthermore, migration from rural areas causes imbalances in the labour supply and skill mix in rural areas. Eventually, it becomes difficult for coastal agriculture to sustain production despite the increased viability of genetic improvements to crop varieties that renders them more suitable for growth in coastal areas.

Climate change scenario in coastal India

Entire coastal region of India is climatically vulnerable and exposed to Nor'wester and cyclonic storms because of its proximity to the sea. IPCC has projected rising trend of Earth's surface temperature of 0.74°±0.18°C over a period from 1906 to 2005. Climate change projections made for India indicate an overall increase in temperature by 1–4°C and precipitation by 9–16% toward 2050s (Krishna Kumar et al., 2011). Another significant aspect of climate change is the increase in the frequency of occurrence of extreme events such as droughts, floods, and cyclones. All of these expected changes will have adverse impacts on climate sensitive sectors such as agriculture, forest, and coastal ecosystems and also on the availability of water for different uses and on human health.

The major natural disaster that affects the coastal regions of the world is tropical cyclone, which causes devastating hazards via high winds, intense rainfall, terrestrial flooding and storm surges. Effects vary depending on the locations of origin and landfall. About 5% of global tropical cyclones originate in the Bay of Bengal, but adjacent countries including Bangladesh, India and Myanmar experience more than 75% of the global casualties.

The East Coast is more prone compared to the West Coast. An analysis of the frequencies of cyclones on the East and West coasts of India during 1891-2000 showed that nearly 308 cyclones (out of which 103 were severe) affected the East Coast. During the same period, 48 tropical cyclones crossed the West Coast, of which 24 were severe cyclonic storms. The west coast of India experiences high wave activity during the South-West monsoon with relatively calm sea conditions prevailing during the rest of the year. On the East Coast, wave activity is significant both during South-West and North-East monsoons. Extreme wave conditions occur under severe tropical cyclones, which are frequent in the Bay of Bengal during the North-East monsoon period (Kumar et al., 2006).

The Bay of Bengal is more turbulent and the epicenter of frequent devastating Bay cyclones, sea surges etc. because of the special nature of the coastline, shallow coastal ocean bottom topography and characteristics of the tide. Their coastal impact in this region is very large because of the low, flat coastal terrain, high density of population, low awareness of community, inadequate response and preparedness and absence of hedging mechanism. People can remember the devastation of cyclone *Aila* during 2009, and recent *Amphan* in West Bengal (Figure 1), super-cyclone in Odisha during 1999 as well as *Tsunami* in eastern coast during 2004 when thousands of hectares of fertile agricultural land and adjoining mangrove forests were turned into a vast wasteland due to ingress of saline sea water.

A recent study of global deltas showed that the entire Bengal delta is sinking at a perilous rate due to sediment compaction from the removal oil, gas and water from the inland delta's underlying sediments, the trapping of sediment in upstream reservoirs, flood plain engineering, and rising sea level (Syvitski et al., 2009). Based on global sea level data and modeling, Eriscson et al. (2006) have estimated that the sea level rise of the Bay of Bengal is the World's highest, at >10 mm year^{-1}. Other studies have confirmed this trend, but with rates ranging from 4.0 mm year^{-1} in the Western zone to 7.8 mm year^{-1} in the Eastern zone (Han et al., 2010; Unnikrishnan and Shankar, 2007). A 2007 report by UNESCO "Case Studies on Climate Change and World Heritage" has predicted 45 cm rise in sea level likely by the end of the 21st century. The Intergovernmental Panel on Climate Change (IPCC) in its fourth assessment report presented similar observational evidence of climate change in the coastal regions and indicated the evidence of increased ocean temperature, changes in precipitation, corresponding upstream river discharge and rising sea level. This will generally lead to higher coastal flooding and increased salinity (Parry, 2007).

Fig. 1. Damage due to cyclone *Amphan* on May 20, 2020 at Sundarbans region.

Mandal et al. (2018) analyzed the PSMSL (Permanent Service for Mean Sea Level) tide data to detect the changes of sea level trend and the results showed that the rate of sea level changes for four stations, namely, Garden Reach, Diamond Harbour, Haldia and Gangra in West Bengal coast are found to be +7.48, +4.27, +3.24 and +2.06 mm yr^{-1} whereas for the stations Mumbai and Cochin in west coast the rate of changes were +0.78 and +2.07 mm yr^{-1} and for Chennai and Vishakhapatnam in the east coast the rate of changes were +1.06 and +1.00 mm yr^{-1} (Figure 2). The rise in sea level in West Bengal coast was more during the last ten years (2005-2014) and it increased by 1.65% of long term average in the region.

Climate change impact on coastal agriculture

Coastal agricultural systems, sandwiched between land and sea, are particularly under threat from increasing climate and non-climate related hazards. Coastal farming is more susceptible to climate change compared to inland farming because, along with the rainfall and temperature changes, they also have to deal with frequent changes in coastal ecosystems. Furthermore, coastal agriculture is experiencing increasing anthropogenic disturbances through land degradation and encroachments of agricultural lands (Gopalakrishnan et al., 2019).

Agricultural production in India is closely linked to the performance of summer monsoon (June–September) which contributes about 75% of the annual precipitation. Apart from the inter-annual variability in summer monsoon rainfall the occurrence of many of the hydrometeorological events is found to influence Indian agriculture at different spatial scales. Apart from the summer monsoon rainfall, India receives about 15% of annual precipitation during the winter months of December–March. This precipitation is very important for rabi (winter) crops. Though rainfall and its distribution have a profound influence on the Indian food grain production, other climatic elements like radiation and temperature are also exerting considerable effect.

Climate change may affect agricultural crops in four ways (Srinivasarao et al., 2016). First, changes in temperature and precipitation will alter the distribution of agroecological zones. Second, carbon dioxide effects are expected to have a positive impact due to greater water use efficiency and a higher rate of photosynthesis. Third, water availability (or runoff) is another critical factor in determining the impact of climate change. Fourth, agricultural losses can result from climatic variability and the increased frequency of extreme events such as cyclones, floods, drought etc.

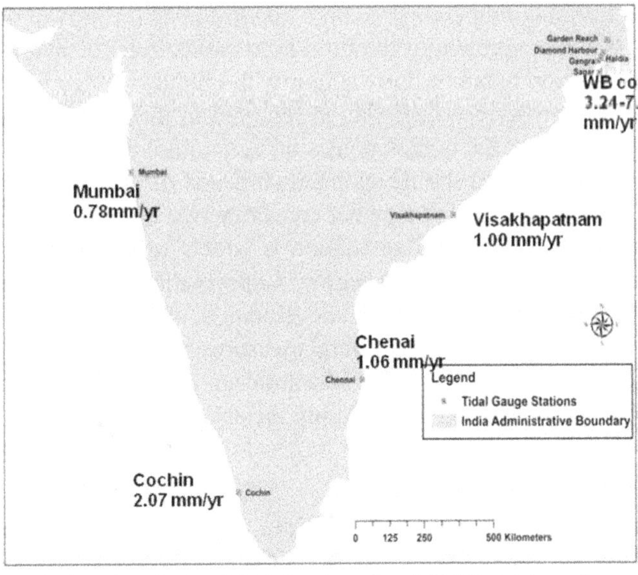

Fig. 2. Trend analysis of tidal data based on PSMSL database for analyzing sea level rise (*Source*: Mandal et al., 2018)

Rice is the predominant form of land use in many coastal and deltaic regions of the tropics (Mandal et al., 2019a). No crop other than rice can be grown under these adverse conditions of unstable water levels and highly saline locations. Changes in precipitation may exert a stronger impact on rice production than temperature changes. Similarly, frequent floods due to heavy precipitation may result in higher yield losses under progressing climate change. Moreover, higher temperature will increase sea level due to (1) thermal expansion of sea water and (2) rapid melting of glaciers and ice caps. As a consequence, fragile coastal and highly productive deltaic rice cultivation areas will be more exposed to inundation and salinity intrusion.

Climate change is modifying the distribution and productivity of marine and freshwater species and is already affecting biological processes and altering food webs. The consequences for the sustainability of aquatic ecosystems, fisheries and aquaculture, and the people that depend on them, are uncertain. Coral reefs will degrade with increasing water temperatures and acidification of the oceans and be more sensitive to the threats of over–fishing, pollution, poor tourism practices and invasive species. This will affect the quantity and type of fish available to coastal communities.

Innovations in livelihood improvement

Coastal regions are rich in natural resources including mangroves, coral reefs, and water bodies which provide a range of services for human well-being. This natural diversity, while providing immense economic opportunities for growth of the regions, also supports a vast segment of the rural households in particular, to earn their livelihoods. However, human actions along with climate change have put these resources at risk with serious degradation reported in different spatial and temporal scales. By and large, the overemphasis on agriculture development focused on irrigated belts overlooked the developmental needs of vast tracts of rainfed areas as well as coastal regions that together constitute two-third of India's cultivable land. Salinity ingress and land degradation is widely reported due to over-extraction of groundwater in the coastal region. Conservation of land and water, therefore becomes the central theme for the development of land and water use in these regions. To address this problem, several measures have already been adopted in the coastal region to reduce salinity and drainage congestion during rainy season and availability of irrigation water during the dry season.

Innovations in land and water management

Implementing innovative land management practices in combination with the productive utilization of opportunities of the coastal areas like excess rainwater and vast brackish water resources could be the best approach to meet the challenges of land degradation like salinization, waterlogging, drainage congestion. Land shaping by definition is the alteration of the surface of the land primarily for harvesting rain water for creating a source for irrigation, reducing the effect of groundwater salinity, reducing drainage congestion and growing multiple and diversified crops round the year (Burman et al., 2015; Mandal et al., 2019b). Harvesting rainwater in farm pond and raising of low land with excavated soil reduced the impact of saline groundwater table on salinity build up of soil, improved drainage congestion of low-lying land and created irrigation resources for irrigation water starved coastal areas.

Following land shaping models are popular in the coastal region in India

Farm pond: About 20% of the farm area is converted into on-farm pond of about 3m depth to harvest excess rainwater (Figure 3). The dug-out soil is used to raise the land to form high land/dike and medium land situations besides the original low land situation in the farm for growing multiple and diversified crops throughout the year instead of mono-cropping with rice in kharif season. The high land is free from water logging in Kharif and with less salinity build up in dry seasons and thus, can be used for multiple & diversified crop cultivation throughout the year (Table 2).

Deep furrow and high ridge: About 50 % of the farm land is shaped into alternate ridges (1.5 m top width × 1.0 m height × 3m bottom width) and furrows (3m top width × 1.5 m bottom width × 1.0 m depth) (Figure 3). These ridges remain free of water logging during Kharif with less soil salinity build up in dry seasons (due to higher elevation and presence of fresh rain water in furrows. The remaining portion of the farmland including the furrows is used for growing more profitable paddy + fish cultivation in kharif. The rainwater harvested in furrows is used for irrigation.

Shallow furrow and medium ridge: About 75 % of the farm land is shaped into medium ridges (1.0 m top width × 0.75 m height × 2.0 m bottom width) and furrows (2.0 m top width × 1.0 m bottom width × 0.75 m depth) with a gap of 3.5 m between two consecutive ridges and furrows.

Paddy-cum-fish: Trenches (3 m top width × 1.5 m bottom width × 1.5 m depth) are dug around the periphery of the farm land leaving about 3.5 m wide outer from boundary and the dugout soil is used for making dikes (about 1.5 m top width ×1.5 m height × 3 m bottom width) to protect free flow of water from the field and harvesting more rain water in the field and trench (Figure 3). A small ditch is dug out at one corner of the field as shelter for fishes when water will dry out in trenches.

Broad bed and furrow: This involves shaping of land for broad beds (4-5 m width and 1 m height) and furrows (5-6 m width and 1m deep) with a provision of

(2 m x 4 m x 1 m) fish shelter at the end of the furrow alternatively in low-lying lands. Raised beds are used for cultivation of vegetables round the year and fish is cultivated in the furrows. This system provided the scope for *in-situ* rainwater harvesting of about 3800 m^3 ha^{-1} and which is used to cultivate the second crop during dry seasons.

Three tier land configuration: In this technique of land shaping degraded low-lying land is shaped into three equal portions as raised land, medium or original land and pond with a depth of 2.5-3 m and dikes of 5 m wide and 1.5 m height. Pond at the lower part of the land is used for harvesting of rain water of about 4500 m^3 ha^{-1} and poly-culture of fish. Paddy in medium (original) land along with vegetables on raised land and dikes are cultivated.

Paired bed technique: In paired bed technique degraded low lying land is shaped into broad furrow of 9 m width x 2 m depth and two beds of 6 m width. In this technique, a nursery pond of 5 m x 9 m size is also created at one end of the furrow for raising fingerlings while broad furrow is used for brooders. Two dikes are created of 2-3 m width at both ends. Broad furrow is used for harvesting of rain water of about 3750 m3ha^{-1}. Vegetables are grown round the year in the raised beds and dikes.

***Ail* cultivation (cultivation on land embankment):** The so-called *Ail* is the land embankment, which is made to demarcate a given land. This demarcation is made to consider the land with individual ownership. Traditionally, the width and height of this *Ail* remain upto 40 cm and 30 cm respectively. In Sundarbans, it is usually noticed that the depth of water on the low land become 2½ ft. to 3 ft. during the rainy season. Only long duration traditional paddy is grown in the low land situation keeping the *Ail* fallow. No vegetable crops are possible in this land situation during the rainy season and as the optimum moisture condition of the soil comes in too late, it is not also possible to grow any vegetable during the winter season. Besides, the source of sweet water for irrigation is minimum to irrigate the winter vegetables. For productive use of the *Ail*, the width and height of this *Ails* was modified upto that level where vegetables can be grown both in rainy and winter seasons. To overcome the inundation problem during the rainy season, the height of the *Ail* was raised upto 1 meter. The width of the top of the *Ail* is made 90 cm where base remains 150 cm. The transverse section of this modified land embankment would be like a trapezium, which is required for strengthening the structure. To make a raised land embankment along the circumference of 1 ha (10,000 sq. meter) land, having 400 running meter *Ail*, it requires 480 cubic meter soil. This volume of soil is collected by making a trench surrounding the *Ail* or by digging a small sized pond of about 24 m length, 10 m width and 2 m depth within the paddy field itself. The trench so developed is utilized for harvesting of rain water for life saving irrigation of the vegetable crops grown on the *Ail* during the winter season. The pond excavated within the paddy field for the construction of

Ail is used for pisciculture also. As paddy-cum- fish culture is a popular practice here, the fish takes shelter in that pond when the water of the paddy field becomes insufficient or dry up. During rainy season, as the height of the *Ail* is more than that of the existing field, there is no chance of damaging the vegetable crops on *Ail* by submergence. Once a raised *Ail* is developed, it lasts for about 8-10 years without incurring additional expenditure to great extent and only periodical repairing of the structure is required as and when necessary. Cultivation on the land embankment (*Ail*) starts from the month of June each year just before the onset of monsoon. During the rainy season, crop like okra, bitter gourd etc. and during winter season tomato, french bean etc. are cultivated on the land embankment.

Brackish water aquaculture pond: There are many areas in the coastal areas particularly near the brackish water river or sea coast remain highly saline throughout the year and not suitable for crop cultivation. These lands are shaped into shallow depth brackish water pond. The pond size varies from 0.13 - 0.4 ha with a depth of 1.0 -1.5 m. The height of the embankment of the pond is determined by the tidal height occurring in the area, generally about 30 cm above the maximum flood level. In general about 1.2 m hight and 1.6 m wide embankment is made on the periphery of the pond. Polyculture system of brackishwater fish farming with tiger shrimp (*Penaeus monodon*) along with brackishwater fish like golbhangon/ bhangon (*Liza tade*) and aansbhangon (*Mugil cephalus*) is practiced in the pond with brackishwater from the nearest river.

About 370 ha of land in disadvantaged areas in Sundarbans and Andaman& Nicobar Islands has been converted from mono-cropped to multi-cropped with integrated crop and fish cultivation through the implementation of different land shaping techniques like farm pond, deep furrow & high ridge, paddy-cum-fish, broad bed & furrow, three tire system, paired bed system and drainage improvement network under GEF funded National Agricultural Innovation Project (CSSRI, NAIP, 2014). About 1943 water harvesting structures were developed under various land shaping techniques with water storage capacity of 13,04,600 m^3 per annum. With land shaping techniques, different land situations like, high land, medium land and low (original) apart from rainwater harvesting structures like farm pond/furrows/ trenches etc. were created in low-lying and degraded farmers' fields (Figure 3). Raising of land and creating water harvesting structures reduced the problem of drainage congestion during kharif season and this provided the scope for growing high value crops like vegetables during this season and it also facilitated early sowing of rabi crops. Salinity build up in the soil of different land situations especially medium land and highland/ridges/dikes in the land shaped area were reduced and, fertility status and biological activities in soil was increased under land shaping techniques (Table 3). The cropping intensity has been increased up to 240 % from a base level value of 100%. Land shaping techniques have increased the employment and income of the households by many times (Table 4). Net income per ha of farm land has been increased from Rs. 22000 to Rs. 1,23,000 in

Sundarbans and from Rs. 22400 to Rs. 1,90,000 in Andaman & Nicobar Islands. By adopting brackish water aquaculture ponds in Sundarbans particularly near the brackish water rivers, farmers were benefitted from this technique with a net income of about Rs. 1,43,000 ha^{-1} of pond area. Land shaping techniques were financially viable and attractive proposition for the coastal salt affected region (Burman et al., 2015).

Table 2. Crops and cropping/farming systems under land shaping techniques

Land shaping models	Land situation created	Crops	
		Kharif season	Rabi/summer season
Farm Pond (FP)	(a) Pond (20%)	Fish	Fish
	(b) High land and dikes (20%)	Vegetables, fruit crops	Vegetables, fruit crops
	(c) Medium land (20%)	HYV Rice	Vegetables, low water requiring field crops
	(d) Original lowland (40%)	Paddy + fish	Low water requiring field crops/ vegetables, short duration rice
Deep furrow & high ridge (DF)	(a) Furrows (25%)	Paddy + Fish	Fish
	(b) Ridges (25%)	Vegetables & fruit crops/ multi-purpose tree species	Vegetables & fruit crops/ multi-purpose tree species (MPTs)
	(c) Original low land (50%)	Rice under paddy + fish	Low water requiring field crops/ vegetables
Paddy-cum-fish (PCF)	(a) Trenches (11%)	Fish under paddy + Fish	fallow
	(b) Dikes (12%)	Vegetables & fruit crops/MPTs	Vegetables & fruit crops/ MPTs
	(c) Original low land (77%)	Paddy + fish	Low water requiring field crops/ vegetables
Control (farmers' practice)	Low land	Paddy	Fallow
Rice-rice	Low land	Paddy	Paddy

Table 3. Effect of land shaping systems on soil properties. (Reference: Mandal *et al.*, 2019b) EC, Electrical Conductivity; OC, Organic C; Av N, Available N; Av P, Available P; Av K, Available K; MBC, microbial biomass carbon. Values with the same lowercase within land uses are not significantly different at $P<0.05$.

Parameters	ECe Winter	ECe Summer	OC	Av N	Av P	Av K	Bulk Density	MBC
	dS m^{-1}	dS m	%		Kg ha^{-1}		Mg m^{-3}	µg g^{-1}
Rice-monocrop	3.82aA	9.77aB	0.762a	122.8a	9.30a	514.6ac	1.502ab	187.6a
Rice-rice	3.28bA	5.33bB	0.913b	124.1a	18.10b	474.5a	1.474a	288.2bc
Farm pond system-lowland	2.1cA	2.64cB	0.855b	114.3a	27.49c	394.5b	1.589b	243.8b
Farm pond system-medium land	2.47cA	2.62cA	0.865b	129.5a	36.93d	479.2a	1.542b	279.1b
Farm pond system-upland	2.57cA	2.82cA)	0.926b	194.2b	49.78e	519.3ac	1.562b	342.2cd
Deep furrow and high ridge system-lowland	2.89cbA	3.59cB	0.793ac	169.3b	24.72c	642.2d	1.584b	291.1bc
Deep furrow and high ridge system-ridge	3.12bA	3.55cB	0.822c	234.8c	46.56e	499.8ac	1.512ab	376.3d
Paddy cum fish system-lowland	2.46cA	2.74cA	0.810c	178.3b	39.22de	442.0ab	1.601b	225.6ab
Paddy cum fish system-dikes	2.88cbA	3.92cB	0.847bc	198.2b	46.12e	549.1c	1.541b	325.9c

Table 4. Enhancement in cropping intensity, employment generation and net income under different land shaping techniques in Sundarbans and Andaman and Nicobar Islands.

Land shaping	Cropping intensity (%)		Employment generation (mandays/hh/yr)		Net return (Rs./ha/yr)	
Technologies	Before land shaping	After land shaping	Before land shaping	After land shaping	Before land shaping	After land shaping
Farm pond	114[a], 100[b]	193[a], 200[b]	87[a], 8[b]	227[a], 22[b]	22000[a], 10000[b]	140000[a], 148000[b]
Deep furrow & high ridge	114[a]	186[a]	87[a]	218[a]	22000[a]	102000[a]
Paddy-cum-fish	114[a], 100[b]	166[a], 200[b]	87[a], 8[b]	223[a], 35[b]	22000[a]	127000[a]
Broad bed & furrow	100[b]	240[b]	9[b]	48[b]	24000[b]	148000b
Three tier	100b	220b	10b	42b	30000b	221000b
Paired bed	100b	240b	9b	54b	24000b	216000b
Brackishwater aquaculture pond			25a	100a		146000a

Note: Costs and returns at the current price of 2012-13, hh means per household (average holding was 0.35 ha in Sundarbans[a], average holding of implementation was 0.20 ha in Andaman and Nicobar Islands[b]. (Reference: Burman et al., 2015).

Layout of farm pond

Layout of deep furrow & high ridge

Layout of Paddy cum fish

Fig. 3. Layout of different land shaping systems (Mandal et al., 2019b)

'DORUVU' technology and skimming freshwater: Along the sea coast of Andhra Pradesh, there are 8.23 lakh ha of sandy soils, of which 1.74 lakh ha have shallow watertable varying from 0.3 to 3.0 m. In the coastal sands, entire precipitation percolates into the soil because of the high permeability of coastal sands and the infiltrated rain water with lesser density form freshwater lenses which floats on the sub- surface saline ground water. In most of these coastal sands, fresh water floating at a depth of 0.5 to 3.0 m below ground level and to a thickness of 3 to 4 m over saline ground waters/clay layers. The waters cannot be extracted in high quantities by conventional tube wells as they are of water table aquifers with poor quality waters underneath. In these soils during summer months the water table fall up to 1.8 to 3.0 m below ground level. The ground water gets recharged during the monsoon season, accumulates and by middle of November the ground water rise up to 0.3 to 0.5 m bgl (below ground level) in most of the areas. As irrigation supplies are not possible either though canal or tube wells in most of the coastal sandy soils, the situation has forced the farmers to search for alternatives and the technique innovated traditionally is to draw water manually that collects in the dugout conical pits locally called '*doruvus*' (Raghu Babu et al. 2009). The water collected from a *doruvu*, that occupies an area of about 200 m^2 is just sufficient for irrigating an area of 800 m^2. Maintenance of 10-12 such *doruvus* is a common practice for each hectare of land. The main drawbacks of traditional *doruvu* techniques are wastage of productive land towards *doruvus*, which is about 20% of the land covered under irrigation; unproductive evaporative loss of fresh water from open water surface of *doruvus*; irrigation process involving manual splash watering requires huge application expenditure and does not permit the use of modern irrigation equipment; more chances of proliferation of vector borne diseases like mosquito breeding in open pits.

To overcome these constraints, the skimming well technology (improved *doruvu* technology) has been evolved by the All India Coordinated Research Project on Management of Salt affected Soils and Use of Saline Water in Agriculture, Bapatla Centre of Acharya N. G. Ranga Agricultural University, popularly known as 'Improved *Doruvu* Technology' to skim the shallow depth fresh water in coastal aquifer without disturbing the hydro-dynamic conditions. The system consists of a collector well with lateral collector lines installed at shallow depth. The collector lines are connected to the sump on either side and imbedded for the collection of lateral flow by digging a trench. Depending on the water table head above the collector pipes, the collectors are continuously charged with subsurface fresh waters throughout their length and water flow into the well under gravity. The recharge occurs due to precipitation, upstream canal seepage and also by recycling of irrigation water. As a result of this recharge, flow lines are sustained. With built-in advantages of skimming well, the technology has become very popular with the farmers of coastal sandy belt in Guntur and Prakasam districts of Andhra Pradesh.

Sub surface drainage for reclamation of saline waterlogged soils: The sub-surface drainage (SSD) technology has been developed to lower the water table in saline waterlogged areas. In SSD system, perforated corrugated/ smooth PVC pipes are installed manually or by laser-controlled trencher machines mechanically at a desired design spacing and depth below soil surface to drain out the excess water and soluble salts from the affected area. These pipes are covered with gravel/ synthetic filter to prevent the clogging. The success of a few manually laid pilot SSD projects during 1980s proved conducive to the commissioning of many large, mechanically installed projects in the states of Haryana, Rajasthan, Maharashtra, Karnataka, Gujarat, Punjab and Andhra Pradesh. Thanks to the availability of adequate funds under different government-sponsored schemes and the entry of private companies in a public–private partnership mode, key installation and operational constraints have been overcome resulting in fast spread of SSD technology in many salt-affected regions. Till date, about 66,500 ha waterlogged saline soils have been reclaimed with SSD in India (Sharma et al., 2016). The reclaimed soils exhibit marked improvements in soil health, crop productivity, cropping intensity and overall environmental quality (Sharma and Singh, 2015). Despite these benefits, smooth establishment and operation of SSD projects suffers from constraints such as high-installation costs and the lack of farmers' active involvement in the maintenance and running of SSD sites. Again, in many cases, problems encountered in the safe disposal of saline drainage water hamper SSD adoption in the landlocked regions. A range of solutions have been recommended to overcome these obstacles. These include provision of evaporation ponds, bio-drainage, conjunctive use of fresh and saline water in irrigation and the use of salt-tolerant cultivars. Of these approaches, the last 3 have received considerable attention to enhance the acceptability of SSD technology at farmers' fields.

Alternate land use systems: A number of salt tolerant agro-forestry and fruit trees, shrubs, grasses and medicinal and aromatic plants have been identified for commercial cultivation in salt-affected soils. Some of the promising agro-forestry species are *Prosopis juliflora*, *Acacia nilotica*, *Tamarix articulata* and *Casuarina* *equisetifolia*. A *Prosopis juliflora-Leptochloa fusca* silvi-pastoral model has been found promising for sustained fuel wood and forage production in high pH soils. When followed for more than four years, the tree and grass cover reclaims sodic soils to such an extent that normal agriculture crops such as *Trifolium alexandrinum* and T. *resupinatum* can be grown successfully. Appropriate planting techniques have been standardized for raising tree plantations in saline (sub-surface planting, ridge-trench method, subsurface planting and furrow irrigation system) and sodic (ridge-trench method, auger-hole method, pit auger-hole method, pit-auger hole and furrow method) soils. Different fruit-based agro-forestry systems, with bael (*Aegle marmelos*), aonla (*Emblica officinallis*) and karonda (*Carissa congesta*) as tree components and cluster bean (in *Kharif*) and barley (in *Rabi*) as subsidiary components, have been found practically feasible and Remunerative with the use of moderate (ECIW between 4.0 to 5.8 dS m^{-1}) to high (ECIW between 8.2-10.5 dS m^{-1}) saline waters. The saline and sodic lands under long-term tree cover exhibit overall improvement in soil quality which is attributed to an increase in soil organic carbon and nitrogen contents, enhanced microbial biomass, microclimate modification under tree cover and nutrient uptake from deeper layers by the tree roots (Sharma and Singh 2015). Different medicinal and aromatic plants such as isabgol *(Plantago ovata)*, aloe (*Aloe barbadensis*) and kalmeg (*Andrographis paniculata*) produce high biomass under saline (ECIW 8.5 dS m^{-1}) irrigation.

Afforestation and agroforestry have tremendous potential for carbon sequestration not only in aboveground C biomass but also belowground C biomass in deeper soil depths. Reforestation of 75 million ha degraded lands with suitable trees and grasses / crops may sequester about 4 Pg of carbon. Cropping systems involving forage or legumes invariably conserve soil organic carbon (SOC) and maintain a great biological nutrient pool. Agroforestry systems and mixed farming systems are superior in terms of SOC improvement as compared to sole cropping. Cultivation of fast growing trees with arable crops under agri-horticultural or agri-silvicultural systems help in improving SOC content and sequestering carbon particularly in highly degraded barren lands.

Bio-drainage to combat waterlogging and secondary salinity: In many canal irrigated regions of the country, waterlogging and the subsequent secondary salinization have turned vast stretches of agricultural lands unproductive. Water seepage from canals and faulty on-farm water management practices together create shallow water table conditions and significant increase in root zone salinity. Conventionally, waterlogged saline lands are reclaimed by installation of sub-surface drainage systems which are expensive, difficult to operate and pose problems in the disposal of saline drainage effluents. These operational difficulties have generated interest in other viable alternatives such as bio-drainage for harnessing the productivity of waterlogged soils.

Bio-drainage refers to the bio-energy driven pumping out of excess soil water and dissolved salts through rapid transpiration by the perennial trees. Analogous to energy-operated water pumps, it is a proven technology to prevent salinity build-up in canal commands when suitable tree species (e.g., eucalyptus, poplar, and bamboo) are raised in the beginning to prevent waterlogging and salinization. Of late, combined applications of bio-drainage and suitable land modifications are being explored to productively utilize the waterlogged salt affected soils (Sharma et al., 2016).

Agronomic approach to increase the cropping intensity in coastal saline soils

To make cropping systems intensification a reality, a number of systems changes need to be made in the coastal zone (Bell et al., 2019) as follows: surface drainage of excess water at the end of the wet season; better understanding and use of the limited fresh groundwater; increased surface water storage capacity; planting of earlier maturity and shorter duration rice varieties during the monsoon season to provide better timeliness of *rabi* crops in relation to water availability; increased use of salt tolerant *rabi* cultivars, use of mulches to reduce *rabi* season soil evaporation and hence reduce salinity build-up in surface soil.

A wide range of field crops and vegetable crops are feasible in the coastal zone, and profitable intensified cropping patterns have been identified. Among the cropping patterns that increased BCR Benefit cost ratio in Sundarbans region are *kharif* rice - potato - mungbean, kharif rice - mustard - mungbean, kharif rice - Garden pea - Mungbean, kharif rice - spinach-mungbean. One of the most promising new crops introduced was zero-tillage (ZT) potato after kharif rice (Maji et al., 2020). The ZT potato is placed on wet soil and covered by a thick layer of rice straw mulch equivalent to 12 t ha^{-1}. The main opportunity involves changing the cropping season to use stored water from the monsoon season in the *rabi* season when the solute potential is low and to harvest crops early enough to avoid crop stress from waterlogging, salinity/ drought, heat and or storms (Bell et al., 2019).

Agronomic research on salt affected vertisols: Salt affected Vertisols (Black soils) cover an area of about 1.21 million hectares in Gujarat. Problem is increasing in Karnataka, Maharashtra and Rajasthan. Due to high clay content, soils are adversely affected even at low salt and exchangeable sodium contents. Further, with the introduction of mega irrigation projects, the extent of salt affliction is increasing at a rapid rate. Restoring the productivity of these lands, once they become salinized is much more difficult as compared to alluvial sandy loam soils of Indo-Gangetic Plains. Researchers made consistent efforts to study the basic, strategic and applied aspects of salt affected Vertisols (Sharma and Chaudhari 2012). Technologies like cultivation of *Salvadora* (tolerates salinity up to 50 dS/m) for the restoration of highly saline (>40 dS/m) soils and *Dill* (tolerates salinity of

4-6 dS/m) have been adopted by the farmers, NGOs and Govt. Institutions like Gujarat State Land Development Corporation (GSLDC) are growing these crops on a large scale in Gujarat.

Dill (*Anethum graveolens*), a non-conventional seed spice crop has been identified as potential crop for cultivation on saline black soils having salinity up to 6 dS m^{-1} in rabi season with the residual soil moisture. It has multiple uses viz., pot herb, leafy vegetable, seeds used as condiments and seed oil for aromatic and medicinal purposes. The herb contains Vitamin-C as high as 121.4 mg/100g. The oil of dill seeds and its emulsion in water (Dill water) are considered to be aromatic, carminative and effective in colic pains and possesses anti-pyretic and anti-helminthic properties. A substantial increase in yield can be obtained by using saline ground water in conjunction with best available surface water. This crop thus would help farmers of the region to go for the second crop in rabi season on lands, which hitherto remain fallow due to water and salinity constraints (Anil et al., 2015).

Salvadora persica L. (Meswak), a facultative halophyte which is a potential source for seed oil has been identified as a predominant species in highly saline habitats of coastal and inland black soils (ECe >30 dS/m). This species is a medicinal plant of great value and its bark contains resins and an alkaloid called Salvadoricine. The seeds are a good source of non-edible oil rich in lauric (C12) and myrestic (C14) fatty acids having immense applications in soap and detergent industries. This species gives economic returns for the highly saline black soils with salinity values up to 50 dS m^{-1}, also provides eco-restoration through environmental greening and thus forms a niche for highly saline black soils.

Innovations in brackishwater aquaculture: Brackishwater farming in India has been a traditional practice for centuries and confined mainly to the bheries (manmade impoundments) in coastal wetlands. Scientific brackishwater aquaculture started in India with tiger shrimp (*Penaeus monodon*) farming initiated during early 1990s. With the introduction of Pacific white shrimp (*Penaeus vannamei*) during 2009, Indian aquaculture industry has grown rapidly. In addition, certain marine/brackishwater fish such as, seabass, mullets, milkfish and pearlspot have shown a lot of promise. Indian Sundarbans located in the south–east end of West Bengal offers congenial environment for growth of variety of fishes and shrimps. Frozen shrimp and live crab are the main export items from brackishwater aquaculture in Sundarbans. As the economic benefit is greater, there is a tendency of Sundarbans dwellers to shift from fishing to aquaculture for better livelihood. There is vast scope for sustainable development of brackishwater aquaculture in Sundarbans to meet the livelihood demand utilizing the unused and underused areas and adopting advanced farming practices. Challenges faced by Sundarbans aqua farmers need to be tackled by appropriate management tools like social mobilization of aqua producers, technology assessment and refinement, participatory planning, and capacity building of key stakeholders.

There are different opportunities to improve brackishwater aquaculture production. Among short-term strategies, improving traditional system of farming, development and dissemination of low-cost farm-made feed, control of diseases, and improved production through good aquaculture practices like, biofloc and periphyton based system, are the best options available (Ghosal et al., 2019). Long-term strategies like, development of organic farming, genetic improvement of shrimp stock through selective breeding, improved and intensive culture methods, adoption of cage farming of fishes, and overcoming the constrains through participatory fisheries management, public-private partnership, refinement of indigenous technical knowledge, can be taken up. Apart from brackish water aquaculture, freshwater sector contributes parallel economy and livelihood security of the peoples living in close vicinity in the eco-region.

Indian coastal line has a potential for seaweed farming. The required manpower for the seaweed farming and processing can be met considerably from the fishermen communities which are vulnerable to climate changes and thereby can enhance their livelihoods. Seaweed farming and its associated preliminary processing can be taken in village or community levels as a strategy of adaptability and resilience in the context of climate change. Presently, commercial scale seaweed cultivation is being done only in the state of Tamil Nadu, India which is at the southern coast of India. But lots of coastal areas are still unexplored so far in term of cultivation activity.

Use of digital science for transforming coastal agriculture

India has made tremendous progress in achieving self-sufficiency in food production and now is one of the leading contributors to the global food basket. Despite this, the major challenges confronting Indian agriculture today are declining total productivity, degrading natural resources, increasing demand for food, stagnating farm incomes, fragmented land holdings and climate change (Seth and Ganguly, 2017). Agricultural productivity and income in the coastal regions are low due to soil salinity and waterlogging problems and is further constrained by the associated risk of crop failures arising from natural calamities. Several technologies (improved crop, soil and water management practices, stress-tolerant and high yielding crop varieties, land shaping technologies, etc.) are available for improving productivity in the stress-prone coastal region. However, many of the farmers are not able to take benefit of these improved technologies for augmenting their farm productivity and income for lack of awareness as well as timely access to advisories. Besides, the market is another factor which drives a farmer to decide what to produce and how much to produce.

The impact of technology adoption on modernizing farmers' production practices, reducing risks of crop failure and increasing crop yields thereby leading to more sustainable income has been well established (Seth and Ganguly, 2017). Access to the right information at the right place and at the right time will help in empowering the farmers and improving the efficiency and viability of small and marginal farms.

Digital science through information and communication technology has the potential in providing technological support by way of assisting in information dissemination related to crop and input selection, pest and disease advisories, weather advisories, credit and insurance, realtime data on market intelligence, post-harvest processing in a need-based and user-friendly mode (Seth and Ganguly, 2017). The application of digital technologies which include remote sensing (satellites and drones), wireless sensors, internet of things (IoTs), geographical information systems for crop and soil health monitoring, livestock and farm management is expected to be the future in providing newer solutions to the problems faced by the farmers (Gill et al., 2017).

With the strong penetration of mobile and internet connectivity in rural areas, there has been an increase in the development of many services and applications in agriculture for the benefit of the farmers. The Government of India has launched its flagship scheme 'Digital India' with a vision to transform India into a digitally empowered society and knowledge economy and it could be instrumental in providing solutions to the farming community. The electronic National Agricultural Market (e-NAM) is a unified market platform that operates on a pan Indian basis by providing effective linkage of the Agriculture Produce Market Committees (APMCs) for improved service delivery and the discovery of the highest price possible for agricultural commodities. Several private business conglomerates have also come forward in building up digital platforms for providing one-stop solutions to the farmers, such as ITCs e-Choupal, a comprehensive digital hub for the farmers, Mahindra and Mahindra's Tringo which is a mobile-based app for custom hiring of tractors and the mKRISHI platform by Tata Consultancy Services offers personalized advisory services in audio and visual formats using mobile phones. Besides, there are also startups, such as Skymet Weather Services involved in monitoring and predicting weather and providing agri-risk solutions, Barrix Agro Sciences offers eco-friendly crop protection methods while Agri Suite by eKutir offers a one-stop solution for all needs of a farmer and train farmers to use their application (Seth and Ganguly, 2017).

The mobile-based applications provide the latest information on agriculture about latest technologies and methods being used, help to identify pests and diseases, provide real-time weather data, early warnings about storms, local markets offering best prices, seeds, fertilizers, etc. Also, via these apps, the farmers can interact and seek guidance from agriculture experts across the country (Bhaskar et al.,

2017). The various apps can be downloaded from https://mkisan.gov.in/ or https://krishi.icar.gov.in/ or Google play store. Some of the apps are Kisan Suvidha, Pusa Krishi, Soil Health Card, Krishi Video Advice, IFFCO Kisan Agriculture, Krishi Vigyan, e-NAM, Agrimarket, riceXpert, Pashu Poshan, etc.

The importance of information technology in the field of agriculture is enormous in today's world of "Information Age" and the growth in the agricultural sector will be information-driven. New information must reach to the ultimate user at the fastest speed to harness its potential benefits. Knowledge-intensive and precision farming techniques will be the guiding lines for sustainable agriculture. Thus, the convergence of IT with agriculture research will add a new dimension in understanding and finding solutions to the problems and constraints we are facing today in Indian agriculture and would help in ushering an "evergreen revolution".

Use of renewable energy resources in coastal agriculture

Renewable energy resources are those generated by natural processes and are continuously replenished by these processes at a rate comparable to or faster than its rate of consumption by humans (Chel and Kaushik, 2011). It includes sunlight, geothermal heat, wind, tides, water and various forms of biomass. Renewable energy is also called "clean energy" or "green power" as is does not pollute the environment (Bharthi et al., 2019).

Being located in the tropics, India receives high solar insolation and there is huge solar potential (Kumar and Majid, 2020). India receives a solar energy equivalent of more than 5000 trillion kWh per year, which is far more than its total annual consumption. The daily global radiation is around 5 kWh m^2 day^{-1} with sunshine ranging between 2300 and 3200 h $year^{-1}$ in most parts of India (Khare et al., 2013). The coastal states of Maharashtra, Gujarat, Andhra Pradesh, Orrisa and West Bengal, and inland states like Haryana, Punjab, Rajasthan, Madhya Pradesh and Bihar, receive 4-7 kWh m^2 $day^{-1}$of solar radiation, which is equivalent to 2300-3200 hours of sunshine every year. Sunlight can be converted directly into electricity using photovoltaic (PV) systems. The photovoltaic generation is a technique of converting solar radiation or photon energy into direct current electricity using a semiconductor material that exhibits a photovoltaic effect. Another way of using solar power is for the generation of heat.

Wind power is the conversion of the kinetic energy of the wind into a useful form of energy, such as using wind turbines to make electrical power, windmills for mechanical power or wind-pumps for water pumping or drainage (Khare et al., 2013). Wind energy leads amongst all the renewable energy sources in terms of installed capacity and energy generation in India. The recent assessment indicates a gross wind power potential of 300 GW gigawatt in the country at 100 m above ground level. Besides, India is blessed with a coastline of nearly 7,600 km with relatively shallow waters near to shore where there are good prospects of harnessing offshore wind energy along the coastline (Chaurasiya et al., 2019). Studies have suggested that offshore wind energy potential along the coast of Gujarat is 36 GW and 35 GW along the coast of Tamil Nadu (Dash, 2019). The coastal states of Gujarat, Maharashtra, Tamil Nadu, Karnataka and Andhra Pradesh have the highest potential for wind energy.

Bioenergy that can be obtained from biomass is another form of renewable energy that has a potential use. As per the estimates by the Ministry of New and Renewable Energy (MNRE), India every year around 500 MT of biomass, including agro-residue and forestry waste, is generated in India. However, only 25-30 % of it, capable of generating 18 GW of power, is available for power generation. At present, a major portion (around 65%) of biomass power is derived from bagasse produced in sugar mills (Pal et al., 2017).

The oceans are the source of thermal energy due to surface heating by the sun as well as mechanical energy arising out of tides and waves. In India the long coastline of 7600 km with estuaries and gulfs producing strong tides are the potential areas for harnessing tidal energy, however, it is the least harnessed renewable source of the country. It is estimated that the total available tidal potential of India is 40,000 MW, whereas the total identified potential is about 9000 MW in the Gulfs of Cambay and Kutch in Gujarat and Sundarbans Delta in West Bengal (Pal et al., 2017).

Solar energy can be used in two ways, first by converting solar energy into electricity and secondly by converting into heat. For converting solar energy into electricity, photovoltaic cells are used. The most common use of the photovoltaic systems is for pumping of water irrigation and livestock uses. The solar-powered drip irrigation system (Figure 4) is most widely used nowadays as an efficient and cost-effective irrigation system. The electricity generated can also be used for lighting the farm households and the buildings and enclosures in animal and poultry farms. The solar power generated can also be used for various other uses, such as grinding of feed, electric-powered egg collection, product refrigeration, livestock or fish feeder, sprayer motors, aerator pumps in aquaculture, electric fencing, etc. (Bharthi et al., 2019). The thermal energy i.e., heat derived from solar energy has application in agro-processing like drying of produce and processed products and heating of water in animal or poultry farms and also in fish farms. As solar power plants with photovoltaic systems compete with agriculture and also bioenergy for land, concerns have been raised about the use of land for food and energy production. To overcome this dispute of land use, the agri-voltaic system has been conceptualized where the same piece of land can be used for photovoltaic system as well as growing crops at the same time. Such systems have been successfully developed for generating electricity from solar energy and for the growing of crops, such as mung bean (*Vigna radiata*), moth bean (*Vigna aconitifolia*), cluster bean (*Cyamopsis tetragonoloba*), isabgol (*Plantago ovata*) cumin (*Cuminum cyminum*), Aloe vera, etc. (Santra et al., 2018) and grapes (Malu et al., 2017). Another advantage of such a system is that the solar panels can be used as a surface for rainwater harvesting which can be stored and used for providing supplemental irrigation and for washing the solar panels. The farmers can further gain income by selling the excess electricity produced over the grid. For an area of one acre, the break-even period of the agri-voltaic system has been estimated to be 9-10 years (Santra et al., 2018).

Like solar energy, wind energy can also be used for generation of electricity and mechanical power. The electric power can be used directly or stored in batteries. The electricity generated from wind turbines can be used for pumping water for irrigating crops and for livestock, fish farms, etc. Wind energy-based aerators and water circulators can also be used in aquaculture. In the wind mills, grains can be ground to produce flour.

Fig. 4. Solar power drip irrigation system used in Sundarbans region.

In the coastal areas, the tidal farms can use the tidal energy to generate electricity. Using paddle wheel fans and turbines the energy of the tide can be used to produce electricity which can be directly used or stored in batteries for future use.

The advantages of renewable energy are that remote areas can be electrified, it is clean as it does not pollute the environment, it is cheaper so helps in reducing production costs making farming more efficient, cleaner and profitable. However, high installation costs and maintenance, lack of connectivity to grid, conflicts in land use and lack of awareness are some of the bottlenecks in large scale adoption of renewable energy.

In comparison to fossil fuel-based technologies, use of renewable energy resources in agriculture have the additional benefit of earning carbon credits and contribute to the clean development mechanism (CDM). Renewable energy technologies in the agriculture sector are being promoted in India through various clean development mechanism (CDM) projects, however, due to high investment costs, such projects would require governmental support (Chel and Kaushik, 2011). Thus, there is a need to promote the use of renewable technologies in agriculture to reduce fossil fuel consumption, mitigate climate change and protect the environment for the benefit of the society as a whole. Further, these technologies will help the farmers in achieving self-sufficiency in energy requirement, reduce input costs and achieve higher income.

Mangrove forests and sustainability of coastal ecosystem

Mangroves (Figure 5) are considered to be highly productive ecosystems (Pramanik et al., 2019), having immense importance in terms of ecological functions and are considered as the national wealth of maritime countries (Kaladharan et al. 2005). They help in the protection and stabilization of the coastal zones and at the same time contribute to the recycling of nutrients in the coastal waters (Pramanik et al., 2019). The mangroves perform many important ecosystem services such as maintenance of coastal water quality by removing pollutants particularly toxic heavy metals, reduction in the severity of natural calamities like Tsunamis, storms cyclones and flood damage, bio resources as well as provide the breeding ground for fish species of commercial importance (Kathiresan and Bingham, 2001; Sahu et al., 2015).

There exists a significant potential of climate change mitigation through carbon capture and sequestration by mangroves (Kathiresan, 2018). As per the studies conducted in India, it has been revealed that Andaman Island possesses the highest biomass carbon stock (118.3 t C ha^{-1}) followed by Tamil Nadu (62.81 t C ha^{-1}) Karnataka (50.40 t C ha^{-1}) and Gujarat (24.57 t C ha^{-1}) (Kathiresan, 2018). The mangroves of the Indian Sundarbans spread over an area of 2,118 km^2, can absorb over 41.5 million tonnes of carbon dioxide daily, valued at around US$79 billion in the international market (Bhatt and Karthiresan, 2012). Therefore, mangroves play an important role in maintaining the ecological balance in the fragile ecosystem and their restoration could be a novel option in climate change mitigation (Sahu et al., 2015).

Fig. 5. A view of mangrove ecosystem in the Indian Sundarbans.

Mangrove ecosystems are very rich in biodiversity and are inhabited by diverse groups of terrestrial and aquatic organisms. It occupies a diversified habitat like core forests, litter-forest floors, mud flats, water bodies (rivers, bays, creeks, etc.) coral reefs and sea grass ecosystems. In the Indian mangrove ecosystems, a total of 4,011 species including 920 plants (23%) and 3,091 animals (77%) species have been recorded, which is highest in the world (Sahu et al., 2015).

According to the State of Forest Report, 2019, in India the mangrove forest cover extends over 4975 km^2 which is about 3% of the total mangrove cover in South Asia. Of the total cover in India, Sundarbans has the largest mangrove cover, occupying 43% followed by Gujarat with 23%. The third-largest mangrove forest in India occupying 13% of the total cover is found in the Andaman and Nicobar Islands (Kathiresan, 2018). It has been observed that during the past 30 years except for Andhra Pradesh and Andaman and Nicobar Islands where the mangrove cover decreased at 0.61% and 0.34% per year respectively, it increased in all other states like Gujarat (168%), Maharashtra (117%), Tamil Nadu (113), Odisha (22%), Goa (26 times), Karnataka (10 times), Kerala (9 times), Daman and Diu (3 times) and Puducherry (2 times). Overall, there was an increase in mangrove cover by 21.63% over 30 years at 0.72% per year in India (Kathiresan, 2018). The increase in the covers is largely attributed to the efforts of the State Forest Departments and Ministry of Environment and Forests, Government of India for initiating plantation as well as restoration and conservation of mangroves (Sahu et al., 2015). Despite the increase in mangrove cover in the country, the open sparse mangrove cover stands at 40% which deserves attention for its transformation into a dense mangrove cover. Further, *Heritiera fomes* and *Sonneratia griffithii* the two globally threatened species of the total 11 species under IUCN Red List are from India (Kathiresan, 2018). The mangrove forests are threatened due to various factors like conversion for urbanization, aquaculture, agriculture, salt farming and other developmental activities such as tourism, mining, refineries, oil pipeline passages, port/harbour, dam and road constructions; changes in hydrological regimes; increasing salinity; coastal pollution; siltation; exploitation of fishery resources; cattle grazing; private ownership and ineffective institutional regimes. Also, there is the growing threat of climate change especially sea-level rise in low lying coastal areas of the country. Mangrove habitats of the east coast of India and Kerala due to low slopes are vulnerable to sea-level rise than others in the west coast (Karthiresan, 2018). As the mangroves are very valuable in maintaining the ecological balance in the coastal ecosystem, the Government of India has been proactively implementing management plans aimed at conservation and rejuvenation the mangroves in the different coastal tracts of the country. Emphasis is also being given for ecological restoration in which mangroves are allowed to grow back naturally by providing the right conditions for growth. Understanding the local ecological conditions and hydrological regimes is necessary to restore the hydrology and remove the barriers to natural regeneration (Karthiresan, 2018).

Mangroves thus play an important role in maintaining the biodiversity, providing feeding and breeding grounds for different fishes and aquatic animals, fuel, timber, honey, medicines and ensuring the livelihood security of the coastal people. Besides offering protection during natural calamities like floods, tsunamis and cyclones they have tremendous potential for carbon sequestration and climate change mitigation. Therefore, for conservation, protection and restoration of the valuable mangrove wetland ecosystems, there is a need for effective governance, better planning for rehabilitation of degraded mangroves and creation of awareness among the local communities (Sahu et al., 2015).

Research Needed

Re-sodification of reclaimed sodic soils: This is a growing concern in the areas where sodic soil reclamation took place before 2-3 decades. Though the causes of re-sodification are identified, yet concerted research efforts are to be made to avoid it (Sharma and Chaudhari 2012).

Need for alternative to gypsum: Gypsum being mineral in nature and mined from limited area, its availability in future will no longer be the same as it was in the past. Already there is a crisis and several judicial and social issues involved in gypsum mining. Therefore there is an urgent need to find out an alternative to mineral gypsum and standardize the agro-techniques for its use on sodic soils.

Work at ICAR-CSSRI showed that marine gypsum could be a feasible alternative to mineral gypsum for reviving the productivity of degraded sodic lands. There has been an increasing interest in using elemental S for calcareous-sodic soil reclamation.

Harnessing synergy from modern science: There are opportunities and avenues of reclamation and remediation of both salt affected soils and poor quality waters; and agronomic research in the area of nano-technology applications in salt affected soils.

Bio- and phyto-remediation: Bio-remediation and phyto-remediation of waste waters / poor quality waters and their use in salt affected soils need to be explored on a major scale. This area has a great potential for agronomic research in future with limiting water availability for irrigation.

Bio-saline agriculture: Agro-forestry offers a very good solution to the bio-amelioration of saline lands. It is possible to go for bio-fuels and bio-energy plantations on moderately degraded and severely degraded salt affected soils. This requires extensive research involving multi-disciplinary teams and addressing core researchable issues

Micro-irrigation for enhanced resource use efficiency: Till now cereal systems are dominant on salt affected as well as reclaimed salt affected soils. These soils are not exposed to high value horticulture production. Since these areas are affected by multiple stresses of salt, water and nutrients. Therefore, research efforts need to be diverted for enhancing the use efficiency of water and nutrients on salt affected soils. For this, micro-irrigation, drip fertigation, drip chemigation and use of liquid bio-fertilizer along with automized irrigation need to be explored.

Resource conservation: Entire Indo-Gangetic Plain and Trans-Indo-Gangetic Plain is engaged in rice-wheat cultivation for several decades. Exhaustive nature of rice-wheat cropping system depleted the natural resources and there is resource crunch disturbing agro-ecology of the area. The problem is aggravated in salt affected soils because of multiple stresses. This compels adaptation of resource conservation technology and therefore conservation agriculture options need to be thoroughly evaluated for reclaimed as well as non-reclaimed salt affected soils.

Multi enterprise agriculture: For small and marginal land holdings, multi enterprise agriculture provides an assurance of regular income as well as on-farm employment. Thus, the multi enterprise agriculture involving subsidiary components with conventional crop components in a synergy based recycling manner needs to be evaluated in different agro-eco systems on salt affected soils.

Integrated nutrient management: It is well proved that higher nutrient application is beneficial in salt affected soils as compared to normal soils. But after reclamation soil pH of surface 30 cm layer is within the range of 7.5-8.5, therefore nutrient supply as per the recommended doses of fertilizer is advisable. However, in many areas, farmers use a very high amount of nitrogen fertilizers than the recommended one. On the contrary, they do not use potassium fertilizers leading to imbalance nutrition for the crop, which affects soil health and quality. These issues need a fresh relook and to be critically addressed by location-specific research. Therefore, integrated nutrient management prescriptions need to be developed for different types of salt affected soils.

The adverse impacts of climate change on agricultural production are increasingly becoming noticeable in different parts of the world. Most of the projected changes-rise in atmospheric temperature, sudden and heavy downpour, drying of streams and reduced water flows in rivers-can prove detrimental to agriculture and resource-poor farmers (Sharma and Singh, 2015). The problems of sea-water ingress and cyclonic storms have significantly increased in the last few years, posing threats to the sustainability of coastal soils. High atmospheric temperature, evaporative loss of water and less rainfall are likely to accentuate salt accumulation in soil. In the last few years, problems of resodification (i.e. reappearance of sodic patches in gypsum- amended soils and resalinization of ameliorated saline soils have become widespread. The growing scarcity of freshwater will necessitate changes in the conventional soil-reclamation strategies where good quality is an indispensable input for salt leaching. In many cases, marginal quality water will be the only viable option for irrigation and soil amelioration. Strong climate variability may lead to the emergence of multiple soil stresses that could prove lethal to even the salt-tolerant genotypes. Development of cultivars having tolerance to two or more soil constraints is thus becoming a priority in genetic improvement programmes (Sharma and Singh, 2015).

Towards self-reliance coastal agriculture: challenges and way forward

At present, one-quarter of the world's population live in coastal areas and most of the largest urban concentrations are on the coast. The current coastal urban population of 220 million is projected to almost double in the next 20 to 30 years. Unless appropriate action is taken by governments and users of coastal resources, population pressure and associated levels of economic activity will further increase the already evident overexploitation of coastal resources and environmental degradation of many coastal habitats. In many developing countries, this trend is further exacerbated by widespread extreme poverty and unemployment. The global economic condition under the COVID-19 pandemic situation has become severely disruptive to all contributing sector including agriculture. Govt. of India is keen to achieve self-reliance in all sectors and to achieve this goal has planned for largest public investment of Rs. 20 lakh crores (10% of India's GDP) into the economy. Out of which 1.5 lakh crores are going to be invested in agriculture sector alone, to boost the primary sector in the country. The expectations are to meet the current and future food and nutritional demand of the country through the domestic production as well as targeting exports of commodities wherever possible. The focus will be boosting agricultural production and productivity in all agro-ecological zones including the coastal zones that spread over 7500 km of coast line in the country.

The coastal regions of India are marked as one of the traditionally backward and disadvantaged areas with low agricultural productivity due to poor quality of soil and water. Enhancing farm income under such fragile environment with utilization of poor quality soil and water resources is a key challenge for the region. The success of improved agro-technologies is dependent on sustainable management of soil and water resources. The management of agricultural land to improve farm income in coastal saline environment is quite challenging and most of the agricultural area is characterized by mono-cropping with low yielding rice varieties during kharif season. Under such fragile environment sustaining the livelihoods of this resource poor farmers become a real challenge both for technology developers as well as policy makers. To increase the farm income under the coastal saline environment, strategies have been taken up to increase the adoption of the salt resistant crop varieties and more importantly water harvesting, storing and appropriate management through different kind of land shaping technologies. Farmers in coastal region view their farming operation as a system where crop and fisheries are integral part of their farming system. These land-shaping interventions were implemented through crop-fish integration for achieving high impact and adoptability among the farmers. Land shaping models have created the land suitable for growing multiple crops and rearing fish (Mandal et al., 2013 & 2015). For out-scaling of these technologies to a greater scale there is a need to address some socio-economic constraints and policy support. Such proposition of crop-fish integration in agriculture through these land-shaping models was highly suitable for pulling the income of intra-agricultural sector in coastal region of India.

Adoption of technologies

Farmers' have to operate their farming operation under diverse socio-economic conditions that affect their decisions to adopt new technologies. Socio-economic factors like input prices, market environment, fragmented and small/marginal land holdings, availability of own or hired human labour, labour wage rates, financial and credit needs, availability and capacity to absorption of credit, risk preferences etc., all these factors affect the adoption behaviour towards new technologies. Resource poor farmers were naturally risk averters and they tend to prefer a lower outcome that is relatively certain to the prospect of a higher return with a greater degree of uncertainty was attached. Farmers preferred stability of output even with a lower return rather than the high-cost-high-return technologies where the instability of output was much higher. In general, the farmers in the coastal states were in distress due to their low income and there was a need of concerted efforts to enhance the farmers' income significantly. The research experiments at farmers' field often indicated encouraging results towards higher cropping system intensification and profitability. However, availability of good quality and quantity of irrigation water and accessibility will determine the extent of adoption of such cropping systems. The evolved options might be technologically sound and profitable in terms of output-input ratios. But due to limited water availability farmers might compel to grow multiple crops in small plots (fragmented land) that have primary objectives of meeting household's level food security with less commercial/business motives. For achieving and continuing with higher cropping system intensification from the existing cropping system, farmers need continuous support like, more capital investment, additional knowledge on crop management and assurance of remunerative price. Also, socio-economic factors like, risk preference, risk bearing abilities, risk mitigation strategies and adequacy of capital often determined the level of adoption of new cropping systems.

Transforming agriculture to agribusiness through sustainable value chain

There was a need to transform the smallholder farming operation to self-reliant framing business through promotion and development of sustainable value chain for agricultural commodities. At grassroots level many success stories emerged during implementing livelihoods improvement project. Out of these around one-fifth of farmers continued after the project was over. Most of these farmers' incomes were more, actually more than double, but real challenge remained with the rest of the farmers who couldn't continue without the support. Among many reasons, most crucial was diverse socio-economic conditions. Most important is farmers needed cash to buy inputs before season and one window solution for their problems. They need decision support from professionals for technology plant protection measures, buying insurance or selling of produce. Agricultural production needed to be demand driven; to produce what consumers want through development of sustainable value chain system. Under this chain all actors, farmers, service provider, retailers and consumers would be interdependent and affecting one's interest will break the chain and affect the business and profitability of entire chain. Transforming agriculture to sustainable agribusiness was the need and agriculture needed to be handled by professionals.

Three new agricultural policies, namely, The Essential Commodities (Amendment) Ordinance 2020, The Farming Produce Trade and Commerce (Promotion and Facilitation) Ordinance, 2020 and The Farmers (Empowerment and Protection) Agreement on Price Assurance and Farm Services Ordinance, 2020 aimed at creation of 'One-Nation-One-Market', selling agri-produce outside the APMC yards or their districts, towards free market. Key expectations from these policies are facilitating One India One Market, Free Trade, rremoval of APMC monopoly competition and multiple buyers, better price discovery and increasing farmers income. These bills are few steps way forward towards making domestic market free for farmers and providing them options to sell their commodities directly to traders. Agriculture is a rural sector, need to connect all farmers (including small and marginal), inclusion of the excluded, through increasing the rural infrastructure (road and connectivity) adequately and need a number of competitive firms in the market to ensure the better tomorrow for the farmers.

Market prices were the key drivers to change the farm economy but we all are aware of the high price volatility in agri-marketing sector that adversely affect the up-scaling of these agro-technologies. Some of the agricultural marketing strategies needed were (1) to delineate the clusters of districts/areas which have favourable resources, climate and socio-economic conditions for growing each of these high value commodities and differentiated strategies could be formulated for triggering the development of the sector; (2) On one hand corporate retailers are consolidating rapidly asking for bulk purchase of agri-commodities, and on other hand producers are becoming more and more fragmented and producing small quantity of marketable surplus, so farmers must be organized to increase their volume of trading to increase their bargaining power to take advantage of these marketing systems. and (3) Developing agri-marketing infrastructure calls for large-scale investment both from public and private players. Public investment is continuing and many have been initiated recently (e.g., promotion of farmer's producers company/FPO and *Krishak* Bazaar) needed to be complemented by the private investment. Allowing Foreign Direct Investment (FDI) would be a good option towards this. As the markets were being de-regulated across the states and also in the country, corporate entry was inevitable – sooner or later. Commodity preference across the varied income groups of people was creating different 'niche' markets and the corporate retailer can meet the demand and expectations of such growing consumers.

Formation of Farmers Producers Organisation (FPO) can be one of the good propositions to ensure the regular input supply as well as better marketing. Farmer Producer Organisations (FPOs) are the recent institutional innovations to provide single window solution for multiple problems of farmers and ensuring better price realization through establishing direct market linkages. Key objective of FPOs is to ensure better income for the producers through an organization of their own. As the small-holder producers do not have the volume of produce individually (both inputs and produce), such institutions provide the benefit of economies of scale as well as better share to consumers' price by reducing the length of marketing channel (no of intermediaries) and lowering the transaction cost. Also the FPOs are expected to have better bargaining power as bulk buyer of inputs and bulk seller of produce. The essentials of FPOs are, it is formed by a group of producers for either farm or non-farm activities; it is a registered body and a legal entity; producers are shareholders in the organization; it deals with business activities related to the primary produce/product; it works for the benefit of the member producers; a part of the profit is shared amongst the producers and rest of the surplus is added to its owned funds for business expansion. As the part of the profits is shared among the members, they are actually the owners of the organisations and therefore successful FPO members' feels like collective winners. Also Govt. of India currently is supporting such farmers company across the country through Small Farmers Agribusiness Consortia (SFAC). As such if small farmers' could be aggregated, the problems related to inputs could be overcome and transactional costs could be reduced and the farmers' could realize better prices (Parthiban, 2015, Nikam et al, 2019). Producer organisations were increasingly becoming important for small producers for securing better access to services through intermediation (Buckley, 2007).In Karnataka FPOs arranged input supply at a price lesser than the market, by procuring inputs on wholesale price, seeds, planting material, chemicals, micro-nutrients, bio-pesticides and custom-hiring services of farm machineries and implements for their producer members (Venkattakumar et al. 2019).

Policy making: The acceptable and efficient distribution of resources by appropriate institutions is a prerequisite to sustainable coastal development. Agricultural development has advanced tremendously since the "green revolution", but environmental concern is more prominent in the planning level than found in the actual implementation. Transformation of coastal agricultural practices by individual farmers without supervised planning may not always be environmentally friendly. For example, sustainable fisheries can define the survival capability of many coastal communities; nevertheless, unplanned and excessive shrimp culture that requires ponds of saline water may not be sustainable and may be a concern regarding salinity intrusion (Gopalakrishnan et al., 2019). In fact, coastal agriculture in India particularly in eastern coast has been undergoing a major shift from cereal production, especially paddy, to vegetable production and shrimp farming because climate change and market demand have changed the economic incentives of different farming practices. Farmers are not willing to stick to the existing practices if changes in farming systems offer a higher income. The sustainable management of coastal agriculture in polluted ecosystems is difficult. Major sources of pollution are agriculture, urban disposals and industrial wastes. The implementation of policy is required to minimize pollution from both agricultural and non-agricultural sources. The stake holders who are directly vulnerable to climate change should take a role in preparing and implementing strategies for adaptation. Local communities are likely to contribute to developing and implementing sustainable agricultural strategies if their immediate needs are fulfilled without damaging the ecological carrying capacity. Individual adaptation is motivated by self-interest that may not be beneficial for all. Therefore, joint adaption is required through government interventions in training and market-based incentives.

Policies concerning the sustainability of coastal agriculture involve several stakeholders, such as administrators, policymakers and farming and non-farming communities living in the coastal areas. Integrated coastal management will be effective regarding governance, the coastal environment and socio-economic aspects. In addition, the raising of awareness to promote coastal management is necessary to render the current patterns of coastal management sustainable.

Education and voluntary incentive-based programs to improve the adoption of conservation practices can perform better than regulatory tools at controlling pollution in the case of industrial pollution in coastal areas and to promote sustainable agricultural land management.

Land-use planning is at the heart of sustainable agriculture. Different countries, as well as different farming communities (e.g., subsistence farming, family farming and capitalist farming) should be recognized and addressed differently in the integrated management policies. Public policies need to diversify on-farm activities and to improve rural services with special attention being paid to the adoption of sustainable agricultural practices by smallholders. This would help to minimize the out-migration of rural farmers.

From the perspective of sustainability the adaptation and mitigation to climate change should address the role of gender in strategic decision-making and in the access to the use of and control over resources. In such a scenario, women's capacity building, along with the integration of participatory approaches in the management of natural coastal resources, is required to promote agricultural sustainability. Besides cultural and economic barriers, the limited access to resources and inadequate policy support are the constraints faced by women in coastal areas that need to be minimised for attaining sustainability. The perception of climate change influences women more than men while selecting adaptation strategies. In such a scenario, ignoring half of the population (women) in coastal planning and management may not help adaptation programs to become successful.

There are several technological options available for its management and productive use of natural resources in coastal zones (Mandal, et al., 2018). By adopting these management options coastal saline soils can be managed sustainably. There is a need for policy support for wider application of these technologies. Some specific recommendations are

- **Launching of dedicated coastal area development programme** - Management of coastal soils needs location specific solution at micro-management level. Often the impact, in terms of loss of production is not realized at the macro level, but serious concern at the regional level. Yield losses are particularly detrimental at a local scale because salt-affected soils are not uniformly distributed and threaten the continued existence of agriculture in coastal regions of the countries. This programme can cover all the coastal districts in India.

- **Incentives and mandatory measures for management of salt-affected soils** - Incentives might be extended to reducing cost of amendments and mandatory measures to check excessive groundwater exploitation. Experiences with technology adoption during implementation of livelihoods projects in the past indicated that capital intensive natural resources management of technologies do not go very far without supportive policy environment. The expectations of return from farming in salt-affected areas are normally much lower as compared to other good land due to substantial production loss. The extent of losses a farmer can afford and still be in the farming depends on the farmers' own available alternative options and his motives of farming. Very often it is seen that farmers are de-motivated and leave farming abandoning the cultivable lands subject to further degradation. Under this circumstances Govt. tries to keep farmers back into his farming business through incentive mechanism (subsidy for amendment, cost sharing) or enforce rules and regulation (limiting groundwater extraction crop choice) to arrest further degradation of soils. Policy interventions for effective management of salt-affected soils need to be a blend of providing incentives and regulating mandatory measures. Incentive measures to be included consists of providing subsidies for buying amendments, initial investment for land shaping/modifications, based on actual costs and need of the farmers and thus mainstreaming them in developmental programmes of State/Central Governments.
- **Effective community engagement** - Agriculture is a state subject and interventions on land management activities are always complex and sensitive. Understanding local community, private bodies and common property resources needs particular focus. Technologies for coastal salt-affected land management have been demonstrated in farmers field in various parts of the country and benefits accrued mainly to the individual farmers. Technologies have been highly successful for increasing the private benefits but by and large could not be made socially inclusive. Adoption of salt-affected management technologies requires sizeable investment on land reclamation or modification. Incentives for such investment on land, particularly for the tenant farmers are not very attractive proposition and thus impeding its large-scale adoption. Effective engagement through collaboration with non-governmental organisations (NGOs) and focussing on higher community engagement would be leading to higher social benefits

and larger adoption. Besides developing technologies effective for private lands, there is a need for community level land management technologies effective for community at large.

- **Prioritization of research based on farmers' need** - Research on problems of salt-affected soils had distinct phases like understanding basic cause-effect relationship of salt and plants productivity, crop variety development with salt tolerance, finding solution and optimization of amendment (both chemical and biological) applications, transferring technologies to problem areas and analysing impacts. Besides, amendment options, development of crop varieties and agronomic management for higher returns have been given top priorities. Often the experimental output showed remarkable increase in crop productivity and best in terms of technology (positive output-input ratio) but actual adoption remained low, indicating even the researchers' best was not good enough for the farmers and failing to translate into societal benefits. Farmers failed to own the technology due to mis-match between priorities of problems as assigned by researchers and those faced by the farmers. Research prioritization needs clear distinction as well as blending of basic, applied and action research. Research priorities needs to be based on target areas (how much area affected?), target population base (or how many households?) and with due consideration of farmers' socio-economic conditions (which categories of farmers?).
- **Creation of real time database and impact assessment** - Assessment of accurate salt-affected coastal areas is one of the major impediments for effective planning and implementation of any policy. Although land degradation is recognized as a serious problem but information available on the severity as also the area affected (real time) by various forms of degradation is limited, highly variable and sketchy. For effective decision making for coastal salt-affected interventions, there is a need for block level intervention planning. Almost every states are having detailed database on land use statistics developed by NRSC, NBSS&LUP or ICAR-CSSRI, however its application in planning has been limited so far. The available information on the extent of affected areas to be simplified and made user friendly so that one can locate the actual salt affected area. Collaboration among these institutes is needed for translating these databases into decision making applicable at grassroots level, taking into consideration the intrinsic character of land, the concerned user groups and future possibilities. Also, there is a need for developing socio-economic database for the affected farmers and collating this information with the existing land use data for identifying particular needs of the farmers and possible implications on technology dissemination at macro level.

Conclusion

In coastal area the land shaping is an innovative technology for addressing the key challenges like land degradation (salinity), drainage congestion and scarcity of fresh water for irrigation and in turn have the potential to enhancing production, productivity, income and employment. This is one of the most important strategies not only to control run-off in the region and soil loss but also contribute to climate change mitigation as well as increased ecological resilience due to improvement of degraded land and water quality and more carbon sequestration due to more plant cover. These techniques are financially viable and attractive proposition for the coastal region. However, major constraints for adoption of land-shaping techniques are marginal land holdings that too divided into several parcels, high initial investment, presence of acid sulphate soils near surface or at shallow depth at places, distance from residential village etc. Though the technologies have been well adopted at farm level, there is lack of information on larger watershed or basin level hydrological impacts such as availability of rain water for downstream flow and groundwater recharge. There is a need to understand and resolve issues on a large scale dissemination of land-shaping technology covering the areas of input-supplies and management, market and marketing environment – the driver of change in cropping pattern and production, credit needs and absorption of the farmers, and the role financial institutions therein. More intensive study, particularly the long term implications of these techniques should be undertaken to address those issues so that the land shaping will be adopted in a large scale for the sustainable agricultural development in the salt affected coastal region.

While the sustainability of coastal agriculture is a very complex issue and is affected by many inputs, both climatic and non-climatic, it needs to be addressed for future societal viability and food security. Integrated coastal resource management and proper land-use planning are of foremost importance. The inclusion of farmers in the planning and implementation of the policies is necessary. However, policymaking is more effective when education and communication work in a synergistic manner. Therefore, the gap between farmer perceptions and scientific records of climatic information must be minimised to increase their participation in adaptation actions concerning self-reliance and sustainability in coastal agriculture.

References

Anil, R.C., Ali, S., Kulshrestha, N., Singh, R.K., Bundela, D.S., Kumar, P., and Sharma, D.K. (Eds.). 2015. Intellectual property management and commercialization of ICAR-CSSRI technologies for management of salt-affected and waterlogged soils of India. ICAR-Central Soil Salinity Research Institute, Karnal-132 001, India, p 62.

Bell, R.W., Mainuddin, M., Barrett-Lennard, E.G., Sarangi, S. K., Maniruzzaman, M., Brahmachari, K., Sarker, K. K., Burman, D., Gaydon, D.S., Kirby, J.M., Glover, M., Rashid, M. H., Khan, M. S. I., Kabir, M. E., Rahman, M. A. and Hossain, M. B. 2019.

Cropping systems intensification in the coastal zone of the Ganges Delta: opportunities and risks. Journal of the Indian Society of Coastal Agricultural Research. 37(2): 153-161.

Bharathi, S., Cheryl, A., Uma, A., Ahilan, B., Aanand, S. and Lingam, R.S.S. 2019. Application of Renewable Energy in Aquaculture. Aqua International March Issue: p. 48, 50, 52 & 54.

Bhaskar, G., Murthy, L. and Sharma, V.P. 2017. Mobile apps empowering farmers. Extension Digest. 1(2): 1-35.

Bhatt, J.R. and Kathiresan, K. 2012. Valuation, carbon sequestration and restoration of mangrove ecosystems in India. In: Sharing Lessons on Mangrove Restoration, Macintosh D. J., Mahindrapala, R. and Markopoulos, M., (Eds.), Mangroves for the future, Bangkok, Thailand and IUCN, Gland, Switzerland. pp 19-38.

Buckley, C. Penrose 2007. Producer Organizations: A Guide to Developing Collective Rural Enterprises. Oxford: Oxfam GB.

Burman, D., Mandal, S., Bandopadhyay, B. K., Maji, B., Sharma, D. K., Mahanta, K.K., Sarangi, S. K., Mandal, U. K., Patra, P., De, S., Patra, S., Mandal, B., Maitra, N. J., Ghoshal, T. K. and Velmurugan, A. 2015. Unlocking production potential of degraded coastal land through innovative land management practices: a synthesis. Journal of Soil Salinity and Water Quality.7: 12-18.

Chaurasiya, P.K., Warudkar, V. and Ahmed, S. 2019. Wind energy development and policy in India: A review. Energy Strategy Reviews. 24: 342-357.

Chel, A., and Kaushik, G. 2011. Renewable energy for sustainable agriculture. Agronomy for Sustainable Development. 31: 91-118. https://doi.org/10.1051/agro/2010029

CSSRI, NAIP, 2014. Final Report of NAIP Sub-Project on Strategies for Sustainable Management of Degraded Coastal Land and Water for Enhancing Livelihood Security of Farming Communities. ICAR-Central Soil Salinity Research Institute, Regional Research Station, (CSSRI, RRS), Canning Town - 743 329, India. p. 104.

Dash, P. K. 2019. Offshore wind energy in India. AkshayUrja. 12(5): 23-25.

Ghoshal, T.K., De, D., Biswas, G., Kumar, P., and Vijayan, K.K. 2019. Brackishwater aquaculture: opportunities and challenges for meeting livelihood demand in Indian Sundarbans. In: The Sundarbans: A Disaster-Prone Eco-Region, Coastal Research Library 30 (Eds. H S Sen), Springer Nature, Switzerland. pp. 407-437.

Gill, S. S., Chana, I., and Buyya, R. 2017. IoT based agriculture as a cloud and big data service: the beginning of digital India. Journal of Organizational and End User Computing. 29(4): 1-23.

Gopalakrishnan, T., Hasan, K.M., Haque, A.T.M.S., Jayasinghe, S.L., and Kumar, L. 2019. Sustainability of Coastal Agriculture under Climate Change. Sustainability, 11, 7200.

Han, W., Meehl, G.A., Rajagopalan, B., Fasullo, J.T., Hu, A., Lin, J., Large, W.G., Wang, J., Quan, X., Trenary, L.L., Wallcarft, A., Shinoda, T., Yeager, S. 2010. Patterns of Indian Ocean sea-level change in a warming climate. Nature Geoscience. 3: 546–550

Islam, M.T. 2014. Vegetation changes of Sundarbans based on Landsat Imagery analysis between 1975 and 2006. Landscape Environment. 8: 1–9

Kaladharan, P., Nandakumar, A, Rajagopalan, M. and George, J.P. 2005. Mangroves of India: Biodiversity, conservation and management. Marine Fish Information Service, Technical and Extension Series No. 183, ICAR-Central Marine Fisheries Research Institute, Kochi, Kerala. pp 8-14.

Kathiresan, K. 2018. Mangrove forests of India. Current Science.114(5): 976-981. doi: 10.18520/cs/v114/i05/976-981

Kathiresan, K. and Bingham, B. L. 2001. Biology of mangroves and mangrove ecosystems. Advances in Marine Biology. 40: 81-251.

Khare, V., Nema, S. and Baredar, P. 2013. Status of solar wind renewable energy in India. Renewable and Sustainable Energy Reviews. 27:1-10.

Krishna Kumar, K., Kamala, K., Rajagopalan, B., Hoerling, M.P., Eischeid, J.K., Patwardhan, S.K., Srinivasan, G., Goswami, B.N., and Nemani, R. 2011. The once and future pulse of Indian monsoonal climate. Climate Dynamics, 36 (11): 2159–2170.pp. 335–406

Kumar, J.C.R. and Majid, M.A. 2020. Renewable energy for sustainable development in India: current status, future prospects, challenges, employment, and investment opportunities. Energy, Sustainability and Society. 10(2). https://doi.org/10.1186/s13705-019-0232-1

Kumar, V.S., Pathak, K.C., Pednekar, P., Raju, N.S.N. and Gowthaman, R. 2006. Coastal processes along the Indian coastline. Current Science. 91(4): 530-536.

Maji, B., Sarangi, S.K., Mandal, U.K., Burman, D. and Balasubramaniam, P. (2020). Climate Change and Management Options for Sustainable Soil Health and Crop Production: Eastern Coast of India. Advanced Agricultural Research & Technology Journal. IV(1): 28-41

Malu, P.R., Sharma, U.S. and Pearce, J.M. 2017. Agrivoltaic potential on grape farms in India. Sustainable Energy Technologies and Assessments. 23: 104-110. doi: 10.1016/j.seta.2017.08.004

Mandal, S., Sarangi, S.K., Burman, D., Bandyopadhyay, B.K., Maji, B., Mandal, U.K., Sharma, D.K. 2013. Land shaping models for enhancing agricultural productivity in salt affected coastal areas of West Bengal – An economic analysis, Indian Journal of Agricultural Economic.Vol.68, No. 3, pp. 389-401.

Mandal, S., Burman, D., Bandyopadhyay, B.K., Mandal, U.K., Sarangi, S.K., Mahanta, K.K., Maji, B., Sharma, D.K., Maitra, N.J., Ghoshal, T.K., Velmurugan, A., Ambast, S. K., Mani, P.K., Mandal, B., Patra, P., Patra, S. and De, S. 2015. Crop-Fish Integration through Land Shaping Models for Enhancing Farm Income under Eastern Coastal Region of India. Agricultural Economics Research Review. 28(Conference issue):47-54. DOI:10.5958/0974-0279.2015.00021.X.

Mandal, S., Raju, R., Kumar, Anil, Kumar, Parveen and Sharma, P.C.2018. Current Status of Research, Technology Response and Policy Needs of Salt affected Soils in India – A Review, Journal of Indian Society of Coastal Agricultural Research, 36(2):40-53.

Mandal, U.K., Burman, D., Bhardwaj, A.K., Nayak, D. B., Samui, A., Mahanta, K.K., Lama, T.D., Maji, B., Mandal, S.,Raut, S. and Sarangi, S.K. 2019b. Waterlogging and coastal salinity management through land shaping and cropping intensification in climatically vulnerable Indian Sundarbans. Agricultural Water Management.216: 12-26.

Mandal, U.K., Maji, B., Mullick, S., Nayak, D.B., Mahanta, K.K. and Raut, S. 2019a. Global climate change and human interferences as risk factors, and their impacts on geomorphological features as well as on farming practices in Sundarbans eco-region. In: The Sundarbans: A Disaster-Prone Eco-Region, Coastal Research Library 30 (Eds. H. S. Sen), Springer Nature, Switzerland. pp. 407-437.

Mandal, U.K., Nayak, D.B., Samui, A., Jana, A. K., Mullick, S., Lama, T.D., Bhardwaj, A.K., Mahanta, K.K., Mandal, S., Raut, S., Sarangi, S.K. and Burman, D. 2018. Trend of sea-level-rise in West Bengal coast. Journal of the Indian Society of Coastal Agricultural Research. 36(2): 64-73.

Nikam, Vinayak., Singh, Premlata., Ashok, Arothy and Kumar, Shiv. 2019. Farmer Producer Organisations: Innovative Institutions for upliftment of Small Farmers. Indian Journal of Agricultural Sciences 89(9):1383-1392.

Pal, K., Yadav, P. and Tyagi, S.K. 2017. Renewable sources in India and their applications. In: Sustainable Biofuels Development in India, A.K. Chandel and R.K. Sukumaran (Eds.), Springer International Publishing. pp. 39-71. DOI 10.1007/978-3-319-50219-9_3AG 2017

Parry, M.L. 2007. Climate Change 2007: Impacts, Adaptation and Vulnerability: Contribution of Working Group II to the Fourth Assessment Report of the Intergovernmental Panel on Climate Change. Cambridge University Press, Cambridge, UK.

Parthiban, R. Shakthi., Main, M.S., Singh, Rashmi., Kumar, Shiv and Chahal, V.P. 2015. Farmers producer organization in reducing transaction costs: A study of Tamil Nadu Mango Growers Federation (TAMFED), Indian Journal of Agricultural Sciences. Vol. 85(10):1303-7.

Pramanik, A., Sengupta, S. and Bhattacharyya, M. 2019. Microbial diversity and community analysis of the Sundarbans mangrove, a world heritage site. In: Microbial Diversity in the Genomic Era, S. Das and H.R. Dash (Eds.), Academic Press, Elsevier. pp. 65-76. doi:10.1016/B978-0-12-814849-5.00005-8.

Raghu Babu, M., Prasad, P.R.K., Ravindra Babu, P., Lakshmi, G.V., and Gupta, S.K. 2009. Technologies for skimming fresh water in coastal sandy soils-Bulletin No. 1/2009. 44p. Saline Water Scheme, Bapatla-522101, India.

Sahu, S.C., Suresh, H.S., Murthy, I.K. and Ravindranath, N.H. 2015. Mangrove area assessment in India: Implications of loss of mangroves. Journal of Earth Science and Climate Change 6 280. doi:10.4172/2157-7617.1000280

Santra, P., Singh, R.K., Meena, H.M., Kumawat, R.N., Mishra, D., Jain, D. and Yadav, O. P. 2018. Agri-voltaic system: crop production and photovoltaic-based electricity generation from a single land unit. Indian Farming. 68(01): 20-23.

Seth, A. and Ganguly, K. 2017. Digital Technologies Transforming Indian Agriculture. In: The Global Innovation Index 2017: Innovation Feeding the World, Dutta, S, B Lanvin and S Wunsch-Vincent (eds.), Cornell University, INSEAD, and WIPO, Ithaca, Fontainebleau and Geneva. pp. 105-111.

Sharma D.K., Singh, A. and Sharma, P.C. 2016. Role of ICAR-CSSRI in sustainable management of salt-affected soils— achievements, current trends and future perspectives 4th International Agronomy Congress. (November 22–26): 91-103.

Sharma, D. K., and Chaudhari, S. K. 2012. Agronomic research in salt affected soils of India: An overview. Indian Journal of Agronomy. 57(3rd IAC Special Issue): 175-185.

Sharma, D. K., and Singh, A. 2015. Salinity research in India: achievements, challenges and future prospects. Water and Energy International. 58, 35-45.

Singh, G. 2009. Salinity-related desertification and management strategies: Indian experience. Land Degradation and Development. 20:367-385.

Srinivasarao, Ch., Gopinath, K.A., Prasad, J.V.N.S., Prasannakumar, Singh, A.K. 2016. Climate resilient villages for sustainable food security in tropical India: concept, process, technologies, institutions, and impacts. Advances in Agronomy. Volume 140: 101-214

Syvitski, J. P. M., Kettner, A. J., Overeem, I., Hutton, E. W. H., Hannon, M. T., *et al.* 2009. Sinking deltas due to human activities. Nature Geoscience 2. doi:10.1038/ngeo629.

Unnikrishnan, A. S., and Shankar, D. 2007. Are sea-level-rise trends along the coasts of the north Indian Ocean consistent with global estimates? Global and Planetary Change. 57: 301–307.

Velayutham, M., Sarkar, D., Reddy, R. S., Natarajan, A., Krishan, P., Shiva Prasad, C. R., Challa, O.,Harindranath, C. S., Shyampura, R. L., Sharma, J. P. and Bhattacharyya, T. 1999b. Soil resource and their potentialities in coastal areas of India. Journal of the Indian Society of Coastal Agricultural Research. 17: 29-47.

Venkattakumar, P., Mysore, Sudha., Venugopalan, R., Balakrishnan, B., Narayaswamy, B., Atheequlla, G.A., Paripurna, Achala., and Reddy, T.M. 2019 Performance of Farmers Producers Organizations (FPOs) and Associated Factors in Karnataka: Producers' Perspectives. Indian Res. Journal of Extension Education 19 (2&3): 7-12.

Yadav, J.S.P. 2004. Coastal agricultural research-present status and future perspective with special reference to value addition. Journal of the Indian Society of Coastal Agricultural Research. 22 (1&2): 1-8.

22

Structural Reforms for Inclusive Agricultural Growth and Self-Reliance of Bundelkhand Region

Sunil Kumar[1], *Gaurendra Gupta*[1], *A.K. Handa*[2] *and P.K. Ghosh*[3]

[1]*ICAR-Indian Grass land and Fodder Research Institute, Jhansi, Uttar Pradesh*
[2]*ICAR-Central Agroforestry Research Institute, Jhansi, Uttar Pradesh*
[3]*ICAR-National Institute of Biotic Stress Management, Raipur, Chhattisgarh*

Introduction

Bundelkhand is an important region of Central India which is often called as heart-land of the nation. It is spread in the geographical area of 7.08 mha, includes 14 districts seven, each of Uttar Pradesh and Madhya Pradesh and supports 1.84 crore (as per 2011 census) of socio-economically and culturally diverse population. The present region of Bundelkhand lies between approximately 23°10' and 26°27' (north) latitude and 78°4' and 810 34' (east) longitude. It lies between the Indo-Gangetic Plain to the north and the Vindhya Range to the south. Bundelkhand is a hot and semi- humid climate with highly variable depending on the region and with gently sloping upland, distinguished by barren hilly terrain with sparse vegetation, although it was historically forested. As with the rest of the Indian sub-continent, Bundelkhand experiences two main seasons, monsoon and dry. The monsoon brings over 90 percent of the annual rainfall between the months of June and September with the highest precipitation occurring in July and August. On an average, the region receives 75 cm to 125 cm of rain each year. The dry plains in the north usually receive less rain compared to the south- eastern part of the region. Minimum temperature varies from around 6-12 °C and maximum temperature varies from 38- 48 °C.

The region is blessed with plenty of natural resources. It had truly been endowed by nature, with beautiful mountains, good perennial rivers, fast flowing seasonal rivulets, large rich forest areas with valuable timber, fruit trees, numerous medicinal plants / herbs in almost each district. The rainfall here had been on average much better than many parts of India. Bundelkhand is a part of the central zone of India with a group of north-bound perennial rivers flowing and paying tribute to river

Yamuna. From west to east these are Sind, Pahuj, Betwa, Dhasan, Ken, Baghein Paisuni and Tons. Along with Chambal and Yamuna these are ten major rivers. Other water resources in Bundelkhand are in form of water stored / conserved on surface termed as ponds, tanks, lakes etc. Bundelkhand has, undoubtedly, been one of the regions in India very much known for its historical rain-water management through man-made ponds, tanks and lakes. Topography here helped to build such structures but more than anything it reflected fore-sight and positive attitude of those rulers towards conservation of nature.

The soils in the Bundelkhand region have developed from Vindhyan rocks abounding in gneiss and granites of the Deccan trap with highly ferruginous beds and often soft lime stone. The soils can be divided into two broad groups, (1) red and (2) black soils. Locally these soils are termed as *rakar, parua, kabar* and *mar*, respectively. The red soils, *rakar* and *parua*, belong to the order Ultisols, Alfisols, and Inceptisols and the black soils are grouped under the great groups, Pellusterts, Chrousterts and Ustochnepts. The *rakar* soils are residual, slightly acidic, coarse grained, shallow and excessively permeable soils occupying higher elevations. The *parua* soils are alluvial, mildly alkaline, very deep soils with free $CaCO_3$ accumulation at lower depths. The black soils, *mar* and *kabar*, are very deep soils confined to low lying landscapes-having fine texture and the remarkable property of shrinkage and swelling on drying and wetting. These soils are normally deficient in organic matter and difficult to work. All available soil-types had been productively used by farmers for growing varieties of indigenous crops, suited to particular soil, slope and local climate.

The most important resource of Bundelkhand is, in true sense, the dense rich forests of this region. These area not only provide valuable timbers, wild-fruits, edible tubers, large number of medicinal plants and herbs, varieties of minor forest produce (MFP) and grasses for millions of cattle grazing here from hundreds of villages situated near and far. The excess rains flowing out of forests as floods used to bring valuable humus to agricultural fields of villages on its way. Forest is the home for large number of indigenous tribes and also enormous wild life. These together created a congenial atmosphere for all living beings through providing healthy air, water and atmosphere. Bundelkhand is also a home to around 450 different herbs/ medicinal plants since time immemorial and have been utilized as medicinal purpose.

Major issues of Bundelkhand

Extreme weather

Agriculture in Bundelkhand is mainly rainfed, diverse, complex and risky owing to increasing vulnerability to the frequent and unseasonal extreme weather conditions, like droughts, short-term rain and flooding in fields. In the recent past,

9 years experienced moderate to disastrous drought in this region, wherein rainfall deficiency ranged between 24 and 60% from normal value. Among the different locations, Lalitpur showed maximum drought probabilities i.e. once in three years. However, the disastrous drought (more than 50% rainfall deficit than normal) probabilities at Banda, Hamirpur, Lalitpur, Damoh, Sagar and Chhatarpur are in between two to five decades. Moderate drought is a recurring feature at Tikamgarh with probability of once in six years.

Water scarcity

In all the districts of Bundelkhand, the non-availability of water for drinking purpose and irrigation is a major problem. Despite of number of major rivers like Betwa, Dhasan, Ken flowing in the region but with the improper management of rain water, and degrading forests, these rivers serve the purpose of only draining the water of Bundelkhand into Yamuna. The recharging of ground water resources through collecting the rain water in traditional ponds is not taking place properly due to degradation of such ponds, encroachment of forest lands, cutting of forests and blocking the catchment areas of watershed regions, as a result wells being used for drinking water and irrigation are not providing enough water to serve the purpose for the whole year.

Low fertile soils

The landscape is undulating with characteristics hillocks soil of the region. Soils of the region exhibit greater heterogeneity with regard to texture depth, fertility (low OC= <0.4%) leading to operational and management problems. In some of the part of this region, there is notable problem of soil erosion and deep ravenous areas due to its topographic gradients & geological setup.

Practice of Monocropping

Bundelkhand region is pre-dominantly occupied by some selected crops as per water resource availability. Recent trends also witnessed a moderate growth of diversification of 1% in all the crops. Due to this reason, the area under mono-cropping or patterns are seen in entire Bundelkhand. Vegetables or oilseed/pulses based diversification is available but that too in pockets.

Low livestock productivity

Livestock sector in this area is characterized by high population of unproductive cattle, indigenous milch animals with low productivity, ineffective input delivery mechanism, and traditional milk marketing systems and lack of adequate feed resources both quantitatively and qualitatively. The traditional practice of "*Annapratha*" in Bundelkhand region, where cattle are let loose for free grazing has been a great setback for crop production even during *kharif* season. Majority of producers are unaware about scientific dairy farming practices and value addition. The major cropping season (*kharif* & *rabi* both) crop cultivation is possible only with some protection of fields through fencing and watch of farmers. Recently, this problem has taken a shape of menace to the farmers and due to roaming animals, growing of the fodder, vegetable and other than food crops has significantly reduced. Lack of fodder and grazing resources on community lands are putting further pressure and making these grazing areas totally unproductive. This is creating now conflicts among village community and government.

Unregulated mining

Mining consumes huge amounts of water and deplete the water table. The lure of commercial profits includes foreign exchange earnings and ruins the chances of agriculture. The present mining operations in Bundelkhand are so indiscriminate that the entire environment has been endangered. Mining starts with deforestation and leads to groundwater wastage, soil erosion, silting and pollution of rivers and streams.

Subsistence-like farming

The agriculture in Bundelkhand region is subsistence-like and characterized by low productivity and profitability and high risk. This insecurity of livelihoods and lack of supportive governance have led to forced large-scale migration of the local population. The reason of poor agriculture development is mainly due to adoption of local varieties with poor plant stand and also due to soil moisture deficit during *rabi* season. The villagers depend upon agriculture mining, employment on development at work and collection of forest produce. People are bound to migrate in search of other alternatives for sustenance due to lack of employment industries, low education status and poverty.

Lack of industrial development

Bundelkhand is known as either state with low industrialization or low urbanization even among the least developed region. The mineral wealth of the Madhya Pradesh part of Bundelkhand is the only industry which is cursed with illegal mining. The existing industries in the region are confined to the Jhansi city as it is easier to export the raw material from Jhansi to the national capital and to the other developed parts of the country.

Lack of support from off-farm employment

The decrease in the number of days of rainfall, frequent droughts, a decline in the quantity of water available for irrigation and often even for usufruct and drinking has affected agriculture most severely, burdening the already struggling small and marginal farmers. Non-farm employment is stagnating and even declining and modern economic opportunities are few and beyond the reach of the skill profile of the average young person born in this region. All this leads to a particularly disheartening economic and employment situation that reflects an inescapable economic cycle, that is unable to expand its horizons and size.

Poor infrastructure to support tourism

There are several tourists' attractions in Bundlekhand ranging from monuments, havelis, natural habitats, religious and spiritual destinations. Despite of rich heritage and tourists' spots exists in the region, the tourism in Bundelkhand does not attract many domestic and foreign tourists. It's mainly because of lack of infrastructure of road, energy, security, amusement and promotional ventures of standard quality.

Inadequate implementation of schemes and multiplicity

During past few years, many programmes and schemes were launched for development of Bundelkhand and bring prosperity to its people. In last decade, Bundelkhand package was announced but its impact need to be further out scaled in larger area. The escalating helplessness and marginalization have made imbalance in the rural economy of Bundelkhand. In these conditions, the villagers' plight are at stake by the service providers, line departments, Panchayati Raj Institutions and also bank functionaries are also not properly serving for the poor. All these factors have forced villagers to migrate and the region further enters in vicious cycle of poverty and backwardness.

Indigenous Technical Knowledge (ITK) of Bundelkhand

There are numerous ITKs available which are being followed by the peoples of Bundelkhand for soil and water conservation, disease and pest management of crops and animals, and other age old and traditional activities related to various strata of social life . Although, most of these ITKs are under documentation still these are popular among the peoples. Therefore, in order to maintain the cultural identity of Bundelkhand and to keep those worthy techniques and traditions alive, the ITKs should find an important place in the policies being drawn in Bundelkhand area.

Some of the notable, popular and age old ITKs of Bundelkhand region are

General

- *Haveli* system of cultivation (rainwater is impounded in bunded fields)
- Stone wall for *Nallah* bank protection in erosion prone areas
- Sand bag structure (*Bori Bandhan*) constructed during peak flow across the drainage line of field with cement bags filled with soil/sand to check the flow of water

- Village pond/ *talab* a common rural rainwater harvesting technology (surface runoff)
- *Talai*- A small water storing structure *Talai* is an indigenous water harvesting technique
- Farm pond (small pond) -generally constructed by a farmers, whose land remain temporarily submerged and after monsoon
- Smaller cross-sectional *Nali* Structure most suitable soil and water conservation techniques for hilly areas moderately sloping land nali of about 30- 45 cm depth are constructed across the slope almost on contours for enhancing in-situ moisture conservation
- ITKs for plant and animal protection from pests and diseases

Specifics

Crop related (Triathi et al., 2020)

- Treatment of ginger and turmeric seed by the solution of fresh cow dung (1 kg cow dung in 5 liter water) for the management of soil and seed borne disease and better germination.
- Sclerotial wilt and root rot in betelvine control using sesame, mustard and neem cake.
- Soil-borne disease and nematode management in vegetables using Karanj cake.
- Wilt disease of chickpea control using 30 kg seed of chickpea with 05 gm Hing (Asafoetida) + 200 gm salt mixed in one liter of butter milk (Whey).
- Growing of marigold with solanaceous vegetable crops for the management of bacterial wilt.
- Seeds of pulses treated with cow urine for the protection of soil borne fungi.
- Control of leaf defoliating insects, thrips, aphids and jassids using 01 kg dried chilli powder with equal quantity of garlic mixed with 200 litres of water.
- Management of insect pests of vegetable crops through *Nicotiana tabacum* (tobacco) + Onion extract specially thrips and white fly.
- Growing coriander on the bunds of cauliflower and cabbage field is a popular practice in the Bundelkhand region for control of Diamond back moth (DBM) and other defoliators.
- Mixing neem leaves @ 4-5 kg / quintal and seed of Sitafal @ 2-3 kg / quintal in stored grains.
- Red pepper and Salt (NaCl) in the bags of rice to protest the grains from stored grain pests. Eucalyptus wood ash is also added with paddy to protect seeds during storage in Tikamgarh district.

- Betel leaf cultivation- The betel leaf (popularly known as *paan*) cultivation is one of the traditional ways of livelihood in the Mahoba, Lalitpur and some parts of Banda districts of Bundelkhand. Around 2000 *barejas* (a kind of traditional greenhouse for growing paan) presently exist in *paan* growing regions of Bundelkhand.
- *Bidi* making from forest plants- In Bundelkhand, Beedi (hand rolled, unfiltered sun dried tobacco flakes, rolled in a *tendu* leaf or *temburni* leaf and then tied with cotton thread) making is an age old industry and one of the traditional job providing sector particularly for women, in the unorganized sector. Tendu (*Diospyros melanoxylon*) is a traditional flora found in the forests of Bundelkhand; andis being used for bidi rolling. However, traditional bidi rolling area workers are facing scarcity of work opportunity and shrink in income level.
- Hard wheat (*kathia gehun*) of Bundelkhand- Hard wheat (popularly called *kathia gehun*) is a reddish coloured special type of wheat which was in practice of cultivation in the river Yamuna belt as well as districts of Bundelkhand mainly, Banda, Chitrakoot and Hamirpur. This wheat was so popular in the region by virtue of its nutritious nature, hardiness to aberrant weather and ability to be cultivated in resource deficit conditions. Eefforts for the conservation of traditional crops and their cultivars are needed to be taken at policy level.
- Minor millets of Bundelkhand- Bundelkhand have been home to the various traditional millet foods such as *Jowar, Bajra, Chea, Quinova, Teff, Kodo, Sawan* etc.. Minor millets are rich in nutrition, *kodo* and *kutki* are gluten-free and non-acid forming foods. However, with the advent of various high remunerative and productive crops such as wheat, sesame, mustard and pulses has sent these millets to the background level.
- Traditional earthen grain storage structure (*Kothi*)-These storage structures are quite common in rural areas of Bundelkhand for the storage of grains and other seeds. The capacity of such storage structures varies from 1 to 50 tonnes. These are made from mud mixed with dung and straw. There are many sizes and dimensions of Kothi made for storing grains.

Animal husbandry related

Treating dysentery: Three pieces of Golmorich (Black pepper, *Piper nigrum*), 5 g Jowan (Bishop's weed, *Trachyspermum ammi*) and 5 g Chirata (*Swertia angustifolia* var. pulchella) are grounded and fed to the animal for 3-4 days (Amitendu et al., 2004).

For arthritis: A luke worm paste is made from Lahsun (Garlic, *Allium sativum*) and ghee and applied on the affected part

For cough and cold: Efflorcence of Tulsi (Holy basil, *Ocimum sanctum*) and *Basak* (*Adhatoda vasica*) leaves are mixed and extract is taken and mixed with ghee, *Adarak* (Ginger, *Zingiber officinale*) and molasses and fed to the cattle

For wound; *Haldi* (Turmeric, *Curcuma domestica*) is grounded and applied topically.

Prioritisation of focus areas in Bundelkhand

Agriculture and livestock sector has vital importance for the upliftment of socio-economic status of peoples of Bundelkhand since majority of its population rely upon agriculture and livestock sector for livelihood and employment. Infact, government is much concerned for the holistic development of Bundelkhand through special purpose schemes and funds for uplifting industry, roads and transport, agriculture, health and education sector. Considering present scenario, the following areas need priority:

Agriculture and allied sector

Agriculture and livestock sector of Bundelkhand is lifeline of its peoples. Therefore, it becomes very crucial for upliftment of this sector via technology overhaul and adoption, assured input and credit supply, crop insurance and better produce price, post-harvest value addition, produce storage facilities and proper transport and marketing facilities. In this regard, some of the specific areas of agriculture and livestock which needs attention are as following:

- Adoption of Integrated Farming System (IFS) so that higher land and other resource use efficiency could be achieved through integration of various agriculture and livestock components. Adoption of IFS will also help in generating round the year income and employment.
- Crop diversification should get importance since it imparts diversity in the farm produce as well as it reduces risk of crop failure due to various weather calamities such as drought and water logging which are major risk factors in Bundelkhand. It will also guarantee minimum production at least for livelihood sustenance under highly unfavourable circumstances.
- Livestock sector of Bundelkhand is also an important sector which needs to have attention by the way of popularising cross breeding for breed improvement, bridging the gap between availability and requirement of progeny tested proven dairy sires, ensuring the availability of quality feed and fodder, giving place to fodder crops in farming system, enhancing milk collection, procurement, transport and marketing facilities (Rathod et al., 2020).

- Fisheries offer much greater opportunities than have been utilized so far. While MP Government has been keen to promote fisheries in their part of Bundelkhand, similar effort must also come for the UP side.
- Focus should be on soil and water conservation by water harvesting through farm ponds, field bunds and haveli structures etc.
- Adoption of agro-forestry will also help in reducing soil erosion and enhancing water harvesting and conservation.
- Imparting awareness and providing technical knowledge to the farmers on various aspects of integrated plant protection (weed, insects and diseases) measures since lack of knowledge on this aspect results into mismanagement.
- In view of future challenges of natural resource management and maintaining soil health, it becomes imperative to infuse the spirit of integrated and precision nutrient management *via* diversity of nutrient sources such as fertilizers, manures, bio fertilizers, green manuring, crop residue incorporation etc.
- Water scarcity is very common due to poor ground water resources and unmanaged water received from rains. Efforts for water harvesting, its distribution, judicious use and recycling has to be attempted seriously,
- Micro irrigation can be boon to water starved areas of Bundelkhand for addressing more crop per drop.
- Bundelkhand has good potential for organic farming due to its typical topography, vegetation diversity, less use of chemicals in dry lands and abundance of forest areas. Development of suitable package of practices of crop cultivation, assured input and credit supply, organic produce certification services and promoting marketing platforms are the need of the day
- Other than above priority areas, setting up of Food Park, agri-market, knowledge dissemination on ICT based agriculture, improving insurance and credit supply structure, post-harvest value addition and food processing, proper facility of agri-produce storage and marketing are also important.

Industry and infrastructure

Industrial development of any region governs its socio-economic status since it is an important sector which generates employment as well as infrastructure facilities to the local peoples in the region. The study reported that in terms of share of total investment in core industry sector across different regions of UP-Bundelkhand, accounts for only less than 2% of the investment against more than twice in the central and eastern regions (also considered backward) and 60% in western UP. This scenario is enforcing the peoples of Bundelkhand to migrate to some other places in search of employment.

Road and transport

Bundelkhand being located in the centre of India, the region needs good connectivity for developing as a logistical hub. If it can generate enough warehousing and transport related infrastructure, its location advantage can be exploited. Adequate schemes promoting investments in this regard already exist and an aggressive effort to make this a logistic hub can develop this sector substantially.

Health

The situation in Bundelkhand is dismal, and reflected in extremely poor usage of government health facilities and the coverage is much poorer, with roughly one primary health care (PHC) per 50 villages in both UP Bundelkhand and MP Bundelkhand. As per a survey, only 10% of people in rural areas who consulted any medical practitioner went to a government health facility, the rest went to a private sector doctor (35%) or a quack or traditional healer (55%). Among poor families in rural areas, the percentage was even lower (7%). Therefore, considering these factors and hurdles, significant improvement is needed at policy and ground level by strengthening work culture in the government hospital, curbing corruption in the service delivery, providing modern and world class infrastructure (Niti Ayog, Human Development report, 2012).

Education

Bundelkhand has been educationally backward as its overall literacy rate (44.2 and 48.5 % for MP and UP Bundelkhand, respectively) and literacy among women is not up to the mark. Bundelkhand is one of the most backward regions and lag far behind on the development indicators. This pathetic condition of education in Bundelkhand is not due to a single factor but due to the interlinked factors of social, psychological, economic and policy related. Education in urban areas is relatively better since there is number of private schools in urban areas. Despite these facilitites, progress in urban education is not significantly better than rural education. Infact , every second child in the UP-Bundelkhand area, and every fifth child in the MP-Bundelkhand area is attending private school.

Tourism

The region has a large number of historical tourism sites, such as palaces, forts, etc. The region has 277 protected monuments and archaeological sites in Bundelkhand (173 in UP and 104 in MP). Bundelkhand houses two of the nation's most attractive tourist destinations and both remain under developed. The first of these two sites is the Khajuraho cluster and the other cluster is around Chitrakoot, one of India's principal centres for religious tourism. In these two sites also, the infrastructure is poor, and there are no facilities for high-spending tourists and almost no supporting activities. The development of tourist attractions of Khajuraho and Chitrakoot can provide a stimulus to the local economies of the region. The potential of Bundelkhand in tourism remains unexplored for its geographical and monumental architectural beauty.

ICTs based accurate weather forecasting for the field operations

Drought or drought like situation is a regular phenomenon of this region. Agriculture is the main livelihood of the inhabitants of Bundelkhand but the weather is certainly the most important factor determining the success or the failure of agricultural crops. ICTs based tools and services need to be upgraded and made effective to address the weather related vagaries, monsoon aberrations, rainfall occurrence intensity & prediction), dry spells, hail storms, untimely rains, heat & cold waves etc. at block or cluster of village levels. Effective follow up of these services will definitely contribute in raising income of farming community of the region, and their risk for crop cultivation could be minimized substantially.

Strengthening of government and civil service organizations

The escalating helplessness and marginalization of masses have made imbalance in the rural economy of Bundelkhand. Practically, the villagers' rights are denied by the service providers, line departments, Panchayati Raj Institutions and also bank functionaries due to lack of awareness and socio-economic disparity. Local institutions such as gram panchayats are in an abysmal condition and not delivering effective good to the rural people. Gender equity and access for people of all communities are required to help people to participate and benefit from services in agriculture, rural schemes benefit, business & trade, market, credit, education and health.

Technologies for transformation of agriculture in Bundelkhand

Bundelkhand has not been benefitted at large scale by the technologies for overcoming its backwardness in the area of agriculture, rural development and social sector. There is huge gap in potential of existing technologies and actual outcome attained with the adoption of these technologies by different categories of the farmers of the region. However with the current initiative and focus of GoI, for doubling of Bundelkhand farmer's income in this region priority should be focus on irrigation with per drop-more crop, quality seed and soil health, investments in warehousing and cold chains, value addition through food processing, creation of a regional farm market and linking to national eNAM, new revolutionary crop insurance scheme to mitigate risks at affordable costs and promotion of ancillary activities like poultry, beekeeping and fisheries.

With an impetus to transform the agricultural scenario in Bundelkhand, various institutions and departments are working in the Bundelkhand region for enhancing agricultural and livestock production and raising farmer's income. These include institutions like ICAR-IGFRI, Jhansi; SAUs under UP & MP jurisdiction (BUAT, Banda, RVSKVV, Gwalior, JNKVV, Jabalpur), KVKs, CAU, state line departments, other central government/state government institutions, NGOs, SHGs, FPOs, various farmers groups and private companies. Several technologies have been developed which are especially suitable under Bundelkhand conditions and notable technologies which deserve prioritised attention are given below:

A. Watershed Technology

The most important environmental problems in Bundelkhand regions are land degradation, loss of soil nutrients, deforestation, ground water depletion, insufficient ground water recharge and diminishing bio-diversity. Considering all these burning issues of Bundelkhand, a holistic and integrated watershed approach is needed which can improve the natural resources condition and ecosystem services for improving livelihood and services. Various watershed management projects are being implemented by different agencies in Bundelkhand area. A water conservation model for the drought prone Bundelkhand region has been successfully developed and tested in Garhkundar-Dabar area by the ICAR - Central Agroforestry Research Institute (CAFRI), Jhansi. The model has shown a promising way to fight drought in the region and so may help to bring in greenery and prosperity there. In this watershed model, construction of check dams, gabions, drainage structures, water spreaders, and field bunds are major components.

B. Integrated Farming Systems (IFS)

Bundelkhand's economy is predominantly agriculture oriented and livestock rearing forms an integral part of rural living. However, the practices adopted by the farmers are conventional and leads to subsistence type of economic gains. Balanced approach with selection of profitable and sustainable practices however, has been demonstrated to enhance the productivity and income as well as reducing risk of climate variability. Therefore, it is the utmost need that the different components (Crop: monocrop, mixed/intercrop, multi-tier crops of cereals, legumes (pulses), oilseeds, forage etc; Livestock: milch cow, goat, sheep, poultry, bees; Tree: timber, fuel, fodder and fruit trees) of livelihood available in Budlekhand to be chosen selectively. The components should have complementary and synergistic effect in recycling of input, reducing cost and enhancing overall profit of farm. Various IFS models have been developed by the different institutes and organizations engaged in the R&D of Bundelkhand. ICAR-IGFRI, Jhansi have developed livestock resource based IFS model (02 rainfed & 03 irrigated). Among the irrigated IFS models, dairy based IFS provides the highest net return per year (Rs. 162593), employment (431 man-days/year) and resource recycling followed by intensive IFS with balanced enterprises (Rs. 108638 and 336 man-days/year, respectively). These IFS models are being validated and up scaled at farmer's field also. In this model allocation of land for the food crops, fruits, vegetables and fodder crops was in the order of 40, 20, 20, 20% percent land along with 3-4 livestock, respectively. In rainfed crop dominated IFS model, higher net return (Rs. 37424) and BC ratio (2.03) can be obtained with emphasis on water harvesting, silvi-pasture and rainfed food-forage crops sustaining 1 livestock and 4-5 goats. Such models are to be promoted and up scaled at large scale through convergence of government schemes like NRMSA, RKVY, NHM and having participation of NGOs, SHGs, FPOs, local communities.

C. Crop Diversification

Crop diversification in Bundelkhand over the period, gained momentum due to overall input assurance, irrigation facility, economic growth, diversification in food basket, better market institutions, credit availability, infrastructure, urbanization, contract farming and improving rural education level. In this region, crop diversification has major positive effect for small millets, sesame and blackgram crop owing to less risk and higher water economy. Along with this, it was also found that area allocation by the farmers under chickpea, pigeonpea, lentil, rice, sorghum, small millet, pearl millet and linseed crops were reduced over the period of time. The area has shifted towards more profitable crops with less infestation of pests & diseases which include blackgram, pea, green gram, wheat, maize, barley, sesame, rapeseed-mustard and groundnut. In region like Bundelkhand the crops requiring more ground water, input and yield plateau have grim future which is reflected in area reduction under wheat, rice and sugarcane. Consequently, increase of area under less water intensive and higher value crops like sesame, blackgram, pea, green gram, maize, barley and oilseed crops has taken place. This is further needed to get encouragement from government in terms of procurement on MSP, value addition & marketing, incentives through various schemes of sustainable agriculture (Kumar et al., 2018).

D. Organic Farming

Bundelkhand is bestowed with plenty of natural resources such as rich forests and fertile soils. This physio-graphic and edaphic ecology offers great potential for organic farming in selected commodities like pulses (peas, chickpea, pigeonpea & blackgram), oilseeds (groundnut, linseed & sesame), millets, turmeric and ginger, other medicinal produce like aswagandha, aloe vera, safed musli, tulsi, forest origin products, and some fruits like citrus, guava, ber, bael, aonla. For promotion of organic farming in Bundelkhand basic requirements would be required *viz.*, standard POP, assured bio-input from reliable sources, credit supply, organic produce certification services and marketing platforms are needed. This can be achieved through cluster farming or through creation of farmers groups (FPOs) which has in-built mechanism for technological package, capacity building, linkages for input supplies and platforms for value addition and marketing. During last few years, government has started making efforts to make Bundelkhand, a hub for organic farming and zero budget agriculture through commissioning of specialized schemes and forming groups.

E. Agro-forestry

Climate change is a reality and Bundelkhand region would be one of the worst hit area witnessing more extreme weather events that include heat waves, droughts, strong winds, and unexpected heavy rains etc. in the coming future. With this climate change, agriculture would be worst affected, because majority of crops are unable to survive more than 15 days dry spell or heavy rainstorm. The situation is worsened when there are very less area under irrigation and resources for supplemental irrigation. Moreover, sustainability in Bundelkhand's agriculture is only possible when woody perennials are included into the crop system because they suffer least from climatic aberrations and withstand weather vagaries on account of woody nature. Further, agro-forestry provides shelter to companion crops against bad weather. Therefore, agroforestry, being climate resilient practice, provides a perfect strategy in Bundelkhand to combat climatic aberrations and at the same time, ensuring better crop produce, income, employment and clean environment which bring overall sustainability and prosperity.

In Bundelkhand multipurpose tree species like teak, neem and *Melia* can provide livelihood support to the resource poor farmers. The tree species like Moringa provides nutritional security. The suitable agroforestry models have been developed by ICAR – CAFRI for adoption in the region. Under Kisan Mitra Project of RKVY, large scale plantation of these tree species on farm boundaries has been undertaken in the seven districts of Bundelkhand in Uttar Pradesh.

F. Fodder Technologies

Profitable livestock rearing may be possible with improved breed and proper feeding and caring of animals. Cost of feed and fodder reaches to the tune of 65-70%. Various fodder production technologies from irrigated lands (intercropping of forages in existing systems, round the year fodder involving seasonal fodder crops (maize, sorghum, pearl millet, guar, cowpea, berseem, oat etc.), perennial fodders (*bajra* napier hybrid, guinea grass, TSH grass, anjan grass etc.), non-traditional fodders (cactus, beet, sugarcane, azolla etc.) and use of tree fodders (subabul, moringa, sesbania, deshi babool, as Anjan tree (*Hardwickia binnata*), Vilayati babool *Prosopis juliflora,* Shirish tree (*Albizia lebbeck*) for fodder supplementation have been developed by ICAR- IGFRI, Jhansi. Multiple crop sequences like sorghum + [cowpea (berseem + mustard)], NB Hybrid+ (cowpea-berseem- mustard), NB Hybrid+ (cowpea (berseem) are promising round the year fodder production systems (270 t/ha green fodder for 8-10 adult cattle units, ACUs) under irrigated conditions of Bundelkhand. Under rainfed conditions of Bundelkhand, perennial fodder production systems are more suitable over the seasonal fodder crops. Suitable fodder crop species for silvi-pasture system are *Cenchrus* grass, marvel grass, Dhaulu grass, blue panic grass, *Desmodium*, purple bush-bean, *Stylosanthes*, Lablab bean. In degraded lands, spineless cactus (protected from direct grazing) can be grown successfully which provide fodder (Crude protein: 6-9 %, water-90-92 %) during dry summer months. Spineless cactus (green cladodes chopped) can be fed to all type of livestock up to 30-35 % mixed with dry fodder. *Azolla* cultivation is also proving a successful technology for producing, nutritious (30-35 percent protein and 10-15 percent minerals on dry weight basis) green fodder under resource (particularly land) deficit condition with least investment and efforts. Apart from these fodder production technologies, proven fodder conservation techniques such as 'silage' and 'hay' are also need to popularise in Bundelkahnd region for ensuring green and quality fodder during lean period.

G. Soil Testing and Integrated Nutrient Management

Bundlekhand soils are not only thirsty but hungry also. The organic carbon content is below 0.5%, low in N & P and majority of soils are deficient in Zn, S, Mo & B. Presently, lot of emphasis is being made towards soil health based nutrient supplementation to crops. The central government scheme "Soil Health Card (SHC)' is a novel approach in this direction. SHC based précised and mechanised fertilizer application and soil organic carbon enrichment techniques (FYM application, green manuring, residue incorporation, legume crops cultivation, agro-forestry) are being recommended for maintaining soil fertility and productivity. Nowadays, in all POPs, INM based recommendations with due emphasis on organics, composting, residue cover, green manuring, and inclusion of legumes in copping system is advocated to farmers which could prove a positive strategy for maintaining soil fertility of Bundelkhand soils.

H. Micro-irrigation

Micro-irrigation has proved to be most effective method for irrigation in rainfed and drier regions having 30-35% more WUE than conventional method of irrigation. In Bundelkhand region, water scarcity in various parts is the reality. In such areas, micro-irrigation technologies are proving a potential solution for addressing prevailing water crisis by the way of enhancing water productivity which is popularly referred as 'per drop, more crop' concept. Implementation of micro-irrigation results in significant benefits which are noticed in terms of crop yields, enhanced water use efficiency (owing to reduced conveyance losses, runoff, evaporation losses, seepage and deep percolation losses significantly), reduction in energy consumption, reduction in the use of chemical fertilizers & pesticides. It also ensures additional benefits like non-exploitation of groundwater, reduction in the cost of weeding and relief from water scarcity induced labour migration.

I. Adoption of Best Management Practices (BMPs)

Most of the small and marginal farmers of Bundelkhand region doing the subsistence farming and have lack of knowledge of good management practices *viz.* quality seeds, time and method of sowing, seed rate, varieties, nutrient, irrigation, weed, pests and disease-management. Major best management practices for different crops in Bundelkhand region are: improved and mechanised tillage, quality seed of improved and disease/pest resistant varieties, seed treatment with fungicides-insecticides-biofertilizer, précised and mechanised sowing, integrated weed management, integrated nutrient management, irrigation scheduling, integrated pest and diseases management, timely and mechanised harvesting, safe post-harvest management and storage. These BMPs are to be popularised through different extension functionaries of state line departments, KVKs, ICAR & SAUs. Models of BMPs to be up scaled taking block of the district as a unit.

J. Breed Improvement of Milch Animal

Low animal productivity is the major factor responsible for less return from animal husbandry in Bundelkhand region. In order to enhance animal productivity of Bundelkhand, breed improvement technologies such as artificial insemination (AI) through sorted semen is required. This technology ensures birth of female calf. In this way, sorted semen based AI technology could give relief from unproductive and burdened male calf as well as could improve animal breed.

K. Use of Digital Science

Agriculture is the main livelihood of the inhabitants of Bundelkhand but the weather vagaries or normalcy determines the success or the failure of agricultural crops (in *kharif* due to abnormal rains/ long dry spells & in *rabi* due to rains or frost/ hails). However, these crop losses can be minimized substantially by adjusting timely and accurate weather forecasts and providing weather based advisory and the SMS services. Various ICT based digital service platforms e-Commerce, e-Governance, e-Chaupal, Digital Green, mFarmer (IFFCO-Airtel collaborative plan for free voice messaging to farmers), mKisan (farmer-extension-expert linkages via text and voice messaging), aAQUA (stands for almost All Questions Answered is a farmer-expert Q&A database), are being promoted in Bundelkhand area. However, the reach of such ICTs is limited to certain pockets and category of farmers. Government call centers, free SMS services by state level agencies, mobile apps etc. are also in place to provide weather related advisories and also to give information on the market selling rates, POP, credits information etc. Effective implementation of these services in large areas covering remote villages/ communities will definitely contribute in raising income of farming community of the region, and their risk for crop cultivation could be reduced substantially. Farmers can easily solve their farming problems related to pest or insect attack or any problems which put them in a difficult situation.

Bundelkhand farmers also rely much upon various Apps related with agriculture and livestock sector, which gives the guideline on weekly basis for adopting the improved farming practices like seed & sowing, nursery & intercultural operations, nutrient and weed management, pest management, precautions in harvesting, storage and post-harvest handling of produce. Some of the Apps which are popular among the farming community and general public of Bundelkhand are: Chara app, Farm Tree, Pashu Poshan, Krishi Gyan, Crop Insurance, Kisan Suvidha, Agri Market, IFFCO Kisan, Agriculture RML, Farmer Pusa Krishi, Agri App, Kheti Badi.

Government Interventions

Conspicuous to frequent climatic and hydrological droughts, the Bundelkhand experiences severe agricultural droughts. In the era of globalization and modernisation and inclusive growth initiative of different regions by GoI, the plight of Bundelkhand peoples is getting attention and now has been included in agenda of development. The brief detail of the plan, programmes, projects & schemes has been outlined as under:

The Bundelkhand Package (Planning for sustainable development of the region)

The state planning department is responsible for implementing the Bundelkhand package with a mandate to bringing prosperity in the region. The Bundelkhand package allocates funds primarily, towards drought mitigation, watershed management, institutional credit, water resource agriculture and animal husbandry development etc. First phase was completed in 2011 and afterwords two more packages in the year 2016 and 2020 has been operational.

Bundelkhand Expressway

Central government in Chitrakoot this year laid the foundation stone for the Bundelkhand Expressway that will come up at a cost of Rs. 15,000 crore. The 296-km long expressway would pass through the districts of Chitrakoot, Banda, Hamirpur and Jalaun, and connect the drought-prone region to the national capital through the Agra-Lucknow Expressway and the Yamuna Expressway. Expressway would alter lives in Bundelkhand, "creating thousands of employment opportunities" and link the region to big cities. It would "speed up" the Defence Corridor that's coming up in the region. Bundelkhand would be known for making India self-reliant in manufacture of arms and ammunition.

Mahila Dairy Yojna

In wake of migration of rural people, in search of job due to drought and reverse migration of the masses due to COVID-19 pandemic, 'Mahila Dairy Yojna' has been launched for the Bundelkhand people under `National Rural livelihood Mission'. The centre has approved the plan which aims to provide financial assistance to 50,000 women including the `women self help groups' in the region for self-employment which will add additional production of 1.60 lakh liters of milk per day. Under the plan dairy industry will be promoted in 22 development blocks of District Banda, Jhansi, Hamirpur, Jalaun and Chitrakoot.

PM-KISAN Scheme

Pradhan Mantri Kisan Samman Nidhi is an initiative by the government of India in which all farmers will get up to Rs. 6,000 per year as minimum income support. PM KISAN is a central sector scheme with 100% funding from Government of India. State Government and UT administration will identify the farmer families which are eligible for support as per scheme guidelines. The fund will be directly transferred to the bank accounts of the beneficiaries. The scheme is intended to support Bundelkhand poor farmers for procurement of inputs (seed, chemicals and other inputs) during crop season.

National Mission for Sustainable Agriculture

Rainfed is the dominant feature of agro-ecology of Bundelkhand since, about 60% area of Bundelkhand still falls under this category. All the schemes being formulated for Bundelkhand gives utmost place to the potential and limitation of rainfed area in order to achieve the goals of such schemes. National Mission for Sustainable Agriculture (NMSA) is such a scheme which has been drafted for Bundelkhand region by sub setting the scheme into component focus areas through plans such as Rainfed Area Development (RAD) plan, Soil Heath Mission (SHM) and Traditional Agriculture Development Scheme (PKVY). It aims to make agriculture of Bundelkhand more resilient to climate change by identifying new varieties of crops (example: thermally resistant crops, capable of withstanding extremes of weather, long dry spells, flooding, and variable moisture availability), alternative cropping patterns and expanding its coverage to rainfed areas for integrating farming systems with livestock and fisheries so that agriculture continues to grow in a sustainable manner in Bundelkhand.

Sub Mission on Agroforestry

Forest department of Uttar Pradesh is the nodal agency to implement Sub Mission of Agroforestry in the state. The main objective of SMAF is to enhance tree cover by encouraging farmers to adopt tree based farming. Incentives are provided to farmers for planting trees along with crops in farm lands or on the farm boundaries. Capacity building of all the stake holders is another important objective of this mission. Farmers and other organizations provided financial assistance in the form for quality planting material.

Based on National Agroforestry Policy 2014, Uttar Pradesh Government has relaxed some important agroforestry tree species from harvesting and transit rules. It will help in promotion of agroforestry particularly in the rainfed areas of Bundelkhand.

National Bamboo Mission

The national Bamboo Mission aims to promote bamboo planation and its utilization. In Bundelkhand there is huge potential for different bamboo species to be grown on farm boundaries. The bamboo cultivation will provide additional source of income to the farmers and raw material to the wood based industries.

Government Interventions for Water in Bundelkhand

Non-availability of water for drinking and irrigation purpose has been a major problem in whole Bundelkhand. There is a big gap between production capacity and actual production of important crops in Bundelkhand region. Major factor responsible for this is acute shortage of water for irrigation purpose since, irrigated area of UP and MP Bundelkhand is hardly 40%. Moreover, whatever rainfall occurs in Bundelkhand, major portion goes to waste due to insufficient infrastructure related to water harvesting and conservation. Therefore, governments of both the states have focused on this area through various plans and schemes such as: Atal Bhujal Yojna, Piped water supply scheme, Bundelkhand farm pond scheme Bundelkhand specific subset of Pradhan Mantri Krishi Sinchai Yojna (PMKSY), National Action plan for Climate Change framework, Accelerated rural water supply program (ARWSP), Rain Water Harvesting and Distribution of Sprinkler set Scheme by UP govt, Construction of small check dams and bunds in major and minor nalas.

Interventions for Promotion of Organic Farming in Bundelkhand

Bundelkhand, by virtue of its plentiful natural resources and dynamic agro-ecology, has immense potential to be harnessed for organic agriculture. Considering this as important aspect, governments and research institutions are focusing sincerely for the promotion of organic farming in Bundelkhand *via* creating access to required inputs for organic farming, strengthening marketing facility of organic produce and facilitating through credit and subsidy support. MP Govt has planned to promote organic farming in MP- Bundelkhand through Madhya Pradesh State Organic Agriculture Policy. The policy entails on "farm – to – fork" approach reassuring abundant supply of "healthy food for all". The policy statement provides concurrent thinking of the state with its futuristic, pragmatic pro farmer initiatives. Similarly, UP govt. is much serious for making Bundelkhand as a hub and role model for organic farming. To achieve this aim, Govt. is promoting organic farming and zero budget agriculture via credit, input and infrastructure support.

Programmes for Energy and Electricity in Bundelkhand

According to census 2011, 33.5% households of UP, Bundelkhand and 55.3% households of MP Bundelkhand had access to electricity. Now-a-days ICT and digital technologies are revolutionising the farming community not only at national level but also in Bundelkhand region. Therefore, with the learning from less achievement in past regarding electricity access and considering the importance of electricity in agriculture and livelihood, governments have focused critically in this sector. In this regard, various schemes such as Rajiv Gandhi Grameen Vidyutikaran Yojana (RGGVY), Remote Village Electrification Programme (RVE), Jawaharlal Nehru National Solar Mission and Livelihood Support Activities are transforming the lives of peoples of Bundelkhand through assuring the electricity supply at deepest and farthest level of society which have been untouched with earlier development.

Strengthening Agricultural Educational & Research Institutions

Government in the recent past has taken steps for strengthening the agricultural education, research and development in the region of Bundelkhand. ICAR-IGFRI, Jhansi; Central Agroforestry Research Institute (CAFRI), Jhansi, Indian Institute of Soil and Water Conservation Research, Regional Centre (IISWC, RC), Datia are the premier institutions which are serving for the well-being of agriculture and livestock in the Bundelkhand region. It has got further impetus with the recognition of Banda University of Agriculture and Technology (BUAT, Banda) as a nodal SAU of Bundelkhand region of UP and the establishment of Rani Laxmi Bai Central Agricultural University (RLBCAU), Jhansi for UP & MP both part of Bundelkhand for overall development of agriculture. The ICAR institutes located in this region are engaged with the development of technology and implementation of their outputs at on-farm through their number of projects like: NICRA -demonstration of technology at farmers field, KISAN MITRA-Project for doubling farmers income in Bundelkhand region of Uttar Pradesh by ICRISAT-IGFRI-CAFRI, Farmers FIRST- The Farmer FIRST Programme (FFP) is an ICAR initiative and MGMG- *Mera Gaon Mera Gaurav* programme implemented by ICAR institutes of the region.

Role of Voluntary Organisations

There are organisations, involved in development activities with the approach to improve the condition of the poor people. Besides small and big NGOs, working in field, there are community based organisations (CBOs) which are activity or identity based groups of the community *viz.* mahila mandals, youth groups, credit groups, forest management groups, watershed development groups etc. Nehru Yuva Kendras(NYK) and cooperatives are involved basically in the formation of youth- groups at village-level and involve them in constructive activities, cultural activities and awareness- generation about the government programmes. Other than the above, the central and state government has come up with numerous plans, programmes and schemes such as National Disaster Relief Fund (NDRF), dedicated contingency plan for drinking water in Banda, Chitrakoot and Mahoba districts, Provision of extending mandays under MNREGS, specialized provisions of National Food Security Act, special bonus in the MSP for Bundelkhand, Pradhan Mantri Fasal Bima Yojana (PMFBY) etc.

Climate Change and Adaptation Measures

Assessment of climate change in Bundelkhand clearly indicates a downward trend in rainfall in the last 30 years (by almost 7-8% from mean 1980 levels) and an increase in mean summer temperature by 0.5 °C. The region has experienced droughts in 18 of the last 30 years, and these have been continuous spells of droughts spanning 3 to 4 years at a stretch. This has led to highly reduced water availability. Surface water availability has been minimal and groundwater water sources have been extensively tapped for agricultural purposes. All these factors put together will increase the water requirement for agriculture resulting in a higher demand of energy for irrigation.

Although, considering the futuristic projections and significance of climate change, various efforts has been taken by public/private/NGOs. MP State Action Plan on Climate Change (MPSAPCC) has been identified as one of the better state action plans because of the consultative approach which was followed for developing it. Some of the key strategies in the water sector, which have been identified to address climate change concerns include: development of comprehensive water data base in public domain; accelerate surface water development activities in the state; promote recharge of groundwater with special focus on over exploited areas; plan for efficient water supply systems and management; encourage water management practices like water auditing; regulated exploration of groundwater, water recycling, etc; enhance basin level integrated water shed management; review of existing water storing structures in view of excess precipitation; restoration of traditional water storing structures as groundwater recharging structures. Similar action plan are need to be implemented for mitigation the climate change implication at present as well as in future also.

Role of CSOs

There are many small civil society organizations (CSOs) at the grass roots level working on water issues, sanitation, agriculture, energy, employment and development of agriculture based entrepreneurship in Bundelkhand region. However, their impact has not been uniform and replicable due to short run models and lacking real field connect with communities and their aspirations. In the absence of a good development model which can influence the mainstream institution is also a challenge, especially in the UP part of Bundelkhand. However, the CSOs lack resources to reach out and impact multitudes of people in the region. Recognizing the precarious nature of Bundelkhand a number of bilateral and multi-lateral agencies have been undertaking developmental and climate change adaptation related projects within the region. These agencies have experienced the climatic reality of the region first hand and have harvested lessons from the projects which they have piloted over the years. One of the CSO namely Sir Dorabji Tata Trust (SDTT), in partnership with local NGOs, launched the Bundelkhand Initiative, to address poverty and inequities in the Bundelkhand region. However, these two have various problem in working; hence, PRADAN, in its effort to reach out to the very poor population in weaker areas, identified Bundelkhand for developing partnerships with existing local NGOs and worked for a common cause of development.

Involvement of FPOs

Farmer Producer Organisations (FPOs) are now seen as one of the potential strategy to organise farmers for adoption of technological model for scaling of agricultural entrepreneurship and aggregation of produce for value addition and marketing. In Bundelkhand, FPOs can be a game changer for the small and marginal farmers to enhance their competitive advantage in terms of better access to markets, better prices, awareness on climate resilient agricultural practices, convergence with government schemes and providing training support. There has been a steady progress in this regard and following FPOs are listed in this region:

S.No.	Name of FPO/FPC	Location	Mandates/activities
1.	Bamaur Farmers Producer Company	Jhansi	Pigeonpea, Chickpea, Pea Lentil
2.	Brahmanand Kisan Utpadak Producer Company Ltd.	Hamirpur	Pigeonpea, Chickpea, Pea Lentil
3.	Mahakaal Kishan Producer Company Ltd.	Banda	Wheat , Rice, Mustard
4.	Neelkanth Kisan Producer Company	Banda	Mentha, Wheat, Rice
5.	Mandakini Kisan Utpadak Producer Company	Chitrakoot	Rice, Wheat, Potato Mentha, Garlic
6.	Asoh Farmers Producer Company Limited	Chitrakoot	Sorghum
7.	Padari Farmer Producer Company, Etura Farmer Producer Company	Jalaun	Chickpea, Pea, Lentil
8.	Veer Bhoomi Kisan Utpadak Producer Company Ltd.	Mahoba	Agricultural and animal husbandry
9.	Lalitpur Krashk Farmer Producer Company Limited	Lalitpur	Agriculture, Horticulture Marketing

S.No.	Name of FPO/FPC	Location	Mandates/activities
10.	Chhatarpur Mahila Kisan Producer Company Limited, Chhatarpur Krishi Chetna Producer Company Limited	Chhatarpur	Agricultural and Animal Husbandry
11.	Damoh Krishi Chetna Producer Company Ltd.	Damoh	Agricultural and Animal Husbandry
12.	Panna Krishak Chetna Producer Company Ltd.	Panna	Agricultural and Animal Husbandry
13.	Tikamgarh Krishi Chetna Producer Company Ltd.	Tikamgarh	Agricultural and Animal Husbandry
14.	Jagriti Dalhan Producer Company Ltd.	Sagar	Pulses
15.	Bundelkhand Krishi Producer Company Limited	Datia	Agricultural and Animal Husbandry
16.	Orchha Farmer Producer Company Ltd	Niwari	Agricultural and Animal Husbandry

NGOs

According to a survey, there are broadly three kinds of NGOs in Bundelkhand. First one is established NGOs with clear perspective and considerable resources (e.g. Developmental Alternatives). Second type is operational in certain regions, basically activist NGOs and labour-unions. But the largest are the small and new emerging NGOs with less developed organisational set up, poor financial resources and programmes with narrow domain. Due to this, there is not a single established NGO with wider implications, in terms of programmes, beneficiary groups and area coverage, except Ekta Parishad which has tried to build upon the Gandhian and Sarvodaya. Ekta Parishad is active in almost all the districts of Bundelkhand particularly in Chattarpur, Tikamgarh and Satna, with its coordination office at Mohalla near Jabalpur. It functions on issues concerning land, water and forests through community mobilisation and group formation. There are more than 100 NGOs each in UP part and MP part of Bundelkhand working for agriculture and rural development and bringing prosperity through innovative technology with community participation.

Bundelkhand Knowledge Platform

Bundelkhand knowledge platform is a forum where ideas and experiences are exchanged leading to sustainable practices in agriculture, livestock, energy and water security; and advocacy of appropriate development strategies and policies for Bundelkhand in the context of climate change. The participants include representatives from Civil Society Organisations, Community Based Organisations, Research Organizations, Krishi Vigyan Kendras, other local distinguished individuals and other stakeholders. Knowledge Platform would bring out material such as policy briefs, case studies, perspective notes etc. which would help the decision makers in taking informed decisions and design sound policy. Apart from this, a compilation of materials available from various partners such as literature, tools, manuals, etc. will be maintained as a common good.

Linkage with Different Sectors for Rural Upliftment

There is a need for linkages with different sectors of rural upliftment like water and food security, industry and water security, forests and agriculture, employment and food security, public-private partnership in the Bundelkhand region. This will usher a new era of overall development with cohesive efforts of inter-departmental coordination and completion of schemes and programmes in mission mode.

Strengthening Seed Sector

Seed is the basic, vital and central input in agriculture and quality seed alone contributes to the total production of about 15– 20% depending upon the crop and it can be further raised up to 45% with efficient management of other inputs (Poonia, 2013). In Bundelkhand, currently quality seeds are met to the extent of 20 per cent only. Farmers themselves meet the balance of 75 per cent through own saved seeds which may be of local variety and poor seed standards. The seed sector needs immediate attention as seed replacement ration (SRR) is very poor in majority of crops grown under rainfed situations. Strong seed network with effective backup to NSC, SSC, research organisations and SAUs is required. Timely availability of quality seed of promising varieties of the region need to be made available to the farmers. Small scale seed industries need to be promoted to minimize movement of seeds over long distances. A village, wherein trained group of farmers are involved in production of seeds of various crops should be accommodated as per the needs of themselves through a "seed village" approach. In Bundelkhand, 'Seed Hub for Pulse crop' can be a novel step to promote seed sector and raise the average productivity of the small & marginal farmers.

Livelihood Improvement

The region is dominated by small & marginal farmers with poor coping management to climate vagaries. The risky farming is often leading to crop failures and meagre income to the farmers. The cropping intensity is also less and diversification of agriculture is not widely practiced, which makes farmers and rural youth unemployed in most part of the year. This necessitates migration to big cities in search of job. There are many avenues for livelihood improvement of farmers of this region:

- Innovations in climate resilient agricultural technologies (watershed based developmental models, agro-forestry, livestock-based IFS, organic farming, nich & commodity-based fruits, vegetables, spices production, forest-based activities).

- Reviving of traditional knowledge and activities
- Commercialisation of agriculture and livestock in agri-business mode
- Industrialisation and digitization
- Improvement in service delivery by government/private functionaries
- Marketing, transport and infrastructure facilities
- Proper health and education facilities
- Inclusive and all-round rural development, and
- Giving pace to tourism sector

Capacity building

Bundelkhand farmers, livestock keepers, rural youth & women folk have very poor knowledge on improved practices related to agriculture and allied activities which may be beneficial for adoption of sustainable climate resilient production system. There is wide gap in awareness and skill updating in almost all spheres of their endeavours. This makes agriculture less attractive and competitive to them which are essential for addressing multipronged problems related to environment input, practices, post-harvest management and marketing. Capacity building of farmers in such situations becomes mandatory to make them understand the worth of improved practices and train them to take up the activities for enhancing productivity and raising their income. Similarly capacity building of NGOs, CBOs ,voluntary groups, SHGs and farmers groups is also important to build their competence and confidence. The Bundelkhand R&D institutions (ICAR institutes, KVKs, SAUs) are engaged in capacity building of different stakeholders through their outreach programmes like institutional extension mechanisms, FLDs, Farmers FIRST, SC-SP programme, Tribal sub plan. State governments are also active for capacity building through ATMA and converging many central & state government schemes. Certain established NGOs like Development Alternative, PRADAN are also engaged in imparting knowledge and skill based training to the beneficiaries of their programmes.

Women empowerment

Gender inequalities in the region are very high as women have very little control over common property resources and household decision making. There is a need for empowering the women in policy issues, planning, executing and monitoring at different stages of the purchase of the livestock and their management. This would enable them to increase the productivity of the animals and system productivity of the farm. To improve their decision making and women empowerment women education level, livestock/land ownership, credit facilities, women's participation in social gatherings like 'kisan gosthi', 'village chaupal' etc. should be encouraged. Women societies, cooperatives, SHGs and organizations should be started which can help in their building for improving their contribution in livestock sector economically. Government interventions through schemes such as Mahila E-Haat and STEP are creating employment opportunities and enhancing the competency. Trainings must be arranged especially for women at community level after assessing their specific training needs through female extension workers extensively and their impact analysis must be done.

Research needed

- Mission mode R&D action plan for sustainable water management for agriculture & non-agriculture uses with emphasis on micro-irrigation and conjunctive use of water.
- Emphasis on sustainable management of soil with emphasis on conservation agriculture, soil health based nutrients application to crops and integrated management of nutrients.
- Developing self-sustaining models for natural/organic farming models
- Diversification of agriculture through fruits, vegetables, medicinal, forest species, livestock & fisheries in each district for risk mitigation and enhancing farmers' income.
- Promotion and economic validation of IFS modules for different category and farming situations.
- Problem of '*Annapratha*' need to tackled urgently involving scientific interventions like sex-stored semen, horti-pasture & silvipasture, development of common pasture resources (CPRs) and creation of fodder banks in the village clusters.
- Development of degraded waste lands through afforestation, reforestation, and restoration for fodder production.
- Use of ITKs of local communities and indigenous peoples for control of natural calamities and conservation for future generation.
- There is need to develop methodology to quantify these ecosystem services to assist the rural population adopting tree based farming.
- Availability of quality planting material of multipurpose tree species is another challenge to enhance the overall productive and better returns.
- Need to develop an extension module for dissemination of suitable agroforestry models for large scale adoption.

Policy reforms needed

For making Bundelkhand self- reliant, many policy reforms are required to address the challenges of socio-economic region in holistic way;

- Water management / inter-basin river transfer
- Watershed based planning for development of agriculture emphasising IFS in region
- Establishment of MSMEs for employment to local peoples and strengthening regional product services
- Evidence based planning, development of balance sheet of project beneficiaries and monitoring
- Specialised agriculture production zone for organics (oilseed, pulses, medicinal & fruit crops, forest produce) and nutri- cereals
- Mission on livestock based models for Bundelkhand
- Seed hub for selected pulses and oilseeds grown in central part of India
- Strengthening FPOs, SHGs for entrepreneurship in agri-foods and value added products
- Promotion of tourism in Bundelkhand
- Allocation to livelihood support activities through district level inter-departmental convergence and coordination

Recommendations

- Eco-regional based planning culminating into various micro agro climatic-zones for overall agriculture development
- Appropriate management of existing resources of soil, water, minerals and vegetation
- Emphasis on natural resource management with scientific and social planning of water use by different sectors
- Cluster based farming for organic and specialised agriculture
- Promotion of boundary plantation "Her Med Per Ped" for natural resource conservation and increase tree cover in the region.
- Provision of credit and financial assistance along with market infrastructure for agro-forestry produce to treat it at par with agriculture produce for all policy matters
- Provision of insurance for tree based farming in line with crop insurance in PMFBY (Pradhan Mantri Fasal Bima Yojana).
- Provision of subsidy for drip and micro irrigation facility for tree based farming.
- Promotion of 'One District- One product'scheme of UP government
- Generation of alternative sources of income like secondary agriculture, local artisans, utilising ITKs

- Creation of various development infrastructures, such as schools, health centers and roads etc.
- Above all there is an urgent need to rethink over the issue of public policy towards drought in the region.
- Establishing small scale-agro based industries to provide jobs to redundant agriculture labourers in agricultural slack season but reduce out migration to other states
- Sound planning for mineral based industry in both part (MP & UP) of the Bundelkhand region
- Promotion of eco-tourism and development of tourist cluster for domestic as well as international tourists

References

Amitendu, D., Arya, H. P.S., Babulal, T and Goswami, A. 2004. Indigenous technical knowledge in animal husbandry. Livestock Research for Rural Development. Vol. 16, Art. #59. http://www.lrrd.org/lrrd16/8/arun16059.html

Kumar, S., Kumar, S., Prakash, V., Chahal, V. P. and Singh, D. R. 2018. Trends and determinants of crop diversification in Uttar Pradesh. Indian Journal of Agricultural Sciences 88(11): 1704–1708.

Poonia, T. C. 2013. History of seed production and its key issues. International Journal of Food, Agriculture & Veterinary Sciences 3(1): 148-154.

Rathod, P. and Dixit, S. 2020. Dairying in Bundelkhand region of Uttar Pradesh: Constraints to realizing the potential. Indian Journal of Animal Sciences 90(1): 3-11.

Tripathi, A. K., Gupta, O. and Gupta, P. 2020. A case study: Traditional methods of insect-pest and plant diseases management in Bundelkhand region of Madhya Pradesh. Journal of Entomology and Zoology Studies 8(2): 1572-1574.

Theme 6: New Science and Speciality Agriculture

23

New Vistas in Secondary Agriculture

Krishnasree, V., Sujatha R., Seema, B.R. and Archana, P.

College of Agriculture, Padanakkad, Kasaragod, Kerala

Introduction

The natural economic movement of a country goes from agrarian economy which is the primary sector to an industrial economy (secondary sector) and then only to the tertiary sector i.e., service economy. But due to various socio-economic factors, India has jumped from an agrarian economy to a service economy in the past and had a low growth in secondary sector.

Though primary sector is vital, there is a natural limit on how much can be extracted from this. The country is witnessing a peak in food grains production (285.21 million tonnes) nonetheless rupees 92,000 crores worth food grains are wasted annually (MOFPI annual report 2018-19). Hence, with an increasing population, decreasing arable land (52.6%), irrigation water resources and with present level of climate change scenario, food processing should be considered as a need to enhance the profitability and sustainability of agriculture in the country (Swaminathan and Bhavani, 2013). Thus, effective post-harvest management is one of the strategies for envisaging an evergreen revolution along with increasing yields. A well-developed food processing industry increases farm gate prices reduce wastages, ensure value addition, promote crop diversification, generate employment opportunities as well as export earnings.

When compared to other industries, food processing sector contributes fourteen percent of gross domestic production (GDP) of the country and is providing employment to 18.54 lakh people directly in registered processing industries. Even less capital investment food producing industries contributes to 14.09% of output among other large industries like refined petroleum products (12.06%), chemical products (8.02%), basic metals (11.53%) etc. (http://mospi.nic.in/). The copious resources in the country is offering a cauldron of opportunities for secondary agriculture that could widen the horizons of food processing sector and open new vistas for doubling farmers income. The major segments of food processing in India are milk and other dairy based products, fisheries, meat & poultry, cereals & other grains, fruits and vegetables. However only 10 percent of the available food resources are utilized for processing, mostly involving milk (35%), marine (23%) and meat (21%) while cereals, millets and horticultural (2%) resources are yet to be fully explored (https://mofpi.nic.in/).

The concept of secondary agriculture came into prominence when the Ashok Dalwai Committee submitted its report on "adding value to primary agriculture and building agricultural enterprises in rural India" through "farm-linked activities and secondary agriculture" in February 2018 with major thrust on four aspects viz., sustainability of production; monetization of farmers' produce; strengthening of extension services and recognizing agriculture as an enterprise. Secondary agriculture involves "all practices and process which add value to primary agricultural commodities by using efficient technologies, market information and consumer preference". It is basically an alternate production and marketing strategy. The surge in urbanization is characterized by increased demands for value added goods like ready-to-eat, ready-to-serve food, nutraceutical and functional food in both international and domestic markets.

Financial support is necessary for any industrial activity to flourish, preferential rates under priority sector lending, 100 % income tax deduction on capital expenditures (warehouse/ cold chain) and concessional customs duties on imported equipment are some key fiscal incentives for promoting food processing operations. Thus food processing sector is expected to grow with sufficient resources, industrial facilities and policy supports.

Trebling Opportunities in the Major Sectors

Cereals & other grains

Indian agricultural basket contains variety of cereals and other grains that worth USD 38.49 billion in the export market with current export rate of 34.36%. The major export destinations for Indian rice are Iran, Saudi Arabia, UAE, Nepal and Iraq (FAO, 2018). As per the fourth advance estimates of food grain production, the projected rice production is 118.43 million tons, followed by wheat (107.59 million tons), coarse cereals (30.33 million tons), pulses (23.15 million tons) and nutri-cereals (17.15 million tons) (Ministry of Agriculture and Farmers Welfare, 2019).

Despite high production, the effective usage of these cereals produced in the domestic market are reduced due to improper storage facilities and unscientific post-harvest handling mechanisms resulting in 4.65%-5.99% wastage. This lacuna in fact opens up the new trajectories in the sub segments of processing and post-harvest handling of these abundant resources. Convenient foods are one of the booming industries where in, many products are cereal based like breakfast cereals, extruded snacks, breads and other baked products with a compound annual growth rate of +0.5% from 2017-25, thus, expanding the market volume of cereal based products *viz*, bread and baked products to 135 MT globally (www.agriexchange.apeda.gov.in).

Likewise, millets namely sorghum, finger millet, pearl millet and other minor millets are attracting much attention in processing sector because of the upsurge demand for healthy replacements in junk foods considering the efficacy of millets in reducing diabetes and obesity (Kalaiselvi and Fathima, 2016). The government initiatives in re-naming millets as nutri-cereals and promoting it for nutritional security also created awareness among the population to enhance the consumption of these micronutrient rich staples. Capacity building for value addition of millets through establishment of centre of excellence for technology transfer, providing entrepreneurship skills and facilitating market linkages were other major aspects covered in this initiative (www.nfsm.gov.in).

As the current global demand for gluten free cereals is increasing, processed millets also have export potentiality which is still unexplored. Hence millet based processed products have immense scope in the international as well as domestic market.

Fruits & Vegetables

The horticultural produce of the country needs urgent attention due to the high perishable nature of the same. Severe losses in key horticultural crops reported by CIPHET in 2015 were as follows: guava (15.88%), mango (9.16%), apple (10.39%), grapes (8.63%), papaya (6.70%) and banana (7.76%).

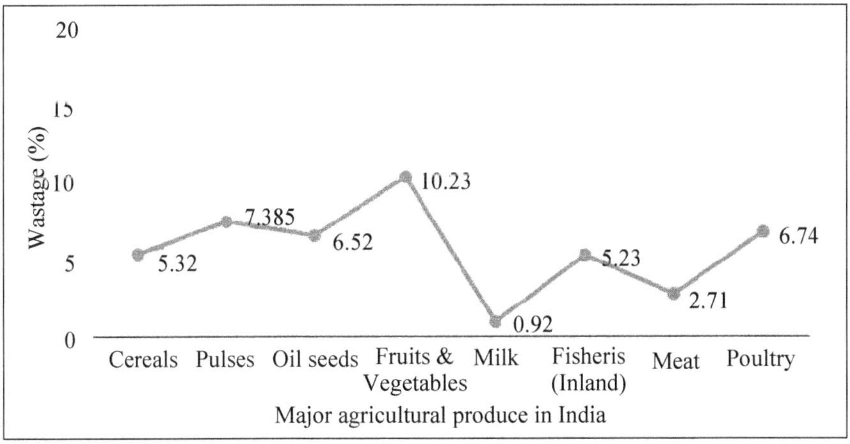

Fig. 1. Percentage loss of major agricultural produce in India

Source: Ministry of Food Processing Industries, Government of India

In an opportunistic way, investment in processing of fruits and vegetables will reduce the wastage and also enhance diversified products in the market. As per 2019-20 export data, India exported fruits worth 668.75 USD millions and vegetables worth 608.48 USD millions to countries viz., Bangladesh, Nepal,

Malaysia, Sri Lanka, UAE, Oman, Qatar, Netherland and UK. Establishment of cold storage facilities and centres for perishable cargoes contributed to the increase in export rates (www.apeda.gov.in).

Effective post-harvest handling of fruits and vegetables could be achieved by adopting scientific management practices viz., utilization of selected varieties with better processing quality, application of pre-harvest sprays for on-tree storage, harvesting at proper stage, systematic grading and sorting, storing at hypobaric storage or zero-energy cool chamber (Nath et al., 2018).

Milk & Other Dairy Based Products

Milk is a prime nutritious food with a higher protein digestibility score (PDCAAS = 100) when compared to other vegan sources like soy (PDCAAS = 91) and wheat (PDCAAS = 42) with per capita availability of 352 grams per day (Hess and Slavin, 2016). India has a tradition of using various value added products from milk viz., ghee, butter, curd, flavored milk and cheese to name a few. As a main probiotic source in Indian diet, milk and its related products are always favoured with a recorded consumption of 78% and 85% among rural and urban dwellings respectively (NSSO, 2011-12). A worldwide increase (1.41%) in milk production is also depicting the sustainability of this sector (FAO, 2018). Global dairy trade is witnessing an import demand spurred for dairy products by countries viz., China, Viet Nam, Oman, Japan, Algeria and Mexico create an export momentum will benefit Indian dairy industry. Nationally also dairying is the major livelihood opportunity securing income to 80 million rural households with small and marginal land holdings. Inclusion of milk in mid-day meal programs and other ongoing schemes of Ministry of Women and Child Development offer tremendous opportunity for enhancing dairy business locally.

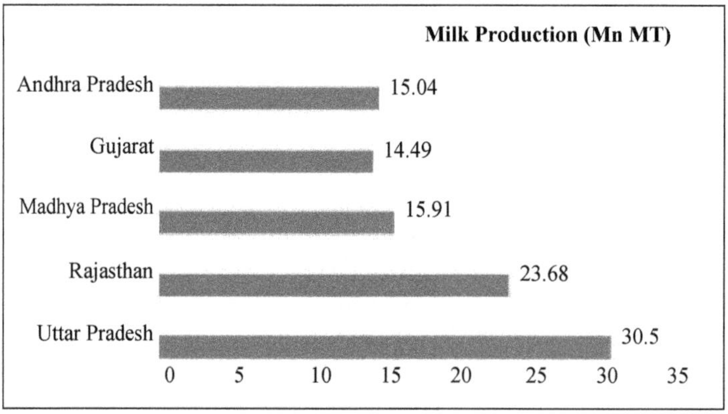

Fig. 2. Major milk producing states in India (2018-19)
Source: Department of Animal Husbandry, Dairying & Fisheries, Ministry of Agriculture and Farmers Welfare, GoI

Apart from milk, India also hold good stocks of its value added products *viz.*, butter (96,000 MT) and skim milk powder (2,13,971 MT) having longer shelf life (FAOSTAT, 2018). Likewise, the import of milk based products and its by-products are reduced over the past five years. Government have initiated various schemes to promulgate the dairy industry in India with some of the notable programs like National Program for Dairy Development (NPDD), Dairy Entrepreneurship Development Scheme (DEDS), Supporting State Cooperative Dairy Federation (SSCDF) and Dairy Processing & Infrastructure Development Fund (DIDF) (https://www.nddb.coop/).

DIDF is the latest central scheme that aims to elaborate the current milk processing sector by establishment of infrastructure with an additional milk processing (126 lakh per day) and manufacturing capacity (59.78 lakh per day) that consists of milk chilling units (140 lakh liter per day) and electronic milk adulteration testing facility. Currently the scheme have implemented 22 sub- projects (https://www.nddb.coop).

Fisheries, Meat & Poultry

Presence of inland and brackish water culture in India makes it the second largest producer of fish with a capture fisheries production of 5.1 MT and aquaculture production of 5.7 million tonnes also exporting 5.5 USD billion worth fish (FAOSTAT, 2018). Among sea foods frozen shrimp contributes to the major share (38%) export. With respect to processing, canned fishes are mostly explored followed by fish pickles, however scope for ready to serve value added fish products through retort pouching, fish sausages, nuggets etc. should also be included (https://foodprocessingindia.gov.in/).

Poultry and pig meat are in demand (16 kg per capita consumption each) followed by bovine (9 kg per capita consumption) and sheep meat (2 kg per capita consumption). India is one among other international players viz., United States, EU and Argentina to meet this global demand in the meat supplies (FAOSTAT, 2018). India currently processes 6 per cent of poultry and 21 percent of meat. Government initiatives in formalizing abattoirs ensure safe and hygiene meat delivery in the country. Under the modernization of abattoirs scheme technical and financial assistance was provided for establishing new abattoirs with chilling facility to prevent microbial contamination in slaughtered meats (https://mofpi.nic.in/).

Supply Chain in Food Processing Industry

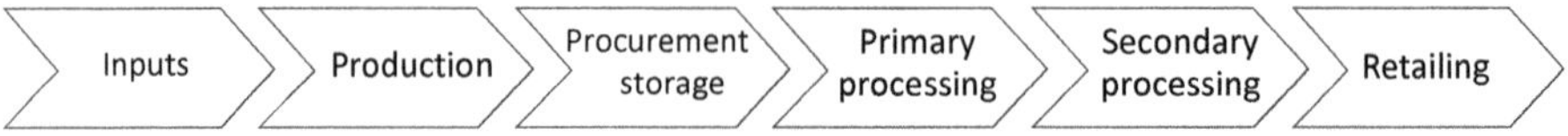

Primary Processing

It involves steps such as cleaning, grading, sorting, packing etc. to convert raw agricultural produce into human consumable form. Food grains are widely cultivated in different regions and climatic conditions to feed the population. In this process, it is necessary to advocate certain codes of practice to reduce wastage and enhance quality of the foods. Proper washing and disinfection technologies using antimicrobial or antioxidant solutions are done to avoid contamination from equipment used in harvesting and preliminary processing (FDA, 2009; Fallon et al., 2011). Different types of preliminary storage facilities are being used for this purpose.

Zero energy cool chambers

Zero energy cool chambers are on-farm low cost storage facility developed by Indian Agricultural Research Institute, Delhi that work on the principle of evaporative cooling for ensuring immediate storage of the produce with in a lesser temperature (10–15°C below ambient) than the outside environment (Devi and Singh, 2018).

Cold storages

Cold storages are generally recommended for bulk handling of highly perishable foods like fruits and vegetables. In cold storage these produce could be stored at low temperature and at optimum relative humidity. The storage temperature will vary for each produce and it is necessary to take caution while deciding the combination of produce to be kept together, for example high ethylene producers (ripe bananas, apples) should never be stored along with ethylene sensitive products (cucumber, carrot, potato) (Krishnakumar and Dayanandakumar, 2002).

Secondary and tertiary processing

It involves advanced processing methods to convert the primary product into new or modified food products. However, the processing and value addition sector in developing countries are in an immature phase in spite of several research being organized in this sector. Most of the unorganized sectors use unscientific processing methods without adhering to quality control and safety measures. The major concerns of Indian food processing sector are lack of on-farm storage facilities, limited infrastructure and machinery for preliminary & secondary processing and improper logistics. The products of basic processing are called functional foods and minimally processed foods.

Functional Foods

It is a kind of food into which one or more new ingredients has been added so that the new product has a new function, often one linked to health promotion or disease prevention. Eg : Foods that are fortified with health-promoting additives such as "vitamin enriched" products.

Minimally processed foods

Any procedure, short of the traditional complete preservation procedures that adds value to the food are called minimally processed foods. Mainly used in fruits and vegetables. It involves minimum preparation procedures, peeling or cutting. Minimally processed fruits and vegetables available are peeled and sliced potatoes; shredded lettuce and cabbage; washed and trimmed spinach; hilled peach, mango, melon, and other fruit slices, vegetable snacks like carrot and celery sticks, etc.

Advanced Processing Technologies

Numerous technological and scientific interventions are applied at various stages of processing and packaging to obtain good quality products.

Edible Coating (Waxing)

This is a concept of using edible films as protective coating for foods. An effective coating is one that forms an effective obstruction to moisture loss, has selective permeability to gases, controls migration of water soluble solutes to retain natural colour pigments and nutrients and incorporates additives such as colors, flavors or preservatives that impart specific functions and properties. The most well-known example is coating the fresh fruits and vegetables with waxes or waxy-water emulsions. Waxing minimizes moisture loss thereby reduce the wilting of leafy vegetables and juicy fruits. Major edible coating is one by carnauba wax, apart from which certain combination of anti-microbial plant extracts like neem, aloe-vera, holy basil etc. are also tried (Nath et al., 2013).

Irradiation (Cold sterilization)

Irradiation refers to the physical means of exposing food products or bulk food to gamma rays. These exposures to radiations cause ionizations and functional and structural changes in exposed molecules and results in cell death in cases of living organisms. Radiation damage to components of cell other than DNA, such as cytoplasmic membrane may also contribute to cell injury.

Non thermal processing techniques

Minimally processed foods are gaining momentum as consumers are looking for product with optimum nutritional and sensory characteristics like a fresh produce. Non-thermal processing aids in processing of foods with high nutritional value and better organoleptic qualities when compared to conventional thermal processing techniques. Major non-thermal processing techniques are conducted through application of high pressure, pulsed electric field, ultrasound, cold plasma etc.

High pressure processing

Hydrostatic pressure at 300-800 MPa (4000-9000 atm) 6000-8000 bar will be applied to foods that are pre-packed in vacuum-packs or flexible packaging material so as to ensure uniform pressure throughout the product to inactivate the microorganisms due to denaturation of proteins while preserving the nutrients. Besides preservation, it improves rheological and functional properties of food and is environment friendly with less affluent problems. Researches have confirmed that application of hydrostatic pressure at 600 MPa for 6 minutes at 31^{0}C will inhibit the pathogenic microbes in *Salmonella enterica, Listeria monocytogenes* and *E. coli* in meat products (Jofre et al., 2009). HHP is also found to be an effective technique for preservation of fruits juices (Silva et al., 2012; Tola and Ramaswamy, 2014).

Pulsed electric field

Application of high voltage pulses for few seconds to foods that are placed or flowing between two electrodes produces a lethal effect on microorganisms thus enhancing the shelf life of the product. Pulsed electric field processing is mostly done in liquid foods like milk, fruits and vegetable juices so that this technology preserves the bioactive antioxidants present in these fruits and vegetables which otherwise will be lost due to conventional thermal processing (Koubaa et al., 2018).

Manothermosonication

It represents an interesting combination of low pressure (0.3 MPa) heat treatment and ultrasonic wave treatment that seems effective for inactivation of microorganisms. The combination procedure generally has the effect of reducing the apparent heat resistance of microorganisms by about 5-20°C depending on the temperature, the organism and its z-value. The combined heat treatment can be used in processing fruit juices and other drinks.

Ultrasound

Ultrasound is a green technology that involves in inactivation of microbes and enzymes in foods without compromising the nutritional qualities. Ultrasound waves of 16 kHz frequency will be propagated through foods, however the technology is not widespread industrially due to its high cost of establishment. Another application of ultrasounds are reduction of viscosity in dairy and foaming in beverages (Patist and Bates, 2008).

Cold plasma

The principle of cold plasma preservation is utilization of ionized gas that comprises of electrons, free radicals, positive and negative charged ions that accumulates on the surface of food and rupture the cell membrane leading to oxidation of lipids, nucleic acids and amino acids of microorganisms (Misra et al., 2011). This technology is still in infancy as further in-depth research is required to understand its impact on the chemical composition of foods with which it is treated.

Cryogenic grinding

It is a method involving the powdering spices or herbs at sub-zero temperatures. The herbs/spices are initially frozen with liquid nitrogen as they have to be ground without causing any change in the chemical composition of plant. The dry, cold, inert atmosphere is generated by a liquid nitrogen vapor blanket, minimizes thermal reaction with the material and decreases the loss of volatile components from spices during processing.

Modified Atmospheric Packaging (MAP) technology

Considering the high respiration rate and other physiological changes in horticultural produce, modified atmospheric storage and packaging is applied. The principle behind MAP is to regulate the respiration rate by providing controlled temperature and optimal O_2/CO_2 concentrations. MAP is a well adopted technology for preservation of fresh fruits and vegetables due to its cost effectiveness (Mangaraj and Goswami, 2011). Studies have established that MAP could reduce the moisture loss in fruits and delay senescence thus ensuring an extended shelf life than conventional storage methods (Singh et al., 2018).

Hurdle technology

It is a combination of numerous preservation factors (hurdles) that the microorganisms present should not overcome, accounting for the final safety and microbial stability of food. It illustrates the complex iteration of temperature, water activity, pH, heat, chemicals, salt, sugar, additives and so on to the microbial stability of food. In hurdle technology, microbial stability of product is achieved by selecting few hurdles which in turn selectively disturbs the homeostasis of microorganisms.

Nanotechnology

It is termed as creation, utilization and manipulation of materials, devices or systems at nanometer scales. It exhibits great potential for food industry. Nano materials are smaller than 100nm and have unique properties due to high surface to volume ratio and physicochemical properties such as color, solubility and thermodynamics. These properties improve various sensory attributes of foods like taste, texture and colour (Yu *et al.*,2018).

Bio-preservation

Use of some antagonistic microorganisms or their metabolic products for the purpose of inhibition or destruction of the undesired microorganisms that may be present in foods to enhance food safety and extend the shelf life.

Group Initiatives in the Sector

Public-Private Partnership (PPP)

This is an arrangement between government and private sector for the provision of public assets and /or public services, where investments are being undertaken by the private sector entity for a specified period of time while retaining the responsibility by the government for providing the services. Private entity is chosen on the basis of open competitive bidding and receives performance linked payments. This is done especially when government faces constraints on borrowing money for important projects.

Agro Processing Centres (APC)

Essential facilities for primary and secondary processing such as storage, handling and drying of cereals, spices, oilseeds, pulses, fruits and vegetables are made available on rental basis to rural people. These are usually managed by individuals / co-operatives / community / organizations / voluntary organizations. Machines and equipment of small to medium capacity are used by these centers so that it will be easy to work and handle. The center meets the processing, preservation, handling and marketing needs of surplus produce available in cluster of villages. Thus, it is a means of providing employment and income to rural people through agro-processing activities of various produce and it creates employment opportunities within sector and more opportunities in service sector. APC in the production catchment has advantages of enhanced income through value addition of the farm produce and reduction in post-harvest losses as a means to provide gainful rural employment (Mann et al., 2018).

Farmer Producer Organizations (FPO)

Collective participation in farming and marketing is essential for viable economic realization from small farmers. Thus formation of farmer producer/ village produce organizations with SFAC (Small Farmers' Agri-business Consortium) as the nodal agency at the national level are necessary for minimization of transaction cost involved in production and for tapping high value market options. FPOs reduce the raw input cost in production, incremental reduction to labour, better post-harvest handling of the produce and enhance market access and connectivity. Currently 741 FPOs are promoted by SFAC and another 339 have been promoted by respective state agencies, resulting in 1080 FPOs in the country. The doubling farmers' income (DFI) committee recommended establishment of a minimum of 7000 FPOs by 2022-23. Financial reforms with respect to income tax exemption for 5 years to FPOs and handholding supports in the form of regular training through incubation facilities will enhance the prospects of FPOs (http://sfacindia.com/). Similarly, women can be given training in the area of processing and can go for value addition through coconut, mango, jackfruit, guava, cashew, banana, rice etc. Development of integrated pack house are required for grading, sorting, packaging of agricultural and horticultural produce.

Implementation of farmer producer organizations have benefited several agricultural community, for example, establishment of Prayag Raj Farmer Producer Company Limited at Allahabad was a boon for the farming community there, who were forced to purchase farming input of spurious quality at high prices. The FPO established agri-input retail outlets ensuring availability of quality inputs at reasonable prices. Rameshwar FPO at Varanasi facilitated the vegetable marketing in that region, in the usual context the farmers have to give 6% commission to wholesalers to sell their vegetables in the big mandis where as Rameshwar FPO facilitated this process by reducing the commission to 3%. Mayyil Rice Producers' Company established at Kannur district of Kerala to enhance the paddy farming of the district which was earlier confined to numerous small land holdings (from 4000 sq.ft to 3 acres) resulting in high production cost and low returns. The company enhanced the paddy production through better practices suggested by agricultural scientists and by forming paddy farming committees at grass root level. The company also initiated preliminary rice processing through providing mini rice mills. Thus the company branded and market their produce to the market without any intermediaries.

Skill Development Centres (SDC) & Agri Business Incubators (ABI)

Agribusiness means a collective of all the business activities that are being performed from farm to fork in horticultural crop processing and packaging. Trained manpower in the production, post-harvest & processing industries is required to cater to the demand for processed horticultural products for domestic trade and export purpose. An entrepreneur requires a continuous flow of funds not only for setting up the business, but also for successful operation as well as regular upgradation/ modernisation of the unit. To meet this requirement, the Government (Central and State) has been undertaking several steps like setting up of banks and financial institutions; formulating various schemes, etc.

Marketing Strategies

Linking farmers with the market through aggregation and assaying/grading of agricultural produce can help them in value enhancement and appropriation. Agribusiness incubation and acceleration, if made marketable, can benefit farm communities and appropriate market linkages can improve farmers' realizations through collectivization. The surplus food production and processing should be entwined with efficient marketing strategies to attain the goal of 'capturing value from every grain, every ounce, every drop of agricultural produce' as envisioned by the doubling farmers' income committee. Current marketing channels for farmers are limited to nearby farm mandis or agents and they are blindsided to other options further up the value chain. During post-independence, many state governments implemented Agricultural Produce Market regulation Act (APMR) with an aim to regulate trading practices in wholesale markets. However, this has restricted the free flow of agriculture produce, further constrained a direct association of farmers with end users *viz.*, processors/exporters/bulk buyers leading to rampant multiple intermediaries in value chain without commensurate benefits to farmers. In this regard, several new marketing strategies have been proposed namely, PRAM (Primary Rural Agricultural Markets), NAM (National Agricultural Markets) and nutrition led marketing.

PRAM (Primary Rural Agricultural Markets)

PRAM centers forms the foundation of the new market architecture that facilitates aggregation as well as flow of farm produce. Thus, PRAM performs the dual function of being retail markets for assisting local purchase and as storage clusters giving farmers a chance to connect with nationwide markets without worrying about the post-harvest management of their produce. Financial assistance for the operation of PRAM will be met from RKVY (Rashtriya Krishi Vikas Yojana), Integrated Scheme for Agricultural Marketing (ISAM), Mission for Integrated Development of Horticulture and MGNREGA (Ministry of Agriculture & Farmers' Welfare).

Nutritional led marketing

Various nutritious products produced in the India are not featured in the plates of majority due to various reasons. Certain products irrespective of market signals should be promoted by creating awareness through rigors marketing and linking them to organized retail portals. According to ATNI report recommendation for reformulation of packaged foods with less sugar, salts and saturated fats will facilitate to redirect the demand to indigenously produced nutritious food items like millets, camel milk, native fruits and vegetables (https://accesstonutrition.org/).

Policy Reforms & Fiscal Incentives

Food processing has been acknowledged with priority status and Ministry of Food Processing & Industries is the nodal agency to support and facilitate the food processing activities in the country. Several reforms are implemented to promote investments, innovation and brining best practices with an aim to elite India as the 'global food basket'.

Pradhan Mantri Kisan Sampada Yojana (PMKSY)

Major reforms in food processing was set up by PMKSY with an allocation of 6000 crores for the period from 2016 – 2020. The program is an extensive package for creating forward and backward linkages, research and development, integrated cold chain and value chain infrastructure, mega food parks, infrastructure for agro-processing clusters, food safety & quality assurance and operation greens.

With a concept of linking farmers with processors and retailers, mega food park project was implemented envisaging minimal wastage of raw produce, creating employment and increasing farmers' income. Currently twenty mega food parks are functioning across the country. Integrated cold chains are necessary for distribution of dairy, livestock products, fisheries and horticultural produce. Hence through PMKSY, cold chain facilities could be established by partnership or proprietorship firms, companies, self-help groups etc. Agro processing clusters are another major highlight of this program which enables implementation of core infrastructure like tetra pack, sorting, grading, Individual quick freezing (IQF) units. Agro-processing clusters could be set up by project execution agency/government organizations/joint ventures having at least 10 acres of land.

PMKSY will also support project implementation expenses like consultation, preparation of detailed project report, engineering/designing etc. The scheme enhanced linkage between farmers and food processors by reducing the avoidable intermediaries. Also it facilitated proper networking between research institutions, farmers and industries.

Start-up India/Standup India/Make in India

These schemes launched in 2014, implemented with an aim to create conducive environment for ease of doing business and to encourage companies to manufacture in India and incentivize dedicated investments into manufacturing. Food sector has given priority status under 'Make in India' scheme. Government also has allocated special funds in National Bank for Agriculture and Rural Development (NABARD) for extending credits to designated food parks (www.makeinindia.com)

Atmanirbhar Bharat

Around 25 lakh food processing enterprises in the country are functioning as unorganized sectors without registration and minimal investment in plant and machinery (7%). Nonetheless, these enterprises contribute to 74% of the employment. Thus, it is necessary to formalize these unorganized food processing units by providing need based support for technology upgradation, branding and marketing. The scheme supports farmer producer organizations (FPOs), Self-help groups (SHGs) with credit linked subsidies. The units will be identified based on "one district one product approach" in which that particular unit is functioning, other potential units will also receive grant-in-aid (https://mofpi.nic.in/).

Food industry capacity and skill initiative (FICSI)

Industry specific skill set is mandatory for the human resource in participating food processing sector. In this regard FICSI, with the financial support from National Skill Development Corporation (NSDC) is offering quality training to food processors in the country. The scheme has developed sector specific modules and curriculum for enhancing the skill set of food manufacturing employees (www.ficsi.in).

Policies for promoting investments

To facilitate foreign investment, 100 percent FDI is allowed in food processing industries under the automatic route. Further 100 percent for retail trading (including e-commerce) that are either manufactured or processed in India. At the backdrop increasing foreign direct investment enhance technology transfer, backend integration and create more employment by more food industries setting up (https://mofpi.nic.in/).

Policy initiatives for food processing sector (2014-15 to 2018-19)

Year	Policy initiative	Description
2014-15	Creation of special fund Rs. 2000 crore in NABARD	Availability of affordable credit to mega food parks and food processing units set up therein
	Launched investors portal	Information on potential and opportunities for investment in the food processing sector and incentives provided by the Central and State Govts were made available to the prospective investors at a single point

Year	Policy initiative	Description
	Strengthened Grievance Redressal system	A committee of three independent monitors was constituted to address the grievances of the applicants whose proposal for mega food parks and cold chains that could not be selected
	Food map of India	It maps the potential food processing in surplus production areas
	Streamlining of monitoring process	Monitoring to ensure utilization of fund
	Excise duty of machinery reduces from 10% to 6%	Reduction in cost of investment in food processing projects and enhancing viability of projects
2015-16	Inclusion of food and agro-based processing unit and cold chain infrastructure in priority sector lending	Availability of additional credit facility
	Service tax on pre conditioning, pre coding, ripening, waxing, retail packaging and labelling of fruits and vegetables exempted in cold chain projects	Encourages more investment in cold chain
2016-17	100% FDI in retail trade including e-commerce of food products manufactured and or produced in India	For encouraging foreign investment in in India
	Online for filing claims for infrastructure development projects	Quicker disposal of claims
	Setting up of investment tracking and facilitation desk of invest India in the Ministry	Desk will identify potential investors and approach tem in focused and structured manner for investment cases by providing hand holding services
	SAMPADA scheme launched restructuring existing schemes and designing new schemes	SAMPADA scheme to target creation of infrastructure and increasing capacity of processing and preservation in the entire supply chain of food processing sector right from farm gate to retail outlets
	Excise duty reduced from 12.5% to 6% on refrigerated containers	Reduction of cost investment in food processing projects and enhancing viability

Year	Policy initiative	Description
2017-18	e-NAM to be expanded from 250 to 585 APMC's to have primary processing facilities and assistance for primary processing to cleaning, grading and packaging	Encourage direct procuring by processing units and retail traders
	Model law on contract farming to be prepared to integrate farmers	Will attract investment in post-harvest management activities
	Dairy processing and development fund of Rs. 8000 crore to be setup in NABARD	Modernise old and obsolete milk processing units in co-operative systems
	National policy on food processing	Policy for holistic development for food processing sector, create conducive environment for the growth of supply chain and promote food processing in the country
	World Food India-2017	Ministry is organising World Food India Expo 2017 from 3-5 Nov 2017, New Delhi
2018-19	Exempting all the processed food items from the purview of licensing under the Industries (Development and Regulation) Act, 1951	
	100% Foreign Direct Investment, under government approval route, for trading, including through e-commerce, in respect of food products manufactured or produced in India	
	Lower GST for raw and processed product; nearby 80% food products are covered in lower tax slab of 0%, 5% and 12%	
	Provision of profit linked tax holiday under section 80IB and investment linked deduction under section 35 AD of Income Tax Act, 1961	
	Cold Chain and Food Parks covered under Harmonised Master List of Infrastructure Sub-sector	
	Simplifying Application Forms of all the schemes and minimizing requirement of documents	
	Assisting creation of skill Infrastructure in Food Processing sector and skill development initiatives through the Sector Skill Council [i.e. Food Industry Capacity and Skill Initiatives (FICSI)]	

Source: Press Information Bureau, GoI, MoFPI

Conclusion

In conclusion, secondary agriculture has all the potential to drive the growth of primary agricultural sector in the country. The strengthening and expansion of secondary agriculture through rural industrialization following a cluster approach aims at providing value addition to primary agricultural commodities through processing to suit the demands of the consumers. By developing processing technologies and new products, the demand push for under-utilized crops are created which may help a large number of farmers.

Agribusiness is very important to sustain the livelihood of millions of farmers in India. It could be noted that India has a vast scope for agribusiness and also a large number of opportunities for doing agribusiness. Developing entrepreneurs in agriculture can immensely benefit Indian economy by generating employment opportunities for rural youth, reducing the burden on agriculture, increasing individual and national income, prevents the pressure on urban cities by reducing migration from rural areas to urban area. The innovative post-harvest technologies, post-harvest goods and services provided to small holder farmers and marketers will allow them to join in modern value chains, improve their potential to reach new markets and help to reduce food losses while increasing incomes. For enhancement of shelf life and value addition, the produce is generally processed by using different techniques. There is a necessity to devise more efficient processing techniques to add value to the crop as well as minimization of the post-harvest losses. This would also be useful to ensure the economic stability to the developing nations.

References

Devi, S. R. and Singh, L. K. (2018). Zero energy cool chamber, low cost storage structure for vegetables and fruits in Churachandpur District of Manipur. Journal of Krishi Vigyan 7(1):216.

Fallon, S. D., Rios, D. V. and Fonseca, J. M. (2011). Scrutinizing the sanitizing process involving aqueous chlorine from field to fresh-cut iceberg lettuce. Acta Horticult. 906:37–42.

FAO. (2018). Food Outlook- Biannual report on global food markets - November 2018. In Global information and early warning system on food and agriculture.

FAOSTAT . FAO; 2018, http://www.fao.org/faostat/en/#data/QC/visualize.

FDA, Food and Drug Administration. (2009). Draft Guidance for Industry: Guide to Minimize Microbial Food Safety Hazards of Leafy Greens. www.fda.gov.

Hess,J., and Slavin, J.(2016).Defining "Protein" foods. Nutrition Today. Doi: 10.1097/NT.0000000000000157

http://mospi.nic.in/sites/default/files/publication_reports/instruction_for_filling_part-I-II_ASI_Schd16-17.pdf

http://nfsm.gov.in/Guidelines/DRAFTguidelines%20of%20Initiative%20for%20Nutritional%20Security%20through%20Intensive%20Millets%20Promotion.doc-1.pdf

https://agriexchange.apeda.gov.in/Weekly_eReport/Cereal_Preparations_Report.pdf

https://foodprocessingindia.gov.in/

https://mofpi.nic.in/

https://www.nddb.coop/

Jofré, A., Aymerich, T., Grébol, N., Garriga, M. (2009). Efficiency of high hydrostatic pressure at 600 MPa against food-borne organisms by challenge tests on convenience meat products. LWT Food Sci. Technol., 42, 924–928

Kalaiselvi, A. and Fathima, L. A. (2016). Awareness and Consumption of Millets by Women – A Study on Coimbatore city. Indian Journal of Applied Resaerch, 6(2): 96-97.

Koubaa, M., Barba, F. J., Bursać Kovačević, D., Putnik, P., Santos, M. D., Queirós, R. P., Moreira, S. A., Inácio, R. S., Fidalgo, L. G., and Saraiva, J. A. (2018). Pulsed Electric Field Processing of Fruit Juices. In Fruit Juices: Extraction, Composition, Quality and Analysis. https://doi.org/10.1016/B978-0-12-802230-6.00022-9

Krishnakumar T. and Dayanandakumar. (2002). Design of cold storage for fruits and vegetables. 10.13140/RG.2.2.14335.82082

Mangaraj, S., and T.K. Goswami . (2011). Modeling of respiration rate of litchi fruit under aerobic conditions. Food Bioprocess. Technol. 4:272–281. doi: 10.1007/s11947-008-0145-z.

Misra, N., Tiwari, B., Raghavarao, K., Cullen, P. (2011). Nonthermal Plasma Inactivation of Food-Borne Pathogens. Food Engineering Reviews, 3(3) 159-170.

Nath, A., K. Barman, S. Chandra, P. Baiswar. (2013). Effect of plant extracts on quality of Khasi mandarin (*Citrus reticulata* Blanco) fruits during ambient storage. Food & Bioprocess Technology, 6(2):470-474.

Nath, A., Meena, L.R, Kumar, V. and Panwar, A. S. (2018). Postharvest management of horticultural crops for doubling farmer's income. Journal of Pharmacognosy and Phytochemistry, 6(2): 2682-2690.

National Sample Survey Organisation. Ministry of Statistics & Programme Implementation, Government of India; 2007. Consumer Expenditure Survey, 2011-2012.

Pandiselvam, R., Subhashini, S., Banuu Priya, E. P., Kothakota, A., Ramesh, S. V., & Shahir, S. (2019). Ozone based food preservation: a promising green technology for enhanced food safety. In Ozone: Science and Engineering. https://doi.org/10.1080/01919512.2018.1490636

Patist, A., and Bates, D. (2008). Ultrasonic innovations in the food industry: from the laboratory to commercial production. Innovative Food Science & Emerging Technologies, 9(2), 147-154.

Sastry, S. K. (2008). Ohmic heating and moderate electric fields processing. Food Science and Technology International, 14, 419.

Silva, F.V.M., Tan, E.K., Farid, M. (2012). Bacterial spore inactivation at 45–65 °C using high pressure processing: Study of *Alicyclobacillus acidoterrestris* in orange juice. Food Microbiology, 32, 206–211.

Singh, S., M. Sharma, and S. K. Singh. (2018). Use of shrink and cling film for modified atmosphere packaging of 'Kinnow' (*Citrus nobilis* L. x C. *delicious* L.) fruits. Vegetos, 31:12–16. doi: 10.5958/2229-4473.2018.00050.2.

Swaminathan, M. S., and Bhavani, R. V. (2013). Food production & availability - Essential prerequisites for sustainable food security. Indian Journal of Medical Research. https://doi.org/10.1142/9789813200074_0025

Tola, Y.B. and Ramaswamy, H. S. (2014). Combined effects of high pressure, moderate heat and pH on the inactivation kinetics of *Bacillus licheniformis* spores in carrot juice. Food Res. Int., 62, 50–58.

Torres, B., Tiwari, B. K., Patras, A., Wijngaard, H. H., Brunton, N., Cullen, P. J., and O'Donnell, C. P. (2011). Effect of ozone processing on the colour, rheological properties and phenolic content of apple juice. Food Chemistry. https://doi.org/10.1016/j.foodchem.2010.06.050

Varghese, K. S., Pandey, M. C., Radhakrishna, K., and Bawa, A. S. (2014). Technology, applications and modelling of ohmic heating: a review. In Journal of Food Science and Technology. https://doi.org/10.1007/s13197-012-0710-3

www.agricoop.nic.in

www.agriexchange.apeda.gov.in

www.makeinindia.com

www.nfsm.gov.in

24

Trends in Hi-Tech Agriculture Sector

Prabhat Kumar[1], S. Gorantiwar[2], S.M. More[2], Awani Kumar Singh[3] and Pinaki Roy[4]

[1&4]National Agriculture Higher Education Project
Indian Council of Agricultural Research, New Delhi
[2]Mahatma Phule Krishi Vidyapeeth, Rahuri, Maharashtra
[3]ICAR-Indian Agricultural Research Institute, New Delhi

Introduction

The protected cultivation aims to create a favorable environment for harnessing maximum potential even in adverse climatic conditions. Farmers in India are facing different challenges including small and fragmented land holdings, climate change, vagaries of weather, poor soil health and lack of knowledge and exposure to modern agricultural technologies. The protected cultivation farming is emerging as one of the better alternatives in agriculture and gaining foothold in India. However, for successful protected cultivation farming, one needs to understand the protected cultivation structures, climate control, production technologies, plant protection, water management, marketing, guidelines and policies. At present, various innovations are adopted under the greenhouse technology *viz*., Soil based cultivation and soil less cultivation. Soil less cultivation consist of solid media based and liquid media based. Solid media include coco-peat, vermiculite, perlite, rice husk, sawdust, mass grass, wheat - paddy straw, sand, gravels, rock wool etc. whereas liquid based media include hydroponic, aeroponic and aquaponics cultivation.

Advantages of Protected Cultivation

- Complete control of all kind of abiotic and biotic stress.
- Enhance quality, quantity, colour, and taste along with nutritional value of the product.
- Crop can be grown as per market demand-off, early, late and main season or crop can be grown year round.
- Create employment opportunity.

- Provide livelihood opportunity in land less or marginal and small farmers those are belonging weaker section, hilly and tribal areas farmers.
- Increase in yield up to 5 to 8 times .
- Significant saving in inputs like water (up to 50%), fertilizers (25%) pesticides and reduced labour cost (75-80%).
- Feasible even in undulating topography and problematic soils.
- Soil less media method is also used for cultivation of crops in the indoor and roof top.
- Media may be re-used as well.

Hi-tech or controlled environment agriculture is a concept of modifying the natural environment for optimum plant growth (Sirohi, 2002). It controls the air and root zone temperature, relative humidity, radiant energy (light), air velocity, atmospheric concentration of carbon dioxide (CO_2), root zone oxygen concentration, nutrient and moisture supply to control crop growth. The aim of protected cultivation is to achieve independence of climate and weather and to allow crop production in areas where the natural environment limits or prohibits plant growth (Lorenzo et al., 2005).

Potential areas for Protected Cultivation

- Urban and peri-urban areas to meet requirements of fresh produce like vegetables, fruits and flowers round the year.
- Areas with limited land and water resources.
- Areas where availability of land for cultivation is restricted due to snowfall and low temperature is prevalent restricting cultivation of crops under open field conditions.
- As an agri-business enterprise for enterprising youth in rural and urban peripherals.

Improvement in yields and quality under protected cultivation

Protected cultivation enables to maintain growing conditions fully or partially *viz.* Environmental (Temperature, Humidity, Light, CO_2, UV and IR protection, Physical protection from hail, storm, frost etc.) and Other like (Media, EC, pH ,Pest and disease, Variety of crops, Irrigation management (EC and pH, Field capacity use of tensiometers), Nutrient management (EC and pH, NPK ratios, Micronutrient) Protection from pest and disease (Sanitation, Preventive care) and Proper support system for crop.

Climatic requirement for flowers and vegetables under protected cultivation

S. No	Name of Crop	Temperature (^{0} C)		Relative humidity (%)	Light intensity (Lux)
		Day	Night		
Flowers					
1	Rose	24-28	18-20	65-70	60000-70000
2	Carnation	16-20	10-12	60-65	40000-50000
3	Chrysanthemum	22-24	15-16	60-65	35000-40000
	Cut flower	23-26	16-19	60-65	35000-40000
4	Gerbera	20-24	18-21	60-65	40000-50000
5	Orchid	22-24	18-20	740-80	25000-30000
6	Anthurium	22-25	18-20	70-80	25000-30000
Vegetables					
7	Tomato	22-27	15-19	50-65	50000-60000
8	Capsicum	21-24	18-20	50-65	50000-60000
9	Cucumber	24-27	18-19	60-65	50000-60000
Nursery					
10	Nursery	22-27	16-19	50-65	50000-60000

(*Source*: www.ncpahindia.com)

Protected Cultivation Technologies

The protected cultivation technologies have a different type of classification on the basis of the covering materials and structure. The following classifications are given below.

a. **Greenhouse group:** This group having a different permanent protected structure like glass house, polyhouse, insect proof net house, shade net house, FRP seed house etc. The cost of the installation of these infrastructure is in the range of E=RS 1000-2000 / meter square.

b. **Tunnel group**: This structure is known as micro temporary green house or short duration green house. Tunnel is making of two types. Low tunnel (height 1 mt, width 1 mt, length according to field) and second is a high tunnel (height 2-2.5 ft, width – 4-5 mt, length -20-25 mt). Both tunnel covering by 30-50 micron transparent UV stabilized polythene and high tunnel covering by 100 micron transparent UV stabilized polythene. Then other covering by nets according to the need of the farmers. Both are the temporary and safety in structure. Low tunnel installation cost Rs. 50000-60000/ per acre along with drip irrigation and mulching system and high tunnel installation cost Rs 30000-40000/ per 100 sq meter. The tunnel technology used for early cucurbits production and off season and around year nursery production of all horticultural crops.

c. **Mulch group:** This group has been two types of mulch. One is plastic mulch and another one is bio mulch (use any bio mass and natural products those are conserving moisture and protects weed infestation and increase organic matter in soil). These technologies are used in other name as surface cover cultivation. Using of surface mulch with 20-30 micron insect proof , net will be cost of Rs. 20000-30000/ per acre.

d. **Staking or vertical group:** Staking technology used in different type of model or structures as a vertical as well as horizontal. This technology used protect against soil borne disease and pests, water logging, weed incidence because all cucurbits and indeterminate tomatoes crop grow vertically more than 1-2 mt in height, single, double or multi row system polyhouse and open conditions with the help of bamboo, wooden and Gipps along with plastic ropes and iron wires with an estimated cost of Rs 25000-30000/ per acre.

e. **Net group:** This group having different type of the net and have various objective like anti hail net to protect hail, anti-bird net to protect bird , anti-animal net to protects animal. (This made by iron base). This net can't be used for to create micro climate rather only protect against natural hazards. This nets are used in hilly areas for the protected of apple orchard and planting materials. And the cost will be near to Rs. 40000-50000/ per acre.

Different type of Greenhouse (Polyhouse)

- Fully climatic control- High tech, full device base
- Semi climatic control/ Evaporative cool poly house- fan pad poly house
- Forced ventilator- Exhaust fan poly house
- Natural ventilator- zero electric and device base poly house

(i) Greenhouse *(Glass house, Polyhouse, Net house, Tunnels)*

A greenhouse stricture group is a framed or an inflated structure covered with transparent material, which permits at least partial control of growing environment to carry out cultural operations. Greenhouse is usually a covered structure of plastic film which allows the solar radiation to pass through it but traps the geothermal radiations emitted by plant inside (Sirohi and Behera, 2000). Greenhouse technology implies production of plants for economic use in covered structure that permits rapid harvesting of solar radiation and modification of microclimate conductive for plant growth and development. It is evident that greenhouse structures are economical for producing high value vegetable crops at a time when no other crops is possible to grow under open condition. The main advantage of greenhouse farming is that the production can be obtained round the year even in adverse climatic conditions. The cultivation of vegetable crops in tropical area is quite difficult under natural conditions due to adverse climatic conditions. Thus, the greenhouse or polyhouses appear to be an alternative in tropical areas for growing of vegetable seedlings or crops, since these structures maximize solar radiation transmission, while permitting control of temperature, humidity and carbon dioxide concentration for maximizing the photosynthetic activity. The carbon dioxide (CO_2) released by the plants during night is also trapped inside which increases the rate of photosynthesis during day time. The evaporation from soil and plants is raised which ultimately increases the humidity inside greenhouse. Occurrence of frost coupled with low temperature during month of December and January cause death of tomato plant when grown in open field conditions. By nearly optimum conditions in greenhouse, 10-15 times total higher yields of vegetable crops can be obtained than in open conditions.

A few recently commercially introduced greenhouses covering material: 200 micron thickness, **5-layer** UV stabilized wide range of polyfilms suitable to various agro climatic conditions is given below.

A few recently commercially introduced greenhouse covering material

Film characteristics	Drip Lock Diffused AV Blue	205 N	Drip Lock clear	Drip Lock Cool	IR 504
Light transmission	77%	88%	90%	87%	87%
Light diffusion	60%	60&	25%	77%	55%
Anti-dust	Yes	Yes	Yes	Yes	Yes
UV blocking	Yes	Yes	No	No	Yes
Anti Sulfur (upto 1500)	Yes	Yes	No	Yes	Yes
Anti Drip	Yes	No	Yes	Yes	No
Thermic	No	No	No	No	Yes

(Ref: www.agriplst.co.in)

Area and crop recommendation

	Drip Lock Diffused AV Blue	205 N	Drip Lock clear	Drip Lock Cool	IR 504
Areas recommended	Recommended for hot areas	Recommended across India	Recommended for cold areas with limited sunlight like Shimla, Kuluetc	Recommended for hot area Reduces temperature by 3-4°C inside polyhouse.	Recommended for cold areas

	Drip Lock Diffused AV Blue	205 N	Drip Lock clear	Drip Lock Cool	IR 504
Crops recommended for cultivation	Tomato, Bell Pepper, Cucumber, Strawberry, Roses, Lettuce, Broccoli, Chrysanthemum	All crops including roses (except bicolour roses)	Carnation, Nursery except roses	All horticulture and floriculture crops including roses	All crops including roses (except bicolour roses)

Source: www.agriplst.co.in

(ii) Net houses

Net houses or shade-net house is a framed structure made of materials such as GI pipes, angle iron, wood or bamboo. It is covered with plastics net (Nets are made of 100% Polyethylene thread with specialised UV treatment) having different shade percentages. It provides partially controlled atmosphere and environment by reducing light intensity and effective heat during day time to crops grown under it. Hence round the year seasonal and off-season cultivation is possible. Shade nets are used to cut solar radiations and protect the crops from wilting/scorching.

Shade-nets are available in different shade percentages or shade factor *i.e.* 15%, 35%, 40%, 50% 75% and 90% (for example 35% shade factor means - the net will cut 35% of light intensity and would allow only 65% of light intensity to pass through the net). Each plant has its individual requirements for sunlight and shade under which it flourishes at its best. To create optimum climatic conditions, selection of the correct percentage of shade factor plays an important role to enhance plant's productivity to its optimum. The shade nests are available in two forms: Tape-nets and Mono-nets.

Insect Proof Nets: Insect nets are manufactured from HDPE monofilaments granules which are UV stabilized by adding suitable UV stabilizer with linear density 42.2 Tex (380Denier). Insect proof nets are available in different mesh from 25 to 60 mesh which are white or milky in colour. The growing of seed crop under insect proof net house fetches higher net returns as compared to open field. Singh et al., (2010) found that insect proof net house are suitable structure for virus free vegetable seed crop that protects the seed crop against major insects/ pests like shoot and fruit borer in brinjal, fruit borer in okra and tomato, fruit fly in cucurbits, leaf miner in cucurbits and solanaceous vegetables, and red pumpkin beetle in cucurbits. Mess size of this net is given below-

Insect net	Type	Mesh	Weight
IS16513:2016	Type I	30	90 GSM
For Agriculture and Horticulture Purpose	Type II	40	105GSM
	Type III	50	430GSM

(Tolerance percentage is ±5% in GSM)

(Protected Cultivation Guideline 2019, MSHMPB, Dept of Agriculture, Govt of Maharashtra)

(iii) Row Cover

Row covers are often used for the early production of vegetable crops in different regions of the world. The purpose of row covers in the temperate regions differs from that in the subtropical regions. In the temperate regions, row covers are used for protection from cold, frost, hail, wind, bird and insect attacks, suppression of transpiration and promotion of germination, growth and yield . On the other hand, their main purpose in the subtropics is to protect the crops from typhoons, pests, and for the attenuation of strong solar radiation and drought. Helbacka, (2002) conducted a study on row covers for vegetable gardens. It was reported that many cucurbits (squash, cucumber, and melons) respond well under row covers with increased yield of as much as 25%. Hochmuth et al., (2000) conducted a study on row covers for commercial vegetable culture in Florida. It was found that row covers are used to enclose one or more rows of plants to enhance the crop growth and production by increasing both air and soil temperatures and reducing wind damage.

Product and purpose of net

Product Name	Description	Colour	Width	Length (m)	Recommended for
Crop shield	17 gsm	White	1.6 m	625	Protection from insect pests, animal, birds attack, sun burn frost injuries on vegetables fruit and nursery plants
	17 gsm	White	2.13m	625	
	17 gsm	White	3.2m	625	
	17 gsm	Red	1.6m	625	Higher vegetative growth, recommended in cold climate,
	17 gsm	Blue	1.6m	625	Promote shorter intermodal distance
	23gsm	White	1.6m	625	Protects from frost injury in cold climate
Banana Shield	17 gsm	White	31.5 inch	200	For bananas for better quality and higher yield protection against sunburn and insect

Product Name	Description	Colour	Width	Length (m)	Recommended for
Fruit Shield	17 gsm	White	6-8 inch	200	For fruit like Pomegranates, Mangoes, Avocado, Peach, Apple, Orange, Guava etc To protect against sunburn and insect

Source: www.agriplast.co.in

(iv) Plastics Tunnels

Plastics tunnel popularly known as low tunnels, hi-tunnel and cloche are short duration temporary miniature structures producing greenhouse like effect. These tunnels facilitate the entrapment of carbon dioxide, thereby enhancing the photosynthetic activities of the plant that help to increase yield. These structures also protect the plants from high wind, rain, frost and snow. Besides being inexpensive, these structures are easy to construct and dismantle. Low tunnels are being used for producing high quality, high valued nurseries and crops such as tomatoes, cucumber, radish, beans and capsicum. With this technology, the farmers can capture the market in the early season and may get good return of the produce. Another advantage of such technology is that low tunnels can be easily dismantled and used in the next year. A portable low plastic tunnel of size (3.3 X1.2 X.1.0) M is put on the nursery bed. Use of low tunnels has been effective in crops such as Tomato, Cucumber, Radish, Beans, Asparagus, Strawberries, Melon and Tobacco etc. Streck et al., (2007) recommended that umbrella-like tunnels can be used instead of conventional low tunnels, which need daily management. This technique allowed lettuce to grow throughout the year and only required lateral adjustment according to the season.

New Innovative Production Techniques are used in Greenhouse Structures

After the installation of greenhouse (polyhouse) structures then growers are used various type of innovative technique with and without soil cultivation *i.e.* soil less production techniques with vertical and horizontal spotting system details are described in below.

A. Hydroponic or Water Culture

The rapid expansion worldwide of hydroponic systems in the last three decades tends to their independence from the soil and its associated problems, *i.e.* the presence of soil-borne pathogens at the start of the crop and the decline of soil structure and fertility due to continual cultivation with the same or relative crop species. Hydroponic appears to be the safest and most effective alternative to soil disinfection by means of Bromomethane. It is therefore becoming increasingly important in protected cultivation – not only in modern, fully equipped glasshouses, but also in simple greenhouse constructions designed to optimize favourable climatic conditions.

Details of Systems, Technique, Methods and Equipment for Hydroponic (water culture) in Greenhouse (polyhouse) Structures

Intense research and experimental activities in soilless cultivation have led to the development of numerous systems characterized by different water volume methods of water supply, nutrition management, size and shape of growing modules, and by the presence or absence of a variety of growing media (substrates).

a) Feeding Plants in Soilless Systems: In all modern soilless systems, fertilization and irrigation are integrated into one system able to supply fertilizers and water at the same time (fertigation). Once it became evident that all nutrients essential for crops (macro- and micronutrients) could be supplied through hydro soluble fertilizer salts, systems were developed with fertilizers dissolved at relatively high concentrations in special stock solutions. Stock solutions are injected and diluted in the irrigation water. Generally, two fertilizer tanks containing the stock solutions are used to separate fertilizers that can interact. A possible combination is a tank "A" containing essentially calcium fertilizers and a tank "B" with essentially phosphate and sulphate fertilizers. In this way, Ca is separated from P and SO_4-S to avoid precipitation of calcium phosphate or calcium sulphate, which is sparingly soluble. A third tank "C" contains a concentrated solution of an inorganic acid which is used to control the pH of the nutrient solution obtained after the injection of the stock solutions into the irrigation water, and to wash the irrigation system and avoid clogging of the nutrient solution emitters.

b) Open-Loop System: In open-loop systems the water and nutrients are supplied as for a conventional on-soil crop and the drained nutrient solution is thrown out of the system. The leachate may be collected and reused to fertilize on-soil crops but in most cases, it is lost causing harm to the environment. Open-loop systems determine the nutrient solution to supply in conjunction with leaching, i.e. the volumetric ratio of the leachate to the applied nutrient solution.

c) Closed-Loop systems: In this system the drained nutrient solution is recovered, replenished and recycled. Compared with the open-loop system, it requires more precise and frequent control of the nutrient solution; technical know-how is needed, as it is more sensitive to operational mistakes, in particular during spring due to the possible increase of nutrient concentration in the solution with increasing temperature and solar radiation. The returned nutrient solution has to be treated to restore its original nutrient element composition and to remove any foreign substances. Moreover, spreading of root-borne diseases may occur, thus sterilization of the solution must be provided to kill pathogens.

d) Deep Water Culture (DWC): DWC, created in 1929 by Professor W.F. Gericke of the University of California, was the first hydroponic method proposed for commercial purposes. It consists of a bucket filled with nutrient solution, covered with a net and a cloth on which thin layer of sand (1 cm) is placed to support the plants; the roots are suspended in the nutrient solution. Alternatively, the bucket may be covered with a lid and the plants, contained in net pots, suspended from the centre of the cover. The main drawback of the system is the hypoxic conditions occurring at root level, due to the limited air-water exchange area, compared with the volume of the solution, and the low diffusion coefficient of oxygen in the water. This constraint has been overcome by means of air pumps oxygenating the nutrient solution or by applying recirculation deep-water culture systems (RDWC) that use a reservoir to provide nutrient solution to multiple buckets. In RDWC, as the water is reintroduced to the reservoir it is broken up and aerated with the use of spray nozzles.

e) Float Hydroponics: Plants are grown on trays floating in tanks filled with nutrient solution. This method has a long history but its use in greenhouse production spread following the introduction of high-density polystyrene or other "ultralight" plastic (e.g. Styrofoam) trays. It was used for the first time by Professor Franco Massantini at the University of Pisa, Italy, in 1976, to grow lettuce, cardoon and strawberry. Nowadays, the technique is principally adopted for the cultivation of fresh-cut leafy vegetables (lettuce, chicory, rocket, lamb's lettuce etc.) and aromatics (basil, mint, thyme etc.).The system appears to be particularly interesting due to the low set-up and management costs and the little automation required for monitoring and adjusting the nutrient solution.

Classic Float hydroponics systems are based on tanks 0.20–0.30 m deep made of low-cost material (concrete, bricks and wooden planks) or directly dug into the greenhouse. Tanks are sealed (e.g. waterproofing with PE film) and filled with nutrient solution (150–250 litres m^{-2}). The large volume of nutrient solution buffers the temperature and reduces the frequency of adjustment and reintegration of the solution. During the growing season, the O_2 concentration in the nutrient solution should range between 5 and 6 mg per litre. The easiest way to oxygenate the nutrient solution is by pumps that drive part of the solution into a pipe onto which a Venturi tube is inserted to insufflate air. However, the airflow should never become very strong, to avoid root damage and recirculation of plant exudates. In the greenhouse, a single-tank or multiple-tank system may be used. The former taking up almost the whole of the span, reduces the incidence of barren areas and allows the automation of certain operations, such as the placement and removal of floating trays; the latter consists of several tanks of $\geq 4\ m^2$ (2×2 m), with spacing of 0.5–1.0 m, and it reduces the risk of operational mistakes and diseases.

f) Nutrient Film Technique (NFT): NFT is a hydroponic technique whereby a very thin layer (film) of nutrient solution flows through watertight channels (also known as gullies, troughs or gutters), wherein the bare roots of plants lie. Channels are on a slope of 1.2–3.0 percent and nutrient solution is applied at the elevated end so that the solution flows down through the channels keeping the roots completely wet. The slope may be provided by the floor itself, or benches or racks may hold the channels and provide the required elevation. The thin water stream (1–2 mm deep) ensures sufficient oxygenation of the roots, as the thick root mat which develops on the bottom of the channel has its upper surface continuously exposed to the air. At the lower end of the channels, the solution is drained to a large catchment pipe, which conducts the solution back to the cistern to be recalculated. Depressions in the channel floors must be avoided because ponds of immobile solution will lead to oxygen depletion and growth retardation.

Flow rates beyond this range are often associated either with oxygenation or nutritional problems: too rapid and the water becomes too deep and oxygenation of the roots inadequate; too slow and the result is lack of nutrients, especially for plants with roots downstream in the channel and exposed to water from which many other plants have already extracted nutrients, especially nitrogen and potassium. The rate of nutrient depletion along the channel also depends on length. As a rule, length should not exceed 12–16 m.

In order to overcome these problems, a modified system called super nutrient film technique (SNFT) has been developed: nutrient solution is distributed by nozzles arranged along the channel, ensuring adequate availability of both nutrients and oxygen near the roots.

The delivery of nutrient solution may be continuous in a 24-hour cycle or intermittent (alternating watering and dry periods to improve oxygenation of the root system). Another possibility – a compromise between these two approaches is the continuous recirculation of the nutrient solution during daylight hours (dawn to dusk) and the automated switching off at night.

Plants destined for use in NFT systems are raised in small pots or plugs or in rock wool cubes and are placed in the channels when a substantial root system has formed.

The main advantages of NFT over other systems are the absence of substrate and the reduced volume of nutrient solution required, resulting in significant savings in water and fertilizers and reduced environmental impact and costs related to the disposal of the substrate.

On the other hand, owing to the low water volume, the nutrient solution is subjected to major temperature changes along the channel and during growing seasons. Moreover, NFT has very little buffering against interruptions in water and nutrient supplies, and there is a considerable risk of the spread of root-borne diseases. Technically most crops could be grown in a NFT system, but it works best for short-term crops (30–50 days), such as lettuce, because plants are ready to harvest before their root mass fills the channel.

Growing Media for Soilless Cultivation: The cultivation of plants in systems without soil *in situ* is defined in literature as "soilless culture" (Gruda, 2009). Many such systems are based on the use of solid rooting media for growing plants. They are usually called "growing media" or "substrates"; however, sometimes terms like "aggregate systems, "supporting media" or "potting soil" are used. With reference to plant cultivation and propagation, "growing media" or "substrates" are defined as all those solid materials, other than soil, which alone or in mixtures can guarantee better conditions than agricultural soil (for one or more aspects). Hence, media of different origin take on the role of soil and provide anchorage for the root system, supply water and nutrients for the plant, and guarantee adequate aeration in the root area (Gruda et al., 2006). Growing media are used in containers (organic substrates, perlite etc.). However, sometimes they are used in the form of prepared cubes (rock wool cubes for seedling and transplant production), bags and slabs (peat-based substrates and rock wool, respectively), mats (polyurethane foam) and troughs (rock wool); these last three are also used for vegetable production in soilless culture systems. In recent years, many innovative cultivation procedures using new growing media methods have been developed, including systems without a solid medium, as well as aggregate systems in which inorganic or organic substrates are used.

Suitable crops grown under Hydroponics Techniques

Type of crops	Name of the crops
Cereals	Rice, Maize
Fruit	Strawberry
Vegetables	Tomato, Bell pepper Chilli, Brinjal, Cucumbers, Melons, Green bean, Winged bean, green Onion, Beet.
Leafy vegetables	Lettuce, Parsley, Celery, Swiss chard, Oregano Atriplex, Spinach, Coriander, Methi, Mint, Sweet basilleaves
Flowers	Roses, Carnations, Chrysanthemum, Marigold
Medicinal	Indian Aloe, Coleus
Fodder	Sorghum, Alfa-alfa, Bermuda grass, Carpet grass

Varieties grown under Hydroponics techniques

Crops	Varieties
Lettuce	Bib: Deci-mionr, Ostinata, Cortina, Rex, Salina, Milou,Vegas, Cortina Lose leaf: Domineer, Black, Seeded Simpson, Grand Rapids, Waldmann's Dark Green Head/Iceberg: Great Lakes 659, Montemar Roamaine:ValamaineCos, Cimmaron, Parris Island Cos
Tomato	Beefsteak: Dombito, Caruso, Larma, Perfecto, Belmondo, Trend, Apollo, Match, Blitz, Quest, Laura Cherry: Pusa cherry-I, Favorita, Conchita

Crops	Varieties
	Tomato on Vine (TOV): GS-600, Himsona, Himshikher, NS-4466, ID-37, Rakshita, Naveen+, Tradiro, Ambiance, Balance, Cronos.
Cucumber	Pusa seedless cucumber-6, Isaitish, Kian, Hilton, Multi star, Sunsrtar, Minianzal, Toska70, Pandex, Uniflora D, Corona, Farona, Marillo, Fidelio, Bronco,Mustang, Exacta,Ventura 1289, Jessica, Optima, Flamingo, Dominica, Accolade, Discover, Milligon
Bell Pepper	Red: Indra, Natasha, Bomby, Delphin, Plutona, Tango, Cubico, Mazurka, Val Valeta Yellow: Swerna, Orobely, Tanivi+, Yellow wonder, Luteus, Goldstar, Samantha, Gold Flame, Kelvin Orange: Wonde, Eagle, Narobi, Fellini Purple: Violetta

Crop Nutrition in Soilless Culture: Soilless culture except Deep Water Culture and Float Hydroponic requires fertigation systems: to mix appropriately water and fertilizers dissolved in concentrated stock solutions; and to uniformly supply feed solution to every plant.

Principles: In soilless culture, all essential plant nutrients should be supplied via the nutrient solution, with the exception of carbon, taken up from the air as CO2. To prepare nutrient solutions containing all essential nutrients, inorganic fertilizers are used as nutrient sources, except for iron, which is added in chelated form to improve its availability for the plants. Most fertilizers used to prepare nutrient solutions in soilless culture are highly soluble inorganic salts but some inorganic acids are also used. In commercial soilless culture, the fertilizers needed to prepare a nutrient solution are mixed with water to form concentrated stock solutions which are then automatically mixed with irrigation water to form nutrient solution.

Optimum range of Ec and pH values for different hydroponic crops

Crops	Ec (dSm^{-1})	pH
Asparagus	1.4-1.8	6.0-6.8
African Violet	1.2-1.5	6.0-7.0
Basil	1.0-1.6	5.5-6.0
Sage	1.0-1.6	5.5-6.0
Bean	2.0-4.0	6.0
Broccoli	2.8-3.5	6.0-6.8
Cabbage	2.5-3.0	6.5-7.0
Cucumber	1.7-2.0	5.0-5.5
Egg Plant	2.5-3.5	6.0
Peppers	0.8-1.8	5.5-6.0
Tomato	2.0-4.0	6.0-5.5
Leek	1.4-1.8	6.5-7.0
Lettuce	1.2-1.8	6.0-7.0
Pak Choi	1.5-2.0	7.0
Parsely	1.8-2.2	6.0-6.5

Crops	Ec (dSm^{-1})	pH
Celery	1.8-2.4	6.5
Spinach	1.8-2.3	6.0-7.0
Rose	1.5-2.5	5.5-6.0
Carnation	2.0-3.5	6.0
Courgettes	1.8-2.4	6.0
Strawberry	1.8-2.2	6.0
Ficus	1.6-2.4	5.5-6.0
Banana	1.8-2.2	5.5-6.0
Rhubarb	1.6-2.0	5.5-6.0

Source: Nisha Sharma et al., 2018

Irrigation management for soilless production

It is a water transport to the root zone and the decisions "when" to irrigate the crops and "how much" to apply. Irrigation scheduling requires good knowledge of the crop's water demand and the substrate's physical properties. Irrigation management is one of the main factors determining the overall performance of soilless culture systems, since both nutrients and water are supplied to the root zone via the irrigation system. One of the most important advantages of soilless culture compared with soil-grown cultures is the accurate control of water availability in the root zone. Moreover, soilless culture systems greatly improve water-use efficiency and water management in crop production. In soilless culture the root zone volume is much smaller than in soil-based cropping systems and thus the total volume of available water per plant is smaller, despite the higher water holding capacity, lower moisture tension and greater hydraulic conductivity in most crops grown on substrates (Schroder and Lieth, 2002). Smaller root volume results in restricted root length and surface area and, therefore, limited capacity of the plant to take up nutrients and water. Therefore "little and frequent" irrigation and fertilization is applied to maximize yield. A standard recommendation in soilless culture is to apply a constant volume of irrigation water at each irrigation event and vary the number of irrigation applications (as opposed to keeping the number of irrigation events constant and varying the volume of irrigation water).

Irrigation scheduling: Irrigation scheduling is the decision related to "when" to irrigate and "how much" water to apply to the crop. It is based either on substrate water status, where the moisture content or potential is measured directly to determine the need for irrigation, or on plant water status, which does not indicate how much water to apply. In soilless culture, little and frequent irrigation is required. Thus, rapidly growing crops in the summer may need 15–20 or even more irrigation events a day. Increasing irrigation frequency reduces fruit defects, such as cracking and blossom end rot.

Water Conservation for Hydroponic: As water becomes scarce and important as a resource, the use of hydroponics and other water saving technologies for crop production is needed now and is poised to popularize in time. Hydroponics uses substantially less water as compared to the soil farming. In soil farming, most of the water that we supply to the plants gets leached deep into the soil and is unavailable to the plant's roots, whereas in hydroponics, plant roots are either submerged in water or a film of nutrients mixed in water is constantly encompassing the root zone, keeping it hydrated and nourished. Water is not wasted in this process, as it gets recovered, filtered, replenished and recycled. Waste nutrient solution can be used as an alternate water resource for crop cultivation under hydroponic system (Choi et al., 2012). Savings in irrigation water, fertilizer and increase in vegetable and water productivity under hydroponic system as compared to conventional agriculture is depicted in Table. NFT based hydroponics can reduce irrigation water usage by 70% to 90% by recycling the run-off water. It is possible to effectively grow high value, good quality vegetables under controlled hydroponic conditions using 85 to 90% less water than traditional soil-based production. Water sources from groundwater or dam/river water commonly contain factors that can influence plant yield and affect plant condition, including salinity, dissolved solids and pathogens.

Irrigation decisions: Generally, common approaches in irrigation decisions entail timer-based, sensor based or model-based irrigation control methods.

- **Electrical timers** specifically designed to control irrigation valves are used. This operation is the cheapest, simplest and easiest approach for triggering irrigation events. However, time-based irrigation needs skilled personnel and knowledge to compile a present irrigation schedule.
- **Sensor-based** control depends on the measurement of the water status either in the substrate (i.e. tensiometer, frequency domain reflectometer) or in the plant (i.e. sap flow meter, thermal sensing).
- **Model-based** control is based on the estimation of plant water loss related to one or more environmental variables (i.e. temperature, solar radiation). Scheduling is computerized to obtain in real time crop irrigation needs based on the data provided online by sensors. Many methods are available for the estimation of evapotranspiration and most of them use either a modified version or a combination of models originally developed by Penman and Monteith. However, the application of these models in commercial practice needs appropriate calibration for each crop.

B. Aeroponic (Spray) Technique

Aeroponic is actually a subgroup of hydroponics, except that it uses no growing medium at all, and the plants grow by misting rich nutrients water. Aeroponic is the growing of plants with the root system suspended in a fine mist of nutrient solution applied continuously or intermittently. Plants are secured in holes on polystyrene panels using polyurethane foam: panels are placed horizontally or on a slope, and fixed over a metal frame, arranging closed containers with a square or triangular section. Water and nutrients are supplied by spraying the plant's dangling roots with an atomized nutrient solution by means of sprayers, misters or foggers inserted in PE or PVC pipes placed in the unit. The flow rate of the sprayers may range from 35 to 70 litres h^{-1}, whereas the spacing depends on the design and size of the cropping modules. As a general indication, they should be placed about 0.50 m apart to assure homogeneous nebulisation all along the Aeroponic unit. Spraying usually lasts 30–60 seconds, and their frequency varies according to species, plant growth stage, growing season and time of day (e.g. in summer during rapid vegetative growth, a crop grown in northeast Italy may require up to 80 sprayings per day). At each nebulisation, the drainage is collected at the bottom of the modules and recalculated. Aeroponic permits a major reduction in water and fertilizer consumption and ensures adequate oxygenation of the roots. However, aeroponically grown plants may experience severe thermal stresses, especially in summer. Another disadvantage is the inability to buffer interruptions in the flow of nutrient solution (e.g. power outages). Aeroponic may be used for small-sized vegetables (e.g. lettuce and strawberries) and medicinal and aromatic plants.

Advantages of Aeroponic Technique

- Aeroponic succeeds more in vertical growing arrangements and using the space efficiently.
- Aeroponic systems can reduce water usage by 98 percent, fertilizer usage by 60 percent, and pesticide usage by 100 percent, all while maximizing crop yields.
- The plants can more easily be transplanted, since they don't suffer from transplant shock.
- Aeroponics also allow you to observe the plants directly without disturbing them, which allows you to adjust the nutrient mix that you're using and cut off any problems that you might be having before they actually have a chance to become a problem.
- Aeroponic systems are favoured over other methods of hydroponics because the increased aeration of nutrient solution delivers more oxygen to plant roots, stimulating growth and preventing pathogen formation.
- Aeroponic techniques have proven to be commercially successful for propagation, seed germination, seed potato production, tomato production leaf crops, and micro-greens.

- Vegetable crops like potato, yams, tomato, lettuce and some of the leafy vegetables are being commercially cultivated in Aeroponic system. Aeroponic appeared to be a highly feasible method for the production of both aerial parts and roots

References

Choi, B., S.S. Lee and O.Y. Sik. 2012. Effects of Waste Nutrient Solution On growth of chinese Cabbage (*Brassica campestris* L.) in Korea. Korean Journal of Environmental Agriculture. 30(2):125-131.

Gruda, N. 2009. Do Soilless Culture Systems have an Influence on Product Quality of Vegetables? J. Appl. Bot. & Food Qual. 82:141-147.

Gruda, N. and W.H. Schnitzler 2006. Wood Fiber Substrates as a Peat Alternative for Vegetable Production. Eur. J. Wood Prod. 64: 347-350.

Helbacka, J. 2002. Row Covers for Vegetable Gardens. King County Coop. Extn. Ser., Washington.

Hochmuth, G.J., R.C., Hochmuth, S. Kostewicz and W. Stall. 2000. Row Covers for Commercial Vegetable Culture in Florida. Publication of Series of the Horticultural Sciences Department, UF/IFAS Extension, http://edis.ifas.ufl.edu.

Kaddi, G., B.S. Tomar, B. Singh and S. Kumar. 2014. Effect of Growing Conditions on Seed Yield and Quality Of Cucumber (*Cucumis Stivus*) Hybrid. Indian Journal of Agricultural Sciences. 84(5): 624-7.

Lorenzo, P., M.C. Sanchez-Guerrero, E., Medran, T. Soriano and N. Castilla. 2005. Response of Cucumbers to Mulching in an Unheated Plastic Greenhouse. J. Hort. Sci. Biotech. 80:11-17.

Sharma, N., S. Acharya, K. Kumar, N. Singh and O.P. Chaurasia. 2018. Hydroponics as an Advanced Technique for Vegetable Production: An Overview, Journal of Soil and Water Conservation. 17(4):364-371.

Schroder, F.G. and H.J. Lieth. 2002. Irrigation Control in Hydroponics. In D. Savvasand H.C. Passam, eds. Hydroponic production of vegetables and ornamentals. 263-298. Embryo Publications, Athens, Greece.

Singh, B., B.S. Tomar, A. Kumar and M. Kumar. 2010. Vegetable Seed Production under Protected conditions. Paper presented In: National Conference on production of Quality Seeds and Planting Material-Health Management in Horticultural crops. (March 11-14, 2010), New Delhi.

Sirohi, N.S. and T.K. Behera. 2000. Protected cultivation and seed production in vegetables. Indian Hort. 45:23-25.

Sirohi, N.S. 2002. Growing Vegetables under Protected Conditions. Proc Intconf on vegetable. pp 162, Banglore, India.

Streck, L., F.M. Schneider, G.A. Buriol, J. Luzza and M.A. Sandri. 2007. A System to Grow Lettuce inside Low Plastic Tunnels. Ciencia Rural. 37, 667–675.

25

Use of Space Technology in Agriculture: Innovations for a Change of Paradigm

Debashis Chakraborty[1], Sunayan Saha[2] and Saurav Saha[3]

[1]ICAR-Research Complex for NEH Region, Umiam, Meghalaya
[2]ICAR-Central Potato Research Institute, Regional Station-Jalandhar, Punjab
[3]ICAR-Research Complex for NEH Region, Mizoram Centre, Kolasib, Mizoram

Introduction

The country's population is growing at fast pace putting pressure on agriculture sector to grow more with declining resources. Proper planning and management become ever more crucial to ensure the sustainability of food production with minimal environmental footprints. Space technologies have unique advantages of providing large area synoptic coverage, with repetitive, objective and scientific datasets essential for monitoring a hugely vast and heterogeneous country like India. Agriculture being very dynamic and highly dependent on different natural resources needs continuous monitoring. Therefore these technologies have immense potential for application in the agriculture sector of the country. Space based technologies provide information on crops and natural resources on a real-time basis, proving unmatched opportunity for planning including in-season manipulation of crop production and management practices improving the resilience of farming against weather aberrations. Advanced estimates of regional agricultural output are critical to anticipate and mitigate the course of food shortage from sudden crop failure.

Incidentally, the first practical application of remote sensing in India was in the field of agriculture, wherein early detection of coconut wilt disease in Kerala was possible using air borne infrared cameras (Dakshinamurty et al. 1971). To explore the potential of space technology through satellites Joint Experiment Programme (JEP) started for defining the appropriate configurations of sensors in Indian Remote Sensing (IRS) satellites during the early 1980s. After exhaustive field experimentation, the potential of remote sensing (RS) technology to estimate crop yield and area to forecast its productivity at district and regional level due to the multispectral, large area and repetitive coverage was established. By that time this technology was being used operationally used by a few advance nations. With the launch of the first IRS satellite IRS-1A in 1988, in India, the Crop Acreage and Production Estimation (CAPE) project was initiated to provide district and state level pre-harvest acreage and production forecasts meeting 90 percent accuracy with 90 percent confidence level accuracy criterion. The crops covered were wheat, rice, cotton, mustard, sorghum and groundnut using RS data along with ancillary ground and weather data (Navalgund et al. 1991). The datasets of IRS-1A and 1B was also used by the National Natural Resources Management System (NNRMS) for the monitoring and management of the natural resources of the country. Since then the satellite sensor based vegetation index (Normalized Difference Vegetation Index- NDVI) was calculated over 11 states to assess and monitor the agricultural drought condition under National Agricultural Drought Management and Monitoring System (NADAMS). This was followed by the mapping of saline, alkali and salt affected soils and further to map the soil of the whole country using RS as well as ground surveys.

Based on the experience of CAPE, synthesizing state-of-the-art in econometrics agricultural meteorology and remote sensing, Forecasting Agricultural Output using Space, Agro-meteorology and Land-based Observations (FASAL) project was initiated for making multiple in-season forecasts at national level. Further with the advent of Synthetic Aperture Radar (SAR) datasets the mapping and monitoring of crops during *kharif* season especially the rice crop could be possible with accuracy over 90% (Chakraborty et al. 1997). Using the in-season multi-temporal images, crop phenology i.e. the crop spectral emergence, maximum growth and length of the growing season was derived. The crucial crop bio-physical parameters like leaf area index (LAI), fraction of photosynthetic active radiation (FAPAR) etc were retrieved. Soil water being one of the most critical factors for crop growth was estimated and further to estimate the water stress evapotranspiration was also derived through energy-balance monitoring. To understand the spatial variability in crop growth, crop simulation model was integrated with remote sensing inputs for deriving farmers' field scale information (Sehgal et al. 2001). Apart from monitoring, remote sensing has also been extensively used for planning purposes like site suitability evaluation of different crops and cropping systems, sustainability of the cropping systems over a region, water resources development, wasteland management and even possibility of crop expansion in certain regions. SAR datasets are nowadays being increasingly used due to their all-weather conditions applicability for even retrieving crop biophysical parameters like crop height, LAI and also to understand the complex phenomenon like crop lodging (Chauhan et al. 2018). Hyperspectral remote sensing, internet of things (IoT) sensors, high speed data processing and intelligent systems are bringing a completely new dimension in the capabilities of RS applications.

Since, the start of the journey in the late 1960s there has been a lot of improvement in the capabilities of the space technology i.e. sensors and satellites point of view as well as their application in different sectors including agriculture. Agriculture sector, being the prime consumer of the natural resources, most economic use of them is must for long term sustainability. Moreover, to help farming in day to day decision making, the country has agro-advisory service systems. In view of increasing variability in the weather system due to climate change, the need of the hour is that farmers adopt climate smart agriculture practices based on weather or agro-meteorological advisories. Nowadays, India is capable of objective monitoring of all its resources in near real time using space technologies. Though there has been much improvement in the technology front, policy makers should promote the mass scale use of these technologies through research-based outcomes so that an ecosystem can be created where these technologies are used to its fullest.

Indian space capabilities

India has a plethora of satellites and sensors working from optical, thermal to micro-wave domain. The following is a list of satellite-sensors and characteristics important for agriculture sector.

Table 1: List of Indian satellites important for agricultural applications

Satellite	Year	Sensors	Spectral bands (µm)	Spatial resolution (m)	Swath width (km)	Repeat cycle (days)
Polar Orbiting Satellites						
IRS-1A	1988	LISS-I, and LISS-II A/B	0.45-0.52 0.52-0.59 0.62-0.68 0.77-0.86	72.5 m LISS-I 36 m LISS-II	148 74 x 2 (swath of 148 km)	22
IRS-1B	1991	LISS-I and LISS-II A/B	same as for IRS-1A		148 74 x 2	22
IRS-P2	1994	LISS-II M	0.45-0.52 0.52-0.59 0.62-0.68 0.77-0.86	32 m x 37 m	66 x 2 (131 km for combined swaths)	24
IRS-1C	1995	LISS-III	0.52-0.59 0.62-0.68 0.77-0.86 1.55-1.70	23.5 23.5 23.5 70	142 142 142 148	24
		PAN	0.50-0.75	5.8	70	24 (5)
		WiFS	0.62-0.68 0.77-0.86	188	804	5

Satellite	Year	Sensors	Spectral bands (μm)	Spatial resolution (m)	Swath width (km)	Repeat cycle (days)
IRS-P3	1996	WiFS	0.62-0.68 0.77-0.86 1.55-1.70	188	804	5
		MOS-A MOS-B MOS-C	0.75-0.77 0.41-1.01 1.595-1.605	1500 520 550	195 200 192	Ocean surface
		IXAE	Indian X-ray Astronomy Experiment			
IRS-1D	1997	Satellite and instruments are identical to those of IRS-1C				
IRS-P4 (OceanSat-1)	1999	OCM MSMR	0.4-0.9 6.6, 10.65, 18, 21 GHz (frequencies)	360 x 236 105 x 68, 66 x 43, 40 x 26, 34 x 22 (km for frequency sequence)	1420 1360	2 2
IRS-P6 (ResourceSat-1)	2003	LISS-IV	0.52-0.59 0.62-0.68 0.77-0.86	5.8 5.8 5.8	70	24 (5)
		LISS-III*	0.52-0.59 0.62-0.68 0.77-0.86 1.55-1.70	23.5 23.5 23.5 23.5	140	24
		AWiFS	0.62-0.68 0.77-0.86 1.55-1.70	56 56 56	740	5
IRS-P5 (CartoSat-1)	2005	PAN-F PAN-A	0.50-0.75 0.50-0.75	2.5 2.5	30 30	2-line stereo camera
IRS-P7* (CartoSat-2)	2007	PAN camera	0.50-0.85	< 1	9.6	
RISAT-2*	2009	SAR	9.590 GHz (X-band)	< 1 to 8	10-100	14 (3-4)
OceanSat-2*	2009	OCM SCAT ROSA	0.40-0.90 8 bands 13.515 GHz GPS occultation	360 x 236 25 km x 25 km	1420 1400	2
ResourceSat-2*	2011	Satellite and instruments are identical to those of ResourceSat-1				

Satellite	Year	Sensors	Spectral bands (µm)	Spatial resolution (m)	Swath width (km)	Repeat cycle (days)
Megha Tropiques* (ISRO/CNES)	2011	MADRAS SAPHIR ScaRaB GPS-ROS	5 channel radiometer Atmos sounder Radiation budget Occultations	40 km x 60 km	1700	2
RISAT-1*	2012	SAR	5.350 GHz (C-band)	< 2 to 50	100 - 600	25 (12)
SARAL* (ISRO/CNES)	2013	AltiKa DORIS Argos-3 LRA	35.75 GHz Ka-band altimeter S/C tracking for POD services Data collection system Satellite laser ranging			35
HysIS* (Hyperspectral)	2018	VNIR SWIR	0.4 – 0.95, 10 nm bands 55 0.85 – 2.4, 10 nm bands 256	30 m	30	
CartoSat-3*	2019	PAN MSP MWIR	0.4 – 0.7 4 bands 5 m	0.28 1 m		
RISAT-2BR1*	2019	Satellite and instruments are identical to those of RISAT-2B				
Geostationary Satellite						
INSAT-3D* (positioned over 82° E Longitude)	2013	6 Channel MSI 19 channel sounder DRT SAS&R	0.52-0.72 1.55-1.70 3.80-4.00 6.50-7.00 10.2-11.2 11.5-12.5	1 km 1 km 4 km 4km 8 km 8 km		30 min
INSAT-3DR* (positioned over 74°E Longitude)	2016	Satellite and instruments are identical to those of INSAT-3D				30 min

Satellite	Year	Sensors	Spectral bands (μm)	Spatial resolution (m)	Swath width (km)	Repeat cycle (days)
Navigation/Positioning Satellites						
IRNSS-1 A to I* (Eight satellites) (positioned over 32.5°E to 129°E Longitude)	2013	A navigation payload CDMA ranging	Navigation signals at L5 and S-band		32.5°E to 129°E Longitude	

*Indicates the satellites which are operational as on date.
ISRO- Indian Space Research Organization; CNES: National Centre for Space Studies, France

Aryabhata was the first Indian Satellite launched in 1975 but operational remote sensing services were started with the launch of first Indian Remote Sensing Satellite (IRS), IRS-1A in the year 1988. This satellite had various sensors (Linear Imagine Self Scanning System- LISS-I and II) to provide varied spatial, spectral and temporal resolution images. The IRS-1C was launched in the year 1995 which had quite significant improvements having three different sensors namely improved LISS-III, panchromatic camera (PAN) and Wide Field Sensor (WiFS). These three sensors improved spatial, spectral as well as the temporal resolution. IRS-P4 or OceanSat-1 was the first satellite designed to study the ocean which had two unique sensors namely Ocean Colour Monitor (OCM) and a Multi - frequency Scanning Microwave Radiometer (MSMR). A quantum jump in the earth observation capabilities was achieved with the launch of IRS-P6 or ResourceSat-1 in 2003. Along with LISS-III it had LISS-IV with improved spatial resolution (5.8 m) and an option of steerability up to 26° for obtaining stereographic imagery to achieve five day revisit capability. It also had an Advanced Wide Field Sensor (AWiFS) which is the improved version of WiFS. IRS-P5 or CartoSat-1 and IRS-P7 or CartoSat-2 had high resolution PAN cameras to provide 2.5 m and 1 m spatial resolution panchromatic images, respectively. With the launch of Radar Imaging Satellite (RISAT-2) in the year 2009, India entered the domain of synthetic aperture radar (SAR). It works in the X-band domain. ResourceSat-2 was launched in 2011, which provides data continuity regarding the sensors onboard ResourceSat-1. Megha Tropiques is a joint venture between Indian and French space agencies to study the water cycle and energy exchanges in the tropical region. It was launched in 2011 and have four very crucial sensors to meet the objectives. Another important achievement is the launch of HysIS or Hyperspectral Imaging Satellite in the year 2018 through which India records datasets in the hyperspectral domain. Apart from these remote sensing satellites, Indian National Satellite System (INSAT) is multi-purpose geostationary satellites providing telecommunication, broadcasting and meteorological services. The Indian Regional Navigation Satellite System (IRNSS), with an operational name of NavIC (Navigation with Indian Constellation) provides real-time positioning information over the region which is very crucial for various applications. With various types of satellites and sensors systems working in different parts of the electromagnetic spectrum for different application purposes, Indian space capabilities are at par with the most advanced nations.

Applications in Agriculture

Crop growth, acreage and yield

Crop health information during the process of development is crucial for understanding the probable economic output. Crop health refers to the state of growth i.e. whether the crop is doing well (healthy) or infected with pests/ diseases or nutrition deficiency. Monitoring of these is very crucial for resource management especially under precision agriculture where fertilizer, irrigation or pest management decisions are to be appropriately taken for maximizing the productivity. Detecting a specific growth phase from satellite remote sensing platforms is difficult and hence, a general descriptor of vegetation cycle called land surface phenology is used for this purpose (Zeng et al. 2020). Vegetation index values can directly provide information about the crop growth over any particular region. Further the comparison of vegetation index of a particular year with that of a good year can provide a glimpse about its status. Vegetation biophysical and biochemical parameters (such as leaf area index-LAI, fraction of photosynthetically active radiation, chlorophyll, water content etc.) are also derived from vegetation indices or inversion of Radiative transfer models (RTM) to understand the growth or health of the crops (Chakraborty et al. 2015). Wheat LAI was retrieved using IRS LISS-III with a considerable accuracy having root mean square error (RMSE) of 0.56 m^2/m^2 (Sehgal et al. 2016).

In case of crop phenology detection, multi-temporal datasets are used. The time-series of vegetation index datasets are at first smoothened and then applied with different phenology extracting methods for its extraction (Zeng et al. 2020). Savitzky-Golay filtering was used for extraction of phenology parameters of *kharif* and *rabi* crops over Indo-Gangetic Plains of India (IGP) (Sehgal et al. 2011). The study reported increase in the peak growth over IGP indicating the increase in productivity of crops over time from 1981-2001. Using TIMESAT software for extraction of wheat crop phenology over North India, earliness in the start of the growing season over Punjab and Haryana while its delay in the eastern part of Uttar Pradesh was reported (Chakraborty et al. 2018). The farmers have started sowing wheat crop early as an adaptation strategy to avoid the terminal heat stress.

The information about crop specific acreage of a state or country is very important for planning purposes. Spectral signature which is the typical characteristics of any feature arising out from its interaction with the electro-magnetic radiation enables the identification of that feature. The uses of radiance/reflectance in specific spectral bands of the satellite sensors or the combination of them to generate different spectral indices are used. Most popular spectral indices are ratio vegetation index (RVI), normalized difference vegetation index (NDVI), enhanced vegetation index (EVI) etc, which are used to classify features into few broad groups like vegetation, built-up area, water bodies etc. Based on ground truth collection and different classification algorithms, crops can be separated from the rest. Multi-date in-season satellite imageries are capable to differentiate crops which are grown in the same seasons like wheat, mustard and potato (in *rabi*). Satellite based study on rice, mustard and potato reveal that classification accuracy between LISS -II (36 m spatial resolution) and LISS-III (23.5m) data increased by about 6% with the inclusion of MIR (Middle Infra-Red) band mainly due to better separability between low land rice and other vegetation. In case of mustard and potato grown in small plots, the classification accuracy increased by 5% to 8% due to higher spatial resolution of LISS-III (Manjunath et al. 1998). Evaluation of potential usefulness of IRS-1C WiFS data for dry land *rabi* sorghum crop discrimination indicated that single date WiFS data gave slightly lower accuracy (79%) compared to the single date LISS-1 data (81%). Overall accuracy for 5 dates WiFS NDVI data were far superior to the single date data of the IRS-1C as well as the IRS-LISS 1. These results indicate that with increase in spatial resolution the accuracy increases but the spectral resolution also plays crucial roles. Compared to the single date data, multi-date data possess better separability even that may be of coarser spatial resolution. Nowadays the satellite remote sensing derived vegetation index (EVI or NDVI) is used for generation of the crop phenology which are used in different machine learning algorithms (random forest, neural network learning etc.) to identify different crop types and their acreage over large areas (Wei et al. 2019).

Better and accurate yield estimation has been one of the most explored areas of space technology applications in agriculture. Attention has been given to the development of yield models, where near real time remote sensing information is the most crucial input. Successful attempts have been carried out in India in the spectral response and yield relationships of different crops such as wheat, rice, mustard, chickpea, groundnut, sorghum, etc. based on field studies. Work on yield modeling using space borne data has received increasing attention with the launching of series of indigenous Indian remote sensing satellites (IRS 1A, 1B, 1C, 1D, P1, P2, P3 and P4). The simplest way of yield estimation is through relationship of yield with spectral indices like RVI, NDVI etc. derived from satellite data of the crop during the maximum vegetative growth phase. Apart from that remote sensing input with weather variables, crop biophysical parameters (LAI, Phenology etc.) derived from remote sensing and dynamic crop simulation models with remote sensing inputs are also used for developing yield models at regional scales (Dadhwal et al. 2003). Nowadays using the information from crop-calendars, weather forecast and remote sensing datasets through machine learning/artificial intelligence accurate crop yield is estimated (Cunha and Silva, 2020).

Horticultural Inventory

The first application of remote sensing in India was in the horticulture sector the detection of coconut wilt disease in Kerala (Dakshinamurti et al. 1971). In India, largely, people grow most of the horticultural crops as part of their homestead. If properly managed, the sector offers huge potential to address malnutrition issue in large part of our population, sustaining rural agro-industries, local employment generation and providing options for crop diversification. The foremost requirement for improving this sector through need based prioritized interventions is preparing inventory. To accomplish that, we need accurate spatial maps on dynamic basis along with standard spatial databases on various themes such as land use & land cover, weather, soil, crop type, species & variety, age, health & stress level, disease severity, management practices, infrastructure & socio-economic indicators. In this sense, space technologies are supplementing the various ongoing programmes of the Govt. such as Mission for Integrated Development of Horticulture (MIDH) for improving the sector. Indigenous satellite data such as Resourcesat-2 LISS III (23.5 m resolution), LISS IV (5.8 m) and LISS IV + Cartosat (2.5 m) merged product are being used depending upon the spatial extent and heterogeneity of the crop cover.

Area estimation and production forecasting

It has been done for some major horticultural crops, *viz.*, mango, banana, citrus, potato, onion, chilli and tomato in 185 districts of 12 states. In case of vegetables and spice crops, major pockets of the country are being covered. However, for potato, area and production estimation are being carried out at district/state/ national level on operational basis.

Horticulture Development and Planning

Cropping intensity maps are being studied to identify areas with low crop intensity along with favourable conditions where horticultural crops could be introduced. Site suitability analysis for horticulture has been done in North-Eastern states where soil degradation/erosion of fertile soil from hill slopes due to indiscriminate Jhum/ shifting cultivation is a major concern. Satellite remote sensing provides unique opportunity to derive the soil (depth, texture and nutrient content), topography (slope, aspect and elevation) and climate information over difficult hilly terrain for development of these plans. Horticulture expansion plan have been proposed as an alternative in seven districts taking one from each state. Similarly, researchers also suggested plans for apple plantations of Arunachal Pradesh and cultivation of grapes in large areas in Mizoram using geomatic tools.

Design plans for optimal post-harvest infrastructure

At present, post-harvest losses in horticulture sector in India are huge (>50%). Inaccessibility of small and marginal farmers to cold storage facility is a major reason for this and such losses can be minimized by establishing cold storage at optimal locations instead of clustering only around the major markets/ consumption centers. An optimal plan for setting up cold storage facilities for potato in Bardhaman district of West Bengal was developed. Such plan was also developed for Bihar state (Ray et al. 2000). These plans were generated by combining satellite derived crop maps with different socio-economic thematic information like settlements, roads, nearest cold store facility etc.

Wetland mapping for Aqua-Horticulture

Some horticultural crops like Makhana/Fox nut are grown in wetlands. In order to identify more areas suitable for such crops, countrywide wetland maps at 1:50,000 scales, derived from satellite data, & GIS database of analogous ecology and economic conditions are being used (Garg, 2015).

Orchard mapping and rejuvenation

Locations of orchards along with some important characteristics like age, species and variety etc. are being generated. Such maps are useful for better management of the Orchards, to identify plantations/orchards that require rejuvenation and to evaluate whether any shifting of orchards is needed. For, example, due to climate change, some apple orchards in Himachal Pradesh need to be shifted to further uphills for proper fruiting. Similarly, possible shift in mango (West Bengal & UP), orange (Darjeeling) and citrus in Arunachal Pradesh have been identified.

Natural Resources Monitoring

Soil resources mapping (Land use & land cover)

Mapping and identifying land use/land cover and its change is the most important and widely researched topic in remote sensing. Land use/land cover analysis is extensively used to derive a number of biophysical variables, such as vegetation index, biomass, and carbon content. The soil reflectance in visible (0.4 - 0.7 μm) near infra-red (0.7-1.1 μm), shortwave infra-red (1.1-2.5 μm) and thermal infra-red (3-5 and 8-14 μm) regions have distinctive patterns playing deterministic role in remotely sensed data acquisition for land use /land cover classification. Although the term 'land use' and 'land cover' is often used interchangeably, their actual meanings are quite distinct. Land use refers to the existing utilization prospect that the land serves such as, recreation, wildlife habitat or agriculture. Land use applications involve both baseline mapping and subsequent monitoring on timely information acquisition for dominant land use pattern and assessment of dynamic changes in land use on year to year basis. Land cover refers to the surface cover on the ground- vegetation, urban infrastructure, water, bare soil etc. Satellite imageries are well suited to classify land use/land cover where land use categories are associated with homogeneous land cover at the sub-pixel level, but aerial photography or ground observations are essential for validation. Landsat 8 OLI satellite image (U.S. Geological Survey), CORINE Land Cover – CLC (European Environment Agency), GlobeLand30 (Geomatics Center of China) etc. are some of the easily accessible dataset with 25-30 m spatial resolution for land cover analysis under variable time scale. The finer spatial resolution of 0.5 m is available from Orthophotos (National Agency for Cadaster and Land Registration under A.N.C.P.I. Bucharest, Romania). Landsat Thematic Mapper imagery, using a pixel-object-knowledge (POK) based classification yielded 80% accuracy under 10 categories of land cover viz. cultivated land, forest, grassland, shrub land, wetland water bodies, tundra, artificial surfaces, bare land, permanent snow and ice. Image resolution and atmospheric condition often adversely affected the accuracy of the land use/ land cover classification algorithms. For complex landscape, integrating satellite images with ancillary data derived from currently available land cover models are essential with minimizing high omission and commission errors like built-up areas and riverbeds or deciduous forests and orchards under GIS environment (Rujoiu-Mare and Mihai, 2016). However, newer approaches such as fuzzy classification, artificial neural network and object-based classification will be useful to formulate a nationwide common framework of land use/land cover monitoring to detect real time changes in regional scenario. Extensive training on analyzing textural measures and human supervision are essential pre-requisites for more accurate land cover/land use classification using fractals variograms, lacunarity and spatial autocorrelation statistics (Lam et al., 2002). Efficient methods for rapid monitoring in textural changes in the acquired remote sensing images viz., band 4 (0.5-0.6 µm) and band 6 (0.7-0.8 µm) of Landsat TM 5, C-band data and dual polarization images Microwave satellites (VV, VH, HH, and HV; Bousbih et al., 2019) are essentially required for the development of real time decision support system on crop water management and periodic risk assessment in integration with GIS tools towards raising farming resiliency in Indian Agriculture.

Mapping problem soil areas

Remote sensing tools are useful for detecting the exact location and assessing the spatial extent of problem soils/wasteland for real time monitoring and effective management of agricultural resources towards achieving sustainable development goals.

Land degradation

Land degradation includes ensemble of surface processes viz. wind erosion, water erosion, soil compaction, water logging etc. Remote sensing tools based on the acquired soil reflectance over any large spatial domain are useful for real time detection, assessing, mapping, and monitoring of land degradation process. Since mid-1970s, several space borne sensors were used for land degradation studies. In India, Indian Remote Sensing Satellites along with Landsat-TM, SPOT are often used for land degradation studies with better spatial and spectral resolution periodically to identify and quantify the Gully lands, water logged, saline areas ravine land and shifting cultivation areas against the traditional field survey. Landsat 7 ETM+ and Landsat 8 Operational Land Imager were highly effective to combat land degradation and conserve land resources in Indo-Gangetic Plain region of North India (Ahmad and Pandey, 2018; Dandapat et al., 2020).

Soil erosion

Satellite data are useful for direct detection of both wind and water erosion processes and their consequences through direct detection of erosion features and eroded areas. These advantages facilitated extensive use of satellite imageries for detailed identification of individual large and medium sized gullies (El Jazouli et al., 2017). Several studies confirmed the potential of different soil reflectance based spectral indices such as form index (FI), coloration index (CI), brightness index (BI), and normalized difference vegetation index (NDVI) for characterizing soil erosion particularly in ravine areas of arid and semi-arid regions (Hill et al. 1994 Haboudane et al. 2002, Barakat et al. 2016). High resolution satellite sensors are also available for precise estimation of soil loss (Vrieling, 2006; Kourgialas et al., 2016). Quantification of soil loss and delineation of degraded areas from water erosion is often relied on the measurements of erosion intensity based on empirical relations for effective conservation planning. The revised version of Universal Soil Loss Equation (RUSLE) is considered as the best model for estimation of surface soil erosion with suitable ground validation (Chatterjee et al., 2014; Tang et al. 2015).

Water logging and salinization

Excess water stagnation because of topography and poor drainage often results in physical deterioration of land quality. The water logging may be surface or sub surface type. The net impact depends on the duration of water logging and it may develop soil salinity particularly with shallow saline below ground water table. For water logged soils, the use of digital elevation model (DEM) is essentially required for construction of artificial drainage infrastructures, mapping of drainage channels and catchment outlets with reasonable accuracy. Aerial photography, satellite imageries and space based radar are often useful for the precise estimation. Hydrological modeling using DEM facilitated the advanced planning for creating drainage network in water logged soils (Yang et al. 2011; Bastawesy and Ali, 2013).

Anthropogenic activities often results in secondary salinization. Rapid urbanization or overexploitation of canal irrigation, resulted in surface accumulation of soluble salts from impeded drainage that have deleterious impact on agro-ecosystems and regional agricultural production (Gorji et al., 2017). The total salt affected soils in India are estimated to be about 6.72 million ha (NRSA, 2008). Remote sensing technology is one of the most effective and economic tools for identification and monitoring of the patches affected by soil salinity in canal irrigated areas (Dehni & Lounis 2012; Scudiero et al., 2015; Taghadosi et al., 2018). The accumulation of salts on the terrain surface can be detected directly either from bare surface with salt efflorescence or crust presence in arid and semi-arid regions of India. The indirect measurements relied on changes in vegetation characters as affected by soil salinity. Arial photography, infra-red thermography, visible/infrared multispectral and microwave imageries can be used to detect the extent. Multispectral satellite imageries are very promising tool to map salinity and generating electrical conductivity (EC) maps (Dwivedi and Sreenivas, 1998). Composite indices based on multiple bands were more efficient for detection of soil salinity than a single band (Allbed et al., 2014; Fan et al., 2015). Terrain features viz. relative altitude of drainage channel, and topographical wetness controlled the extent of responses. Evaluation of vegetation indices using high spatial resolution satellite imageries can also be effectively used to serve the purpose (Setia et al. 2013). Several remote sensing based indices such as the salinity index, the normalized difference salinity index and the brightness index are often used in addition to normalized difference vegetation index (NDVI) for getting effective estimates of soil salinity mapping based on blue, green, red and near infrared bands over the salt-affected lands with sparse vegetation cover (Jiapaer and Bao, 2011; Elhag, 2016).

Insolation

The term "Insolation" refers to the incoming solar radiation received on ground on a horizontal surface. Solar radiation is the most important driver for crop production. For insolation estimation, both empirical/statistical and physical methods have been developed. The first attempts were based on regression relationship between the top of atmosphere (TOA) reflectance/transmittance derived from satellite data and ground based pyranometer data of insolation at the time of satellite overpass. Like, visible band data were used first to estimate cloud amount and then based on cloud cover different regression equations were used. In case of physical approach, there are two types, viz. "top down" and "bottom-up" and both uses various radiative transfer equations. In the former, computation starts with shortwave radiation at the TOA and from this attenuation due to atmosphere and cloud are subtracted stepwise till it reaches the ground. In the bottom-up case, surface albedo is determined from planetary (earth plus atmosphere) albedo and then surface insolation is arrived through inversion. The above approaches have been applied to both geostationary and polar satellite data like GOES, METEOSAT, KALPANA-1 (METSAT), GMS, NOAA-AVHRR, LANDSAT-TM etc. Most operational solar radiation products worldwide including in India are generated utilizing geostationary satellite data through semi-empirical approach. Nowadays, the accuracy of such products at daily time scale often match that of ground based pyranometers.

Soil Moisture (SM)

Depending on the application needs various spectral bands have been used to generate products on the top soil (about 5-10 cm depth) moisture content where active roots of most crop plants are present. Due to higher specific heat of water than the soil particles, higher amount of water in soil shall delay changes in its temperature i.e. exhibit some kind of inertia to the change of state. High water content also causes darker appearance of soil due to lowering of albedo. These properties can be used to compute apparent thermal inertia (ATI), as an indicator of soil wetness, by utilizing surface albedo from optical (visible) band and day-night soil brightness temperature difference from thermal infrared (TIR) bands. Using suitable combination of optical and thermal bands field scale resolution SM products can be generated but only under clear sky conditions. By using time series ATI data from geostationary satellites Kalpana-1 (VHRR) and Meteosat the soil moisture saturation index (SMSI) could be computed and soil moisture information derived using soil specific higher & lower moisture limits (Bhattacharya et al. 2008).

Microwave sensors are capable to generate SM products under all weather conditions by using L or C bands, however, at the cost of spatial resolution. Between passive & active, the former produces coarser spatial resolution (of the order of 50 km or so) product whereas the latter can provide relatively better spatial resolution but has limited coverage per scene. SM estimation using optical bands mainly depends on the relationship between SM and the surface reflectance or vegetation index. Whereas, in case of thermal bands, the relation between SM and the surface temperature or variations of surface temperature /vegetation index are explored. For microwave remote sensing, estimation of soil moisture is based on the principle that any changes in soil moisture content also bring changes in its dielectric properties. Dielectric constant in turn changes the emissivity of soil. In passive microwave remote sensing system, the electromagnetic radiations (emitted by the soil surface) received by the radiometer is converted into microwave brightness temperature (Tb) which in turn is co-related to soil moisture. On the other hand, in active remote sensing, any dielectric changes causes change in the microwave backscattering coefficient (Bruckler et al. 1988; Calla et al. 2008; Zhang & Zhou, 2016). Satellites like RISAT-1, RADARSAT-2, SMOS, SMAP, MODIS have been used for soil moisture estimation. Some products are in the form of index values instead of direct values of soil moisture content. Soil Wetness Index (SWI) is one such. It is derived utilizing SMAP L-band radiometer data, normalized to the extreme values of 0 and 1, corresponding to the dry and saturated soil wetness conditions respectively. We expect in coming days, by suitably combining the optical and or thermal images with microwave data, more relevant SM products at small field scale and at optimal temporal frequency can be produced. Such product will have the potential to replace tedious & costly soil moisture monitoring methods using ground based instruments for irrigation scheduling.

Consumptive water use

Consumptive water use of a crop is basically the actual evapotranspiration (AET). Direct field measurement of AET is quite cumbersome and involves costly equipments such as lysimeter and eddy covariance systems. Single source models such as SEBAL, SEBI, S-SEBI and SEBS are nowadays being widely used for spatial AET estimation using satellite data. These models are based on the residual methods of energy balance, use both empirical relations as well as physical parameterization equations and require minimum ground based data. Insolation, surface albedo, land surface temperature and vegetation index (VI) maps are needed to run these models. VI maps are used to derive land surface emissivity, roughness parameters and ground heat flux. Sensible heat flux is computed through flux inversions with respect to two extreme types of pixels i.e. dry (no evaporation) and wet (total energy to be spent for latent heat flux). Then, latent heat flux which is equivalent to AET is approximated as the residual of energy balance. In India, national scale AET products at 8 Km resolution using indigenous geostationary satellite Kalpana-1 data are nowadays available. Globally MODIS-based AET data are also available.

Disaster Monitoring

Drought

Drought is a phenomenon that principally occurs due to abnormality of rainfall. About 68% of cropped area in India is vulnerable to drought. One-third of this, where mean annual rainfall is less than 750 mm, is highly susceptible and regarded as chronically drought-prone. Space technology offers unique advantages in drought monitoring. The main approach followed here is based on the assessment of the crop /vegetation condition at a place in a given time of the year in comparison to the corresponding normal condition. To start with, vegetation indices like NDVI or Normalized Difference Water Index (NDWI) are computed by utilizing the red (R) along with near infrared (NIR) or shortwave infrared (SWIR) band data of satellite images. Then, relative deviation of the current year NDVI or NDWI from that of normal (average of at least three recent normal years data) is used as criteria for inferring the crop condition/occurrence of drought and its severity. As an alternative to relative deviations of NDVI or NDWI, Vegetation Condition Index (VCI) method has also been developed. In case of VCI approach, long-term (at least ten years) data of NDVI or NDWI are used to create a normalized scale by assigning the minimum & maximum extremes to 0% and 100%, respectively. Position of the current year NDVI or NDWI values is then determined on that scale to infer crop condition/drought severity. There is another useful index, TCI i.e. Temperature Condition Index, that can be computed using thermal bands. There are other important approaches wherein Geospatial techniques like GIS with or without satellite images are also used to generate maps on drought situation. These are for example, i) SPI i.e. Standardized Precipitation Index (based on rainfall deficiency), ii) deviation in crop sown area, iii) soil moisture based indices like Moisture Adequacy Index (MAI) & Percent Available Soil Moisture (PASM) computed using agro-meteorological models.

Locust

Desert locusts are one of the most harmful insects. These are grass-hoppers with short horns having marked polymorphism and most importantly highly migratory with voracious feeding behavior. They form swarms (adult' congregation) and hopper bands (nymphal congregation) which can travel long distances and cause devastation to the natural as well as cultivated vegetation. Desert locusts inhabits in the semi-arid to arid regions starting from West Africa to North West of India covering about 16 million km^2 over 30 countries. In India the locust upsurge has occurred several times during the 19th and 20th century causing huge losses to the cultivated areas but it decreased gradually after 1970s. These locusts generally intrude western districts of Rajasthan and Gujarat. In the year 2020, during the month of April, locust swarms entered India from the western border and covered almost the entire Rajasthan and spread to the adjacent districts of Madhya Pradesh and Uttar Pradesh. In India, Locust Warning Organization (LWO), Directorate of Plant Protection Quarantine and Storage, Ministry of Agriculture & Farmers Welfare, Department of Agriculture, Cooperation & Farmers Welfare is responsible for monitoring, survey and control of Desert Locust (http://ppqs.gov.in/divisions/locust-control-research), while the Desert Locust Information Service (DLIS) of FAO, Rome coordinate the activities for the whole world (http://www.fao.org/ag/locusts/en/activ/DLIS/dailyphotos/index.html). The locust population increases very fast over space and time under favourable bio-climatic condition. Therefore, monitoring of the bio-climatic condition of the breeding grounds remains the key for early detection of the locust swarm. Remote sensing technologies are used for deriving this information from inaccessible desert grounds. Satellite data provides the large area accessibility with high temporal resolution for generation of reliable datasets on the bio-physical variables which are then modeled under GIS environment with the locust physiology and historical migration behavior to get the probable region where it can spread in the due course (Fig. 1; Dutta et al. 2003).

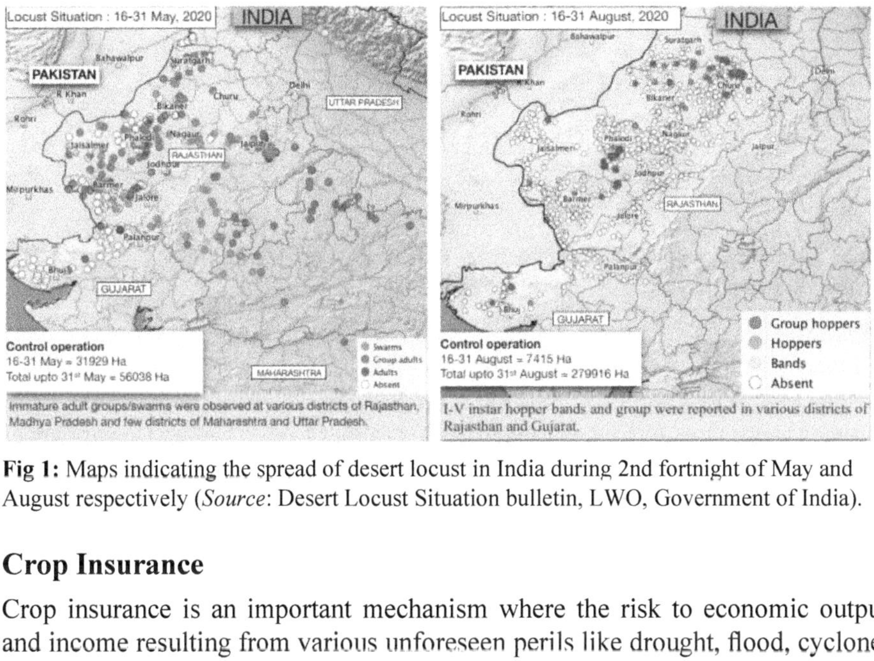

Fig 1: Maps indicating the spread of desert locust in India during 2nd fortnight of May and August respectively (*Source*: Desert Locust Situation bulletin, LWO, Government of India).

Crop Insurance

Crop insurance is an important mechanism where the risk to economic output and income resulting from various unforeseen perils like drought, flood, cyclone, and heat stress which are beyond the control of the farmers is managed. Through insurance the crop production risk is transferred to the insurer at the cost of premium. Crop insurances are basically of two types: indemnity and index based. In case of indemnity based insurance, the insurance amount is paid based on the measurement of actual crop loss obtained through crop cutting experiment. However, it is having certain limitations like slow claim settlement and high administrative cost. In index based insurance, a robust relationship is developed between yield and any weather based or remote sensing derived index and then, insurance is paid based upon aberrations of the index parameters. Hence it is objective, very fast and incurring low administrative cost.

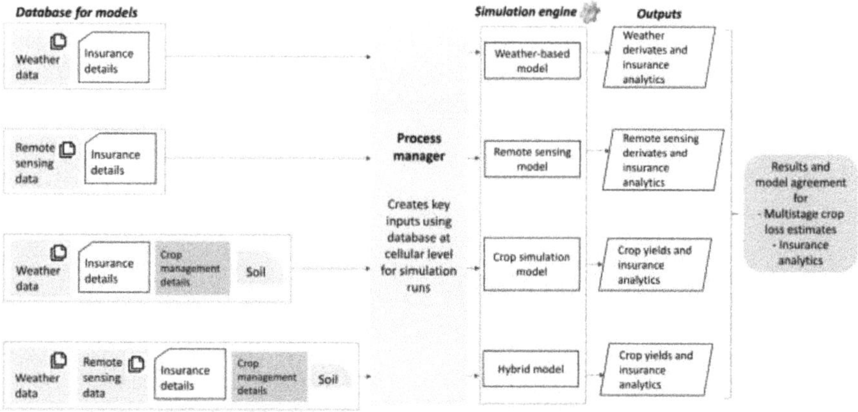

Fig 2: The processes and modules of the Crop-Loss Assessment Monitor, CAM (*Source*: Aggarwal et al., 2020).

Remote sensing and geospatial technologies have many applications in crop insurance sectors like crop mapping & area discrepancy, prevented /failed sowing risk, mid-season adversaries, crop loss due to natural calamities, crop cutting experiments (CCE), localized risk, insurance coverage and compliance, rate making and designing and even post-harvest damage assessment. The use of remote sensing and weather datasets along with the historical yield value of a particular crop over a particular place has been demonstrated for scientifically designing of index insurance products (Chakraborty et al., 2018). CCE is one of the most crucial factors for settlement of the insurance. Firstly, using advanced geo-tagging these can be transparently monitored. High resolution satellite datasets and statistical sampling techniques can improve the quality of the CCEs and reduce the number of CCEs to be conducted for getting reliable estimates of losses. Further remote sensing inputs in consonance with crop simulation and statistical models become very handy for analysis of near-real time crop growth monitoring, crop loss estimation etc. which are crucial for crop insurance. Crop-Loss Assessment Monitor (CAM) is such a system which incorporates multi-models and work under multi-stage of the crop to enable the decision makers take appropriate decisions (Aggarwal et al., 2020). The processes and modules of the CAM are illustrated in the figure 2. With the advancement of remote sensing, crop modeling and different data assimilation and analytics tools, lots of possible applications in the field of crop insurance is possible.

Fisheries (Potential fishing zone)

Identification of potential fishing zones (PFZ) is very crucial for increasing the Catch Per Unit Effort (CPUE) of the fishermen community. Ocean colour i.e. chlorophyll content along with the sea surface temperature (SST) play critical roles for determination of the PFZs. It is a well known fact that good fishing grounds over the world oceans are located in the regions where upwelling occurs and it contributes to about 50% of the total annual fish production. Hence, timely and reliable information on different meteorological as well as oceanographic variables such as winds, wave-circulation, SST along with ocean colour is required by the fishermen for increasing their effectiveness. India with 7500 kms of coastline covering 10 states and 3100 villages with fishing activities, this information is highly sought for. The real time availability and synergistic analysis of data from Ocean colour sensors and thermal sensors of IRS-P4 OCM chlorophyll and AVHRR SST could explain the variation in the surface features of the oceans. SST and chlorophyll features have coincidence in particular locations and those are also found to have good fish catch. To standardize the methodology Dwivedi et al. (2002) and Solanki et al. (2001) have also carried out experiments along with validation by the fish catch data.

Climate Change Studies

For some time now, we have been coming across news pieces through various media, sources, scientific studies or technical surveys that there have been some perceptible changes on ground in one or other parts of the globe which can be attributed to climate change. Be it early blossoming of fruit trees, appearance of grasses in Antarctic, changes in cropping pattern from wheat to maize in Bihar shifting of temperate fruits toward higher latitude like apple in Himachal Pradesh or coral bleaching, such list goes on and on. It is now evident that the earth is facing ever increasing challenges due to the compounding effects of climate change that we are clearly witnessing since the past few decades. Due to global tele-connections of the climatic events, it was very much important to know the spatial and temporal pattern of the aforesaid changes. And there was no better technology than space based remote sensing in combination with appropriate modeling techniques and geospatial tools to achieve this. The applications so far, can be broadly put into three groups as described below:

Mapping and monitoring of climate change indicators

Examples include

- Maps of changes in Coral Reef habitat & coral bleaching/degradation (caused due to rising sea surface temperature and increased acidity of ocean water prepared using LISS-II and LISS-III data, Bahuguna et al. 2008).
- Glacier inventory mapping at various scales & estimation of retreating rate of many glaciers in Himalayas have been done using IRS-LISS-II & LISS-IV data. Likewise, changes in sea ice cover in polar region could be monitored quasi-continuously during past 30 years using Passive Microwave Radiometers like SSMR (Nimbus-7) and SSM/I (DMSP) (Navalgund & Singh, 2011)

Mapping of global change forcing agents

Concentration of greenhouse gases (GHGs) in the atmosphere (e.g. CH_4, CO_2, H_2O, N_2O), aerosols, cloud type & amount are some of the important drivers of climate change. Any changes in them cause corresponding changes in the quantum of radiative forcing and thus associated with global warming/deeming and the climate change. GHGs emission inventories like methane emission from paddy fields & livestock or from wetlands have been done using integrated approach. Spatial and temporal distribution mapping of carbon dioxide and methane over India by utilizing satellite data in different combinations have been demonstrated (Goroshi et al. 2011; Prasad et al. 2014). The figure 3 indicates average columnar methane concentration during the *kharif* season of 2004 and 2005.

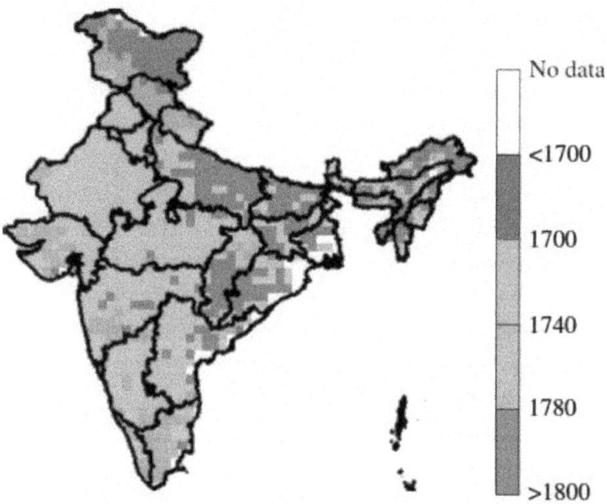

Fig 3: The average concentration of columnar atmospheric CH_4 during the *kharif* seasons of 2004 and 2005 over 0.5° grid (*Source*: Goroshi et al., 2011).

Long-term impact assessment

A warmer or cooler year, in association with other atmospheric parameters like humidity, rainfall, atmospheric CO_2 concentration and sunlight availability typically affects crop productivity due to changes in photosynthetic gain vs respiratory losses or magnitude & type of disease-pest. By removing the effect of technological interventions such as new variety, irrigation, fertilization etc. the impact of climate change on agriculture can be studied. Such studies can throw light on the behavior of an ecosystem in terms of net source/sink of carbon as a function of climate. Studies have been conducted using long-term satellite images of past years to know the impact on vegetation productivity (measured in terms of

GPP i.e. Gross Primary Productivity or NPP i.e. Net Primary Productivity or NEP i.e. Net ecosystem productivity) of ecosystems including agro-ecosystems in India (Nayak et al. 2011 and 2013).

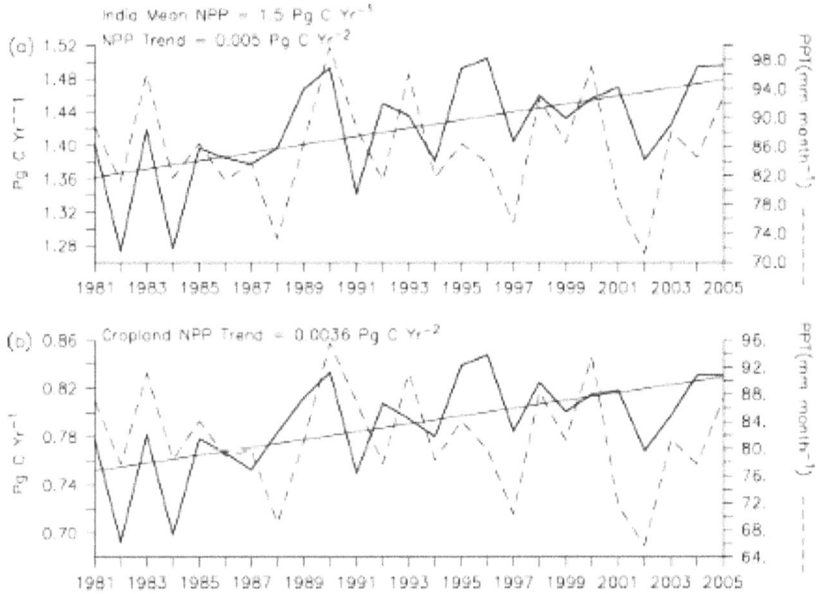

Fig 4: The annual total NPP (solid line) and precipitation (dotted line) over India (a) and its cropland (b) (*Source*: Nayak et al., 2013).

Government programmes relating to applications of space technologies in agriculture

The usefulness of space technologies was conceived by the Indian governments from early 1960s with the establishment of ISRO. Consequently, applications in several sectors like agriculture, oceanography, atmospheric science, geology etc. have started and advanced. Ministry of Agriculture and Farmers Welfare started collaboration with the Department of Space since long through different application based projects and their commitment for application of remote sensing in agriculture. This has led to the establishment of Mahalanabis National Crop Forecast Centre (MNCFC) in 2012. This Centre under the Department of Agriculture, Cooperation and Farmers Welfare (DAC&FW), Ministry of Agriculture and Farmers Welfare is dedicated for application of the remote sensing technologies in the field of agriculture in India through different projects.

FASAL (Forecasting Agricultural output using Space, Agrometeorology and Land based observations)

This project is for forecasting the acreage and production of crops through integration of different inputs from remote sensing, weather-based models and field observations. Nine crops namely, jute, *kharif* rice, *rabi* rice, cotton, rapeseed & mustard, *rabi* sorghum, weheat, *rabi* pulses and sugarcane are covered. It uses optical (AWiFS /LISS-III/Sentinel 2/ Landsat-8) and microwave (RISAT/Sentinel 1A) datasets for issuing different pre-harvest multiple in-season forecasts over different spatial scales (National/State/District). The programme encompasses a host of strategies starting from the use of semi-physical remote sensing based models, spectral index based yield models, district level statistical weather based agromet models, crop simulation models to the crop cutting experiments datasets from field levels.

CHAMAN (Coordinated Horticulture Assessment and Management using geoiNformatics)

Understanding the importance and scope of horticulture for the growth of Indian agriculture, government has launched the CHAMAN project during 2014 under the Mission for Integrated Development of Horticulture (MIDH). The main aim of the project is to generate reliable and timely statistics using geo-spatial technologies and field observations. The focus has been in the area assessment and production forecasting of major horticultural crops with the application for horticultural management and planning such as orchard rejuvenation, site suitability assessment, crop intensification, post-harvest infrastructure, aqua-horticulture, GIS Database Creation etc. Further this project envisages developing detailed scientific methodology for crop identification, yield estimation, pest-disease identification and associated loss assessment through field experiments in collaboration with different agencies. Till date it has estimated the area of seven major horticultural crops (Citrus, mango, banana, potato, chilli, onion and tomato) over major horticultural states (12 states, 185 Districts). Further it is also involved in developing horticulture expansion plan for the Jhum area of northeast India and other parts of India, horticultural crop maps and methodologies for area and yield estimation of other crops such as coriander, cumin, mentha, grape, pomegranate, apple, coconut etc. Management of high value horticultural crops through precision farming using hyperspectral remote sensing and unmanned aerial vehicle (UAV) are also on its future objectives.

NADAMS (National Agricultural Drought Assessment And Monitoring System)

This project provides near-real time information regarding the prevalence, persistence and severity levels of the agricultural drought over different spatial scales (state/district/sub-district). The original methodology of NADAMS was developed by National Remote Sensing Centre (NRSC), Hyderabad. Presently it covers seventeen states of India, which are predominantly of agrarian nature and prone to drought (Andhra Pradesh, Assam, Bihar, Chhattisgarh, Gujarat, Haryana, Jharkhand, Karnataka, Madhya Pradesh, Maharashtra, Odisha, Punjab, Rajasthan, Tamil Nadu, Telengana, Uttar Pradesh and West Bengal). It uses rainfall datasets for computation of different indices (rainfall deviation, dry spell, standardized precipitation index-SPI), district-wise soil moisture models and remote sensing derived vegetation indices. The crop sown area progression i.e. changes in normal crop calendars, hydrological data, extreme weather events along with ground observations are also incorporated. Drought warning is issued from the datasets of June, July and August months in the form of normal, alert and watch. Whereas, drought declaration is done after incorporating datasets during September and October months in the form of mild, moderate and severe category.

KISAN (C[K]rop Insurance using Space technology And geoiNformatics)

Started in the year 2015, the project aims to develop methodologies to use high spatial resolution remote sensing including UAV datasets in order to optimize the CCE and improving yield estimation accuracy for application in crop insurance. It also envisaged to do research for identification and evaluation of remote sensing based index crop insurance. During *kharif* season, rice and cotton and during *rabi* season wheat, rice and *rabi jowar* crops are covered in four states of Haryana, Karnataka, Madhya Pradesh and Maharashtra. The optimization of CCE is being done based upon various remote sensing derived indices such as sowing/ transplanting time, leaf area index, biomass or crop health etc.

Crop Intensification

This project had the objective of mapping and monitoring of *post-kharif* rice fallow lands over east and northeast India using satellite remote sensing and GIS technologies suitable for *rabi* crop area expansion. It was initiated under National Food Security Mission (NFSM) during 2016-17 covering six states of Chhattisgarh, Bihar, Jharkhand, Odisha, West Bengal and Assam where there is presence of *post-kharif* rice fallow area. It uses remote sensing, climate (rainfall, temperature etc.) and soil datasets to find out the *rabi* crop suitability over the region.

Future Plans and Perspectives

There are several satellite missions already planned by the ISRO to mainly ensure the continuity in the datasets of the thematic satellites like ResourceSat & RISAT (Water and Land), Cartosat (Cartography), Oceansat (Oceanography and Atmospheric) and INSAT (Meteorology). Apart from that technological enhancements in the sensors for meeting different application demands are also taken care for future planning.

Table 2: List of future Indian satellites important for agricultural applications.

Satellite	Year Planned	Sensor	Spectral Resolution	Spatial resolution (m)	Applications
GISAT (Geo Imaging Satellite or) {Geostationary with very high temporal resolution}	2020	Multispectral (VNIR) Hyperspectral (VNIR) Hyperspectral (SWIR)	0.45-0.875 in 6 bands 0.375-1.0 in 158 bands 0.9-2.5 in 256 bands	42 318 191	Near real time imaging of natural resources and disaster management
NISAR (NASA-ISRO Synthetic Aperture Radar) {First Dual frequency Radar}	2022-23	L-Band SAR S-Band SAR	1260 MHz 3200 MHz	5 to 10 meters (4 to 6 times a month)	Global coverage of earth's biomass, cryosphere for surface studies

India is one of the experts in space technology over the world. The country possesses a plethora of satellite sensors having capabilities to monitor its resources. But it has been seen that in the remote sensing arena of the world, the satellite data of India is not used much as these are not open source. Though these datasets are of very good quality and resolution having very high potential for application but its accessibility is restricting the wide spread use. The more the datasets are used, the better analytical techniques and algorithms with these datasets can be developed by the researchers. Moreover, some of the datasets (LISS-III & IV) are of sufficiently long duration which can be used for different application in natural resources management especially in agriculture. Government policy should be modified from both the supply side (policies regarding availability of Indian satellite data) as well as the demand side (the use of satellite dataset for different applications) to create an ecosystem for higher use of space technology in various applications over the country. Already some steps have been taken to use space technology as much as possible in different applications by the various departments of the government. Concerted efforts should be taken by the governments for the application of these technologies in agriculture in the fields like finding the suitable places for cultivation of high value crops, crop intensification/diversification and disaster planning with management etc. Promotion of use of these technologies with development of skilled manpower and availability of Indian datasets will usher a new era of effective management with space technology in the country.

References

Aggarwal, P., P. Shirsath, S. Vyas, P. Arumugam, S. Goroshi, S. Aravind, M. Nagpal and M. Chanana. 2020. Application note: Crop-loss assessment monitor–A multi-model and multi-stage decision support system. Computers and Electronics in Agriculture. 175:105619.

Ahmad, N. and P. Pandey. 2018. Assessment and monitoring of land degradation using geospatial technology in Bathinda district, Punjab, India. Solid Earth. 9:75–90.

Allbed, A., L. Kumar and Y.Y. Aldakheel. 2014. Assessing soil salinity using soil salinity and vegetation indices derived from IKONOS high-spatial resolution imageries: Applications in a date palm dominated region. Geoderma. 230:1-8.

Bahuguna, A., S. Nayak, N. Ray Choudhury, S. Sharma, R.J. Bhanderi, M.V.M. Wafar, M. Venkatesh, C.N. Pandey, B.H. Patel, S.A. Chauhan and D. Kumaraguru. 2008. Assessment of coral reef health using satellite data. Scientific Note, Space Applications Centre, Ahmedabad. SAC/RESA/MESG/MCED/SN/62.

Barakat, A., M. El Baghdadi, J. Rais, B. Aghezzaf and M. Slassi. 2016. Assessment of spatial and seasonal water quality variation of OumErRbia River (Morocco) using multivariate statistical techniques. Int. Soil Water Conserv. Res. 4(4):284–292

Bhattacharya, B.K., K. Mallick, R. Nigam, N.K. Patel, N. Padmanabhan, S. Mahammad, R. Ramakrishnan and J.S. Parihar. 2008. A study on land surface radiation budget parameters using KALAPNA-1 VHRR and INSAT-3A CCD data for agrometeorological applications; ISRO Scientific Report ISRO. ISRO Scientific Report.

Bousbih, S., M. Zribi, C. Pelletier, A. Gorrab, Z. Lili-Chabaane, N. Baghdadi, N. Ben Aissa and B. Mougenot. 2019. Soil Texture Estimation Using Radar and Optical Data from Sentinel-1 and Sentinel-2. Remote Sens. 11:1520

Brucklcr, L., H. Witono and P. Stengel. 1988. Near surface soil moisture estimation from microwave measurements. Remote Sensing of Environment. 26(2): 101-121.

Calla, O.P.N., D. Bohra, R. Vyas, B.S. Purohit, R. Prasher, A. Loomba and N. Kumar.2008. Measurement of soil moisture using microwave radiometer. In 2008 International Conference on Recent Advances in Microwave Theory and Applications. IEEE. pp. 621-624.

Chakraborty, D. 2018. Weather And Remote Sensing Based Crop Modeling For Wheat Crop Insurance (Doctoral Dissertation, Division Of Agricultural Physics ICAR-Indian Agricultural Research Institute New Delhi).

Chakraborty, D., V.K. Sehgal, R. Dhakar, D.K. Das and R.N. Sahoo. 2018. Trends and change-point in satellite derived phenology parameters in major wheat growing regions of North India during the last three decades. Journal of the Indian Society of Remote Sensing. 46(1):59-68.

Chakraborty, D., V.K. Sehgal, R.N. Sahoo, S. Pradhan and V.K. Gupta. 2015. Study of the anisotropic reflectance behaviour of wheat canopy to evaluate the performance of radiative transfer model PROSAIL5B. Journal of the Indian Society of Remote Sensing. 43(2):297-310.

Chakraborty, M., E.S. Panigrahy and S.A. Sharma. 1997. Discrimination of rice crop grown under different cultural practices using temporal ERS-1 synthetic aperture radar data. ISPRS Journal of Photogrammetry and Remote Sensing. 52(4): 183-191.

Chatterjee, S., A.P. Krishna and A.P. Sharma. 2014. Geospatial assessment of soil erosion vulnerability at watershed level in some sections of the Upper Subarnarekha river basin Jharkhand, India. Environmental Earth Sciences. 71(1): 357-374.

Chauhan, S., H.S. Srivastava and P. Patel. 2018. Wheat crop biophysical parameters retrieval using hybrid-polarized RISAT-1 SAR data. Remote sensing of environment. 216:.28-43.

Cunha, R.L.D.F. and B. Silva. 2020. Estimating crop yields with remote sensing and deep learning. arXiv preprint arXiv:2007.10882.

Dadhwal, V.K., V.K. Sehgal, R.P. Singh and D.R. Rajak. 2003. Wheat yield modelling using satellite remote sensing with weather data: Recent Indian experience. Mausam. 54(1):253-262.

Dakshinamurti, C., B. Krishnamurthy, A.S. Summanwar, P. Shanta and P.R. Pisharoty. 1971. Remote sensing for coconut wilt. In Proceedings of the Seventh International Symposium on Remote Sensing. 1: 25-29.

Dandapat, K., R. Hazari, G.S. Bhunia and P. K. Shit. 2020. The Potential Gully Erosion Risk Mapping of River Dulung Basin, West Bengal, India Using AHP Method. In P. K. Shit (eds.), Gully Erosion Studies from India and Surrounding Regions, Advances in Science, Technology & Innovation. doi: 10.1007/978-3-030-23243-6_6.

Dehni, A. and M. Lounis. 2012. Remote sensing techniques for salt affected soil mapping: application to the Oran region of Algeria. Procedia Engineering. 33:188-198.

Dutta, D., S. Bhatawdekar, B. Chandrasekharan, J.R. Sharma, S. Adiga, D. Wood and A. McCardle. 2003. Desert locust monitoring system–remote sensing and GIS based approach. Satellite Remote Sensing and GIS Applications in Agricultural Meteorology. p.401.

Dwivedi, R.M., H.U. Solanki, D. Gulati, M.E. John, V.S. Somvanshi and S.R. Nayak. 2003. Exploration of living marine resources in the Arabian Sea through integration of ocean colour with SST. International Archives Of Photogrammetry Remote Sensing And Spatial Information Sciences. 34(7/B):1404-1409.

Dwivedi, R.S. and K. Sreenivas. 1998. Image transforms as a tool for the study of soil salinity and alkalinity dynamics. International Journal of Remote Sensing. 19(4): 605-619.

El Bastawesy, M. and R.R. Ali. 2013. The use of GIS and remote sensing for the assessment of waterlogging in the dryland irrigated catchments of Farafra Oasis, Egypt. Hydrological processes. 27(2):206-216.

El Jazouli, A., A. Barakat, A. Ghafiri, S. El Moutaki, A. Ettaqy and R. Khellouk. 2017. Soil erosion modeled with USLE, GIS, and remote sensing: a case study of Ikkour watershed in Middle Atlas (Morocco). Geoscience Letters. 4(1):1-12.

Elhag, M. 2016. Evaluation of different soil salinity mapping using remote sensing techniques in arid ecosystems, Saudi Arabia. Journal of Sensors.. Article ID 7596175, 8 pages, 2016. https://doi.org/10.1155/2016/7596175

Fan, X., Y. Liu, J. Tao and Y. Weng. 2015. Soil salinity retrieval from advanced multi-spectral sensor with partial least square regression. Remote Sensing. 7(1):488-511.

Garg, J.K. 2015. Wetland assessment, monitoring and management in India using geospatial techniques. Journal of environmental management. 148:112-123.

Gorji, T., E. Sertel and A. Tanik. 2017. Monitoring soil salinity via remote sensing technology under data scarce conditions: A case study from Turkey. Ecological Indicators. 74:384-391.

Goroshi, S.K., R.P. Singh, S. Panigrahy and J.S. Parihar. 2011. Analysis of seasonal variability of vegetation and methane concentration over India using SPOT-VEGETATION and ENVISAT-SCIAMACHY data. Journal of the Indian Society of Remote Sensing. 39(3):315-321.

Haboudane, D., F. Bonn, A. Royer, S. Sommer and W. Mehl. 2002. Land degradation and erosion risk mapping by fusion of spectrally-based information and digital geomorphometric attributes. Int J Remote Sens. 23:3795-3820

Hill, J., W. Mehl and M. Altherr. 1994. Land degradation and soil erosion mapping in a Mediterranean ecosystem. In Imaging spectrometry-a tool for environmental observations. Springer, Dordrecht. pp. 237-260..

Jiapaer, G., X. Chen and A. Bao. 2011. A comparison of methods for estimating fractional vegetation cover in arid regions. Agricultural and Forest Meteorology. 151(12):1698-1710.

Kourgialas, N.N., G.C. Koubouris, G.P. Karatzas and I. Metzidakis. 2016. Assessing water erosion in Mediterranean tree crops using GIS techniques and field measurements: the effect of climate change. Natural Hazards. 83(1):65-81.

Lam, N.S.N., H.L. Qiu, D.A. Quattrochi and C.W. Emerson. 2002. An evaluation of fractal methods for characterizing image complexity. Cartography and Geographic Information Science 29(1):25-35.

Manjunath, K.R., N. Kundu and S. Panigrahy. 1998. Evaluation of spectral bands and spatial resolution of LISS II and LISS III sensors on-board IRS satellites for crop identification. Journal of the Indian Society of Remote Sensing. 26(4):197-208.

Navalgund, R.R. and R.P. Singh. 2011. Climate change studies using space-based observation. Journal of the Indian Society of Remote Sensing. 39(3):281.

Navalgund, R.R., J.S. Parihar Ajai and P.P. Nageshwara Rao. 1991. Crop inventory using remotely sensed data. Current Science. 61(3 & 4):162-171.

Nayak, R.K., V.K. Dadhwal, N.R. Patel and C.B.S. Dutt. 2012. Inter-annual variability of net ecosystem productivity over India. ISPRS - International Archives of the Photogrammetry, Remote Sensing and Spatial Information Sciences. XXXVIII-8/W20. 125-128. 10.5194/isprsarchives-XXXVIII-8-W20-125-2011.

Nayak, R.K., N.R. Patel and V.K. Dadhwal. 2013. Inter□annual variability and climate control of terrestrial net primary productivity over India. International journal of climatology. 33(1):132-142.

Prasad, P., S. Rastogi and R.P. Singh. 2014. Study of satellite retrieved CO_2 and CH_4 concentration over India. Advances in Space Research. 54(9):1933-1940.

Ray, S.S., N. Kundu, S. Dutta, and S. Panigrahy. 2000. A GIS and remote sensing-based approach for siting cold storage infrastructure for horticultural crops: A case study for potato crop in Bardhaman district, West Bengal, India. Journal of the Indian Society of Remote Sensing. 28(2-3):171-178.

Rujoiu-Mare, M.R. and B.A. Mihai. 2016. Mapping land cover using remote sensing data and GIS techniques: A case study of Prahova Subcarpathians. Procedia Environmental Sciences. 32:244-255.

Scudiero, E., T.H. Skaggs and D.L. Corwin. 2015. Regional-scale soil salinity assessment using Landsat ETM+ canopy reflectance. Remote Sensing of Environment. 169:335-343.

Sehgal, V.K. 2001. Regional wheat yield estimation by crop simulation model using remote sensing inputs and geographic information system. (Doctoral Dissertation, Division of Agricultural Physics ICAR-Indian Agricultural Research Institute New Delhi).

Sehgal, V.K., D. Chakraborty and R.N. Sahoo. 2016. Inversion of radiative transfer model for retrieval of wheat biophysical parameters from broadband reflectance measurements. Information Processing in Agriculture. 3(2):107-118.

Sehgal, V.K., S. Jain, P.K. Aggarwal and S. Jha. 2011. Deriving crop phenology metrics and their trends using times series NOAA-AVHRR NDVI Data. Journal of the Indian Society of Remote Sensing. 39(3):373-381.

Setia, R., M. Lewis, P. Marschner, R. Raja Segaran, D. Summers and D. Chittleborough, 2013. Severity of salinity accurately detected and classified on a paddock scale with high resolution multispectral satellite imagery. Land Degradation & Development. 24(4):375-384.

Solanki, H.U., R.M. Dwivedi, S.R. Nayak, V.S. Somvanshi, D.K. Gulati and S.K. Pattnayak.., 2003. Fishery forecast using OCM chlorophyll concentration and AVHRR SST: validation results off Gujarat coast, India. International Journal of Remote Sensing. 24(18):3691-3699.

Taghadosi, M.M., M. Hasanlou and K. Eftekhari. 2019. Soil salinity mapping using dual-polarized SAR Sentinel-1 imagery. International Journal of Remote Sensing. 40(1):237-252.

Tang, Q., Y. Xu, S.J. Bennett and Y. Li. 2015. Assessment of soil erosion using RUSLE and GIS: a case study of the Yangou watershed in the Loess Plateau, China. Environmental Earth Sciences. 73(4):1715-1724.

Vrieling, A. 2006. Satellite remote sensing for water erosion assessment: A review. Catena. 65(1):2-18.

Wei, Y., X. Tong, G. Chen, D. Liu and Z. Han. 2019. Remote detection of large-area crop types: the role of plant phenology and topography. Agriculture. 9(7): 150.

Yang, L., X. Meng and X. Zhang. 2011. SRTM DEM and its application advances. International Journal of Remote Sensing. 32(14):3875-3896.

Zeng, L., B.D. Wardlow, D. Xiang, S. Hu and D. Li, 2020. A review of vegetation phenological metrics extraction using time-series, multispectral satellite data. Remote Sensing of Environment. 237:111511.

Zhang, D. and G. Zhou. 2016. Estimation of soil moisture from optical and thermal remote sensing: A review. Sensors. 16(8):1308.

26

Role of Digital Solutions in Agricultural Sector

Gopal, U. Shinde[1], Prabhat Kumar[2] and P.K. Ghosh[3]

[1]*National Agricultural Higher Education Project-CAAST-DFSRDA*
Vasantrao Naik Marathwada Krishi Vidyapeeth, Parbhani, Maharashtra
[2]*National Agricultural Higher Education Project, ICAR, New Delhi*
[3]*ICAR-National Institute of Biotic Stress Management, Raipur, Chhattisgarh*

Introduction

The digital farming solutions by various advanced tools and technologies became a prime important to see the future smart agriculture in India with enhanced productivity and farmer's lifestyle upliftment. The agriculture industry facing challenges from farm to market by climate interfacing and energy conservations practices during farming operations. The use of innovative industry 5.0 is targeting a farming society to achieve high productivity by hardware devices like Robots, Drones, Automated Guided Vehicles (AGVs) working on the basis of Artificial Intelligence (AI), Machine Learning (ML), Deep Learning (DL), Big-Data, IoT and Cloud-based software tools and interactive software programs.

The use of tractors, combine harvesters, power tillers etc. like advanced machineries are practiced in present farming practices by youth farmers; however the digital tools with an automation of existing advanced farm machinery became an attraction to see human comfort. The custom operative design and development of digital tools are essentially becomes a need to innovate and exercise in finding best technologies for Indian farming society. The agricultural society spent several years and now also synthesizing an operational description for digital farming. There are several ways of opting solutions by using new digital logistic tools with software and hardware engagements. As we know agricultural production is influenced by climate, weather, soil properties, plant genetic profile, plant peripheral environment, plant protection, plant nourishing, harvesting and post harvesting like operational tasks. All such a challenging task can be handled by advanced digital tools and technologies.

The climate resilient situations, optimum use of resources, skilled labor scarcity, and optimum time management like challenges can be managed by digital tools. So, the digital agriculture concerns to tools that digitally gathers, stores, shares by analysis in to electronic data and creates optimum information with value aided chains for modern farming society. Digital Farming Solutions (DFS) is shown in Fig.1, and relative nomenclatures for smart agriculture practices as mentioned in Table 1.

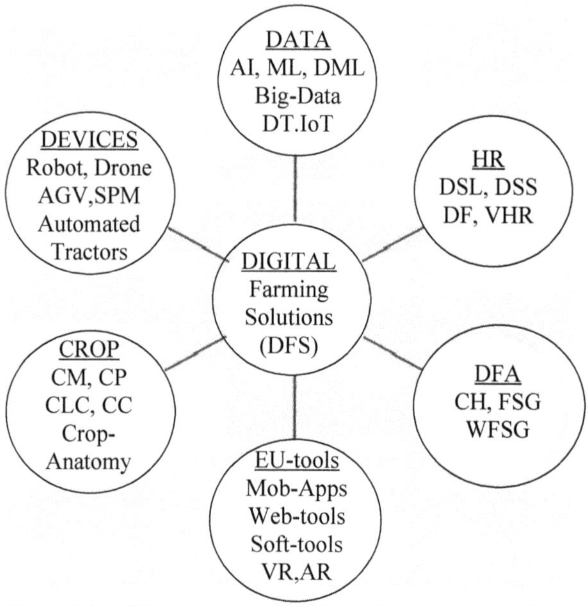

Fig. 1. Digital Farming Solutions by Robots, Drones and AGVs (DFSRDA)

The term digital agriculture, e-Agriculture, precision agriculture, smart agriculture, digital farming, smart farming, e-Farming, modern farming, etc. like many more terms that represents digital technology applications in agricultural operations however as our challenges to grow the demand for higher quality food, climate change, water scarcity, pollution, reducing biodiversity, significant population growth, wealth driving higher value nutrition. Hence Digital Farming is one key answer to these challenges.

Agriculture 1.0 to 6.0 and Digital Farming

The development in agricultural sector can be discussed as Agriculture 1.0 to 6.0 just like stages of industrial growth and development be designated as Industry 1.0 (Yr-1760) to Industry 5.0 (current-day scenario),

- **Agriculture 1.0:** Agriculture from ancient time till 1920, labor intensive agricultural operations.

- **Agriculture 2.0:** Agriculture from 1920 to 2010 Machines, fertilizers, pesticides and seeds with agronomic practices introduced an agribusiness or industrial culture.
- **Agriculture 3.0:** Agriculture entered in to high-tech sensors, cloud computing, specialized software, Internet of Things (IoT), Machine Learning (ML),Deep Machine Learning(DML)etc. in farming and farmer's data by traditional successful practices in to methodic tools applied to efficient use of resources such as seeds fertilizers, pesticides , water energy by smart farm equipment and machineries.
- **Agriculture 4.0:** Present global market and new challenges of scarcity problem can equip by three general trends with technology disruptions.

1. Use of Hydroponics, algae feed-stock, Desert Agriculture & sea water farming.
2. Use of technologies for food chain supply by vertical and urban farming,3-D printing precision equipment for food products, cultured meat and genetically modification and sea water culture.
3. Precision agriculture for efficiency and productivity by developing several agricultural tools around the globe. Data analytic, Nano Technology, AI, Block Chain, Drones, Robots, AGVs, Sensors, ML, DML, DL, Data mining, Big-Data, Food Sharing and crowd farming will high takeover and smart machines will no longer a science fiction.

The youth farming society needs to develop their skills in handling new advanced digital technologies with an interest of updating and continuing a state of art to develop a digital human resource for Indian Agriculture.

- **Agriculture 5.0:** Economic and Profitable decision tools, when data driven farms in corporates robotics, AGVs, Drones, Automation with AI algorithms to their systems, Use of AR (Augmented Reality) and VR (Virtual Reality) is referred to as Agriculture 5.0.
- **Agriculture 6.0:** The creation of a mechanism that synthesizes the art of digital technologies in agriculture with those factors identified in this work such as cultural, educational and institutional elements, at the stage of basic elements and enablers, considering the economic, social and environmental impact. This could involve further development of a Digital Agriculture Readiness Index, expanding on previous work.

Such an index would help provide context for the development of future digital agriculture strategies for the sensitizing countries to the concept of digital agriculture and the importance of digital technologies for the agri-food sector, and continues with steps towards the digital agriculture transformation process with local to global impact.(See Fig.2)

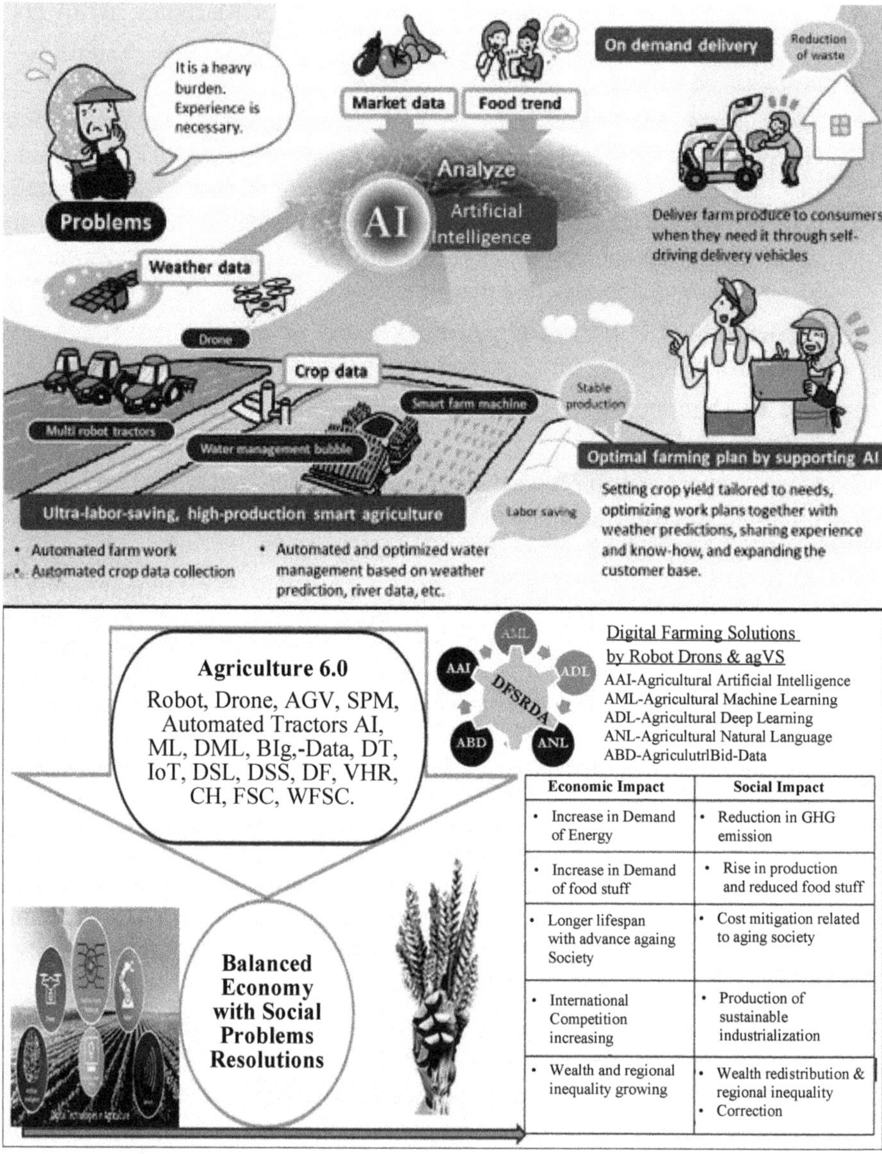

Fig. 2. Agriculture up to 6.0 with Economic and Social impact
(*Source*: https://www8.cao.go.jp)

Smart Digital Applications in Agriculture

Advanced Data Management in Digital Agriculture by specialized terminologies (AAI, AML, ADML, and ABD)

Agricultural Artificial Intelligence (AAI) is an artificial think tank created by agri-experts in which the successful repetitive agricultural field operations can be electronically stored and a logical interpretation is prepared for hardware devices to work like a human by software programming. **AAI**-derives the solutions that will not only enable the farmers to do more with less, but it will also improves a quality with an assurance to be a faster enter in the market for crops. The agricultural domains are facing many challenges such as disease and pest infestation, climate change, improper soil treatment, inadequate drainage and irrigation, etc. The field of artificial intelligence with its rigorous learning capabilities by experts are being developed to assist the agricultural experts for better solutions throughout the world.

AAI potentials: Agricultural Natural Language Processing (ANLP), Agricultural Robotics, Agricultural Machine Learning (AML), Agricultural Automated Reasoning, Agricultural Knowledge Representation, Agricultural Expert Systems, Computer Vision, Speech Recognition, Automated Data Analytics, Virtual Reality, Augmented Reality, Agri-Internet of Things (Agri-IoT), Cloud Computing, Statistical Computing, Deep Learning etc. are some of the major subareas of AI with huge potential in solving complex problems of agriculture.

Agricultural Machine Learning (AML) and Deep Learning (ADL): Agricultural Machine learning **(AML)** is Process of making a agricultural machine to train for automatic operations with an ability to learn from "experience" from experts data without being strictly programmed to perform a task. **AML** is being applied in more and more agricultural scientific fields.

Agricultural deep learning **(ADL)** has also recently entered the domain of agriculture with applications of ADL in agriculture is image recognition, to overcome many obstacles as observed in agriculture. Applications of AML in agricultural production systems can be categorized as crop management, yield prediction, disease detection, weed detection crop quality and species recognition, livestock management, water management, soil management.

AML has been applied in multiple agricultural applications such as crop management, yield prediction and disease detection (Liakos et al., 2018). However, traditionally the driving intelligence behind AAI was the AML method, which determines the decisions that are made by AAI technologies and discovers the hidden patterns or trends that can be used to for paste predictions (Pierson, 2017). This will help farmers to make data-driven decisions for different cropping system.

AML and ADL: Applications and Techniques

1. Species management

a. **Species Breeding :**Species selection for specific genes for water nutrients according to climate change

b. **Species Recognition:** Genes detection, classification by colour and shape of leaves, leaf vein analysis, leaf morphology recognition.

2. Soil management

a. **Water Management:** hydrological and climatological impact, agronomical balance, water management schedule for irrigation systems, temperature, weather phenomena and evapotranspiration and evaporation estimation.

b. **Soil Properties:** soil characteristic, soil resistance, stresses in soil, soil erosion, and soil environment.

3. Crop management

a. **Yield Prediction:** yield mapping and estimation, matching of crop supply with demand, and crop management. Computer vision technologies for crops, weather, and economic conditions.

b. **Crop Quality:** crop quality characteristics, comparative evaluation for meaningless data.

c. **Disease Detection:** Both in open-air and greenhouse conditions, the pest detection and disease control in plants.

d. **Weed Detection:** Weed characterisation, mechanical weeding or chemical weeding decisions, Computer vision by AML algorithms.

4. Livestock management

a. **Livestock Production:** weight prediction and health detection in cattle and eggs production.

b. **Animal Welfare:** Animals behaviour classifiers, chewing signals unhappiness and exhausted by energy, diseases, weight gain and production. Chat-bots application tools.

c. **Models Behind:** most popular models in agriculture are Artificial and Deep Neural Networks (ANNs and DL) and Support Vector Machines (SVMs). ANNs by the human brain functionality and biological neural network imitate the complex functions such as pattern generation, recognition, learning, and decision making.

These above explained advanced terms in digital agriculture, specializes to focus on agricultural work domain for its application oriented data management in design and development of smart tools and equipment. The logical interpretation systems for the terms such as AAI, AML, AD, ABI, ANLP, ABD explained below in Fig.3.

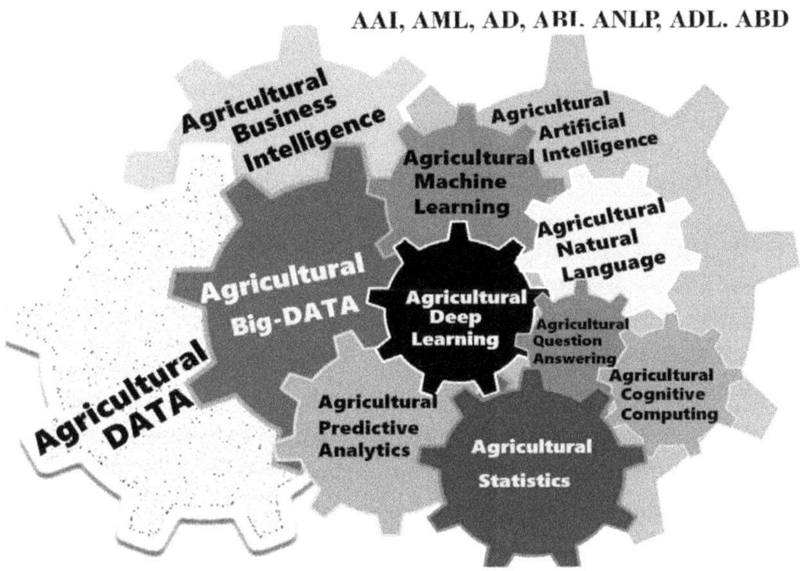

Fig. 3. Advanced Data Management in Digital Agriculture

Present Status of Digital Farming

- The overall productivity of the state is expecting to grow 60% more by 2030. However, the medical care of crops and post-production, value-added products with the potential to sustain within the national or international market.
- Digital technologies interventions in agriculture thanks to an epidemic situation as raised two difficulties like food security and provide chain disruptions in India, while challenging the increase within the income of smallholder farmers. the govt is preparing for online and software based links of agricultural produce supply directly from the farm to fork.
- The field-grown or home-grown agri-products like Ninjacart app. developed for fruits and vegetable supply from farmers to your doorsteps to increase in supply chain efficiency with enhanced market linkages.
- Now Indian farmers are inspired in handling digital technologies like Agri-Drones, Agri-AGVs and will enter in to Agri-bots like mobile technologies.
- Farmers can use the mobile Apps, Web-links, Chat-bots to share the sort, quantity, and price of crops they want to sell by keeping updates with local and global market links. Using such technology to match supply and demand of agricultural resources and food by operating on thin margins
- India is at a critical juncture when it involves the size of adoption of advanced digital technologies within the food and agriculture sector. Government policy is focussing to double-up the farmers' income.

- The role of the Make-in-India plan, *Atmanirbhar* plan are mean targeting to develop our own technology and become self-dependant agri-entrepreneur.
- Digitisation of agriculture is already showing a transformative impact, although it's still at a nascent stage.

Agricultural Robots, Drones and AGVs Technologies

I. Agri-Bots/Agricultural Robots/AgBots

Application of robot is reaching to various developing countries however India is one where in work ethics and methodologies need to be developed as field farming robots will be the future machinery with an intellectual operational task for small and marginal farmers of India.

- The use of basic big data is essential for such operational task where use boundaries with too many variables are constrained .
- The ultimate aim is to improve the productivity and manage the production activities efficiently such that farmers will make quality of life by boosting their business and products.
- Climate change, labour scarcity, labour cost working with less resources and increased population are some of the important issues where in dynamic work environment, robot is a device created by mechanical electrical electronics computer information technology discipline to work for hazardous repetitive tedious , non capacity of human race where in hardware and software skeleton is capable of handling work with enhanced productivity.
- There are several opportunities and new way to work in Agro-Industries involving youth farmers of India .As we discuss the labour shortage due to poor labour service charges and limited educational and technoical skill to work in advanced automation system.
- The youth will turn their interest to work with advanced digital machinery and equipments due to comforts and less physical efforts
- The use of mobile technology in overall Indian Farming Society has initiated to take intrest in Digital platform so youth farmers will develop their digital technology handling skills.
- The golbal cropping system review and technical specification of Agribots can be studied for Indian cropping System.
- The world population may expected to reach 9 billion by 2050. Robot employes may be hired by industries for customer service, packaging and shipping, manufacturing, and transportation.

- Several global agricultural robot manufacturers are listed in Table 3.who are introducing various designs and applications for digital farming with productivity enhancement solutions see Fig.7 and 8.

Table 3. List of top leading Agricultural Robot Manufacturer in the world

Ecorobotix	Agribotix	Energid	Agrobot E-Series
Blue River LettuceBot2(BRT)	Naio Technologies	Small Robot Company	Weighnigan University
Boston dynamics	HRC	Sydney University	Vision Robotics
Robotnik	Clearpath	Ubiquity	ISO group
Abundant Robotics	Agrobot	American Robotics	Bear Flag Robotics
Rabbit Tractors	Rowbot	PrecisionHawk	Soft Robotics

- Agriculture robots with sustainable design to work hard in huge land work envelope which is challenge to consitantly work by robots .The task that is repetative ,time consuming or hazardous can be performed even day or night by robots.
- In many of the developed countries, increasing labor costs and unavailability especially during the harvesting seasons is driving the need for agricultural robots. The driverless tractors, automated farm machinary are is expected to grow with a robotic applications and sustain with consitant working capacity.
- The driverless tractors ,retrofitted automated steering systems, GPS,NDVI, Wireless technology applied for farming operations are experienced by japanese ,korean,chaineseamerican,australian etc like many more countries however Indian startups and manufactureres are thinking of small prototype digital technology based automated machinaries for small farmers.
- Most of the owners of smaller farms, as well as a few owners of medium sized farms are expected to lease tractors ,Combined harvesters ,robots ,drones, autonmated vehicles from service providers or by custom hiring systems, Whereas larger sized farms are expected to directly purchase various types of agricultural robots including drones, weeding robots, and driverless tractors.
- New Design Innovations in Agribots for Digital Farming are developed by top world ranking universities and manufacturer comapanies as mentioned in Fig. 4

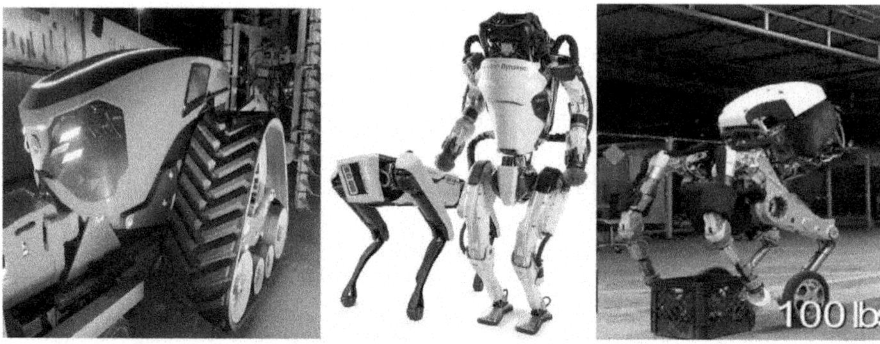

John Deere: Robotic Maintenance of Tractors and Nimble robot by Boston Dynamics

Material handling humanoid and animoids models of robots by Boston Dynamics

RIPPA robot for Pests and Weeds by Sydney University Nio-Robotics

Killer robots, Tom and Dick Hybrid-electric Agribot, Eco-robots for Weeding,Spraying

Fig. 4: New Design Innovations in Agribots for Digital Farming

Bonirob for Weed control Scouting Vine-bots Cotton picker Robots

Fig. 4: New Design Innovations in Agribots for Digital Farming
(*Source*: Bostan Dynamics, Naio, USA,Ecorobotix, Naio,Robtnik, TerkRecoms, Sydney University etc)

The autonomous agricultural robots for nursery , horticultural plants are designed and developed by Gummi-Arm robot, TREX RC Robot Lawn Mower, Interesting-Engineering (IE) as per IDTechEx Reports, are becoming new popular technologies to vegitable,floriculture or horticultural growers on large scale production basis as shown below in Fig. 5.

Fig. 5: Autonomous Agricultural Robots for Nursery, Horticultural Plants (Ref.: Gummi-Arm robot, Interesting-Engineering (IE), IDTechEx Reports, TREX RC Robot Lawn Mower).

II. Agri-DRONES / Agri-UAVs (Agricultural Drones/ Agricultural Unmanned Aerial Vehicles)

- Unmanned aerial vehicles (UAVs) are also termed as Unammaned Aierial Systems (UAS) or Drones, becoming a valuable digital tool to collect field data in a variety of contexts. Their use in agriculture is particularly suitable, as those areas are often vast, making ground scouting difficult, and hazardous sometimes.
- The use of UAVs for crop orchards,and forests monitoring for identifying water ,nutrients,tempreture in plant environment for its monitoring and control operation.
- The new hybrid designs with combinations of robot and UAVs such as landing on ground will act as robot where as when take of will work like a UAV for the management of water, diseases, nutrition deficiencies, and pests.
- AAI,ADL,AML,AgriBig-Data or Internet of Things (IoT) software tools can be used to a provide a decision support with significant potential for Smart and Precision Agriculture with a real-time environmental data.

Importance of Drones / UAV technology for agriculture

- Speeds up the field operation with realtime climate change situations.
- Faster speed and optimistic Spraing operations.
- Quick detection of diseceses and paste management operations .
- Keen observative plant health condition monitoring
- Optimum plant population counter .
- bird fire or wild animal thretening sound for crop care.

Applications

- Weed mapping and management
- Vegetation growth monitoring and yield estimation
- Vegetation health monitoring and diseases detection
- Irrigation management
- Corps spraying

It is used for time saving and faster decisions in field operations to support the four pillars of precision agriculture, i.e., apply the right practice, at the right place, at the right time and with the right quantity.

Challenges

1. Selection of types of UAVs for right field operations
2. The efficient use of UAVs power storage and pilot skills
3. Any component failure causes an accidents during operation.
4. Short flight duration due limited power increases no. of operation cycles.
5. Requires all the time permissions to fly from DGCA by network link.

Crop features that can be monitored with UAVs.

Following are the design features required to consider while design and development of any kind of agricultural drones or UAV in table 4.

Table 4. Field features required for UAV design and developments

Vegetation	Nitrogen status	Moisture content	Vegetation color
	spectral behavior of chlorophyll	Spatial position of an object	Plant anatomy
	biomass	Temperature	Vegetation indices
Soil	Moisture content	Temperature	Electrical conductivity

Top 10- Drone Companies in the world – 2020

Ranking of Drone and Drone Accessories Hardware manufacturers

1. **DJI:** A revolutionary **Mavic Mini**, packing high quality drone features into a tiny drone that weighs only a half pound. It is intermediate pilots and videographers and the simultaneous.
2. **Parrot: ANAFI** with camera drone is for smartphones and tablets usful for military reconnaissance aircraft.
3. **Yuneec:** Aerial video quadcopter ,ready-to-fly fixed wing RC airplane Typhoon **H520** (2017) voice-controlled drone, **Mantis Q**(2018).
4. **Kespry:** Popular for capturing, viewing and analyzing aerial imagery and survey data.
5. **Autel Robotics:**Simple solutions by photography /fiming and imaging.
6. **Skydio: Model R1**(2014),autonomous drones for consumers and Skydio 2(2019) built around the Skydio Autonomy Engine – the most advanced flying artificial intelligence system in the world.
7. **Insitu:** Applied for surveying and reconnaissance ,also for oil or gas pipeline inspection and force protection.
8. **Delair**: Model-**UX11** offerings combine high performance, long range fixed-wing UAV hardware with sophisticated analytics and operational services like mining, construction, agriculture, utilities, transportation, oil and gas.

9. **EHang: GhostDrone 2.0,** smart algorithms,dual sensor's support .EHang 216 two seats passengers. It's first US trial took in early 2020.
10. **Aerialtronics: Altura Zenith** is flat, compact and lightweight design for dual vision cameras to gas sniffers to radiation detectors .Aerialtronics systems for Safety & Security, Inspection, Surveying & Mapping, Agriculture and Research.The different design features of UAVs used by Defence organisation in the world

Indian Drone Regulations by Government of India

- Ministry of Civil Aviation Government of India derived rules and regulations by National Aviation Authority of India.
- Foreigners are not allowed to fly drones in India.
- Unique Identification Number (UIN) and UAOP from DGCA is required to obtain for All drones except those in the Nano category .
- The Nano category flown below 50 feet and those in the Micro category flown below 200 feet are permitted. maximum flying vertical height up 400 feet.
- Drone pilot training from authorised traing centre approved by DGCA authority is required.
- Drone pilots must maintain a direct visual line of sight at all the times.
- Permission to fly in controlled airspace can be obtained by filing a flight plan and obtaining a unique Air Defense Clearance (ADC)/Flight Information Center (FIC) number.

Drone classifications by weights in India

Registration is required for all but the Nano category.

i. Nano: ≤ 250 grams (.55 pounds)
ii. Micro: 250 grams (.55 pounds) to 2kg (4.4 pounds)
iii. Small: 2kg (4.4 pounds) to 25kg (55 pounds)
iv. Medium: 25kg (55 pounds) to 150kg (330 pounds)
v. Large: >150kg (33 pounds)

The drone industries are grownup like local custom market now a days as there are several thousand of cutom designs and shapes which can be manufactured by using 3-D printer technology see as in Fig. 6.

Fig. 6: Different commercial designs of Drones / UAV by 3-D printers

The agricultural drones are targeting specifically for crop mapping ,plant poulation counter,paste detection and management,sprayers scouting or crop monitoring like functions.

The following equipments are required for UAV setup in India

GPS; Return-to-home (RTH) ;Anti-collision light ; ID plate; A flight controller with flight data logging capability; RF ID and SIM/No Permission No Takeoff (NPNT).

- Indian UAV market poised to grow at a CAGR of 18% during 2017–23 in terms of revenue. (6W research)
- According to Stockholm International Peace Research Institute (SIPRI) indicates that with 22.5%5 of the world's UAV imports, India tops the list of drone-importing nations.
- Again, these are numbers primarily for military purposes, and commercial drones are showing healthy growth as well.

The drone business will grow up by 35% oin dian up to 2030. It will create smart agribusiness by lot of investment to shape out smart agriculture.

Comparison of Satellite, Aircraft and UAV

Satellite and Aircraft based technologies are the conventional methods of remote sensing which have their benefits and hindrances in the form of Coverage area, spectral, spatial and temporal resolution with onboard sensors. Along with less revisit frequency, satellite suffers from cloud cover conditions where information beneath the clouds is shadowed entirely.

Aircraft systems may sustain with higher payloads and speed, but the hovering at a place and maintaining high and low speeds are the challenges Several studies were performed by various researchers to compare the three technologies such as NDVI survey of larger farmland, soil and crop mapping, Land classification. ***Unmanned Aerial Vehicles*** somehow bridges the gap of conventional technologies regarding hovering capability, maintain low and high Speed, low elevation and more importantly higher spatial and temporal resolution of images.

The agriculture drones market is expected to grow from USD 1.2 billion in 2020 to USD 5.7 billion by 2025 at a CAGR of 35.9% from 2020 to 2025.

A few key factors driving the growth of this market are pressure on global food supply due to growing world population and increase in venture funding for the development of agriculture drones.

III Agri-AGVs (Agricultural Automated Guided Vehicles)

Agricultural Automated Guided Vehicles (Agri-AGVs) have transformed the way agricultural operations or produced materials that can be handled within the field and shopfloor food production activities or poly house with controlled atmospere. In a agricultural production environment where several operations are carried out simultaneously, these vehicles ensure a predictable and reliable farm operation such as spraying ,scouting ,paste detection plant counting ,goods tranport like many more applications within closed loop program in operation cycle.It can also transfer a raw materials as well as manufactured products from one point to another point within the facility, thereby eliminating any potential disruption in production. These vehicles can operate safely around structures, machinery, and employees, as they are equipped with accessories, such as camera vision and LiDAR sensors, which help in detecting junctions, identifying floor signs, and avoiding collisions with any obstacle.

The rural craftman are taking intrest in small portable digital tools or machinery manufacturing ,however a lack of accessories or digital components nearnes to market like difficulties are minimized by internet marketing.So the AgriAGVS for middle and small farmers can be possible to develop their agribusiness.as shown in Fig. 7.

Automated AGV farmer's cart

Automatic Guided Portable Field-Car

Fig. 7: Commercial smart Agri-AGVs manufactured by rural craftman

The navigation based autonomous mobility technology in agriculture has also evolved past rigid infrastructure-dependent automation.In parallel, the advent of unmanned mobility can give rise to agricultural vehicles with new form factors. In particular, large fleets of small, light, slow agrobots may come to replace a few large, heavy and fast traditional agriculture vehicles. These new vehicle form factors are possible because autonomous mobility eliminates the wage bill overhead per vehicle. In this case, a few persons will remotely monitor and control the operation of a large fleet. Here, the productivity of individual units may be lower than traditional powerful vehicles such as tractors but the overall productivity- at the level of the fleet- could be higher.

The survey for Agricultural Robots and Drones 2018-2038: Technologies, Markets and Players, we provide technology and market analysis/forecasts for AGVs in dairy farming, autonomous tractors (level 3 to 5) and unmanned agricultural robots for data scouts, weeding and multi-platforms see Fig.8.

Fig. 8: Agri-AGV models for field operations and process industries
Source: mdpi.com, FarmWise.com, UPV Valencia, naiyim_garage

International Market forecast up to 2025 after COVID-19 Impact Analysis for Agribots, AgriDrones and AgriAGVs.

The consitant rising population in India reached to 140 Crore up to October 2020. The contenous natural calamity and uncertain challenges raising the food production difficulties as a results farmer's of suciding group is nervous and dipressed due to seek agricultural production and yield for improving lifestyle.

The advanced digital technologies applications in agriculture needs an Information Technology(IT) Infrastructure establishment in rural sector. The use of robots ,drones and AGVs in agriculture will grow as the international manufacturers taking intrest to develop for agriculture operations.

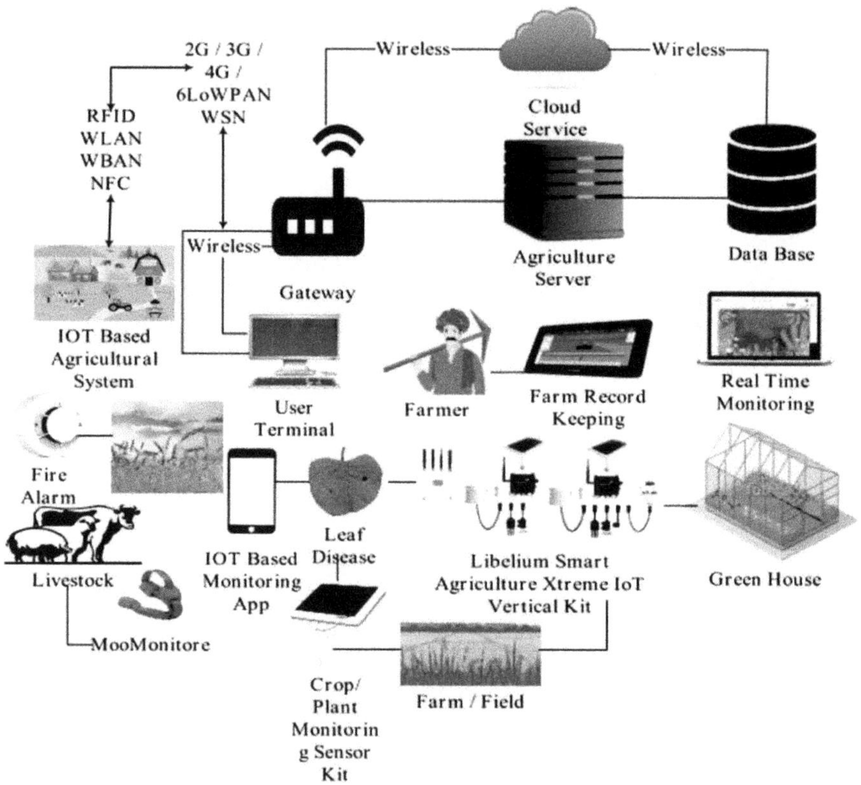

Fig. 9: Overall IT –infrastucutre for Digital Agricultural System

Future Policy Reforms and Future Challenges in Digital Farming

Digital Farming Solutions in India Project programs and policies are introduced by Government of India since Year 2018.The project has three core components, viz. ***digital infrastructure, digital services and digital literacy***. Mobile phone tool introduces **m-*Governance, m-Services, m-Agriculture and m-Gram Bazar,*** like many more services. Following points are constrained for future policy reforms.

1. Development of innovative digital technologies for Indian crop production.
2. Use of advanced digital tools for agribusiness communications.
3. IT-infrastructure development for digital agriculture in India.
4. Skill development of youth farmers in agricultural industries of India.
5. Indian Farmer's assurance system in doubling income of agribusiness.
6. Virtual agriculture market and strong agribusiness networking structure.
7. Climate resilience and Soil, Crop Health Card Scheme.
8. Indian Crop insurance scheme.
9. Farmer's Lifestyle upliftment by overall Productivity enhancement.

Following activities are needed to be covered in digital agriculture policy implementation

a. Assessment of the impact by demographic factors of the farmers on their level of awareness on Digital technologies in Indian agriculture
b. Analysis of approach and practices in fostering practices under Program.
c. Finding the association between the farmers level of satisfaction on the authorities approach towards the implementation of Digital India Program.
d. Deriving the impact of experience and age of farmer's awareness and knowledge on digital India initiatives.
e. Awareness creation regarding digital India and its benefits to the farmers.

Government Programmes and Policy Reforms for Digital Agriculture

First, how can government policies and programmes appropriately facilitate the adoption of digital technologies by the agriculture and food sectors? Policymakers will need to consider potential benefits, costs and risks, and to understand the factors affecting technology uptake so that interventions can be targeted to where there is a market failure, or a public interest.

How can governments make use of digital technologies to design and deliver better agricultural policies? This requires understanding how technology can help in different components of the policy cycle, and may require government bodies to expand their skillsets, invest in technology and training, or partner with other actors (both government and non-government).

How might digital technologies change the roles of government? On the one hand, digital technologies may create new roles or responsibilities for governments, including to enable the digital infrastructure (is there a case for governments to be a provider or a rule maker of new digital infrastructure, and under what circumstances); but on the other hand, if technology can reduce information asymmetries and transactions costs, less government intervention may be needed.

For policymakers, the challenge will be to shape policy and regulatory settings so that they facilitate opportunities offered by digital technologies. At the same time, and not unique to the agriculture sector, digital technologies raise questions about privacy, interoperability, and even potential liability issues, all of which will need careful consideration.

Reaping the benefits of digital technologies in agriculture requires the participation and co-operation of farmers, researchers, private sector, non-profits and government

OECD work is examining the benefits and challenges of using technologies for policy in agriculture, with particular insights drawn from agri-environmental policies, and for agro-food trade. Further issues include how the regulatory environment can influence sustainable and inclusive uptake of digital technologies; how technologies might affect skills needs in the sector; and how tracebility technologies can improve agricultural value chain transparency, enhance food safety and combat fraud; and potentially enablee new demand-side policies to promote sustainability.

For more information on how digital technologies are impacting food growth and distribution, you can download presentations and background information from our 2018 Global Forum on Agriculture.

Future Challenges

As there are several challenges and frequent change in climate the farming operations on production battleground the use of intelligent system is helpful to work with automated high precision devices such as Agribot, Agri-drones and Agri-AGVs to use optimum resources and developing a short crop production cycle with IoT supports.

References

Agricultural Market report by https://www.marketsandmarkets.com/ for 2020-2025

Anderson, et al. (2011). Recasting robotics challenges as experiments. IEEE Robotics and Automation Magazine,18(2),10-11.

Baumüller, H. 2015. Assessing the role of mobile phones in offering price information and market linkages: the case of m-farm in Kenya,EJISDC.(68)

Bhattacharjee, S. et al. 2016. Social Media: Shaping the Future of Agricultural Extension and advisory Services. GFRAS Interest Group on ICT4RAS discussion paper, GFRAS:

Doubling the Farmers Income (DFI) — Volume 14

Downes, L. 2009. The Laws of Disruption: Harnessing the New Forces that Govern Life and Business in the Digital Age. Basic Books. European Commission. 2013. i. food security to farmers income security by 2025 http://www.ipsnews.net/2020/07/digital-agriculture-linking-indian-farmers-consumers-impact-food-security/

Jacoff A., et al. 2012. Emergency Response Robot Evaluation Exercise. Proceedings of the Workshop on Performance Metrics for Intelligent Systems, Pages, 145–154.

Jan, N., et al.(2018). The IWSLT 2018 Evaluation Campaign. In International Workshop on Spoken Language Translation (pp. 2-6).

Kitano, H., et al. (1997). Robocup: The robot world cup initiative. In Proceedings of the first international conference on Autonomous agents (pp. 340-347). ACM. Pratt, G., & Manzo, J. (2013).

Lindau, Switzerland. Coldiretti, 2018. Report for the agri-food forum of Cernobbio 2018. Trieste: Instituto Ixe Srl [In Italian].Disrupt Africa. 2018. African tech start-ups funding report 2018. Disrupt Africa (available at: http://disrupt-africa.com/funding-report/#)

Mckinsey report- Harvesting golden opportunities in Indian agriculture: From NASSCOM — Agri-tech in India: Emerging Trends in 2019

Schuller, B. W., et al (2018). The Interspeech 2018 computational paralinguistics challenge: Atypical & self-assessed affect, crying & heart beats. Proceedings of the INTERSPEECH International Conference, Hyderabad, India.

Tee, K.P., et al.(2017). The ICRA 2017 Robot Challenges[Competitions]. IEEE robotics and automation magazine, 24(3), 15-21.

The DARPA robotics challenge [competitions]. IEEE Robotics and Automation Magazine, 20(2), 10-12.

27

Waste to Wealth: Innovative Ideas and Technological Initiatives

Shiv Dhar[1], Renu Singh[2], Manoj Shrivastava[2] and Seema Sangwan[3]*

[1]Division of Agronomy, ICAR-Indian Agricultural Research Institute New Delhi
[2]Centre for Environment Science and Climate Resilient Agriculture (CESCRA) ICAR-Indian Agricultural Research Institute, New Delhi
[3]Division of Microbiology, ICAR-Indian Agricultural Research Institute New Delhi

Introduction

More than 380 million urban population living in town and cities generates 62 million tones of municipal waste every year out of which 75% is recyclable but only 30% is recycled and reused whereas less than 15% of municipal solid waste is treated. Generally municipal authorities of the country dump the waste outside the city to keep the urban area clean. According to World Bank report, by the year 2025, waste generation will reach 377,000 tonnes per day in India (CPCB 2020). The waste generated in industrial and urban area of the country is tangible and concerning. This is the high time to address this issue with a great sensitivity and planning. Ragpickers works as the informal waste collection and recycling sector which plays an important role in waste management and leads to lesser waste to the landfills (Gupta et al., 1998).

Waste management is the process of treatment and disposal of solid, organic and hazardous waste. It can be categorized based on the source sector i) urban waste; ii) electrical waste iii) medical waste; iv) waste produced by disasters; v) agricultural waste vi) industrial waste etc. The type of waste has variable potential harms for human health, social implications, policies and the environment (FICCI, 2011). *Agriculture itself is a big source of waste generated at different paths of* food supply chain that occurs as follows:

- Crop cultivation: losses during manual/mechanical harvesting and sorting
- Post harvest handling and storage: losses due to spillage and degradation during handling, storage and transportation from farms to market.

- Marketing: losses in the market system
- Processing: losses during industrial or domestic processing, e.g. washing, peeling, slicing and boiling, juice production, canning and bread baking.
- Consumption: losses and waste generation during consumption at user end.

Management of the waste differs depending on its profile (solid, liquid or gaseous, toxic, dangerous). The method of collection of waste will be decided based on the nature of waste at the source itself to minimize contact and further contamination in the environment., The sorting and selective collection of waste consisting of separating and recovering waste according to their nature, at the source, to minimize contact and contamination. It allows to be given a "second life", by reuse and recycling, thus the wealth generation and therefore reduction of ecological footprint. Recycling reduces the volume of waste, and preserves the natural resources. This is the major economic activities of the consumer society. Few processes are simple and inexpensive but, some are complex, expensive, and may be unprofitable.

To make India self-reliant, recently established Prime Minister's Science Technology and Innovation Advisory Council (PM-STIAC) launched the waste to wealth mission project. This will ensure the integration of effective recycle, reuse and resource recovery of waste at different platform where stakeholders can work together. The objective of this project is to implement various technologies available with industries, academic institutions, research agencies and other platforms. It encourages the site-specific development, installation and demonstration of waste management technologies in India. Long term goals, strong collaboration and waste to wealth programme will help in evolving solutions and policies related to waste management and minimization. Processing, treatment and disposal related technologies are required in India so that the volume of waste not only can be reduced but also used to convert into wealth. Conversion of waste into manure, energy, recycled materials and other value-added products are the real pathways towards waste to wealth instead of dumping the waste in the landfill sites. Production of compost from waste and biomethanation methods are common but can be utilized more for waste management and decreasing loads on landfill sites (Bhide and Sundaresan, 2013).

Major issues of waste management in India

Production, collection, storage, treatment, transfer and utilization are basic functions of any waste management systems. Waste management is a challenge in India due to various reasons related to rapid urbanization, massive waste generation, lack of site and source specific guidelines, improper waste collection and segregation, insufficient policies and inappropriate treatment and recycling systems. The major issues are municipalities are not managing all the areas properly, lack of 360-degree waste management plan, waste to energy plants are very few and waste segregation strategies are not sufficient. Along with this, e-waste is the real challenge before the society as it poses threats to environmental safety (Sharma et al., 1996).

Different mobile phone-based applications (apps) and websites for waste to wealth initiatives

For the reduction of waste and maximization of its recycling, various informative and innovative apps are available in the country. These apps are majorly focused towards the type and waste managed pathways to ensure its recycling, to locate the waste collection and pick up points, larger coverage of waste related news, chatting with representatives and innovations and to increase insight, knowledge and hands on trainings for waste handling and management. These apps are useful for residents, institutions, organizations, industries and related departments.

Name of App	Developer	Target
Waste to Wealth DESKU ENVIS Team	Dept. of Environmental Science, University of Kalyani.	Mobile apps for providing industry solutions for both residents and industry leaders
Waste is wealth TechLearn Institute		Diverse ways of handling waste to create wealth
Yo-Waste Yo-Waste LL		Zero waste goals by connecting independent haulers & recycling businesses
IVRI- Waste Management Guide App:	IVRI-Waste Management Guide App is designed and developed by ICAR-Indian Veterinary Research Institute, Izatnagar in collaboration with ICAR-Indian Agricultural Statistics Research Institute, New Delhi.	To impart information and knowledge to graduating veterinarians, field vets, general public, farmers and other stakeholders about management of waste originating from agriculture, livestock and household activities.
Pom Pom Trash to Cash	Deepak Sethi and Kishore Thakur, Delhi	Sell anything and set a price per kilogram
Ecolekt	JoGiv, Bengaluru	e-waste collection, disposal, recycling and awareness
Sellixo	Anupama N, Sanjana Vasanth, Mahima Mehendale, Navyasree B and Swasthi P Rao, Bengaluru	Sell dry waste (such as paper, metal or plastic)
Encashea	Encashea, Bengaluru	Tie up with vendors who buy scrap, and lists the prices on the app
Swachh Delhi & I Clean India	NGO Janaagraha	

Name of App	Developer	Target
Toter	Waste Ventures India	Waste picker and customers interface
Recycler	Freya	Uber of Recycling
ECS e-waste mobile App	ECS Environment Pvt Ltd	e-waste collection, disposal, recycling and awareness
Karma Recycling		Trade-in operator and redistributor of electronics

So many websites are working in the field of waste collection, disposal, recycling and value added product development. Websites with idea about best out of waste are easily available on internet which says unlimited creative ideas and concepts for the development of products where children can also be involved. These websites not only makes the fun out of waste but also generates awareness among the children and youth towards environment. These ideas sometimes can be materialized into big projects for the generation of employment and money. The concepts of best out of waste are gaining global acceptance and catching the market at the fast speed. The commonly used materials in such projects are newspaper, plastic bottles, old cloths, electric bulbs, used furniture etc. People of all age groups may participate depending on their mental level and project demand. Do It Yourself (DIY) based videos are providing hands on training to the views. DIY apps and websites are very popular and fastest growing creative field among the children and women. Indigenous innovative technologies for waste to wealth (https://www.agnii.gov.in/innovation-listing)

1. Vengurla municipality Sindhudurg district (Maharashtra) used the segregated households waste and used wet waste to generate biogas, producing 30 units of electricity per tonne. The United Nations Development Programme (UNDP) provided one biogas powered plastic shredding machine having the capacity of 180 kg of light plastic crushing every day. This plastic is being used for road building in turns of which Vengurla has 12 km of plastic roads. They also earned @INR 15 per kg of plastic from contractors for road-building in adjoining areas.
2. GPS Renewables-Bangalore based enterprise founded by Mainak Chakraborty and Sreekrishna Sankathatr is working to solve the problem of urban organic waste with the rule of Zero waste. The company launched a pilot project named BioUrja, is a compact plug-and-play system that users can install anywhere with minimal civil work, works on kitchen and other organic waste and is good for biogas production. The system processes around 600 kgs of kitchen waste every single day and solves the issues of poor installations and cost. The company has now expanded beyond India into Bangladesh, Malaysia, and Sri Lanka.
3. A Noida-based company named Code developed by Vishal Kant and Naman Gupta is working on discarded after smoking cigarette butt. They are collecting these cigarette butts @ INR 700 for every kilogram of cigarette

waste and Rs. 80 for every 100 grams from the people who smoke and selling cigarettes. They are using this waste in plantations and nurseries more over the recycled polymer material is used to make cushions, garlands, small stuffed toys, accessories, key chains, among other things. Cellulose Acetate, the recycled polymer from cigarette butts has many potential applications due to its high filtering properties and resistance to heat properties. They are in the process of developing air-purification systems for kitchen chimneys, and other more useful by-products.

4. Tea Wine is a product which addresses the waste coming from tea. Its distinctive aroma and flavor and are rich in antioxidants and this wine can be made sweet or dry as per demand and the alcohol content can also be customized.
5. Waste Polymers to Energy technology can convert waste plastics to fuel which can replace gasoline in industries and power generation sector. The output fuel is superior to the gasoline in terms of colour, viscosity, density etc.
6. Methane from waste water is a process which can produce methane gas from running waste water. It's a novel source of renewable energy where drainage system can be used for methane gas production.
7. 100% recyclable and reusable Kerbstone can be produced from waste material which provide the solution for the management of three types of waste namely, concrete, plastic and rubber. This product is hiny, water resistant and the use of reflectors on the kerbstone may prevents road accidents by providing improved visibility.
8. Bio based polymer developed from agricultural crop waste based from biopolymers such as poly (lactic acid). This product not only solves the problem of plastic disposal problem but also evokes commercial interests worldwide as it has the potential to meet environmental and health requirements.
9. Two bioactive chambered Ecopot is coming as a solution of unprocessed organic waste that may cause diseases and pollution. It is an effective organic waste management system which can reduce the fertilizer cost involved in vegetable cultivation. It has a inbuilt composting unit which produces manure and a pot for growing plants.
10. Dry waste processing of muncipal solid waste to value added products through a technique which is a combination of gasification and pyrolysis model is the solution of urban waste that can be a feedstock for energy generation. This is about the conversion of dry waste into various value added products; fuel oil from plastic waste and electricity generation by using a updraft gassifier under an integrated approach. This type of plant

is scalable from 50 kg/hour to 50 tons/day. This model is initiated in few municipal bodies for effective waste management and time deriving value-added products.

11. Xaper is a hybrid solid waste sorting cum processing machine which can segregate dry waste, process segregated wet waste to compost or can simultaneously segregate waste in three size fractions and process wet waste from mixed fractions of municipal solid waste. This innovation can solve the problem of waste segregation and enables the user to utilize the segregated waste for different wealth generation options.
12. Agringenium is the innovative idea for addressing twin problems of environmental pollution and health. This innovation converts agricultural residues into biodegradable packaging materials and solves the problem of indiscriminate application of plastic products. The feedstock used in Agringenium is rice husk, sugarcane bagasse, straw and stalks of agricultural produce. This inexpensive product can be used for packaging food and non-food commodity products

Involvement of startups and Non-Government Organizations (NGO) in waste to wealth initiatives

The Swachh Bharat Mission aiming at more public-private partnerships with innovative startups and NGOs working in the field of waste management is growing at a rapid pace. With the objective of reduce, reuse and recycle, these NGOs from different Indian cities manage their waste. Waste management may be a huge problem in urban India thus startups are tapping the large opportunity to make wealth out of waste. (https://yourstory.com/2017/11/waste-management-startups).

1. Saahas Zero Waste started as an NGO in 2001, and evolved into a for-profit business. This Bengaluru-based startup, supported by Indian Angel Networks and Upaya Social Ventures, caters to clients from Bengaluru and Chennai. This enterprise has diverted more than 15,000 tonnes of waste from reaching landfills.
2. HasiruDala Bengaluru-based HasiruDalahelps over 25,000 households to manage their waste and gives the solutions to apartments, corporate offices, events, and also helps in designing, set up and maintenance of urban gardens.
3. Namo e-waste Started by Akshay Jain, from Delhi, processes over the 20 tonnes of e-waste daily across 12 states and union territories in the country. This startup is in line of strategic partnerships with leading electronic companies, by providing solution for collection of e-waste.

4. Delhi-based GEM Enviro Management Incorporated established in 2013 collects polyethylene terephthalate (PET) waste from factories, offices hotels, and institutes, and recycles it into products such as T-shirts, caps and bags. The barns name is Being Responsible of products with a clientele of Bisleri, Cipla, IDFC Bank, Pepsi Sun Pharma and Google.GEM Enviro Management has processed over 5,000 tonnes of waste.
5. The Bengaluru-based startup Citizengage - waste-to-resource technology has built an end-to-end waste management system to help communities and businesses in their waste at source. This Tech30 startup has recycled about 4,000 tonnes of waste. by converting waste into biogas and compost, and also diverted the segregated waste to piggeries.
6. Chennai-based organization Paperman runs over 250 waste paper marts across Chennai to help citizens in clearing trash at their doorstep. Citizens are also getting value generated from the trash.
7. Kolkata-based startup - Vital Waste provides recycling solutions to residential communities, corporate offices and schools in the city.
8. The social enterprise - ExtraCarbon caters 8 cities, including Delhi Gurugram, Noida, Greater Noida, Ghaziabad, Manesar, Lucknow and Ludhianaprovide on-demand kabadiwala service to collect recyclable waste from commercial areas and homes.

Community participation

The concept of common waste treatment facility (ENVIS Newsletter, December 2010) is being promoted and accepted as manages waste as a feedstock by using it as a fuel or raw material in manufacturing processes. This gives the way to Public Private Partnership (PPP) models in waste management which has immense opportunity for business establishment in waste management (TERI 1997).

Amolak Singh a farmer in Punjab making money also helping in keeping clean the air in cities just by selling agricultural waste to Sukhbir Agro- a rice mill at a mutually negotiated price. This is not the only case in the village as they are now generating money from waste that can be used in annual farm labour payments. SuhkbirAgro has 40 existing paddy centres to collect straw after the harvesting season in districts of Muktsar, Bathinda, Mansa and more in neighbouring states. The agro waste collected is used in company's biomass plant to generate green electricity around the year and surplus can be sold to the Punjab government at a price higher than thermal power. Over a dozen such biomass plants will be in coming few years with generation power of 200 MW of power and disbursing approximately Rs. 50,000 crores to farmers for waste every year. Timarpur Okhla Waste Management Co. Ltd. installed first commercial waste-to-energy project in India with the aim of converting 1/3rd of city waste into electricity to serve 600,000 homes. Surat treats 57 million litres of sewage per day, which provides 40 million litres of potable water that saves huge money. Jamshedpur in Jharkhand converts waste into biogas at bio-gas stations (NIAM 2012).

Maharashtra's Sitara district is showing twin benefits from sugarcane molasses by generating electricity for the grid and fertiliser for the crops. German federal technical agency GIZ and Sugarcane Farmers' Cooperative is giving new sustainable development model by putting a unit that would ensure the free agro waste transportation and generation of electricity. Protoprint, a Pune based enterprise has collaborated with the waste pickers and is converting plastic waste into filaments for 3D printing. Siddhant Pai founder of Protoprint partnered with SWaCH (Solid Waste Collection and Handling, or SWaCH Seva Sahakari Sanstha Maryadit started buying plastics filaments and developed process of 3D printing.

Measures taken by Agricultural Produce Market Committee (APMC)

APMC Mumbai making efforts to utilize the agricultural waste generated from fruit and vegetable market yards for power generation and biogas production in association of B.A.R.C Mumbai. Agriculture Marketing Board Koyambedu Chennai is producing 250KW electricity with the association of RAMkY Group, by using 150metric Ton agriculture waste on daily basis. Kolkata Agriculture Trader Association is proposing to produce electricity by paddy husk, whereas APMC Delhi is utilizing agro waste to produce compost power. APMC Jalandhar is generating wealth by using sunflower waste for ignition purpose ((MOFPI 2020).

Role of engineering and machinery in waste management

In such an important field of waste management, machinery and engineering have crucial role to play. Comfortable, healthy and sustainable waste management is possible with site and waste specific machineries. sensors based automation in waste collection and sorting (plastic, metallic and organic) are chief approaches in urban waste management. Garbage bins fittest with ultrasonic sensors for level indication directs garbage collection drivers via SMS on phone. In India such sensor-based garbage bins would be useful in automated uploading to get rid of unhygienic conditions. These technologies such as LCD display, driver circuit, robots, infrared (IR) and ultrasonic sensors also helps in effective recycling.

RVM Recycle Pvt Ltd. has created a product based on reverse vending technology where the fully automatic waste integrated collection and segregation machine can process heterogeneous waste such as plastic, steel and aluminium. They also reward locals for their contribution in Swachh Bharat Abhiyan by providing them gifts discounts coupons, currency benefits and free of cost Wi-Fi facilities. these machines are sensor-based which indicates full load limits so that cost of transportation can be reduced.

The concept smart bins to robotic road in smart cities can reduce carbon footprint. They have 3D laser scanners, weight-measurement sensors, touchsensors, spectrometers, metal detectors and Visual sensors that's why they can help in collection, segregation and recycling of waste. Deploying robots in waste management and recycling can be a success after proper awareness among people. Finland-based company Zenrobotics in 2011, is the first commercial supplier of robots for waste sorting. Robots powered by artificial intelligence (AI) for garbage sorting are also developed by a Spain-based company. Thus, these robots are promising technology as they are capable of reducing the cost of trash-bots; can curtail efficient waste-management systems and fits in the concept of smart cities.

Cost reduction in waste management

Some possible waste management options are waste minimization, material recycling, waste processing, waste transformation and sanitary land filling. Cost of any waste management plant is based on few components such as: land price, scale, term, quantum of waste, rate of waste recovery & recycling, energy consumption and revenue generation (Appasamy and Nelliyat 2007).

To make waste management profitable, the concept of social costing must be applied properly. The summation of development cost and positive/negative externalities comprises the social cost. Table no.1 is showing simple explanation of positive and negative external cost associated with some waste management processes. Environmental cost related to pollution can also be quantified by using standard methodologies (Saleem et al., 2016).

Table 1. Negative and positive external cost associated with waste management processes

Waste management process	Negative external costs	Positive external costs
Recycling	Energy consumption	saves Green House Gas (GHG) emissions and water consumption; minimizes waste quantity; direct monetary benefits
Landfilling	Emissions of methane; risk of surface- and groundwater contamination; odors, noise problems to the residents	Irreversible use of land
Composting/ anaerobic digestion	odor, noise and gaseous emissions	Production of biogas Compost can substitute chemical fertilizer
Incineration	Toxic and GHG emissions	Energy can substitutes fossil fuel based electricity.

Cost reduction in waste management can be achieved by adoption integrated operation which is technically sound, eco-friendly, financially stable, supported by sturdy machinery and local community and long-term planning. Measures such as commodity tax, bans on landfilling of recyclable materials, mandatory waste disposal charge can make waste management profitable. Approaches like willingness to pay is also very acceptable nowadays where people are willing to pay for services related to hygiene, hence such services are sustained. Intelligent charging mechanisms and models are also been developed to overcome the issues related to willingness to pay. Variable waste charging from people requires strategic planning having following points to ponder:

- Estimated waste generation
- Rationale behind the accountability of waste generator
- Unit pricing of collected amount or availed services for individual charging
- Creation of environmental taxes
- Solid waste user rewards or landfill fees
- Provision of municipal bonds for capital investment
- Inter-municipal arrangements to capture economies-of-scale
- Private sector inclusion for operating new facilities

Agricultural crop residues waste management intiatives taken by Government of India

Agricultural waste is the product of the agriculture operations on farm and off the farm. It starts from on from to the post-harvest to marketing to processing to house-hold. These days due to the mismanagement of the waste that is generate from the agriculture produce has led to many environmental issues, that has affected the society in the form pollution and the mis-management of these waste has affected farmers and economy of the country by investing in other control measures that has formed due to some practices that costed in other forms .So government of India along with other state governments has proposed some schemes and polices in order to manage this in a right manner.

Some of the policies and schemes initiated by government of India will help the farmers who are the backbone of our country. But the government usually fails to apply the policies and schemes in a proper manner. So seeing from side of economy of the country the farmers are burning the resource biomass due to the economic and social problems such as lack of awareness and the government fails in the application part of the policies in an efficient manner. Collection of crop residue by government and efficient utilization is required. But the new initiatives taken by the government will help the farmers and economy of our county if it is correctly implicated.

In the past few years solid waste has given much more attention by government of India and state government in India. One such scheme proposed by government India is "programme of energy from urban, industrial, agricultural wastes/

industrial, municipal solid waste". In this project they have used some agricultural waste in such a way to promote Biomass Gasifier for feeding power into the grid or meeting captive power and thermal needs of rice mills/other industries and villages. This was an project initiated by government of India in the form of waste to energy concept. Currently there are 56 biogas-based power plants operational in India, the majority of them are in the states of Maharashtra, Kerala and Karnataka. The recent development in studies has shown the possibilities of rice straw and other crop residue instead of dung and vegetable waste. Small biogas type has been introduced to some rural villages where 1-10m^3 biogas can produced per day. In some private enterprises they are producing a 12 MW rice-straw power plant can consume 120,000 tons of stubble collected from nearly 15,000 farmer .These private enterprise are giving Rs 500- Rs1000 to farmers for per ton of straw. In India in the past few years the crop residue burning has become a major problem due to the inadequate management practices. This practice is mainly under go in the month of October and November mainly for the field of rabi crop, i.e. wheat and potato. This is practiced mainly in Punjab and Haryana both together in 2018-19 has produced 28.1 million tons of paddy straw. In that 10.1 million tons of paddy straw has been burnt. Government of India has undertaken so many initiatives for creating awareness among the farmers, a new Central Sector Scheme on "Promotion of Agricultural Mechanization for In- situ Management of Crop Residue" in the States of Punjab, Haryana, Uttar Pradesh and NCT of Delhi' for the period from 2018-19 to 2019-20 was proposed .Mass awareness campaigning's has been done through KVK's among the farmers they have performed some on-farm techniques how to manage the residual waste. The government is providing financial assistance to the individual farmers by giving them 50% of the cost for buying the crop residue management-equipments/machine. Happy seeder, Rotary mulcher, super straw, Hydraulic Reversible M.B. Plough, management system are some of the machinery/equipment comes under this scheme .Due to this scheme and the initiation taken by the KVK's in 2017 Punjab was able to convert 25 villages into zero stubble villages that has been increased to 76 villages in 2018-19.

National Thermal Power Corporation (NTPC) has been directed by government of India to mix crop pellets (10%) with the coal for the purpose of power generation. This will help the farmers to curb the crop residue burning and in-turn will get profited by Rs-5500 per ton of the straw, the policy is yet not in action. but it is sure that this will help the farmers .The national policy Some of the laws that are to curb the crop residue burning are: The Section 144 of the Civil Procedure Code (CPC) to ban burning of paddy; The Air Prevention and Control of Pollution Act 1981; The Environment Protection Act, 1986; The National Tribunal Act 1995; and The National Environment Appellate Authority Act, 199.These laws are very strictly followed by Rajasthan, Haryana, Punjab and Uttar Pradesh. Few initiatives of bio-composting have been taken by the government of India by teaching farmers the agro-waste conversion and bio composting. Around 456 farmers from Uttar Pradesh has learned this and the Rashtriya Krishi Vikas Yogna (RKVY) has done on-farm demonstrations to the farmers of different villages of Azamgarh and Marinath Bhanjam districts.

The Ministry of Agriculture of India recently developed a National Policy for Management of Crop Residue (NPMCR) with some specific objectives such as to curb the crop burning with the help of new technologies for the in-situ management and optimum utilization, promoting the use if machineries such as happy seeder using the technology of Remote Sensing for the management of the of the crop residue using National Remote Sensing Agency (NRSA) and Central Pollution Control Board (CPCB).Other than energy extraction from agricultural residue, it can be used for the purpose of construction and rice husk can be used in cement husk and sugarcane waste is used in paper industry, husk and bagasse ash has been used in mushroom cultivation too.

Conclusion

Along with these startups many more small and big organizations are helping Indian communities in the battle with the complex problem of waste management. Growing awareness among citizens and more innovative solutions we can hope to see lesser garbage go to waste in future. The unique economic and social development must be focused on improving efficiency of waste management through controlling waste generation and achieving maximum rates of recycling while utilizing waste as the commodity that can be converted into another value-added product which are additional resources to finance waste management.

References

Appasamy P. P. and P. Nelliyat. 2007. Financing Solid Waste Management: Issues and Options, Madras School of Economics, Chennai. Proceedings of the International Conference on Sustainable Solid Waste Management, 5 - 7 September 2007, Chennai, India. P.537-542.

Bhide, A. D. and B. B. Sundaresan. 2013. Solid Waste Management in Developing Countries. Indian National Scientific Documentation Centre. New Delhi. 222.

Central Pollution Control Board (CPCB). 2020. Pollution Control Implementation Division. http://www.cpcb.nic.in/divisionsofheadoffice/pci3/Profiles.pdf.

Federation of Indian Chambers of Commerce and Industry (FICCI). 2011. Resource Assessment for Livestock and Agro-Industrial Wastes – India The Global Methane Initiative.

Gupta S., K. Mohan, R. Prasad, S. Gupta and A Kansal. 1998. Solid waste management in India: options and opportunities Resources. Conservation and Recycling. 24:137–154.

https://www.agnii.gov.in/innovation-listing

Ministry of Food Processing Industries (MOFPI). 2020. Government of India. htp://mofpi.nic.in/.

NIAM 2012. Research Study Handling of Agricultural Wastes in APMCS Research Report 2011-12. National Institute of Agricultural Marketing (A Govt. of India organization) Jaipur, Rajasthan pp. 1-83.

Saleem W, A. Zulfiqar, M. Tahir, F. Asif and G Yaqub. 2016. Latest technologies of municipal solid waste management in developed and developing countries: A review. International Journal of Advanced Science and Research. 1(10):22-29.

Sharma V.K., P.V. Beukering and K.V. Ramaswamy. 1996. Conclusions and Policy Recommendations in International Trade and Recycling in Developing Countries: the case of waste paper trade in India. In: Van Beukering P., K. S. Vinod, editors. Mumbai: Institute for environmental studies, Netherlands and Indira Gandhi Institute of Development Research. 147–155.

TERI. 1997. Looking Back to Think Ahead: Green India 2047. India: Tata Energy Research Institute, New Delhi. 346.

28

Post-harvest Management Through Convergence of Innovation and Technology

Sudheer, K.P.[1]*, Sreelakshmi, K. Unni*[2] *and Vithu Prabha*[3]

[1]*Kerala Agricultural University, Thrissur, Kerala*
[2]*KAU RAFTAAR Agri-Business Incubator, Department of Agricultural Engineering, Kerala Agricultural University, Thrissur, Kerala*
[3]*NAHEP – CAAST Project, KAU Agri-Business Incubator, Kerala Agricultural University, Thrissur, Kerala*

Introduction

The post-harvest and marketing system comprised of a series of interconnected activities which starts after the harvesting process and continues until the distribution of food products to the consumers. An effective post-harvest system makes sure that the food product reaches the customer in a short time without sacrificing volume, quality, and safety of the food product (Gardas et al., 2018). Increasing population growth in the last few centuries and the requirements from industrialization forces the need for sustainable food production and better food processing technologies. There is a requirement of balance between sustainable food production practices, a healthy environment, and a healthy population due to the rapid changes in the climate and population. According to the Food and Agricultural Organization (FAO), there will be a requirement of a 70% increase in food production in 2050 due to the growing population across the world (FAO, 2017).

The challenges such as improper land use, decrease in biodiversity, aquatic eutrophication by over-fertilization, global warming, water shortages due to improper irrigation etc. imposed awareness among the stakeholders to bring more efficient approaches in agricultural and food production practices. This brings the idea of terms such as recyclable and eco-friendly packaging, buy local, whole foods, free-from, and fair-trade (Boye and Arcand, 2013). Recent trends in climate and changes in the environment may pose challenges to achieve the required food production in the future. In addition to the steps taken for increasing food production, the steps to reduce food loss are important. In 2011, it was calculated that 11.3 billion tonnes of food was wasted in a year, which is almost one- third of the total food production (FAO, 2011).

Food loss also leads to scarcity of water, wastage of agricultural land, and the increase in greenhouse gas emission. Therefore, reducing food loss is critical to achieve food safety as well as sustainability concerning sources of land, water etc. (Kummu et al., 2012). Developing countries lack advanced harvesting techniques, storage and transportation facilities and these drawbacks lead to the food loss during production and processing stages of food (Lundqvist et al., 2008). About 45% of global food loss and waste-related greenhouse emission is contributed by industrialized Asia and South and South East Asia (Porter et al., 2016).

India is the world's largest producer of milk, rice, wheat, vegetables, and fruits. India is also one of the leaders in producing spices, fish, poultry and plantation crops. Even though India is a leading food producer, the country also acts as a potential waste hotspot (Hiç et al., 2016). Being a developing country, the majority of food loss and waste is occurring at food production and post-harvesting stages (Kashyap and Agarwal, 2020). Therefore, efficient post-harvesting technology and management are necessary to counter the issue of food loss, especially post-harvest loss. Food loss is the mass of food lost in the part of the supply chain and it is measured only for products that are meant for human consumption, excluding feed and parts of products which are not edible. Food loss can be calculated as

$$F^i_{loss} = \alpha^i \times F^i_{production}$$

where F^i_{loss} is the mass of food loss (mass) of the ith food item, $\alpha^i$is the proportion of total loss in farm operations and storage stages and $F^i_{production}$ is the total food production directed towards human consumption (Kummu et al., 2012; Kashyap and Agarwal, 2020a). In 1974, the World Food Conference held in Rome pointed out the importance of reducing post-harvest food loss to increase food availability.

Post-harvest loss (PHL) and the factors affecting the post-harvest loss

Raut et al. (2018) listed the critical factors which affect post harvesting loss. Some of the factors are related to the lack of infrastructures such as insufficient storage facilities, processing facilities, and packaging facilities. A significant number of critical factors are related to the improper communication or linkages between different stakeholders such as the insufficient linkage between farmers and processing units, lack of linkages between industry and government, improper record-keeping, tracking, and traceability facilities, lack of interaction between farmer and consumer, lack of information related to marketing to small scale farmers and their units etc. Insufficient dissemination and creation of knowledge also lead to a significant amount of food loss such as lack of effective post-harvest technologies, misunderstanding present market scenarios, deviations in the prediction of climate weather conditions etc. (Raut et al., 2018).

Gardas et al. also evaluated critical causal factors for post-harvest losses (PHL) in the fruit and vegetables supply chain in India using decision making and trial evaluation laboratory (DEMATEL) method. This study determined the cause-effect relationships among the identified factors and found that lack of proper packaging facilities, lack of proper storage facilities and insufficient infrastructure were the most significant factors which lead to food loss. This study also recommended that policymakers should give attention to these factors to introduce new policies to bring down food loss (Gardas et al., 2018). The reduction in post-harvest loss will improve all pillars of food security, through an increase in food availability improvement of food access due to reduction in prices, an increase of food utilization owing to the decrease of qualitative PHL, and higher stability obtained by an increase in stored food (Bendinelli et al., 2020).

In less developed countries, biological spoilage of food is common. Lack of proper refrigeration system leads to spoilage of livestock products, fish, fruit and vegetables. But in the case of products such as grains and roots, inefficient post-harvest handling of the products leads to the spoilage. Spoilage is mainly for scattering and crushing of cereals for cereals during their handling (Golob et al., 2002). In addition to the issues created by handling, storage conditions also reduce the quality of cereals and grains. Weather also determines the quality of the food products as well as food loss. Poor quality food not only leads to food loss, but it can also impart health issues to the people (Hodges et al., 2011).

Present post-harvest loss scenario

As per FAO, percentage of food loss is highest for roots, tubers and oil-bearing crops (>25%), followed by fruits and vegetables (>20 %). FAO estimates that 13.8 percent of the total food produced in the world is lost between farms and up to, but excluding retail. These estimates also considered an economic weightage of the food items (FAO, 2019). Lipinsky et al. listed potential direct causes and indirect drivers of the food loss and wastage. Direct causes are inferior production and agronomic practices and choices, improper packaging, poor stock management failure of equipment, improper harvest management and logistics management. Indirect drivers of food loss are not utilizing or collecting the crop due to poor quality, due to a drop in the prize, lack of proper storage and variable demand for perishable food items etc. (Lipinsky et al., 2013). ICAR-Central Institute of Post-Harvest Engineering and Technology (CIPHET) assessed quantitative harvest and post-harvest losses of major crops and commodities in India. They have found that food wastage occurs in many stages of the supply chain in the country. The post-harvest losses according to this study which is carried out during the years 2012-2015 are listed in Table 1. The estimated annual value of the losses is calculated as Rs. 92651 crores.

Table 1. The estimated annual post-harvest losses

S. No.	Crops	Post-harvest loss (in %)
1.	Cereals	4.65-5.99
2.	Pulses (gram, tur & others)	6.36-8.41
3.	Oilseeds	3.08-9.96
4.	Fruits	6.70-5.88
5.	Vegetables	4.58-2.44
6.	Plantation crops	4.17-4.91
7.	Spices	1.18–6.51

(Jha, et al. 2015, ICAR-CIPHET)

Suggestions made by the CIPHET to reduce harvest and post-harvesting loss were as follows,

a. Standardization of harvesting and threshing practices and their refinements.
b. Popularization of appropriate techniques and infrastructure for short-term storages.
c. Right processing techniques, value addition to the products and proper storage of the surplus food products.
d. Skill development, entrepreneurship development, and promoting incubators.
e. Spread the awareness of the right post-harvest technologies and the introduction of favourable policies.
f. Investment in post-harvest infrastructure and mega Food Parks (Jha et al., 2015)

It is projected that the production of food grains in India will be 315 million metric tonnes by 2030 and the food loss may lead to a water loss of 32.3 km^3/ yr, land usage loss of 5.24 Mha and 59.8 million metric tonnes CO_2eq GHG emissions (Kumar et al., 2016). Climate change in the immediate future will have an important role in food loss. Extreme weather conditions will affect food storage and food spoilage. An outbreak of pests and diseases associated with climate change can seriously affect food storage. Inefficient post-harvesting techniques and reduced resources may induce a food scarcity in future. Therefore, this issue should be addressed, and solutions should be developed to tackle food scarcity issues (Kashyap and Agarwal, 2020).

Strategies and challenges to reduce post-harvest loss

Reduction in Food loss and waste (FLW) improves food security, nutrition and economic growth. Reduction of post-harvest loss also leads to benefits such as

a) Expenditure for food by the consumer will be reduced
b) Increased profit for food suppliers and food producers

c) Improved food security and nutrition

d) Reduction of Green House Gases and less usage of natural resources

Collective food production and processing approaches

The collective action of farmers performs an escalating role in agriculture and rural development. Many collectives of farmers try to invent alternatives to the present food production and processing approaches by adopting co-operative initiatives in production, processing, and marketing. This enables them to improve their access to the market and the income and support regional economic and employment developments (Knickel et al., 2008).

Problems faced by farmer collectives

The major issues faced by farmers collectives related to post-harvest loss are a) Unavailability of the suitable processing technique due to the lack of interaction between processing industries and farmers, b) Insufficient backward-forward integration and vertical integration from farmer to customer leads to increase the risk, c) Lack of direct interaction between the market and farmers, negatively affects the supply chain and decreases the sales, d) Inaccessibility and unavailability of advanced post-harvesting techniques lead to considerable food losses and decrease the efficiency of the supply chain e) Inaccurate forecasting of climate and market lead to the wrong realization of the commodity prices, heavy losses, and delay in delivery (Gardas et al., 2017).

In recent decades, the role of farmers' associations in developing countries has been increasing due to several projects promoted by local and international governmental and non-governmental institutions and organizations (Orsi et al., 2017). Compared to other developing countries, little has been said in the Indian context about what role farmer organizations can play to help smallholder farmers specifically, improve their position in those emerging value chains. Wide education of farmers on the main causes of losses, infrastructure improvements to connect small farmers to markets, more effective supply chain with access to microcredit to improve the technological level and public-private partnerships (PPPs) sharing investment costs and risks in the market targeted interventions (Bendinelli et al., 2020).

Role of Farmer Producer Organizations (FPOs), Self Help Groups (SHGs) and Farmer Interest Groups (FIGs)

Farmer aggregations play an important role in enhancing direct market linkages, and in providing training and services. Implementing new technologies related to post-harvesting is relatively easy with the help of farmer collectives. Different types of aggregation models are operating in India such as FPOs, Self Help Groups and Farmer Interest Groups, and Agricultural Entrepreneurs etc. In addition to these models, Primary Collection Centres, Contractual Agreements and Digital Platforms are also being operated by public and private sectors. FPOs are legally registered collectives of farmers which focus on enhancing farmer incomes. Co-operative societies and farmer producing companies are two different types of FPOs. Initially, FPOs operate with the guidance of civil society organizations (CSOs). FPOs enables the farmers to access the post-harvesting technologies which may not be accessed if farmers operate independently. For example, FPOs promoted by the Association for Social Advancement (ASA) collaborate with companies like Mahindra and Mahindra's Trringo and arrange the equipment for lease to the farmers. FPOs also enables the farmers to access knowledge on good agricultural practices developed by R&Ds across the globe. Government organizations must educate farmers about the requirement and benefits of creating FPOs, and the legal aspects of FPOs. To attract equity investments and to facilitate the cash flow, FPOs should make a joint venture with other FPOs who work in similar sectors.

SHGs and FIGs are voluntary collectives of a small number of farmers who work collectively towards a common goal which is established through discussion and mutual consensus. SHGs and FIGs do not necessarily require external support for the formation. But with the help of NGOs, they have the potential to become FPOs. Due to the informal structure, SHGs and FIGs possess a good amount of cohesiveness and common purpose. SHGs and FIGs can avail mechanization and tools by leasing from other entities to reduce post-harvest loss. Presently, the Government of India promotes these aggregations by giving loans of low-interest rates. The major problem faced by SHGs and FIGs is the lack of proper leadership, competition from big private industries, and the incapability to raise huge investment.

In the future, FPOs and SHGs focused innovations are required to reduce PHLs significantly. These innovations should be in the field of farm mechanization and information and communication and technology (ICT), crop selection and decision analytics. Encouraging policies such as relaxation of FDI norms, modification of contract farming law and Agricultural Produce Market Committee act are necessary to enhance the significance of FPOs and SHGs. Actions should be taken to bridge the gap between on-farm and near farm storage and crop protection infrastructure, which ensures the farmers to sell their product at a better price in the market. Initiatives like the FDI and APMC reforms and e-NAM, along with the private sector thrust on direct procurement, will continue to open new avenues and strengthen linkages with existing markets for farmers to sell their products (Shiferaw et al., 2011; Trebbin, 2014 and Ganesh et al., 2019)

Role of public and private sector

The role of the public sector in the agricultural sectors of developing countries has varied significantly over the last few decades. In most of the countries, the public extension has been relatively less effective to fulfil the needs of small-scale, marginal farmers due to insufficient resources and the lack of suitable expertise. To meet the requirements of the farmers, alternative organizations have emerged in many developing countries. These alternative organizations provide the required technical and organizational needs of the farmers. Private sector firms and nongovernmental organizations (NGOs), and other collectives of farmers including co-operative sectors are examples of alternative organizations which provide training, processing, production and organizational assistance facilities to farmers. Private firms have a superior infrastructure for research and development activities and the outcomes of the research help farmers in the field of selection of feeds, production and processing practices, and post-harvesting techniques. Nongovernmental organizations conduct different programs such as human resource development and social capital development programs for farmers and their small collective organizations (Swanson and Samy, 2002). In India, several NGOs work in the field of uplifting rural women by giving supports for them to enter different agricultural sectors.

Establishment of traceability

Modern postharvest systems should have traceability as an attribute. Each step related to products from the producer to the final consumer should be well documented and coordinated to achieve traceability. Traceability is necessary to provide information to consumers and to gather feedback from the consumers to improve the performance or quality of the food products. The drivers for traceability are categorized into hard and soft traceability. Hard traceability requirements include compliance with regulatory or trade treaty requirements, food safety, and production standards. Soft traceability requirements are related to supply chain, market survey and customer needs (Bollen et al., 2006). Traceability provides information about good agricultural practices followed by each step in the food production and marketing. Traceability also provides evidence of the sustainability of the processes followed by the manufacturers. As per ISO 9000:2005, product traceability is the ability to identify the product at any stage in the supply chain. In traceability for the postharvest system requires tracing of products as well as the processes. To make the traceability more functional, identifiable units (IUs) which adhere to the uniqueness requirements with additional attributes are required. Components of traceability of a system are identification technologies such as alphanumeric ID, Bar Code, RFID etc and the information systems. Other uses of traceability in post-harvesting systems are the improvement of feedback tools cool chain quality management etc. (Bollen and Emond, 2014).

Supply chain management and cold supply chain management

Supply chain loss in the post-harvest management of agriculture products is one of the major factors which determines the food problem in most of the developing countries. Food loss and wastage lead to both material loss in terms of proteins, nutrients etc. as well as the economic loss. According to literature, one of the major reasons behind the food loss is the lack of cold storage for fruits, vegetables and meat. Controlling the levels of waste is beyond the capability of individual farmers or consumers. The problem is wider and involves various factors in the supply chain of agro-products (Chandrasekaran and Ranganathan, 2017). One of the important steps to solve this supply chain loss is temperature management in the supply chain. Perishable foods require cold environment throughout the storage and supply. Cold environment protects perishable foods from decay and reduces the growth of foodborne pathogens. Any failure in the cold supply chain leads to spoilage of food and compromise the safety and quality of food. The supply chain of cold food starts in the postharvest stages and ends in the refrigerators of retailers or consumers. Cold supply chain management is a multidisciplinary area which involves different stakeholders and requires modern technologies.

A study conducted by BIO Intelligence Service for the Global Food Cold Chain Council predicted that the cold chain could save up to 550 million tonnes of CO_2 equivalent globally. In developing countries, an effective cold supply chain requires government policies as well as partnerships between the public and private sectors (PPPs). Several studies conducted to reduce the postharvest losses by multiyear projects for strengthening the PPPs such as the study conducted in the Asia-Pacific region from 2013 to 2018. PPPs can build infrastructures for efficient cold chain systems, including precooling amenities, refrigerated trucks, cold storage services, and advanced distribution centres. PPS can also play an important role in developing national food grading standards, inspection, and information exchange and contributes to related research. To reduce food loss through cold supply chain management, holistic approaches including managerial, technological, legal, and regulatory aspects are required. Technological advances improve the efficiency and transparency of the food cold chain. Establishing these legal requirements can improve the cold chain supply system by making the process from a voluntary to a compulsory basis. Collaboration between different regions as well as in partnerships between the PPPs can result in a better cold supply system which will reduce the food loss and wastage (Ndraha et al., 2020).

Entrepreneurship and skill development programs

Agri-entrepreneurship can be defined as sustainable, community-oriented, directly-marketed agriculture. Agri-entrepreneurship can also be defined as the formation of a novel economic organization for the intention of growth under risk and uncertainty in agriculture (Bollen et al., 2006). In India, agriculture still employs about 44 percent of the workforce (in 2019; ILO, 2019) and an average consumer spends nearly 45 percent of her expenditure on food (in 2011; NSSO, GoI, 2014). Most of the farmers have very small landholdings. The average size of landholding in India is just 1.08 ha (Agriculture Census, 2018). Challenges faced by small scale agri-entrepreneurs in developing countries (Trebbin, 2014; Kucukvar et al., 2019 and Fischer, 2020) are listed below.

- Lack of skilled and managerial manpower
- Lack of infrastructure
- Inadequate marketing
- Inaccessibility of technology
- Higher capital cost
- Inadequate government policies

The only way to feed the rising population with increasing per capita incomes as well as augment incomes of farmers, from a very low base, would be to raise productivity by using modern inputs (such as high yielding seeds, fertilizers, pesticides, farm machinery, precision farming practices, etc.) sustainably. However, the use of these modern technologies requires skills that may be different from those necessary for traditional agriculture. This requires policy attention and institutional support to provide appropriate skills and vocational training to the people who are engaged in not only production activities in agriculture but all along the agricultural value chains and food-processing sector. The skill level and educational attainment of the workforce determines the productivity, income levels as well as the adaptability of the working class in a changing environment. Skill-based training enhance the knowledge base and efficiency of the farmer on the use of technology, thereby increasing the production and productivity of crops. Vocational and professional training in horticulture and allied areas are important and it will help in employment generation and entrepreneurship development in the farm sector (Skill development of rural entrepreneurs, 2016). Government of India has a significant number of initiatives for the development of agri-entrepreneurship and few of them and their objectives are listed below.

a. Initiative for Development of Entrepreneurs in Agriculture (IDEA).

This initiative is to promote Agri-business ventures in the North East Region and to assist those ventures to achieve profitability. This initiative also aims to generate more employment and make availability of input supply and services (IDEA, 2016).

b. Capital Investment Subsidy Scheme for Commercial Production Units for Organic/ biological Inputs.

The major objectives of this initiative are to promote organic farming in the country by making available of bio fertilisers, biopesticides, waste compost and increase agricultural productivity while maintaining soil health and environmental safety (NABARD, 2020).

c. Agriclinic and Agribusiness Centres Scheme

The scheme is implemented by Ministry of Agriculture and Farmers' Welfare, Government of India, with NABARD acting as subsidy channelizing agency. The major objective of this scheme is to supplement efforts of public extension to the farmers on payment basis or free of cost as per the business model of agripreneur. This scheme also aims for self-employment opportunities to unemployed agricultural graduates, diploma holders etc (NABARD, 2020). Figure 1 visualises a structure model fore role of agribusiness incubators according to Smilor (1987) and Wiggins and Gibson (2003).

d. National Livestock Mission

It is an initiative of the Ministry of Agriculture and Farmers' Welfare and NABARD is the subsidy channelizing agency. The components of the National Livestock Mission are Poultry Venture Capital Fund, Integrated Development of Small Ruminants and Rabbit (IDSRR), Pig Development (PD) and Salvaging and Rearing of Male Buffalo Calves (SRMBC) (NABARD, 2020)

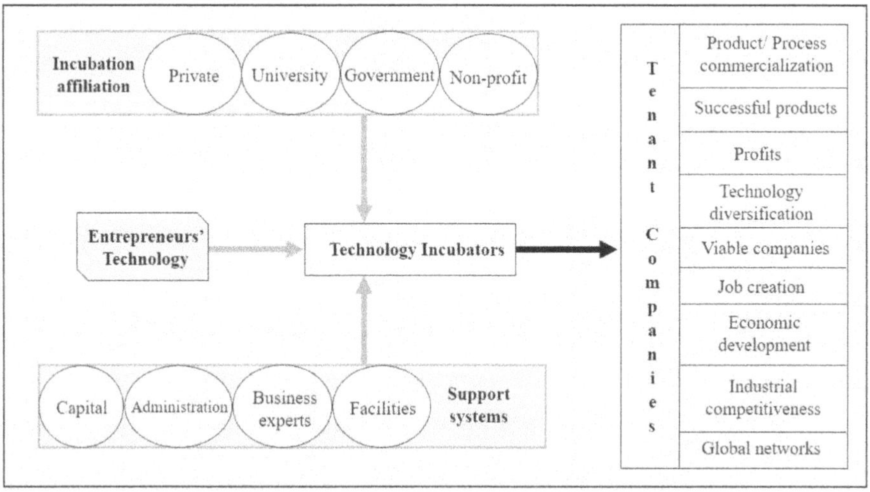

Fig. 1. Technology business incubator model for food enterprises
(*Source*: Modified from Smilor (1987), and Wiggins and Gibson (2003))

e. Credit Linked Capital Subsidy Scheme for Technology Upgradation

The scheme aims to facilitate technology up-gradation in MSEs by providing capital subsidies for the introduction of well-established and advanced technology and upgrade the plant & machinery with state-of-the-art technology (Development Commissioner Ministry of Micro, Small & Medium Enterprises, 2020)

f. NRLM / NULM (Deendayal Antyodaya Yojana)

DAY-NRLM (Deendayal Antyodaya Yojana - National Rural Livelihoods Mission) is a scheme of The Ministry of Rural Development (MoRD), Government of India for promoting poverty reduction through building strong institutions of the poor, particularly women. This scheme facilitates institutions to access a range of financial services and livelihoods services (NABARD, 2020).

g. Startup Agri-Business Incubation

Main Objective of this initiative is to seed support to aspiring incubates by enabling minimum viable product (MVP) to the marketable stage and scale-up the product and business.

h. Agriculture Grand Challenge

The Ministry of Agriculture & Farmers Welfare started this challenge to give opportunities for Agri-tech start-ups to solve defined challenges faced by in the Agriculture sector in India.

i. Other government of India schemes

- Pradhan Mantri Fasal Bima Yojana (PMFBY):
- Pradhan Mantri Krishi Sinchayee Yojana (PMKSY):
- Paramparagat Krishi Vikas Yojana (PKVY):
- Agricultural Technology Management Agency (ATMA):

State governments also introduced several schemes to uplift the agri-entrepreneurship. For example, Government of Kerala has numerous schemes in the field of Crop Husbandry, Agricultural Research & Education,Central Sector Schemes, Agricultural Marketing & Quality Control, Rashtriya Krishi Vikas Yojana (RKVY), Rural Infrastructure Development Fund Projects etc.

Quality and safety management systems

One of the biggest challenges in the food industry is to maintain the postharvest quality of food products. Chemicals were used in the past to avoid the spoilage of food. But these chemicals are being phased out and new advances in the field of postharvest management give attention to the concern of consumers and the environment. The changes in the laws related to food safety and the consumer's presumptions forced food industries to develop or innovate new technologies in the logistics of postharvest handling and the value chain of food products. For example, a drastic change occurs from the use of synthetic pre- and post-storage treatment to eco-friendly products.

Another example of emergent technology in the field of postharvest technology to maintain the food quality is modified atmosphere packaging (MAP) which maintain the nutritional value and enhances the shelf-life of the fruits and vegetables. Food safety and quality are the prime concerns in the food industries. In the present world of globalization, both the food and agricultural firms adapted systems which work in a way that food standards and international trade work together to achieve superior food quality and safety. Food industries follow several regional and international food safety procedures to get certification from the government and other agencies to survive in both the domestic and international markets (Ryan, 2014).

The essential components in the food safety include handling, preparation and storing of the food products to ensure the food quality and to avoid food-borne diseases. The unwanted substance in the food product included by an accident is known contaminants and the unwanted substance added to the food product to reduce the cost or to improve the other properties are known as adulterants. Contaminants and adulterants may lead to producing microbiological pathogens and toxins which induces serious concerns over the health of the consumer (Jha, 2016). Food safety is an aspect of food quality as it refers to hygiene and harmlessness of the food that reaches consumers (Luning et al., 2009). According to the Food and Agriculture Organization (FAO) of the United Nations, food safety is a basic human right due to its economic and social implications. According to Codex Alimentarius Commission (CAC), 1997, "food safety is the assurance that food will not cause harmto the consumer when it is prepared and/or eaten according to its intended use" (Xu, 2016).

Food safety systems focus on safety, quality, efficiency, reliability, interchangeabality and environmental friendliness along with economic factors (Wentholt et al., 2009; Panghal et al., 2018). A food safety management system (FSMS) is defined as a set of interrelated or interacting elements to establish policy and objectives and to achieve those objectives used to direct and control organizations with regards to food safety. The FSMS programme helps food business operators (FBO) to ensure that food is safe for human consumption. Food Safety and Standards Authority of India (FSSAI) has chalked out the FSMS programme in the regulations and FBOs are required to follow it.

Quality is a broad term which can be defined according to various conditions. A common definition of quality is "the total of features and characteristics of a product or service that bears on its ability to satisfy given needs". Progressive organizations now define quality as "quality is meeting or exceeding customer expectations". A quality management system can be defined as "a set of coordinated activities to direct and control an organization to continually improve the effectiveness and efficiency of its performance". Quality control refers to the operational techniques and activities, which are used to fulfil the requirements for quality. Individual quality and /or safety management systems can be classified according to the extent of activities as,

- Basic safety systems such as GAPs, GMPs etc
- Advanced safety systems such as HACCP
- Integrated food safety management such as ISO22000
- Basic quality management systems such as ISO 9001 (Rotaru et al., 2005)

FAO defines good agricultural practices are a "collection of principles to apply for on-farm production and post-production processes, resulting in safe and healthy food and non-food agriculture products, while taking into account economic, social and environmental sustainability". Good hygiene practices (GHP) measures are necessary to ensure the safety and suitability of food at all stages of the food chain. Good manufacturing practices (GMP) are required to conform with codes of practice, industry standards, regulations and laws concerning production, processing, handling, labelling and sale of foods decreed by industry, local, state, national and international bodies to protect the consumer from foodborne disease, product adulteration and fraud (CAC - Codex Alimentarius Commission, 2003).

Continuous improvement should happen not only in the quality of the product or process also on the working environment and in safety. Total Quality Management (TQM) is a customer-oriented continuous process of increasing the quality of the product or services by eliminating waste and the non-value adding activities in the system. TQM ensures participation of all the collaborators of the organization is necessary to achieve continuous improvement with the help of a global management strategy. It comprises of elements such as continuous improvements in customer focus, employee empowerment, use of quality tools, product design, process management and managing supplier quality. One of the key features of TQM is the organization's focusing on customer requirements, which indicates that quality is customer-driven (Gould and Gould, 1991). Total Quality Management can be defined by understanding each of three words in Total Quality Management and the three words are explained in Figure 2.

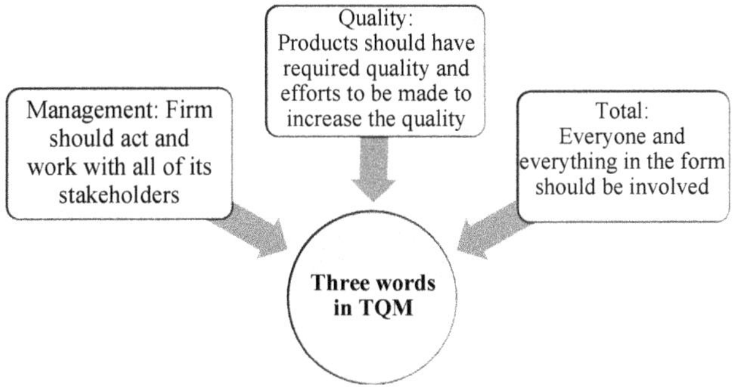

Fig. 2. Three words in Total Quality Management (*Source*: Gould and Gould, 1991)

The TQM strategy is different from organization to organization; however, a set of basic primary principles such as customer focus, leadership and structure of the organization, the involvement of people, process approach, strategic approach, continual improvement, and decision making. The elements which lead to an effective TQM programme can be listed below.

- An efficient and clear quality policy
- Effective organization for quality management
- A better understanding of performance indices and quality costs
- Efficient use of computers, process control, problem-solving techniques, benchmarking and quality improvement teams
- An environment that encourages continuous improvement to satisfy the customers' needs

Technical interventions and novel technologies

The foods produced annually to meet the requirements of food industries and export trade need to be managed appropriately to maintain their freshness and quality attributes. The present scenario shows a huge post-harvest loss for cereals, roots and tubers, and fruits and vegetables, annually. Table 2 presents the basic considerations to get a consistent supply of more quantity of food from the existing level of productions as per Kiaya (2014). The knowledge of technological advances in post-harvest management is significant for scientific and farmers community to increase profitability. A few of the established innovative technologies in post-harvest storage, processing and packaging are comprehended in the subsequent sections. Despite, the choice of appropriate approach depends on the type of crop and the product to be prepared, the scale of production and feasibility of use by the farmers (Kitinoja and Gorny, 1999; Jat, 2010). Figure 3 visualises the potential of recent technologies in the management of food from the field to the fork.

Farm ⟶ **Fork**

Pre-processing practices	Processing & value addition	Packaging solutions
• Smart harvesting systems	• Low temperature drying	• Report pouching
• Machine vision system	• Hybrid dryers	• Vacuum packaging
• Non-destructive quality analysis	• Vacuum frying	• Intelligent packaging
• Modified atmosphere storage	• Non-thermal processing	• Antimicrobial pacakaging
• Controlled atmosphere transit	• Extrusion processing	• Nitrogen flusing
• Ethylene control	• 3D printing	

Fig. 3. Technological advances in post-harvest management of agricultural commodities

• Smart farm practices

The integration of advanced technologies in post-production of food helps to preserve the freshness of biological materials. An approach known as 'smart farming' aims at increasing the net quantity of the produce apposite for further use, processing or marketing. This involves the application of electronics and, information and communication technologies such as precision equipment, the Internet of Things, sensors and actuators, geo-positioning systems, big data, drones and robotics to surge the quantity and quality of products (Wang et al., 2006; Saiz-Rubio and Rovira-Más, 2020). A few such systems that serve a link to reduce the loss or wastage of food materials after harvest are as below:

- Drone-based crop protection systems
- Crop maturity sensors
- Robotic harvesting systems
- Heavy metal detection sensors
- Non-contact quality analysis devices
- Mobile applications for crop nutrient information

Post-production, and during handling and transportation thereof, food commodities require precooling, blanching or drying operations to prevent the loss in quality. As discussed in former sections about the cold supply chain, refrigerated conditions offer stable conditions for horticultural produce during storage and transportation by controlling the physiological and microbial activities. However, the use of environmentally-hostile refrigerants and the release of greenhouse gases involved in this technology is demanding new alternatives. In this context, prefabricated underground cellars (*Groundfridge*, developed by *Weltevree*, Netherlands) can be in-field zero-energy cool chambers to keep the fresh foods throughout the year at below 12 °C. Apart, liquid nitrogen engine-based refrigerated transport systems have also been developed for food transport with zero-emission (Garlo et al., 2002).

• Machine vision systems

Automatic inspection of fruits and vegetables in the food supply chain is promising with machine vision systems. They are rapid, repeatable and reliable as compared to the manual grading and sorting methods, which depend on several factors including the psychological conditions of the workers to obtain correct results. Machine vision system or computer vision systems acquire the spectral and spatial features from the food product (in the form of an image), which are then processed and analysed in digital platforms. Artificial intelligence tools are employed to process this information and to correlate and predict the objective quality attributes. The size, shape, colour, appearance of the product and external damage, and eventually the nutritional and sensory characteristics can be commercially inspected this way followed by image analysis methods (Du and Sun, 2006).

The technology is simple and less tiresome for industrial quality control. However, electronic vision systems pose problems in bulk quality analysis. An intricate digital platform capable of executing complex algorithms is necessary since the accuracy of the classification system is governed by the number of features extracted. This would further complicate the image processing stage when commercial applications expect fast outcomes (Blasco et al., 2003; Saldaña et al. 2013). Other than the technical difficulties, high capital investments owing to the expensive image sensors and electronic accessories for data processing also poses a difficulty for budding food processors.

Machine vision system works by acquiring digital information from the products. Therefore, depending on the physical stimuli involved in the system (spectra varying in wavelengths, or sound, chemical and electric impulses), various technologies have been developed so far. Since these approaches obtain the quality information of the food products without any particular sample preparation or analysis standards, they are commonly known as 'non-destructive technologies'. A few other similar technologies used in the quality control of food products in a continuous mechanical food supply line are presented in Figure 4.

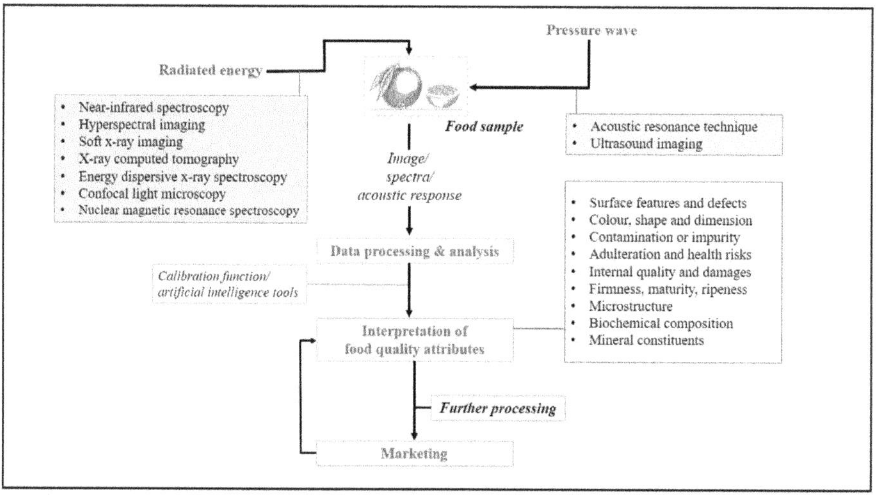

Fig. 4. Quality control in the food supply chain using Non-destructive techniques

Table 2. Basic strategies for reducing immediate food loss in the food supply chain

Stages	Grains	Roots and tubers	Fruits and vegetables
Harvesting	• Harvest at proper maturity stage (moisture content is 20-25%) and peak quality	• Harvest without bruises or damages • Use proper tools and equipment	• Harvest at the proper time (early morning) • Do not expose the product to the heat of the sun • Shade drying to minimize product weight loss and wilting
Handling/ Drying/ Threshing/ shelling	• Lower moisture content unsupportive to mould growth during storage (<13-15%) • Reduce delay in threshing/ (millet and sorghum) and shelling (maize) before storage	• Movement of products without peeling off the skin • Removal of skin initiates physiological deterioration, dehydration and microbial attack	• Handling avoiding mechanical injury • Reduce the number of handling times
Winnowing/ sorting/ cleaning	• Screening to remove stones, sand, and extraneous organic matter using grain cleaners	• Separate poor produce to minimize bulk handling • More marketability based on quality fractions	
Packaging/ Transportation		• In-field packaging • Need cushion against impact and compression • Arrange packages compact with proper aeration	• Pack in smooth and ventilated containers • Transport during the cool part of the day • No watering before loading to avoid decay
Storage	• Dry and well-vented storage conditions before complete drying before storage	• On-farm storage for fresh produces (pits, mounds, or clamp) with locally available ventilation and insulation	• High-quality crops (mature, undamaged) stored at an optimal temperature, relative humidity and environmental conditions.

(*Source*: modified from Kiaya (2014))

• Modified/ controlled atmosphere storage

The atmosphere surrounding the fresh and respiring produces has been altered (low oxygen and high carbon dioxide levels) to slow down respiration, retard maturation, lag ripening and senescence and to control pathogenic attacks. This technology can extend the shelf life of perishables to twice or thrice to that stored at normal atmospheric conditions. When there is a lower degree of control of the gas concentration, it is referred to as modified atmosphere storage or packaging. Here, the atmosphere condition established initially in the storage or package remains transient and maintains according to the interaction with commodities' physiology.

For controlled atmosphere storage/ packaging, the levels of gases are continuously monitored and adjusted to maintain the optimal concentration (Zagory and Kader, 1988; Yahia, 2006). It necessitates, a high degree of atmosphere regulation, particularly temperature, pressure and gas compositions, which is capital intensive. Researchers claim that the approach is cost-effective only for exotic foods amenable to long-term storage. The optimal gas concentration is significant to extend the shelf life of the commodity without compromising its biological value. It varies with the type of product, physiological age, temperature and duration of treatment. Beyond the optimal tolerable levels of gas concentrations, the technology can result in detrimental changes in the products. For exotic fruits like berry, for example, the optimal oxygen level should be above 2% and that of carbon dioxide at a maximum of 20% (Harb and Streif, 2004; Dziedzic et al., 2020).

• Ethylene controlling technology

The delay in ripening and senescence of fruits and their preservation is critical in dry and high-temperature areas. Ethylene gas is the activator of ripening in some fruits, particularly climacteric fruits. Active packaging solutions that scavenge ethylene from the surrounding of fruits, or by producing compounds that facilitate a neutral storage condition have been proposed. A few of the approaches reported in this area are *e+® active Ethylene Remover* and *SmartFresh* (quality enhancer based on 1-methylcyclopropene). They work by limiting the production of ethylene from stored fruits or than in transit and thus slow down the ripening. Furthermore, they have added advantages such as retaining firmness and greenness (Di Terlizzi, 2016).

• Low-temperature drying by heat pump

Heat pump drying is a novel dehydration method applicable to process heat-sensitive biological materials at low temperatures. Principally, the dryer consists of components of an air-conditioning refrigeration system. The food materials are dried using hot dehumidified air. The dehumidification of the dried air occurs in the evaporator section, which is then reheated to desired temperatures in the condenser section and is passed over the materials to be dried. The benefit of this technology is that the drying air temperatures can be controlled from −20 °C to 70 °C with appropriate refrigerant and varying the compressor capacity and air flows within the system (Patel and Kar, 2012). Further, around 60–80% less energy is consumed in this method as compared to the conventional dryers operating at the same temperature (Strommen et al., 2002; Colak and Hepbasli, 2009).

Since the heat is recoverable, this technology has been recognised as energy-efficient. However, the requirement of high-grade energy (electricity) is a drawback. Likewise, it has also been regarded as a slow and high capital intensive process as compared to the conventional drying (Eisa, 1996; Fayose and Huan, 2016). Other than heat pump drying, 'hybrid dryers' have also been developed for food application, which uses a combination of one or more technologies. This way, drying and energy efficiency can be improved, and drying time can be reduced. A few of such dryers that have been developed for various agricultural commodities are,

- Solar based air dryers
- Vacuum air dryer
- Microwave-assisted air/vacuum dryer
- Radiofrequency assisted air/vacuum dryer
- Ultrasound-assisted air dryer
- Infrared air dryer

• Vacuum impregnation

Vacuum impregnation is a promising technology for incorporating physiologically active compounds in a liquid phase into the porous structure of fruits and vegetables. It can replace the conventional practices in the preparation of confectionary foods, which depends on natural osmosis, a slow and lengthy mass transfer process. In this technology, the hydrodynamic mechanism promoted by pressure change transfer the external liquid (mostly, sugar syrup or brine) into the pores of the product. Initially, vacuum pressure (50-100 mbar) created inside the closed tank for a short period promotes the expansion of the product and outflow of the internal gas and native liquid. Afterwards, the pressure is restored to atmosphere conditions, which promotes the flow of external liquid into the porous structure of the product (Fito et al., 2001; Duarte-Correa et al., 2020).

The advantage with this technology is that it is a fast mass transfer process and is outstanding to retain the natural colour, aroma and other quality attributes of the fruit or vegetable used. Researchers suggest that vacuum impregnation is suitable for dry salting of meat, fish and cheese probiotic enrichment in dried fruits and fortification in chips (Chiralt et al., 2001; Betoret et al., 2003; Duarte-Correa et al., 2020). However, the high capital investment and high operating costs owing to the machinery associated with the regulation of pressure conditions is a factor to be considered while employing this technology. Further, the necessity of timely inspections to assure leakage-free condition in the chamber and processing line is also a concern for continuous processing.

- **Vacuum frying**

Vacuum frying is a novel deep-fat frying technology that offers multitude advantages over conventional frying that works under atmospheric conditions. It employs frying at low temperatures and minimal exposure to oxygen, thus prevents the degradation of nutritional quality of the product. Dehydration of the products is achieved rapidly under reduced pressures (~8 kPa) without excessive darkening or scorching of the product. This is due to the reduced boiling points of the oil and the product at vacuum conditions (Shyu et al., 1998; Mariscal and Bouchon, 2008). The advantage of this technology is that the vacuum fried products possess lower oil contents (35% less) and acrylamide contents (97% less) than those from the conventional fryers (frying temperatures above 150 °C). Likewise, the fried foods offer lighter colour and better texture, and thus high marketability. Moreover, the degree of heat-induced changes in frying oil is also very less in vacuum frying (Garayo and Moreira, 2002; Granda et al., 2004; Belkova et al., 2018).

The vacuum fried products are sold at higher prices than conventionally fried products. However, setting up of a vacuum frying unit is highly capital intensive. The main requirements for the consistent production of quality fried products are the machinery for frying, machinery for conveying oil and vacuum-air, and pre-treatment facility (such as deep freezer). Also, systematic utilisation of fried oil is a factor to be considered.

- **Non-thermal inactivation**

Preservation of foods without compromising the quality and safety is a challenge. With thermal processing, the microbial safety and stability of the products may be ensured, but heat-sensitive nutrients, vitamins and polyphenols may be destroyed. Likewise, the development of undesired and harmful substances can also happen at high processing temperatures and time (Hellwig and Henle, 2014; Lima et al., 2014). Non-thermal processing technologies have been introduced in food processing to preserve the food products with minimal or no changes in quality and essential nutrients (Rawson et al., 2011; Zhang et al., 2019). The sterilisation or processing by these methods is devoid of heat application even though a considerable amount of heat can be produced in some methods. Figure 5 presents the major non-thermal microbial inactivation methods proposed so far.

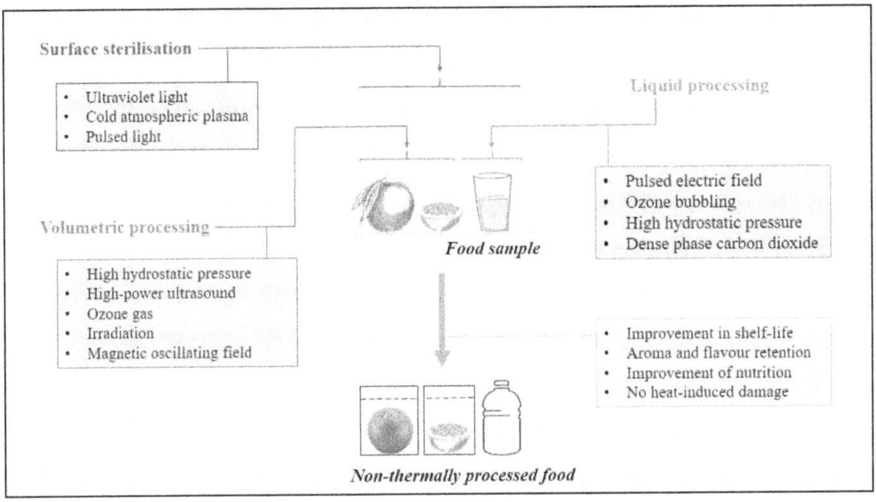

Fig. 5. Non-thermal inactivation methods in food supply chain
Source: Modified from Rawson, et al., 2011; Zhang, et al., 2019

Non-thermal processing methods affect the cell membrane of microorganism in general, thereby changing its regulatory function. Also, the structure of protein chains in the genetic material of microorganism may be affected, causing metabolic disorders and eventual death. The main advantage with these preservation techniques is that microbial inactivation or processing is possible at ambient temperatures thus reducing negative effects on food nutritional and organoleptic characteristics. Apart, many of these technologies are used as pre-treatments to extraction, drying and other mass transfer processes as they may change the structure of cell membrane of the biological materials (Tiwari et al., 2009; Zhang et al., 2019).

Researches show that each non-thermal processing methods have their advantages and disadvantages in the food industry. The selection of an appropriate method is based on the type of the product, degree of microbial log reduction and the further processing requirement of the products. However, the expensive machinery required for a closed and sealed system, and operating costs and personnel safety are important considerations in non-thermal processing units.

• Extrusion processing

Extrusion is precision nutrition introduced to produce low density expanded snack foods of desired shape and size such as puffed cereals. When the food material is heated to above 100 °C, the process is known as extrusion cooking (or hot extrusion). In this technology, the set of mixed ingredients are subjected to high temperature, pressure and shearing, which convert them into a plasticised rheological status enabling to be forced through an opening in a perforated plate or die. The size of the extruded product may be adjusted in the extruder. The advantage of this preservation method is that the high temperature-short time process reduces water activity levels (0.1–0.4), thus lowering microbial contamination and enzyme activity of the extruded snacks. Also, the variability of shapes such as sphere, rods, strips and shells can be formed in the extrusion process (Bordoloi and Ganguly, 2014; Leonard et al., 2020). A few of the related technologies to extrusion processing are mentioned below.

- Cold extrusion (at ambient temperature): pasta and shaped meat products
- Low-pressure extrusion (temperatures below 100 °C): liquorice, fish pastes and pet foods

Extrusion processing has been recognised in producing nutritional food snacks, pet foods and by-products processing (Riaz, 2000; Tran et al., 2008; Leonard et al., 2020). The main constraint in extrusion processing unit is the selection of the extruder machinery and its cleaning and maintenance. Also, the end product texture and quality are dependent on the feed composition. Therefore, optimal ingredient mix, moisture content and machine parameters such as extruded speed, temperature and time are important.

• 3D Food Printing

Three-dimensional (3D) food printing is a novel precision nutrition technique. It encompasses a digitally controlled layer-by-layer food manufacturing process to create 3D food constructs with desired microstructure and tailored nutritional contents. By this technique, the conventional food supply chain may be simplified; complex-shaped food designs which cannot be achieved by conventional moulds may be easily prepared using 3D printing (Sun et al., 2015; Yang et al., 2017). The food customisation is controlled by customising the ingredient compositions (or printing ink) and by regulating the eating quality of food according to consumer requirement.

Processed foods, which are cheaper and profitable alternatives to the ultra-processed foods rich in nutritional contents may result in the occurrence of food-related illnesses and nutrient deficiency. Food customisation by 3D printing enables the production of special-purpose foods. The nourishment for individual uniqueness by 3D printing has been recognised in the following areas,

- Foods for people with dietic preferences, such as a vegetarian diet
- Foods for people with specific food-related illness, such as diabetes and obesity
- Foods for people with special ailments, such as dysphagia
- Foods for a special category like infants, feeding mother and sportsmen

Researches on this technology are ongoing. A limiting factor is the requirement of subtractive processing methods for some 3D food construct designs to trim or cut the additional material sticking on them. Moreover, the machinery cost is high compared to conventional mould based or extrusion methods. Apart, the effectiveness of the 3D printing process is high for slow printing.

• Retort pouching

Retort pouches have been recognised as a lightweight and easy-to-pack replacement to conventional tinplate and aluminium cans. They are flexible and laminated packages that can withstand thermal processing temperatures, and retorted foods are storable at room temperatures. The technology combines the advantages of both metal cans and plastic packages. Besides, it lacks the metallic taste of the product and discolourations as reported can packed products. The laminates in retort pouches: polyester, nylon, aluminium and polypropylene, offer gloss and rigidity, puncture resistance, gas barrier and sealing and bonding property, respectively (Shankar et al., 2002; Pal et al., 2019; Khedkar and Khedkar, 2020).

Retort pouches have been recognised widely as a breakthrough in safe storage of ready-to-serve ethnic food products. However, there are few limitations in retorting method of food preservation to scale up in the food supply chain (Williams et al., 1981; Jha, 2001). A few of them are concisely given below.

- Large capital investment for the unique machinery
- Slow pouch filling facilities
- The necessity of leak detection and container integrity evaluation using special equipment such as burst/ tensile tester
- The requirement of continuous monitoring of critical processing parameters (such as residual air, pouch thickness, pouch-pouch contact) for effective thermal processing
- Need for overwrapping before distribution to prevent puncturing of pouches

• Vacuum packaging

The quality aspects of rapidly respiring products and their microbial safety situation are critical deciders in the food supply chain. Vacuum packaging facilitates the transport and storage of fresh perishable commodities. This technology involves withdrawing the headspace air from the package and sealing. The altered atmosphere in the package contains very low oxygen, limiting the respiration of the product and thereby the microbial growth. Also, prevention of oxidation of lipids, discolouration, and fading and protection of vitamins have been reported. Variants of vacuum packaging machines are available (Yamaguchi, 1990); the important types are as below:

- Nozzle-type (footswitch operated)
- Chamber-type (vacuum-seal chamber)
- Skin-type (films packed skin-tight over food)
- Deep-draw type (automated film sealing)

Vacuum packaging has reported successful in extending the shelf life of respiring fruits, meat and fish products, and preservation of quality (Gorris and Peppelenbos, 1992; Goulas and Kontominas, 2007; Sun et al., 2019). However, the selection of impermeable films and its relative cost decelerates the adoption of this technology to a great extent. In this context, the degree of oxygen and vapour barriers of the vacuum-able packaging materials (< 15 cm^3/m^2 /24 h atm and 15g/m^2/24h, respectively) are crucial. Apart, the potential risk of faster spoilage of the product if the packaging gets damaged is also a concern. In such cases, unpleasant odours and flavours due to bacterial activity are susceptible (Seideman and Durland, 1983; Chen et al. 2013; Patil et al., 2020).

• Intelligent packaging system

As already discussed in previous sections, technologies such as time-temperature indicators (TTI) and radio frequency identification (RFID) tags are recognised widely to estimate the condition of perishable foods during the long transit. TTI is available in the form of self-adhesive labels that experiences irreversible changes (colour) when subjected to undesired conditions. It gives information about the freshness of the product, and about the shelf life. However, new designs need to be proposed to reduce the added costs required in such packaging systems. Likewise, RFID tags use wireless technology to understand the storage air, food product temperatures and relative humidity of the fruit supply chains (Gander, 2007). However, acceptance of the technology in the food supply chain is limited, owing to the cost factors.

• Antimicrobial active packaging

Semitropical climates enhance microbial growth that can affect the quality and life of perishable products. Antimicrobial packages or coatings are commonly based on polysaccharide and protein-based biopolymers such as chitosan. They work by migrating the antimicrobial contents to the foods and prevent microbial growth and food spoilage. A system named *BioSwitch* contains antimicrobial that responds to change in inside pH or temperature or when the package is exposed to ultraviolet light (De Jong et al., 2005). Researches in this area suggest that the suitability of conventional polymers for this application need to be investigated to come up with environmentally friendly options (Kerbellec et al., 2008).

• Nitrogen flushing

Flushing the package with inert gases reduces the oxygen levels. Nitrogen flushing is mostly employed to increase the shelf-life of dehydrated and fried foods. With the reduction of oxygen concentration by incorporating nitrogen gas, oxidation of oily products can be slowed down. The formation of intermediate oxidation products such as peroxides and hydrogen peroxides are reduced, which prevents the further breakdown of food products that owe to rancidity and repulsive off flavour. Further, nitrogen flushing stabilises the stored product structure (Lloyd et al., 2009; Marasca et al., 2016).

Researchers reported that nitrogen flushing in commercial applications substantially reduces lipid oxidation rates during storage. However, the effectiveness of nitrogen flushing is evident mostly at very low oxygen concentrations (below 0.01 atm). Also, chances of sulphurous off-flavour have been reported with nitrogen flushing (in heat-sterilised meat). Apart, in commercial applications for fried and snack foods, procurement, storage and handling of nitrogen gas may also pose practical difficulties (Paik et al., 1994; Langourieux and Escher, 1998).

Way forward to mitigate post-harvest losses

Inadequate quality of the food production pipeline and its resulted food waste have a deep negative effect on food availability, productivity and the environment.Indian agricultural sector has to pay more attention to the importance of the processing of foods and protection of the food from spoilage using modern post-harvest technologies. Eventhough the use of types of machinery increased in the past few decadesin India, problems associated with proper drying, storage and delivery still exist. The 21st century is witnessing an enhancement in the commercialization of agriculture. Future post-harvest management should be focused on new cooling systems and temperature control, sustainable packaging to reduce food waste, time-temperature indicator technologies, active, smart and intelligent packaging systems, use of nanotechnologies in different stages of post-harvest management, information technologies such as genomics and computer modelling etc. Policy development related to post-harvest management will play a huge role and these policies should aim to reinforce research and developments, training and investment. Efforts also to be made in theupgrading of local knowledge on food supply chains and food uses, strengthening supply chain integration, transferring superior technologies, building skills and capacity in the food supply chain, and improving small and medium-scale focused entrepreneurship opportunities.

Conclusion

In developing countries, the reduction of post-harvest food losses is important not only to avoid waste but also because of the cost associated with the production of additional food products to counter the food loss occurred. There is a clear need for a more holistic and integrated approach when dealing with postharvest losses. Postharvest innovations, coupled with sustainable development and strategies are required to prevent/reduce food loss to a greater extent. Along with the research & development processes, innovation techniques, the efforts of policymakers in the field of post-harvest management will influence food loss and food securityin the future. Knowledge of novel and modern technologies to mitigate the post-harvest losses and encourage value addition measures is necessary to assure food safety in the future. Community participation, the intervention of government and non-government organisations, policy reforms, skill development programs, sustainable food production, processing and export are essential steps to reduce food loss and ensure food safety.

References

Belkova, B., Hradecky, J., Hurkova, K., Forstova, V., Vaclavik, L., & Hajslova, J. (2018). Impact of vacuum frying on quality of potato crisps and frying oil. Food chemistry, 241, 51-59.

Bendinelli, W. E. et al. (2020) 'What are the main factors that determine post-harvest losses of grains', Sustainable Production and Consumption. Elsevier B.V., 21, pp. 228–238. doi: 10.1016/j.spc.2019.09.002.

Betoret, N., Puente, L., Dıaz, M. J., Pagan, M. J., Garcıa, M. J., Gras, M. L., & Fito, P. (2003). Development of probiotic-enriched dried fruits by vacuum impregnation. Journal of food Engineering, 56(2-3), 273-277.

Blasco, J., Aleixos, N., & Moltó, E. (2003). Machine vision system for automatic quality grading of fruit. Biosystems engineering, 85(4), 415-423.

Bollen, A. F. and Emond, J. P. (2014) 'Traceability in Postharvest Systems', in Postharvest Handling: A Systems Approach. Elsevier Inc., pp. 485–504. doi: 10.1016/B978-0-12-408137-6.00017-X.

Bollen, A. F., Riden, C. P. and Opara, L. U. (2006) 'Traceability in postharvest quality management', International Journal of Postharvest Technology and Innovation, 1(1), pp. 93–105. doi: 10.1504/IJPTI.2006.009186.

Bordoloi, R., & Ganguly, S. (2014). Extrusion technique in food processing and a review on its various technological parameters. Indian Journal of Scientific Research and Technology, 2(1), 1-3.

Boye, J. I. and Arcand, Y. (2013) 'Current Trends in Green Technologies in Food Production and Processing', Food Engineering Reviews. Springer, pp. 1–17. doi: 10.1007/s12393-012-9062-z.

CAC - Codex Alimentarius Comission (2003) 'General Principles of Food Hygiene. CAC/RCP 1-1969 Rev. 4', Codex Alimentarius Comission.

Chandrasekaran, M. and Ranganathan, R. (2017) 'Modelling and optimisation of Indian traditional agriculture supply chain to reduce post-harvest loss and CO_2 emission', Industrial Management and Data Systems. Emerald Group Publishing Ltd., 117(9), pp. 1817–1841. doi: 10.1108/IMDS-09-2016-0383.

Chen, Q., Anders, S., & An, H. (2013). Measuring consumer resistance to a new food technology: A choice experiment in meat packaging. Food Quality and Preference, 28(2), 419-428.

Chiralt, A., Fito, P., Barat, J. M., Andres, A., González-Martınez, C., Escriche, I., & Camacho, M. M. (2001). Use of vacuum impregnation in food salting process. Journal of Food Engineering, 49(2-3), 141-151.

Colak, N., & Hepbasli, A. (2009). A review of heat pump drying: Part 1–Systems, models and studies. Energy Conversion and Management, 50(9), 2180-2186.

Development Commissioner Ministry of Micro, Small & Medium Enterprises, (2020). Available at: http://www.dcmsme.gov.in/schemes/credit_link_scheme.htm, (Accessed: 28-8-2020)

Di Terlizzi, B. (2016). Enhancing Knowledge for Food Security. Mediterra 2016. Zero Waste in the Mediterranean. Natural Resources, Food and Knowledge/International Centre for Advanced Mediterranean Agronomic Studies (CIHEAM) and Food and Agriculture Organization of the United Nations (FAO)–Paris: Presses de Sciences Po, 2016., 363.

Du, C. J., & Sun, D. W. (2006). Learning techniques used in computer vision for food quality evaluation: a review. Journal of Food Engineering, 72(1), 39-55.

Duarte-Correa, Y., Díaz-Osorio, A., Osorio-Arias, J., Sobral, P. J., & Vega-Castro, O. (2020). Development of fortified low-fat potato chips through Vacuum Impregnation and Microwave Vacuum Drying. Innovative Food Science & Emerging Technologies, 64, 102437.

Dziedzic, E., Błaszczyk, J., Bieniasz, M., Dziadek, K., & Kopeć, A. (2020). Effect of modified (MAP) and controlled atmosphere (CA) storage on the quality and bioactive compounds of blue honeysuckle fruits (Lonicera caerulea L.). Scientia Horticulturae, 265, 109226.

Eisa, M. A. R. (1996). Applications of heat pumps in chemical processing. Energy conversion and management, 37(3), 369-377.

FAO (2011) Global food losses and food waste – Extent, causes and prevention. Rome. Available at: http://www.fao.org/3/a i2697e.pdf(Accessed: 31 August 2020).

FAO (2013), Food Wastage Footprint: Impacts on Natural Resources, Rome, FAO. Available at: http://www.fao.org/3/i3347e/i3347e.pdf (Accessed: 31 August 2020).

FAO (2017) Save food for a better climate converting the Food loss and waste challenge into climate action. Available at: www.fao.org/publications (Accessed: 31 August 2020).

FAO (2019) The State of Food and Agriculture 2019. Moving forward on food loss and waste reduction. Rome. Available at: http://www.fao.org/3/ca6030en/ca6030en.pdf(Accessed: 30 August 2020).

Fayose, F., & Huan, Z. (2016). Heat pump drying of fruits and vegetables: Principles and potentials for Sub-Saharan Africa. International Journal of Food Science, 2016.

Fischer, J. (2020) 'Green and/or brown: Food production and politics in India', Research in Globalization. Elsevier BV, 2, p. 100017. doi: 10.1016/j.resglo.2020.100017.

Fito, P., Chiralt, A., Barat, J. M., Andrés, A., Martínez-Monzó, J., & Martínez-Navarrete, N. (2001). Vacuum impregnation for development of new dehydrated products. Journal of Food Engineering, 49(4), 297-302.

Ganesh U. et al. (2019) Reducing Post-harvest Losses in India: Key Initiatives and Opportunities, Intellecap, The Rockefeller Foundation, USA. Available at: https://www.intellecap.com/wp-content/themes/intellecap/pdf/Public-Facing-Report.pdf(Accessed: 31 August 2020).

Garayo, J., & Moreira, R. G. (2002). Vacuum frying of potato chips. Journal Food Processing Engineering, 55(2), 181-191.

Gardas, B. B., Raut, R. D. and Narkhede, B. (2017) 'Modeling causal factors of post-harvesting losses in vegetable and fruit supply chain: An Indian perspective', Renewable and Sustainable Energy Reviews. Elsevier Ltd, pp. 1355–1371. doi: 10.1016/j.rser.2017.05.259. Gardas, B. B., Raut, R. D. and Narkhede, B. (2018) 'Evaluating critical causal factors for post-harvest losses (PHL) in the fruit and vegetables supply chain in India using the DEMATEL approach', Journal of Cleaner Production. Elsevier Ltd, 199, pp. 47–61. doi: 10.1016/j.jclepro.2018.07.153.

Garlov, R., Saveliev, V., Gavrylov, K., Golovin, L., & Pedolsky, H. (2002). U.S. Patent No. 6,345,509. Washington, DC: U.S. Patent and Trademark Office.

Golob, P., Farrell, G. and Orchard, J. E. (eds) (2002) Crop Post-Harvest: Science and Technology, Volume 1, Crop Post-Harvest: Science and Technology, Volume 1. Oxford, UK: Blackwell Science Ltd. doi: 10.1002/9780470751015.

Gorris, L. G. M., & Peppelenbos, H. W. (1992). Modified atmosphere and vacuum packaging to extend the shelf life of respiring food products. HortTechnology, 2(3), 303-309.

Goulas, A. E., & Kontominas, M. G. (2007). Effect of modified atmosphere packaging and vacuum packaging on the shelf-life of refrigerated chub mackerel (Scomber japonicus): biochemical and sensory attributes. European Food Research and Technology, 224(5), 545-553.

Gould, W. A. and Gould, R. W. (1991) Total Quality Assurance for the Food Industries, Total Quality Assurance for the Food Industries. doi: 10.1533/9781845696153.

Government of India (2018). Agriculture Census 2015-16. Phase 1. Provisional Results. All India Report on Number and Area of Operational Holdings. Agriculture Census Division. Department of Agriculture, Co-Operation & Farmers Welfare.

Granda, C., Moreira, R. G., & Tichy, S. E. (2004). Reduction of acrylamide formation in potato chips by low-temperature vacuum frying. Journal of Food Science, 69(8), 405-411.

Harb, J. Y. and Streif, J. (2004). Controlled Atmosphere Storage of Highbush Blueberries cv. 'Duke' European Journal of Horticultural Science, 69 (2). S. 66–72.

Hellwig, M. & Henle, T. (2014). Baking, ageing, diabetes: a short history of the Maillard reaction. Angewandte Chemie International Edition, 53, 10316–10329.

Hiç, C. et al. (2016) 'Food Surplus and Its Climate Burdens', Environmental Science and Technology. doi: 10.1021/acs.est.5b05088.

Hodges, R. J., Buzby, J. C. and Bennett, B. (2011) 'Postharvest losses and waste in developed and less developed countries: Opportunities to improve resource use', Journal of Agricultural Science. doi: 10.1017/S0021859610000936.

Initiative for Development of Entrepreneurs in Agriculture (IDEA) | North Eastern Development Finance Corporation Ltd (2016). Available at: https://www.nedfi.com/node/97 (Accessed: 28-8-2020)

Jha, A. (2001). Retort Processing for Shelf Life Extension of Foods. Advances In Preservation Of Dairy and Food Products, 74.

Jha, S. N. (2016) 'Food Safety and Quality', in Rapid Detection of Food Adulterants and Contaminants. Elsevier, pp. 1–24. doi: 10.1016/b978-0-12-420084-5.00001-9.

Jha, S.N. et al. (2015) Assessment of quantitative harvest and post-harvest losses of major crops and commodities in India. Ministry of Food Processing Industries (Govt. of India), pp.64-67.

Kashyap, D. and Agarwal, T. (2020a) 'Food loss in India: water footprint, land footprint and GHG emissions', Environment, Development and Sustainability. Springer, 22(4), pp. 2905–2918. doi: 10.1007/s10668-019-00325-4.

Khedkar, D., & Khedkar, R. (2020). New Innovations in Food Packaging in Food Industry. In Emerging Technologies in Food Science (pp. 165-185). Springer, Singapore.

Kiaya, V. (2014). Post-harvest losses and strategies to reduce them. Technical Paper on Postharvest Losses, Action Contre la Faim (ACF), 25.

Knickel, K. et al. (2008) 'Limiting and enabling factors of collective farmers' marketing initiatives: Results of a comparative analysis of the situation and trends in 10 European countries', Journal of Hunger and Environmental Nutrition. doi: 10.1080/19320240802244041.

Kucukvar, M. et al. (2019) 'Assessing regional and global environmental footprints and value added of the largest food producers in the world', Resources, Conservation and Recycling. Elsevier B.V., 144, pp. 187–197. doi: 10.1016/j.resconrec.2019.01.048.

Kumar, P., Joshi, P. K. and Mittal, S. (2016) 'Demand vs supply of food in India - Futuristic projection', in Proceedings of the Indian National Science Academy. doi: 10.16943/ptinsa/2016/48889.

Kummu, M. et al. (2012) 'Lost food, wasted resources: Global food supply chain losses and their impacts on freshwater, cropland, and fertiliser use', Science of the Total Environment. Elsevier, 438, pp. 477–489. doi: 10.1016/j.scitotenv.2012.08.092.

Langourieux, S., & Escher, F. E. (1998). Sulfurous Off Flavor Formation and Lipid Oxidation in Heat Sterilized Meat in Trays. Journal of food science, 63(4), 716-720.

Leonard, W., Zhang, P., Ying, D., & Fang, Z. (2020). Application of extrusion technology in plant food processing byproducts: An overview. Comprehensive Reviews in Food Science and Food Safety, 19(1), 218-246.

Lima, G.P.P., Vianello, F., Corrêa, C.R., da Silva Campos, R.A. & Borguini, M.G. (2014). Polyphenols in fruits and vegetables and its effect on human health. Food and Nutrition sciences, 5, 1065–1082.

Lipinsky, B. et al. (2013) 'Reducing food loss and waste. Working paper. Installment 2 of "Creating a sustainable food future".', World Resources Institute.

Lloyd, M. A., Hess, S. J., & Drake, M. A. (2009). Effect of nitrogen flushing and storage temperature on flavor and shelf-life of whole milk powder. Journal of Dairy Science, 92(6), 2409-2422.

Lundqvist, J., Fraiture, C. De and Molden, D. (2008) 'Saving Water: From Field to Fork Curbing Losses and Wastage in the Food Chain', SIWI Policy Brief.

Luning, P. A. et al. (2009) 'Systematic assessment of core assurance activities in a company specific food safety management system', Trends in Food Science & Technology. Elsevier, 20(6–7), pp. 300–312. doi: 10.1016/J.TIFS.2009.03.003.

Marasca, E., Greetham, D., Herring, S. D., & Fisk, I. D. (2016). Impact of nitrogen flushing and oil choice on the progression of lipid oxidation in unwashed fried sliced potato crisps. Food chemistry, 199, 81-86.

Mariscal, M., & Bouchon, P. (2008). Comparison between atmospheric and vacuum frying of apple slices. Food chemistry, 107(4), 1561-1569.

Mariscal, M., & Bouchon, P. (2008). Comparison between atmospheric and vacuum frying of apple slices. Food chemistry, 107(4), 1561-1569.

NABARD (2020), Available at : https://www.nabard.org/content.aspx?id=592(Accessed: 28-8-2020)

NABARD (2020), Available at:https://www.nabard.org/content1.aspx?id=591& catid=23& mid=530, (Accessed: 28-8-2020)

NABARD (2020). Available at: https://www.nabard.org/content.aspx?id=582, (Accessed: 28-8-2020)

NABARD (2020). Available at: https://www.nabard.org/content1.aspx?id=595& catid=23 & mid=530. (Accessed: 28-8-2020)

Ndraha, N. et al. (2020) 'Challenges with food waste management in the food cold chains', in Food Industry Wastes. Elsevier, pp. 467–483. doi: 10.1016/b978-0-12-817121-9.00022-x.

Orsi, L. et al. (2017) 'The role of collective action in leveraging farmers' performances: Lessons from sesame seed farmers' collaboration in eastern Chad', Journal of Rural Studies. Elsevier Ltd, 51, pp. 93–104. doi: 10.1016/j.jrurstud.2017.02.011.

Paik, J. S., Shint, J. I., Kimt, J. I., & Choit, P. K. (1994). Effect of nitrogen flushing on shelf-life of packaged potato chips. Packaging Technology and Science, 7(2), 81-85.

Pal, U. S., Das, M., Nayak, R. N., Sahoo, N. R., Panda, M. K., & Dash, S. K. (2019). Development and evaluation of retort pouch processed chhenapoda (cheese based baked sweet). Journal of Food Science and Technology, 56(1), 302-309.

Panghal, A. et al. (2018) 'Role of Food Safety Management Systems in safe food production: A review', Journal of Food Safety. John Wiley & Sons, Ltd (10.1111), 38(4), p. e12464. doi: 10.1111/jfs.12464.

Patel, K. K., & Kar, A. (2012). Heat pump assisted drying of agricultural produce—an overview. Journal of Food Science and Technology, 49(2), 142-160.

Porter, S. D. et al. (2016) 'A half-century of production-phase greenhouse gas emissions from food loss & waste in the global food supply chain', Science of the Total Environment. doi: 10.1016/j.scitotenv.2016.07.041.

Raut, R. D. et al. (2018) 'Modeling the drivers of post-harvest losses – MCDM approach', Computers and Electronics in Agriculture. Elsevier B.V., 154, pp. 426–433. doi: 10.1016/j.compag.2018.09.035.

Rawson, A., Patras, A., Tiwari, B., Noci, F., Koutchma, T. & Brunton, N. (2011). Effect of thermal and non thermal processing technologies on the bioactive content of exotic fruits and their products: Review of recent advances. Food Research International, 44, 1875–1887.

Riaz, M. N. (Ed.). (2000). Extruders in food applications. CRC press.

Rotaru, G. et al. (2005) 'Food Quality and Safety Management Systems: a Brief Analysis of the Individual and Integrated Approaches', Agroalimentary Processes and Technologies.

Ryan, J. M. (2014) Guide to Food Safety and Quality During Transportation: Controls, Standards and Practices, Guide to Food Safety and Quality During Transportation: Controls, Standards and Practices. doi: 10.1016/C2012-0-05957-1.

Saiz-Rubio, V., & Rovira-Más, F. (2020). From smart farming towards agriculture 5.0: a review on crop data management. Agronomy, 10(2), 207.

Saldaña, E., Siche, R., Luján, M., & Quevedo, R. (2013). Computer vision applied to the inspection and quality control of fruits and vegetables. Brazilian journal of food technology, 16(4), 254-272.

Seideman, S. C., & Durland, P. R. (1983). Vacuum packaging of fresh beef: A review. Journal of Food Quality, 6(1), 29-47.

Shankar, C. R., Gopal, T. S., & Vijayan, P. K. (2002). Studies on heat processing and storage of seer fish curry in retort pouches. Packaging Technology and Science: An International Journal, 15(1), 3-7.

Shiferaw, B., Hellin, J. and Muricho, G. (2011) 'Improving market access and agricultural productivity growth in Africa: What role for producer organizations and collective action institutions?', Food Security. Springer, 3(4), pp. 475–489. doi: 10.1007/s12571-011-0153-0.

Shyu, S., Hau, L., & Hwang, S. (1998). Effect of vacuum frying on the oxidative stability of oils. Journal of the American Oil Chemists' Society, 75, 1393-1398.

Skill development of rural entrepreneurs (2016) Government of odisha, agriculture department, Available at http://odihort.nic.in/sites/default/files/Skill%20Development%20under%20RKVY-2016-17.pdf (Accessed: 28-8-2020)

Smilor, R. W. (1987). Managing the incubator system: critical success factors to accelerate new company development. IEEE transactions on Engineering Management, (3), 146-155.

Strommen, I., Eikevik, T. M., Alves-Filho, O., Syverud, K., & Jonassen, O. (2002, August). Low temperature drying with heat pumps new generations of high quality dried products. In Proceeding in 13th International Drying Symposium (IDS2002) (pp. 27-30).

Sun, J., Peng, Z., Zhou, W., Fuh, J. Y., Hong, G. S., & Chiu, A. (2015). A review on 3D printing for customized food fabrication. Procedia Manufacturing, 1, 308-319.

Sun, Q., Sun, F., Zheng, D., Kong, B., & Liu, Q. (2019). Complex starter culture combined with vacuum packaging reduces biogenic amine formation and delays the quality deterioration of dry sausage during storage. Food Control, 100, 58-66.

Swanson, B. and Samy, M. (2002) 'Developing an Extension Partnership among Public, Private, and Nongovernmental Organizations', Journal of International Agricultural and Extension Education. Journal of International Agricultural and Extension Education, 9(1). doi: 10.5191/jiaee.2002.09101.

Tiwari, B. K., O'donnell, C. P., & Cullen, P. J. (2009). Effect of non thermal processing technologies on the anthocyanin content of fruit juices. Trends in Food Science & Technology, 20(3-4), 137-145.

Tran, Q. D., Hendriks, W. H., & van der Poel, A. F. (2008). Effects of extrusion processing on nutrients in dry pet food. Journal of the Science of Food and Agriculture, 88(9), 1487-1493.

Trebbin, A. (2014) 'Linking small farmers to modern retail through producer organizations - Experiences with producer companies in India', Food Policy. Pergamon, 45, pp. 35–44. doi: 10.1016/j.foodpol.2013.12.007.

Wang, N., Zhang, N., & Wang, M. (2006). Wireless sensors in agriculture and food industry—Recent development and future perspective. Computers and electronics in agriculture, 50(1), 1-14.

Wentholt, M.T.A. et al. (2009) 'The views of key stakeholders on an evolving food risk governance framework: Results from a Delphi study', Food Policy. Pergamon, 34(6), pp. 539–548. doi: 10.1016/J.FOODPOL.2009.06.002.

Wiggins, J., & Gibson, D. V. (2003). Overview of US incubators and the case of the Austin Technology Incubator. International Journal of Entrepreneurship and Innovation Management, 3(1-2), 56-66.

Williams, J. R., Black, J. R., & Steffe, J. F. (1981). A Economic Analysis of the Feasibility of the Retort Pouch for Packaging Fruit and Vegetable Commodities (No. 1091-2016-87400).

Xu, W. (2016) Functional nucleic acids detection in food safety: Theories and applications, Functional Nucleic Acids Detection in Food Safety: Theories and Applications. Springer. doi: 10.1007/978-981-10-1618-9.

Yahia, E. M. (2006). Modified and controlled atmospheres for tropical fruits. Stewart Postharvest Review, 5(6), 1-10.

Yang, F., Zhang, M., & Bhandari, B. (2017). Recent development in 3D food printing. Critical reviews in food science and nutrition, 57(14), 3145-3153.

Zagory, D., & Kader, A. A. (1988). Modified atmosphere packaging of fresh produce. Food Technol, 42(9), 70-77.

Zhang, Z. H., Wang, L. H., Zeng, X. A., Han, Z., & Brennan, C. S. (2019). Non-thermal technologies and its current and future application in the food industry: a review. International Journal of Food Science & Technology, 54(1), 1-13.

Theme 7: Research and Development in Allied Sectors

29

Research and Development, Innovations in Livestock and Fodder Sector

Sunil Kumar[1], *B.G .Shivakumar*[2], *S.K. Mahanta*[1], *Deepak Upadhyay*[1] *and P.K. Ghosh*[3]

[1]*ICAR-Indian Grassland and Fodder Research Institute, Jhansi, Uttar Pradesh*
[2]*ICAR-Indian Grassland and Fodder Research Institute*
Regional Research Station, Dharwad, Karnataka
[3]*ICAR-National Institute of Biotic Stress Management, Raipur, Chhattisgarh*

Introduction

Animal husbandry has been an integral component of Indian agriculture since time immemorial. It is still continuing as one of the major avocations in India contributing nearly 4.9% to the national GDP and 28.4% of the agricultural GDP. Livestock population in the country as per Livestock Census 2019 is 535.78 million which is an increase of 4.6% over Livestock Census 2012 (BAHS, 2019; DAHD, 2019). Cattle, buffaloes, sheep and goats are the major livestock in the country. Today, India is the global leader in milk and milk products with a production of 187.7 million tonnes (MT) and the availability of 394 ml milk per capita per day. The per capita availability of milk has increased substantially during last four decades. The anticipated milk demand is expected to be 266.5 MT by 2030 based on per capita consumption of 468 ml/day (Table 1). It has been made possible with the active involvement of farming community and by creating awareness about high yielding breeds of livestock, improved rearing methods and milking, management of pests and diseases and most importantly, the feeding of the animal. The crop residue constitutes a major chunk of fodder to the livestock in many parts of the country. However due to their low nutritive values and poor palatability, it is not helping to realize the full milking potential of animals. A huge opportunity exists in livestock sector as there is a growing demand for animal products particularly milk and meat (NDDB, 2019). In addition, for promoting organic farming particularly the need of farm yard manure has necessitated the farming community to keep livestock as a complementary component of his farming. It provides draught power and organic manure to crop sector and hides, skin, bones, blood and fibers to the industrial sector. Therefore, the importance of livestock goes beyond its food production function, and has become even more significant (Birthal et al., 2002).

Table 1. Estimated total and per capita consumption of milk and milk products in India by 2030

Sector	Total consumption (MT)	Per capita per day (ml)
Rural	152.2	404
Urban	114.4	592
Total	266.5	468

In spite of booming demand for the livestock products, the bourgeoning of fodder scarcity has a significant negative impact on the productivity and remunerativeness of the livestock keeping, pushing livestock keepers to greater economic stress. At present, the country faces a net deficit of 35.6% green fodder, 11.0% dry crop residues and 44.0% concentrate feed ingredients. The demand of green and dry fodder will reach to 1012 and 631 MT, respectively by the year 2050 (Table 2). At the current level of growth in forage resources, there will be 18.4% deficit in green fodder and 13.2% deficit in dry fodder by the year 2050. To meet out the deficit, green forage supply has to grow at 1.69% annually (IGFRI, 2013). Hence, there is an urgent need for overhauling the situation to make the livestock keeping more profitable through improved fodder production, value addition and conservation technologies and also bridging the gap between the availability and demand for fodder resources.

Table 2. Demand and supply estimates of dry and green fodder

Year	Demand		Supply		Deficit		Deficit	
	Dry	Green	Dry	Green	Dry	Green	Dry	Green
	MT						%	
2020	530.5	851.3	467.6	590.4	62.85	260.9	11.85	30.65
2030	568.1	911.6	500.0	687.4	68.07	224.2	11.98	24.59
2040	594.9	954.8	524.4	761.7	70.57	193.0	11.86	20.22
2050	631.0	1012.7	547.7	826.0	83.27	186.6	13.20	18.43

Source: Vision 2050, ICAR-Indian Grassland and Fodder Research Institute, Jhansi

Major Issues

The livestock sector is continuously expanding over the last few decades since independence in terms of number, although the rate of expansion is unstable. Likewise, the fodder production has not achieved the level of expectation to address the fodder demand of the livestock sector.

Issues pertaining to livestock

India is a global leader with regards to livestock diversity and number, but Indian breeds are less productive, although tolerant to many biotic and abiotic stresses.

Efforts have been made to improve the breeds through breeding programmes and genetic improvement, but the pace of change from indigenous to improved breeds has been very slow. The livestock health is another important issue confronting the country. Our animals are afflicted with a variety of pests and diseases making them less productive (FAO, 2005). The national vaccination programmes for diseases like black quarter (BQ), foot and mouth (FM), hemorrhagic septicemia (HS), anthrax, fascioliasis, brucellosis, tuberculosis, rabies etc. have although contained the problem yet the total control is far away affecting the animal health and leading to low productivity and high mortality rates in the livestock.

In India, the animal rearing is still mainly by open grazing. The lack of natural grazing lands, poor grass cover and overdependence on the crop residues has adverse effects on animal health. The decline in the soil fertility over the period has led to micronutrient deficiency in fodders in many areas. Thus feeding of area specific micronutrients to animals is also being talked over. The very high cost of traditional animal feeds like ground nut cake, cotton seeds, de-oiled cakes, grain mixtures have pushed farmers to discontinue the practice of feeding these items to the animals. The livestock keepers still follow the age-old practices of animal keeping and dairying. These are more manual labour consuming, prone to contamination and less remunerative. There is a need for small machineries for cleaning, milking, and processing at livestock keepers' level.

Issues pertaining to fodder production

Residues of food crops constitute a major source of fodder to the livestock in many parts of the country. Crop residues are low in nutrients, less palatable due to over maturity and contain high crude fiber. These are not good source of food for the animals. Agriculture continues to be a major enterprise and livestock keeping is only subsidiary to it, and often gets neglected.

In addition, field crops with food and commercial value are preferred over fodder crops under assured irrigation. Only 3% of the area continues under the fodder crops since long and is mostly under rainfed conditions with scant care for plant nutrients, irrigation, plant protection etc., and do not meet the green fodder requirement of the livestock. Often animals tend to rely on dry fodder and concentrate feeds for their nutrition. Although there has been marginal improvement in the fodder production, it is still at miniscule and is confined only to intensive dairy farming areas. The cultivation of fodder crops is largely confined to traditional annual fodder crops *viz.* jowar, maize, *bajra*, oats, cowpea, berseem etc. The cultivation of high yielding perennial fodder crops *viz. bajara napier* hybrid, guinea grass and lucerne is still very less. . There is a need for educating the farming community about the high yielding fodder crops for both irrigated and rainfed areas and round the year fodder production. This will ensure the availability of green fodder to the livestock every day.

A large proportion of livestock keepers are small land holders or landless farmers. The available land is either too small or no land is available for fodder cultivation. They are dependent on others for fodder. With disappearing of community lands and no entry into forest areas or land holders' fields for fodder, they often face difficulty in meeting the fodder demands of livestock. This calls for establishment of regular fodder markets. Besides, farmers who grow fodder in excess of their own requirement will also get an opportunity to sell the excess fodder to the needy livestock keepers. The excess green and dry fodders can also be safely conserved and value added in the form of bales, silage and hay making without losing the nutritive value and palatability. The silage and hay can substitute green fodder during lean seasons like summer and winters, when the availability of fresh fodder declines. Hence, there is a need for both value addition and conservation to mitigate the fodder related issues.

Innovations

In view of the low livestock productivity and fodder availability, a number of new innovations and technologies have been tested, and found to be better in addressing the issues of livestock and fodder sector. Few of them have already become popular and others need to be popularized.

A. Livestock based innovations

Cross breeds of livestock

The native breeds of cattle, buffaloes, sheep and goats are hardy, can tolerate the harsh climatic conditions, resistant to many pests and diseases and yields even with poor feeding and management. However, this is no longer attractive as the cost of livestock keeping is escalating while the returns are not attractive. Hence breed improvement through crossing native breeds with the exotic ones is becoming the need of the hour to improve the productivity and monetary returns from livestock keeping (DAHD, 2013). A major systematic effort in this direction is required if an all-round sustainable genetic improvement is sought in the country.

Need based stocking

Forage production and stocking rate records are critical in making timely management decisions. No other single management practice affects profitability of livestock more than stocking rate. In Indian conditions, keeping all sorts of animals under one roof is very common norm. This often leads to less productivity due to even spread of resources and management practices to both productive and non-productive animals within a category or across types of animals. As the productivity of animals depends on the climatic conditions, the most suitable animals prevailed in a given area should be given priority over others to enhance the productivity and monetary returns.

Stall feeding

Techniques like stall feeding after fodder chopping and bypass protein use, have long been in practice in many parts of the country. However, main constraint to large-scale adoption of nutrition technologies in general has been the lack of information to users. The stall feeding reduces the losses of fodder resources and provides scope for individual animal-based feeding leading to more judicious use of fodder resources and avoid the exposure of animal to spread of communicable diseases. It also cuts down the energy of the animal in the process of open-grazing and the same may be translated into productivity. Nowadays many young entrepreneurs are setting up cattle, buffalo and goat/sheep units using scientific rearing practices with stall feeding under semi-intensive and intensive management. However, high demand of milk and milk products and present insufficiency in meat production and growing demand for animal protein open up avenues for the progressive/ commercial farmers and small-scale entrepreneurs to take up 'Commercial Dairy, Sheep/Goat Farming' as a profession and profitable revenue.

B. Fodder based innovations to augment fodder availability

Collection and conservation of crop residues

Proper care for quality crop residues and their collection and conservation will ensure the dry fodder availability to the livestock. The burning and destroying of paddy straw in north India besides causing environmental hazard also adversely affect the fodder availability to the animals. The tops of sugarcane, haulms of potato, and stubbles of pulses could also be collected and conserved as fodder for feeding animals as dry fodder. It gives insurance against the shortage of fodder for the livestock particularly during lean periods.

Popularization of agro-climate based fodder crops

A large of number of fodder crops and varieties have been developed by ICAR-IGFRI, Jhansi and many SAUs over the years. However, the information about suitable crops/varieties for a given piece of land under a particular climate is limiting. Thus farmers are not using suitable fodder crops and varieties according to the climatic conditions prevailing, soil types, cropping system etc. The transfer of suitable package of practices for production of fodder crops in a given area will greatly help the farmers and also help to increase the productivity and monetary returns from the fodder cultivation. The programme, NIFTD (National Initiative on Fodder Technology Demonstration) was launched through network of 100 KVKs, where promotion of appropriate varieties with suitable package of practices was successfully made in different farming situations. This innovative model needs to be replicated in entire country through all 716 KVKs.

Year-round fodder production

ICAR-IGFRI, Jhansi has perfected the techniques for year around fodder production in different agro-climatic conditions. These need to be transferred to the needy farmers and made them adopt for meeting the fodder requirement for their livestock. For example, the system comprises of raising seasonal fodder crops, inter-planted with perennial grasses (hybrid *napier*/guinea grass). Hybrid *napier* based cropping system (hybrid *napier* + (cowpea - berseem + mustard) has green fodder production potential of 273 t/ha and dry fodder potential of 44.3 t/ha per year under assured water supply (water requirement 1090 mm). The system is suitable for large scale dairy farmers in peri-urban and Milk shed areas of whole India except tropical region. Technology can supply round the year good quality fodder (cereal:legume, 67:33) which can sustain 8-10 Adult Cattle Uunit per ha (1 ACU= 350 kg bodyweight). The perennial crops like BN hybrid, guinea grass or *Setaria* based system may be adopted in different regions with seasonal crops like cowpea, horse gram, ricebean berseem and lucerne. This technology has also great potential for commercialization and adoption with market led dairy development in different niches.

Utilization of orchards for fodder production

There are various alternate land use (ALU) systems which provide fodder such as silvi-pasture, horti-pasture and agri-horti-silvipasture. There is ample scope and many opportunities for introducing fodder crops in existing orchards. Horti-pasture system integrates pasture (grass and/or legumes) and fruit trees to fulfill the gap between demand and supply of fruit, fodder and fuel wood through utilizing moderately degraded land. *Ber*, *aonla*, guava, *beal*, moringa, mulberry etc. based horti-pasture systems have shown potential for enhancing forage productivity . Perennial grasses *viz,* NB hybrid, TSH, *Setaria*, guinea grass and other temperate grasses are the potential fodder types, which could be integrated in existing orchards. According to an estimate, approximately 19.0 and 17.9 MT of green fodder could be obtained from mango and coconut orchards of the country (30% area of the total orchard areas). From other orchards (guava, sapota, litchi and apple) contribution could be 4.0 million tonnes of green fodder. Such model can be implemented through convergence of different government schemes like National Horticulture Mission, Coconut Development Board and National Mission on Oil seed and Oil Palm.

Fodder from non-competitive lands

In this system, area specific forage grasses are planted on bunds, terrace risers, steep uncultivable slopes and around bamboo grooves etc. for dairy and goat farmers. The grasses grown on bunds do not require extra input like fertilizer and irrigation, besides, it helps in conserving natural resources. All cultivable area, along the irrigation channels, farm boundaries, backyard, threshing area, animal complex, community land, etc. can be utilized for this purpose. Among different perennial cultivated grasses, *napier bajra* hybrid is suitable for bunds of irrigated areas and tri-specific hybrid (TSH), guinea grass, anjan grass and nandi grass are most suitable under rainfed conditions. Farmers with marginal and small land holdings and livestock keepers with limited forage requirement can effectively grow forages on bunds. Forage on bunds/farm boundaries can produce 7 to 11 q green fodder per 100 m bund length. Moreover, it also works as a guard crop for the main crop, reduces runoff of water and controls soil erosion. As per projection, if 10 % farm bunds are utilized in India, additionally 17.8 MT of green fodders could be produced annually. This can go long way to solve the problem of fodder deficits in different regions during lean period.

Exploitation of non-conventional fodder resources

Non-conventional fodder sources like sugar beet, *lathyrus*, spineless cactus and moringa could be an important fodder resource to all categories of livestock. These sources supplement good quality fodder, energy and nutrients in lean period while grown in marginal or degraded or less competitive lands. Fodder beets contain about 16-22% dry matter with 85% digestibility in ruminant animals. It also contains 7-8% crude protein on dry matter basis. Fodder beet can be cultivated in many parts of the country except high hills with a duration of 140-150 days. Spineless cactus has become an important alternate fodder crop for degraded lands and drier regions. The crude protein content in cactus is 5-10%, and cattle and goats can consume 50 to 70 kg and 6 to 8 kg of fresh cladodes per day, respectively. However, in any case sole feeding of cactus cladodes is not advisable and normally recommended for its feeding up to 35% in mixed rations with dry fodder. *Lathyrus* (grass pea) as a forage (protein: 14-15%, Low ODAP: 0.07 to 0.20 %) has a unique potential to be grown in rice fallows as well as in marginal soils of semi-arid regions on residual moisture.

Hydroponics

Hydroponic fodder production is a boon for the landless, urban livestock keepers to meet the requirement through rapid green fodder production through this method. The hydroponic fodder has good palatability and intake of fresh hydroponics maize fodder by dairy cows may be up to 25 kg/animal/day along with limited concentrate mixture and *jowar* straw. The feedback from the coastal region farmers of Goa and Maharashtra states revealed that milk yield could be enhanced up to 2.5 litre/animal/day. Although this is high initial investment demanding, recent methods using locally available inputs can cut the cost substantially and produce the quality fodder. The cost of production of the hydroponics fodder was about Rs. 2.0 to 3.0/kg fresh fodder when seeds were home-grown. However, when seeds were purchased from the market, the cost of production could be Rs. 3.0 to 3.5 per kg. This is more suitable for farmers maintaining fewer animals with own manual labour and home-grown grains (Nayak *et. al.*, 2013). Production of hydroponics fodder in low cost greenhouses is also an effective solution for fodder scarcity and a very promising technology for sustainable livestock production in different regions of India particularly in deserts, snowfall areas, undulated terrains etc.

Bale and feed block making

Dry fodder is bulky in nature. It occupies more space and pose transport problem from one place to another in case of scarcity. Often the cost of transport becomes more than the fodder itself. To overcome such problems, it is advised to go for bale or block making to reduce the volume and space to store them. Maize, rice straw, cowpea, pea, sorghum, groundnut leaves etc. are of more than 6" size suitable for baling. The volume is reduced up to 20% of the original, making it easy for handling and transportation. This is expected to be very useful for livestock keepers who have limited space. It may also reduce the incidence of pest and disease infestation under storage.

Azolla production

Azolla is an emerging source of green fodder in many areas. On a dry weight basis, azolla has 25-35% protein, 10-15% mineral content, and 7-10% comprising a combination of amino acids, bio-active substances and biopolymers. During lean/drought period it provides sufficient quantity of nutrients and acts as a feed resource. Due to ease of its cultivation, many livestock keepers have started azolla production. It can successfully be grown round the year in southern states like Karnataka, Kerala, Goa, Tamil Nadu and Maharashtra and during monsoon and summer months in states of Punjab, Haryana, Delhi, U.P., Bihar, M.P., Rajasthan, Gujarat, and Chhattisgarh. In hilly states, where winter is pronounced, azolla can be grown in monsoon and summer months. Azolla doubles its biomass in 3 to 10 days, depending on conditions and it can yield up to 37.8 t fresh weight/ha (2.78 t dry matter/ha). This is a cost effective and quality fodder source, provided livestock keeper involves himself and family members in its production. It requires less space and may be undertaken in the kitchen garden, threshing yard, animal shed with minimum additional expenditure.

Leaf meals

Leaf meal made from leaves of legume crops/trees/shrubs are rich in protein essential amino acids, carotene and minerals and could act as a replacer of feed concentrate for livestock to save the valuable feed grain and for providing nutritious diet to the livestock throughout the year. *Leucaena, Acacia, Ficus* sp., Neem, *Grewia* sp., *Celtis* sp., *Sesbania, Prosopis cineraria, Ziziphus* sp., *Flacourtia indica* are used traditionally as fodder and are rich in crude protein (12-26%). Crops like *Stylosanthes* (14-16% CP), Lucerne (18-20% CP), *Sesbania sesban* (18-25% CP) are suitable for leaf meal preparation. They can also be crushed and made blocks using binding agents. All these leaves are rich in nutrients and minerals, and form well nutritious feed supplements to the animals particularly milking and small ruminants, thereby contributing to increase the yield and productivity of the animals.

Fodder banks

The term 'Fodder bank' is becoming very popular in government programmes to meet the fodder shortages owing to failure of monsoon and crop-based fodder availability. It involves purchase of fodder from farmers/growers, storing and distribution to needy farmers/animals in areas of scarcity to meet the fodder demand temporarily. Many state governments have effectively used this concept to address the fodder problem by establishing 'Community Fodder Banks'. Although this is only temporary, this will be useful for meeting the fodder demand in a given situation and prevents starving of livestock for want of fodder.

Forages from forests

The grassland area in the country in terms of permanent pastures covers about 3.7 % of land. The pasture resource in India is spread over an area of 12.04 million ha (mha). Other grazing lands are available under tree crops and groves (3.70 mha), on wastelands (1.50 mha) and on fallow lands (2.33 mha). Forests occupy vast area in many states. The natural grasses grown in the vicinity of the forest areas also form grazing resources in many places for the livestock. However, owing to infestation of several obnoxious plants viz. *Eupatorium, Stroblianthes, Lantana* etc. the natural grass covers is declining. This has led to the loss of habitat for the herbivores in and adjacent to the forest areas and the decline in their population. Hence, there is a need to protect, restore and conserve our grasslands ecosystems in order to sustain livestock as well as wildlife in a sustainable manner.

C. Participatory fodder seed production

The major hindrance for fodder production in many areas is non-availability of quality fodder seeds and planting materials. The scarcity of quality fodder seeds is mainly due to lesser interest in seed production owing to lack of knowledge and unorganized market for fodder seeds and planting materials. If appropriate platform is created and educated, the farmers are ready to take up fodder seed production. However, as there is an uncertainty in the fodder seed marketing, the farmers need assurance of price and market for the fodder seeds produced. This can be ensured with the participatory fodder seed production on buy-back arrangement under pre-agreed price. This will address the fodder seed availability, and will also ensure quality seeds and alternative avenues of earning revenue for the farming community. This has been successfully demonstrated at Dharwad for *Stylosanthes* seed production. Similarly, berseem and oats fodder seeds were produced in which farmers were provided technical knowhow. The success of the model has attracted several farmers to go for fodder seed production and earn higher returns as there is market for the sale of green fodder. The quality seed so produced is sold to many livestock keepers for fodder production. This can be made a self-sustaining mechanism through bringing the fodder seed producers and buyers face to face and creating a win-win situation for both stakeholders. It is also observed that the farmers are often forced to grow other field crops and use it as fodder crops in case of exigencies.

D. Use of digital science

The virtual world of digital science can be gainfully used in livestock and fodder production. Both livestock and fodder production are biological systems, have limited shelf-life, and perishable quotient attached to them. Often livestock keeping and fodder production are happening at different geographical location and they need to be integrated for synergy and complementarities. The digital science plays an important role in linking the clientele for both livestock keepers for their inputs like feeds, medicines and medical care, machineries and market. There are portals, apps, knowledge networks, professional groups, social media and many more. Farmers and livestock keepers can access information on issues around the world at one click (of mouse, and one touch in case of touch-screen gadgets). The stakeholders include the public and private agencies, civil society organizations as well as small groups of individuals and individual farmers. About 45% of the world's ICT projects are implemented in India. Asia's largest number of information kiosks is also implemented across rural India. However, most of the rural ICT projects are implemented in the socio-economically developed states of south and north India. Making available information in internet more usable remains a challenge till today and working on these lines is the need of the hour. Since the digital network provide real time information, both livestock keepers and fodder producers can be benefited to make well informed decision regarding their activities.

Many big dairy owners will have a vast network of livestock keepers for procurement of milk and also buyers of milk so as to make the dairying enterprise a profitable one. The livestock keepers too can be in touch with the feed manufacturers, fodder producers cum sellers, pharmaceutical companies for bulk purchase of medicines, mineral mixtures, machineries such automated milking machines, chillers, transportation cans etc. They can also understand the routine of milk collection pattern by the dairy owners, demand in the open markets etc. This information will help them to plan better. The digital information on the milk yield of animals, heat pattern, medical history and forewarning system will keep them aware of the steps to be taken by them. ICT is expected to be a game changer in animal husbandry and dairy sector. The Department of Animal Husbandry and Dairying is bringing in technological innovations like using genomics which will be helpful in increasing productivity through breed improvement and addressing other related issues.

E. Use of Apps in technology up-scaling

A large number applications (acronym: *apps*) are available in the play stores for the benefit of livestock keepers as well as fodder producers. These apps are giving information on livestock breeds, advance dairy practices, physiology of animals feeding for profit, and information on local dairies, local livestock market feeds and feed markets etc. Likewise, apps are providing information on fodder crops cultivation, availability of fodder seeds, fodder market, value addition in fodder, fodder conservation etc. to make the needy farmers fully equipped with technology and make use of this for the scientific and cost effective methods of production for higher profitability. The ICAR organizations like IGFRI, Jhansi (Chara App, Forage India, Fodder and Range grasses App); NDRI, Karnal (TEAM-CD); NIANP, Bengaluru (Feed Chart); IVRI, Izatnagar (Animal reproduction); etc. have developed several apps which are quite useful to livestock keepers and fodder producers.

F. Public-private-partnership (PPP) for rejuvenation of community grassland

India has vast tracts of grazing land, most of which has fragmented or degraded due to lack of appropriate policy interventions and management inputs. The natural grasslands, which were once the major source of fodder and grazing resources in India, have gradually deteriorated in their quality and use for other activities, thereby minimizing their role and importance in animal husbandry. The active growth of grasses in grazing land occurs only during monsoon season. The grazing intensity in the country is as high as 12.6 adult cow unit (ACU)/ha compared with 0.8 ACU/ha in developed countries. Removal of unwanted vegetation like obnoxious weeds *viz., Eupatorium, Parthenium, Typha, Datura, Lantana* etc. reducing the grazing pressure, cooperative sharing of the fodder etc. will ensure availability of palatable fodder for the village livestock and meet the fodder requirement of landless animal owners. The PPP model of Common Property Resources (CPR) management is envisaged to put in place community based mechanisms aimed at sustainable management of grazing lands, especially around the forest fringe villages. The scheme also aims at arresting the degradation and fragmentation of grazing lands and reversing the degradation process through active management interventions towards developing fodder species on these lands. Fortifying PPP through various groups, individuals and organizations will result in enhancing productivity and create new cooperation opportunities as well as meaningful interaction in animal husbandry sector also.

G. Involvement of FPOs, NGOs, and their linkages

It is difficult for individuals to cope with modern developments in the field of livestock keeping and fodder production. The new technologies are cost intensive and out of individuals reach owing to initial heavy investment. In additions, individuals will have difficulty in accessing modern dynamic technologies and still be profitable enterprise. Hence the involvement of Farmers' Produce Organizations (FPOs), Non-Governmental Agencies (NGOs) and their linkages with vast clientele groups give access to bigger and wider markets to tide over the difficulties faced by the individuals. Several farmers' groups, co-operatives and NGOs have successfully established and are in many parts of the country. Some successful FPOs, NGOs etc. involved in livestock and fodder production include Anand Milk Union Limited (AMUL), Anand, Gujarat, Karnataka Milk Federation (KMF), Karnataka. FPOs were formed under various initiatives of the Government of India, State Governments, NABARD and other organizations over the last 8-10 years, presently numbering around 5000 FPOs (including FPC's). Currently there are 853 SFAC promoted FPOs and these have been registered throughout the country till 30th June, 2020. Apart from these, 339 numbers of non-SFAC promoted FPOs have also been registered in India. In Union Budget 2019-20 a plan has been chalked out to form 10,000 FPOs over next five years which eventually will help livestock and fodder sector.

H. Capacity building

The vast network of NARS (ICAR institutes, KVKs, SAUs) are engaged in capacity building of different stakeholders through their outreach programmes like institutional extension mechanisms, FLDs, FARMERS FIRST, SC-SP programme and Tribal sub plan. Similarly, all the states, including NER states may avail the benefits of the multiple components and the flexibility of choosing them under NLM for a sustainable livestock development. The emergence of new technologies and practices require linkages between stakeholders and this sub-mission enabling a wider outreach to the farmers. To strengthen the capacity building the government of India announced Rs 1 lakh crore Agri-Infrastructure Fund for farm-gate infrastructures for farmers. Furthermore, Rs 10,000 crores scheme for Formalization of Micro Food Enterprises (MFE), Rs 20,000 crores for fishermen through Pradhan Mantri Matsya Sampada Yojana (PMMSY) and Rs. 15,000 crores for Animal Husbandry Infrastructure Development have been announced to make India the self-reliant (GOI, 2020). Promotion of herbal cultivation and beekeeping initiatives have also been considered for capacity building.

I. Women empowerment

Women are deeply engaged in livestock rearing and allied activities, but the contribution in terms of magnitude is difficult to realize because of a high degree of variation across the country. In India, women contribute 32% of their time to agricultural activities and constitute about 69% of workforce engaged in livestock sector (Singh and Sengupta, 2009). There is a need for empowering women in policy issues, planning, executing and monitoring at different stages of the purchase of the livestock and their management. This will enable them to increase the productivity of animals as well as system productivity of the farm. Women's names should be recorded as cultivators in revenue records, for family farms where women operate land that is registered under male ownership. Women societies, cooperatives, SHGs and organizations should be initiated which will help in organizing capacity building program to improve their contributions in livestock sector economically. To improve their decision making and women empowerment, the women education, livestock/land ownership, credit facilities, women's participation in social gatherings like 'Kisan Gosthi', 'village chaupal' etc. need to be encouraged. Trainings must be arranged especially for women at community level after assessing their specific training needs through female extension workers extensively and their impact analysis must be done. Government interventions through schemes such as 'Mahila e-Haat' and Support to Training & Employment Programme for Women (STEP) are creating employment opportunities and enhancing the competency of women.

Policy Reforms

To make livestock and fodder production enterprises more profitable, sustainable and attractive to the practitioners, Followings need due attention:

- There is a need to optimize the number of livestock in the country with the understanding that whether sector has to grow based on resource-driven approach (feed, fodder, breeding and health services) or supply-driven approach (number-driven growth).
- Generating authentic statistical data base on feed & fodder resources, economics of livestock production system and other emerging trends of market, finance & trade needs etc. for a robust and viable livestock sector, supporting agrarian economy.
- Initiation of a national programme in mission mode approach for accelerating production and effective utilization of fodder resources across the regions in the country.
- Formulation of national grassland and grazing policy and rejuvenation of degraded pastures.
- Formulation of a land use policy for protecting sustainable area of different crops and to preserve the freed non-arable land for intensive livestock management.
- Promotion of entrepreneurship and exploiting opportunities in commercial venture.

- Mechanism/legislation for protecting the interests of large number of smallholder population while facilitating in scaling up commercial dairy production system.
- Strengthening of fodder bank, marketing structure and its distribution mechanism in the country.
- Convergence of livestock sector programmes with other mission schemes like horticulture mission, MNREGA and regulation bill on animal feed.
- More investment in livestock sector particularly on feed and fodder development as well as their marketing.
- Service sector needs to be modernized and there should be an organized fodder production dedicated to dairy (diverting land from other crops).
- Orientation and support to various stakeholders, government, NGOs, research organizations towards Media/HRD/ICTs for extension services.
- Introduction of fodder security bill and policy on fodder market should get top priority.
- Policy promoting women empowerment and successful models of women–based livestock management can be replicated in India as well.

Conclusion

There is an urgent need to infuse new ideas and innovations for rejuvenating the livestock and fodder sector. This sector is still dominated by age-old practices and is oblivious to the modern practices in vogue elsewhere in the world. Indeed, a large number of new and potential innovations and practices have been developed by Indian researchers but due to lack of adequate publicity and awareness, they have not resulted in quick adoption and progress. Of late there is renewed interest in making livestock and fodder sector more developmental oriented and forward looking. The efforts of both public (Government) and private sectors have unveiled a bouquet of options available to these sectors.

For addressing the issue of livestock development along with feed and fodder issues, single sectoral approach will not work, rather there is a need to develop comprehensive approach of action. Self-reliant India- the dream in livestock and fodder sector is expected to get boost through-

- Modern breeding, feeding and health approaches to all categories of livestock
- Development of feed and fodder database involving various working groups
- Area-specific– ecoregional development approach of livestock and fodder sector
- Focus on CPRs rejuvenation (particularly through silvi-pastures)
- Tapping export potential of Indian livestock products

- Entrepreneurship development, commercialization and value addition to farmers groups and other private stakeholders
- Application of ICTs and modern tools for extension services including vaccination
- Addressing the issue of insurance of milch animals
- Harnessing the benefits from National Livestock Mission (having feed & fodder as sub-plan)

The recent initiatives of government in the livestock and fodder sectors by creating necessary infrastructural support, technical backstopping, financial support etc. through policies and programmes have made these sectors more vibrant and viable in terms of productivity and profitability. The private sectors both cooperative and limited companies have also contributed in popularizing the new innovations and kick starting the livestock and fodder sectors. With the availability of modern innovations, efforts to make them more affordable and providing the necessary financial backup, both livestock and fodder sectors will scale new heights in days to come.

References

BAHS. 2019. Basic Animal Husbandry Statistics. 2019. Department of Animal Husbandry and Dairying, Ministry of Fisheries, Animal Husbandry and Dairying. GOI,New Delhi.

Birthal, P.S. 2002. Technological change in India's livestock sub-sector: evidence and issues. In: P.S. Birthal and P. Parthasarathy Rao (eds). Technology options for sustainable livestock production in India. National Centre for Agricultural Economics and Policy Research, New Delhi, http://www.icrisat.org/Text/pubs/digital_pubs/J144_2002.pdf.

DAHD. 2013. National Livestock Policy-2013. Department of Animal Husbandry and Dairying, Ministry of Agriculture, GOI, New Delhi.

DAHD. 2019. Basic Animal Husbandry Statistics-2019. Department of Animal Husbandry and Dairying, Ministry of Fisheries, Animal Husbandry and Dairying, GOI, New Delhi.

FAO. 2005. Livestock Sector Brief-India. Food and Agriculture Organization of the United Nations. Rome, Italy.

GOI. 2020. Cabinet approves establishment of animal husbandry infrastructure development fund (press release). Cabinet Committee on Economic Affairs (CCEA), GOI, New Delhi.

IGFRI. 2013. Vision Document 2050. Indian Grassland and Fodder Research Institute, Jhansi, India.

Nayak, P.K., S.P. Gaikwad, M.J. Gupta, R.B. Dhuri, G.M. Dhumal and N.P. Singh. 2013. Success story, low cost devices for hydroponics fodder production. Indian Dairyman. pp. 68-72.

NDDB.2019. Study on demand and milk and milk products in India. Final Report. National Dairy Development Board, Anand, Gujarat.

Singh, R. and R. Sengupta. 2009. The EU FTA in agriculture and likely impact on Indian women. Consortium for Trade and Development (Centad), Heinrich Boll Foundation, India.

30

New Developments in Fisheries Sector

Gopal Krishna

ICAR-Central Institute of Fisheries Education, Yari Road, Versova Andheri West, Mumbai, Maharashtra

Introduction

The Indian fisheries sector is set in a unique and diverse set of resources ranging from the pristine waters of the Himalayas to the sprawling Indian Ocean. The fisheries biodiversity of the country encompasses a wide spectrum of physical and biological components that support the livelihoods of millions of people. Fisheries resources are set in different ecosystems. With growing population and the increasing demand for fish protein, the need for sustainable development of aquatic resources is now felt much more than ever before.

Fisheries and aquaculture remain important sources of food, nutrition, income and livelihoods for hundreds of millions of people around the world. Global capture and culture fish production during 2019 were 97 and 82 million tonnes, respectively with India being second largest producer of fish in the world after China. From 0.75 million tonnes in 1950-51, total fish production of India touched 12.59 million tonnes in 2018-19 with a contribution of 8.90 million tonnes (70.60%) from inland sector. The sunrise fisheries sector contributes about 0.96 % to National Gross Value Added (GVA) and 5.37% to agricultural GVA (2018-19) while engaging over 14.5 million people (2003 census) at the primary level and almost twice the number along the value chain. During 2018-19 the export of marine products that includes production from aquaculture reached 13,77,244 tonnes valued at INR 46589 crore (US $ 6.93 billion). The inland sector grew at a compounded annual growth rate (CAGR) of nearly 6.5% between 1979 and 2018, while overall fish production witnessed 4.5% growth, which is double the growth rate of food grains production in India.

The total fisheries potential of India has been estimated at 22.31 million metric tons (in 2018), of this, the marine fisheries potential stands at an estimated 5.31 million metric tons and the inland fisheries potential has been estimated at 17 million metric tons. In the recent years, the fish production in India has registered an average annual growth rate of more than 7%. The fish production in the country has shown continuous and sustained increments since independence. The total fish production in the country rose from 0.752 million metric tons in 1950-51 to 13.42 million 3 metric tons (provisional) during 2018-19. Of this, the marine fisheries contributed 3.71 million metric tons and the inland fisheries contributed 9.71 million metric tons. During 2018-19, 71% of marine fisheries potential has been harnessed and the inland fisheries potential harnessed during the same period stands at 58%.

Trends show that over the years there has been a significant shift in terms of increasing contribution from inland fisheries sector propelled by freshwater aquaculture and fluctuations in brackish water aquaculture. Aquaculture production has increased to 6.2 million metric tons in 2017-18 from 1.9 million metric tons in 2000-01. About 88% of the farmed fish comes from freshwater aquaculture. In spite of its rapid growth, aquaculture today is beset with several issues impacting production and productivity such as disease, absence of species diversification and genetic improvement, poor brood and seed, high input cost, lack of access to institutional credit, credit guarantee and insurance, environmental sustainability etc.

Major Issues

Marine Sector

The marine fisheries sector is dominated by the socio-economically backward artisanal and small scale fishers whose lives are closely intertwined with the oceans and seas. However, 75 per cent of the total marine fish production comes from the mechanized sector, 23 per cent from the motorized sector and only 2 per cent from the artisanal sector. In spite of stagnation in the growth of marine capture fish production over the years, the dependency of traditional marine fishers on marine capture fisheries for livelihoods has been increasing and it is only imperative that sustainable alternative livelihood opportunities like mariculture activities are developed and promoted. Major constraints impacting the growth of marine capture fisheries include limited scope for expansion due to overcapacities in territorial waters, weak regulation, inefficient management and prevalence of traditional fishing practices. Inadequate infrastructure especially fishing harbours, landing centers, cold chain and distribution systems, poor processing and value addition, wastage, traceability and certification, non-availability of skilled manpower, etc. are some of the other factors constraining the growth of the capture fisheries.

Inland Sector

Although inland fisheries and aquaculture have grown in absolute terms, the development in terms of its potential is yet to be realised. The vast and varied resources, that are unutilised and underutilised, in the form of 191,024 km of rivers and canals, 1.2 million ha of floodplain lakes, 2.36 million ha of ponds and tanks, 3.54 million ha of reservoirs and 1.24 million ha of brackish water resources offer great opportunities for livelihood development. Overall, the public and private investments have been found to be minimal for developing it as an instrument of growth. As the sector is extremely diverse and quite dynamic, there is a need to streamline policies and programs to take R&D benefits to the farmers and fishers for orderly and sustainable resource utilisation

In inland capture fisheries, seasonal nature of fishing operations, depleted stocks in natural waters, issues related with tenure and lease rights, use of obsolete technology for harvesting coupled with low capital infusion are some of the significant limiting factors. Specific problems negating the growth of culture fisheries include poor physical condition of resources (specially the water quality and quantity) low input culture system, lack of diversity in culture practices and species, lower productivity, inadequate regulatory mechanism, lack of cohesion among various institutions, increased incidents of disease, low levels of investment, inadequate access to institutional credit and high cost of credit, inadequate infrastructure for pre-production, production, post-harvest and processing facilities, low adoption of technologies and shortage of skilled manpower in aquaculture and extension services.

Present status, innovations needed in research and new technologies

Despite growth in the aquaculture sector, India presently produces 10 times lower than China (leading fish producer) and stands around 136th-rank in per capita fish consumption with 9kg amongst 160 countries of world. India invests more on research and technological advancements to improve fish production and meet projected demand, 15mt fish by 2020 with 8% annual growth rate. Asian carps including Indian (*Catla catla, Labeo rohita, Cirrhinus mrigala*) and Chinese carps (*Cyprinus carpio, Ctenopharyngodon idella, Hypophthalmichthys molitrix*) besides catfish (*Pangasius pangasius*) and cichlid (*Tilapia* sp.) in warm water and trout (*Salmo trutto fario, Oncorhynchus mykiss*), mahseer (*Tor putitora*) and Indian trout (*Schizothorax richardsonii*) in cold water; shrimps (*Penaeus monodon, P. indicus and P. vannamei)*, seabass (*Lates calcarifer*) in coasts and cobia (*Rachycentron canadum*), pompano (*Trachinotus mookalee*) and groupers (*Epinephelus* sp.) in marine environments are some of the species that are currently cultivated/promoted. Effective utilization of unutilized/under-utilized resources through modern farming systems, mobilizing farmers/stakeholders, technological innovations and policy/support mechanisms are some of ongoing thrusts.

Aquaculture is faced by many challenges, notably, combating diseases and epizootics, brood stock improvement and domestication, development of appropriate feeds and feeding mechanisms, hatchery and grow-out technology, as well as water-quality management (Subasinghe et al., 2003).

Technological innovations in aquaculture

Selective breeding

Aquaculture generally trails far behind plant and farm animal industries in utilizing selective breeding as a tool to improve the biological efficiency of production. It is estimated that at present less than 10% of aquaculture production is based on genetically improved stocks, despite the fact that annual genetic gains reported for aquatic species are substantially higher than that of farm animals. With an average genetic gain in growth rate of 12.5% per generation, production may be dramatically increased if genetically improved animals are used. Importantly, animals selected for faster growth have also been shown to have improved feed conversion and higher survival, implying that increased use of selectively bred stocks leads to better utilization of limited resources such as feed, labour, water, and available land and sea areas (Gjedrem et al., 2012). The highest number of breeding programs is reported for tilapia (27), followed by Atlantic salmon (13) and rainbow trout (13), while among the important most reared group of species (cyprinids) only common carp (8) and rohu carp (1) are represented. This is the group of species where genetic improvement would benefit the most people as production is extensive (consisting of many small farms).

Transgenic fish

The production of transgenic fish has since expanded to research focused on five general applications: to enhance traits of commercially important species, to develop fish as bioreactors to produce bio-medically important proteins, to enhance the utility of fish as indicators of adverse health effects associated with exposure to toxicants in aquatic environments, to develop new non-mammalian animal models for comparative biomedical research and for functional genomics studies (Dunham and Winn, 2014). The great potential of transgenic technology was demonstrated in transgenic salmon, with an average 11-fold increase of growth compared to non-transgenic controls (Devlin et al., 1995). Transgenic alteration of the nutritional characteristics of fish could be beneficial for consumers, and it is now possible to directly alter body composition *via* transgenesis. Zebrafish transfected with B-actin-salmon desaturase genes had enhanced levels of omega-3 fatty acids, docosahexaenoic acid (DHA), and eicosapentaenoic acid (EPA) in their flesh (Alimuddin et al., 2005, 2007). The two- to four fold increase in bacterial disease resistance exhibited by cecropin transgenics is much better than what is obtained through selection for disease resistance in channel catfish, which is sporadically successful in some, but not all strains of channel catfish (Dunham et al., 1993; Waters, 2001). It is likely that transgenic fish and gene-edited fish will be widely used in the future. The best genotypes will be developed by using combinations of traditional, biotechnological, and transgenic approaches.

Monosex production

Monosex refers to the culture of either all-male or all-female populations, a sought after approach in aquaculture. Monosex population culture is desired for several reasons. In most species there are sexually dimorphic patterns of favorable traits that are more accentuated in either males or females. In general, most fish species show dimorphic growth rates. Males grow faster in some species (like tilapia and catfish), while females grow faster in other species (like grass carp, several salmonids, and cyprinids). Monosex population culture of the faster growing sex therefore increases the production rate. While culturing single sex, the culture conditions are tailored to better suit the energetic demands of the cultured sex. Also, since there is no unwanted breeding, the genetic breeding program is managed in separate compartments and is more controlled. Unplanned reproduction in grow-out ponds would result in crowding and higher densities than intended and lead to energy wastage due to sexual activity on the expense of growth. The end product of monosex culture will therefore be more uniform as a consequence, a crucial marketing advantage (Ventura, 2018). In order to produce monosex populations in vertebrate aquacultured species, usually synthetic estrogens or androgens are administered to induce sex reversal. This method is not environmentally friendly since these compounds end up in waterways (Megbowon and Mojekwu 2014). In several species temperature manipulation proves useful to some extent, although it does not hold for most species. It is important to know the process of sex determination and sexual differentiation in species of interest and using biotechnological tools sex reversal could be achieved. For example gynogenesis and production of super males in tilapia and manipulation of IAG which is considered to be the key regulator of sexual differentiation in crustaceans. The case study of the giant freshwater prawn serves as a model for sex reversal through AG manipulation or IAG gene silencing. These methodologies are environmentally friendly since the intervention is species-specific, and unlike the practice common in vertebrate species, it does not include hormonal treatment that might affect other organisms. Some of the key vertebrate factors like amh and Dmrt1, could be utilized in a species-specific way to develop novel techniques for monosex population in aquaculture biotechnologies.

Molecular markers

Molecular techniques also show significant promise for aquaculture application, in that they help provide more accurate information on the genetic diversity of natural stocks and allow genetic tagging of animals in breeding programmes (Subasinghe et al., 2000). Effective breeding programmes need to identify and track the pedigrees of individual organisms. Physical tagging of early life-history stages of many aquatic species is difficult, thus, non-invasive, genetic markers using microsatellite DNA, and AFLP's (amplified fragment length polymorphisms) need to be developed to track pedigrees and provide linkage maps to identify quantitative trait loci (QTL's – genes coding for characters that have production value, such as growth rates, disease resistance or cold tolerance) (Garcia et al., 1996, Benzie, 1998, Moore et al., 1999; Agresti et al., 2000).

Egg production using environmental manipulation

In most fish species, reproduction occurs during limited periods of the year, when environmental conditions are more suitable for offspring survival (Vazzoler et al., 1997; Martínez, 2000; Bromage et al., 2001). On the other hand, one of the prerequisites to produce farm animals is the capacity to control all the phases of their life cycle, especially reproduction (Mylonas et al., 2010). Thus, the restriction on the production of eggs and larvae to the natural breeding season is a major problem for fish farmers, since it results in disruption of fingerlings production (Bairwa et al., 2013). In this context, the adjustment of the reproductive period through environmental manipulation is a highly applicable tool in getting out-of-season eggs, providing greater control of fingerling production and reducing off-season (Ramos et al., 2002; Duncan et al., 2013).

Disease Management

Production of specific pathogen free (SPF) and specific pathogen resistant (SPR) stocks

SPF stocks are produced by selecting animals shown to be free of specific pathogens, using them as brood stock, and raising their offspring under strictly controlled sanitary conditions. SPFs are of value for trade to countries or areas that are free of the specific disease agent, or for restocking ponds following a disease outbreak and disinfection. By contrast, SPR stocks are developed through selective breeding of individuals that have survived challenges/infections by specific pathogens. These, therefore have great potential to enhance production in waters endemic for the specific diseases, but are inappropriate for use in non-endemic waters, as they may carry sub-clinical infections of the pathogen in question. The development of SPF stocks will be possible with the simultaneous development of the necessary infrastructure, in terms of biosecurity, diagnostic methods and trained personnel, to successfully select founder populations of candidate SPF stocks from wild or cultured stocks and domesticate those stocks and frame the guidelines for the same. The development of SPF stocks of *P. vannamei*, *P. stylirostris*, *P. monodon*, and other penaeid species has become central to the sustainability of modern shrimp farming. Further development through selective breeding of SPF stocks for disease resistance and other desired characteristics is very likely to be a principle characteristic of the industry for the indefinite future (Lightener, 2011).

Vaccine technology

Despite multiple approaches to innovative therapy, fish diseases remain a major economic issue in commercial aquaculture worldwide. Although antibiotics or chemotherapeutics may be implemented for disease treatment, there are some clear drawbacks, such as drug resistance issues and safety concerns (Sneeringer, 2019). Vaccination, as an effective method of preventing a wide range of bacterial and viral diseases, and contributes to environmental, social, and economic sustainability in global aquaculture (Ma et al., 2019). Since the first reports in the 1940s of fish vaccination for disease prevention (Snieszko and Friddle, 1949), there have been many vaccines developed that significantly reduced the impact of bacterial and some viral diseases in fish (Gudding and Goodrich, 2014). Millions of fish are vaccinated annually, and in some areas of the world there has been a transition away from antibiotics and toward vaccination. For example, there has been a dramatic reduction in the use of antibiotics in Norwegian salmon farming since the introduction of vaccines (Rodger, 2016) and vaccination has become the most cost-effective and sustainable method of controlling infectious fish diseases (Horzinek et al., 1997). Alternative vaccine technology that includes sub unit vaccines, nucleic acid vaccines (DNA and RNA based vaccines) need to be propagated for sustainable aquaculture.

Biosensors for evaluating fish health

Methods that can be used to check the health of fishwhich determine their physiological condition and detect abnormalities at an early stage, have been attracting recent attention. The health of fish can be determined by the assessment of stress responses, prediction of spawning time, and highly sensitive detection of bacteria that cause diseases in fish, etc. However, it can be difficult to accurately measure these parameters due to factors such as sampling errors, uncontrollable aquaculture conditions and unsound methodological approaches. Also, the methods currently used are time-consuming and may involve complicated processes. Thus, there is a high demand for novel approaches for the assessment of fish health. The development of a biosensor to measure target factors of fish health by combining biocatalytics and electronics technology can greatly improve the assessment of the health of fish produced for consumption. Biosensors take advantage of molecular elements of biological processes such as enzymes and antibodies to identify target substances. When biocatalysts react with a target substance, they can serve as discriminating elements that can cause minor changes in parameters such as current, resistance, and heat due to the generation and consumption of chemical substances. Biosensor system allows for real time monitoring of fish stress, and provides both numerical data and visualized results (Endo and Wu, 2019).

Prebiotics, probiotics and immune-stimulants

Prebiotics, probiotics, and synbiotics are considered natural functional food ingredients and an alternative feeding strategy for immunomodulation and antibiotic eradication in aquaculture (Hasan et al., 2019). Different types of immune-stimulants have been used as feed additives aquaculture to upregulate innate immunity and disease resistance. Although both immune-stimulants and prebiotics have positive effects on host immunity, they differ in their immunomodulatory mechanisms.

Feed Technologies

Find ways to replace fish meal with vegetable protein from more sustainable sources

Bioencapsulation

Research into bioencapsulation and use of live feed as a means of oral delivery for compounds to enhance survival and fitness of larval stages of aquatic organisms merits further research priority.

Bioremediation

Bioremediation is another promising biotechnological approach for degradation of hazardous waste to environmentally safe levels using aquatic microorganisms, or other filtering macro-organisms (Rao and Sudha, 1996).

Cell-Based Fish: A Novel Approach to Seafood Production

Cellular agriculture is defined as the production of agricultural products from cell cultures rather than from whole plants or animals. With growing interest in cellular agriculture as a means to address public health, environmental, and animal welfare challenges of animal agriculture, the concept of producing seafood from fish cell- and tissue-cultures is emerging as an approach to address similar challenges with industrial aquaculture systems and marine capture. Cell-based seafood—as opposed to animal-based seafood—can combine developments in biomedical engineering with modern aquaculture techniques. Biomedical engineering developments such as closed-system bioreactor production of land animal cells create a basis for the large scale production of marine animal cells. Aquaculture techniques such as genetic modification and closed system aquaculture have achieved significant gains in production that can pave the way for innovations in cell-based seafood production (Rubio et al., 2019).

Artificial Intelligence (AI) in Aquaculture

AI, real-time sensors and IoT (Internet of Things) have many advantages. They can identify water quality changes at the initial stage and detect changes in the consumption and growth pattern of animals or help farmers take preventive measures before a disease outbreak. Collecting most of their information from sensors, many aquaculture technology companies are harnessing the power of AI to improve decision-making. The Yield, an Australian company that provides technologies for all types of agriculture, uses its Sensing+Aqua technology to create predictive analytics for enhanced data-driven decision-making. A robotic fish known as Shoal uses AI, or swarm intelligence (SI), to detect pollution underwater. The robots are sent out as a group and must be able to navigate their environment, avoid obstacles, including those of other robotic fish, recharge themselves at charging stations and generally make decisions autonomously of humans. The introduction of AI can greatly reduce overexploited fish species through camera and data collection systems that use AI to identify species and enable greater accountability of harvesting practices. The Seafood Innovation Cluster launched the AquaCloud platform, which aims to help managers, researchers and scientists gain new insights through its massive data collection and analysis. Particularly focused on sea lice management, the platform then uses AI to aid in the monitoring of infestations' development and spread within the environment, effectively promoting more effective area management systems for the control of pathogens.

Innovations in Production Systems

Fish holding systems

Fish can be held in holding tanks if they are going to be sold within 2-3 days. The design and operation of a live fish holding system plays an important role in maintaining fish health, preventing worker illness, and in the prevention of foodborne illness. Proper design will facilitate cleaning and sanitizing of the equipment. In addition, the hydraulic design of the unit is important to assure an adequate quantity and quality of water for the intended purpose.

In addition open sea cage farming, sea ranching, re-circulatory aquaculture system (RAS) are some more technologies gaining momentum

These (listed above are some of the technological interventions that can be further taken up to enhance the fish productivity and facilitate the growth in a sustainable way

Use of digital science and its preparedness

Information is critical for the efficient functioning of markets. When information is limited or costly, agents are unable to engage in optimal arbitrage. Excess price dispersion across markets can arise, and goods may not be allocated efficiently. In this setting, information technologies may improve market performance and increase welfare of fish farmers. Many studies have revealed that adoption of mobile phones by fishermen and wholesalers was associated with a dramatic reduction in price dispersion, the complete elimination of waste, and near-perfect adherence to the Law of One Price. Both consumer and producer welfare increased (Jensen, 2007). In India, fish markets are dispersed, and communications infrastructure is poor. Producers and traders often have only limited information, perhaps knowing only the price in a handful of nearby villages or the nearest town, so the potential for inefficiency in the allocation of goods across markets is great. By improving access to information, ICTs have immensely helped poorly functioning markets work better and thereby increased incomes and/or lower consumer prices.

Use of social media platforms by fish farmers has been playing a great role. Active use of WhatsApp and Facebook by fish farmers made them to access to all the information from seed stocking to harvest of fish. Now most of fish farmers' queries are answered in the WhatsApp by veteran fellow farmers and that to correct solutions. In addition hundreds of mobile apps are available on google playstore/ IOS platform which provide deeper insights into various aspects of fisheries and aquaculture and all stakeholders are encouraged to use the same. Data reveals that Apps providing information on marine fisheries are high. It is expected that the software market for precision agricultural tools will grow and there is a need for fisheries organizations to partner with Information Technology providers and come out with solutions which will be visionary for both sustainability and profitability.

In the field of academics and research amidst this pandemic, Virtual teaching and learning has been a game changer. Earlier the use of virtual platform like Zoom, Google Meet, Cisco WebEx and GoToMeeting etc. in fisheries sector was meagre. Many were not even aware such things exist but now as they say necessity is the mother of invention all these digital hubs have spurted up and now almost all the fisheries colleges and research institutes are using these platforms for teaching and conducting webinars. The use of virtual platform has not been restricted to academicians and research scholars but now fish farmers also actively use these platforms to gain the knowledge. It has opened new venues to learn new skills and in future also will cater to the needs of the sector.

Recent initiatives as per policy reforms introduced by Govt of India, Artificial Intelligence (AI) technologies will be used through suitable platforms and apps to improve efficiency and sustainability of fisheries sector.

Role of public and private sector in fisheries

Till date the public sector led growth has fulfilled the requirements of the fisheries sector deciphering a top-down approach focusing on the areas of production and culture which is not sufficient to meet the ever-increasing consumers demand for fish. Achieving the goal of a world free of hunger and malnutrition by 2030 – as defined by the UN Sustainable Development Goal (SDG 2) – requires a food system transformation that will effectively and sustainably use all its production components (HLPE, 2017). Fish being an essential and nutritious food component, has the potential to become a food, eradicating the problems of food security, especially nutritional security. However, the government of India has envisioned doubling the fish production by 2025, whereas the present rate of production scenario reveals that it is merely a dream seeing the present conditions. To fulfil the goal of doubling fish production, the fisheries sector necessitates radical revamping in its extension systems for fostering intensification of the sector (Ojha and Dey, 2019). The public sector led growth has paved the way for maintaining constant growth rate in the fisheries sector but to foster the missions of blue revolution and blue economy the involvement of private sector is sinaquanon. The private sector agencies like feed companies, seed companies, NGOs and community-based organisations has started mushrooming in the fisheries sector and their involvement in the sector is promulgating towards prosperity of the sector. In order to make the fisheries sector more self-sustaining and self-reliant, the promotion of private sector in aquaculture and fisheries needs to be emphasized (Ojha et al., 2020), so that higher investments can be curbed towards the sector resulting in the development of a business ecosystem. Promotion of PPP and PPCP (Public Private Community Partnership) led technology adoption models may lead to the growth and prosperity, resulting into intensification of the fisheries sector.

Investment shared by private and public sector

Investments are backbone of any economy and it triggers the demand supply cycle boosting growth. Union government and State/UTs need to coordinate with financial institutions like- NABARD, NCDC, and external Agencies - World Bank, ADB, JICA, etc. to ensure increased flow of institutional credit to fisheries sector. Business friendly environment must be made to attract private investment across the value chain based on demand rather than incentives or subsidies. Business models are to be developed in consultation with IIT/IIM/ICAR etc. for replication in the sector. A robust system of public private partnership under various policy frameworks have been envisaged to be developed where the private sector, industry, farmers, communities, government, research institutes and civil societies will be part of it. Access to institutional credit as priority lending on the lines of crop sector especially to small and marginal fishers and fish farmers and unemployed youth will be promoted.

Involvement of FPO and NGOs

Establishment of FPOs in the fisheries sector will be a step towards boosting the fishermen's income as formation of FPO will increase their remunerations and thereby provide plethora of services like marketing (input supply, output supply and marketing information), financial (savings, loans and other forms of credit), technological (education and extension), educational (business skill and health) and welfare services which are essentially required by fish farmers to reach the consumers easily and effectively. The Government of India is also focusing on formation of fisheries based FPOs which will receive financial assistance through Small Farmers' Agri-Business Consortium (SFAC) under its venture capital assistance (VCA) scheme. The primary benefits of forming an FPO includes reduction in price of inputs, production cost, help in promoting cooperative marketing mechanisms, disseminating market information and technological innovations, help in packaging, branding, labelling and standardization of the produced products (NABARD, 2015). Fisheries sector involves a lion share of small-scale farmers having a small land holding capacity. In this regard, formation of such farmer led organizations may foster better income and growth of the small-scale fish farmers as well as the fisher women who lacks the opportunity in selling their produce at the wider markets.

NGOs are mainly non-governmental bodies following a bottom-up and participatory approach towards addressing the problems in a sector. They play a significant role in developing countries where there is often a lack of political will for action plans to tackle important issues that require prior attention and may also complement government approaches towards undertaking work on policy issues (Shariff, 1996). In fisheries sector NGOs has proven to be effective in motivating target fish farmer/producers to improve aquaculture practices as well as the fish producer's access to credit, fish seed, technology, markets, and the water resource itself (Blueplanet Matters, 2012).

Thus, promoting community-based organisations like NGOs and FPOs may lead to prosperity and growth of the sector.

Linkages needed

Multi-institutional and multi-stakeholder nature of the fisheries sector in India envisages the importance of linkages and consensus for the efficient utilisation of common use resources and maintaining the sustainability. Multiple conflicts on the use of resource has created lot of issues especially in managing the off-shore fisheries and reservoir fisheries. In many recent policy frameworks (NIFAP, 2018 and NPMF, 2017) it has been stressed that fisheries sector will be recognized as a partner in determining the water policy, river management, watershed management and rural development. Inter-sectoral coordination among users *i.e.* irrigation, hydroelectricity, industries or any other purpose would be essential to protect the integrity of aquatic ecosystem. Union MoF, AH&D and States/UTs will take up with nodal departments to ensure that Department of Fisheries is fittingly represented in the all the committees of various levels entrusted with the management and usages of water resources. Institutional mechanism will be strengthened to preserve and improve the quality of fishery products at various stages. Laboratory network for testing and cold chain facilities will be developed. Use of undesirable chemicals will be banned. Food Safety Standards Authority of India (FSSAI) regulations will also be applicable for domestically traded fish to protect consumer interests. Brand promotion based on the inherent strength of the country will be encouraged to give unique identity for the Indian products.

Programs focused to fill the technological gaps in various fields related to production, processing, value addition, food safety, quality control and conservation etc. must be initiated through involvement of user groups namely State DOFs, farmers/fishers and their organizations and industry in prioritizing research agenda of ICAR- Fisheries Institutes, Fisheries Universities and other agencies. 'Center of Excellence' needs to be established entailing state of art facility for different activities at appropriate place to supply services and demonstrate the technical feasibility, economic profitably and commercial viability of fisheries activities.

Skill Development

Fisheries sector, comprising Aquaculture and Capture Fisheries, has shown a continuous growth of 5 to 6% over the years and there is a dire need for skilling in this sector for doubling the production and farmers' income. Skilled manpower is inevitable as this sector is moving from subsistence farming to commercial activities. Earlier, the fisheries and aquaculture industry was mostly, traditional in nature. In the last two or three decades, a lot of scientific and technical developments have taken place, giving it the commercial status, particularly in Aquaculture where new developments are taking place throughout the world.

While initiating the skilling programs in fisheries one of the challenges is that unlike in Agriculture and Animal Husbandry, many training partners in this sector are not available. Traditionally, all fisheries capacity building and extension programs are being handled either by the state fisheries' departments or through the eight fisheries ICAR institutions, Fisheries Universities and Colleges under the SAUs.

Agricultural Skill Council of India (ASCI) is in the process of entering into MoUs with ICAR Institutions, State Agricultural and Fisheries Universities to take forward the skill development programs in fisheries sector. Since many private players as training partners are not available, they are trying to establish linkages with many NGOs and corporates having presence in fisheries and aquaculture.

Union Government programme on SKILL INDIA Initiative post-harvest segment programmes will be implemented to upscale and improve the skill sets of the workers. In the harvest/post-harvest/processing and value addition sector, trained human resources and support facilities to move up the value chain and meet international standards is the need of the hour. Skilling will act as a catalyst for blue revolution and hence more investments need to be diverted to prioritise the skill development programmes in fisheries sector to meet the skilled manpower requirement.

Capacity Building

Capacity building is a cross cutting theme which is one of the major elements for sustainable development. It is suggested that capacity exists, and needs to be strengthened at, four levels: in individuals, in organizations/institutions, in sectors and networks, and in the overall environment in which the first three function. Capacity building has to be done keeping a holistic perspective leading to all round development. The primary capacity development areas in fisheries may be in fisheries science and research (ecosystem sciences, fisheries sciences, fisheries technology, post-harvest utilization and marketing, Aquaculture Technology, Data collection & Information management, social sciences, communication technology), fisheries sector management (policy development, sectoral planning and management, regulation of fishing capacity, Monitoring control and surveillance, Fisheries law and law of the sea, cross-sectoral integration, fisheries economics, fisheries socio-economics and livelihoods) and general societal skills and knowledge (community mobilization, management and administration, conflict management and problem solving, good governance, environmental awareness, sustainable trade, information & communication) (Macfadyen and Huntington, 2004). Capacity development in the aforementioned areas in the fisheries sector will help in strengthening fisheries research and growth among all of its stakeholders ranging from scientist to fish farmers.

Community Participation

Fisheries mainly comprise of Common Property Resources (CPR) like reservoirs lakes, community ponds, beels, etc which are managed by the local communities and Government entities. In this regard, community participation needs to be encouraged for sustainable maintenance and effective management of the common fisheries resources. In addition to this, there are studies which emphasize that how local resource users can become more and better involved in the management of their natural resources (Jansen et al., 2000). The areas for community participation encouragement may be in the harvesting sector of the fish resources the processing, distribution and consumption sectors of the fisheries (Jansen et al. 2020). Public participation is a key ingredient of good governance, aimed at (a) engaging the public,(b) resolving conflicts and supporting implementation, and (c) improving decision quality (Coffey, 2005). Engaging community participation in fisheries may lead to the involvement of the small-scale fishers, women traders youths who are mostly ignored from managing community resources due to the involvement of contractors and political issues. Besides this, strengthening community participation will also address the problems of food security, poverty employment generation among the rural areas fulfilling the global goals of SDGs. A community-based fisheries management system shares responsibility over the resource between the fishing communities and the government, which may help in reducing the problem of common property by allocating exclusive fishing rights to the fishing communities in their respective areas or villages through CFMA (Fatuma, 2012).

Women Empowerment

Women's engagement in fisheries can be viewed from social, political and technical perspectives, all of which show that the role of women is often underestimated. This inadequate recognition of women's contributions hampers the sustainable development process, resulting in increased poverty and food insecurity. Women constitute about half the population involved in fisheries development activities. In aquaculture, women often carry out most of the work of feeding, harvesting and processing fish and shellfish. They can become managers of small household enterprises, such as fish ponds, and thus improve their families' income and nutrition. However, women tend to have limited control over ponds and inputs for aquaculture, so they rarely participate in production beyond satisfying their household needs. These women, estimated at approximately 90 million, are often invisible to policy-makers who have traditionally assumed – mistakenly – that fisheries are largely a male domain (FAO, 2016).

Policies and development plans obscure the economic contribution of the pre- and post-harvest subsectors, where women are most active. In addition, reported statistics largely fail to capture youth and children employed in the sector, and the limited data available are rarely sex-aggregated. Gender issues in the fisheries sector are seldom examined, and women's important role is often not adequately considered. In view of their important role in the post-harvest subsector, women must be allowed to participate in decision-making processes, and be provided with the access to physical and capital resources for developing their industry and meeting their needs and aspirations. Both women and men must be accorded equal rights and be able to participate in the development process, in order to ensure that their interests and needs are adequately protected and fulfilled. It is necessary to provide them with training and formal education to improve the efficiency, profitability and sustainability of their activities; to ensure adequate infrastructure, equipment, technologies, and access to markets. This will support their enterprises, increase their income potential and reduce their marginalization. It is also urgent to give women equal control in the value chain and profit margins. This is particularly relevant as women often still work in low-status, less-skilled and low-paid jobs, and on informal, casual and temporary contracts that disqualify them from receiving social benefits. The increasing demand for fish and fisheries resources is leading to overexploitation, and there is a need to better utilize present global catches while reducing waste and losses. Therefore, fisheries strategies should take gender-specific divisions of labour into account. To avoid placing additional time burdens on women and to promote broad social and cultural acceptability, the opportunities offered by aquaculture should be assessed from a gender perspective, using a participatory approach. Strategies must ensure that aquaculture can contribute to women's empowerment. To achieve a sustainable impact on the social and economic outcomes of communities, it is important that men and women equally benefit from any fishing project and from technical and financial inputs, such as boats, fridges, educational materials, technical instruments and credit.

"Investing in women capabilities and empowering them to exercise their choice is not only valuable itself but also the surest way to contribute to the economic growth and development" (UNDP, 2008).

Entrepreneurship

Fisheries sector with its important role played in the socio-economic development of the country has become a powerful income and employment generator, and stimulates the growth of a number of subsidiary small, medium and large-scale industries. Sir Peter Drucker has also elucidated that "aquaculture, not the Internet, represents the most promising investment opportunity of the 21st Century".

The current utilization pattern of inland water resources for fish farming and the productivity of our aquaculture systems are far below the global average. Conversely, the share of employment in capture fisheries is stagnating or decreasing, and increased opportunities are being observed in the aquaculture sector (FAO, 2010; FAO 2018). Since aquaculture requires both scientific capacity and business skills, hence promoting aquapreneurship development through skilled professionals can be a viable option in exploiting and expanding the aquaculture potentials of India. Aquapreneurship is a successful marriage of entrepreneurship with aquaculture. Aquapreneurship turns the farm into an aqua business. It mainly relates to establishing aqua enterprises that may reap favourable benefits for the farmers. Aquaculture employs large numbers of people either directly in fish farming activities (for example, fish pond/fish pen/fish cage operators, caretakers, construction workers, pump tenders, vehicle/ machine operators, harvesting aides) or indirectly as employees in related or ancillary industries (as net manufacturers, boat-makers, fry gatherers, bamboo suppliers) (Baluyut, 1989). Entrepreneurial possibilities in Indian aquaculture sectors are still under- exploited (Mohanty and Sajesh, 2018). Entrepreneurial possibilities exist across the value chain in aquaculture, from production to value addition and marketing. Technological options for various subsectors like harvest, post-harvest, aquaculture production, by-product utilization is being generated, and the process is still going on in research institutes and fisheries universities (Mohanty and Sajesh, 2018).

The fisheries and aquaculture sector entails a plethora of sectors associated with it like inland culture, capture, ornamental fisheries, brackish water aquaculture marine fisheries which has opened up a bunch of opportunities for people to dive into the sector. The prospective rural aquaculture enterprises can be ornamental fisheries and its accessories enterprises, fish seed/ hatchery enterprises, fish marketing and processing enterprises, Inland aquaculture system enterprises, waste utilization enterprises, transportation/ facilitation enterprises, sea weed and algae extraction enterprises etc. (Dey and Chinmaya, 2020). However, entrepreneurship development requires a push from the government and other institutions and it is encouraging to see some incubation centers coming up in this sector since the last five years. Though there is provision for technical support from the industry financial support to sustain incubation of Startups and help them accelerate and grow is required. When it comes to handling risks in this sector, if the business becomes sustainable, the risks get mitigated.

Policy Reforms

Various policies have been formulated by Government of India in collaboration with various ICAR research institutes, multiple stakeholders' representatives as part of core committees. Most essential, fishers/farmers and their opinions have been also taken into consideration while formulating policies to manage fisheries sector in India viz; National Policy on Marine Fisheries 2017, National Inland Fisheries and Aquaculture Policy (NIFAP) 2018 and National Fisheries Policy 2020. Holistic strategies have been formulated to address the identified problems/ issues inherent to the sector and proper logical action plans have been prepared to cater to the need of the sunrise sector. Recently, the Pradhan Mantri Matsya Sampada Yojana (PMMSY) has been digitally launched. Several other initiatives in the fisheries and animal husbandry sectors have also been launched on the occasion. PMMSY is a flagship scheme for focused and sustainable development of the fisheries sector in the country as a part of the Aatmanirbhar Bharat Abhiyan.

It aims at

- Adopt 'Cluster or Area-based Approaches' and create fisheries clusters through backward and forward linkages.
- Focus especially on employment generation activities such as seaweed and ornamental fish cultivation.
- Address critical gaps in fish production and productivity, quality, technology, post-harvest infrastructure and management, modernisation and strengthening of the value chain, traceability, establishing a robust fisheries management framework and fishers' welfare.
- Consolidate the achievements of Blue revolution and bring new interventions such as fishing vessel insurance, support for new/up-gradation of fishing vessels/boats, integrated aqua parks, e-trading/marketing, etc.

Targets

- **Enhance fish production** by an additional 70 lakh tonne and increase fisheries export earnings to Rs.1,00,000 crore by 2024-25.
- **Double the incomes** of fishers and fish farmers.
- **Reduce post-harvest losses** from 20-25% to about 10%.
- **Generate** an additional 55 lakhs direct and indirect gainful **employment opportunities** in the fisheries sector and allied activities.

Budget and Time Period

- An estimated investment of **Rs. 20,050 crore for a period of 5 years from financial year (FY) 2020-21 to FY 2024-25** in all States/Union Territories.
- This investment is the **highest ever in the fisheries** sector yet.

Implementation: It will be implemented as an umbrella scheme with two separate components namely:

- Central Sector Scheme: The project cost will be borne by the Central government.
- Centrally Sponsored Scheme: All the sub-components/activities will be implemented by the States/UTs and the cost will be shared between Centre and State.

PMMSY, aims to plug critical gaps in the fisheries value chain, including modernization, traceability, production, productivity and post-harvest quality control if executed immediately. In addition to this, Interim Budget FY20 had created a separate ministry for animal husbandry, livestock, and fisheries, and Union Budget FY20 allocated Rs 770.25 crore for fisheries for a comprehensive development. The budgetary allocation has further increased to Rs 825 crores in the current budget for FY 21. This budgetary support has recognised the significance of 'blue economy'. These financial packages need careful tinkering and strategic roll-out so as to achieve desired results in a sector that is severely battered by the prolonged COVID 19-induced lockdown and associated multiple ramifications.

Conclusion

The bottom line to this whole discussion is that India has immense potential in terms of resources and youth to fulfil the dream of Honourable Prime Minister to become self-reliant in fisheries sector and achieve Blue revolution. The dire need is to put concerted efforts in terms of strategic implementation of policies, budget allocation, linkages among institutions, smooth credit flow, research, innovations, development of new technologies, community based extension activities, entrepreneurship development, better market infrastructure information system, proper regulations and skilled manpower to roll the wheels of the fisheries economy. A multi-pronged approach will result in a self-reliant sector.

References

Agresti, J.J., S. Seki, A. Cnaani, S. Poompuang, E.M. Hallerman, N. Umiel, G. Hulata, G.A.E. Gall and B. May. 2000. Breeding new strains of tilapia: development of an artificial centre of origin and linkage map based on AFLP and microsatellite loci. Aquaculture, 185: 43-56.

Alimuddin, Y.G., Kiron, V., Satoh, S., Takeuchi, T., 2005. Enhancement of EPA and DHA biosynthesis by over-expression of masu salmon delta 6-desaturase-like gene in zebrafish. Transgenic Res. 14, 159-165.

Alimuddin, Y.G., Kiron, V., Satoh, S., Takeuchi, T., 2007. Expression of masu salmon delta5-desaturase-like gene elevated EPA and DHA biosynthesis in zebrafish. Mar. Biotechnol. (NY) 9, 92-100.

Bairwa, M.K., Saharan, N., Rawat, KD., Jakhar, J.K. & Bera, A. 2013. Photoperiod, melatonin and its importance in fish reproduction. Central European Journal of Experimental Biology, 2(4): 7-15.

Baluyut, E.A., 1989. Aquaculture systems and practices: a selected review. United Nations Development Programme & Food and Agriculture Organization of the United Nations. Rome.

Benzie, J.A.H. 1998. Penaeid genetics and biotechnology. Aquaculture, 164: 23-47.

Blueplanet Matters, 2012. Role of NGOs in promotion and development of fisheries sector. Available at:https://blueplanetmatters.blogspot.com/2012/09/role-of-ngos-in-promotion-and.html#:~:text=NGOs%20can%20be%20effective%20in,and%20the%20water%20 resource%20itself. (Accessed on 20.09.2020)

Bromage, N., Jones, J., Randall, C., Thrush, M., Davies, B., Springate, J., Duston, J. & Barker, G. 1992. Broodstock management, fecundity, egg quality and the timing of egg production in the rainbow trout (Oncorhynchus mykiss). Aquaculture, 100: 141-166

Coffey, C., 2005. What role for public participation in fisheries governance?. In Participation in fisheries governance (pp. 27-44). Springer, Dordrecht.

Devlin, R.H., Yesaki, T.Y., Donaldson, E.M., Du, S.J. and Hew, C.L., 1995. Production of germline transgenic Pacific salmonids with dramatically increased growth performance. Canadian Journal of Fisheries and Aquatic Sciences, 52(7), pp.1376-1384.

Dey, S. and Chinmaya, N., 2020. Aquapreneurship: a viable option for the next generation of entrepreneurs. Aquaculture Magazine, August -September Issue, 46(4), pp. 16-20.

Duncan, N.J., Sonesson, A.K. & Chavanne, H. 2013. Principles of finfish broodstock management in aquaculture: control of reproduction and genetic improvement. In: Allan, G. & Burnell, G. (Eds.). Advances in aquaculture hatchery technology. Woodhead Publishing, Cambridge, pp. 23-75.

Dunham, R.A. and Winn, R.N., 2014. Production of transgenic fish. In Transgenic Animal Technology (pp. 305-334). Elsevier.

Dunham, R.A., Brady, Y. and Vinitnantharat, S., 1994. Response to challenge with Edwardsiella ictaluri by channel catfish, Ictalurus punctatus, selected for resistance to E. ictaluri. Journal of Applied Aquaculture, 3(3-4), pp.211-222.

FAO, 2010. The State of World Fisheries and Aquaculture 2010. Rome, p. 19.

FAO, 2016. Promoting gender equality and women's empowerment in fisheries and aquaculture. Available at: http://www.fao.org/3/a-i6623e.pdf.

FAO. 2018. The State of World Fisheries and Aquaculture 2018 - Meeting the sustainable development goals. Rome.

Fatuma, S., 2012. Community Participation in Fisheries Management in Tanzania. IIFET Proceedings

Garcia, D.K., A.K. Dhar, and A. Alciva-Warren. 1996. Molecular analysis of PAPD marker (B20) reveals two microsatellite and differential mRNA expression in Penaeus vannamei. Mol. Mar. Biol. Biotech. 5: 71- 83.

Gjedrem, T., Robinson, N. and Rye, M., 2012. The importance of selective breeding in aquaculture to meet future demands for animal protein: a review. Aquaculture, 350, pp.117-129.

Gudding, R.; Goodrich, T. The History of fish vaccination. In Fish Vaccination, 1st ed.; Gudding, R., Lillehaug, A., Evensen, Ø., Eds.; John. Wiley & Sons, Inc.: New. York, NY, USA, 2014; pp. 1–11.

Hasan, M.T., Je Jang, W., Lee, J.M., Lee, B.J., Hur, S.W., Gu Lim, S., Kim, K.W., Han, H.S. and Kong, I.S., 2019. Effects of immunostimulants, prebiotics, probiotics, synbiotics, and potentially immunoreactive feed additives on olive flounder (Paralichthys olivaceus): a review. Reviews in Fisheries Science & Aquaculture, 27(4), pp.417-437.

HLPE, 2017. Nutrition and food systems: A report by the High-Level Panel of Experts on Food Security and Nutrition of the Committee on World Food Security, Rome. High Level Panel of Experts on Food Security and Nutrition of the Committee on World Food Security, Rome.

Horzinek, M.C.; Schijns, V.E.C.J.; Denis, M.; Desmettre, P.; Babiuk, L.A. General description of vaccines. In Veterinary Vaccinology; Pastoret, P.P., Blancou, J., Vannier, P., Verschueren, C., Eds.; Elsevier Press: Amsterdam, The Netherlands, 1997; pp. 132–152.

Jansen, E.G., Abila, R.O. and Owino, J.P., 2000, June. Constraints and Opportunities for 'Community Participation' in the management of the Lake Victoria Fisheries. In Forum for Development Studies (Vol. 27, No. 1, pp. 95-133). Taylor & Francis Group.

Jensen, R., 2007. The digital provide: Information (technology), market performance, and welfare in the South Indian fisheries sector. The quarterly journal of economics, 122(3), pp.879-924.

Krueger, K., Rubio, N., Datar, I. and Stachura, D., 2019. Cell-based fish: a novel approach to seafood production and an opportunity for cellular agriculture. Frontiers in Sustainable Food Systems, 3, p.43.

Lightner, D.V., 2011. Status of shrimp diseases and advances in shrimp health management. Diseases in Asian aquaculture VII. Fish health section, Asian fisheries society, Selangor, Malaysia. P, 385, pp.121-134.

Ma, J., Bruce, T.J., Jones, E.M. and Cain, K.D., 2019. A review of fish vaccine development strategies: Conventional methods and modern biotechnological approaches. Microorganisms, 7(11), p.569.

Macfadyen, G. and Huntington, T., 2004. Human capacity development in fisheries. Food and Agriculture Organization of the United Nations.

Martínez, E.A. 2000. Reproducción en peces teleósteos. Estrategias reproductivas e influencias de los factores ambientales en los ciclos reproductivos. In: Navarro, S.Z., Díaz, B.A. & Hernández, M.P.G. (Eds.). Acuicultura I: Biología marina. Reproducción y desarrollo. Servicio de Publicaciones, Universidad de Murcia, Murcia, pp. 49-62.

Megbowon I, Mojekwu TO (2014) Tilapia sex reversal using methyl testosterone (MT) and its effect on fish, man and environment. Biotechnology 13:213–216

Mohanty, A.K. and Sajesh, V.K., 2018. Fishpreneurship development–A New Direction in Enhancing the Farm Income. ICAR-Central Institute of Fisheries Technology.

Moore, S.S., V. Whan, G.P. Davis, K. Byrne, D.J. S Hetzel and N. Preston. 1999. The development and application of genetic markers for the Kuruma prawn Penaeus japonicus. Aquaculture, 173(1-4): 19-32.

Mylonas, C.C., Fostier, A. & Zanuy, S. 2010. Broodstock management and hormonal manipulations of fish reproduction. General and Comparative Endocrinology, 165: 516-534.

NABARD, 2015. Farmer Producer Organizations. Available at: https://www.nabard.org/demo/auth/writereaddata/File/FARMER%20PRODUCER%20ORGANISATIONS.pdf. (Accessed on 20.09.20)

Ojha, S.N. and Dey, S., 2019. Revamping Fisheries and Aquaculture Extension Strategies: Doubling the Fish Production by 2025 in India. Souvenir book. Fourth PAF Congress on Increasing Aquaculture Production in India through Synergistic Approach between Multinational Industries, Domestic Entrepreneurs and Aquaculturists, ICAR- CIFA, Bhubaneswar, Odisha, India, pp.14- 27.

Ojha, S.N., Dey, S. and Babu, S.C., 2020. A bottom-up approach for a private fisheries extension system: A framework and action plan for an aqua-chamber of commerce in India (Vol. 1931). Intl Food Policy Res Inst.

Ramos, J., Rodríguez, L., Zanuy, S. & Carrillo, M. 2002. Influencia del fotoperiodo sobre la aparición de la primera madurez sexual, comportamiento reproductivo y calidad de puestas en hembras de lubina Dicentrarchus labrax (Linnaeus, 1758). Boletín del Instituto Español de Oceanografía, 18: 175-182.

Rodger, H.D. Fish disease causing economic impact in global aquaculture. In Fish Vaccines, 1st ed.; Adams, A., Ed.; Springer: Basel, Switzerland, 2016; pp. 1–34

Shariff, M., 1996. Role of non-governmental organizations (NGOs) in fish health management in the Asia-Pacific. FAO Fisheries Technical Paper, pp.51-59.

Sneeringer, S. and Clancy, M., 2019. Incentivizing New Veterinary Pharmaceutical Products to Combat Antibiotic Resistance. Applied Economic Perspectives and Policy.

Snieszko, S.F.; Friddle, S.B. Prophylaxis of furunculosis in brook trout (Salvelinus Fontinalis) by oral immunization and sulfamerazine. Prog. Fish-Cult. 1949, 11, 161–168.

Sriniwasa Rao, P.S. and P.M. Sudha. 1996. Emerging trends in shrimp farming. Fishing Chimes, 16(3): 25-26.

Subasinghe, R. and Bernoth, E.M., 2000. Disease control and health management: aquaculture development, health and wealth.

Vazzoler, A.E.A.M., Lizama, M.L.A.P. & Inada, P. 1997. Influências ambientais sobre a sazonalidade reprodutiva. In: Vazzoler, A.E.A.M., Agostinho, A.A.A. & Hahn, N.S. (Eds.). Planície de inundação do alto rio Paraná: aspectos físicos, biológicos e socioeconômicos. EDUEM, Maringá, pp. 267-280.

Ventura, T., 2018. Monosex in Aquaculture. In Marine Organisms as Model Systems in Biology and Medicine (pp. 91-101). Springer, Cham.

Waters, P.J., 2001. Response and correlated response to selection of channel catfish, Ictalurus punctatus for resistance to Edwardsiella ictaluri, Flexibacter columnaris, low levels of dissolved oxygen and toxic levels of nitrite. M.S. thesis. Auburn University, AL.

Wu, H., Fujii, Y., Nakano, T., Arimoto, T., Murata, M., Matsumoto, H., Yoshiura, Y., Ohnuki, H. and Endo, H., 2019. Development of a Novel Enhanced Biosensor System for Real-Time Monitoring of Fish Stress Using a Self-Assembled Monolayer. Sensors, 19(7), p.1518.

Theme 8: Agricultural Education, Extension and Socio-economic Reforms

31

Innovations and Reforms in Agricultural Marketing

R.S. Pundir[1], Ganga Devi[2] and Rachana Bansal[2]

[1]International Agribusiness Management Institute, Anand Agricultural University, Anand, Gujarat
[2]Department of Agricultural Economics, Bansilal Amrutlal College of Agriculture Anand Agricultural University, Anand, Gujarat

Introduction

India is predominantly an agricultural economy and about half of the workforce is engaged in agriculture sector. Around 48 per cent of households in the country are agricultural household and they derive the monthly income from crop cultivation. India is leading in production of many agricultural crops but finding a market for marketed surplus and getting fair prices have always been a major challenge. Long time ago, traditional market system was existed in India. Marketing implied greater magnitude as yields and production started rising after farmers embraced with advanced technologies such as high-yielding varieties, fertilizers, chemical inputs, block chain development, farmers business network, market linkages, diversification towards high value crops, integrated farming system approach, precision farming including digital agriculture, drones, GIS technology, remote sensing, soil health management, crop and livestock insurance, mobile apps, web portal, machine learning and many mores. These catered the need of efficient regulated marketing system to support the farmers and keep away from them getting manipulated in the market.

India has gained self-sufficiency in the agriculture production. There is need to translate this production into better remuneration for the farming community by focusing into agricultural marketing system. The agricultural marketing system is characterized by various short comings like heavy sale of commodities at primary level immediately after the harvesting period, absence of on-farm grading of produce, poor packaging, insufficient marketing infrastructure, long marketing channels, existence of various malpractices in the marketing of agri-produce, non-transparent price discovery mechanism, lack of market information system, low marketable surplus especially of small and marginal farmers, *etc.*

Rising population and increasing demand for food are prominent factors driving the digital farming market. Moreover, the growing demand to efficiently deal with pests, weeds, and other diseases is encouraging farmers to digitize their farm. In developing regions, cheaper internet data and strong penetration of smart devices (especially smartphones) are driving the digital farming market. Active participation of governments in digitizing farms through investment is also driving the digital farm market.

Market Reforms

Time to time the Government has taken many initiatives for improving agricultural markets because in earlier time the main focus is given to the production technologies, according to the new guidelines there is a need to strengthen the marketing system for effective implementation of the technologies. On production side India is the global leader in many crops and agricultural enterprises, but marketing is the key problem which faced by the farmers. Generally the marketing channels or the route which follows by the farmers are dominated by the middle men and they exploit the farmers by offering the lower price for their produce as farmers need the immediate cash for fulfil their family and other requirements. Many times the farmers are not aware about the market situation like price, demand supply *etc.* may be due to the lack of market information and knowledge also and farmers has to forcefully sale their produce to the middlemen's which results into the distress sale condition. This will make the advantageous for the middlemen's and commission agents, they form the cartel to put the downward pressure on the farmers sale price and as a result of this, the farmers share in consumer's rupee was low. But gradually farmers shared increased over the last decade due to many market reforms in agriculture sector (Reddy, 2016). Therefore, marketing is as critical as to better performance in agriculture market and hence, reforms and marketing systems improvement ought to be an integral part of policy and strategy for agricultural development.

Agricultural market regulation in India has move towards a very long way since the first regulated Karanjia Cotton Market was established in the year 1886 by the British rulers under Hyderabad Residency Order. In the year 1928 the Royal Commission on agriculture stated out the problems of traditional marketing such as high marketing cost, unauthorized deductions, and prevalence of various malpractices. In 1935, the establishment of the Directorate of Marketing and Inspection, enactment of the Act for grading and marking of agricultural produce in 1937 and the conduct of commodity market survey and the establishment of regulated markets in the states were some of the other measures taken before independence of India to improve the marketing situation.

In Early 1960's the Agricultural Produce Market Regulation Act (APMR) was circulated and took almost 16-18 years for the major states to enact in the country. It was considered as the first phase of reforms with main objectives *viz.*, creation of physical infrastructure facilities in primary markets, to establish farmers-dominated management committees, to put in place certain marketing practices, standardize market charges and creation of some amenities in the market yard. At the same time, there are many problems identified under the effective implementation of APMR like there is no timely payments was made, barriers to entry of firms in the market, no new trading license are provided, more concern given to the collection of fee amount, inefficient and lees amenities in the yard which results inefficient marketing system for needy farmers. Looking to the challenges in APMR act several review committees are strong-willed to the second phase of reforms were commenced in early 2000's.

The Government of India decided and encourages all the states in the country to implement the Model APMC Act, 2003/ Model Rules, 2007, Model Rules 2010 and now recently the Model Act, 2017 and the Ordinance, 2020. The Model Agricultural Produce Market Committee (APMC) Act, 2003 operates mainly on two principles.

1) It ensures that farmers are not exploited by the different market functionaries/ intermediaries/money lenders who force farmers to sale their produce at the farm gate and very less price.
2) It was bring into the notice that first all the food produce should brought to a market yard and then sold through auction method which would be helpful to the farmers for getting right price for their produce.

The major indicators which representing different areas of reforms in APMC markets are;

- **Direct marketing:** Producers sale their produce directly to different organizations/exporters/bulk byers/processing industries *etc.*, which eliminate all the market middlemen and malpractices existing in the indirect markets. Direct marketing helps many folds to increase the producer share in consumer's rupee.
- **Contract Farming:** Contract farming is imperative to reduce the marketing and other risks associated with production, price, institutional, environmental *etc.* It also essential to promote food processing industries, provide technical and financial support, quality inputs to smallholders *etc.*
- **De-link the provisions of compulsory requirement of shop/space for registration of traders / market functionaries:** The provision of compulsory licensing of commission agents/traders in the regulated markets has led to the monopoly of these licensed traders and they acting as a main entry barrier in existing APMCs for new entrepreneurs and prevent the perfect competition in the market. This practice of licensing should be replaced by simple registration at nominal fee with no requirement of ownership of a shop or godown in the marketing area.

- Take fruits and vegetables out of APMC Act: There should be a provision with producers to sale their produce to anyone they wish like FPOs, integrators, cooperatives, processors or in APMC mandi.
- Setting up of private mandis
- Single point levy on agriculture produce
- e-trading
- Single license for traders valid in all the markets across the state
- Rationalization of taxes on agricultural commodities
- Implementation of e-NAM in all states

The Ministry of Agriculture and Farmers Welfare in the year 2017 come out with the new APMC model act "Model Agricultural Produce and Livestock Marketing (Promotion & Facilitation) (APLM) Act, 2017" which replace the existing APMC act. The APLM Act provides better price realisation to the farmers as they can sale their produce according to their choice to the buyers and consumers also benefited by paying the lower price for agricultural products.

The major intention of amendment of APLM act was declaration of whole State/UT as one unified market area, establishment of special commodity market yard and Market yard of National Importance (MNI) of specified agricultural produce and livestock's, establishment of farmer- consumer market yard (direct marketing; farmers can directly sale the produce to consumer in retail), provision of geographically restriction-free trade transaction, integration of supply chain provision of scientific storage structures, e-trading, single point levy of market fee across the state and unified single trading licence *etc.*, with these key features the APLM act was enacted.

In spite of the best intentions and efforts made by the government, many of the APMCs has many short falls, the conception of a notified market area creates the problem with the farmers those who are falling under the notified area as they are bound to sale the produce at the selected APMCs. Lack of transparency in price discovery was found due to the cartel among traders of APMC markets. The provision of compulsory license of traders led to the monopoly and they act as a main entry barrier in existing APMCs for new entrepreneurs and prevent the perfect competition in the market and lack of scientific storage infrastructure and amenities in the yards. Therefore, inefficient infrastructure results in losses during different marketing practices *viz.,* handling, cleaning, grading and transporting. There is no provision in most of the APMCs for handling of the bulk volume of perishable commodities. With respect to all these short falls in the existing market reforms Government of India come out with the new frontier agricultural marketing concept "one nation one market" in the year 2016.

National Agriculture Market (*e*NAM): Uttam Fasal, Uttam Inaam

National Agriculture Market (NAM) is a pan-India electronic trading portal networks the existing APMC *mandis* to create a unified national market for agricultural commodities which provides a single window service including arrivals and prices, buying and selling trade offers, provision to respond to trade offers and other services both, at state and National level and promotes uniformity, streamlining of procedures across the integrated markets. It was launched by Ministry of Agriculture, Government of India across the country on *14 April 2016 by Prime Minister of India, Narendra Modi*. The Portal is managed by Small Farmers' Agribusiness Consortium (SFAC) and NFCL's, iKisan division helps in providing the technical knowhow.

The foremost objectives and benefits of trading on e-NAM for various stakeholders including farmers, traders, buyers, processers, exporters and consumers are transparent trading activities, accurate information sharing at right time with all stakeholders, real time price discovery based on demand and supply, better price realization by farmers, low transaction cost for buyers, reduction of wastages, stability in the prices for consumers, quality certification, warehousing and logistic, more efficient supply chain, guarantee of payment and delivery, single point levy of market fees, one license for a trader across the state, lack of errors in the reporting of transaction and it enhanced the accessibility to the markets round the nation.

Current Status of *e*NAM

About 1,000 markets located in 18 states and 3 union territories are integrated on *e*NAM by the year 2020. While 585 *mandis* were integrated in the first phase and 415 *mandis* were integrated in the second phase. *e*NAM has registered a user base of 1.66 crore farmers, 1.31 lakh traders, 73,151 commission agents and 1,012 farmers producers organisations (FPOs) in last four years. The transacted value over Rs 1 lakh crore on the eNAM platform with a trade volume of 3.43 crore tonnes of commodities, and 38.16 lakh bamboo and coconuts were generated. Presently, 150 commodities, including food grains, oilseeds, fibres, fruits and vegetables, are traded on *e*NAM platform (https://economictimes.indiatimes.com/news). Moreover, 25 more commodities are included on *e*NAM platform by 29th August 2020. At present total 175 agricultural commodities are traded (enam.gov.in).

Many initiatives has been taken by Government to handle the present situation Covid-19 for smooth running of ***e***NAM such as development of FPO module for direct trading options from warehouses and collection centres, development of logistic module to provide the transport facilities, development of Kisan Rath mobile app to support and provide the logistic services across the country to all the stakeholders and implementation of *e*-NAM call centre for the solution of constraints faced by farmers specially logistic and others.

Major challenges in implementation of *e*-NAM

Despite of the progress and growth several key issues and challenges appears in the implementation of *e*-NAM is inadequate marketing infrastructure, lack of coordination between the institutions and poor market information. Infrastructure includes poor quality of rural road, lack of scientific storage, warehousing, very few cold storages and refrigerated vans, limited access of agriculture produce markets due to huge variation in the density, very limited markets having the facility of assaying and grading, inadequate facility and equipment to deal with high volume of agricultural commodities in the peak season, diverse standards for agricultural commodities, disjointed APMCs, lack of synergy between marketing organizations and service providers, less involvement of traders in the marketing of agricultural produce, poor internet connection, scarcity of computers, interrupted power supply etc. The other foremost bottlenecks are lack of skilled manpower and trained traders for trading online platform, lack of awareness among the farmers for the benefits of *e*-NAM other procedure and formalities for e-trading (Bisen and Kumar, 2018).

In spite of the challenges, there are way-outs to cope with these issues being faced in effective implementation of *e*-NAM. The accurate and timely market information and market intelligence system needs to be addressed and revitalized. Subsequently, there is a need to focus on the development of required infrastructure related to cold storage, scientific storage structures, refrigerated vans, coordination among the institutions, channelize and integration of the existing supply chain, capacity building programmes to create the awareness among all the stakeholders about benefits of *e*-NAM trading and farmers producer organization to offer vital link to market.

Recently, the Government of India introduce the three reform package through the ordinance route. First, has amended the Essential Commodities Act to remove the existing restrictions on stocking food produce. Second, it has introduced a new law the i.e. Farmers' Produce Trade and Commerce (Promotion and Facilitation) Ordinance, 2020 or the FPTC Ordinance to end the monopoly of the APMC and allow anyone to purchase and sell agricultural produce. Third, the Farmers (Empowerment and Protection) Agreement on Price Assurance and Farm Services Ordinance, 2020 or FAPAFS has been enacted to legalise contract farming, so that big businesses and companies can cultivate vast swaths of land on contract.

The Farmers' Produce Trade and Commerce (Promotion and Facilitation) Ordinance, 2020: It was promulgated on June 5, 2020. It seeks to provide for barrier-free trade of farmers' produce outside the markets notified under the various state agricultural produce market laws (state APMC Acts). The Ordinance allows intra-state and inter-state trade of farmers' produce outside: (i) the physical premises of market yards run by market committees formed under the state APMC Acts and (ii) other markets notified under the state APMC Acts such as private market yards and market sub-yards, direct marketing collection centres, and private farmer-consumer market yards. Such trade can be conducted in any place of production, collection, and aggregation of farmers' produce including: (i) farm gates, (ii) factory premises, (iii) warehouses, (iv) silos, and (v) cold storages.

Ordinance allows farmers, farm producer organisations as well as anyone who buys farmers' produce for: (i) wholesale trade, (ii) retail, (iii) end-use, (iv) value addition, (v) processing, (vi) manufacturing, (vii) export, or (viii) consumption, to engage in such intra-state or inter -state trade.

Farmer Producer Organisations (FPOs)

To promote and support the small and marginal farmers the concept of FPOs was initiated in the year 2011-12. A producer organisation is an authorized body formed by primary producers, *viz.,* farmers, milk producers, fishermen, weavers, rural artisans, craftsmen. It can be a producer company, a co-operative society or any other legal form which provides for sharing of profits/benefits among the members. In some forms, like producer companies, institutions of primary producers can also become member of producer organization. It registered under the special provisions of the Companies Act, 1956. With respect to the current status around 5000 FPOs including farmer producer companies are in existence and it formed under various initiatives including SFAC, state governments, NABARD, as well as other organisations over the last 8-10 years by the Government of India. But majority of them in growing stage and many are struggling to survive. Recently, the Government announced in the Union Budget 2019-20, for formation of 10,000 new FPOs to ensure economies of scale for farmers over the next five years. They aimed at providing a platform to small farmers collectively tackle their challenges right from accessing various agricultural inputs to marketing their produce. These FPOs will turn farmers into businessmen.

The important activities under taken by the FPOs for the benefit and development of small and marginal farmers are procurement and financial facilities for inputs, dissemination of market information, technology and innovations, aggregation and storage of produce, primary processing like drying, cleaning and grading, brand building, packaging, labelling and standardization, quality control, participation in commodity exchanges, export *etc*. This resulted the low cost of cultivation, high producer share in consumer's rupee and better bargaining power of producers.

In promotion and development of FPOs many bottlenecks are embedded *viz.,* difficulty and delay in mobilisation of farmers, the organizational and management capacity is limited, limited networking and handholding support from various stakeholders, problem of raise necessary equity, limited capability to autonomously invest in primary/ secondary processing, storage and custom hiring facilities, difficulty in access institutional credit sans collateral and there are several challenges at the policy level due to poor information asymmetry and understanding as a result of which FPOs are unable to reap their real benefits to survive and development (SFAC, 2019). Hence, there is a need of hand holding support from all the stakeholders to address the challenges being faced by the FPOs by envisaged the interventions through capacity building programmes, there is a need to create the awareness among the producers and other stakeholders, need of competent Cluster Based Business Organisations (CBBOs), by increasing the farmers equity share, need to be provided input license, as well as APMC license and dealerships on priority basis, need to make available required infrastructure and technical facilities, credit guarantee scheme, by spreading the successful cases of FPOs. These may be facilitated to FPOs for their effective implementation, growth and development.

Agricultural Market Infrastructure

According to report of task force on National Infrastructure Pipeline (NIP), there is requirement of agricultural reforms broadly in three areas: e-market infrastructure, storage and processing, and research and development (R & D). The report of the task force headed by Economic Affairs Secretary Mr. Atanu Chakraborty, pointed out that insufficient market infrastructure, including grading and certification facilities, ineffective cold chain management and low level of processing of agriculture produce as main challenges in agriculture and food processing industry. Due to these challenges, India's post production wastage levels are high, leading a loss of Rs 44,000 crore annually.

The existing marketing infrastructure in the form of rural primary markets, regulated wholesale and assembling markets, grading and quality control systems, retail markets, storage including cold chain infrastructure, infrastructure required for linking the commodity futures with the farmers, perishable cargo centres, rural farm road infrastructure, market information infrastructure, infrastructure for livestock markets, poultry and livestock meat markets, slaughter house facilities and quality assurance infrastructure of various agricultural commodities is far below the desired / required levels both in terms of capacity as well as quality of the facilities. This infrastructure is also inadequate to realize the potential competitiveness of multiple commodities for taking them to the global markets.

Agriculture Infrastructure Fund (AIF)

Government of India on August, 2020 launched the Rs 1 lakh crore Agriculture Infrastructure Fund (AIF) to be used over the next four years. This allocated fund will be used to build post-harvest storage and processing facilities, largely anchored at the FPOs but also be availed by individual agri-entrepreneur. This fund will also be used to provide loans at concessional rates to FPOs or individual entrepreneurs through primary agriculture credit societies (PACs) and this initiative will be taken NABARD in association with the Ministry of Agriculture and Farmers Welfare.

There are three features of the AIF. First, the banks and financial institutes who want to avail of interest subvention and credit guarantee for their borrowers will have to sign MoU with NABARD and Department of Agriculture Cooperation and Farmers Welfare (DAC). Second, the projects will be eligible for a grant under existing schemes of Union or State Governments. Third, the scheme provides indicative targets for the States and UTs.

Agricultural Marketing Infrastructure Facilities

1. **Storage:** It is an important link in the whole system of procurement to distribution. Three-tier structure *i.e.* Central Warehousing Corporation (CWC) at National level with 99.79 lakh MT storage capacity, State Warehousing Corporations (SWCs) and the state governments at centres of States and district level with 340.82 lakh MT storage capacity and cooperatives at the primary and marketing society's level (village/taluka), are providing warehousing facilities. Financial assistance for construction of godown's at the rural level in states comes under the scheme of National Grid of Rural Godowns, which stands transferred to State Governments with effect from April 1, 1992. The Food Corporation of India (FCI) constructs godowns for storage of food grains to maintain buffer stock. It constructs storage capacity at certain nodal points, keeping in view its requirements. It is the single largest agency which has a capacity of 37.62 million tonnes.
2. **Cold storage:** India produces 134.5 million tons of fruits and vegetables and the output is likely to go up to during the next 10 years. More than 30 per cent of fruits and vegetables produced in the country are lost due to lack of proper handling, storage and processing facilities. As on 31st March, 2001, a total of 4199 cold storages are exists in the country with a total storage capacity of 15.38 million tons. Foreseeing the future requirements of the fresh/precooked/frozen fruits and vegetables and their products as well as anticipated change in the food habits in favour of processed food, the capacity requirement for post-harvest management of perishables is estimated at more than five times the available capacity. In the next 10 years, 15000 additional cold storage units with a capacity of 45 million tons should be created. The additional capacity requirement would need

an investment of the order of Rs 27,000 crores. In future, there would be a need for multi-chamber type of cold storage units for various perishables and other products.

Cold Storage industry at a glance

Projected investment in domestic capacity	Rs 16,000-21,000 crore
Existing capacity	37-39 million tonnes
Total number of cold storages	7,645
Cold storage occupancy by top commodity potato	68%
Projected growth of industry during 2019-23	13-15% CAGR
Private sector ownership of cold storages	95%
Cold storage capacity of top state Uttar Pradesh	14 million tonnes

Sources: Crisil Research/Industry, March 22 2019

3. **Cleaning, Grading and Packaging:** At present, the grading facility is available only in 1321 markets out of total number of 7127 regulated markets. The quantity graded at producer level is negligible. There is need to create facilities for cleaning, grading and packaging not only at primary level but also in the village from where produce is brought to the market sale. In absence of such facility at the village level, the kind of pollution and congestion created at market yards during the peak arrivals is well known. The APMCs should encourage private entrepreneurs to promote such units in or around the yard/sub-yards. There is need to promote proper packaging after grading so that further chances of adulteration or temptation may not be there. Therefore, there is strong need to educate the farmers for proper packaging and grading before they bring the produce to the market. Scientific packaging should be encouraged at the farm level through subsidy support. The expert committee feels that an investment of Rs 2000 crores should be earmarked for this purpose during the next 10 years.
4. **Transportation:** Efficient transportation helps in the expansion of markets by reducing the time and costs. Roads in movement of produce are just like the arteries in human body for blood circulation. Public investment in rural roads had a significant impact on access to markets by farmers, will improve the supply chain and marketing efficiency. The Indian road network is the largest in the world. According to one study, rural road investments contributed 36-68 per cent reduction in transport expenses, 27 per cent increase in agricultural wages, 5 per cent decrease in fertilizer costs, and 4 per cent increase in output prices. Railway wagons are also used for transportation of agricultural commodities from wholesale markets to consumption centers. The existing rail facilities in the country are highly inadequate as it is not connecting some of the districts in the country. Existing air cargo facilities available in limited states are in poor condition and much below the international standards.

5. **Agricultural Produce Markets:** Actual buying and selling of agricultural commodities takes place in market yards, sub-yards and rural markets/ haats spread throughout the length and breadth of the country. Agricultural produce regulated markets have been playing a major role in the smooth distribution of food grains, oilseeds, fibre crops and fruits and vegetables to meet the supply and demand needs of the farmers, traders, processors and consumers.
6. **Commodity Markets/ Futures Markets:** The risk management tools like- forward, future and options markets are used to avoid the risk and uncertainties due to wide fluctuations in prices of agricultural commodities. Forward and future contracts enable price discovery, which helps in making economic decisions as to which commodities produce, how much to sell and at what prices, how much to store and for how long and further it enhances the share of farmer in consumer rupee. The major agricultural commodities traded in futures exchanges are guar seed/gum, gram, soya oil, sugar, raw cotton, gur, castor seed, rubber, wheat, rice and other oilseeds/edible oils traded through six National Commodity Exchanges.
7. **Agricultural Credit Facility:** Agricultural credit is available to the farmers in the form of short-term credit (financing crop production programs) and in the form of medium-term/long–term credit (financing capital investment in agriculture and allied activities). Loans are also available for storage, processing and marketing of agricultural produce. Agricultural credit is disbursed through a multiagency network consisting of Commercial Banks (CBs), Regional Rural Banks (RRBs) and Cooperatives.
8. **Market Information System (MIS):** Right from planting to marketing, farmers need accurate market information at right time and place. It helps the farmers in comparing the prices offered by different firms in different markets and also in the selection of alternative outlets available. The MIS reduces business risks of farmer, sellers and traders. There are MIS centres in the country. Wholesale prices of important agricultural commodities from selected markets are collected daily by these centres and are transmitted to Head Office for further transmission to TV and AIR stations. The use of electronic medium in agricultural markets for transmission of information is relatively low. Markets of some states are linked with National Information Network (NIC-NET) to provide the speedy and timely dissemination of market information to the growers.
9. **Specialised Markets:** There is need for developing specialised markets for fruits and vegetables. It has been assessed that there are at least 241 such places in the country where fruits and vegetables markets should be developed. The infrastructure required for such markets depends on the volume of arrivals which in turn depends on the size of population to which

these markets cater. The investment requirement for fruits and vegetables markets in the country is around Rs 970 crores.

10. **Rural Connectivity:** Rural roads constitutes one of the most important marketing infrastructure which reduce the cost of production and marketing by providing external economies to farmers, traders and public at large. The status of road in India indicates that only 47.83 % villages were covered with roads till mid-90s. Considering average road length as 4 km, to connect each village with the main road or public yard, the total length of rural roads required connecting to remaining 3.7 lakh villages come to around 14.8 lakh km and the investment is requirement is Rs 74000 crores.

11. **Rural Haats/ Rural periodic market:** The haats or weekly bazaars which constitute first contact point for the producers-sellers have not been provided with amenities and facilities required for operational, technical and pricing efficiency. Improvement in the efficiency of the rural markets has direct impact on farmers' level of income. There are in all 27294 rural periodic markets including those for livestock, in the country. The investment requirement for developing these primary rural market places is estimated at Rs 2146 crores. These markets need improved services for users to facilitate marketing of the local produce, creating an element of market security for the growers. The rural market can also be used for effective credit, input marketing and procurement activities. Haats need to be provided with mobile banks on haat days by Gramin Banks.

12. **Export zones and Food Parks:**With a view to taking advantage of new global trade environment, there is a need to encourage export of high value products grown in India. Fruits and vegetables having export potential in form of raw as well as processed and packed spices like cumin, fennel, coriander *etc.* for which there is significant demand in international market. However, there is a need to educate and train the growers of these crops in producing, grading and packing for global markets and creates necessary infrastructure. A scheme of creating Export Oriented Agri-Zones (EOAZ) has been announced by the government of India which should be promoted by providing institutional and physical infrastructure in each of these as per the needs of the specific commodity. In some of the EOAZs, there is also a need to establishment food parks. In these parks, some common facilities like electricity and warehouse should be created with central government assistance which will help in attracting investment by the private sector and the state government. The estimated public investment is Rs 200 crores and private investment of around Rs 400 crores on fifty such EOAZs. In identification of EOAZs and food parks, the Government of India, through APEDA and DMI should take an active stance, rather than leaving it to the state governments.

Digital Agricultural Market

Technology imparting both tangible and intangible benefits, provide a more convenient, efficient and faster way of doing business transactions and acts as an enabler for positive progress. With the advancements in technology, businesses changed the way they transact. Most of the transactions which were traditionally handled by the skilled manpower are taken over by the technology. The scale of operations and need for transparency, use of technology have become inevitable. Use of Information and Communication Technology (ICT) and intelligent automation became necessary to position the organization in the emerging market economy across almost every sector.

The Digital technology can process large amounts of data faster, speed up the clearance process and facilitate risk management and can exchange the information in real time across the globe. Therefore, trading of commodities through web-based applications is one of such initiatives where technology has started playing a significant role. eAuctions replaced the traditional auctions and facilitated in providing larger participation, fast transactions and transparent dealings.

e-NAM is one of the most important policy initiative of the government to make potential use of ICT for the maximum benefit of the all stakeholders with particular focus on farmers. In this context, e-trading initiative for dairy cooperative by National Cooperative Dairy Federation of India (NCDFI) is worth mentioning. NCDFI eMarket is a convenient platform which is accessible from anywhere with limited hassle. The increased transparency and reduced time circle facilitate easier and faster commodity transactions through NCDFI eMarket. Large data base of sellers/buyers help members to procure and sell their products effectively and get higher returns for their produce. Considering its performance till date, and huge potential of the eMarket scenario, it is expected that most of the future trading of dairy/agri commodities would be through eAuctions. In an emerging market economy like India, NCDFI eMarket provides an excellent opportunity for the cooperatives to bring in more efficiency and transparency in their operations through etrading. It is therefore, envisaged that the cooperatives should contemplate the changing business environment, and utilize the NCDFI eMarket services for positioning themselves in a better condition.

The global digital agriculture market size is estimated to grow from USD 5.6 billion in 2020 and is projected to reach USD 6.2 billion by 2021, recording a CAGR of 9.9 %. The increase in demand for agricultural food products, shift in consumer preferences to higher standards of food safety and quality, and non-availability of labour during COVID-19 are some of the driving factors for the market.

The market for digital agriculture, by region, has been segmented into Asia Pacific, Europe, the US, and Rest of the World. The Asia Pacific region segment is to be the fastest-growing market through 2021, by value, in the digital agriculture market. Asia Pacific accounts for the most populated countries such as China and India with increasing demand for agricultural products. These countries are also the most two of the most affected countries during COVID-19. The government policies supporting the digitization of farm processes and the need for efficient usage of natural resources, along with decreasing availability of migrant labourers, are expected to increase the adoption of digital agriculture techniques (Research and Market, 2020).

Increased farm mechanization and developing digital agriculture infrastructure are expected to increase the adoption of digital agriculture among the farmers. The COVID-19 impact on the market is estimated to be positive. Labour shortages and supply chain disruptions are expected to raise the need for digital agriculture, globally (Markets and Markets, 2020).

Adoption of variable rate, remote sensing, and guidance technologies

The adoption of technologies such as VRT, remote sensing, GPS, GIS, and guidance technology in precision farming has led to an increase in productivity and a decrease in labor overheads. Likewise, the use of cutting-edge technology-based agricultural equipment has enabled the optimum utilization of resources. Precision farming equipment based on advanced technologies not only adds to the bottom-line savings by reducing the labor cost but also improves farm operations to a great extent (Markets and Markets, 2020).

Tasks performed before harvesting, such as weeding, application of fertilizers, and site inspections, account for about 50–70% of farm operating expenses. In precision farming, drones equipped with cameras are used for spraying, weeding, pruning, and harvesting applications. These drones take hyper spectral images, which help farmers to monitor processes at farmlands remotely. In the next 10 years, precision farming technology is expected to gain momentum, reducing the need for manual labor in farms, which would result in a drastic reduction in the labor cost (Markets and Markets, 2020).

Integration of smartphones with hardware devices and software applications

The smartphone is among the most widely used technological inventions in history, and with the availability of multiple user-friendly farming applications (apps), smartphones are becoming powerful and easy-to-carry tools for farmers. Most of the smartphone apps provide crucial information related to farming, such as weather and climate conditions and which effectively help farmers to make smart decisions.

Smartphones can be integrated with several hardware devices such as sensors, high-resolution cameras, and GPS receivers for various activities such as sample collection, aerial imaging, and record-keeping (a task recommended by experts). Some farming apps can load field information, and then with the help of software tools, grid it acre-wise (Markets and Markets, 2020).

Guidance Technology

GPS-based guidance technology allow growers to reduce overlapping of equipment and tractor passes, thus saving fuel, labor, time, and soil compaction. The main components needed for a guidance technology based system or device are a GPS/GNSS, DGPS receiver, antenna, controller, guidance & steering, handheld devices and display, along with the accompanying wiring harness. Other optional items include a data logger, sound device, or visual display. Guidance technology has been categorized into GPS/GNSS based guidance technology and GIS based guidance technology (Markets and Markets, 2020).

Market for weather tracking and forecasting

Increasing demand for climate information services owing to their benefits that enable farmers to deal with climate change and variability to improve decision-making in agriculture is the major reason behind the high growth of the market for weather tracking and forecasting. Moreover, the increasing demand of cloud platform for climate services is a major factor propelling the growth of the precision farming market for this application. Irrigation management is expected to witness an attractive rate of growth during the forecast period. Irrigation management involves the use of technologies such as weather-based controllers, rain sensors, temperature sensors, network elements, water meters, and sensor-based controllers. The increasing demand for water conservation and food production, and reduction in the wastage of resources are the major driving factors for the growing adoption of precision tools for irrigation management applications (Markets and Markets, 2020).

Digital Agriculture Service

The Digital Agriculture Service aggregates granular, real-time data from a variety of sources, including environmental sensors in the field, NDVI images from UAVs (that show crop stress before it's visible to the naked eye), and sensors mounted on field equipment, weather forecast data, and soil databases. By combining telemetry from these different sources and leveraging a proprietary decision support engine encoded with crop-specific business rules, the Digital Agricultural Service can provide practical recommended actions for farmers to take to improve their farms' economic output. The Digital Agriculture Service can be connected to the farm's work management system, which enables the platform to automatically schedule the relevant machinery and people when the farmer accepts specific recommendations.

Conclusion

To create double digit growth in agriculture, investment in infrastructure is indeed. Some push drivers of infrastructure recognized in this changing environment are: opening up of railways infrastructure to 100 % foreign direct investment, rural electrification, driving the construction of new airports and sea ports across India *etc*. The creation of adequate infrastructural facilities in a liberalized and market driven economic environment is necessary particularly in rural area for minimizing economic disparities between rural and urban area and it is also justified for reducing the migration of people from rural to urban centre.

Small farmers cannot hold stocks for long as they have urgent needs to meet family expenditures. Hence, the value of storage facilities at the FPOs level could be enhanced by a negotiable warehouse receipt. In this case FPOs can give an advance to particularly small and marginal farmers up to 80 per cent of the value of their produce at the current market price. But situation is that, FPOs will need large working capital to give advances to farmers against their produce as collateral. Therefore, handholding by the government in this area is required to strengthen the market infrastructure.

For encouraging private entrepreneurs in cold storage business, there is a need to provide subsidy to make the units viable for some initial years. There is need to provide incentives in reducing current expenses such as tax relief in electricity.

Rural periodic markets in general, and the tribal markets in particular, remained out of its development ambit. It was envisaged that physical markets with facilities and services would attract the farmers and the buyers creating competitive trade environment thereby offering best of the prices to the producers. Also, there is direct need for maximum use of ICT in agricultural marketing.

References

Bisen, J. and R. Kumar. 2018. Agricultural marketing reforms and e-national agricultural market (e-NAM) in India: a review. Agricultural Economics Research Review, 31 (Conference Number): 167-176.

Government of India (GoI). 2016. Agricultural statistics at a glance. Ministry of Agriculture and Farmers Welfare, New Delhi.

https://dea.gov.in/sites/default/files/Report%20of%20the%20Task%20Force%20National%20 Infrastructure%20Pipeline%20%28NIP%29%20-%20volume-ii_1.pdf

https://www.iimb.ac.in/turn_turn/managing-agricultural-supply-chain.php

Markets and Markets, 2020 : https://www.marketsandmarkets.com/Market-Reports/covid-19-impact-on-digital-agriculture-market-222616344.html

Markets and Markets, 2020 :https://www.marketsandmarkets.com/Market-Reports/precision-farming-market-1243.html

National Bank For Agriculture & Rural Development (NABARD).2018. All India rural financial inclusion survey 2016-17. National Bank for Agriculture and Rural Development, Mumbai.

National Cooperative Dairy Federation of India (NCDFI). 2020. NCDFI eMarket: A Revolutionary eTrading Initiatives for Dairy Cooperatives, National Cooperative Dairy Federation of India, Anand, Gujarat.

Reddy, A.A. 2016. Status of Market Reforms in India. Indian Farming. 66(8): 33-37;

Research and Market, 2020: https://www.globenewswire.com/news-release /2020 /07/03 /2057417/0/en/Global-Digital-Agriculture-Market-Expected-to-Grow-from-5-6-Billion-in-2020-to-6-2-Billion-by-2021-Recording-a-CAGR-of-9-9-Revised-to-Reflect-the--Implications -of -the-COVID-19-Pande.html

Research and Markets, 2020: Field mapping only : https://www.globenewswire.com/news-release/2020/07/03/2057417/0/en/Global-Digital-Agriculture-Market-Expected-to-Grow-from-5-6-Billion-in-2020-to-6-2-Billion-by-2021-Recording-a-CAGR-of-9-9-Revised-to-Reflect-the-Implications-of-the-COVID-19-Pande.html

Small Farmers' Agribusiness Consortium (SFAC). 2019. Strategy Paper for promotion of 10,000 Farmer Producer Organisations (FPOs) New Delhi, Pp 1-32.

32

Innovations in Infrastructure Development for Agricultural Growth

Pawanexh Kohli[1], Raka Saxena[2], Ritambhara Singh[3] and Vinita Kanwal[2]

[1]National Centre for Cold-chain Development and Ministry of Agriculture & Farmers Welfare, New Delhi
[2]ICAR-National Institute of Agricultural Economics and Policy Research New Delhi
[3]International Agri-Business Management Institute, Anand Agricultural University, Anand, Gujarat

Introduction

The food production in India has been on a rising trajectory and the country is among the top producers of several crops. It is the highest producer of milk and second highest producer of fruits and vegetables. Total food grain production in the country is estimated at record 291.95 million tonnes (MT; Second Advance Estimates for 2019-20) which is 6.74 MT higher over 2018-19. While enhanced field productivity is an essential component of agricultural sector, improved post-harvest handling, marketing and processing is equally essential to reduce the waste, particularly in a country where such losses are substantial and add stress on limited farming resources, appropriate agrilogistic is needed. Post-production management and marketing infrastructure is therefore, not only vital but a priority. Such infrastructure and the services they provide are essential not only for the efficient performance of various marketing functions, but also for effective expansion of the size of markets which allows for growth.An efficient post-harvest logistics network not only ensures that produce will reach destination markets in quality and quantity, but should also trigger a steady reverse flow of market information to the source.

Inadequate market infrastructure leads to higher marketing costs and results in low share of producer in consumer's rupee (Bala, 2009). At the same time, mere availability of a good market infrastructure is not enough. The modern day circumstances demand that farmers need not be forced to transact at the first probable markets near their farms, but should avail other post-harvest logistics services so that they can transact with any buyer, spatially and over time. With such a provision, farmers are better integrated with agricultural supply chains and build the capacity to supply any of the primary users of their produce. Once empowered with such connectivity options, with various market channels made accessible, farmers will have the option to undertake market-led crop diversification, to vertically integrate as a value chain component of existing processors and/or horizontally integrate with multiple wholesale markets.

Studies have reported that markets in India are devoid even of basic infrastructural facilities particularly for post-harvest handling of perishable and high value commodities. It has been reported that market arrivals have increased at a much higher rate than the growth in production, indicating a widening gap between the increase in marketed surplus and the number of markets (Chand, 2012). At the same time, end consumers are also relying on imports, which were growing in volume. This situation requires policy changes and calibration to empower the farmers to convert the higher surplus into income growth. Higher priority needs to be placed on developments that aim to increase the marketing radius of domestic farmers. This will help them first their sales transaction at optimal markets and improve the monetisation of their enhanced marketed surplus.

These infrastructure facilities are commonly delineated as infrastructure platforms for aggregating, transporting, storing and merchandising, and an IT platform for flow of information. Their components may be differentiated, depending on produce and supplies cycles, but are necessary for the agricultural market ecosystem to prosper and grow. A choice of modern technologies is available for adoption, to make the systems cross functional, dynamic and energy efficient. Infrastructure plays a strategic role in producing large multiplier effects in the economy with growth in agriculture (Mellor, 1976). If a gradual trend towards commercialization and diversification of agriculture is to be sustained and promoted, rural infrastructure supporting trade in farm products and inputs and processing of the produce must be strengthened with an emphasis on its quality (Chengappa et al, 2012). To make agriculture more viable, by enhancing incomes and making it sustainable, there is a need to develop holistic post-production management of farm produce, to enable market connectivity for perishable and semi-perishable produce, cull unmarketable lots early in the supply chain, direct the movement of produce in demand-linked manner, divert culled produce to gainful uses so as to capture value from every grain, drop and ounce produced and be efficient in supplying agro-processing industry (both food and non-food).

The real empowerment of farmers shall come by developing their horizontal integration with multiple market channels. Such horizontal integration of producers with users is most beneficial in case of perishables produce as the high quality produce can rapidly downgrade into non-saleable discards or depletion in quantity. It also ensures a wider market reach, especially where the primary produce is in demand and is directly consumed by the end consumers (as in fruits, vegetables meats, milk, etc.). In case of produce which can only be consumed after processing the farmers need to vertically integrate with the value chain of processors, and their growth is directly linked to the growth of the processors. The marketing system must play an important role in opening up markets for farmers, beyond the customary near-farm mandis, and need to function not merely as a place to sell, but as a platform that enables transactions at distant locations. For this, the system must provide suitable services in the form of logistics, IT and to facilitate and monitor the eventual sales in the One India market. Infrastructure and post-harvest services stimulate economic growth and contribute to capacity creation. In light of above facts, this chapter discusses the status and processes involved in post-harvest management of agricultural commodities, on infrastructure and its innovations in India.

Emphasis on Post-Production Activities

Farmers are skilled in production, which is their core activity. Post-production, the farmers are ingrained to seek to dispose the produce as soon as possible and prepare for the second cycle. Generally, they seek to evacuate their produce to the nearest primary market, which is their established avenue for monetisation. This is also necessitated as the scale of most output is small lots, and the focus on economy of scale in the post-production supply chain was missing. It was considered appropriate to increase the market density such that a farmer did not have to travel far to make a sale. Having to sell at markets near growing areas, the farmers began to face an inverse relation between production and price; higher the output, lower was the price in production zones. Meanwhile, the price dispersion between farm-gate and terminal user was typically in multiples, higher than the cost of intermediary function of connecting logistics. The marketing system was only serving to strengthen opportunistic intermediary trading and weaken the producer-owner. There were no services developed or made available, such that the local farmers could consolidate their small lots into a larger aggregate load, prepare it for logistics and dispatch in their ownership to a buyer elsewhere in the country. The farmer, as producer, was no longer the owner when the produce reached the next step in its life cycle, i.e. processing or fresh consumption. Typically, the market system merely facilitated to delink the farmer from produce very early in the post-production supply chain. Hence, left constrained and limited in their selling range, the farmers thereby have no further direct role in the overall value system. These methods became entrenched in the agri-marketing practices, approach and policies.

The desired situation is when the producer remains the owner as long as possible in the supply chain. The supply chain should be producer-owner driven, whether the terminal user of the produce is the consumer, or another sector or industry. This will also ensure the value of produce is affixed (priced) closer to demand centres. The necessary infrastructure to initiate this transformation is the aggregation platforms at first-mile. The ability to aggregate lots from farms as early in the supply chain as possible has multiple benefits. It brings decision making to village level, improves traceability of produce, ensures good post-production handling reducing losses adds relevant scope for waste-to-wealth interventions, etc.

Lack of suitable logistics connectivity creates a situation that markets get farther away from the reach of most villages, and therefore, the small and medium farmers do not find it feasible or economically viable to directly access optimal markets. Currently, as farmers lack any other option, intermediaries step in to close the logistics link for farmers, at times even up to the first level assembly markets (local *mandi*).In case of food grains, the government procurement system is an important mode of monetisation, wherever implemented. Further growth in this model is linked to the capacity of the exchequer to continue to spend on such procurement and develop a robust disposal mechanism like public distribution system, sale in the markets, etc.

In reality, farmers do have more than one avenue to sell into, provided they have options at hand to connect to each opportunity and depend upon the type of crop grown (Fig 1). From any farmer's perspective, the production is of value to him/her in its primary format and at its primary demand. The subsequent changes in value, unless under ownership of the farmer, is irrelevant. As indicated in the figure, the primary producer sees demand either at agro-processors or at end-consumer depending on the produce category. However, between the farmer and each user of the primary produce, there exists the need to aggregate/pre-condition/prepare the farm-produce for the transaction. This is where development efforts where relaxed and the necessary post-production activities were not organised. A more informal and fragmented approach resulted, with losses incurred on the harvested produce passed on to the harvest owner.

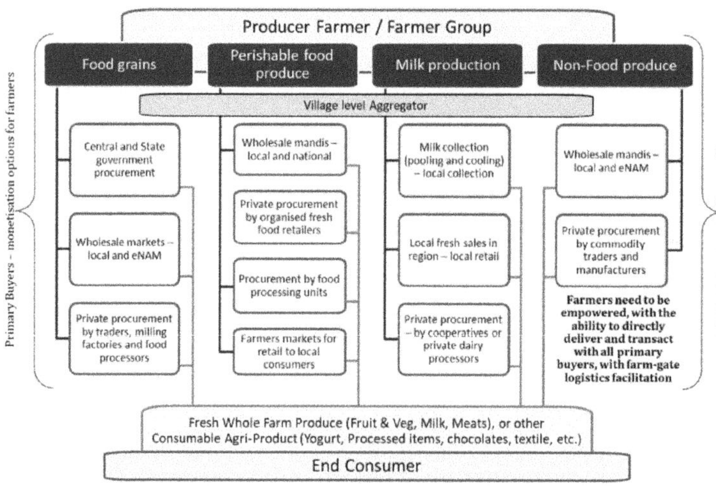

Fig. 1. Produce-wise primary channels to monetise output
Source: Authors

First-mile aggregation systems, when developed as a service, empowers farmers with viable options for post-production logistics. Currently, the farmer is not able to directly link with his/her primary buyer and this transaction is subject to multi-layered interfaces between the farmer and the primary buyer. There is also the option for farmers to directly connect with end-consumer though peri-urban farmers markets and similar, but these only provide a short-term gain on price and do not fully contribute to overall growth that can be harnessed by connecting and expanding their reach to multiple markets. Holistically, farmers need to be empowered to deliver to every market channel possible for monetising their production.

To transform farmers' income growth, it will also be important to strengthen business and operational models that encourage them integrating horizontally with their user markets. In practice, farmers would partake in both, i.e. vertical and horizontal integrations with their market opportunities. Horizontal integration expands links with multiple markets offering choice, builds capacity for market expansion for future growth, encourages entrepreneurship and market competition, mitigates risk from localised demand fluctuations, offers other earning modes, and results in greater transparency in agricultural trade; but requires associated development to organise suitable logistics capabilities and services commencing at village level. Vertical integration with a single buyer-processor, provides an assured off-take, mitigates risk from price fluctuations from external factors, promotes fixed specified quality for processors, and builds long term business interdependences, but limits farmers' growth options, as it is then immediately linked to that of the buyer.

Food Losses

Globally accepted reports of the Food and Agriculture Organization (FAO) of the United Nations state that worldwide, about 1.3 billion tonnes of food incurs loss and goes waste. Among those the highest share is in the case of fruits, vegetables and tuber crops. With greater understanding of food loss and food waste, a harmonised interpretation has emerged. Across all countries, food loss is much higher and is suffered in handling the produce until it is merchandised, while the waste is that which is incurred after monetisation in the hands of consumers. Mitigating food loss in the supply chain is a first priority and requires technology inputs. Food waste requires behavioural changes in consumers and in India, it is barely 9%of the total food loss and waste. Food losses must also be understood in light of frequent reports of unfulfilled demand of certain fruit and vegetables in large cities, while the same crop is discarded alongside farms, for want of effective market connectivity. Coincidentally, most high perishable crops are also high nutrition foods and comprise the bulk of high-value-agriculture (HVA).

The Ministry of Food Processing Industries (MoFPI) had commissioned the Central Institute of Post-Harvest Engineering and Technology (CIPHET) of ICAR to evaluate the food loss in India and the most recent report was published in 2015. The CIPHET study indicates that the post-harvest loss incurred, in per cent of production, in cereals is in the range of 4.65 to 5.99, in oilseeds &pulses 3.08 to 9.96, in spices 1.18 to 7.89, in livestock produce (milk, meats, fish) 0.92 to 10.52, and in fruits & vegetables at 4.58 to 15.88(Table 1).However, in subsequent evaluations, it was observed that the study did not consider the losses that occur in the course of long haul transport to terminal markets, having assessed only the first mile transport.

Table 1. Post-harvest Losses of food produce

Agricultural Produce	Overall Total Loss (in % of production)
Milk	0.92
Meat	2.71
Marine Fish	10.52
Inland Fish	5.23
Egg	7.19
Poultry Meat	6.74
Cereals	4.65-5.99
Pulses	6.36-8.41
Oilseeds	3.08-9.96
Fruits & Vegetables	4.58-15.88

Source: CIPHET (2015)

The losses reported in India by CIPHET (2015) are much lower than those reported globally by FAO. However, the Indian Council of Agricultural Research (ICAR) reports that the Indian agriculture sector incurs 18 to 25 per cent losses in the entire supply-chain. This is much higher than the losses reported by CIPHET, and such variance is also observed in studies by other organisations. The Small Farmers' Agri-Business Consortium (SFAC) had conducted a value chain assessment[1] in the North Eastern region in 2012,reporting higher per cent loss in the vegetables produced. Undertaking a primary field survey, the losses reported ranged from 9 per cent for potato to 32 per cent for chick pea. Other crops like cucumber, onion, chilli, ginger, pumpkin and bittergourd were also reported with losses in the range of 20-25 per cent of production. The National Centre for Cold-chain Development (NCCD) also undertook a continuous sampling survey tracing the same load across its transit from farm to wholesale markets, measuring at each stage the produce was handled. This was conducted with Amity University for 29 fruits and vegetables, in three northern Indian states over a two year period, and the report[2] demonstrated that food loss was much higher, in the range of 7 to 44 per cent.. Produce that reaches points of demand is less likely to result in food loss. Lack of a delivery system, is the leading cause for recurring losses.

[1]Value Chain Analysis of Select Crops in the North Eastern States, SFAC

[2]Study and Documentation of Produce Specific Protocols for Post-harvest and Cold-chain Management, NCCD 2017

Infrastructure status

Required physical logistics and the effective market linkages can reduce the food losses. The primary development focus for agricultural post-production infrastructure has been in the form of warehousing and cold stores, for holding inventory for extended durations. There is still lack of conducive market infrastructure facilities for the sale of agricultural produce. Several markets are still found to be poorly equipped to handle the agricultural produce; it is unfortunate that the existing numbers of regulated markets are not sufficient enough to cater to the expanding agricultural production of the country (Saxena et al., 2015). The infrastructure needed to connect with markets after the storage phase may not have found strategically linked policy support.

Warehousing for non-perishable produce

Storage infrastructure forms a necessary part of agricultural markets. Lack of inadequate scientific storage facilities causes heavy losses to farmers in terms of huge wastage of quantity and quality of crops in general and of fruits and vegetables in particular (Jairath, 2004). (Ramesh, 1999) reported that high wastage and value loss are due to lack of storage infrastructure at the farm level. The country has established widespread godowns and storage for food grains, including cereals and pulses. The Warehousing Development & Regulatory Authority (WDRA) estimated that the storage capacity of 126.96MT was available in India (Table 2).

Table 2. Status of available storage capacity in warehouses

S. No.	Organisation/sector	Storage Size (in million tonnes)
1.	Food Corporation of India (FCI)	35.92
2.	Central Warehousing Corporation (CWC)	11.72
3.	StateWarehousingCorporations (SWCs) and Stateagencies	45.28
4.	Cooperative Sector	15.07
5.	Private Sector	18.97
	Total	**126.96**

Source: Annual Report (2015-16), Warehousing Development and Regulatory Authority

ICAR-NIAP estimated that food grain will reach 281 MT by 2020-21 (179 MT for direct household consumption and 102 MT in indirect demand like fodder, seed, industrial use, etc.), a need for approximately 196 MT of warehousing (about 70 per cent of production) is frequently projected. However, interventions to improve inventory turnover ratios will free warehousing space, and should be kept in mind when planning new capacity. In addition to capacity under FCI, Central and State agencies,65.9 MT of new capacity has been sanctioned since 2001, under the Integrated Scheme for Agricultural Marketing (ISAM), of which about 58MTisthe new capacity created as of 31 March 2017. Anestimated 7MT capacity remains under construction. Including estimations by WDRA, that about 18.97 MT capacity is with the private sector, it can be concluded that the current available storage capacity is about 185 MT, almost equal to the capacity required in 2021 (projected need of 196 MT).

A large share of the warehousing capacity is for use of central and state procurement agencies. The storage capacity includes storage of type 'Cover and Plinth' (CAP), besides covered warehouses and/or silos. CAP storage is more liable to incur losses and upgradation is required. The storage with FCI, and a part of warehousing capacity with the Central Warehousing Corporation (CWC) and the State Warehousing Corporations (SWCs) is used for storage of food grains procured for Central Pool. The capacity under FCI comprises 15.43 MT owned by FCI, and the balance is hired from private sector, CWC, SWC and state agencies. As on February 2017, the total capacity available for storing Central Pool Stocks was 77.625 MT, with average utilisation of 66%of capacity. This comprises covered godowns of 62.608 MT and CAP storage of 15.017 MT. The infrastructure in the form of transport and storage for food grains is well recognised and future development is to modernise and follow advancements in storage technology. The required capacities may be reassessed, incorporating stock liquidation cycles.

Surplus grain stocks still need to be linked to end-user markets to make the inventory viable. Too often, the infrastructure model relies on rentals alone. The infrastructure is to be a tool for enabling the trade and not an end unto itself. Holding inventory, only to counter surplus supply, does not transform the trading status and is normally linked to existing demand patterns. As a result, growth in the trade can only be incremental in nature, depending largely on incremental change in population (demand). To change the situation, the need is to open trade with new markets, especially exports. Losses also need to be minimised by modernising the existing infrastructure. Future policies may look at modernising such infrastructure.

In case of food grains, each harvest cycle should trigger an inventory replacement process instead of adding new storage to allow larger unfruitful procurement. This only leads to non-productive inventory and cost overruns for the exchequer. Managing inventory-turns in Strategic Reserves will mitigate loss of inventory and safeguard value. The predetermined quantity held in reserve should be cycled into consumption, so that aged inventory need not be discarded or disposed at a total loss. A first-in-first-out (FIFO) procedure can be implemented. The available storage capacity (CAP storage, closed Warehouses or Silos) can be optimised by improving inventory distribution and management.

Introduction of "Food grain Stocking Norms for the Central Pool" for better management of food grains stocks while meeting the requirements for food security, monthly release of food grains for supply through the TPDS and other welfare schemes, for price stabilisation in the open market and to combat emergency situations arising out of unexpected crop failure, natural disasters etc. Procurement of food grains is done through FCI and State agencies at Minimum Support Price (MSP) declared by Government of India for the marketing season. Currently, MSPs are announced for 23 commodities, but price support is effective only alongside procurement, which is primarily for wheat and rice, in selected States. For the first time, procurement of pulses was also initiated under the Price Stabilisation Fund (PSF), to check price increase with consumers' interest in mind. With the farmers' interests in mind, purchase under PSF was carried out at MSP. This along with procurement made under Price Support Scheme (PSS) led to a stock build-up of 2 MT of pulses in the year 2016-17.

Market Infrastructure

Mere regulation does not help in improving market performance, if the required infrastructure is not there. Regulatory barriers constrain the investments in development of storage and processing which hampered the development of effective market institutions, and lowered the capacity of agricultural producers to be internationally competitive (Government of India 2011). The benefits received by the farmers by sale of agricultural produce in main and sub-yards of regulated markets vary from one area to another (Acharya, 2004). Market infrastructure is an important tool to facilitate the trade of harvested produce. Today, agricultural marketing is serviced through a network of regulated market yards, from 286 at the time of India's independence to 6,615 (as of31.03.2017) comprising about 2339 principal markets with remaining 4276 being their appended sub-market yards (Fig 2). There are more than 22,000 Rural Periodical Markets (RPMs), i.e. the *haats*.

The studies have reported that only two third of the regulated markets are equipped with covered and open auction platforms; only one fourth of the markets have common drying yards (Government of India, 2011). There is considerable gap in terms of the facilities created in these yards, even auction platforms are not available in one-third of these markets and common drying facilities are not available in 74 per cent markets (Acharya, 2004).

The majority of the markets have godowns or warehouses and very few have modern facilities to handle perishable produce. Scientific assaying, packaging and pre-conditioning of produce, is not readily possible at these centres. The situation is more acute in case of perishable produce, which are also traded at these markets. Milk is one of the produce types not handled at such markets. The market being a channel to the end-consumer, a more dynamic understanding on the essential infrastructure, as a tool to link with the consumer is needed. For example, majority of perishable produce such as fresh fruits and vegetables need handling similar to fast moving goods, as is visible in case of fresh milk. There is, therefore, the need to revisit the concepts and policies that were previously formulated with a non-differentiated understanding of the post-harvest management of farm produce.

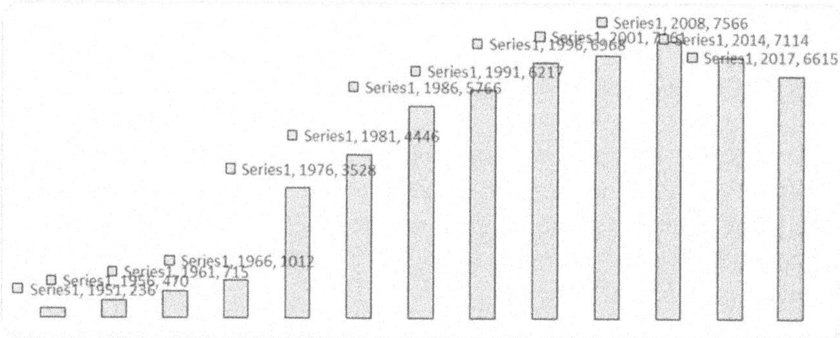

Fig. 2. Regulated Market yards in India (in numbers) since independence (*Data includes APMC markets plus their sub-market yards*)
Source: Directorate of Marketing & Inspection

The Department of Agriculture, Cooperation & Farmers' Welfare (Ministry of Agriculture & Farmers' Welfare, Govt. of India) thereafter formulated the Model Agricultural Produce and Livestock Marketing (Promotion & Facilitation) Act, 2017 (APLM), which incorporates changes to reflect the agenda of a unified national market for agriculture, besides facilitating alternate market channels, including alternate online marketing platform. This Model APLM Act recommends, inter-alia, establishing of markets in the private and cooperative sectors, direct marketing and farmers markets and not only provides a level playing field to existing stakeholders, but opens the marketing arena to new stakeholders. These initiatives will introduce competition for fair play and transparency in price discovery.

Cold-chain for Market Access

Horticultural and livestock produce are perishable and therefore, the marketing range of the farmers is inherently limited to a small radius around the place of harvest. The solution lies in developing cold-chain systems (Kohli, 2019). The cold warehouse is a middle-stage component in the supply chain, designed to buffer the delivery system. In the past, cold storage was understood to be synonymous with the cold-chain. All government subsidies had mainly focused only on supporting the development of refrigerated warehouses. The cold store, and its function, actually varies depending on its positioning in the overall movement of perishable products. A well-designed cold-chain provides produce owners time, which must be used fruitfully, to connect the produce with markets, much before its final expiry. Cold-chains ensure that fresh whole produce reaches gainful end-use and there by mitigates food loss, there in minimising the negative impact on depleting natural resources. Efficient agri-logistics, enhances market linkage and justifies efforts to produce more and brings overall growth to agriculture. In case of perishable food crops, the involved logistics chain is known as the cold-chain which comprises various allied infrastructure components. Refrigeration on its own cannot stop food loss, but when used for market connectivity, it allows the produce under care to reach gainful use. Mere refrigerated cold stores cannot fulfil this purpose for the larger basket of fruits and vegetables. Just as milk needs to undergo rapid chilling prior to subsequent market movement, produce like meats, fish and most horticultural crops also require first-stage pre-conditioning to prepare them to connect to market channels.

The post-production link to markets for horticulture and livestock requires cold-chain as a facilitator especially when growth is desired through market expansion. Therefore, integration with refrigerated transport is important – India mainly built capacity in storage only, lagging severely in refrigerated transportation.

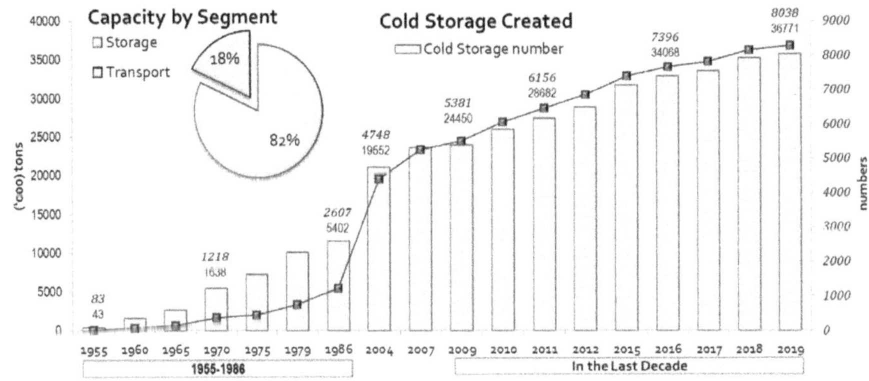

Fig 3. Infrastructure status for cold-chain (2019)

An estimated 36 MT of storage capacity in cold stores has been created (as of March 2019), but with allied development of only about18,000 refrigerated transport units (Fig 3). The transport units are exclusively as trucks, as there are no multi-modal reefer containers for domestic shipment of temperature sensitive fresh produce or processed products. A comprehensive assessment on cold chain infrastructure, conducted by NCCD in 2015, demonstrated that in relation to large capacity created in cold storage, the associated capacity in actively refrigerated transport was only about 12% (Table 3). The data in 2019 indicates that the share of reefer transport capacity may have improved to about 18%, however, this mismatch between storage and transport capacity results in incomplete solutions and associated business models; in addition, infrastructure necessary for safe handling of fruits and vegetables (such as pack-houses) to precondition the fresh produce for long distance travel were also not a focus area until 2014. As integrated pack-houses are minimal, this in turn limits the availability of preconditioned horticultural goods and discouraged any growth in reefer transport capacity. The incongruity in capacity integration, is related to the fact that the majority of the cold stores were designed to service crops like dried chillies, potato and seeds, which do not necessitate prior precooling or onwards cold transport to reach markets. As the entire chain was not addressed, it resulted in a vast preponderance of fruits and vegetable finding no recourse to expand their market linkages, which is the primary value gain enabled by the cold-chain.

Table 3. Cold-chain infrastructure – Status & Gap (2015)

Type of Infrastructure	Infrastructure Requirement	Infrastructure Created	All India Gap	% share Shortfall
Integrated Pack-house	70,080 nos.	249 nos.	69,831 nos.	99.6
Reefer Transport	61,826 nos.	<10,000 nos.	52,826 nos.	85
Cold Storage (Bulk)	341,64,411 MT	318,23,700 MT	32,76,962 MT	10
Cold Storage (Hub)	9,36,251 MT			
Ripening Units	9,131 nos.	812 nos.	8,319 nos.	91

Infrastructure in number, refers predefined unit size; in MT denotes metric tonnes
Source: NCCD(2015)

NCCD recommended the creation of modern pack-houses at village level as a thrust area, along with refrigerated transport units and ripening units. The report emphasised that the lack of allied infrastructure components, left the cold stores for the use of a limited number of produce types, and that the associated inefficiencies in the supply chain meant that the majority of perishable produce could not avail the benefits possible from the cold supply chain. Notwithstanding holding the world's largest capacity in refrigerated warehouses, India was falling far short of integrating the cold-chain.

The success of a dairy cooperative is most apparent when it functions with efficient first mile logistics, in the form of village level procurement or pooling systems. Therefore, cold-chain requires a series of activities where a common chain of custody is maintained of the produce being handled. Any activity in isolation will not typically meet the necessary objectives. Cold-chain therefore is spoken of as an integrated chain of logistics activities. The critical missing links are modern pack-houses at village level to aggregate and precondition the fresh produce for onwards market connectivity.

Agro-Processing

Agro-processing is a necessary intervention, especially in case of farm produce that cannot be consumed in its natural format. Similar mediation of a processing unit is seen in case of oil extraction, the milling units that process grains into consumer ready flour, the units that harvest and make meats and fish into consumer ready food products, and others. The processing industry, is not only an intermediary between farmer and end-consumer, but becomes the producer/initiator of a new value chain, as manufacturer/producer of a new/different product. The raw farm-produce changes form and value, and the farmer-producer is delinked after having supplied the raw produce. The processor is the terminal consumer in the farmers' value cycle. Competition among agro-processors ensures that the farmer as raw material supplier gets demand linked price for the supply of produce. In India, a very large bulk of agricultural food produce undergoes food processing. The major bulk of food grains and pulses undergo dry or wet milling processes (de-husking, pitting, powdering or polishing, etc.)and are converted into flour or other consumer ready products. Food processing is not always organised in large commercial scale, and homestead units for jaggery, honey, pickling sorbets, *kulfi*, etc. are common. Small scale food processing for pickling, jamming, dehydrating are suited methods to regain value from leftover produce and optimally attached to modern pack-houses. Increasingly, food processing is aimed at keeping the product as natural as possible. Juice making has also progressed to cold press techniques which are perceived to safeguard the inherent nutritional value.

In case of food items, the Indian consumers have a comparatively frequent buying cycle, and near daily purchase of fresh food items are common. This also stems from the fact that India is the largest concentration of vegetarians in the world. Though the bulk of food consumed is food grains, there is a growing preference for fresh fruits and vegetables. In the fruits and vegetable segment, the food processing industry, is therefore subject to steady competition from the fresh format market, besides the consumers' predilections for a specific taste or brand of processed products. The consumer's preference for fresh is an opportunity for farmers to directly connect with terminal wholesale markets, where possible.

The infrastructure needs of individual processing units are both crop and procedure dependent, while the common intervention of warehouses and transport, either dry or refrigerated, are universally shared tools. Various industrial processes are undertaken in processing units which may include multiple activities for milling, cooking, manufacturing, weaving or those that are preservative in function.

Integration in the logistics chain

Over the years, a number of institutional mechanisms have been established with a mandate to accomplish one or more areas of the agricultural marketing system such as procurement, storage and warehousing, credit, co-operative marketing, exports, food processing, agricultural prices, marketing training, research and extension. In infrastructure terms, special focus was given to creating storage capacities both dry and refrigerated as well as market yards. The most essential activity, of physically transporting the farm produce to buyers' destination, was largely unguided, left to individual commercial interests, which has then developed in a fragmented fashion. Neither was attention given to provisioning of farm-gate or village level centres, to aggregate and prepare the produce for subsequent post-production market linkage. The private sector participation in agribusiness trade also developed, given the opportunities from government's initiatives as well as the near perpetual demand for food and agri-based products. However, the various control orders to regulate and manage the market, did not allow for larger private enterprises to organise and develop more holistic and appropriate infrastructure for all types of agricultural produce in the country.

The primary produce is not consumed in its natural format but undergoes processes to make it ready for consumers. This includes treatment to grain, cereal, pulse, extracting oils, and other transformative processes. The 'difficult' crop types -are those that are more perishable and require further care in handling, -need specialised logistics, and this may have deterred equal interest from large players, except a few in the last decade. Other than the success in managing and marketing of milk, the bulk of other perishable farm produce remains subject to poor post-harvest management. The scale of private sector involvement has been slow to develop in post-production activities in perishables. Yet, consumer demand for fresh produce is seeing rapid growth in the country. From nutrition perspective, fresh food formats are considered far superior and increasingly preferred over their processed form. To strengthen the marketing reach of the farmers producing perishable food, cold-chain is needed, lack of which is a major inefficiency. The cold-chain has an empowering impact if developed as an agri-logistics system, in allowing the farmers to safely convey the value as harvested into a choice of markets. The cold-chain is a market channel that allows the harvest to access far away demand, and greater development of the relevant infrastructure is needed.

Market expansion is vital if productivity gains on farms are to translate into gainful productivity. In case of perishables, widening the market radius requires using technologies that temporarily offset perishability, so that the extra time is available to bridge the distances. The cold-chain empowers producers by allowing for cross-geography access or distance-price arbitrage, as well as time arbitrage or cross-seasonal trading. Without the cold-chain, the consequence is a local glut, where production in surplus to near-farm demand has a negative impact on price, or there are no buyers at all. In an era of technology-driven agriculture, any increase to farm-level productivity must be met with a matching multiplier in its consuming base (Kohli, 2019).

Farm-gate Aggregation for Agri-logistics

Within the country, spreading the silo and storage systems closer to high density population centres will ease the distribution bottlenecks and allow for a more streamlined supply system and help minimise losses. There is need to explore and link the stored inventory with the demand outside of the country/region. This requires greater marketing effort for export of surpluses, with suitable support provided to processing industry (food and non-food) and marketers. The support would need to be directed towards increasing the global level competitiveness of the products and to making the final products more acceptable. The perishable produce has a shorter post-harvest holding life and this category requires pre-conditioning of the food item before it can be dispatched to distant markets. Pre-conditioning is defined as activities that prepare the produce for market, without changing its essential characteristics of the produce, i.e. it remains agricultural produce (whole food) and not a manufactured product. The activities of pre-conditioning involve trimming, cleaning, pre-cooling, waxing, retail packing, labelling of fruits and vegetables and ripening if needed. These preparatory procedures extend the marketable life of the produce, for subsequent travel to consumers, including short- and long-term storage to buffer the supply.

In the case of milk, the success of this sector can be attributed to the fact that the raw milk supply chain was developed by prioritising on establishing the first level pooling systems, to function as the preliminary collection or aggregation centres. Individual farmers pool their milk at the aggregation or milk collection centres for onwards linkage, getting their share of the value ascertained at the milk receiving or processing facility – the value is not locally determined at the near-farm pooling point, but is evaluated on the basis of demand and supply at the remote processing unit. As such, the farmer is able to tap into value that is directly linked to the wider market demand. A similar aggregation system for the larger basket of perishable fruits and vegetables is needed and is in deficit in the country. Modern pack-houses receive produce directly from farms, and area vital first step in pooling and organising the post-production handling of perishable farm produce. These facilities have been developed in the country in case of grapes and bananas, and have boosted trade including exports, which in turn have been aiding greater and sustained productivity. In case of apples too, these pack-houses are increasingly seen to be assimilated into the near-farm cold storages, specially designed for apples.

It is recommended that pack-houses at village level be promoted as a priority infrastructure, to receive farm output. Pack-houses should be created with a small sized storage, only to stock about two or three days of harvested quantities, so that onwards market connectivity using transport is stimulated. If the transport arrival to a pack-house and dispatch to market is expected to be delayed, the pack-house in turn signals the farmer to defer that day's harvest, which mitigates post-harvest distress. Pack-house units therefore, can serve as signalling centres to decide the harvesting activities depending on market linkage. This signalling can be made robust by application based information messaging to the registered farmers. At such facilities, the produce is assorted into market lots by quality, physically graded for packaging purposes to ease transportability, packed for onwards transport and eventually pre-cooled to extend the freshness. Thereafter, the produce is staged in suitable lots for onwards transport to wholesale markets. Since produce is initially assayed by quality, the pack-house is the gateway that decides which destination will generate the most viable returns. Integrated pack-houses are created at village level also called farm-gate, and area necessary requirement if the farmers are to extend their reach into markets. A modern pack-house is actually the nerve centre of the fruit and vegetable supply chain. This food segment shows faster growth in demand, fed by fast changing food preferences and growing affluence in country and shows scope to transform the economic situation of farmers.

Successful deployments of modern pack-houses would have impact on farmers' incomes and provide impetus to productivity (higher yields). Created at village level, they also bring near farm jobs. In India, common examples are seen with grapes, banana, apple and floriculture. Initially, pack-houses were used in case of exports, but increasingly domestic market is also showing preference for good quality fresh whole produce. The main enabler for growth in imported fruits and vegetables is that source farms abroad, have recourse to modern pack-houses, to prepare and initiate the produce for the long travel to the importing countries. Unlike food grains, which have a simpler curing procedure at farm level, the perishable food items require more comprehensive pre-conditioning. Countering perishability has to be matched with equal logistics, to connect produce with markets well before it eventually perishesand slips into loss. Use of technology to extend holding life is not sufficient, as without a market, the goods will still perish and turn into loss. Market linked transport remains important.

Transport and storage for Agri-logistics

In case of long holding produce (e.g. cereals, food grains, etc.), the aggregation point itself can be co-located with the godown or warehouse. These dry produce types can catch onwards movement on ordinary trucks with or without any elaborate packaging. Any form of packaging or bagging is for the purpose of segregation and handling of inventory. The destination after the storage phase is normally another handler or processor where the produce is treated, extracted or milled into a final product before it undergoes retail level packaging. Modern movement of grain can also happen in conveyors or pipelines when loose bulk is handled. In most such cases, the farmers would have completed their participation in the market chain by having off loaded their produce for revenue at the first stage, in the hands of a primary user. The linkage to the end consumer is managed by the processor, or marketing agency. Commodity traders also partake as the produce is easier to handle and hold in comparison to perishable crops.

When the linkage with demand is well established through distribution channels, the produce can be forwarded in quantities that suit the processing capabilities and capacities, as well to suit dynamics of markets. For long holding produce, the physical connectivity is accessible on existing transportation modes, unlike perishables.. In effect, the storage and transport technologies for long holding produce exist. Modernisation and scaling up is required.

Major Initiatives under *AtmaNirbhar Bharat* (Self-Reliant India)

To overcome the challenges posed by the COVID-19 pandemic, the Indian Prime Minister Shri Narendra Modi gave a clarion call to become self-reliant (*AtmaNirbhar*), to bring back the economy on wheels and drive sustainably. There are five pillars to *AtmaNirbhar Bharat* and they focus on economy, infrastructure, system, vibrant demography and demand. The announcements under the *AtmaNirbhar Bharat* were made in five phases namely for Businesses including MSMEs, Poor including migrants and farmers, Agriculture, New horizons of growth, and Government reforms and enablers. To become self - reliant in Agriculture, the Government has decided to pump in Rs 1 lakh crore funds to establish Agriculture Infrastructure projects at farm-gate and aggregation points (Primary Agricultural Cooperative Societies, Farmers Producer Organisations, Agriculture entrepreneurs, Start-ups, etc.) for efficient post-harvest management of crops. It is well established fact that inadequate cold chain and Post-Harvest Management Infrastructure near the production centres has led to the increased post-harvest losses. The fund is expected to provide an impetus to create affordable and financially viable post-harvest management infrastructure

- Rs 10,000 crores has been proposed for Formalisation of Micro Food Enterprises (MFEs) to promote the vision 'Vocal for local with Global outreach'. The fund will be used to technically upgrade the MFEs to attain FSSAI food standards, building brand and marketing to help the stakeholders realise higher profits, and promote quality value addition which creates demand, both nationally and internationally. This fund will support not only the MFEs but also the Farmer Producer Organizations, Self Help Groups and also the cooperatives. A Cluster based approach as for example saffron in J&K, Bamboo in the North East will be adopted to create value chain linkages. This fund is expected to establish health and safety standards, integration with the retail markets, and the same time improve incomes of everyone involved in the value chain. Not limited to the local or national boundaries, this is also expected to reach the untapped export markets further creating income and employment opportunities.
- Rs 20,000 crores are granted to the Fishermen through PM Matsya Sampada Yojana to address critical gaps in fisheries value chain. The scheme is meant for integrated, sustainable, and inclusive development of marine and Inland fisheries. Out of Rs 20,000 crores, Rs 11,000 crores are for activities in marine, inland fisheries and aquaculture while Rs 9000 crores has been reserved for infrastructure creation like fish harbours, cold chain, markets etc. In Infrastructural set up new fishing vessels, cage culture, laboratory network etc, are describes as the key activities. This huge investment is pumped to achieve an additional fish production target of 70 lakh tonnes over 5 years, creating employment for 55 lakh persons and double the exports to Rs 1,00,000 crores
- Rs 15,000 crore is reserved for creating Animal Husbandry Infrastructure Development Fund aiming at primate investment in dairy processing, value addition and cattle feed infrastructure. The Government also aims at providing incentives to establishing plants for export of niche products.
- A fund of Rs 4000 crores was announced to promote herbal cultivation. The scheme proposed to bring 10,00,000 hectares under herbal cultivation in next two years generating Rs 5000 crores for farmers involved. The fund will also be used to set up mandi network for medicinal plants. A corridor of 800-hectare area under medicinal plants will also be developed along the banks of river Ganga.

- Rs 500 crore is proposed for beekeeping and related infrastructure. The infrastructure will be developed to establish integrated beekeeping Development centers, collection, marketing and storage centers, post-harvest and value addition facilities, etc. This will also help in implementation of standards and developing traceability mechanism. It is expected to increase income of around 2 lakh beekeepers and quality honey to consumers.
- Rs 500 crores worth Operations Greens is proposed for strengthening supply chains of all fruits and vegetables. A 50 per cent subsidy is proposed on their transportation from surplus to deficient markets and another 50 per cent subsidy on storage, including cold storages. The scheme will be on pilot basis for six months and later will be extended and expanded. This is expected to bring in better price realisation to farmers, reduce wastages, and bring in affordable products for consumers.

Modernising Infrastructure for Agri-logistics

Modernising of the food handling infrastructure will also aid compliance with the country's food safety regulations and will help ensure competitiveness at a global level. In addition, it promotes efficiency, integrity and safety of the individual operations at enterprise level. In relation to farmers' income, the availability of appropriate infrastructure, as the medium to connect with markets is important and a matter of precedence. The supply chain for non-perishable agricultural produce has the opportunity of a large storage capacity and associated transport linkages, coupled with longer holding life of the produce. Given the surpluses in stock and trends in consumption, gross level growth in this segment is intrinsically linked to population growth and with efficiency and optimisation of the existing supply chain. Another option is to foster links with international demand, for which support from industry and exporters must be garnered. However, developing exports requires long term changes to cultivation and handling practices to meet the quality norms of export markets. These interventions are ongoing and will continue to over the longer term.

To bring about the targeted doubling of the farmers' income, the identification and connectivity with domestic consumers needs to be prioritised. The consumer's preference for fresh fruits and vegetables is a decisive factor for further prioritising efforts to develop infrastructure for such connectivity. Consumption trends will indicate the sectors that show potential for immediate growth (fruits, vegetables, floriculture, dairy, fish and meats). Reinforcing, to commence any logistics chain, at first instance there is need to build a viable load for the carrier or transport. Therefore, aggregation centres are imperative to serve as loading and dispatch facilities at village or block level. It is visible in case of food grains and other cash crops, where the handling yards or warehouses have become logistics hubs for onwards delivery to users. The dry goods storage system for long term holding crops can benefit most from modernisation of existing infrastructure and improved inventory management. However, for perishables, the modern pack-house units that concentrate the harvest into market linked loads are in shortage and warrant new creation of such infrastructure. Lack of such units, allows fragmented players and traders to step in, causing multiple handling and aggravates the risks. The post-production supply chain for milk commences, at first instance, with aggregation at village level and this model can be suitably emulated and adapted for other produce types.

It is important that besides the ongoing efforts in developing of warehousing and food processing facilities, a high priority initiative be undertaken to develop the modern pack-houses and other associated components in cold-chain. Horticultural production alone is estimated at >295MT in 2016-17, and only about 300-350 modern pack-houses are developed as of now to handle such produce. This is far short of the estimated numbers required to handle the current consumer demand in a scientific manner. This is a potential area for future investments. Future development of the food processing units will bring an associated demand for skills and the industry is currently more focused on building competitiveness and compliance with food safety norms. Consumers have shown increasing preference for food items, which have lesser additives / preservatives / sugar, cold pressed, and similar. These variations in consumer mind set, is also seen in textiles, with biodegradable and natural fibres rebounding. However, consumers may not always make decisions on basis of comprehensive information but are fickle and sway depending on generic media reports. Private sector inputs indicate that this industry may be getting saturated and undergoing a plateau stage in some areas, and is in greater need of support for upgrading of processing technology and to build global level competitiveness. From the farmers' angle, the linkage with processors allows them yet another option to directly sell their produce to primary users. In accordance, promoting food processors to grow their backwards linkage for direct sourcing from farmers and to partner for quality assurance is preferred.

There is need to focus on strategies for development of an agricultural marketing system in the country with thrust on infrastructure creation, efficient flow of produce, access to market information and reduced food waste. Each aspect of strategy must aim to help farmers to organise the aggregation into viable loads, transport and sell more of what they produce, and thereby, leading to growth in income and in turn motivate further increase in production.

The small and marginal farmers fail to establish linkages with the market-domestic or international. In many cases it is found necessary to organize farmers into groups or work with existing groups, associations and cooperatives in order to supply markets successfully and this also applies to links to export markets (FAO, 2007). Farmers producer organizations could be an effective way of linking farmers to the market. There have been notable examples of co-operatives from across the world, but these are very limited in their success. It has been found that despite of being an independent entity, cooperatives have ended up as quasi Government bodies, with poor management skills. These organizations have been raised through the support and have been found that they struggle for the survival and suffer difficulties once the support is withdrawn. Farmers may directly link to the FPOs or through the groups. The advantages FPOs provide are both in establishing backward and forward linkages. They provide inputs, technical assistance, etc. They also assist in crop marketing, packaging, grading, storage and sometimes processing activities also. Furthermore, the FPOs create potential for farmers to sell large volumes and achieve economies of scale. There have been several examples all over of successful FPO's. The Report on Doubling Farmers' Income also emphasized the role of aggregation in marketing of agricultural goods, both perishable and non-perishables. The report suggested that aggregation helps in displacing the village trader based intermediation (common in fruits and vegetables) and increases the marketing efficiency of the goods. It is easier to take the aggregated sorted, graded and packaged at aggregation centers only to be transported further to market destinations. The product can also be traded online. The report also suggested the facilitation of aggregation and movement of goods from farmer to consumer via establishment of Farm-Produce Marketing Organizations (FPMOs), a cooperative or a company performing marketing only.

References

Acharya, S.S. 2004. State of the Indian farmer, A millennium study on Agricultural Marketing, Department of Agricultural and Cooperation, Ministry of Agriculture, and Academic Foundation, New Delhi.

Annual Report 2015-16. Warehousing Development and Regulatory Authority, Govt. of India. 63 p.

Bala, B. 2009. Marketing efficiency of cooperative marketing societies in Himachal Pradesh: A case study of vegetables. Indian Journal of Agricultural Marketing.23:99-108.

Chand, R. 2012. Development policies and agricultural markets. Economic and Political Weekly47: 53-63

Chengappa, P.G., A.V., Manjunatha, V. Dimble and K. Shah. 2012. Competitive assessment of onion markets in India. Agricultural Development and Rural Transformation Centre Institute for Social and Economic Change, Nagarabhavi, Bangalore-560 072

FAO. 2007. Approaches to linking producers to markets. Available at: http://www.fao.org/3/a-a1123e.pdf, accessed on September 15, 2020

GOI. 2020. Agriculture: AtmaNirbhar Bharat, Available at https://cdnbbsr.s3waas.gov.in/s3850af92f8d9903e7a4e0559a98ecc857/uploads/2020/05/2020051736.pdf, accessed on September 16, 2020

Government of India,2011. Report of the working group on agricultural marketing infrastructure on secondary agriculture and policy required for internal and external trade for the twelfth five-year plan 2012-17. Agriculture division, Planning Commission, New Delhi.

Government of India (2018). Vol III of report of the committee on doubling farmers' income on post production agri-logistics: maximizing gains for farmers.

ICAR-CIPHET (2015). Assessment of Quantitative Harvest & Post-harvest Losses of Major Crops and Commodities in India

Jairath, M.S. 2004. Agricultural marketing infrastructural facilities in India. Indian Journal of Agricultural Marketing, 18: 52-61.

Kohli, P. 2019. Public-private knowledge partnership - redefining the cold-chain. Agriculture for Development, 36 11-15. DOI: 10.6084/m9.figshare.12318740

Mellor, J.W. 1976. New economics of growth: Strategy for India and the developing world. Cornell University Press, Ithaca, New York, USA.

Ramesh, A. 1999. Priorities and constraints of post-harvest technology in India, In JIRCAS International Symposium Series No. 7, 5th JIRCAS International Symposium, Tsukuba, Ibraki, pp33-43. 9-18 September, 1999, Japan.

Saxena, R., S. Pavithra, R.K. Paul, S. Rana and K. Pal. 2015. Market intelligence: Need and relevance. Manual on Price Forecasting Techniques, pp 1-7SaxenaRaka, Pavithra S, Paul R K, Chayal S and ChaurasiaS. (Eds). ICAR-National Institute of Agricultural Economics and Policy Research, New Delhi.

33

Innovations on Extension Models for Self-Reliant India

P. Mooventhan[1], *R. Roy Burman*[2] *and Souvik Ghosh*[3]

[1] *ICAR-National Institute of Biotic Stress Management, Raipur, Chattisgarh*
[2] *ICAR-Indian Agricultural Research Institute, New Delhi*
[3] *Visva-Bharati University, Bolpur, West Bengal*

Introduction

Agricultural Extension is as ancient as agriculture itself. When it moves in agricultural trends, the importance on extension has also been transformed. Firstly, agricultural extension was the component of the research-extension-farmer approach. In the Training and Visit (T&V) period, communication of innovation was the crucial concept and the role of extension was to successfully disseminate the technological interventions. However, after the green revolution, the gap between resource rich and resource poor farmers widened with information and extension reaching only a few of them. With the conceptualization of new models, like Agricultural Knowledge and Information Systems (AKIS), the focus of extension has changed to formation of groups and collective farming. In time, the private sector has also entered into the agricultural arena, introducing the concept of pluralism in extension. Emphasis on mass media has also increased. More recently, Agricultural Innovation Systems (AIS) is introduced within the innovation systems framework. It focuses on the actors, their interactions, the institutions, and the policy environment to support or hinder it. Communication for innovation is the key AIS concept tried to identify the actors in systems, their core activities and bring them into an innovation platform for effective service and successful innovations (Saravanan, 2017).

Non-availability of appropriate technologies and relevant information to all categories of farmers is the major hurdle for achieving sustainable farm production. Many research report indicated that our farmers knew only 40 percent of the developed farm technologies and hence there exists a wide knowledge gap. Moreover, it is noted that, not all the technologies transferred are known and used by majority of clients. Access to information and improved communication is a crucial requirement for sustainable agricultural development. Modern communication technologies when applied to conditions in rural areas can help improve communication, increase participation, disseminate information and share knowledge and skills. However, it is observed that the rural populations still have difficulty in accessing crucial information in order to make timely decisions. There is a concern that the gap between the information rich and information poor is getting wider. New information and communication technologies are generating possibilities to solve problems of rural poverty, inequality and giving an opportunity to bridge the gap between information rich and information poor and support sustainable development in rural and agricultural communities. It is essential that information availability is demand driven rather than supply driven. The challenges are not only to improve the accessibility of communication technology to the rural population but also to improve socio-economic status and its relevance to local development (Balit et al., 1997).

Major Issues in Extension Models

While all the models tried in recent times have been theoretically sound, the problem lies with the implementation. Extension was once merely a mouthpiece of research. It ignored resource-poor farmers during the green revolution era, mostly because it was not equipped to communicate on a mass scale. While T&V was an acclaimed approach, reaching the grassroots became very costly and due to lack of fund, it has to be abandoned. In the AKIS model, the last mile is still unreached and collectivization is also happening at a very slow pace, further marginalizing marginal farmers. With pluralistic extension, the duplication of efforts has occurred because of a lack of communication among stakeholders. In the AIS model, convergence remains a lacuna because of the lack of a common platform for research, innovation and dialogue. Effective communication and reaching the last mile are still major hurdles in extension models, which make it vitally important to focus on appropriate technologies and available forms of ICTs to reach, communicate, converge and improve on extension to reaching to farmers instead of investing in the development of new models of extension. Models like the Agricultural Technology Management Agency in India, implemented with the aim of increasing convergence and reducing the duplication of efforts in agricultural development, have failed to achieve the desired success. To a great extent, this could have been avoided with the proper use of ICTs and creating a knowledge network (Saravanan, 2017).

Issues in Technology Adoption

About 70% of the people in rural are engaged in agriculture, hence activities designed to address the vulnerability of these rural poor are often geared toward educating agricultural practices. The following factors are considered important for technology adoption (Parvan, 2019).

S.No.	Factor influences adoption of technologies	Nature of influence
1.	Farm Size	Farm size is one of the important factors measured when modeling adoption processes due to its diversified holdings and location specific nature.
2.	Risk and Uncertainty	Almost all the technological adoption decisions carry with them some mixture of subjective risk such as variations in rainfall, pests, diseases and other blights, and the timely access to critical inputs.
3.	Human Capital	It is based on individual or community characteristics like education, human health indicators, age and gender demographics, and their relationship to technology adoption.
4.	Labor Availability	The labor market affects technology adoption based on the net labour shortage or surplus; seasonal availability adds further more. In addition, whether the proposed technology is labor-saving or labor-intensive. Logically, higher labor supply is associated with higher rates of adoption of labor-intensive technologies.
5.	Credit Constraints	Access to credit is an important indicator to adopt any new technologies. Especially, it significantly affects small and marginal farmer's adoption preferences.
6.	Tenure	Mainly tenure incorporates issues addressed are farm credit constraints, risk and uncertainty
7.	Commodity Market Access	New technological interventions often required recurrent and steady use of new agro inputs such as fertilizers and pesticides. Farmers needs to dispose their increased output. If there are no market facility available their entire efforts will be in vein. Further, their investment in new agricultural technologies will be of no use.

Indigenous Technologies and National Research System

As per NARS, indigenous technologies would bring significant relevance to research domain, particularly at regional levels. The strongest point of indigenous techniques is relevance to research and when scientific community use this as the initial point of their own methodical research efforts, relevance would be built in their research programmes. Indigenous technologies which are arise from within this mileage are, by definition, fully reliable and it has been totally imprudent on our part to ignore this logic (Jha, 2008).

Blending of Indigenous and Scientific Knowledge

Agricultural research system has been highly discipline-oriented in our nation. Customary scientific approaches generate packages, while resource-poor farm families involve in farming as a nonstop performance. Institutionalised technologies are focused mainly on achieving high yield of targeted crops. The introduction of modern scientific technologies through the application of chemical fertilizers, agrochemicals, machinery, and irrigation in developing countries was a deviation from our traditional agriculture and greatly contributed to pollution and land degradation. Subsequently, lack of relevance to small and marginal farm conditions was found to be one of several constraints in the research institute technologies. In addition, while developing the technologies farmers are usually not considered as a source of innovation. In contrast, farmers experiment on alternative coping approaches to avoid extreme climatic conditions diversified cultivation techniques like intercropping and border cropping in order to widen their food and fodder requirements, and altering their sowing and harvesting periods to meet the local market demand are commonly ignored (Prakash *et al.*, 2008).

Emerging Information and Communication Technological Innovations in Agriculture

Mobile apps in technology transfer

Timely information access is a crucial requirement for decision making in agriculture and allied sectors. ICTs are easing faster access and exchange of targeted information. Among ICT gadget usage, there has been increasing use of mobile phones with a number of services providers. Mobile phones have entered in rural India linking the digital divide era, resulting in development of a number of mobile applications by various government and non-government agencies. Many mobile apps which are specifically aims to address farmers need and farmers are using these application to find out the market price of commodities, weather, plant protection and other advisory services. With the increase number of mobile application farmers' got more awareness on the specific information and services provided by these agro apps.

The key advantages of mobile application for farmers are, easy to access information on their mobile phone. The information can be stored in the mobile itself for easy access. Wherever the content is dynamic, such as weather information, market details, advisory services, the mobile application requires connectivity to get the data from the back-end server databases. The mobile services, predominantly the SMS service is only a one-way information provider to the farmers. The farmer needs two-way-communication for day-to-day farming and some of the application like Kisan Suvidha, Pusa Krishi, Soil Health Card App, Crop Insurance app, Plantix, Digital Mandi, e-NAM, IFFCO Kisan Agriculture, APEDA Farmer Connect, Shetkari Masik Android App, riceXpert, Pashu Poshan, Cattle Expert System, mKrishi, Fisheries App, Fisher Friendly Mobile Application (FEMA), RML Farmer, MyAgriGuru, Rythu Nestham, etc. are providing dynamic information.

Need of Mobile ICT Intervention for Agriculture

Agricultural sector role in economic development is essential in developing countries. To progress the welfare of the majority of people is by helping to rise, (i) the farmer's productivity increase by growing food and cash crops, (ii) the market prices they getting for those crops produced. Though, it is an essential condition, increasing agricultural output is not by itself adequate to attain an increase in rural welfare (Perkins et al, 2007). Telecommunications and ICTs, as a means of sharing information, is not simply a connection between people, but a link in the chain of the development process itself (Hudson, 1995). To reduce the information gap between the farmers and to build productive and competitive market, different ICT tools providing to make rural and under-developed markets become efficient and productive. Due to the knowledge gained through ICT gadgets, farmer can effectively manage with the middle men or the market forces to get better price for his produce.

Agro-information services through Mobiles

The internet and mobile networks have the potential to deliver agro-information that are easly affordable, relevant content (timely and customized information), searchable contents and up-to-date information (Ramamritham *et al,* 2004). The mobile technology offers electronic capabilities (battery backup, processing power and memory), reach to mass customer, privacy of user, and anytime/anywhere, no physical contact services (Helen Nierinck, 2008). Keeping these issues and needs of farmers in mind, numerous applications and services provided by mProjects for dissemination of farm information. The Value Added Services (VAS) focused on crop advisory, market intelligence, forecasting, micro-finance and finally local social networking at rural settings.

Mobile based projects' overview for rural India

The following initiatives are some of the important mobile based projects implemented in the country to serve the various stockholders engaged in agriculture.

i. **Aqua Mini**: This initiative offers real-time decision-support tools (aAQUA) to progressive farmers and organizations (Bahuman and Kirthi, 2007). This project is working on the basis of revenue generating business model. The information services provided are largely to farmers include, localised, remote crop diagnostic solution, audio prompted guide applications (in English/Marathi/Hindi), disease diagnostics, micro-weather info, SMS enabled register & query mechanism, online poll, and service is available on GSM and CDMA networks. The aAQUA eAgriService is a problem-solving system dedicated to find solutions to problems posed by Indian farming community. aAQUA was developed at the Developmental Informatics

Lab at IIT Bombay and this lab involved in various projects funded by Media Lab Asia, Ministry of Communications and IT (GoI), Development Gateway Foundation (GoI, World Bank) and Pan Asia Networking (UNDP, IDRC, APDIP, ISOC, APNIC and Microsoft).

ii. **Fisher friend** – The Fisher Friend project of the M.S. Swaminathan Research Foundation (MSSRF) in Tamil Nadu and Puducherry leverages mobile technology to provide vital livelihood information to fisher folk. MSSRF partnered with Qualcomm, Tata Teleservices and Astute Technology Systems for developing the Fisher Friend Mobile Application. The tool was designed after a thorough needs assessment of the fisher communities and incorporation of feedback from central stakeholders. Upon sending a single-button-click request from an icon-based software module on mobile, fishermen gain access to vital updates on wave height, wind speed and direction, potential fishing zones, news, government schemes and market prices. All content is displayed in the local language - Tamil. This unique application is helping fisher folk make better choices and avoid hazardous situations. It is enabling them to conduct their livelihood operations in a safe and profitable manner. The services provided are broadly to fishing farmers include, weather forecasting; market price; sea wave heights information; fish shoals location information (via satellite scan data); and service is available on CDMA (pre-configured handset) networks only, but not on GSM networks.

iii. **mKRISHI** – Tata Consultancy Services' (TCS) Mobile Agro Advisory System (mKRISHI) connects farmers with an ecosystem that empowers them to make effective decisions about agriculture, drive profits, and conserve the environment. This allows the farmer to make a query in a local language from a mobile phone and receive personalised advice or relevant information on the same in local language (Horvath, 2008). This is the project working on private partnership based revenue generating business model in Maharashtra and Uttar Pradesh states at present. It is testing to test its sustainability with Indian farmers' needs. The services provided are broadly to farmers include, crop disease diagnosis; sensors based remote land & crop property recording (grape, cotton, soybean and potato); micro-weather Information (temp, cloud cover, precipitation) and service is available on CDMA networks only, but not on GSM networks. The mKRISHI application enable farmers to send queries, comprising of text, voice and pictures, specific to their land and crop to agricultural experts, using their mobile phones.

iv. **Reuters Market Light (RML)** - provides Indian farmers up-to-date farm information, local and customised commodity pricing details, news and weather updates (Amit Mehra, 2007). The project is working on public private partnership (PPP) revenue generating business model in Maharashtra and Punjab states. The extensive services provided to farmers which includes, localised - commodity pricing (Onion, Cotton, Soybean, Pulses, Pomegranate); weather updates; news (agriculture & general) and service is accessible on GSM networks only, but not on CDMA networks. RML is a pioneering mobile phone-based highly-personalised professional information service particularly designed to support the farmer community.

v. **IFFCO Kisan Sanchar (IKSL)** - The idea is to make use of IFFCO's extensive reach and establish a low cost telecom delivery channel through the network of cooperative societies. To achieve the task IFFCO tied up with Airtel to build and provide a platform for the farmers through the cooperative society network. The distinctive venture offers the farmer the much desired inputs on real time basis which is going to assistance him on agri-related issues and would guide him for his day to day chores. The project is working on public-private-NGO partnership based revenue generating business model across major states covering in two stages. The services are largely to farmers include, telecom products and services of Airtel; free daily voice updates on VAS platform (mandi prices, farming techniques, weather forecasts and fertilizer availability) and dedicated helpline for farmers to answer their queries

vi. **Life Tools** – It has a range of agriculture, education and entertainment services designed specifically for the consumers in small towns and rural areas of the emerging markets. This project going to work on Private Partnership (PP) based revenue generating business model in our country. The services provided specific to farmers include, information on seeds, fertilizers, pesticides, weather (temperature, rainfall, wind conditions) and prices in English, Marathi and Hindi language option and prevailing market prices, education service in dual language display option.

vii. **CERES** – This initiative aims to assist farmers by providing exhaustive information covering all areas in timely and customized manner to meet specific local needs to increase the overall productivity of agricultural practices (Anurag at el., 2008). This is the project working on private partnership based revenue generating business model in Gujarat state covering 78 villages in Vadodara district. The services provided specific to farmers include, information on seeds, fertilizers, pesticides, disease and farming input; market prices and weather (micro-climatic, rain/storms, temperature, humidity, precipitation, wind speed) on weekly and monthly basis.

viii. **KISSAN Kerala** – is an integrated, multi-modal Agricultural information system, which provides several dynamic information and advisory services for the farming community across the state. The core achievements of the project are integrated multi-component, multi-modal delivery of Agriculture Information Services system that is accessible anywhere/anytime by all concerned. The project has adopted an approach of providing right information to the right people and empowers the farmers with adequate knowledge, which helps them to take appropriate decision. The project also solves the problem of content gaps by providing the authentic agricultural information through various delivery methods like Television, Internet, Telephone, and Mobile. The farmers can choose any medium to seek the relevant information. The project offers the following major services through the effective integration of ICT systems and tools to reaching out to the farming community.

(a) **Online Agri advisory service:** The dynamic, multi-lingual web portal based online Advisory services for the farmers (www.kissankerala.net) - This dynamic portal provides several interactive contents and advisory services to the farming community. The farmers can seek the advisory from the expert scientists through online and get better scientific advisory for their problems.

(b) **Kissan Krishideepam:** The production and telecast of agriculture based weekly television program - in local language through satellite channel.

(c) **Online Agri video Channel:** The project has launched the country's first online video channel in Agriculture in collaboration with Google and YouTube with more than 150 telecast quality videos are available for mass access.

(d) **Tele Advisory Services:** This project also provides telephonic based Agri advisory services through a dedicated telephone number to provide real-time information and advisory; A dedicated telephone connection has been established and backed up by Agricultural scientists to meet out the farmers need.

(e) **The mobile based Agri Advisory services:** The project provides text, voice and video content based Agri information services through mobile phones. A dedicated PAN India based mobile service has been established to provide text, voice and video based information services. It offers several agro-services like crop advisory, weather forecast, soil test information etc through farmers mobile.

ix. Sanchar Shakti - This scheme has been implemented by the Universal Service Obligation Fund (USOF) of Department of Telecommunications (DoT), Ministry of Communications & IT towards the empowerment of rural women. "Sanchar Shakti" is a programme of pilot projects aimed at facilitating SHGs' access to ICT enabled services.

x **Engaging Farmers, Enriching Knowledge: Agropedia - ICAR initiative** Dynamic content availability and its intelligent organization continues to be a serious challenge in Indian agriculture scenario. This prevents offer of meaningful and efficient advisory and allied services to farmers and other stakeholders. **Agropedia** is an effort to infuse semantic and social networking technologies into agriculture information management system to alleviate this kind of farming related issues. In its short span (less than 2 years) of time; it has developed several path breaking concepts and demonstrated their feasibility:

- Crop knowledge models (KMs)
- Use of KMs for tagging content and people; useful for searching information and locating people with similar interest
- Content management platform for storing and searching agriculture related information.
- Agropedia deployment options – It offers appliances for off-line/on-line access where the connectivity is poor
- Social networking platform – it gives like wikis, blogs, chat rooms for interconnecting agriculture community.

Thc utility of agropedia and its worth as a concept has been demonstrated in Phase I. The agenda for Phase II is to translate these elements and strengths into a community product that can gain widespread acceptance amongst the varied constituencies in Indian agriculture, and which in so doing be able to quickly react to the emerging challenges faced by the sector. In the next phase the work on issues related to the form and nature of content, and its distribution has been taken up. Since the information is hosted in the electronic format it is also possiblc to make it *active* and *personalized contents*. A simple example of such content, is the package of practices: an electronic version of package of practices is an *app* (short form for application) sitting either in the server or the Farmer/Extension workers (cell phone) which can 'wake up' every day and figure out any intervention/care is needed. Such an *app* requires the content to be time-line-based, the farmer be registered (what crop, soil characteristics) and the interventions in the crop so far (like "fertiliser applied on day 16").

Phase II mainly concentrates on the creation and development of new focus areas, such as content streams, user groups and partnership opportunities. These can be summarised under the following activities, which are then elaborated on:

Activity 1: Extension Content Mobilisation

Activity 2: Evangelisation, Capacity Building

Activity 3: Social Network Development, Impact monitoring/assessment

Activity 4: Extension 3.0 / vKVK (voice KVK)

Activity 5: Knowledge models 2.0

Activity 6: Platform Enhancement

Activity 1: Extension content mobilisation

Agropedia, in its present form is a well organised book rack in need of books to fill its shelves. There is a need for good quality content to populate agropedia. In this context, UASR will play the role of extension content provider for existing 4 crops (Paddy, Pigeonpea, Chickpea and Cotton) and new crops/topics selected namely Pomegranate, Chilli, Inland Fisheries/Livestock Management (Cows and buffaloes). They would build capacity of KVK personnel and subject matter specialists as well as develop relevant e-extension compatible content – text, photo, video and audio for selected crops/ topics.

Providing SMS services to the selected progressive farmers of 79 districts where the Krishi Vigyan Kendras are operational in the state of Uttar Pradesh (66) and Uttarakhand (13). Further, timely alerts on emergent information and contingencies for adversaries like flood, drought or outbreaks of insect-pests and diseases also offered. The content development and its regular updation are taken up on agro-climatic zone basis. There are nine agro-climatic zones in Uttar Pradesh and one in Uttarakhand and accordingly, zone specific seasonal messages are developed for its transmission to the selected farmers of different districts. Video and audio clippings are also added. All this information is aggregated, resulting in a valuable resource with potential for commercial distribution. Commercial potential of advisories are evaluated by IIM (Indian Institute of Management) Kolkata, while looking at the business models followed by Reuters and IFFCO (Indian Farmers Fertilizers Cooperative Limited).

ICRISAT also participates with UASR (University of Agricultural Sciences, Raichur) and select SAU partners in mobilising upload of validated documents such as POP's as information objects into Agropedia. In order to facilitate this, ICRISAT will offer MSc and PhD level fellowships which proved a success in the previous phase. ICRISAT will also organise workshops with MoA (Ministry of Agriculture)/GoI (Government of India) officials concerned and with select States Directorates of Agriculture to obtain extension content that is certified.

Activity 2: Evangelisation, capacity building

All partners participate in this activity, with participation in seminars and conferences as well as organising joint workshops, participating in Krishi melas etc. IIM Kolkota leverages the strengths of agropedia to create a fabric of knowledge that will serve new needs and expected challenges. This fabric is embodied in the vKVK and Extension 3.0. This involves capacity building, user-centric demonstrations, developing incentive compatible mechanisms to catalyse and sustain interest in user-driven knowledge generation modules such as wikis and blogs. ICRISAT participates in the efforts to popularise the use of agropedia as a platform including its use as software as a service (SaaS). It would aim at having at least one university having its entire extension portfolio on Agropedia SaaS.

Activity 3: Social Networking, Impact monitoring/assessment

In this phase, the primary is on issues related to the form and nature of extension (a fuller explanation follows), and its delivery mechanisms through electronic means. The technology platform allows the formation of a network in which extension personnel are able to provide crop knowledge to farmers (and their representatives); in which formal and experiential knowledge is exchanged. IIM Kolkota will undertake to monitor the agropedia website - in particular usage patterns, needs and pathways - to ensure that the development aims of agropedia Phase II are met. To do this, use of the agropedia platform, features and tools must be consistently high at all levels of user expertise - a requisite for content creation from the community. Focus is maintained on key events (related to the weather or the crop calendar for example), and on accessibility of information objects (documents or media) and ease of use.

A guiding premise for Phase II is that agriculture extension can benefit substantially by using emerging social networking technologies like blogs, twitter and special interest groups. agropedia can form the basis of a modern extension mechanism (Extension 3.0) creating virtual communities of practice and knowledge exchange. A proposed feature of this activity is the building of a social networking platform specially tuned for the agriculture community and the evangelising of its use.

Activity 4: Extension 3.0 / vKVK

A voice KVK (vKVK) is a set of scientific advisors (KVK scientists, extension scientists) and peers (lead farmers) connected through mobile and internet technologies to help farmers. In the vKVK, unlike in a traditional KVK, the interaction between the two parties can be entirely on electronic mode. The agropedia platform acts as a 'middle ware' for this interaction providing amplification (one-to-many), persistence (messages are stored and that can be searched, retrieved also), monitoring and other utilities which are possible when the content is electronically stored and semantically indexed. Sub-systems are developed to address the needs of vKVK scientists, farmers and the middle ware to deliver advisory services, alerts and Q&A services over SMS, voice and Web. The farmer's default interface would be a mobile phone and the KVK Scientist and the Extension worker would interact through a web-client running on an ipad-like light portable device or even a desktop.

ICRISAT has a research interest in documenting and assessing the way new information linkages and flows influence the adoption of research-derived technologies and practices in farming. Thus, ICRISAT works with the UASR and ZPD in extension 3.0 activities, primarily looking at changes from a baseline scenario over the project duration. Standard methods in research on development communication will be followed in setting up the baseline. ICRISAT, as an auxiliary activity, will promote Extension 3.0 in the cluster of villages in Addakal Mandal in AP (Andhra Pradesh) State, with a community-based woman micro-credit/SHG federation as a special process.

IIM Kolkota leads the effort to develop and deploy tools to gauge the progress of vKVKs as well as of alternate delivery modes. It would also develop training modules and train extension scientists on the use of Extension 3.0.

Activity 5: Knowledge models 2.0

The knowledge models in agropedia at the moment are primarily intended for tagging. They can be extended to more sophisticated levels to support services based on the knowledge model itself – like diagnostic and recommendation systems.

ICRISAT makes use of institutionally available expertise on the five crops above and carry out changes and refinements in the available crop KM's. The refined KM's will be far more amenable to extension information in different formats (including audio and video and animation). These will be validated with extension and agricultural communication experts through crop-specific workshops and published via Agropedia. On a trial basis, in collaboration with the IIT-Kanpur, ICRISAT will build inferencing layers in support of extension requirements.

UASR will develop Knowledge Models on Pomegranate, Chilli, Inland Fisheries/ Livestock Management (Cows and buffaloes), Providing content for Pomegranate, Chilli, Inland Fisheries/Livestock Management (Cows and buffaloes). Four KVKs namely Saharanpur, Kanpur Dehat and Pratapgarh from Uttar Pradesh and Dehradun from Uttarakhand will be included in the programme for taking up special activities like Knowledge Model development by ZPD.

Activity 6: Platform Enhancement

The agropedia platform turned out to be robust and has not suffered a single crash since it was launched. With an increased understanding of the needs of the stake holders and the increased *demands* from them, the platform needs to be extended and enhanced. SMS and Voice based workflows for delivering extension services need to be developed. A KVK net platform for social networking based features of agropedia has been generated.

Mobile Governance policy

The governments around the globe have long recognized the need and potential of ICTs tools to make government services available to all the residents at their door step. As a result, e-Governance has emerged as a potential option to deliver government services around the world. However, e-Governance in an implementation sense is restricted primarily to the use of computer based internet access to deliver multidimensional services. In countries where the availability of computers and internet is comparatively low, such as in India, there is an apprehension that the reach of e-Governance may be limited and it considered as one of the bottleneck.

The limited reach of e-Governance has made governments think of new technologies, such as mobile phones, to reach the residents and deliver public services. The e-government is generally defined as the conventional government services made available for all users through electronic means such as computers connected to the internet. Now the mobile government or m-government is defined as the overall strategy and processes for delivering various public services through all kinds of wireless and mobile technologies, applications and devices to all users, including residents, government departments and agencies and businesses.

Potential of Mobile Governance

The mobile phones have marvellous potential to increase the access to and reach of public services in India. The swiftly expanding subscriber base of mobile phone users in India can help in accelerating the use of ICTs for enhancing governance and ushering in comprehensive development. The relevance of mobile devices as a medium for delivery of public services is also obvious when we compare the subscriber base of mobile phones to that of the internet connectivity. Extensive access to mobile phones in the country has made it an ideal platform for Government to resident interface, especially in the rural India.

Policy Goals

In order to attain the vision and mission as specified above, the mobile governance policy of Government of India aims to:

a) Build an enabling mobile service delivery infrastructure system which consis of a Mobile Service Delivery Gateway (MSDG) that is fully integrated with the existing infrastructure created under the National e-Governance Plan (NeGP).

b) Formulate the relevant standards for applications for mobile governance to safeguard seamless interoperability of services across multiple service providers and multiple Government departments and agencies as well.

c) Develop a suitable regulatory mechanism for mobile governance to safeguard proper coordination among manifold stakeholders, ensure compliance with the standards for applications and ensure seamless interoperability of services and implementation of short and long codes for public services across multiple service providers.

d) Develop suitable interface to allow users to pay for public services through mobile phones.

e) Identify vital public services for delivery though mobile platform through stakeholder consultations.

f) Formulate a Project Assessment Framework for the Government departments and agencies to safeguard compliance of services planned under the mobile governance policy.

g) Create a state of the art knowledge portal as well as various toolkits and gadgets for deployment of mobile governance. The purpose will be to make India a thought leader as well as a global role model in the domain of mobile governance services.

h) Develop and implement innovative Public Private Partnership (PPP) and Multi Stakeholder Partnership (MSP) models for design and delivery of mobile governance services as well as encouraging development of cloud based implementation models and use of light technologies.

i) Deployment of a suitable capacity building framework to enhance delivery and absorption capacity for m-Governance services.

References

Amit Mehra (2007) "Reuters market Light now available in local post offices across Maharastra", Press statement on December 20, available at: http://www.reuters.com.

Andrei Parvan (2019) Agricultural Technology Adoption: Issues for Consideration When Scaling-Up. Cornell University. Available at https://blogs.cornell.edu/policyreview/2011/07/01/agricultural-technology-adoption-issues-for-consideration-when-scaling-up/

Anurag P., Vivek B. and Sasank T. (2008) "CERES Information Services", Press statement, available at: http://www.ceres.co.in/.

Balit, S., Calvelo Rios, M. and Masias, L. 1997. Communication for development for Latin America: A regional experience. FAO. Rome, Italy.

Bahuman Anil and Kirthi R. (2007) "aAqua Mini", available at: www.agrocom.co.in

Danielson Anders (2001), "When do the poor benefit from growth, and why?" Poverty and Growth workshop, Lidingo, August 8th.

Dayanatha Jha (2008) Indigenous Technology and Agricultural Research System. Agricultural Economics Research Review Vol. 21 January-June 2008 pp 1-4

Helen Nierinck, (2008), "Mobile proximity payments: Scenario for market development. Analysis Mason", 2008/6, Table 1.1.

Robert Horvath (2008) "Innovation-Mobile Services", CII/GIS 2008 proceedings, May 28th 2008.

Hudson E. Heather, (1995), "Economic and Social Benefits of Rural Telecommunications: A Report to the World Bank", available at: www.usfca.edu/fac_staff/hudson/papers/BenefitsofRuralCommunication.pdf

Misra Satyan, (2005), "Prospecting Rural ICT business–Drishtee Model", Online: visited on August, 2005, http://digitallearning.in/.

N. Prakash, S. S. Roy and S. V. Ngachan (2008) Role of ITK in Conservation Agriculture: Blending Indigenous and Scientific Knowledge. Available at http://www.kiran.nic.in/pdf/publications/ITK.pdf

Perkins, Radelet, Snodgrass, Gillis, Roemer, (2007), "Economics of Development", Fifth edition, W.W.Norton & Company, Inc publication.

Parvan, A. Agricultural Technology Adoption Issues for Consideration When Scaling-Up. Cornell Policy Rev. 2011, 1, 495–504.

Ramamritham K. (2006), "Innovative ICT tools for information provisioning via agricultural extensions",1st IEEE/ACM International Conference on ICT4D, Berkeley.

Ramamritham Krithi, Bahuman Anil and Duttagupta Subhasri, (2004), "aAqua: A database-backended multilingual, multimedia community forum", ACM Conference Proceedings.

Rao, Kasina V., Sonar R.M. (2009), "Farmers Needs and the Use of Mobile Services and Applications: A Study on Indian Pilots", Proceedings of the International Conference on Knowledge Management and Intellectual Capital, Institute of Management Technology, Ghaziabad, India.

Saravanan, R. and Suchiradipta, B. 2017. Agricultural Innovation Systems: Fostering Convergence for Extension. MANAGE Bulletin 2 (2017), National Institute of Agricultural Extension Management, Hyderabad, India.

Saravanan R. 2017. Modernising extension: New models or better communication? Retrieved from https://spore.cta.int 05th December, 2020.

Thomas K. Thomas., (2007), "Qualcomm mapping out 4 rural connectivity projects", Business line magazine, online edition, available at: http://www.thehindubusinessline.com/2007/09/10/stories/2007091051010200.htm.

34

Innovations on Agricultural Education and Policy

Prabhat Kumar[1], *P.K. Ghosh*[2]

[1]*National Agriculture Higher Education Project, ICAR, New Delhi*
[2]*ICAR-National Institute of Biotic Stress Management, Raipur, Chattisgarh*

Historical Perspective of Agricultural Education Policy in India

Pre-Independence Era

During the reign of the British Crown, vocational education and training suffered a great setback and there was a lack of emphasis on agricultural education, even though the Indian economy, to a great extent, depended on agriculture. In 1901, only five institutions were imparting theoretical and practical instruction in agriculture, but that education was not used in extension or fieldwork—it served only as an opening for government white-collar jobs (Crane, 1965).

Post-Independence Era

Prior to the establishment of agricultural universities, university-level education in agriculture, animal science, horticulture, agricultural engineering, and allied subjects were largely conducted by colleges affiliated with general universities that had no special commitment to the farming community. The University Education Commission of 1947, among other things, examined the state of agricultural education in India. The report of this Commission emphasized the need for having what was termed "Rural Universities." In 1954, the First Joint Indo-American Team on Agricultural Research and Education fully endorsed the recommendations of the University Education Commission and strongly recommended the immediate establishment of agricultural universities in a few selected states. The U.S. agricultural advisors who began to arrive in India in the early 1950s were products of the land-grant university system, and many of them naturally believed India could benefit by creating a similar system (Goldsmith, 1988). The Government of India accepted this recommendation and the first agricultural university in Uttar Pradesh now named as Govind Ballab Pant of Agriculture & Technology was established in 1960, patterned on the land-grant university model of the United

States. This was a significant turning point in the history of agricultural education in India. The Second Joint Indo-American Commission (1959) reiterated the recommendations of the First and recommended that at least one agricultural university be established in each state. Agricultural universities have now been established in 21 of the 28 states and the concept of the agricultural university now forms an essential part of the national policy on education.

Current Scenario

India currently has a very strong agricultural education system consisting of three Central Agricultural University, 64 State Agricultural Universities (SAUs), and four National ICAR Institutes having the status of "Deemed University": the Indian Agricultural Research Institute (IARI), established a century ago and given the status of Deemed University in 1958; the Indian Veterinary Research Institute (IVRI), Izatnagar (Uttar Pradesh); and the National Dairy Research Institute (NDRI), Karnal and the Central Institute of Fisheries Education (CIFE), Mumbai, which cater to quality education in animal sciences and the dairy and fishery sectors, respectively. The SAUs are spread over the entire country (Department of Agricultural Education and Research, 2008).

Some Steps Taken by the Government for Agriculture Education

- The government of India established a full-fledged department of Agricultural Research and Education in 1973 to strengthen the linkage between the centre and States Governments to enable the ICAR to deal with international agencies.
- The period between 1966 to 1974 witnessed revolutionary changes in the agricultural education policy and structure in India.
- ICAR established Krishi Vigyan Kendras for vocational training of farmers and fishermen and the first KVK was established in 1974 in Tamil Nadu.

Present National Agricultural Policy

1. Agriculture is a way of life, a tradition, which for centuries has shaped the thought, the outlook, the culture, and the economic life of the people of India. Agriculture therefore is and will continue to be central to all strategies for planned socio-economic development of the country. Rapid growth of agriculture is essential not only to achieve self-reliance at national level but also for household food security and to bring about equity in distribution of income and wealth resulting in rapid reduction in poverty levels.
2. Over 200 million Indian farmers and farm workers are the backbone of India's agriculture. Despite having achieved national food security the well-being of farming community continues to be a matter of grave concern for

the planners and policy makers of the country. The establishment of an agrarian economy which ensures food and nutrition to India's one billion people, raw materials for its expanding industrial base and surpluses for exports and a fair equitable reward system for the farming community for the services they provide to the society, will be the mainstay of reforms in the agricultural sector. For this all Agriculture is a vital element.

3. The National Policy on Agriculture seeks to actualize the vast untapped growth potential of Indian agriculture, strengthen rural infrastructure to support faster agriculture development, promote value addition, accelerate the growth of agro-business, create employment in rural areas, secure a fair standard of living for the farmers and agricultural workers and their families, discourage migration to urban areas and face the challenges arising out of economic liberalization and globalization.

During the next two decades it aims to attain

- A growth rate in excess of 4% per annum in the agriculture sector
- Growth that is based on efficient use of resources and conserves our soil, water and biodiversity.
- Growth with equity.
- Growth that is demand driven and caters to domestic markets and maximizes benefits from exports of agricultural products in the face of challenges arising from economic liberalization and globalization.
- Growth that is technologically sustainable, environmental friendly and economically viable. For this education at farm is a must.

4. Sensitization of farming community with the environmental concerns should receive high priority. Balanced and conjunctive use of biomass, organic and inorganic fertilizers and controlled use of agro chemicals through integrated nutrient and pest management (INM & IPM) will be promoted to achieve the sustainable increase in agricultural production. A nationwide programme for utilization of rural and urban garbage, farm residues and organic waste for organic matter repletion and pollution control will be worked out. For this task, extension workers need to be properly educated ,
5. The history and traditional knowledge of agriculture particularly of tribal communities relating to organic farming and preservation and processing of food for nutritional and medicinal purposes is one of the oldest in the world. Concerted efforts should be made to pool, distill and evaluate traditional practices, knowledge and wisdom and to harness them for sustainable agricultural growth.
6. The Government will endeavor to create a favourable economic environment for increasing capital formation and farmer's own investments by removal of distortions in the incentive regime for agriculture, improving the terms of

trade with manufacturing sectors and bringing about external and domestic markets reforms, backed by rationalization of domestic tax structure. It will seek to bestow on the agriculture sector in as many respects as possible benefits similar to those obtaining in the manufacturing sector and development of effective delivery systems and feeing movement of agro produce.

- In recent decades, India has significantly improved the well-being of its people. With over half a century of development, agricultural education and research have been instrumental in ushering in a Green Revolution in the country (Singh, 1999). The green revolution with its impressive social and economic impact would not have been possible without the significant contributions made by the Agricultural Universities both in the form of trained scientific manpower and the generation of new technologies.

Major Landmarks

- 1952 Indian Council of Agricultural Education (ICAE) set up (worked up to 1964 under ICAR).
- 1958 First `Deemed University` status bestowed on IARI.
- 1959 Chief of Agricultural Education and ex-officio Secretary to ICAE appointed.
- 1960 Emergence of SAUs, starting with Pantnagar, based on the recommendations of Joint Indo-American Teams.
- 1965 Standing Committee on Agricultural Education replaced the Education Panel
- 1965 ICAR reorganization with four Divisions including Agricultural Education.
- 1966 ICAR developed Model Act for Agricultural Universities in India.
- 1973 Second reorganization of ICAR with the establishment of Department of Agricultural Research and Education (DARE) to provide greater 6 autonomy to ICAR, and Regional Committees to take care of regional needs, and creation of Agricultural Research Services (ARS) and Agricultural Scientists Recruitment Board (ASRB).
- 1974 Norms and Accreditation Committee (NAC) replaced Standing Committee on Agricultural Education. 1988 G.V.K. Rao Committee recommendations to revamp the functioning of ICAR including its role in Agricultural Education.
- 1995 Agricultural Human Resource Development (AHRD) project, with World Bank Support, launched (ended in 2001).
- 1996 Accreditation Board established replacing NAC.

- 2007 Fourth Deans Committee revised UG course curricula and syllabi, and norms, standards and academic regulation.
- 2009 National Core Group revisted PG (Master's and PhD) course curricula and syllabi and the common academic regulations.

Systemic Reforms

Reforms in Governance of AUs and increasing System's internal efficiency

1. Organizational reforms: All SAUs must adopt the Model Act for Agricultural Universities in India (2009) developed by the ICAR in consultation with all Vice Chancellors of the AUs in order to achieve uniformity in structure and functioning of the universities for enhanced efficiency and inter-institutional acceptability.
2. e- Governance in AUs: Currently, most AUs are functioning in traditional ways. However, to make it efficient and vibrant, e-governance will be introduced by providing necessary hardware and software and training for operationalization. This will be achieved by utilizing internal experience and out-sourcing. The aim would be to make the office work paper less.
3. Financial Management System: FMS software with necessary capacity development will be put in place to enhance the financial management and its efficiency.
4. Project Implementation and Monitoring Unit: Project Implementation and Monitoring Unit will be created at the Education Division, ICAR with deployment of core manpower and contractual resource persons and support staff for implementation and monitoring the activities of the project

Strengthening of and Compliance with Quality Assurance Policies and Mechanisms

In 1996, ICAR established Accreditation Board with three sectoral committees and developed comprehensive accreditation system to regulate the quality of agricultural education in the country. The new process of accreditation has been welcomed by all AUs and so far many universities have been accredited and others are in the process of accreditation. Through implementation of the new process for over ten years, many of the short comings have come to force. Need has been felt to develop a new system of accreditation which will be objective oriented with much transparent and enable periodic monitoring of quality improvement efforts made by the academic institutions. There is also need for convergence of accreditation system among different national and international agencies. The Accreditation Board needs to be further strengthened and provided power on the pattern of NAAC, so that it may lead to not only compliance for quality improvement but also a mechanism for continuous improvement in the quality of education. There is also need to develop policy papers in respect of academic regulations, norms and standards as well as personal policies for sustaining quality of higher agricultural education. Developing the capacity of an institution for continuous internal quality evaluation is critical and will be addressed.

Human Resource Need Assessment– Strategic Planning for 2030

Presently human resource development is largely based on ad-hocism in total disregard of man power need assessment in different sectors of Agriculture for example, large State like Bihar has a very limited intake in Agriculture compared to Maharashtra which admits disproportionately high number of students. Sectorally, there is a high demand for competent human resource in the area of Veterinary and Animal Sciences, Dairy Technology, Food Technology, Horticulture, Fisheries and Agri Business Management but the number of graduates coming out in these disciplines is lower than the requirement. In fact, there is no creditable system of human resources need assessment. In order to have proper and adequate development of human resources, it is necessary that human resources planning are undertaken on scientific basis so that enrollment in disciplines is regulated. This would require need assessment for professionals in specific subject areas, review of their training needs for sustainable performance and career advancement. It would also be important to consider knowledge and skill profile needs of prospective job market. Therefore, man power need assessment and organization of workshops sectorally at national level and involvement of different stakeholders in review and assessment of skilled human resources will be undertaken.

Attracting Young Talent to Agricultural Education

In order to attract talent in higher agricultural education and research, an innovative ASPIRE Scholarship programme is being recommended. This may attract 10+2 pass outs from colleges to go for higher agricultural education and enable them to face the emerging challenges of agriculture and food security in an imaginative and business like way. The scheme aims to offer 2500 scholarships every year to talented youth in the age group 17-22 years, for undertaking Bachelor and Masters level education in agriculture and allied Sciences with following Criteria for Eligibility for ASPIRE Scholarships:

a. Based on Performance in Board Examinations: The candidates should obtain aggregate marks in the top 2% in the Board examination at XII levels.
b. Based on Performance in Competitive Examinations: Performance within the top 10,000 ranks in Joint Entrance Examination of IIT, AIEEE (Engineering) and top 5000 in CBSE medical examination and joining degree level courses in Agriculture or allied sciences in ICAR-AU system will get preference.
c. Value and Type of Support: Eligible candidates will receive annually scholarship @ total value Rs. 120,000 per candidate. The cash value payable

to the ASPIRE scholarship holder is Rs. 90,000 per annum. A summer time attachment fee of Rs. 30,000 will be paid to the mentoring institution for undertaking summer time project.

d. Duration of the Support: Selected candidates will get support for a maximum period of six years starting 1st year course in agriculture and allied science, except Veterinary Science for which the maximum duration may be 7 years, and integrated course leading to Master's degree or the completion of the course, whichever is earlier.

New Strategies and Institutional Reforms to meet the Challenges

The basic objective in redefining agricultural education in the changing scenario is to promote development and delivery of educational programmers that would enhance the employment potential and build up an easy, accessible and cost-effective knowledge intensive information system. The guiding principles, approach and strategies are as follows:

Guiding Principles

- Initiatives should have long term impact.
- The focus should be on ameliorating quality, relevance and reach.
- Linkage with the market requirements and demand of stakeholders.

(a) Approach

- Performance and output oriented.
- Optimal utilization of existing infrastructure and resource.

Strategies for Education

- Shaping up academic environment through total quality management at all levels
- Entrepreneurship development and self-employment orientation in agricultural education should receive high priority through infusion of vocational courses (certificate/diploma levels).
- Periodic review and revision of curriculum consistent with national and global scenario, market trends, self-employment avenues and industries requirements.
- Devising of Qualification Framework.
- Make Agricultural Education an independent subject at School level and integrate it with the higher education system.
- Emphasis on "Distance Education" in Agriculture with extensive use of innovations of IT, digitalization of course content, virtual universities, e-learning and video conferencing.

- ICAR needs to play a more pro-active role in initiating, implementing, reviewing and monitoring reforms in education system.
- Distinguish agricultural education from training for public service, thereby, fostering training aimed at meeting the needs of the private sector and preparing students for entrepreneurship.
- The educational system today suffers from lack of accountability of the teachers and institutions. The quality of individual institutions be critically assessed, and the accreditation system should be made more stringent to enforce accountability.
- Facilitate, promote and support agreements, cooperation and coordination at inter-institutional and international levels.
- Promote integration of population, environment and sustainable development themes into agricultural education and extension programme. Include specialized training in degree programmes.
- Develop well trained globally competitive human resource for doing research in advanced areas of science to become internationally competitive.

For Research

- Need a paradigm shift from single discipline orientation to multi-disciplinary approach.
- Encourage privatization, planning, monitoring, evaluation and assessment as core component of research management process.

For Extension

- Harness benefits of new scientific advances such as bio-technology, cloning, remote sensing, modeling, information and IT for farmers and rural communities. Internet aided extension networks have to be materialized.
- KVK's have to be strengthened as nerve centre for dissemination of technical knowhow.
- Extension programmes to be based on rural participatory approach.
- Formation of commodity specific extension kiosk on indigenous food, horticultural crops, livestock, poultry, rabbitry, bee keeping, mushroom, medicinal and aromatic plants, sericulture, tea and value addition enterprises.

  Distance education for farmers and small agri-entrepreneurs need to be explored and strengthened through television, radio, interactive audio and video systems, besides print and programmed learning materials which would be the distinguishing feature of extension teaching and learning process.

(b) Institutional Reforms

- Effective partnership linkages with advanced centres of education in the country and abroad.
- Greater autonomy to academic institutions, SAUs and statutory power to ICAR.

During the last 4 years following new initiatives have been taken up to improve the quality of education by:

- Training the Faculty at national and international levels
- Infrastructure development at participating universities
- To conduct All India level Exam for admission of 15% seats at UG and 25% at PG level to avoid inbreeding
- About 65 disciplines have been initiated across the universities.

To achieve the goal of excellence in agricultural education, the following points need special attention at appropriate level:

1. The present structure of agricultural education needs to be practical oriented, self-employment generating, production oriented, environment friendly, innovative in approach, commercial and industry based one.
2. Emphasis should be given to provide various job-oriented certificate/ vocational courses at high or higher secondary level with the purpose of obtaining first-hand knowledge on a particular agro-business or agricultural production self-earning aspects.
4. Agricultural degree programme should be offered to the students in two modules. Module-1 should offer the terminal degree in agriculture with objective of producing generalist and Module-ll should offer higher education on specialised fields to serve as specialist, such as scientist, researcher or teachers on a very selective basis. For this purpose, course-curriculum should be reviewed from time to time to keep pace with the changing agricultural scenario and developing frontier of knowledge. In doing so the opinion/views of the stake holders should also to be looked into.
5. Agricultural education should be sufficiently oriented to commercial agriculture in conformity with the globalisation and liberalisation of economic policies in the post GATT era. Knowledge on international agriculture is also very much relevant in this regard. Our agricultural education must incorporate aspects to understand the emerging market forces, which will help effective transit in product mix and input uses. The aspects of agro-business management should be thoroughly incorporated into the agricultural education curriculum with proper emphasis on entrepreneurial development

6. To achieve excellence in agricultural education there is an urgent need to train the academic staff in pedagogy. Our teachers have academic distinction in the subjects but not pedagogy. Hence this area of weakness should be very seriously addressed.
7. Agricultural education should be such that its products are hot favourite in the job market. To prepare the students for job markets, domestic, national or international the role of students guidance and counseling cell play vital role. Hence a training and placement cell is necessary for all educational institutions. The success of students in job markets is amply focused on the quality of education in that particular organization.
8. For quality improvement and also to give better exposure to the teachers outside their own institute, tie-up with other national or international centres of excellence is very much beneficial for improving the quality of agricultural education. Teachers training should form a part of the education policy for quality improvement.
9. Effective accreditation system should be evolved for monitoring the quality of degree programmes in the line of Indian Medical Council of VCI, etc.
10. Accountability of teachers in the SAU system should be worked out to improve the quality of education.
11. Release of fund more particularly the ICAR Development Grant should be directly made available to the colleges with a freedom to the Deans for better utilization of the fund in the areas of specific need.
12. There is an urgent need for improving the present examination evaluation system. A transparent appraisal system is a must for quality improvement in agricultural education.

Some Key Elements of the New Policy

a. New human resources for preparing India for the agriculture of the 21st century - more productive, environmentally safe, more dependent on renewable rather than non-renewable resources of energy input, and competitive (taking into consideration the WTO provisions).
b. Shifting the focus from self-reliance to greater interaction with the world scientific community, using new information technology such as videoconferencing and greater opportunities for travel and joint projects.
c. Greater integration of basic sciences and agricultural sciences
d. Catching up with the biotechnology revolution
e. Removing all barriers through an open door policy
f. New mechanisms of more stable funding support

g. New governance mechanisms for improved and efficient management of resources
h. Fewer but better managed institutions

References

https://www.naae.org/whatisaged/
https://www.icar.org.in/files/NAEP-Projec-documentt.pdf
https://pdfs.semanticscholar.org/c8f9/c5b387a7c59ca46799c43b4900f412161a43.pdf
http://naasindia.org/Policy%20Papers/Policy%2031.pdf
https://blog.forumias.com/redefining-agricultural-education-in-changed-scenario/
https://shodhganga.inflibnet.ac.in/bitstream/10603/114067/9/09_chapter%203.pdf
https://www.expresscomputer.in/features/indian-agriculture-goes-hi-tech-with-new-technologies-like-ai-ml-and-iot/45432/
http://agropedialabs.iitk.ac.in/openaccess/sites/default/files/WS%2014.pdf

35

Structural Reforms of KVK for *Atmanirbhar Bharat*

Subrata Kumar Roy[1], Jyotirmayee Udgata[2] and Partha Pratim Pal[1]

[1]*ICAR-Agricultural Technology Application Research Institute, Kolkata West Bengal*
[2]*Krishi Vigyan Kendra, Jharsuguda, Odisha*

Introduction

India is considered as one of the emerging superpowers of the world. This potential is attributed to several indicators, the primary one being its demographic trends and a rapidly expanding economy. In 2015, India became the world's fastest growing economy with a 7.5% estimated GDP rate. But in the year 2020, the pandemic lockdown situation created adverse impact on the global economy. At this juncture, there is an urgent need to focus on native and indigenous resource mapping and judicious utilization of it to be a self-reliant. The COVID-19 pandemic has necessitated a new set of reforms to strengthen the economy. The first mention of this came in the form of the *'Atmanirbhar Bharat Abhiyan'* or 'Self-Reliant India Mission' by the Government of India during the announcement of the coronavirus pandemic related economic package on 12th May 2020. Almost 70% of India's population depends directly and indirectly on agriculture for their survival. This means the country cannot have a development model that ignores agriculture.

It's time to take a look at India's agriculture economy. Over the last 10 years, (2008-19) India has been a net exporter of many farm produce. In fact, in 2018, the government has come up with a new agricultural export policy aimed at doubling farm income by 2022. However, in 2019-20, agri-exports were just $36 billion and the agri-trade surplus was $11.2 billion. The net agri-trade surplus has been falling over a period of time, making it difficult to achieve doubling of farmers' income and doubling of agriculture exports by 2022as well.

India is not yet self-reliant in various essential agriculture commodities and depends heavily on imports. India imports around 15 million tonnes of edible oil every year — more than 9 million tonnes of palm oil and about 2.5 million tonnes each of soybean oil and sunflower oil. It buys palm oil from Indonesia and Malaysia, and soy and sunflower oil from Argentina, Brazil, Ukraine and Russia. There is a need to create self-reliance in theses commodities enhancing productivity of Indian farms to create competitive advantage for domestic producers. India can step up cultivation of sunflower, groundnuts, mustard and oil palms. Presently, India has about two million hectares of land that is suitable for oil palm cultivation.

Governor of Reserve Bank of India in his speech at the CII National Council emphasized on five dynamic shifts which are underway in the Indian economy. One of the five themes in the speech was "fortunes shifting in favour of the farm sector" hus gives a policy direction for the future. The "Shifting the terms of trade in favour of agriculture is the key to sustaining this dynamic change and generating positive supply responses in agriculture. Terms of trade will be in favour of agriculture if prices of agricultural goods are rising at a faster pace than that of non-agricultural goods. This needs not be consistent with reasonable food prices for consumers. When the value of the terms of trade index is less than 100, it means that the relative prices are against farmers. The data shows that both these indices were rising in the last decade and have stagnated or fallen since. The farmer versus non-farmer terms of trade in 2018-19 was 96.4, while the agriculture versus non-agriculture index was at 106.8. The trend is worse when it comes to the farmers versus non-farmers index. This is also the period when food price inflation growth has moderated. Food inflation shared the upward trajectory of terms of trade for agriculture when the latter was rising. This suggests that maintaining terms of trade in favour of agriculture and ensuring reasonable food prices for consumers might not be compatible (Chart 1)

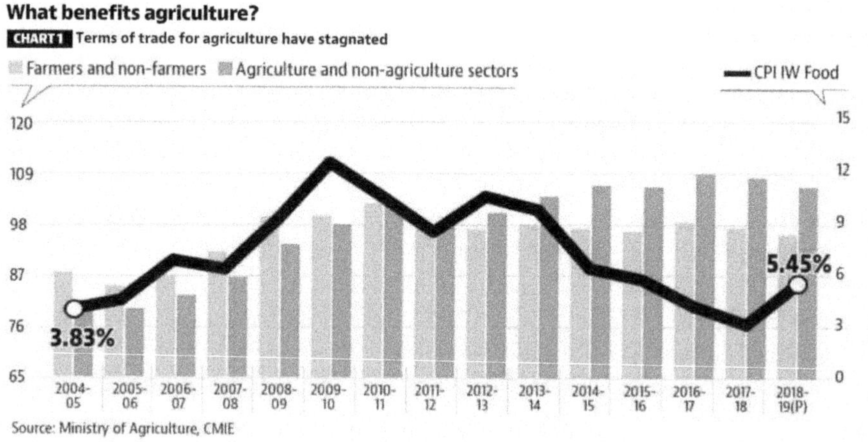

The policy of government procurement at Minimum Support Prices (MSP) has been a key pillar of India's food security. Guaranteed remunerative prices have encouraged farmers to grow more rice and wheat. Government's food stocks, which are built through MSP procurement, come in handy during periods of crisis such as the current one. To be sure, the policy has created problems, especially regarding ecologically unsustainable farming practices in India's original green revolution belt. The policy has also created a regional imbalance. According to latest procurement data from Food Corporation of India - 2018-19 for rice and 2019-20 for wheat - 47% of total procurement was from the states of Punjab and Haryana. As per Centre for Monitoring Indian Economy (CMIE) database, the share of these states in total wheat and rice production in 2017-18 was 22%. These problems notwithstanding, the share of procurement in total rice and wheat production has been increasing continuously. Any sudden withdrawal of this policy is bound to create a huge disruption(Chart 2).

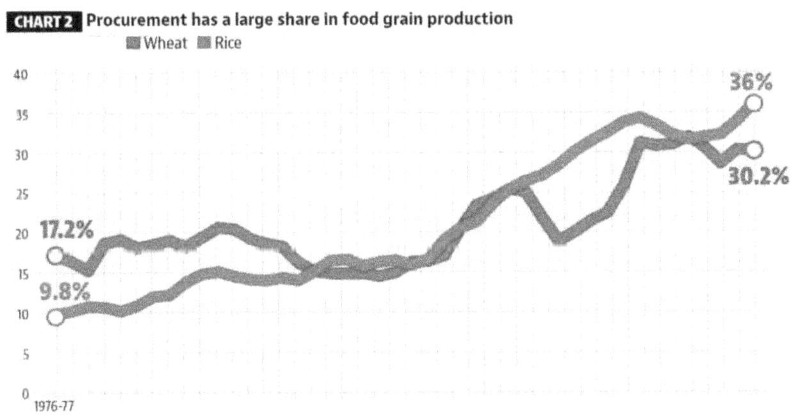

Source: CMIE 2019-20

As the 1.37 billion population in India are dependent on agriculture, the sector must attain self-reliance. A specific long-term strategy is needed with commodity wise plans that factor in comparative advantages, available market, production, productivity, consumption, subsidy and incentive. With the growth of agriculture, the associated industries should also bloom. The entrepreneurship development in agriculture and allied sectors will create an avenue for employment of rural youth. The development of small and medium enterprises in agriculture and allied sector will attract rural youth, which will create a difference in our economy and control labor migration. The innovative technologies can be utilized in agriculture to make it more lucrative and remunerative. The government has also set a vision for doubling farmers' income by the year 2020 to make agriculture profitable and sustainable.. The network of Krishi Vigyan Kendra (KVK) is playing a key role in making the agricultural land of India prosperous and self-reliant utilizing the vast human and natural resources.

Krishi Vigyan Kendra (KVK): The Lighthouse for Farmers

For the sustainable development of agriculture, ICAR generated the idea of establishing KVK (Agricultural Science Centres) as innovative institutions for imparting vocational training to the practicing farmers, school dropouts and field level extension functionaries. The ICAR Standing Committee on Agricultural Education meeting held in August, 1973, observed that the establishment of KVKs was of national importance which would help in accelerating the agricultural production and also in improving the socio-economic conditions of the farming community. The first KVK, on pilot basis, was established in 1974 at Puducherry (Pondicherry) under the administrative control of Tamil Nadu Agricultural University, Coimbatore.

At present there are 720 KVKs across the countryfall under the jurisdiction of 11 Agricultural Technology Application Research Institutes (ATARIs) throughout India. The growth in number of KVKs during various plan periods is graphically represented (Fig. 1)

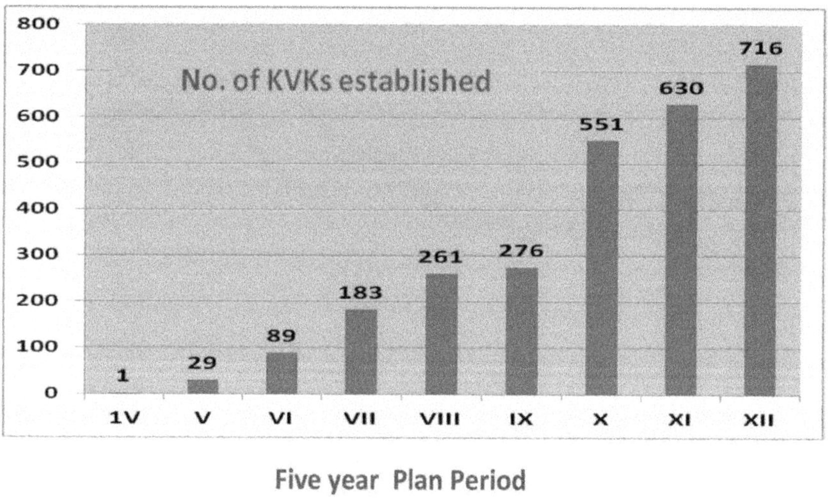

Fig. 1. Number of KVKs in different five year plan period

All KVKs are working towards reducing the time lag between generation of technology at the research institution and its application to the location specific farmers' field for increasing production, productivity and net farm income on a sustained basis with the following mandate. The mandate of KVK is Technology Assessment and Demonstration for its Application and Capacity Development. To achieve this, each KVK is concentration on :On-farm testing to assess the location specificity of agricultural technologies under various farming systems front line demonstrations, training of farmers and extension personnel to update their knowledge and skills in modern agricultural technologies work as resource and knowledge centre of agricultural technologies,. produce and make available technological products and organize extension activities to create awareness about improved agricultural technologies .

Developmental Projects Operating in KVKs

Along with the mandatory activities the KVKs are implementing various projects , the few such programmes are Focused programmes under Knowledge Systems and Homestead Agriculture Management in Tribal Areas (KSHAMTA) in 125 districts; and Nutri-sensitive Agricultural Resources and Innovations (NARI) as well as Value Addition & Technology Incubation Centres in Agriculture (VATICA) in 100 districts have been initiated by KVKs. Since 2016-17, the KVKs are organizing National Skill Qualification Framework aligned skill development training programmes related to agricultural and allied sectors with financial support from Department of Agriculture, Cooperation & Farmers' Welfare. Automatic Weather Station (AWS)is established at 121 KVKs under National Innovations in Climate Resilient Agriculture (NICRA) Project and 100 being established with support from India Meteorological Department

Establishment of 121 Climate Smart Villages under National Innovations in Climate Resilient Agriculture, Integrated Farming System models at 100 KVKs for capacity development of farmers, Establishment of Pulses Seed Hubs in 97 KVKs, 100 KVKs are implementing Attracting and Retaining Youth in Agriculture (ARYA) to attract and empower the youth in rural areas for taking up various agriculture and allied sector enterprises, Conducting Skill trainings for farmers, farmwomen and rural youth of 200hrs funded by Agriculture Skill Council of India (ASCI) are also being carried out.

To achieve the vision of Doubling Farmers' Income (DFI) by the year 2022 and making India self-reliant in agriculture some innovative approaches and models are to be extrapolated. Development of new approaches in agricultural extension in India and worldwide is a continuous process with its focus on increasing productivity and profitability. While the call for demand-driven agricultural extension has existed for several decades now, new modes of reaching out to farmers could have significant impact in India, as they might better reflect the local information needs of farming communities. The diverse nature of the Indian subcontinent, with its wide variety of agro-climatic regions and broad range of socio-economic conditions in the rural population, calls for agricultural extension approaches that are context-specific and situation-specific. Extension organizations in general have been using a wide range of methods for reaching individuals, groups and the wider public in rural areas with new information/ knowledge. Approaches to extension also vary widely from top-down and supply-led to bottom-up, demand-led and participatory. Advances in information and communication technologies (ICTs) have also provided new opportunities for extension to reach more farmers in a short span of time (Sulaiman et al. 2012).

Innovative Models and Approaches

Public Private Partnership (PPP)

The importance of Public Private Partnership (PPP) in agriculture is understood in terms of a shared mechanism among partners for input, resource, market, risk, technology and benefits. In addition, review of various studies indicated the visibility of PPP in various facets of knowledge management, capacity building of women and youth, development of high-end technologies, processing and market promotion and gender mainstreaming in agriculture. The partnership approach apart from developing certain technologies also empowered farmers in terms of enhanced access to technology and market in India through organized farmers groups. Farm women from difficult areas were enabled to empower themselves through gender mainstreaming techniques. The limitations of PPP such as focus mainly on high end technologies, high profit margin areas and crops, perceived mistrust and lack of transparency and non-adherence to agreement among partners could be overcome through appropriate working mechanism and policy support. Establishing PPP cell at research and development organizations would spearhead the growth of PPP and thereby sustainable agriculture and livelihood of millions of poor farm families in India. The seed and quality planting material production in PPP mode will be a great boon in making agriculture self -reliant.

Multi Stakeholder Convergence

The convergence among all the stake holders in the agricultural extension system at district level help in all round development of the farming community. The public extension system although continues to be the leading extension delivery mechanism, NGOs, farmers organizations, State Departments of Agriculture and Animal Husbandry, Panchayati Raj Department , NABARD and CSR team of Corporates have been playing significant roles in extension service delivery. India has a large number of NGOs with varying capacity and strength, which implement a wide range of programmes. Forming farmer interest groups and federating them at block and /or district levels is an important strategy of the ATMA extension approach implemented by the Government of India. The idea is to encourage farmer groups to organize different types of services for themselves, including input supply, credit, technical services and marketing that would increase their incomes, while decreasing their dependence on the government. When there is convergence between research and extension then the gap between technology generation and adoption is reduced. The KVKs are conducting monthly Research-Extension Interface meeting involving the Regional Research Stations of SAUs, KVK, district State agriculture line departments, Lead bank, NABARD, input dealers, CSR department of corporates and progressive farmers. The agenda of the interface is to discuss the various developmental programmes carried by different stake holders, the scope for convergence especially for doubling Farmers' income, to take the information generated in the interface to the grass root level extension functionaries to percolate to the end users, i.e. the farmers.

Farmer to Farmer Extension (Master farmer/Farmer promoter model)

Farmer to Farmer extension approach in agriculture is a systematically utilization of community leadership and informal communication between farmers. This approach aims at strengthening the information flow and enhance the agricultural production. As a generic term we use it as 'farmer trainer', even though we recognize it by different names e.g. lead farmer, farmer-promoter, community knowledge worker may imply different roles. Farmer to farmer approach helps in building the effective, farmer-centered extension systems and empowering farmers as change agents for improving livelihoods in their communities. Farmer to farmer model reduces the extension cost and workload of extension functionaries in a large country like India where extension worker and farmer ratio is very wide. Moreover, this approach needs community as well as government support for sustainability and scalability purpose of the farmer promoters. Farmer promoters facilitate the flow of knowledge, information, and materials (e.g., seed) among farmers leading to increased adoption and productivity, identify and help weak and disadvantaged groups to participate and take benefit, help t o strengthen capacities of their own local communities to access information, innovate and solve problems and help in needs assessment , implementing, monitoring and evaluating the approach and performance of different programmes and schemes.

Farmer promoters or Master farmers act as an extended arm of the KVK, help KVK scientists in selecting site and farmers for carrying out different on-farm testing, demonstrations. In need assessment of the village they play the leading role in identifying problems, prioritizing and providing solutions relating to agriculture including the social needs. They also in many cases help in providing input and market support and linkage to other farmers / groups in the village. KVK organizes Master Farmer Development Programmes has and prepares different module for Master Farmers Development such as Potato seed production Technology, Ground nut seed production technology etc. Long duration courses are conducted in phases as per the availability and need of farmer as well as course content. Good numbers of progressive, knowledgeable and educated farmers aretrained thoroughly on different aspect like Technology, Credit & Marketing and developed Master Farmer.

Custom Hiring Service Centres (CHSCs)

Custom Hiring Service Centre is basically a unit comprising a set of farm machinery, implements and equipment meant for custom hiring by farmers. Mechanization of farm activities is the need of hour to increase production and productivity. Farm mechanization provides the technology to facilitate agricultural growth through optimal use of farm inputs, natural resources, completing different operations, which gives the crop more time to mature, allows the farmer to be more flexible in his farming operations and facilitates multi cropping. Adoption of mechanization ensures timeliness of agricultural operation which reduces cost of production as well as reduces drudgery in carrying out various agricultural operations. Farm mechanization has been helpful to bring about a significant improvement in agricultural productivity. Though subsidy is being provided for farm machinery, due to prohibitive cost of farm machinery all farmers may not come forward to own them. Hence, establishment of CHSC is a boon for farmers especially for the small and marginal farmers. This helps in timely agricultural operations and mitigating the problem of labour availability in peak production time.

The CHCs are established in the NICRA project operated villages in climate vulnerable districts under KVKs. The CHCs are managed by Village Climate Risk Management Committee (VCRMC) formed by the villagers. Through the project the situation and crop specific farm implements have been provided to the CHCs for use of the farmers. By custom hiring the centers are being able to generate income. These CHCs are operating well at village level.

Custom Hiring Centre at village level by KVK, Jharsuguda

Development and Strengthening of Value chain

A value chain is a set of linked activities that work to add value to a product. It consists of actors and actions that improve a product while linking commodity producers to processors and markets. Value chains work best when their actors cooperate to produce higher-quality products and generate more income for all participants along the chain, as against to the simplest kinds of value chains, in which producers and buyers exchange only price information — often in an adversarial mode. Value chains differ from supply chains, which refer to logistics: the transport, storage and procedural steps for getting a product from its production site to the consumer. A value chain encompasses the flow of products, knowledge and information, finance, payments, and the social capital needed to organize producers and communities. Value chains may include a wide range of activities, and an agricultural value chain might include: development and dissemination of plant and animal genetic material, input supply, farmer organization, farm production, post-harvest handling, processing, provision of technologies of production and handling, grading criteria and facilities, cooling and packing technologies, post-harvest local processing, industrial processing, storage, transport, finance, and feedback from market. This can be established for major crops and commodities of the specific region for sustainable production and self-reliance.

Sustainable Local Seed Hub System

A sustainable seed system will ensure that high quality seeds of a wide range of varieties and crops are produced and fully available in time and affordable to farmers and other stakeholders. Pulse seed hub created by KVK/state agency is a good example. However farmers must get benefit of using quality seed by solving the issues including inefficient seed production, distribution and quality assurance systems and access to credit for inputs. Seed production by the farmers and establishment of seed bank at village level will empower the farmers for sustainable and quality production.

ICT Initiative in Technology Dissemination

Information and Communication Technologies (ICTs) have emerged as one of the potent tool for facilitating the agricultural extension. Most of the developing countries are lagging behind in using more advanced ICTs such as for database management, modeling, GIS and remote sensing, knowledge based systems and for agricultural research management. The applied dimension of ICT is Cyber extension which is the extension of agricultural technologies with the help of ICT tools utilizing cyber space.. The major tools of cyber extension include Email, Expert systems providing information on pests and diseases, Internet browsing for extension information, Video conferencing, Call centres and Satellite, communication networks and News and Discussion groups. The ICT initiatives taken up by KVKs are of paramount importance in establishing self-reliance among farmers. Krishi Vigyan Kendra Knowledge Network Portal, dynamic portal having read and write facility, launched on 8 July 2016 to provide information and advisories to the farmers and monitoring. Through mKisan Portal, 90 lakh farmers are being provided advisory by the KVKs

During COVID-19 pandemic the digital platforms adopted by KVKs like virtual meetings, on-line trainings, farmers advisories through Whatsapp group, short videos of technologies and events sent through emails and Whatsapp, web casting and webinars proved to be effective in the transfer of technologies as well as in creating awareness. Multi Location Audio Conference in collaboration with

Reliance Foundation is helping farmers to have direct audio contact with the KVK scientists and other experts and providing real time solution. These methods can be extensively used to connect the farmers to the experts at various places at low cost. The only policy interventioneeded is to provide digital aids to the farmers in subsidized rate so that small and marginal farmers can get benefitof the technology.

KVK online Agri-Mart

Market information for agricultural commodities is available through various Agri Portal. But participatory approach is required to reduce middlemen intervention. Mobile based online marketing platform can efficiently link buyers with seller. Farmers can get better price of their produce, at the same time buyers can get produce in cheaper rate directly form producer with reduced transportation cost. Popularization of such Apps is required through Govt. agencies and agriculture extension functionaries. KVK, Darjeeling developed as android based application for efficient marketing of farmers' produce which is projected below and also available in Google Play Store.

Similar App was developed by KVK on Citrus e-Clinic for monitoring orchards of Mandarin orange, which has very high export potential and one major source of livelihood of Darjeeling district farmers.

Technology Partnership and Networking

Technology partnership between the KVK Rathindra and reputed Governmental Organizations (GOs) and Non-Governmental Organizations (NGOs) have been very effective in technology development, transfer and market development. Rathindra KVK and its associated Networks consist of a group of institutions or associations with the aims of enhancing the capacity to conduct research and improving training and education through interaction thus forming a network to improve access to new ideas, methods, and information sharing and materials exchange. This initiative on partnership and networking can foster technological upgrading and improvement through quick and more efficient extension at a much lower cost to each of the partners thus creating a Win-Win situation for all the partners.

Success Story/Case study Developed by KVKs

i) KVK, Hooghly has developed Leadership Development programmes in collaboration with NABARD wherein, the leaders of farmers club of this district are trained on new technologies, technologies assessed/ refined by this KVK etc which are disseminated to others farmers through the interaction in the monthly meeting of their Farmers' Club .

ii) Food and nutritional security in the farm families especially among the vulnerable groups of women and children is an important indicator of development and many KVK train many farm families on crop diversification, homestead vegetable garden, weaning food etc. KVK, Nimipth have developed a Diet Diversity Score (DDS) card for counting of food groups that a household or an individual used to consume over the past 24 hours for assessing the usual diet of household/individuals and overall food security round the yearThe process involves enquiring and individual his/her diet in the last three days and after distributing them in different food groups, scores are obtained. To achieve this end, one DDS chart consisting of nine (9) food groups from balance diet chart

   a. Rice/Puffed rice/Pressed rice, 2. Dal, 3. Potato, 4 .Green leafy vegetable, 5. Fish/Meat, 6. Egg 7. Milk, 8. Fruit, 9. Oil) is made as below. Tick mark (√) is used instead of "0"or "1" value during interview considering illiterate people. This methodology, developed by the KVK, have been shared in National and International platforms and at present is being used by various NGOs, like, Prava, CWS , Avibyakti foundation (Jharkhand), DRCSC, Pradhan (West Bengal), Jansahas FES, IGSSS (Madhya Pradesh), Living Farm (Orisha), RNN (Nepal) engaged in estimation of food security in the country,

iii) Community Score Card (CSC) is a tool developed by KVK, Nimpith for qualitative monitoring at local level for performance evaluation of services, projects and even government administrative units by the communities themselves. The CSC process is an instrument to exact social and public accountability and responsiveness from service providers. However, by including an interface meeting between service providers and the community that allows for immediate feedback, the processes boost up empowerment.

iv) KVK, Nimpith along with 40 other NGOs from South Asia, East Africa, Philippines and Germany participated to form a Forum – NGO IDEAS

Corporation to develop participatory Impact Assessment Mechanism specifically for women empowerment in the name of "Impact Tool Box" where target groups will play a key role in assessing the impact. The Tool Box consists of 4 tools *viz.* PWbR (Participatory Wellbeing Ranking), SAGE (Situation Analysis and Goal Establishment),

PAG (Participatory Appraisal of Group) and PIAR (Participatory Impact Analysis and reflection). PWbR is designed to involve the community in classifying the households into different categories of well beings SAGE is used to identify the individual goals and to appraise changes at the individual and house hold level of the members of the group or CBO members.. PAG is used to identify goals reached by the group members can measures the extent of their group achievement. PIAR serves an in-depth analysis of outcomes and impacts of the NGOs / GOs works. It sets out from the method of self assessment of the changes among individual / household or CBO, elaborating their results with support of a number of additional element used at different level of PWbR, SAGE. PAG. The four tools are organically linked with each other and at the same time each tool can be applied individually according to the structure of the project or the type of participating organization

v) In West Bengal, there is a serious dearth of qualified veterinary para-professionals in the remote villages., Therfore, ARD, Govt. of West Bengal initiated action of empowering farm women as Para-vets with the rationale that women usually takes care of the homestead animals and birds and this goes in harmony with the objective of gender mainstreaming in agriculture. KVK, Burdwan has empowered 29 such rural women to vaccinate animals and birds in the villages who are now earning to the tune of Rs. 2000 to Rs. 5000 per month.

Model KVK for Becoming Self-Reliant

The model KVK has been proposed here to cater the present need of the farmers for effective technology transfer and enable the farmers to adopt. In this perspective, for a model KVK, the very essential features are good infrastructure facility and a well-developed instructional Farm. A reference lay out plan for a 20 ha KVK farm area is given here in the sketch-1.

Infrastructure development: The basic infrastructure facilities *viz.*, adequate administrative building, trainees' hostel, staff quarters and well planned demonstration units are requisites for strengthening the KVKs.

Demonstration units: In KVK, the Integrated Farming System (IFS) model is being promoted with various crop and livestock components as per the Agro Climatic Zone which can be promoted as Agro Eco-Tourism model. The demonstration units are very necessary for enhancing the practical exposure to the farmers of the district with Nutritional garden, Medicinal garden, Ornamental fish unit, Crop cafeteria, Progeny orchard (Mango, guava), Seed production plots, Polyhouse, Shade net house,Mushroom production unit, Mushroom spawn production unit, Apiary, Fodder unit , Azolla unit, Poultry unit (Brooding, rearing and Hatchery unit, Dairy unit,Goat unit (With improved breed of buck for breed upgradation), Fish hatchery unit, Composting unit, Organic input production unit

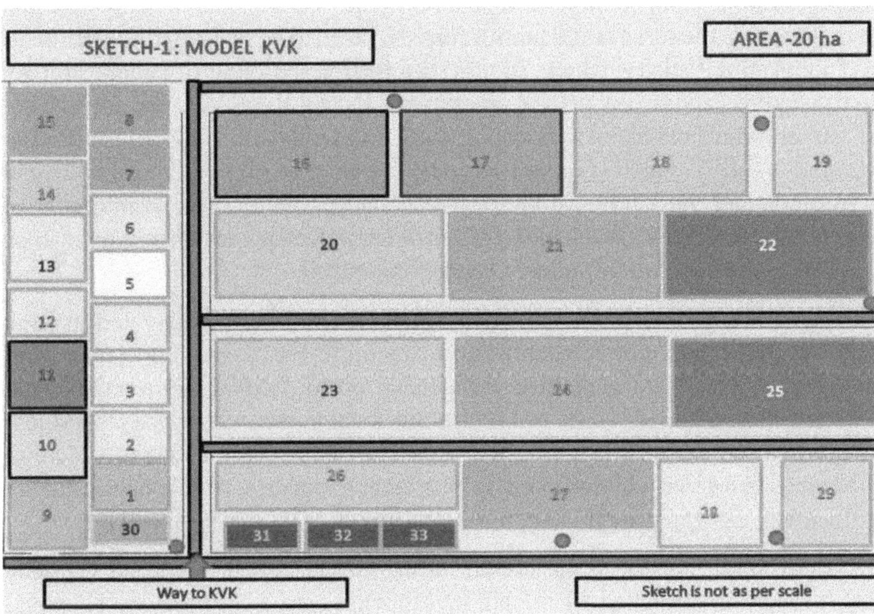

Administrative Building (1)	Organic input unit (11)	Seed Production unit
(21,24,27)		
Farmers Hostel (2)	Poly House (12)	Fruit Progeny
Orchard (22,25)		
Medium Garden (3)	Shade net (13)	Agro Forestry Unit
(28)		
Nutritional Garden (4)	Azolla Unit (14)	RWHS-2 (29)
Ornamental Fish Unit (5)	Quail unit (15)	Lawn and aesthetic
garden (30)		
Exotic Vegetable garden (6)	Dairy unit (16)	Vehicle Shed (31)
Apiary unit (7)	Poultry unit (17)	Godown (32)
Duck Unit (8)	Fodder unit (18)	Threshing Floor (33)
Staff Quarter (9)	RWHS -1 (19)	Borewell ●
Mushroom Unit (10)	Crop Cafeteria (20,23,26)	
Road ▬▬▬	Drainage line ------------	

Future Strategy

Making India prosperous, the task of achieving self-reliance and self-sustainability in agriculture is of utmost importance. The first set of measures rely on branding of local farm products to be sold globally, the branding helping to tide over price spirals and making farmers quality-conscious. There must be standardization and promotion of indigenous technology knowledge (ITK) in agriculture, where techniques are dependent on local resources in dealing with nutrition, disease and pests. These are the sort of rural innovations that need to be registered at a national-level innovation registry, which, in turn, can help in the filing of patents and get micro-venture capital support for enterprises. In fact the KVKs are identifying the farmer innovators and documenting these and sending to National Innovation Foundation (NIF), which is India's national initiative to strengthen technological innovations and outstanding traditional knowledge wealth. The compilation of these farmer innovations are made by ICAR-ATARI, Kolkata which in convergence with SAUs and NIF can be validated and extrapolated.

Entrepreneurship in the farm and agri-business sector is to be promoted in large scale. As SHGs now get collateral-free loans up to Rs 20 lakh and micro food enterprises (MFEs) are supported with funds up to `10,000 crore, connecting these organisations with FPOs at all relevant focal points will spur employment growth in rural areas. Integrating the efforts of these organizations to identify niche areas like value addition, rural infrastructure, logistics, warehousing, quality certification, etc. can provide livelihood opportunities for migrant workers.

The digitization process in agriculture is to be accelerated. Seth and Ganguly (2017) report an estimated 34 private investment ventures in India are receiving US$295 million, compared to US$427 million in China. China had 665 million Internet users in 2014; India had 227 million, partly due to higher Internet costs and smaller physical coverage. Two major constraints in promoting digitization,are lower access to physical capital and human capability. Insufficient connectivity in rural areas, high costs for services, and lack of basic computer knowledge and literacy hinder rapid development of e-agriculture. Substantial investment is needed in physical infrastructure, power, broadband, and transportation. The real need is for direct application of digital and other technologies to increase farm productivity and address the increasingly visible climate change impacts. Remote sensing (via satellites), GIS, crop and soil health monitoring, and technologies for livestock and farm management are the common areas. A digital laboratory with all accessories is to be set up in each KVK which will act as digital incubation center.

Given the magnitude of the challenges, coherent, climate-smart, integrated, national- and state-level agricultural policy and strategy frameworks are needed, beyond the mission mode, as recommended by NABARD (National Bank for

Agriculture and Rural Development) and ICRISAT (NABARD, 2016) with the emphasis on reformed extension systems. To achieve self-reliance in agriculture there must be convergence of all the stakeholders in agriculture application research and extension units at district level. The effective implementation of all developmental activities and innovative extension models can be achieved through convergence of KVK, Regional research stations of SAU, district level state line departments, NABARD, Lead bank, NGOs, Corporate Social Responsibility (CSR) of local corporate bodies. To make India a happy and prosperous , the farmers are to be empowered with enhanced income and a prestigious living in the society, for which KVK plays a pivotal role.

References

Anonymous. 2020. Innovative Technologies developed and strengthened by KVKs in Zone-V, ICAR-ATARI, Kolkata

Basu, K.2016. 'India's digital transformation', The Hindu, 12th February 2016.

https://economictimes.indiatimes.com/news/economy/agriculture/how-india-can-script-a-self-reliance-success-story-in-agriculture/articleshow/75984276.cms?from=mdr

https://www.hindustantimes.com/india-news/india-s-agriculture-policy-dilemma/story-UvdtxBJqQMVFwt1QZ5PN5J.html

https://www.ideasforindia.in/topics/agriculture/how-new-technologies-can-raise-farm-productivity.html

Seth, A and K Ganguly.2017. 'Digital technologies transforming Indian agriculture', in Dutta, S, B Lanvin and S Wunsch-Vincent (eds.), The Global Innovation Index 2017: Innovation Feeding the World, 105–111, Cornell University, INSEAD, and WIPO, Ithaca, Fontainebleau and Geneva.

Sulaiman R, A. Hall,N.J.. Kalaivani, K. Dorai,. and T.S.V. Reddy . 2012. Necessary, but not sufficient: Critiquing the role of Information and Communication Technology in putting knowledge into use, The Journal of Agricultural Education and Extension. 18:331-346

Wani, S.P., D.Bergvinson, K.V. Raju, P. Gaur and R.K. Vashney.2016.Mission India for Transforming Agriculture (MITrA), Research Report IDC-4, ICRISAT, Patancheru.

Theme 9: Governance and Policy Reforms

36

Governance and Policy Reforms

P.K. Ghosh[1], P.N. Sivalingam[1], Debashis Mandal[3], Debashis Chakraborty[2] and Prabhat Kumar[4]

[1]*ICAR-National Institute of Biotic Stress Management, Raipur, Chattisgarh*
[2]*ICAR-Indian Agricultural Research Institute, New Delhi*
[3]*ICAR-Indian Institute of Soil & Water Conservation, Dehradun, Uttarakhand*
[4]*National Agricultural Higher Education Project, ICAR, New Delhi*

Introduction

India is transforming from the phase of revolution to commercial mode of agriculture, a step ahead of self-sufficiency. It further evolves to reach a goal of self-reliant on agriculture and allied sector. The present phase demands several technological innovations, their application and implementation in the market. Meanwhile this phase needs due attention for handling the issues on conservation of natural resources, especially germplasm resource, biodiversity conservation, socio-cultural-economic values of our country, etc. The whole process requires good investment on research and development, various economic measures and implementation. It is only possible with vibrant governance and policy implementation of a Government. Already Government of India has taken several measures to transform agriculture. Recent analysis showed that Indian agriculture is the silent ever green revolution. Food production is kept pace with the growing population since independence. Further, it needs several innovations on technologies coupled with policy support to fill the demand. India is placed 49th position in the world on innovations. Analysis suggested the huge potential for innovation due to its huge human capital, which perhaps need for the country like India to meet the food demand of 333 million tonnes by the year 2050. These innovations must be focussed for growth to efficiency, employment generation, food security to nutrition and health, shortage management to surplus management, input intensive to knowledge intensive agriculture, climate change and sustainability, production and producers, policy interventions, regulations and reforms. The Government emphasize usage of e-trading platforms to sell the farmers' produce of their choice rather using *mandis*. This will enable contract farming and inter-state trade. However, major innovations and investment must be implemented on transportation, logistics, and digital access and payments to enable widespread prosperity. Thus, focus must be on the complete agricultural value chain. In fact, focus needed on value addition, rural infrastructure, logistics, warehousing, and quality certification of agriculture produce. India has greatest challenges to face in the future for food and nutritional security, adaptation and mitigation to climate change and sustainable use of essential and scared resources like water, land and energy.

The proven technology is to be transferred to farmers through demonstration for its adaptation and bringing into production system. Due to obligation of being a member country of WTO a lot of agricultural products of processed and convenient foods with concomitant increasing urbanization and rising income levels are replacing the conventional food products. This shift is being diversified and hence increasing demand for a large number of products. Therefore, several regulative measures need to be developed to ensure quality of supply. The major areas of technological innovations need regulation are vertical farming artificial intelligence technologies particularly automation and robotics, block chain technology, modern greenhouse practices, precision agriculture, and more recently the Clustered Regularly Interspaced Short Palindromic Repeats (CRISPR) and other gene editing technologies. Another important concern of present-day agriculture is the food safety. It needs safety recalls, supply chain efficiency and food traceability of current food system to avoid contamination and may bring transparency in premium products in the market. Climate smart crops multiple disease resistance, bio-fortification of crop plants, hybrid technology hybrid vegetable production under protected conditions, accelerated breeding role of advanced genomic tools, genetically modified crops, marker based genetic purity testing, skill development are going to revolve around us in the shake of increase production and quality improvement. These advance technologies need regulations for their proper usage. Farmers should get proper and timely advice on land uses and anticipates meteorological and marketing trends. To get benefit of the knowledge in farm management, technology delivery system needs urgent attention, structural reforms and revitalization. Presently, the Government has taken initiatives, namely, *Make in India*, *Skill India*, *Start-up-India* and *eNAM* etc. to build new National Agricultural Innovation System that would achieve assured success. Capacity building and Women empowerment in India is the need of hour. Women empowerment in India is mostly dependent on many factors that include geographical location (urban/rural), educational status, social status (caste and class) and age. The policy decisions related to women empowerment have to be taken at administrative/political level. There are significant gaps between policy advancements and actual practice at the community level even after multiple efforts being tried. As the women has important role in the agriculture it needs a special attention. Public research and development should be equipped well to encourage basic research and applied research can be privatized. This step will help in reducing the number of research institutes under public sector.

The institutes engaged in basic research can muster more funds for each institute, popularization of agricultural products input industry should be totally privatized so that the development activities should be independent and encouraged and Tax exemption for the income generated on technology transfer to farmers should be considered. The priority of the Government is to enhance public sector investment in agricultural research and concerted efforts are needed to support the effective transfer of technology. To keep pace in agricultural productivity, Government should enhance the investment to at least 2 or 3 times from 0.6% at present which is insufficient compared to China, Brazil and Japan and even less than many under developed countries. This will foster the growth of innovations. The governance regulations and policy needs to be focused in several sectors are given below.

Crop Improvement and Production

- Adoption of new breeding techniques, utilization of molecular tools, IT, AI are needed for acceleration of improving productivity of crops. Therefore, Government should invest more on research and development on this line.
- Policy on the regulatory system of seed to encourage the seed industry and protecting the farmers' interests are proposed to make self-reliant seed system in our country. The seed act of 1966 needs update and new seed bill 2004 need be formed into an act. The new seed bill needs guiding and seed advisory committees on scientific knowledge, procedures and standards in line with the international systems that will help in making movement of seed to other countries.
- The seed quality testing and assurance system should be liberalized. The scope of Organisation for Economic Co-operation and Development (OECD) seed schemes shall be extended to a greater number of varieties, development of seed hubs in flower, vegetable, and forage seed sectors, several orphan and neglected crops shall be brought under seed chain, and these seed chain are linked to different Government schemes for an assured market and development of organized production centres both at the public and private level.
- Standard policy on GM seeds of prioritized crops along with specified production zones for GM crops, labelling of GM seed and their testing on a large-scale regulation of the GM seeds are needed. The informal seed sector shall be brought under quality assurance system.
- The development of seed farmers' organizations and compulsory certification faster compensation provisions for financial losses at various stages, efficient regulatory system to check illegal seed trade adoption of producer compensation in place of subsidies, simultaneous development of cold chains for seed storage, reliable and robust database is required to integrate the IT and AI in this sector, policy reforms in public seed sector,

formulation of policies should consider this sector as service sector, are essential in this sector. Mere duplication of private industry or privatization of the public sector will deceive the purpose of social responsibility and will be detrimental to Indian agriculture.

- Out of 7,000 edible plant species that are either cultivated only about 150 species have gained significant commercial importance. Pollination, pest control, nutrient cycling ensures greater adaptive capacity to global production challenges including climate change, emerging disease risk and changing consumption patterns. It needs greater research and development efforts to harness available potential using new science.
- Funding in research, capacity development and training, especially of rural youth and women for conservation and rational use and enhance agro-biodiversity needs special attention.
- India has established five National Bureaux under the ICAR to conserve and facilitate utilization of genetic resources of plants, animals, insects, microorganisms and fish. Trade-off for increasing food grain production has contributed to decline and degradation in another ecosystem.
- The demand for edible oil in the country is rising due to enhanced per capita consumption, growing urbanization and steadily increasing affluence levels. India relies on imports to meet over 70% of its vegetable oil requirements; almost 60% of requirement is met through palm oil. Low and unstable yields of most oilseed crops are mainly due to cultivation under rainfed conditions. Growing oil palm entirely under irrigated conditions in which developing hybrids, latest developments in IT in water and nutrient management are essential for higher productivity from unit area of land. Policy is needed for location specific, viable cropping / farming system models, shifts its cultivation to suitable areas and policy support from Governments to ensure price stability and to provide logistic support for achieving success in oil palm.

Natural Resource Management

- Topsoil erosion, depletion of soil fertility, soil acidification, salination, excessive water extraction for irrigation, heavy use of fossil fuel energy for synthesis of fertilisers and pesticides caused severe damages to the ecosystem. A good land management policy is essential for sustaining the productivity of agriculture, forestry, fisheries and hydrology. Advanced knowledge on the hydrological cycle, improved agricultural practices and new tools for mitigating the impacts of chemical pollutants and managing wastewater now offer a set of knowledge-rich solutions to reduce environmental impact.
- Conservation is not something that can be controlled exclusively by legislation. The embedded behaviour of civilized people must be changed.

Policy on precision irrigation should be drafted to manage water efficiently. The high initial cost of the systems and infrastructure development, initial lack of technical aspects and operational know-hows should be in place. Equally important issues are conservation of natural resources in the fragile ecosystem. Policy reforms are needed on utilization and conservation of fragile ecosystem for its sustainable and eco-friendly utilization.

- The arid regions in India occupy 38.7 million ha. At present, hot arid region is devoid of community pasture and is having very low per capita availability of fodder. Therefore, a strong policy intervention is necessary to regenerate and revive community pasture at panchayat level. Policy interventions are urgently required to utilize urban and industrial zones which have huge quantity of grey water after proper treatment for fuel and fodder production in nearby areas. Land degradation and wind erosion is the major land degradation process. There is a strong need for policy intervention to identify the degradation-prone areas and to implement technologies to combat it. Promotion of small sized solar PV pumping system which will be highly beneficial to small and marginal farmers.
- Climate change affects crop productivity and its quality primarily due to changes in growing environment and indirectly due to shifts in the geography and incidence of agricultural pests and diseases, impacts on soil fertility and biological activities. There is a need to integrate these resilient practices developed under the scheme "National Innovation for Climate Resilient Agriculture (NICRA)" into the developmental programmes and policies to reach a large number of farmers in the country. There is an urgent need to integrate these resilient practices with the policy paradigm of the State and Central Governments. A wider refinement in the existing technologies and concerted efforts for research and their adoption by farmers are needed for adaptation and mitigation. The climate action plan/ policy must be incorporated into core National policies and implementation of GHGs mitigation technologies.
- Stringent Government plans and actions regarding the use of eco-friendly and organic techniques with the efficient use of the natural resources will alleviate the threat. Achieving carbon neutrality in Indian soils is a major challenge. It is very difficult to increase soil organic carbon because climatic condition prevails with a high rate of decomposition. Government should provide financial incentive or subsidies to the farmers for adopting different practices which enhance carbon sequestration, implementation of intensive plantation programme, create a large-scale awareness programme against crop residue burning and promotion of alternate crop residue management practices, promotion and adaptation of different soil and water conservation technologies in hilly regions. In fact, resource conservation technologies need to be adopted for sensitive ecosystems on priority basis.

- There is also a need to establish "Green Research Fund" at national level for the development of advance land management technologies and strengthening research on adoption, mitigation and impact assessment of climate change. This is essential to establish climate smart villages, undertake people centred watershed development programmes, convert barren, degraded, non-arable waste land and areas adjacent to road and railway line with planting of herbs, shrubs and trees under "Green India" programme with PPP mode, provide incentive to prepare a stream of "National Green Corps (NGC)" by involving students of schools and colleges, and protect degradation of our valuable natural resources by strict enforcement of regulatory measures for safe disposal of solid and liquid wastes by public-private investment.
- Building of series of food-storage structure and prevention of wastage of crop-harvest could reduce GHGs emission up to 30%. In this aspect, policy intervention and large-scale Government and Non-Governmental financial support are necessary. An up-gradation and value-added advisory should be provided with weather forecasting. Mapping of vulnerable disaster-prone-blocks and biotic-stress-hotspots in GIS-Remote sensing platform are needed. Interlinking of automatic weather stations with data analysis and dissemination throughout the country is the pre-requisite. One of the important points in disaster management is to build on local capacity and initiative to meet the disaster. So, capacity building of local people will be highly beneficial.

Biotic Stress Management

- The biotic stress caused by different agents causes losses to the worth of Rs. 2,25,000 crores annually. The experts of plant protection in India have made a roadmap including policy issues to the Government of India. There is an urgent need for national Policy on Agro-chemicals with an emphasis on safe use of pesticides, fast track transparent time bound on-line registration system, guidelines for registration of newer molecules, re-registration of pesticides (take a mandatory period of 10 years after registration), complete check on sale and use spurious pesticides.
- Establishment of testing laboratories for high quality control accredited national pesticides across the country, generation of bio-safety data for new molecules preferably through notified National Accreditation Board for Testing and Calibration Laboratories (NABL) only, identification of pesticides of biological origin, strengthening and encouraging manufacturing of new molecules under "Make in India" initiative and capacity building and skill development in assessing unregistered pesticides in imported commodities need special attention.

- There is an urgent need to move forward towards innovations in biopesticides, a centre of excellence (CoE) on agro-chemicals, development of low-cost technologies for mass production and bulk availability of biocontrol agents and biopesticides and establishment of national referral laboratories for quick identification/diagnosis and containment, etc.
- Artificial Intelligence for monitoring, tracking breeding sites of pests, migration of movement of locusts and aerial application of pesticides for site-specific management of trans-boundary quarantine pests, exploitation of advanced breeding techniques, including agricultural biotechnology to introduce insect and disease resistance into major crop plants. A nationwide programme on robust monitoring and advisory system for quarantine, invasive, transboundary biotic stresses need to be developed. Robust forewarning systems for timely management of invasive, quarantine pests are to be developed in tune with changing climate, Special emphasise needs to be given on the technological investment to manage biotic stresses in the emerging production system.

Agro-ecosystem Development

- Integrated Farming System is a potential option, where about 70% of the resources should be generated and recycled in the field. The adoption of IFS is the right approach in this direction and should be supported through institutional, extension, policy and marketing interventions in a system approach. Increasing production of vegetables, flowers and post-harvest produce, introducing agro- forestry activities in degraded land, improving dairy sector to yield higher milk, improving backyard poultry, fisheries, farm mechanization and lac sectors are essential. Branding of local farm products, promotion of indigenous technology knowledge in agriculture is needed. These are the sort of rural innovations that need to be registered at a national-level innovation registry. Infrastructure supporting these allied activities must be encouraged as a standalone enterprise, especially among the tribal farming communities.
- Hill agriculture needs special attention by the Central and State Government on seed production, nutri-cereals, IFS for small holders, water harvesting technique, high-tech automated farming, certification of organic farming, secondary and specialty agriculture, processing, value addition and packaging, Use of digital and different Apps in agriculture are available in the other places need to be modified as per the local situation in the hill region. All these can be prompted following "one commodity-one district concept" to promote indigenous niche items for increasing their commercial value and trades. Secondary agriculture, specialty agriculture along with emphasis on farm mechanization will enhance farmer income and

create additional employment opportunities. Adequate skill development programme and hand holding will be required to implement and achieve the vision of self-reliant North-East with a focus on self-reliant India.

- Coastal agriculture in India particularly in eastern coast has been undergoing a major shift from cereal production to vegetable production and shrimp farming. Major sources of pollution are agriculture, urban disposals and industrial wastes. The implementation of policy is required to minimize pollution from both agricultural and non-agricultural sources. The stakeholders who are directly vulnerable to climate change should take a role in preparing and implementing strategies for adaptation. Local communities are likely to contribute in developing and implementing sustainable agricultural strategies if their immediate needs are fulfilled without damaging the ecological carrying capacity. Joint adaption is required through Government interventions in training and market-based incentives. Policies concerning the sustainability of coastal agriculture involve several stakeholders, such as administrators, policymakers and farming and non-farming communities living in the coastal areas. Public policies need to diversify on-farm activities and to improve rural services of smallholders.
- The inclusion of farmers in the planning and implementation of the policies is necessary. However, policymaking is more effective when education and communication work in a synergistic manner. For example, for making Bundelkhand a self- reliant, many policy reforms are required involving farmers with due consideration of their socio-economics condition to address challenges in holistic way. Major recommendations for this region would be; Eco-regional based planning culminating into various micro agro climatic-zones for overall agriculture development, appropriate management of existing resources of soil, water, minerals and vegetation etc.
- Emphasis should be given on natural resource management with scientific and social planning of water use by different sectors, cluster based farming for organic and specialised agriculture, provision of credit and financial assistance along with market infrastructure for agro-forestry produce to treat it at par with agriculture produce for all policy matters, provision of insurance for tree based farming in line with crop insurance, provision of subsidy for drip and micro irrigation facility for tree based farming and promotion of eco-tourism and development of tourist cluster for domestic as well as international tourists.

Technology Advancements

- Secondary agriculture in India needs several reforms particularly on investments, innovation and bringing best practices with an aim to elite

India as the 'global food basket'. The program is an extensive package for creating forward and backward linkages, research and development, integrated cold chain and value chain infrastructure, mega food parks, infrastructure for agro-processing clusters, food safety & quality assurance and operation greens.

- Food sector has given priority status under 'Make in India' scheme. Government also has allocated special funds in National Bank for Agriculture and Rural Development (NABARD) for extending credits to designated food parks. It is necessary to formalize these unorganized food processing units by providing need-based support for technology upgradation, branding and marketing.
- Secondary agriculture has all the potential to drive the growth of primary agricultural sector in the country. By developing processing technologies and new products, the demand push for under-utilized crops is created which may help a large number of farmers.
- New initiatives taken by the Government will help the farmers and economy of our county if it is correctly implicated. One such scheme proposed by Government of India is "programme of energy from urban, industrial, agricultural wastes/industrial, municipal solid waste". Small biogas type has been introduced to some rural villages. In India in the past few years the crop residue burning has become a major problem due to the inadequate management practices. The Government is providing financial assistance to the individual farmers, and introducing new measures to contain these issues are welcome. Along with many start-up programmes many more small and big organizations are helping Indian communities in the battle with the complex problem of waste management. The unique economic and social development must be focused on improving efficiency of waste management through controlling waste generation and achieving maximum rates of recycling while utilizing waste as the commodity that can be converted into another value-added product which are additional resources to finance waste management.
- Indian agricultural sector has to pay more attention to the importance of the processing of foods and protection of the food from spoilage using modern post-harvest technologies. In India, problems associated with proper drying, storage and delivery still exist. Future post-harvest management should be focused on new cooling systems and temperature control, sustainable packaging to reduce food waste, time-temperature indicator technologies, active, smart and intelligent packaging systems, use of nanotechnologies in different stages of post-harvest management, information technologies such as genomics and computer modelling etc. Policy development related to post-harvest management will play a huge role and these policies should

aim to reinforce research and developments, training and investment. There is a clear need for a more holistic and integrated approach when dealing with postharvest losses. Along with the research & development processes, innovation techniques, the efforts of policymakers in the field of post-harvest management will influence food loss and food security in the future. Knowledge of novel and modern technologies to mitigate the post-harvest losses and encourage value addition measures is necessary to assure food safety in the future. Community participation, the intervention of Government and non-Government organisations, policy reforms, skill development programs, sustainable food production, processing and export are essential steps to reduce food loss and ensure food safety.

- Government policy should be modified from both the supply side (policies regarding availability of Indian satellite data) as well as the demand side (the use of satellite dataset for different applications) to create an ecosystem for higher use of space technology in various applications over the country. Already some steps have been taken to use space technology as much as possible in different applications by the various departments of the Government. It should be used on high value crops, crop intensification/ diversification and disaster planning with management etc.
- Government shoud make use of digital technologies to design and deliver better agricultural policies. It may create new roles or responsibilities for Government, including enabling the digital infrastructure and reducing information asymmetries and transactions costs, less Government intervention. For policymakers, the challenge will be to shape policy and regulatory settings so that they facilitate opportunities offered by digital technologies. Reaping the benefits of digital technologies in agriculture requires the participation and co-operation of farmers, researchers, private sector, non-profits and Government.

Allied Sectors of Agriculture

- There is a need to initiate a national programme in mission mode approach for accelerating production and effective utilization of fodder resources across the regions in the country is needed. Formulation of national grassland and grazing policy and rejuvenation of degraded pastures, formulation of a land use policy for protecting sustainable area of different crops and to preserve the freed non-arable land for intensive livestock management are also essential.
- Promotion of entrepreneurship and exploiting opportunities in commercial venture, mechanism/ legislation for protecting the interests of large number of smallholder population while facilitating in scaling up commercial dairy production system, strengthening of fodder bank, marketing structure

and its distribution mechanism in the country, convergence of livestock sector programmes with other mission schemes like horticulture mission, MNREGA and regulation bill on animal feed, more investment in livestock sector particularly on feed and fodder development as well as their marketing, service sector needs to be modernized and there should be an organized fodder production system dedicated to dairy, orientation and support to various stakeholders, Government, NGOs, research organizations towards Media/ HRD/ ICTs for extension services.

- Introduction of fodder security bill and policy on fodder market should get top priority along with policy for promoting women empowerment and successful models of women–based livestock management in India.
- Fishers/ farmers and their opinions have been also taken into consideration while formulating policies to manage fisheries sector in India *viz*; National Policy on Marine Fisheries 2017, National Inland Fisheries and Aquaculture Policy (NIFAP) 2018 and National Fisheries Policy 2020.
- Several initiatives in the fisheries and animal husbandry sectors have also been launched on the occasion. PMMSY is a flagship scheme for focused and sustainable development of the fisheries sector in the country as a part of the *Aatmanirbhar Bharat Abhiyan*. Focus especially on employment generation activities such as seaweed and ornamental fish cultivation, addressing critical gaps in fish production and productivity, quality, technology, post-harvest infrastructure and management, modernisation and strengthening of the value chain, traceability, establishing a robust fisheries management framework and fishers' welfare, and consolidating the achievements and bringing new interventions such as fishing vessel insurance, support for new/up-gradation of fishing vessels/ boats, integrated aqua parks, e-trading/marketing, etc. are included in this new development.
- There is a re-need to put concerted efforts in terms of strategic implementation of policies, budget allocation, linkages among institutions, smooth credit flow, research, innovations, development of new technologies, community based extension activities, entrepreneurship development, better market infrastructure information system, proper regulations and skilled manpower to roll the wheels of the fisheries economy.

Marketing and Infrastructural Development

- Time to time the Government has taken many initiatives for improving agricultural markets and some push drivers of infrastructure recognized in this changing environment are: opening up of railways infrastructure to 100% foreign direct investment, rural electrification, driving the construction of new airports and sea ports across India etc.

- The creation of adequate infrastructural facilities in a liberalized and market driven economic environment is necessary particularly in rural area. FPOs can provide an advance to particularly small and marginal farmers up to 80 per cent of the value of their produce at the current market price. But the fact is that, FPOs will need large working capital to give advances to farmers against their produce as collateral. Therefore, handholding by the Government in this area is required to strengthen the market infrastructure.
- For encouraging private entrepreneurs in cold storage business, there is a need to provide subsidy to make the units viable for some initial years. Physical markets with facilities and services would attract the farmers and the buyers creating competitive trade environment.
- Also, there is a need for maximum use of ICT in agricultural marketing. Appropriate agri-logistic, post-production management needs marketing infrastructure. Markets in India are devoid of basic infrastructural facilities particularly for post-harvest handling of perishable and high value commodities. These infrastructure facilities are commonly delineated as infrastructure platforms for aggregating, transporting, storing and merchandising, and an IT platform for flow of information. The infrastructure in the form of transport and storage for food grains is needed.
- There is a need to revisit the concepts and policies that were previously formulated on post-harvest management of farm produce. The post-production link to markets for horticulture and livestock requires cold-chain infrastructure. The critical missing links are modern pack-houses at village level to aggregate and pre-condition the fresh produce for onwards market connectivity. Policies on integration in the logistics chain, transport and storage, modernising infrastructure for Agri-logistics are needed.
- To bring about the targeted doubling of the farmers' income, the identification and connectivity with domestic consumers needs to be prioritised. Private sector is in greater need of support for upgrading of processing technology and to build global level competitiveness. There is need to focus on strategies for development of an agricultural marketing system in the country with thrust on infrastructure creation, efficient flow of produce, access to market information and reduced food waste.
- The report on Doubling Farmers' Income also emphasized the role of aggregation in marketing of agricultural goods, both perishable and non-perishables and aggregation helps in displacing the village trader based intermediation (common in fruits and vegetables) and increases the marketing efficiency of the goods.

Agricultural Extension

- All the extension models tried in recent times have been theoretically sound but the problem lies with their implementation. Non-availability of appropriate technologies and relevant information to all categories of farmers is the major hurdle for achieving sustainable farm production.
- Modern communication technologies when applied to conditions suits to rural areas can help improve communication, increase participation and disseminate information and share knowledge and skills. An enabling mobile service delivery infrastructure system consisting of a Mobile Service Delivery Gateway (MSDG) that is fully integrated with the existing infrastructure created under the National e-Governance Plan (NeGP) should be built up. There is a need to develop and implement innovative Public Private Partnership (PPP) and Multi Stakeholder Partnership (MSP) models for design and delivery of mobile governance services as well as encouraging development of cloud-based implementation models and use of light technologies and deployment of a suitable capacity building framework to enhance delivery and absorption capacity for e-Governance services.
- There must be standardization and promotion of indigenous technology knowledge (ITK) in agriculture, where techniques are mostly on local resources base in dealing with nutrition, disease and pests. These are the sort of rural innovations that need to be registered at a national-level innovation registry, which, in turn, can help in the filing of patents and get micro-venture capital support for enterprises. In fact the KVKs are identifying the farmer innovators and documenting these and sending to National Innovation Foundation (NIF), which is India's national initiative to strengthen technological innovations and outstanding traditional knowledge wealth.
- Integrating the efforts of organizations to identify niche areas like value addition, rural infrastructure, logistics, warehousing, quality certification, etc. can provide livelihood opportunities for migrant workers. The digitization process in agriculture is to be accelerated. Two major constraints in promoting digitization are lower access to physical capital and human capability. Insufficient connectivity in rural areas, high costs for services, and lack of basic computer knowledge and literacy hinder rapid development of e-agriculture are another constraint. Substantial investment is needed in physical infrastructure, power, broadband, and transportation.
- Remote sensing (*via* satellites), GIS, crop and soil health monitoring, and technologies for livestock and farm management are the common areas. A digital laboratory with all accessories is to be set up in each KVK which will act as digital incubation centre. Given the magnitude of the challenges,

coherent, climate-smart, integrated, national- and state-level agricultural policy and strategy frameworks are needed, beyond the mission mode, as recommended by NABARD (National Bank for Agriculture and Rural Development) and ICRISAT (NABARD, 2016) with the emphasis on bringing reforms in extension systems.

- To achieve self-reliance in agriculture there must be convergence of all the stakeholders in agriculture application country farmers are to be empowered with enhanced income and a prestigious living in the society, for which KVK plays a pivotal role.

Agricultural Education

- Strengthening and compliance with quality assurance policies and mechanisms, man power need assessment and organization of workshops at national level are now needed in the education re-structuring.
- Involvement of different stakeholders in review and assessment of skilled human resources attracting young talent to agricultural education, developing much needed globally competitive human resources for preparing India for the agriculture of the 21st century are also essential.
- Focussing more productive, environmentally safe renewable rather than non-renewable resources of energy input should get priority.
- Shifting the focus from self-reliance to greater interaction with the world scientific community, using new information technology such as video conferencing and greater opportunities for travel and joint projects need formulation of policy. Greater integration of basic sciences and agricultural sciences, catching up with the biotechnology revolution, removing all barriers through an open door policy, new mechanisms of more stable funding support, new governance mechanisms for improved and efficient management of resources are also required.

For Product Safety Concerns and Information please contact our EU representative GPSR@taylorandfrancis.com
Taylor & Francis Verlag GmbH, Kaufingerstraße 24, 80331 München, Germany

www.ingramcontent.com/pod-product-compliance
Ingram Content Group UK Ltd.
Pitfield, Milton Keynes, MK11 3LW, UK
UKHW020131200926
475702UK00005B/697
9781032156989